RNA Technologies

Volume 11

More information about this series at http://www.springer.com/series/8619

Stefan Jurga • Jan Barciszewski
Editors

The Chemical Biology of Long Noncoding RNAs

Springer

Editors
Stefan Jurga
Nanobiomedical Center of the Adam Mickiewicz University,
Poznań, Poland

Jan Barciszewski
Nanobiomedical Center of the Adam Mickiewicz University,
Poznań, Poland

Institute of Bioorganic Chemistry of the Polish Academy of Sciences,
Poznań, Poland

ISSN 2197-9731 ISSN 2197-9758 (electronic)
RNA Technologies
ISBN 978-3-030-44742-7 ISBN 978-3-030-44743-4 (eBook)
https://doi.org/10.1007/978-3-030-44743-4

This Springer imprint is published by the registered company Springer Nature Switzerland AG.
The registered company address is: Gewerbestrasse 11, 6330 Cham, Switzerland

Preface

Introduction

Long Non-protein Coding RNAs

Over the last few decades the development of high-throughput technologies has changed the perception of the Crick's Central Dogma of Molecular Biology stating that genetic information inscribed in DNA is transcribed into RNA and afterwards translated into proteins (Fig. 1). Only ca. 2% of the transcripts can encode proteins, but majority of the transcripts are non-protein coding sequences accounting for 98% of the human genome (ncRNAs).

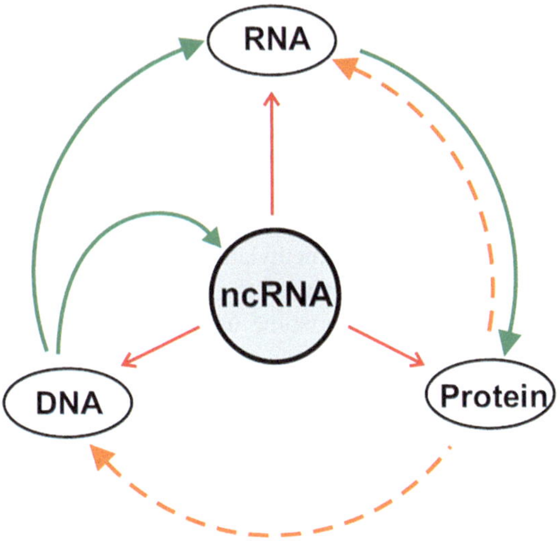

Fig. 1 The central dogma of molecular biology

Non-protein coding RNAs (ncRNAs) family includes structural RNAs (ribosomal RNAs and transfer RNAs) as well as regulatory RNAs. Furthermore, regulatory RNAs can be divided into three classes:

(a) Small non-coding RNAs, 20–50 nucleotides long, include microRNAs (miRNAs), small interfering RNAs (siRNAs), piwi interacting RNAs (piRNAs), cis-regulatory RNAs (cisRNAs), and telomere specific small RNAs (telsRNAs)
(b) Medium non-coding RNAs, 50–200 nucleotides long, include small cytoplasmic RNAs (scRNAs), small nuclear RNAs (snRNAs), small nucleolar RNAs (snoRNAs), and transcription initiation RNAs (tiRNAs)
(c) Long non-coding RNAs (lncRNAs), longer than 200 nucleotides, without or with limited protein-coding potential

For a long time, the DNA sequence of lncRNAs was considered as "junk DNA." Since early 1990s, thousands of lncRNAs have been discovered and investigated, but their exact number and role in human physiology and pathology is still being elucidated.

The lncRNAs take part in many biochemical processes as regulations of genes and proteins, RNA splicing, and maintenance of structure cellular particles.

LncRNAs act at the transcriptional, post-transcriptional, and epigenetic level. At the transcriptional level, they have several functions to disrupt the binding of transcriptional factors with promoters of target genes, altering the localization of transcriptional factors in the genome, competing with endogenous RNA, and forming scaffolds with DNA and proteins. They work as molecular signal transducers, guides for ribonucleoprotein complex, and as sponge to sequester miRNAs. As signalling molecules, they serve as spatiotemporal indicators of gene regulation that reflect the biological effects of transcription factors or signalling pathways. As decoys, they sequester transcription factors and other proteins away from chromatin or into nuclear subdomains. As guides, they recruit RNA binding proteins to target genes, and as scaffolds, they bind several proteins to form complexes with specific biological roles. At the post-transcriptional level, they modulate directly or indirectly the effects of micro-RNAs on target genes and regulate the alternative splicing of mRNA.

At the epigenetic level, they interact with proteins involved in histone modifications, regulate DNA methylation in promoter regions, and interact with chromatin modification complexes.

LncRNAs possess some unique characteristic features. They are found to be abundant mainly in the nuclei of organisms. Their fraction is comparatively lower than the protein coding mRNAs in cells and tissues. However, unlike mRNAs most of the lncRNAs are cell or tissue specific.

LncRNAs can be also classified into five different categories:

(a) Long intergenic ncRNAs (lincRNAs), which do not overlap in close proximity to protein coding genes
(b) Antisense lncRNAs, transcribed from the antisense strand of a protein-coding gene

(c) Sense-overlapping lncRNA, which overlap with one or more intron/exon of different protein-coding genes in the sense RNA strand
(d) Intronic lncRNAs
(e) Bidirectional lncRNAs, transcribed from a promoter of a protein-coding gene, yet in the opposite direction

LncRNAs are found to be sharing many of their structural and functional features, common characteristics with protein coding RNAs (mRNAs). They are transcribed by RNA polymerase II enzyme, many of them indicate a 5′ cap, undergo post-translational modifications, splicing, and are polyadenylated at their 3′ ends. LncRNAs are conserved in a cell type-specific manner and can vary in response to environmental stimuli and during development. They are poorly conserved in sequence, whereas the secondary structure of lncRNAs seems to be conserved across different species. Trying to assign individual functions to individual lncRNAs one has to remember that a single 1000 bases lncRNA is enough to carry out a large number of functions with perhaps different subsets of these functions being active in different tissues and at different stages of development.

In general, the functions of lncRNAs can be split into three groups:

(a) Chromosome modification and of gene transcription regulators (e.g., Xist, as a *cis*-response element maintains the inactivation of X chromosome by binding to X chromosome and attracting Polycomb repressive complex 2)
(b) Protein binders, regulate the activity of proteins, form nucleic acid–protein complexes as structural components, and change the cellular localization of the proteins
(c) Carriers of peptides coding function (some of the lncRNAs are found to have small open reading frames which encode short peptides, e.g., containing 34 amino acids named DWORF mainly located in the sarcoplasmic reticulum)

Majority of lncRNAs have essential roles upon the onset of various diseases and are potential biomarkers for disease diagnosis, treatment, and prognosis evaluation.

The dysregulation of lncRNAs and RNA toxicity with repeat expansion are related to various neurological disorders including Alzheimer's disease (AD), Parkinson's disease (PD), Huntington's disease (HD), frontotemporal dementia (FTD), autism spectrum disorder (ASD), schizophrenia (SCZ), spinocerebellar ataxia (SCA), Fragile X tremor/ataxia syndrome (FXTAS), and amyotrophic lateral sclerosis (ALS).

The number of lncRNAs is huge. In the human genome ca. 144,000 loci for lncRNAs but, and ca. 126,000 in the mouse genome have been found. 28,000 human lncRNA genes show their genuine 5′ transcriptional start site.

So far, most of the works in the field of lncRNAs have been centered on their genetic aspects and development of databases pertaining to lncRNAs. Until now, no structural works are available that deal with the characterizations of the peptides encoded by some lncRNAs.

The book consists of 20 chapters which discuss various aspects of long non-protein coding RNAs at molecular and cellular levels, their targets, functional annotation, and disease association in different biological context. Diversity in form and function of lncRNAs, their localization, biogenesis, and occurrence are discussed. Structural biology analysis of lncRNAs and proteins binding to them reveal mechanisms of their actions. Many chapters cover specific lncRNAs involved in cancer and other diseases and their involvement in development, progression, and treatment.

This book will provide the reader with examples of important roles played by lncRNAs.

NanoBiomedical Centre of Adam Mickiewicz University, Poznań, Poland — Stefan Jurga

NanoBiomedical Centre of Adam Mickiewicz University, Poznań, Poland — Jan Barciszewski

Institute of Bioorganic Chemistry of the Polish Academy of Sciences, Poznań, Poland

Contents

Long Non-coding RNAs Diversity in Form and Function: From Microbes to Humans

Gabriela Toomer, Huachen Gan, and Joanna Sztuba-Solinska

Contents

G. Toomer · H. Gan · J. Sztuba-Solinska (✉)
Department of Biological Sciences, Auburn University, Auburn, AL, USA
e-mail: gzc0025@auburn.edu; hzg0021@auburn.edu; jzs0165@auburn.edu

S. Jurga, J. Barciszewski (eds.), *The Chemical Biology of Long Noncoding RNAs*, RNA Technologies 11, https://doi.org/10.1007/978-3-030-44743-4_1

Abstract Long non-coding (lnc) RNAs are vital regulators of a plethora of biological processes. They have been shown to play major roles as transcriptional and post-transcriptional regulators, with some being involved in imprinting genomic loci, shaping chromosome conformation and allosteric regulation of protein activities. Highly specific patterns of lncRNA expression coordinate cellular differentiation, pluripotency, and development; while the overexpression, deficiency, or mutation of lncRNA genes have been implicated in numerous disorders. Given the diversity and number of lncRNAs, it seems likely that their functions are as numerous as those of proteins, yet, only a small fraction of these transcripts have been structurally and functionally characterized. Although the vast majority of information regarding lncRNA structure and function is derived based on the studies of human lncRNAs, in recent years, these transcripts have been identified in plants, bacteria, and viruses. This chapter aims to overview the salient features, biogenesis, and biological importance of these molecules in various biological systems. We discuss a few prominent examples of lncRNAs expressed in humans, plants, bacteria, and viruses, highlighting their structural properties, physiological functions, and modes of action. There are still many questions that need to be addressed in this relatively new research area. Addressing them in the future research will enhance our knowledge of RNA plasticity as a biological control mechanism and potentially discover novel targets to control the biological outputs of cellular processes in response to different stimuli.

Keywords Viral lncRNA · Bacterial lncRNA · Plant lncRNA · Human lncRNA · RNA structure-function · lncRNA biogenesis

1 Introduction: Identification, Classification, and Properties of LncRNAs

Analyses of the international human genome sequencing results converged that over 90% of the human genome is transcribed, forming an intricate network of non-coding (nc) RNAs. This discovery evident the beginning of a chapter in the understanding of molecular mechanisms governing life. Also, it proved the central

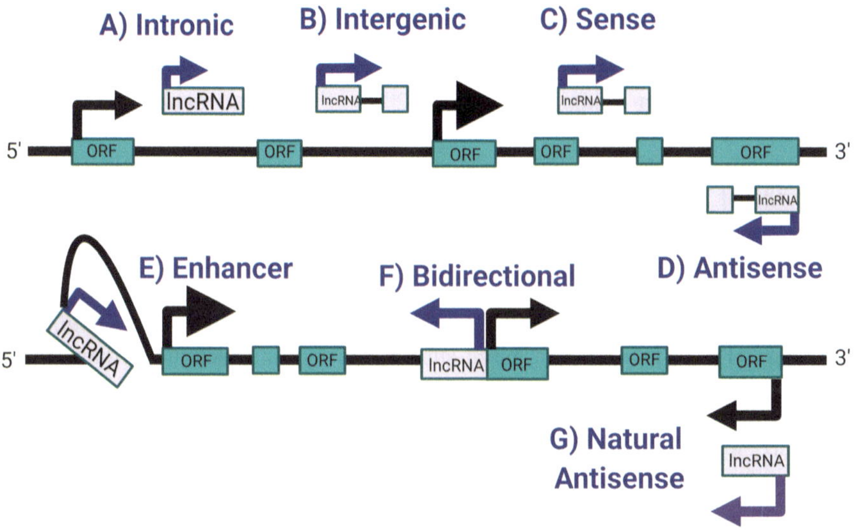

Fig. 1 Classification of lncRNAs based on their genomic localization in relation to the protein-coding genes. Purple boxes represent lncRNAs and their orientation, green boxes correspond to the open reading frames (ORFs) and the direction of the transcript. (Created with biorender.com)

dogma of biology, assuming that DNA functions as a storage medium, whereas RNA exists as an inert molecule used merely for protein production, was an enormous oversimplification. Among, so-called "dark matter" of the human transcriptome, a class of ncRNAs referred to as long non-coding (lnc) RNAs have gained much attention as a crucial layer of cellular regulation, now representing the new frontier in the chemical biology of complex organisms.

LncRNAs are an assorted class of transcripts that engage in numerous biological processes across every branch of life. Although lncRNAs were initially discovered as messenger (m)RNA-like transcripts that do not encode proteins, the recent boom of experimental research has brought to light their unique features which further distinguish them from coding transcripts. LncRNAs are on average longer than 200 nucleotides (nt) (Cao 2014), with some achieving incredible lengths extending 90 kb ("macroRNAs"). Examples contain the 108 kb AIR and the 91 kb KCNQ1OT1 (Korostowski et al. 2011). Also, shorter ncRNAs, near 200 nt, frequently are categorized as lncRNAs, mainly because their biogenesis does not follow the pathways specific for microRNAs.

Based on their genomic location, lncRNAs are classified as intronic, intergenic (sometimes also called intervening or lincRNAs), sense, antisense, or enhancer. Except for lincRNAs, which are located in the intergenic region between two protein-coding genes, most lncRNAs show some overlap with nearby open reading frames (ORFs) (Fig. 1). For example, intronic lncRNAs are transcribed from the introns of protein-coding genes, and NATs are transcribed from the complementary strands of protein-coding genes. Antisense (AS) lncRNAs are particularly common, and up to 72% of genomic loci in mice show evidence of divergent transcriptions leading to the generation of antisense lncRNAs. Besides, bidirectional promoter-associated ncRNAs (pancRNAs), which are oriented head-to-head with a protein-coding gene within 1 kb, account for ~20% of the total number of lncRNAs (Hamazaki et al. 2017).

Typically, lncRNAs show poor sequence conservation across species, as more than 70% of lincRNAs cannot be traced to homologs in species that diverged ~50 million years (Diederichs 2014; Hezroni et al. 2015). Yet, some human lincRNAs have homologs with similar expression patterns and segments of conserved sequences near their 5′ end (Hezroni et al. 2015), surrounded by largely unconstrained regions (Ponjavic et al. 2007; Guttman et al. 2009). Conventionally, a constraint is estimated from the nt substitution rate in functional sequence as a proportion of the rate in neutral, unconstrained, sequence. For proteins, on average, this ratio is ~10%, whereas, for the set of 3122 full-length, and mainly intergenic, ncRNAs, including lncRNAs, it has been estimated to ~90–95% (Ponjavic et al. 2007). Intriguingly, the promoters of the lncRNAs and the binding sites for transcription factors localized within these promoters are well-conserved, suggesting that conserved regulatory mechanisms control lncRNA transcription.

An exact assessment of the number of lncRNAs is quite challenging due to their cell-, tissue-, developmental stage-, and disease-specific expression profiles that restrict their modes and sites of action. Only a minority of lncRNAs are expressed across all tissues or cell types, such as TUG1, taurine upregulated 1 lncRNA or MALAT1, metastasis-associated lung adenocarcinoma transcript 1 (Li et al. 2015a; Ward et al. 2015). The GENCODE V7 (April 2011) placed the number of lncRNAs in humans at ~15,000 and that list has been recently expanded by additional 3574 loci in human and 561 in mouse (Frankish et al. 2019). NONCODE v5.0 includes ncRNAs from 17 species, including model systems like *Arabidopsis thaliana*, fruitfly, and yeast (Fang et al. 2018). Also, more than 120,000 lncRNAs have been annotated and gathered in the Green Non-coding Database (http://greenc.sciencedesigners.com/) (Paytuví Gallart et al. 2016), while such estimates are not yet available for bacterial or viral lncRNAs.

In contrast to the broad genomic annotation of lncRNAs, far fewer have been characterized for their absolute abundance. In general, they tend to be poorly expressed and present at a concentration approximately one order of magnitude below mRNAs (Derrien et al. 2012), with many being quantified at less than one copy per cell (Seiler et al. 2017). Yet, XIST lncRNA, a key regulator of X chromosome inactivation (XCI) in placental mammals, has been estimated to concentrate at 50–100 copies per chromosome, MALAT1 (Tripathi et al. 2010) and NEAT1, nuclear paraspeckle assembly transcript 1, are estimated at ~3000 copies per cell, while HOTAIR, HOX transcript antisense RNA accumulates at ~100 copies per cell (Dodd et al. 2013). Kaposi's sarcoma-associated herpesvirus-encoded (KSHV) polyadenylated nuclear (PAN) RNA involved in the regulation of viral lytic reactivation breaks the record and accumulates at 5×10^5 copies per cell similarly to Epstein-Barr virus-encoded RNAs EBER1 and EBER2.

When it comes to lncRNAs stability, their half-lives vary over a wide range, comparable to, although on average less than that of mRNAs. Clark et al. (2012) used ncRNA array to analyze the half-lives of approximately 800 lncRNAs and ~12,000 mRNAs in the mouse neuronal cell line. Also, Tani et al. (2013) developed a new technique called 5′-bromo-uridine immunoprecipitation chase followed by deep sequencing (BRIC–seq), in which RNAs are labeled by pulse chasing with 5′-bromo-uridine to survey RNA half-life in HeLa cells. Their analyses revealed that intergenic and *cis*-antisense RNAs are more stable than those derived from introns, as are spliced lncRNAs compared to unspliced transcripts. Also, lower stability has been associated with nuclear localization of lncRNAs. Here, one of the least stable lncRNAs is *NEAT1* ($t_{1/2} < 2$ h), suggesting *its* instability contributes to the dynamic nature of nuclear paraspeckles. On the other side of that spectrum, ZFAS1, zinc finger antisense 1 lncRNA, displays no degradation over a 16-h time course, while MALAT1, according to two independent estimates, falls right in between with $t_{1/2} \sim 7$ h. Based on these datasets, an interactive online resource has been created that catalogs lncRNA stability profiles and provides a comprehensive annotation of ~7200 mouse lncRNAs (http://stability.matticklab.com).

The absolute requirement for being non-coding has invited quite the controversy into the lncRNA field, with some studies indicating lncRNAs association with ribosomes, while others demonstrating that lncRNAs do not encode proteins. Bánfai et al. (2012) correlated tandem mass spectrometry data with RNA sequencing (RNA-seq) data and concluded that over 90% of GENCODE lncRNAs are unlikely to encode a peptide. A similar approach by Gascoigne et al. (2012), further strengthen these conclusions. Yet, the steroid receptor RNA activator (*SRA*), a well-characterized lncRNA involved in the nuclear receptor-mediated regulation of gene expression, does both, codes for a polypeptide but also have coding-independent functions. Thus, it seems that a reasonable approach to lncRNAs definition should focus on a coding-independent role of the untranslated transcript, especially, in the light of recent discoveries showing that short peptides encoded by some lncRNAs may perform essential regulatory functions. For example, Matsumoto and coworkers in 2017 uncovered a small regulatory polypeptide of amino acid response (SPAR), encoded by lncRNA LINC00961, which suppresses activation of the mammalian target of rapamycin complex 1 (mTORC1) in response to amino acids stimulation (Matsumoto et al. 2017). Also, HOXB cluster antisense RNA3 (HIXB-AS3) encodes a conserved peptide, which inhibits tumorigenesis and subsequent metabolic reprogramming in colon cancer cells (Huang et al. 2017). To complicate the matter more, some lncRNAs have been shown to express other ncRNAs such as circular (circ) RNAs, tRNAs, miRNAs, and small nucleolar (sno) RNAs. A prime example for this complexity is the growth arrest-specific 5 (GAS5) lncRNA, which hosts 10 C/D box snoRNAs, five of which can be further processed to piwi (pi)RNAs.

2 Discovery of LncRNAs

Development of methodologies that aid the discovery of new lncRNAs has been a fundamental step in the RNA biology field. Initially, scientists were using differential hybridization screens of cDNA libraries to clone and study genes with tissue-specific and temporal patterns of expression. Through this approach, the first non-coding gene discovered in the human was the imprinted maternally expressed transcript H19, although it was initially classified as an mRNA. Soon realized the absence of a long and conserved ORF within H19 and lack of ribosomal interaction prompted the conclusion that it is a non-coding transcript. Following studies led to the discovery of XIST in 1992 (Brockdorff et al. 1992; Brown et al. 1992), AIRN responsible for imprinting of the *IGFR2* gene in 2002 (Sleutels et al. 2002), MALAT1 in 2003 (Ji et al. 2003), and from 2006, numerous lncRNAs encoded within *HOX* gene clusters that regulate gene expression, like HOTAIR (Rinn et al. 2007). These pioneering studies revolutionized our view of non-protein-coding gene functions and the biological relevance of lncRNAs in general. The identification of new lncRNAs in the last decade continues to increase and, as anticipated, largely exceeds that of protein-coding transcripts. Here, we provide an overview of methodologies that were used in the past and have been developed recently, aiding the identification of novel lncRNAs (Fig. 2).

2.1 Tiling Arrays

The tiling array is a classical lncRNA discovery method (Fig. 2a). A typical microarray consists of thousands of spots containing multiple fragments of the same DNA oligonucleotides, known as probes. The probes on the microarray are hybridized to a labeled RNA sample, and the array is subsequently washed. This results in the labeled sample only remaining at the spots where the sample hybridized to probes. The signal intensities on the microarray are used as a measure of the relative abundance of each probe. Tiled microarrays can contain probes that cover non-repetitive sequences of a specific chromosome or, potentially, the entire (Johnson et al. 2005; Weile et al. 2007).

The very high-resolution tiling array with 400,000 overlapping probes, each 40 nt in length, has been applied for the identification of transcripts from four human *HOX* gene clusters. One such lncRNA expressed in an antisense manner within the *HOXC* gene, and hence was named HOTAIR. This lncRNA has been found to repress transcription across 40 kb of the *HOXD* locus on chromosome 2 *in cis* and to interact with polycomb repressive complex 2 (PRC2) (Rinn et al. 2007). Remarkably, tiling arrays can be designed at even higher resolutions. For example, tiling arrays of 25 nt oligonucleotides spaced every 5 bp have been used to analyze RNAs produced from 10 different human chromosomes. As a result, each had a 20 bp overlap with its neighboring probes providing the high-resolution mapping of the transcripts (Cheng

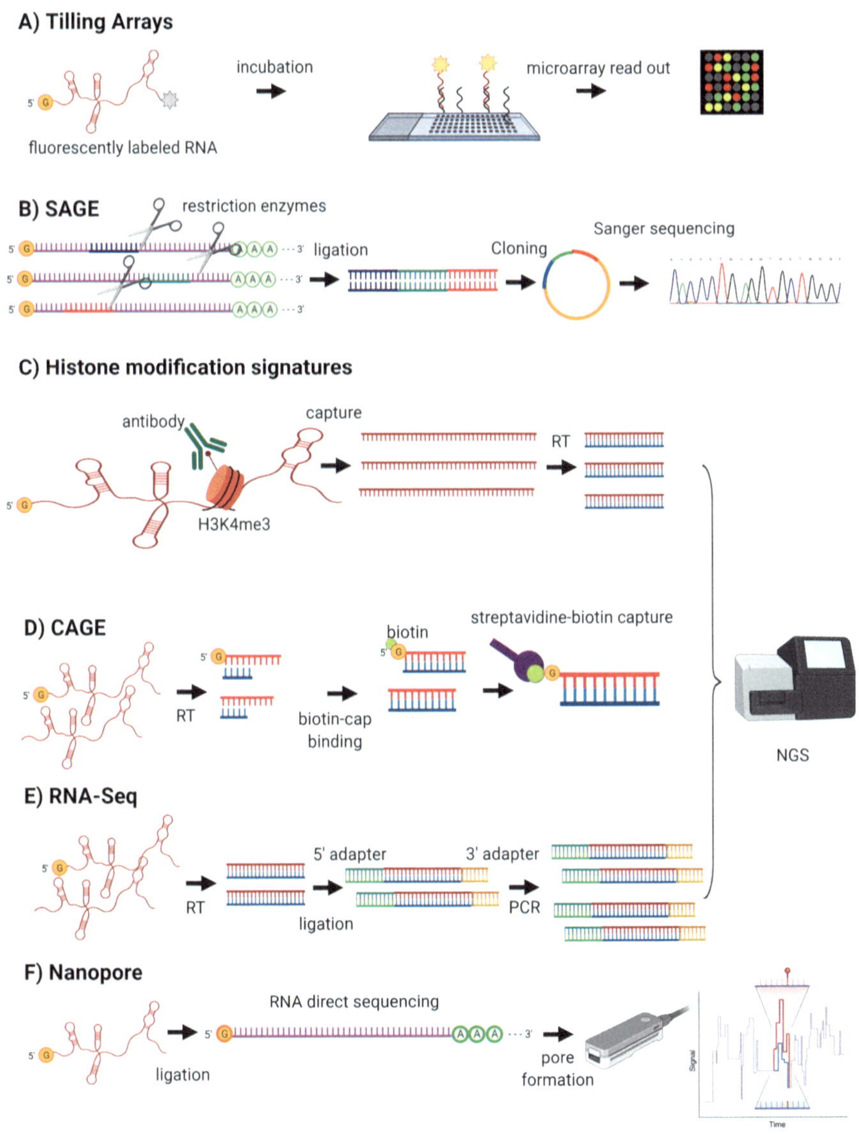

Fig. 2 Schematic overview of methodologies that aid the discovery of new lncRNAs. (**a**) During tilling array, the probes on the microarray are hybridized to a fluorescently labeled RNA sample, and the signal intensities on the microarray are used as a measure of the relative abundance of each probe; (**b**) SAGE relies on the generation of short tags containing restriction sites that are cloned and subjected to Sanger sequencing; (**c**) Histone modification signatures are used to identify lncRNAs by the combination of chromatin immunoprecipitation and NGS; (**d**) CAGE maps and quantifies the expression of 5′-cap-trapped lncRNAs; (**e**) RNA-Seq, the next-generation sequencing of a wide range of caped/uncaped, poly(A)/not poly(A) lncRNAs. (**f**) Nanopore sequencing allows for direct RNA sequencing and identification of epitranscriptomic modifications. (Created with biorender.com)

2005). Also, the separation of polyadenylated and non-polyadenylated RNA fractions from the nuclear and cytosolic compartments has led to the discovery that a large number of unannotated transcripts exclusively accumulate in the nucleus. More than 40% of all RNAs are non-polyadenylated, and among that fraction, more than half of nuclear sequences have been derived from intronic regions (Cheng 2005). Besides, tailing arrays have been successfully used for transcriptome analysis and detection of lncRNAs expressed by herpesviruses, as well as host lncRNAs that are dysregulated during viral infections (Zhang et al. 2015; Gallo et al. 2017).

At a practical level, the microarray procedures are robust, efficient, and technically simple. Also, open-source software are available for data analysis, for example, GenePattern (www.broadinstitute.org/cancer/software/genepattern), MultiExperiment Viewer MeV (www.tm4.org/mev.html), Chipster (chipster.csc.fi), and R/Bioconductor packages (www.bioconductor.org). Yet, soon after the emergence of next-generation sequencing (NGS) technologies, tiling arrays have lost their popularity, mainly due to their inherent problems related to potential noise due to weak-binding or cross-hybridization of transcripts to probes, which hampers the study of repetitive sequences, and also a high cost-benefit ratio when searching through entire genomes.

2.2 High-Throughput Sequencing Serial Analysis of Gene Expression (SAGE)

In 1995, Velculescu et al. described a method for the overall gene expression analysis without pre-existing knowledge of the entire genome (Hu and Polyak 2006). This method is used to generate a library of short sequence tags, referred to as SAGE tags, containing recognition sites for restriction enzymes at the 3′ end of transcripts. SAGE tags are concatenated before being cloned and subjected to Sanger sequencing. This method also allows for both the quantification of pre-annotated transcripts and the identification of new ones, a feature not feasible with microarrays (Fig. 2b).

The SAGE technology has been applied for multi-tissue, cross-cancer lncRNA expression profiling study using 272 SAGE libraries, representing 26 non-malignant human tissues and 19 human cancer types. The analysis has revealed that lncRNAs are broadly distributed across all human chromosomes (Gibb et al. 2011). Only ~1% of the lncRNAs have been shown to be universally expressed, supporting the idea that most function in a tissue-specific manner. The most highly and frequently expressed lncRNAs have been previously associated with critical biological processes, including NEAT1, MALAT1, and sno RNA host gene 6 (SNHG6), which function in RNA modification. Also, some lncRNAs, i.e. MEG3, a maternally expressed 3, have been shown to be aberrantly expressed in more than one type of cancer.

Since the invention of SAGE technology, numerous libraries representing a diversity of human and animal, normal and malignant tissues and cell lines have been providing further insight into differences in global lncRNAs regulation between tissues and normal and cancer cells. NGS technologies replaced this sequencing-based method, mainly because they profile a larger number of transcripts in greater depth.

2.3 Histone Modification Signatures

Guttman et al. (2009) developed a strategy that uses global histone modification signatures to identify novel lncRNAs. Using chromatin immunoprecipitation (ChIP) in combination with NGS, the authors have generated the genome-wide histone modification maps, focusing on signatures associated with Pol II transcription, i.e., trimethylation of lysine 4 of histone H3 (H3K4me3) at the promoter and trimethylation of lysine 36 of histone H3 (H3K36me3) along the length of the transcribed region (Fig. 2c). Their approach has led to the identification of ~1600 large multi-exonic RNAs across four mouse cell types. Codon substitution frequency (CSF) (Clamp et al. 2007), a measure of coding potential, has been assessed for all intergenic transcripts in all three ORFs to confirm that most of the identified transcripts lack coding potential. Also, the authors have assessed the extent of sequence conservation in the identified lncRNAs by modeling the underlying substitution rate across 21 mammalian genomes and finding that lncRNA exons show clear conservation when compared to other intergenic regions.

Additional study has demonstrated that histone modification signatures at the transcriptional initiation region may distinguish functionally distinct classes of lncRNAs, i.e., those arising from enhancer-associated (elncRNA) or promoter-associated (plncRNA) elements (Marques et al. 2013). These classes differ with respect to their evolution, tissue-specificity, expression and co-expression levels with their neighboring genes, suggesting that, if they influence gene expression, they may do so through different mechanisms. Also, Chao D. et al. have applied ChiP-seq and found that the expression patterns of many plant lncRNAs, particularly the non-polyadenylated fraction, are stress-regulated (Di et al. 2014).

2.4 Cap Analysis of Gene Expression (CAGE)

CAGE is another NGS-based method, which simultaneously maps and quantifies the expression of 5′-cap-trapped lncRNAs. This methodology provides a useful tool for the identification of transcriptionally active promoter regions, and POL II-driven transcription start sites (TSS) (Fig. 2d). Yet, as it relies on 5′-capping of the transcript, non-capped RNAs are omitted. CAGE often requires a relatively large amount of starting material, yet, an alternative approach, referred to as nanoCAGE, allows sequencing of as little as 10 ng of total RNA and it has been successfully employed for single-cell transcriptional profiling. This technology also has a paired-read variation called CAGE-scan which counterweighs for CAGE's inability to detect the 3′ end of transcripts by reading in the vicinity of random-priming sites close to the middle of the transcript and providing the insights into the architecture of transcript (Plessy et al. 2010).

Fort et al. (2014) have performed CAGE on human- and mouse-induced pluripotent stem cells, embryonic stem cells, and differentiated cells, identifying more than 4000 novel transcripts in humans and 8000 in mice, that were named non-annotated stem transcripts (NASTs). A large fraction of mostly nuclear NASTs have been shown to be transcribed from long terminal repeats, which constituted either the promoters or the enhancers for these RNAs. Further characterization of NASTs has shown that these RNAs are mainly species-specific, overlap repetitive regulatory regions, and associate with retrotransposon elements. More recently, Hon et al. (2017) have applied CAGE to generate an atlas of 27,919 human lncRNA genes. Genomic and epigenomic classification of these lncRNAs has revealed that most intergenic lncRNAs originate from enhancers rather than from promoters. The authors have also found that 13,896 lncRNAs have conserved exons and 13,228 have conserved transcription initiation regions. By integrating their findings with genomic data, the authors have discovered that 1970 lncRNAs that overlap the single nucleotide polymorphism (SNPs) associated with traits and disease are specifically expressed. 3166 lncRNAs belonging to 5264 lncRNA-mRNA pairs have been shown to overlap the SNPs of expression quantitative trait loci (eQTL) and have co-expressed alongside the mRNA, suggesting a functional role in transcriptional regulation.

2.5 RNA Sequencing (RNA-Seq)

RNA-seq is currently the predominant method used not only for lncRNA discovery, but also to provide top-up information on alternative splicing isoforms, SNPs, gene fusion events, and new splice junctions. RNA-seq is based on the conversion of transcripts into a cDNAs library, that will constitute the sequencing library (Fig. 2e). Initially, RNA-seq libraries have been prepared with oligo (dT) primers to enrich for polyadenylated transcripts. However, oligo (dT) priming simultaneously excludes non-polyadenylated or partially degraded transcripts. Thus, various methods using random primers for cDNA synthesis on rRNA-depleted transcripts have been developed and standardized. Currently, a single RNA-seq run can yield up to a billion reads, with an individual read length of about 10–300 bp to longer read length of 10–20 kb, providing a quantitative as well as the descriptive scenario of the complex cellular content. Using programs such as Cufflinks (Necsulea et al. 2014) or Scripture (Guttman et al. 2010), entire transcriptomes of cells extracted from various species can be reconstructed using only RNA-seq reads and the genome sequence.

RNA-seq has been applied for the analysis of mouse embryonic stem cells, neuronal precursor cells, and lung fibroblasts revealing the presence of more than 1000 mouse lincRNAs, the majority of which have not been previously identified (Guttman et al. 2010). These lncRNAs are polyadenylated, multiexonic, and have an average length of 859 nts and very low coding potential. The authors have also found that the ratio of expressed protein-coding to non-coding genes in these cell types was ~10:1 but that the total number of transcript molecules is more biased toward the protein-coding fraction (~30:1). In another study, the application of RNA-seq has yielded more than 8000 putative lincRNAs, of which 57% form a "stringent" set for which at least one lincRNA isoform has been reconstructed in at least two different tissues or by two assemblers in the same tissue (Cabili et al. 2011). The authors have characterized each lincRNA by structural, sequence, and expression features as an initial step toward fine categorization.

Bhatia et al. (2019) harnessed RNA-seq for different tissues of grapevine at various developmental stages and identified over 50,000 putative lncRNAs. Additionally, the authors investigated relationships among the elements of the non-coding genome, like lncRNAs and microRNAs, and the protein-coding mRNAs.

2.6 *Direct RNA Sequencing (Oxford Nanopore)*

Unlike traditional RNA-seq, long-read nanopore RNA sequencing allows accurate quantification and full-length sequencing of native RNA or cDNA without fragmentation or amplification. During this process, RNA is translocated at a controlled rate through a membrane-bound protein complex, i.e., nanopore under ionic current (Fig. 2f). The changes in an electric current are detected and used to identify each nucleotide and its potential modifications. This technology has reached the mainstream market with MinION technology, which is capable of returning up to five million reads per flow cell.

Recent report describes direct sequencing of RNA from a variety of samples, with read lengths of up to 7.5 kb and sequencing accuracy of 80% (Garalde et al. 2018). Besides, Oxford Nanopore's nanopre sequencing has been successfully used to directly map RNA epitranscriptomic modifications alongside nucleotide sequence. Liu et al. have provided proof of principle for the use of base-calling "errors" as an accurate and computationally simple solution to identify m^6A modifications in human transcriptome with accuracy approaching 90%. Their m^6A-modified and unmodified datasets can be employed to train different machine learning algorithms, and thus obtain improved methods to detect RNA modifications (Liu et al. 2019). Parker et al. have used long-read nanopore direct RNA sequencing to map m^6A in *A. thaliana* epitranscriptome. Besides, the authors have identified previously unannotated antisense lncRNAs (Parker et al. 2019). Also, Hardwick et al. (2019), used targeted RNA capture in combination with nanopore long-read cDNA sequencing to construct an atlas of lncRNAs in the human brain that can be used to connect neurological phenotypes with gene expression.

3 Biogenesis of LncRNAs

LncRNA biogenesis frequently follows the pathways typical for protein-coding transcripts, and it requires canonical factors assisting the RNA polymerase machinery, such as the pre-initiation complex, transcription elongation complex, and specific transcription factors. LncRNA promoters, as mentioned before, are as evolutionarily conserved as mRNA promoters in humans and mice. However, they have specific characteristics that distinguish them from protein-coding gene promoters, i.e., they contain A/T mono-, di-, and trinucleotide stretches and reduced CG skews (Alam et al. 2014). Also, frequently they overlap with regions located between the initiation and elongation histone marks, chromatin states associated with insulators, elongation, weak transcription, and heterochromatin. When it comes to histone modification signatures, lncRNA gene promoters are significantly depleted of almost all of them, except for H3K27me3, H3K9me3, and H3K27me3 (Alam et al. 2014). H3K27me3 is specific for transcriptionally poised regions combining activating and repressing histone marks (Bernstein et al. 2006; Voigt et al. 2013), suggesting that the activation of lncRNA promoters could be regulated under specific conditions. H3K9me3 marks transcriptional repression (Khare et al. 2012), but is also found in certain transcribed regions (Vakoc et al. 2005). H3K36me3 is a mark of transcriptional elongation (Lachner 2003; Vezzoli et al. 2010). These histone signatures, as discussed previously, have been successfully employed to identify novel genes encoding lncRNA (Guttman et al. 2010) and they indicate active, yet distinct chromatin organization of lncRNA promoters.

3.1 Processing of LncRNAs

In mammals, the majority of lncRNAs are produced by highly processive RNA-polymerase II (POL II), while RNA POL I and III are commonly limited to the transcription of housekeeping RNA transcripts (Quinn and Chang 2016). In plants, two additional specialized RNA polymerases, POL IV and POL V, transcribe some lncRNAs (Ariel et al. 2015). The lncRNAs transcribed by POL V are involved in the process of RNA-directed DNA methylation (RdDM), which is a plant-specific de novo DNA methylation mechanism. Also, specific siRNAs, which are combined into ARGONAUTE (AGO), have been shown to pair with lncRNAs transcribed by POL V, which facilitates the recruitment of AGO to define the particular genomic loci (Holoch and Moazed 2015; Böhmdorfer et al. 2016).

Subsequently, the pre-mature lncRNA becomes 3′-polyadenylated and capped at the 5′-end with methyl-guanosine (Losko et al. 2016). The alternatives to these two processes further expand the diversity observed among lncRNAs. For example, lncRNAs that are processed from longer precursors, i.e., intronic lncRNAs, do not include 5′ cap (Yin et al. 2012). Also, POL III-transcribed, ncRNA transcripts, such as *BC200* (Mus et al. 2007) and *asOct4-pg5* (Hawkins and Morris 2010), are not polyadenylated. Interestingly, MALAT1 and NEAT1 both feature a tRNA-like structure at their 3′ end, which is subjected to RNase P cleavage, resulting in a stable RNA triple-helix at the 3′ end of both lncRNAs, compensating the lack of poly (A) tails. Furthermore, many lncRNAs exhibit alternative polyadenylation sites upstream of the most 3′ exon (Derti et al. 2012; Hoque et al. 2013). Since it has been estimated that 15–45% of the conserved elements in lncRNAs are located behind the first polyadenylation site, this process further expands lncRNAs versatility.

Once synthesized, 98% of lncRNAs undergo splicing (Derrien et al. 2012), with more than 25% of them undergoing alternative splicing by having at least two transcripts per gene locus. For example, splicing of GNG12-AS1 lncRNA results in 38 different isoforms with up to 10 exons (Niemczyk et al. 2013). Also, the alternative splicing of HOTAIR, whose unspliced form scaffolds PRC2 and the histone demethylase LSD1 (Askarian-Amiri et al. 2011), leads to the loss of the PRC2-binding domain, potentially changing the functionality of this lncRNA (Rinn et al. 2007; Gupta et al. 2010).

Some lncRNAs may also originate from atypical processing of RNA transcripts. For example, circRNAs, type of lncRNAs thought to antagonize the actions of microRNAs, originate from back-spliced exons. Additionally, circular intronic lncRNAs (ciRNAs) regulating the POL II transcription machinery originate from lariat introns (looped structure generated when the cut end attaches to the conserved branch point downstream, through pairing of guanine and adenine from the 5′ end and the branch point) that escape from debranching (Liang and Wilusz 2014). Also, sno lncRNAs are processed on both ends by the snoRNA machinery, but retain the sequences linking the snoRNAs, leading to the production of lncRNAs flanked by snoRNA sequences on either side, but lacking 5′-caps and 3′-poly(A) tails (Yin et al. 2012). Thus, lncRNA biogenesis occurs through multiple distinct mechanisms, which may direct specific functional outcomes. Intriguingly, for some low abundant lncRNAs, the act of transcription seems to be more important than the transcript itself. Engreitz et al. (2015) have found that loci of some lncRNAs influence the expression of a neighboring gene *in cis*, yet none of these effects required the specific lncRNA transcripts themselves, instead involved general processes associated with their production.

Furthermore, post-transcriptional modifications of lncRNAs further expand their structural and functional versatility (Fig. 3) (Dinescu et al. 2019). For example, N6-methyladenosine (m^6A), N1-methyladenosine (m^1A) and pseudouridine (ψ) signatures have been identified in MALAT1, with m^6A being reported as altering lncRNA secondary structure and regulating the binding of heterogeneous nuclear ribonucleoprotein C (HNRNPC) (Liu et al. 2015; Zhou et al. 2016). N5-methylcytosine (m^5C), ψ and m^6A modification are present on XIST, with m^5C affecting XIST binding to the PRC2 complex (Patil et al. 2016). Also, telomerase RNA component (TERC), antisense non-coding RNA in the INK4 locus (ANRIL), GAS5, and NEAT1 have all been shown to carry various epitranscriptomic signatures (Squires et al. 2012; Schwartz et al. 2014). Nonetheless, the molecular mechanisms triggered by epitranscriptomic modifications in lncRNAs

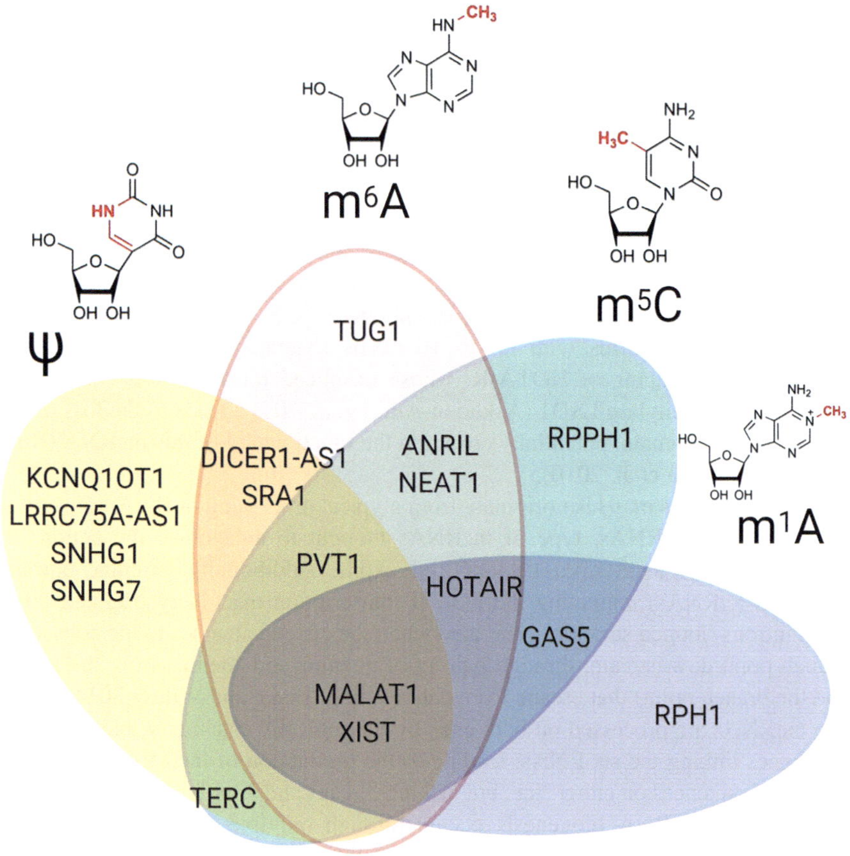

Fig. 3 Venn diagram representing the overlap between different lncRNAs carrying ψ, m^6A, m^5C, and m^1A epitranscriptomic modifications. (Created with biorender.com)

that lead to changes in their function are not well understood and need further characterization to complete another "piece of the puzzle" in the interplay between lncRNAs biogenesis, and regulation of biological processes.

3.2 *Regulation of LncRNA Transcription*

LncRNAs demonstrate a high level of specificity, that is, they are expressed in a cell type, tissue, developmental stage, or disease state-specific. Cabili et al. (2011) have estimated that while 65% of protein-coding genes are ubiquitously expressed, only 11% of lncRNAs are present in all body tissues. Additional data indicate that as many as 78% of lncRNAs are tissue-specific compared to only 19% of coding genes being tissue-specific.

The quantitative differences in the transcriptional regulation of lncRNAs and mRNAs are suggestive of lncRNAs being specifically regulated as a class. It has been shown that the knockout of *DICER1*, a gene responsible for generating miRNAs in mouse embryonic stem cells, results in the lower expression of hundreds of lncRNAs. Also, the transcriptional elongation and initiation of lncRNAs seem to be sensitive to MYC dosage (Zheng et al. 2014). These results indicate that the miRNA circuitry and MYC are important for activating and sustaining lncRNA expression in a manner that is decoupled from mRNA regulation. Also, a genetic RNA interference screen in yeast has identified four distinct chromatin remodeling complexes, i.e., SWR1, ISW2, RSC, and INO80, as global repressors of ncRNA transcription, including lncRNAs that overlap ORFs (Alcid and Tsukiyama 2014). Disruption of these complexes has led to the derepression of antisense lncRNAs and decreased the expression levels of their overlapping mRNAs.

3.3 Degradation of LncRNAs

The turnover rate of lncRNAs has a major role in shaping their subcellular repertoire. In eukaryotes, the cytoplasmic decapping enzyme DCP2 has been implicated in removing the 5′ cap from lncRNA, preparing the transcript for degradation (Geisler et al. 2012). After the decapping, lncRNAs have been shown to undergo cleavage by RAT1, a prominent mRNA 5′ to 3′ exonuclease located within the nucleus. Also, RNA binding proteins (RBPs) have been shown to regulate the lncRNA degradation process by recognizing specific sequences and either increasing or decreasing the efficiency of degradation (Yoon et al. 2015). Specifically, HUR and AUF1 have been shown to enhance lncRNA degradation, while PABPN1 and IGF2BP1 promoted their stability. Yet, other studies suggested that the deficiency of PABPN1 results in the accumulation of a significant number of lncRNAs, implying that PABPN1 rather than promoting lncRNAs stability enhances their turnover via a polyadenylation-dependent mechanism (Beaulieu et al. 2012).

An increasing number of studies have also demonstrated that miRNA interactions with lncRNAs trigger their decay or repress their function. For example, in situ hybridization and confocal microscopy have revealed that miR-9 targets MALAT1 through its binding site in an AGO2-dependent manner in the nucleus (Leucci et al. 2013). Similarly, miR-192 and miR-204 target HOTTIP, HOXA Distal Transcript Antisense RNA, and inhibit proliferation of hepatocellular carcinoma (Ge et al. 2015). Also, *H19* levels have been shown to decline after microRNA let-7a overexpression in an AGO2-dependent manner, although it remains to be determined if microRNAs directly affect *H19* stability, as a reporter construct containing *H19* sequences is selectively inhibited after let-7b overexpression (Kallen et al. 2013).

In addition to the abovementioned mechanisms, it has been shown that about 17% of lncRNAs are targets of the nonsense-mediated decay (NMD) (Wery et al. 2016). For example, GAS5 is sensitive to the NMD protein UPF1 (Tani et al. 2013). Also, mutating enzymes in the NMD pathway in *A. thaliana* results in the global upregulation of both protein-coding genes and mRNA-like lncRNAs, but with more significant effects on the non-coding fraction (Kurihara et al. 2009). NMD has also been associated with an RNA quality surveillance mechanism for the regulation of non-coding pseudogenes in *Caenorhabditis elegans*.

3.4 LncRNA Subcellular Localization

LncRNAs have complex spatially regulated distribution within the cells, which serves as a fundamental regulatory mechanism of their mode of actions (Fig. 4). Using RNA fluorescence in situ hybridization (FISH) for direct visualization of

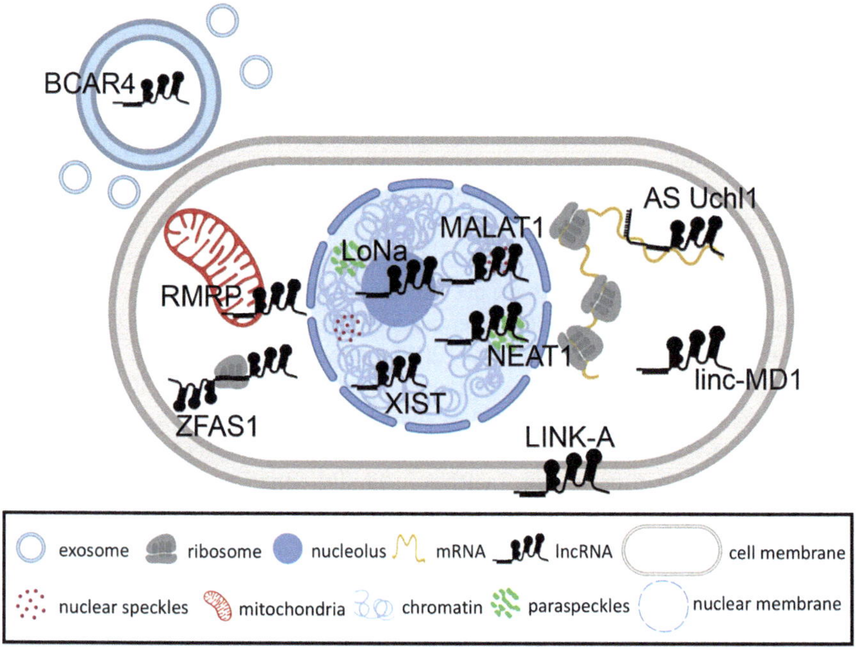

Fig. 4 The overview of lncRNA subcellular location on exemplary lncRNAs, i.e., breast cancer anti-estrogen resistance 4 (BCAR4) localizes to exosomes; RNA component of mitochondrial RNA processing endoribonuclease (RMRP) localizes in mitochondria; ZNFX1 (Zinc Finger NFX1-Type Containing 1) antisense RNA 1 (ZFAS1) co-localized with ribosomes; long nucleolus-specific lncRNA (LoNa) localizes in nucleolus; X-inactive specific transcript (XIST) localizes in nucleus; Metastasis-Associated Lung Adenocarcinoma Transcript 1(MALAT1) localizes in nuclear speckles; Nuclear-Enriched Abundant Transcript 1 (NEAT1) localizes in paraspeckles; long intergenic non-coding RNA for kinase activation (LINK-A) localizes to cell membrane; AS Uchl1, antisense to Uchl1 (Ubiquitin C-Terminal Hydrolase L1) associated with translating mRNA; Long Intergenic Non-Protein Coding RNA, Muscle Differentiation 1 (linc-MD1) localizes in cytoplasm. (Created with biorender.com)

RNA, Cabili et al. (2015) revealed that lncRNAs exhibited a wide range of localization patterns, including not only distinct patterns of nuclear localization but also a nonspecific location in both the nucleus and cytoplasm.

The authors tracked subcellular localization of 34 lncRNAs from 3 different cell lines and divided them into five main categories forming: (1) large nuclear foci, (2) large nuclear foci with single molecules scattered through the nucleus, (3) predominantly nuclear without foci, (4) cytoplasmic and nuclear, and (5) predominantly cytoplasmic. Furthermore, they have observed a strong bias toward nuclear localization for most lncRNAs, stating that 95% of them had higher nuclear fraction than mRNAs (Cabili et al. 2015). Other studies suggest that if the absolute number of RNA molecules in the nucleus and cytoplasm is considered, the median of lncRNAs is cytoplasmic (Carlevaro-Fita and Johnson 2019).

Nuclear lncRNAs can be further detected in specific subnuclear structures, including nucleoli (non-bounded nuclear organelles involved in ribosomal biosynthesis) (Bernard et al. 2010), chromatin speckles (nuclear substructures containing proteins associated with transcription and mRNA maturation), and paraspeckles (heterochromatin domains involved in mRNA nuclear retention) (Nishimoto et al. 2013). On the other hand, cytoplasmic lncRNAs can be localized in mitochondria (Rackham et al. 2011; Dong et al. 2017), ribosom (Ingolia et al. 2011; Zeng et al. 2018), and exosomes (Dragomir et al. 2018).

An individual lncRNA can also be expressed in both the nucleus and cytoplasm and have a different function in each compartment. In general, nuclear lncRNAs detected in the chromatin fraction are more likely to regulate transcription (Mondal et al. 2015) or modulate nuclear architecture (Hacisuleyman et al. 2014), while cytoplasmic lncRNAs more frequently influence translation, mRNA stability, or regulate miRNA functions (Shukla et al. 2011).

Since lncRNAs share structural features with mRNAs, the mechanisms governing their cellular distribution rely on similar export pathways. In the nucleus, proteins from the transcription-export complex (TREX) and the nuclear RNA export factor 1 (NXF1) facilitate the export of ribonucleoprotein complexes (RNPs) to the cytoplasm through the nuclear pore complex (NPCs) by explicitly binding the CAP binding complex, exon-junction complex, and other processing factors (Hautbergue 2017). Viphakone et al. (2019) have identified lncRNAs interacting with both NFX1 and TREX complexes. Interestingly, nuclear lncRNAs showed, on average, less NXF1 binding compared to cytoplasmic ones, suggesting that the lack of binding to the nuclear export adaptor may lead to their nuclear retention. Also, the lncRNA RMRP, RNA component of mitochondria RNA-processing endoribonuclease uses the same protein-export receptor (CRM1) as some mRNAs (Williams et al. 2018; Noh et al. 2018). RMRP binds the HuR protein, which shuttles from nucleus to cytoplasm in a CRM1-dependent manner, mobilizing RNA molecules in the process (Noh et al. 2016; Williams et al. 2018).

Some lncRNAs display distinct localization patterns because of RNA:protein, RNA:DNA, and RNA:RNA interactions that may actively anchor lncRNAs to specific regions, impeding the binding of export factors, or differentially targeting lncRNA stability. For example, the matrix protein hnRNP U is required for the accumulation of XIST on the inactive X chromosome and the *in cis* accumulation of functional intergenic repeating RNA element (FIRRE) (Hacisuleyman et al. 2014), which is essential for their functions. Embryonic stem cells lacking hnRNP U expression fail to form the inactive X chromosome (Hasegawa et al. 2010) and knockdown of hnRNP U even leads to a translocation of FIRRE into the cytoplasm (Hacisuleyman et al. 2014), indicating that hnRNP U is an important factor for both focal localization and nuclear retention of FIRRE. Bacterial lncRNA ornate, large, extremophilic RNAs (OLE) diffuses throughout the whole cell when expressed alone; however, the expression of OLE-associated transmembrane protein (OAPA) anchors the lncRNA to the cell membrane (Block et al. 2011). Also, the binding of cytoplasmic poly(A)-binding protein C1 (PABPC1) and KSHV-encoded ORF57 to viral PAN lncRNA has been shown to relocate PABPC1 to the nucleus, and facilitate shutoff exonuclease (SOX)-mediated accumulation of PAN in the nucleus (Lee and Glaunsinger 2009; Borah et al. 2011; Massimelli et al. 2013).

If lessons learned from mRNA are also binding for lncRNAs, then motifs recognized by RNA binding proteins (RBPs) should be a crucial localization-regulatory mechanism (Martin and Ephrussi 2009). Zhang et al. identified an AGCCC motif responsible for nuclear localization of BORG lncRNA, BMP/OP-Responsive Gene (BORG), whose expression directly correlates with aggressive breast cancer (Zhang et al. 2014). Also, Miyagawa et al. (2012) described several large regions of MALAT1 that promote nuclear enrichment, with region E and M conferring its localization to speckles. Also, a local repeating RNA domain (RDD), present in eight exonic copies in human FIRRE, localizes the lncRNA to the nucleus (Hacisuleyman et al. 2014). Another study implicated an inverted pair of *Alu* elements in nuclear retention of *lincRNA-P21* (Chillón and Pyle 2016).

The subcellular localization for many lncRNAs is a highly dynamic and regulated process that responds to various stimuli. For example, NORAD and SNHG1 lncRNAs display both nuclear and cytoplasmic distributions in human HCT116 colon cancer cells. However, upon DNA damage stress, both lncRNAs are retained in the nucleus (Shen and Corey 2018; Munschauer et al. 2018). The mouse antisense lncRNA, UCHL1-AS1 under rapamycin treatment shuttles from the nucleus to the cytoplasm, where it interacts with UCHL1 mRNA and recruits ribosomes, promoting translation (Carrieri et al. 2012). Also, the epitranscriptomic modifications have been shown to regulate RNA nuclear-cytoplasmic shuttling. For example, m^6A has been reported to promote the nuclear export of mRNAs through the action of the YTHDC1 protein, which directly binds to the modified base and helps to recruit nuclear export factors to the mRNA (Roundtree et al. 2017). Also, the Aly/REF export factor (ALYREF) and a reader for m^5C promotes selective mRNA export from the nucleus (Yang et al. 2017). Whether similar mechanisms apply to lncRNAs remains to be defined.

Various bioinformatics tools have been developed to predict lncRNA localization based on the frequency of short motifs, achieving relatively high prediction accuracy (~87%), and underlining the fact that *cis*-encoded signals contribute to lncRNA compartmentalization (Su et al. 2018). Also, based on 93 strand-specific RNA-seq samples of nuclear and cytosolic fractions from multiple cell types a deep learning algorithm has been used to identify differentially localized lncRNAs (Gudenas and Wang 2018). This approach has the accuracy of 72.4%, the sensitivity of 83%, the specificity of 62.4%, and the authors have concluded that primary sequence motifs are a major driving force in the subcellular localization of lncRNAs. These findings chime with the recent interest in the relationship between transposable elements (TEs) and lncRNAs localization (Carlevaro-Fita et al. 2019). TEs are DNA elements capable of copying and inserting themselves into genic and intergenic regions (Fedoroff 2012). The majority of lncRNAs (~83%) have been shown to carry at least one exonic TE element (Kelley and Rinn 2012) that are evolutionarily repurposed as functional domains that interact with DNA, RNA, or proteins (Kapusta et al. 2013; Johnson and Guigó 2014). In a recently published study to map functional TEs, the correlation has been used to discover a link between certain TE types and the nuclear localization of their host transcript (Carlevaro-Fita et al. 2019). These studies support a "zipcode model" similar to mRNAs, where both short and long elements, contribute to specifying lncRNA localization through *trans*-acting proteins.

3.5 LncRNA Structure and Structure-Mediated Interactions

Recent estimates of evolutionary conservation assessed that ~14% of the human genome shows evidence of purifying selection on RNA structure (Smith et al. 2013), suggesting that much of the non-coding human genome encodes function at the RNA structure level. One such example is represented by HOTAIR, which exists only in mammals and shares 58% of homology between human and mouse (Bhan and Mandal 2015). Covariance analysis across 33 mammalian sequences of HOTAIR revealed a significant number of covariant base pairs, which maintained a consensus structure regardless of the varying sequence, especially in the context of protein-binding regions in lncRNAs (Somarowthu et al. 2015). On the other hand, low sequence conservation that induces changes in structure can drive gain of function and specialization of lncRNA, as it has been shown for human accelerated region 1 (HAR1)-derived lncRNAs, which show unusually high sequence diversity between human and chimpanzee (Beniaminov et al. 2008).

Overall, the determination of the secondary structure of lncRNAs is challenging and often hindered by the fact that the number of probable structures expands exponentially with the increasing length of the target transcript. It has been estimated that for a 2.2 kb long lncRNA, there are over ten thousand possible secondary structures (Novikova et al. 2013). Also, lncRNAs tend to be structurally heterogenic, frequently comprising of distinct base-paired regions and segments structurally less constrained and dynamic (Busan and Weeks 2017). However, recent developments in biochemical RNA probing in combination with NGS technologies provide insight into the structure-to-function relationship of many lncRNAs. Techniques, such as dimethyl sulfate sequencing (DMS-Seq) (Rouskin et al. 2014; Ding et al. 2014), selective 2′-hydroxyl acylation analyzed by primer extension sequencing

(SHAPE-Seq) (Watters et al. 2016), SHAPE and mutational profiling (SHAPE-MaP) (Siegfried et al. 2014; Smola et al. 2016), fragmentation sequencing (FRAG-Seq) (Underwood et al. 2010), and parallel analysis of RNA structure (PARS) (Kertesz et al. 2010) have been successfully applied to analyze lncRNAs structure in vitro (Somarowthu et al. 2015), in living cells (Smola et al. 2016; Sztuba-Solinska et al. 2017), and within virions (Sztuba-Solinska et al. 2017). These studies have revealed that various lncRNAs are organized into modular functional domains that are capable of coordinating RNA:RNA, RNA:protein and RNA:DNA interactions, similar to how proteins are organized into functional subunits but instead encoded at the RNA level. Evidence of this domain-level strategy that scaffolds intermolecular contacts is widespread (Novikova et al. 2012; Somarowthu et al. 2015; Sztuba-Solinska et al. 2017) and further elucidated by various methodologies addressing lncRNA-mediated interaction with effector molecules. For example, RNA antisense purification (RAP) utilizes different crosslinking reagents to address direct and indirect RNA:RNA interactions (RAP RNA) and direct RNA:protein contacts (RAP-MS) (Engreitz et al. 2014). The application of RAP RNA has revealed that U1 snRNA directly hybridizes to 5′ splice sites of its nascent RNAs and MALAT 1 has been found to interact with pre-mRNAs indirectly through protein intermediates (Engreitz et al. 2014). Capture hybridization analysis of RNA targets (CHART) utilizes a similar concept, and it involves the purification of the crosslinked protein, RNA and DNA complexes (Simon et al. 2011). RNA: protein association can also be addressed by RNA immunoprecipitation (RIP), which has been used to identify proteins that interact with XIST (Augui et al. 2011). Also, chromatin isolation by RNA purification (ChIRP) provides insight into associations between lncRNA and chromatin. Chu et al. isolated HOTAIR-associated DNA regions, demonstrating that HOTAIR commonly nucleates at GA-rich DNA regions. Furthermore, ChIRP has also been optimized to elute proteins, which are further analyzed with mass spectrometry (Chu et al. 2015). This approach identified 81 new proteins bound to the XIST lncRNA.

In addition to the secondary and tertiary structure of lncRNAs, the order of events and kinetics of the lncRNA-mediated systems is essential for their mechanistic understanding. Rapid kinetics studies define the overall order of events, while single-molecule studies help elucidate the mechanism of transitions between states (Blanchard 2009). A combination of structural and kinetic information would be necessary to unlock the exact mechanisms governing lncRNA modes of action. For example, MALAT1 has been shown to associate with the PRC2 components (Guil et al. 2012; Biswas et al. 2018) to recruit the epigenetic factors to the chromatin, drive histone methylation, and regulate gene transcription (Wang et al. 2016; Amodio et al. 2018). MALAT1 structure and function are also regulated by m^6A epitranscriptomic modification (Zhou et al. 2016). Thus, the possible order of events involving MALAT1 may be: (1) co-transcriptional lncRNA folding, (2) epitranscriptomic protein binding to the lncRNA, (3) epigenetic protein binding to the chromatin, and (4) action of the epigenetic protein, i.e., histone methylation. Each of these steps would have its time scale and rate-limiting kinetics. Unraveling these details would generate substantial insight into the mechanism of lncRNA action.

4 Human LncRNAs

In this section, we provide a broad overview of prominent examples of human lncRNAs, discussing their structural features, physiological mechanisms, as well as molecular functions. This list is not, by any means complete, and for further details regarding human lncRNAs, we would recommend LncBook website: http://bigd.big.ac.cn/lncbook, that features a comprehensive collection of human lncRNAs and their systemic curation by multi-omics data integration, functional annotation, and disease association (Sun and Ma 2019).

4.1 X-inactive Specific Transcript (XIST)

XIST is a 17–19 kb long, spliced, polyadenylated nuclear transcript that plays a key role in the silencing of one of the X chromosomes in females to compensate the X-linked gene dosage between different sexes. This process is referred to as X-chromosome inactivation (XCI) and is specific for placental mammals. During the XCI, XIST transcript wraps itself around the X chromosome that undergoes inactivation (X_i) and abolishes the expression of most genes by recruiting proteins that govern DNA methylation and chromatin modifications (Sahakyan et al. 2018). The Stochastic Optical Reconstruction Microscopy (STORM) performed in mouse cells has estimated that approximately 50 hubs of XIST RNA occur on the X_i in the maintenance phase, corresponding to 50–100 transcripts per X_i (Sunwoo et al. 2015). XIST lncRNA does not share a lot of sequence similarity between mouse and human; however, it has been shown to maintain a well-conserved structural organization that includes six tandem repeats, named A to F (Brockdorff 2018; Lee et al. 2019). A and F repeats, each hundreds of nts long, form two small domains; B, C, and D repeats build one large 10 kb domain, while E repeat and its surrounding sequence that take up to 1 kb, compose a medium-size domain (Fang et al. 2015). The B repeat is highly repetitive and GC-rich, and together with a part of C repeat, they have been shown to recruit PRC1 and PRC2 by heterogeneous nuclear ribonucleoprotein K (hnRNPK) to exert XIST mediated gene silencing (Pintacuda et al. 2017). The E repeat is believed to anchor XIST RNA to the X_i through CDKN1A-interacting protein CIZ1 (Sunwoo et al. 2017). Smola et al. noted that the RNA-binding proteins TARDBP, CELF1, PTBP1, previously implicated in numerous functional nuclear pathways, can bind to the E-repeat, although these factors are redundant for (McHugh et al. 2015).

Recently, XIST secondary structure has been resolved by application of different RNA structure probing techniques, i.e., DMS-seq (Fang et al. 2015), SHAPE-MaP

(Smola et al. 2016), and SHAPE in combination with DMS probing (Lu et al. 2016) providing the insight into its structure-to-function relationship. It has been shown that the 5′ end of XIST repeat A contains 7.5 copies (in humans 8.5 copies) of 26-mer separated by U-rich linkers required for XIST silencing activity. The A repeat domain forms stem-loop structures in an inter way (base pairs within repeat) not in the intra manner (base pairs with other repeats), each including the AUCG tetraloop. This basic unit of the inter-repeat binding has also been predicted in the XIST secondary structure models presented by Fang et al. and Lu et al. (Fang et al. 2015; Lu et al. 2017). Interestingly, Lu et al. (2017) did not find a single solution to the A repeat region and suggested that XIST likely has a dynamic conformation. Furthermore, the application of reversible psoralen crosslinking for global mapping of RNA duplexes with near base-pair resolution in living cells (PARIS) has shown that the inter-repeat stem-loops flanked by U-rich motifs create binding sites for the silence factor SPEN (Msx2-interacting protein, also called MINT), which recruits histone deacetylase HDAC3.

XIST is also extensively post-transcriptionally modified. It includes 78 putative m^6A residues with some located at or near A repeat, 5 m^5C residues within human XIST A-repeat (Amort et al. 2013) and pseudouridine at position U11249 (Li et al. 2015b). It has been shown that RBM15 and RBM15B are essential silence factors that participate in XIST m^6A methylation by recruiting methyltransferase like 3 (METTL3) (McHugh et al. 2015; Moindrot et al. 2015; Monfort et al. 2015; Patil et al. 2016). Both proteins may bind to the same region as SPEN and compete against it (Lu et al. 2016). The XIST m^6A signature is bound by YTH domain containing 1 (YTHDC1) and that interaction promotes XIST-mediated gene repression. Besides, the knockdown of RBM15, RBM15B, or methyltransferase METTL3 has been shown to inhibit XCI (Patil et al. 2016). The m^5C sites have been shown to affect XIST binding to the PRC2 complex, indicating that post-transcriptional modifications can modulate XIST:protein interactions (Amort et al. 2013). The functionality of the other epitranscriptomic modifications is currently unknown.

4.2 *MALAT1 (Metastasis-Associated Lung Adenocarcinoma Transcript 1)*

MALAT1, also referred to as the nuclear-enriched abundant transcript 2 (NEAT2), is approximately 8 kb long in human, and 6.7 kb long in mouse. It is one of the most abundantly expressed and extensively studied lncRNAs (Sun and Ma 2019) that displays exceptional sequence conservation in mammals (Hutchinson et al. 2007). It has been first identified through a screen for transcripts associated with metastasis and patient survival in non-small cell lung cancer (NSCLC) (Ji et al. 2003). The observed overexpression of MALAT1 in human breast, pancreases, lung, colon, and prostate carcinomas has suggested its important role in tumorigenesis (Lin et al.

2007). Besides, its strong downregulation has been reported in different types of cancer.

The primary transcript of MALAT1 is generated in a tRNA-like mechanism, during which the 3′ cloverleaf-like structure (four-way junction) followed by the poly(A) tail is recognized and cleaved by RNase P. The transcript is further processed by RNase Z, and as a result, MALAT1-associated small cytoplasmic RNA (mascRNA) is produced (Wilusz et al. 2008).

Immunofluorescence assay indicated that MALAT1 primarily localizes to nuclear speckles (NSs, or splicing speckles), sites for splicing factor storage and mRNA maturation (Hutchinson et al. 2007). However, in vivo experiment with MALAT1 knockout mice showed it may not be an essential component of NS (Nakagawa et al. 2012). Furthermore, RNA coimmunoprecipitation (RNA IP) analysis confirmed the interaction between MALAT1 and several of splicing factors (Tripathi et al. 2010). The same report has shown that the depletion of MALAT1 leads to the changes in their phosphorylation status, followed by mislocalization to the NS. MALAT1 has also been reported to regulate gene expression on the transcriptional level. The capture hybridization analysis of RNA targets (CHART) has revealed MALAT1 localizes to the actively transcribed gene loci (West et al. 2014). Subsequent research has indicated the indirect interaction between MALAT1 and pre-mRNA via protein intermediates, which localizes MALAT1 to chromatin at active genes (Engreitz et al. 2014). MALAT has also been reported to work as a scaffold to recruit transcription factor SP1 to the promoter of latent transforming growth factor beta (TGF-β) binding protein 3 (LTBP3), which is an extracellular matrix glycoprotein responsible for efficient secretion, folding and activation of TGF-β (Li et al. 2014).

The X-ray crystal structure showed the 3′ end of MALAT1 and its internal poly (A) tract forms a unique triple helical structure, similar to the triple helix formed in KSHV PAN lncRNA (*discussed below*). Two segments build MALAT1 triple helix: the first one including five consecutive U•A-U triples, which are interrupted by one C^+•G-C triples (C^+ denotes protonated cytosine) and the sequester C-G doublet; and the second one containing four consecutive U•A-U triples. The whole structure is flanked by two stems and one GC dinucleotide bulge with the 3′ terminus buried inside of U•A-U, which produces a 3′-blunt end. The tripe-helix has been shown to promote MALAT1 subcellular stability, contributing to its unusually long half-life (9–12 h) (Brown et al. 2014). Specific triplex-disrupting mutations have been shown to cause MALAT1 degradation (Brown et al. 2012; Wilusz et al. 2012) and reduced levels of the lncRNA significantly inhibit tumor formation, migration, and metastasis (Gutschner et al. 2013; Amodio et al. 2018). Therefore, the triple helix motif is an important target for therapeutic discovery. More recently, the SHAPE and DMS probing have been used to analyze the secondary and tertiary structure of MALAT1 3′ end region, further supporting prior findings (Zhang et al. 2017).

The adenine residue at position 2577 on MALAT1 is reversibly methylated at N6 position, and the modification has been shown to affect RNA structure and structure-mediated protein binding through so-called "m^6A-switch" mechanism (Zhou et al. 2016; Liu and Pan 2016). The nuclear magnetic resonances (NMR) and Förster resonance energy transfer efficiencies (FRET) have shown the m^6A weakens the

Watson-Crick base pairing between A2577 and one U from the downstream U_5-tract, and predisposes its conformation to the binding of HNRNPC, an abundant protein responsible for pre-mRNA processing (König et al. 2010). Also, three sites U5160, U5590, and U3374 in MALAT1 have been reported to be pseudouridnilated, while another methylation, m^1A, occurs at position 8398 of MALAT1, yet their function requires further elucidation (Gutschner et al. 2013; Li et al. 2015b).

4.3 HOX Transcript Antisense RNA (HOTAIR)

HOTAIR is 2.2 kb long polyadenylated lncRNA transcribed in an antisense manner from *HOXC* gene, which has been shown to regulate the transcription of homeobox gene D (*HOXD*) *in trans* (Rinn et al. 2007; Somarowthu et al. 2015). Sequence analysis revealed HOTAIR is expressed only in mammals, and it displays high sequence conservation in primates. Besides, the application of EvoNC, a program detecting selection in non-coding regions of nucleotide sequences, revealed HOTAIR evolves faster than *HOXC* genes (He et al. 2011).

The chromatin immunoprecipitation followed by tiling array analysis (ChIP-chip) have shown the 5′ end of HOTAIR works as a scaffold for PRC2, which in turn mediates histone H3 lysine-27 (H3K27) trimethylation and represses transcription of specific genes (Rinn et al. 2007). Furthermore, its 3′ end binds a histone H3K4-demethylase lysine-specific demethylase 1 (LSD1) complex, another crucial gene silencing factor (Tsai et al. 2010). Both PRC2 and LSD1 complexes are responsible for epigenetic modifications of many tumor and metastasis suppressing genes, whose expression is reduced in the HOTAIR-positive cells. This effect indicates the direct involvement of HOTAIR in processes associated with cancer development (Dinescu et al. 2019).

Also, HOTAIR exerts scaffolding function in protein ubiquitination. It associates with E3 ubiquitin ligases DZIP3 (DAZ interacting zinc finger protein 3) and MEX3B (mex-3 RNA binding family member B) bearing RNA-binding domain, to facilitate ubiquitination of Ataxin-1 and Snurportin-1 and promote their final proteolysis (Yoon et al. 2013). It has been noted that cytoplasmic localization of Ataxin-1 and Dzip3, and nuclear and cytoplasmic localization of Shurportin-1 and Mex3b, indicate that HOTAIR promotes ubiquitination and protein degradation in both cellular compartments. HOTAIR has also been reported to regulate the inflammatory and immune response. In macrophages, lipopolysaccharide (LPS) treatment has been shown to induce the expression of HOTAIR, interleukin 6 (IL-6) and iNOS (inducible nitric oxide synthase). On the other hand, the knockdown of HOTAIR reduces the expression of IL-6 and iNOS in a nuclear transcription factor NF-κB-dependent way (Obaid et al. 2018).

The secondary structure of HOTAIR has been resolved by SHAPE, DMS, and terbium chemical probing (Somarowthu et al. 2015) several years after the publication of the very first low-resolution structure (He et al. 2011). HOTAIR is highly structured, with more than 50% nt being paired. Four independent domains,

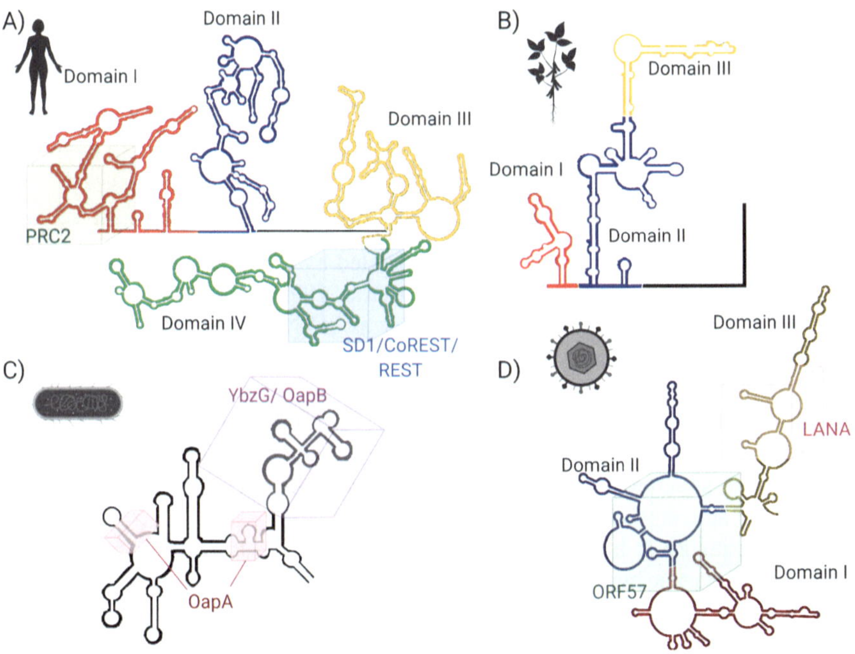

Fig. 5 The outline of secondary structures for representative lncRNAs in (**a**) human – HOTAIR (Somarowthu et al. 2015), (**b**) plants – COOLAI (Hawkes et al. 2016), (**c**) bacteria – OLE (Harris et al. 2018), and (**d**) virus – PAN (Sztuba-Solinska et al. 2017). The interacting protein and functional domains are color-coded. HOTAIR has four domains, with domain I being reported to associate with PRC2, while domain IV binds SD1/CoREST/REST. Plant antisense COOLAIR has three with undetermined functions. Bacteria OLE has two OapA binding sites, while it's 3′ end binds YbzG/OapB. (**d**) Kaposi's sarcoma-associated herpesvirus (KSHV)-encoded PAN folds into three domains that bind viral ORF57 (mRNA transcript accumulation, Mta), LANA (viral latency-associated nuclear antigen), and many more. (Created with biorender.com)

designated to D1–D4, are distinguished, with each being over 500 nt long and consisting of helices, terminal loops, and junctions (Fig. 5a). The deletion experiments have shown the domains D1, and D4 scaffold main protein interactions (Tsai et al. 2010), and the structural elements that surround these sites are evolutionarily conserved. Covariant analysis, including 33 mammalian sequences, has shown that a significant number of base pairs in helices are covariant or half-flips (Somarowthu et al. 2015).

HOTAIR includes m^{5}C epitranscriptomic signature at position nt 1683, in the 3′ end LSD1 binding region and the level of this modification is remarkably high in five tested cell lines, with three of them being cancerogenous with high HOTAIR expression (Amort et al. 2013). Also, m^{6}A-specific methylated RNA immunoprecipitation with next-generation sequencing (MeRIP-Seq) has shown that HOTAIR is m^{6}A modified in HEK293T cells (Meyer et al. 2012). No m^{6}A signal has been detected in HepG2 cells or human brain tissues (Dominissini et al. 2012). The lack of methylation-dependent functional studies hampers the understanding of their significance in the context of full-length HOTAIR lncRNA.

4.4 Steroid Receptor RNA Activator (SRA)

SRA is 0.68 kb long intergenic lncRNA encoded by a well-conserved gene *SRA1* located on human chromosome 5q31 (Lanz et al. 1999; Sheng et al. 2018). *SRA1* encodes multiple lncRNA variants, as well as mRNA translated into a highly conserved among chordates protein, SRAP. SRAP has been shown to associate with multiple transcription factors and act as a transcriptional repressor (Chooniedass-Kothari et al. 2004; Hubé et al. 2011). Similarly to the yeast spliceosome protein PRP18 (Pre-mRNA-splicing factor 18), SRAP includes a five-helix bundle structure in the carboxy-terminal domain, suggesting it may also be involved in splicing (McKay et al. 2014).

The in situ RNA hybridization techniques have shown that SRA lncRNA is mainly cytoplasmic with small fraction localizing in the nucleus in a distinctly speckled pattern (Zhao et al. 2007). Its primary function is to act as a transcriptional coactivator of steroid receptors (SRs), including the receptors for androgen (AR), estrogen (ER), glucocorticoid (GR), and progestins (PR). SRA has been shown to scaffold various proteins containing RNA recognition motifs (RRMs), such as SMRT/HDAC1-associated repressor protein (SHARP), a corepressor of the nuclear receptors (NRs), SRA stem-loop interacting RNA binding protein (SLIRP), a corepressor of NRs or myogenic differentiation, and skeletal muscle-specific basic helix-loop-helix transcription factor (MYOD) (Arieti et al. 2014). A robust human SRA expression in a transgenic-mouse model has been shown to stimulate proliferation as well as apoptosis in vivo. However, overexpression of SRA is not enough to induce tumorigenesis (Lanz et al. 2003), although several reports have shown its expression is upregulated in many human cancers (Murphy et al. 2000; Kim et al. 2018).

The low-resolution SRA structure has been proposed based on comparative sequence analysis, free energy calculations, and site-directed mutagenesis. Six distinct RNA motifs and 11 topological sub-structures have been identified as responsible for the SRA-mediated co-activation of nuclear receptors (Lanz et al. 2002). 10 years later, the full-length SRA secondary structure including four domains (I–IV) comprising of 25 helices (H1–H25) has been revealed through a combined chemical and enzymatic probing, including SHAPE, in-line, DMS and RNase V1 (Novikova et al. 2012). The domains I-III, with the most conserved sequences, have been proposed to be SRA core region; while domain IV showing low sequence conservation, has been regarded as "variable", despite its highly structured conformation. A comparative sequence analysis across vertebrates has shown that evolutionary pressure maintains the structural core in domains III and IV of SRA, rather than its translational product SRAP. This suggests that the primary functions of *SRA1* relate to its lncRNA (Novikova et al. 2012; Sheng et al. 2018).

SRA carries pseudouridine modification at position 206, which is located in the STR5, small hairpin structure in conserved core sequence, and which has been shown to play a pivotal role in a coactivator/corepressor switch for NR signaling (Zhao et al. 2007). Two different pseudouridine synthases (PUSs) PUS1P or PUS3P have been reported as responsible for SRA pseudouridylation at two different positions. However, the only pseudouridylated site confirmed in vivo is U206. The mutation of U206 to A has been shown to lead to hyper-pseudouridylation of SRA (Zhou et al. 2007; Zhao et al. 2018). Interestingly, a synthetic STR5 oligonucleotide introduced into cells can inhibit PUS1P-dependent pseudouridylation of SRA to the same degree as U206A mutation, which may be used as a potential anti-cancer therapeutic strategy (Ghosh et al. 2012).

4.5 *Maternally Expressed 3 (MEG3)*

MEG3 is a 1.6–1.7 kb lncRNA, encoded by the gene in the imprinted DLK1–MEG3 locus (*DLK1*: Delta like homolog 1, a paternally imprinted gene; *MEG3*: Maternally expressed gene 3, homolog in mouse called GTL2 (Wylie et al. 2000)). Although *MEG3* gene includes several small ORFs, the lack of consensus Kozak sequences around AUG region suggests it functions as lncRNA.

There are at least 12 different MEG3 isoforms transcribed from the 10 exons of *MEG3* gene through alternative splicing, all of which can stimulate downstream p53-mediated transactivation and suppress DNA synthesis to a different extent (Zhou et al. 2007; Zhang et al. 2010). It has been widely reported that MEG3 expression is significantly decreased in the primary cancer tissues, including breast cancer (Lu et al. 2016), lung cancer (Wu et al. 2018), prostate cancer (Luo et al. 2015), and hepatocellular cancer (Zhuo et al. 2016). The reduced expression of MEG3 is also observed in various cancer cell lines, and ectopic expression of MEG3 in these cells promotes apoptosis, inhibits tumor growth through p53-dependent or independent way (Al-Rugeebah et al. 2019). The in situ RNA hybridization and nuclear-cytoplasmic RNA fractionation experiments have indicated that the primary location of MEG3 lncRNA is in the nucleus, which is compatible with its function as a gene transcription regulator (Mondal et al. 2015; Sherpa et al. 2018).

The initial prediction of its secondary structure using mfold software has shown MEG3 folds into three major motifs: MI, MII, and MIII, with the subsequent deletion analysis verifying their functionality in p53 activation (Zhang et al. 2010). Further application of ChRIP-seq (chromatin RNA immunoprecipitation (ChRIP)) followed by high-throughput sequencing and modified ChOP (chromatin oligo affinity precipitation) have shown MEG3 associated with PRC2 to regulate a collective set of genes including those of the transforming growth factor-β (TGF-β) pathway and some of the *TGF-β* pathway genes are the direct targets of MEG3 (Mondal et al. 2015).

Recently, the MEG3 lncRNA secondary structure has been resolved by in vitro and ex vivo SHAPE-MaP (Sherpa et al. 2018). MEG3 is intricately branched and highly structured with a flexible center junction connecting 50 double-stranded helices (H1-H50) including 61.1% of nts involved in Watson–Crick or wobble base-pairs and a long-range interaction between its 5′ and 3′ termini. The comparison of the ex vivo and in vitro MEG3 lncRNA models has revealed five conserved structural motifs, designated M-I to M-V, the functional importance of which has been strongly supported by earlier studies (Zhang et al. 2010; Mondal et al. 2015). Previously, it has been indicated that MEG3 M-I is responsible for mediating the formation of RNA-DNA triple helix contacts within *TGF-β* genes (Mondal et al. 2015), while M-II is responsible for the suppression of DNA synthesis (Zhang et al. 2010). Successively, MIII has been identified as a site of PRC2 binding (Mondal et al. 2015; Sherpa et al. 2018), and Motif M-IVb is required for p53 activation (Zhang et al. 2010). SHAPE-MaP footprinting has confirmed that the main PRC2 binding site on MEG3 is within M-III domain, yet their identified site differed somewhat from previous results, which were based on site-specific photocrosslinking (Mondal et al. 2015). A very recent study has reported a pseudoknot between two evolutionary conserved regions of MEG3, which play a key role in MEG3-dependent p53 activation. Disruption of this tertiary interaction leads to the misfolding of MEG3 and impairs the downstream p53 stimulation (Uroda et al. 2019).

5 Plant LncRNAs

Accumulating evidence shows that plant lncRNAs are essential regulators of gene expression in diverse biological processes, such as gene silencing, flowering time control, root organogenesis, photomorphogenesis, abiotic stress responses, and reproduction (Mattick and Rinn 2015; Wang et al. 2017b). Alike human lncRNAs, most of them are transcribed by RNA POL II. Also, two plant-specific RNA polymerases, namely POL IV and POL V, have been identified, and they are required for siRNA-mediated gene silencing of transposons and other repeats. These two multisubunit nuclear enzymes transcribe intergenic and non-coding sequences, thereby facilitating heterochromatin formation and silencing of overlapping and adjacent genes (Wierzbicki et al. 2008).

5.1 Cold of Winter-Induced Non-coding RNA (COLDAIR)

In plants, the developmental transition from vegetative growth to flowering is controlled by the floral repressor flowering locus C (*FLC*). The prolonged exposure to cold temperatures triggers a process called vernalization, during which PRC2 mediates the epigenetic silencing of *FLC* (Song et al. 2012). COLDAIR is a 1098-nt long, 5′ capped and non-polyadenylated intronic lncRNA transcribed from FLC intron 1, in the same direction as *FLC* starting from the vernalization response (VRE) element.

The expression of COLDAIR has been shown to increase in response to cold treatment, and knockdown mutants display delayed flowering without vernalization (Heo and Sung 2011). It has been shown COLDAIR physically associates with PRC2 by a modular 39-nt motif, predicted by in vitro RNA binding assay and bioinformatic analysis as a bifurcated stem-loop. The mutations within the motif that reduce the COLDAIR:PRC2 complex formation result in vernalization insensitivity. The mutations have been presumed to disrupt the double stem-loop of this region as predicted by bioinformatic analysis (Kim et al. 2017). The repression of *FLC* is further mediated by increased enrichment of PRC2 and subsequent trimethylation of H3K27me3 (Csorba et al. 2014).

5.2 *Cold-Induced Long Antisense Intragenic RNA (COOLAIR)*

The RNA-induced heterochromatic region at the end of *FLC* locus has been identified as responsible for the expression of a set of lncRNAs, referred to as cold-induced FLC antisense (COOLAIR). COOLAIR lncRNAs are alternatively spliced, 5′ capped and polyadenylated lncRNAs that belong to two classes I and II, that are respectively ~417-nt, and II ~750-nt long. COOLAIR lncRNAs participate in *FLC* transcriptional repression through increased H3K27me3 and decreased H3K36me3 levels in response to cold temperatures. FCA binding of COOLAIR and SSU72 (an RNA polymerase II C-terminal phosphatase domain) is critical for PRC2 enrichment and H3K27me3 deposition. The class I COOLAIR transcript enhances HK4me2 demethylase activity and reduce *FLC* transcription, resulting in the positive feedback that reinforces proximal adenylation and low expression of *FLC*. On the other hand, class II is associated with high expression of *FLC* (Csorba et al. 2014).

The COOLAIR GC rich promoter region (Jean Finnegan et al. 2005) is regulated by the formation of a three-stranded nucleic acid structure referred to as an R-loop. During this process, the nascent RNA transcript invades the dsDNA generating an RNA-DNA hybrid and impacting gene expression. R-loops have been shown to support the unmethylated DNA state at promoters rich in CpG islands, correlating with transcriptional silencing activity in mammals (Ginno et al. 2012). Also, the stabilization of the R-loop by the homeodomain protein NDX1 has been shown to reduce COOLAIR transcription (Sun et al. 2013), while senataxin helicase is able to resolve R-loops, assisting transcription termination and Pol II release (Yu et al. 2003; Skourti-Stathaki et al. 2011).

Recently, SHAPE biochemical probing has been applied to determine the secondary structure of the COOLAIR transcripts in *A. thaliana* (Fig. 5b) (Hawkes et al. 2016). It has been found the *COOLAIR* transcript is highly structured, with numerous secondary structure motifs, an intricate multi-way junction, and two unusual asymmetric 5′ internal loops (right-hand turn [r-turn] motifs). Part of this structure can be altered by a single non-coding SNP that has been shown to confer functional *cis*-regulatory variation to a naturally occurring *FLC* haplotype. Also, the secondary structure has been used to predict *COOLAIR* exonic sequences in a range of evolutionarily distinct *Brassicaceae* species.

5.3 Cold of Winter-Induced Non-coding RNA from the Promoter (COLDWRAP)

COLDWRAP is a ~ 316 nt, 5′ capped lncRNA that is transcribed in the sense direction with its transcription start located 225 nt upstream from the *FLC* mRNA. Both COLDAIR and COLDWRAP work cooperatively to form a repressive intragenic chromatin loop at the *FLC* locus during vernalization (Kim and Sung 2017).

A putative motif present in the 5′ part of COLDWRAP has been identified to be necessary for the association with PRC2, and the mutation within that region has been shown to impair the PRC2 binding in vitro and in vivo (Kim and Sung 2017). Although clear associations of COLDAIR and COLDWRAP with PRC2 during vernalization have been noted, it remains to be investigated how PRC2 is recruited to *FLC* locus. One probable model involves the co-transcriptional association of lncRNA and PRC2. However, both COLDWRAP and COLDAIR are capable of restoring a vernalization response when they are expressed *in trans*. Thus their co-transcriptional association seems to be not necessary. Another model, consistent with the *trans* effect, suggests the presence of direct DNA-RNA interaction between target loci and lncRNAs.

5.4 Auxin-Regulated Promoter Loop RNA (APOLO)

APOLO, previously known as NPC34 (Ben Amor et al. 2008), participates in chromatin structure remodeling of its neighboring *PINOID* locus. *PINOID* gene encodes a regulatory kinase determining the polar localization of the auxin transporter PIN-FORMED (PIN) 2 in root cells (Ariel et al. 2014). APOLO is transcribed by two enzymes, POL II and V in response to auxin, a phytohormone controlling numerous facets of plant development (Bazin and Bailey-Serres 2015). This dual mechanism of APOLO transcription influences local chromatin topology and directs dynamic auxin-controlled developmental outputs on neighboring genes. It has been shown that altering APOLO expression affects chromatin loop formation, whereas RNA-dependent DNA methylation, active DNA demethylation, and PRC complexes have been shown to control loop dynamics. This mechanism likely underscores the adaptive success of plants in diverse environments and may be widespread in eukaryotes (Ariel et al. 2014).

6 Bacterial LncRNAs

Bacterial genomes encode a plethora of ncRNAs, including house-keeping RNAs, such as 16S and 23S rRNA (Espejo and Plaza 2018), RNAs with catalytic functions, like group I and group II introns, RNase P ribozymes (Mann et al. 2003), and transfer-messenger RNAs (tmRNA), that work as a ribosome-rescue system (Janssen and Hayes 2012). These transcripts create a diverse in structure and biochemical functions molecular space that rivals the activities of proteins. The last decade also brought the discovery of new classes of longer ncRNAs that have not been observed in bacteria before (Harris and Breaker 2018). Although bacteria harbor far fewer of them than eukaryotes, they perform essential roles in the core processes of information transfer, metabolism, and physiological adaptation (Harris and Breaker 2018). Most of them have been discovered through comparative sequence analysis based on a series of computer-aided search strategies, such as Breaker Laboratory Intergenic Sequence Server (BLISS; http://bliss.biology.yale.edu) (Puerta-Fernandez et al. 2006; Weinberg et al. 2009).

6.1 Ornate, Large, Extremophilic RNAs (OLE)

OLE RNAs are ~610 nt long transcripts expressed by extremophilic gram-positive eubacteria (Puerta-Fernandez et al. 2006; Harris et al. 2018), which levels are estimated to exceed that of mRNAs during the exponential growth phase (Block et al. 2011). Direct downstream of *OLE*, a gene encoding a 21-kDa OLE-associated transmembrane protein called OAPA (formerly OAP) has been localized, and its product has been shown to bind OLE lncRNA in vitro. The complex has a 2:1 OAPA:OLE RNA stoichiometry, suggesting that the protein might function as a dimer. The fluorescence in situ hybridization (FISH) microscopy has shown OLE RNA localizes to the cell membrane, but only in the presence of OAPA, which is predicted to be a transmembrane protein and bind OLE RNA (Block et al. 2011; Harris et al. 2018). The position of the tandem-arranged *OLE* and *OAP* genes immediately downstream of the *ISPA* gene, related to cell membrane biochemistry and upstream of the *DXS* gene that codes for a key enzyme in the isoprenoid biosynthesis pathway is conserved in almost all of the bacterial genomes expressing OLE RNA, suggesting the role of OLE RNA might be related to these processes (Julsing et al. 2007).

OLE RNAs are expressed in a wide range of species in anaerobic extremophilic *Firmicutes* (Block et al. 2011). It has been speculated that together with their protein partner, OAPA, they may play an important role in protecting the species from the extreme environment (Wallace et al. 2012). Also, the expression of OLE is elevated when *Bacillus halodurans* C-125 is exposed to the short-chain alcohols at a near-lethal concentration. The mutant strains without *OLE* gene disrupted OAPA expression or both have been shown to exert a high sensitivity to the ethanol as compared to the wild-type strain, and most importantly, the rescue strains have been shown to restore the resistance (Block et al. 2011). Subsequent research in an *OAPA* mutant led to the discovery of the second OLE lncRNA-binding protein, OAPB (also called YBZG), which selectively binds OLE RNA in vivo and forms the functional OLE RNP complex in *B. halodurans* (Harris et al. 2018). OAPB is an essential factor for ethanol resistance originated from the bacterial OLE ribonucleoprotein complex (Harris et al. 2018).

The comparative sequence analysis of 116 bacterial genomes has identified 15 OLE lncRNAs, which are highly conserved with half of their sequences exhibiting at most one mutation or none (Puerta-Fernandez et al. 2006). Larger scale analysis has indicated that one-third of OLE lncRNA nts are more than 90% conserved (Block et al. 2011). Also, the sequences alignment of OLE RNAs has been used to generate a prediction of their secondary structure (Fig. 5c). It has been predicted that OLE lncRNA folds into 14 conserved base-pairing regions, denoted P1 to P14, which are interspersed with complex and variable substructures of bulges and loops (Puerta-Fernandez et al. 2006). The in-line probing of the 5′ end (1–291), two 3′ end segments (449–608 and 284–677) of OLE, and a full-length structural probing via Pb^{2+} cleavage confirmed that OLE folds according to the initial predictions (Puerta-Fernandez et al. 2006; Block et al. 2011). A more recent work based on mutual information (MI) analysis of 52 sequenced genomes and 156 unique sequences from environmental DNA samples, which provides a measure of the dependence of one variable on another, such as covariance of two nucleotides, further supplemented the secondary structure of OLE with three additional paired region P4a, P14a, and P15 (Wallace et al. 2012).

6.2 Giant, Ornate, Lake- and Lactobacillales-Derived (GOLLD)

GOLLD has been initially identified during the metatranscriptome analysis of the environmental samples from Lake Gatun, Panama (Weinberg et al. 2009). Subsequently, eight cultivated organisms distributed among three bacterial phyla have been shown to express it, including *Bacteroides*, *Firmicutes,* and *Verrucomicrobia*. With an average of 800 nts, the GOLLD is the third-largest bacterial lncRNA discovered to date.

Gene encoding GOLLD lncRNA is located near the tRNA genes, but the functional relationship between them is poorly understood (Weinberg et al. 2009; Harris et al. 2018). Studies also reported the *GOLLD* gene localizes in a prophage in *Lactobacillus brevis* ATCC 367, and the increased expression of *GOLLD* in

L. brevis correlates with elevated production of the bacteriophage particles during the lytic cycle with the mitomycin C induction (Raya and H'bert 2009). The termini determined experiments using 5′-RLM-RACE (5'-RNA-ligase-mediated rapid amplification of cDNA ends) and 3'-RACE (rapid amplification of cDNA ends) has confirmed the presence of the full-length GOLLD with the propagated bacteriophage particles in lytic cycle, which suggests GOLLD may exert a helpful function for phage reproduction. *Lactobacillus* bacteriophages and prophages often have large non-coding regions surrounding their tRNA genes (Puerta-Fernandez et al. 2006). They also exhibit high rates of host-parasite recombination and both horizontal and vertical genetic transfer. Thus, it is possible that GOLLD lncRNA is not relevant to the bacteriophage life cycle, which is supported by the fact that a mutant bacteriophage without *GOLLD* gene does not show any phage-propagated defect (Harris and Breaker 2018).

GOLLD lncRNA secondary structure prediction based on the multiple sequence alignment has shown it includes the total of 20 hairpins, 5 of which potentially form pseudoknots, 3 others fold into tetraloops, and remaining 12 contain highly conserved nucleotides (Weinberg et al. 2009; Harris and Breaker 2018). Most of the highly conserved nucleotides and all the predicted pseudoknots localize to the 3′ half of the sequence; while the 5′ half is far less conserved (Harris and Breaker 2018).

6.3 HNH Endonuclease-Associated RNA and ORF (HEARO RNA)

HEARO RNAs are ~350 nt long, highly structured transcripts carrying an embedded ORF that encodes HNH endonuclease (Weinberg et al. 2009; Harris et al. 2018). HNH is a homing endonuclease, which initiates a transfer of mobile genetic elements, i.e., group I and group II introns, by generating DNA double-strand breaks (Stoddard 2011). Also, relatives of the *HEARO* include the ORF associated with *IS605* selfish genetic elements known to exploit small structured DNA motifs as part of their replicative cycle (He et al. 2013). Yet, experiments seeking self-splicing activity of HEARO RNAs have yielded no positive results (Weinberg et al. 2009).

HEARO RNAs are expressed by species from 10 different bacterial phyla, predominantly *Firmicutes, Proteobacteria, Cyanobacteria*, and *Actinobacteri*, with some of them encoding dozens of *HEARO* genes. Although HEARO RNAs do not show high sequence conservation, the covariation analysis provides strong support for their structural conservation. The multiple sequence alignment-based secondary structure prediction suggests HEARO RNAs fold into seven hairpin, one pseudoknot, and few internal loops. These data have been supported by the in-line probing experiments of in vitro synthesized HEARO RNA from *Arthrospira maxima* (Weinberg et al. 2009).

7 Viral LncRNAs

Similarly, to their host cells, many viruses express lncRNAs, with gammaherpesviruses being the absolute champions when it comes to their vast repertoire (Chavez-Calvillo et al. 2018). Since viral genomes have limited coding capacity, these non-coding transcripts are produced only when they directly contribute to viral fitness. Most of them target cellular pathways regulating infection, e.g., the cellular immune response, the transition between replication and latency, leading to the establishment of the equilibrium between the host-virus interactions. With variable length, structure, abundance and binding partners, viral lncRNAs show specificity and diversity concerning the time of expression during the viral infection. Here, we review the prominent examples of viral lncRNAs, their structure, structure-mediated interactions, and cellular functions.

7.1 Adenovirus Virus-Associated RNAs (VA RNAs)

Virus-associated RNAs (VA RNAs) are abundant (~10^8 copies/cell), heterogeneous, cytoplasmic transcripts of varying length (~150–200 nts) expressed by most human adenovirus serotypes (Ma and Mathews 1996). They are transcribed throughout the replication process in the host cells; thus, they are abundant during the late phase of viral infection, and their transcript levels depend on the copy numbers of the viral genome. Although both VA RNAs are initially transcribed at similar levels, following AdV genome replication, RNA POL III becomes limited and as a result, VAI outcompetes VAII for transcription during late infection (Bhat and Thimmappaya 1984).

Although the nucleotide sequences of VA RNAs are poorly conserved between adenovirus, the main function of VA RNAs is to interfere with cellular miRNA biogenesis through (1) competition with cellular RNAs for export from nucleus, (2) saturation of DICER endoribonuclease, and (3) direct interference with RNA-induced silencing complex (RISC) assembly and function (Vachon and Conn 2016). It has been proposed that following transcription by cellular POL III, VA RNAs saturate the nuclear export protein EXPORTIN 5 and the cellular DICER, interfering with pre-micro (mi)RNA export and miRNA biogenesis, respectively. DICER-processed VA RNA fragments are incorporated into the RISC complex as viral microRNAs or “mivaRNAs”, where they may specifically target cellular genes (Vachon and Conn 2016). Also, a deletion study has shown that VAI may perform the role of bi-functional RNA by being processed by DICER at the terminal stem to saturate the RNAi system, while also generating a shortened VAI RNA that remains active to inhibit RNA-dependent protein kinase R (PKR). It has been shown that an RNA mutant lacking the entire terminal stem is still able to bind PKR and inhibit its activity (Wahid et al. 2008). VAII RNA, even though present at a lower concentration than VAI, is the preferred substrate for DICER (Xu et al. 2007) Also, a fraction of the VAII-derived mivaRNAs has been shown to associate with polyribosomes, indicating a possible role in regulating translation of cellular mRNAs (Xu et al. 2007).

Although the nucleotide sequences of VA RNAs are poorly conserved between adenovirus serotypes, their secondary structure has been predicted to be well maintained, and can be divided into a terminal stem, a panhandle apical stem, and a more structured central domain (Ma and Mathews 1996). It has been shown that the apical stem forms highly stable hairpin with T_m of >80 °C, which is also resistant to denaturation, even in 6 M urea (Wahid et al. 2008). This latter feature causes VA RNAI to migrate at ~220-nt RNA in denaturing polyacrylamide gel electrophoresis (PAGE) (Wahid et al. 2008; Clarke et al. 1994). The central domain has been predicted as the most structurally complex of the three (Nicot et al. 2005), and takes the form of a three-helix junction, linking the apical and terminal stems, and containing the two universally conserved, complementary tetranucleotide sequences. Additionally, a formation of a pseudoknot, a higher-order tertiary structure, has been proposed based upon sequence complementarity and mutual insensitivity to single-strand specific ribonucleases of two-loop regions within the central domain (Ma and Mathews 1996). Recently, the structure of VAI has been refined based on constraints derived from SHAPE and DMS RNA chemical probing and small-angle x-ray scattering (SAXS) measurements (Launer-Felty et al. 2015). VAI adopts an extended conformation with the apical and terminal stems emanating from a central bulge. This model shows how the apical stem and central domain assemble into an extended duplex that is tuned to bind a PKR monomer with high affinity, thereby inhibiting activation of PKR by viral dsRNA.

7.2 *Kaposi's Sarcoma-Associated Herpesvirus (KSHV) Polyadenylated Nuclear RNA (PAN)*

KSHV belongs to the family of Gammaherpesviruses, and it is the causative agent of several human cancers and lymphoproliferative disorders (Ganem 2006). During the lytic stage of infection, KSHV produces polyadenylated nuclear (PAN) RNA, that has also been found in the cytoplasm (Massimelli et al. 2011), associating with ribosomes (Arias et al. 2014), and within viral particles (Sztuba-Solinska et al. 2017). PAN has been shown to regulate almost every stage of viral and cellular gene expression, cell cycle, pluripotency, modulation of host-pathogen interactions, and the production of infectious viral particles (Rossetto and Pari 2011, 2012; Rossetto et al. 2013b). PAN actively participates in chromatin remodeling by recruiting the protein components of PRC2, as well as the histone methyltransferase MLL2 and the demethylases UTX and JMJD3. Using chromatin isolation by RNA purification (ChIRP-Seq), Rossetto et al. demonstrated the great extent to which PAN manipulates viral and host gene expression programs (Rossetto et al. 2013b). Eighty-four cellular gene promoters that are involved in the regulation of inflammatory and antiviral responses (IFNγ, IL-18, IFNA16, and RNase L), cell death (TRIM68, RAD52, INPP5E, EPHB2, PAX2) and development (HIST1H4A, HIST3H3, PAX6, PAX5, CDKN2B), and 35 viral gene promoters involved in direct regulation of KSHV lytic gene expression (i.e., PAN, orLyt-L, K14, ORF4, ORF64, ORF50, ORF74) were shown to be directly recognized and regulated by PAN RNA. PAN also interacts with PABPC1, which relocates to the nucleus during the lytic phase of KSHV infection. This relocation is directly caused by the shutoff exonuclease (SOX) protein, which downregulates the expression of host mRNAs and upregulates levels of PAN RNA. Therefore, PAN RNA acts downstream of SOX, further contributing to viral manipulation of gene expression. In addition, multiple viral proteins have been shown to associate with PAN RNA. The interaction with ORF26 likely facilitates PAN packaging into virions, while the ORF59 facilitates the recruitment of PAN RNA to the viral episome. PAN RNA has also been shown to regulate the function of the latency-associated nuclear antigen (LANA) protein (Campbell et al. 2014). During latency, LANA wraps around the KSHV episome and silences the expression of lytic genes. Lytic reactivation is marked by an abundance of PAN RNA, which sequesters LANA away from the episome, thereby relieving the repressive activity and facilitating the expression of lytic genes.

PAN RNA reaches up to 80% of total polyadenylated transcripts and accumulates at astonishing numbers of 5×10^5 copies per cell (Conrad 2016). Its abundance has been proposed to be the outcome of two structure-related strategies: (1) the interaction between mRNA transcript accumulation protein (MTA or ORF57) and the 5′ MTA responsive element (MRE, or ORF57-responsive element) that has been proposed to prolong PAN stability and maintain its mainly nuclear localization, and (2) the sequestration of PAN 3′ poly(A) tail within the expression and nuclear retention element (ENE) forming a triple helix structure, that protects PAN from deadenylation and decay. Secondary structure prediction and analyses of mutants using cellular assays suggested that the ENE forms a hairpin containing a U-rich internal loop flanked by short helices (Conrad and Steitz 2005). The mode of interaction between the ENE and PAN poly(A) tails has been furthered through crystallography. It has been revealed that the ENE internal loop interacts with A9 oligonucleotide that simultaneously engages both sides of the internal loop in an extended U-A•U major-groove triple helix through Watson-Crick and Hoogsteen interactions. Hydrogen bonds are also observed between the A9 phosphate backbone and the ribose hydroxyl groups of the Hoogsteen U strand. Additionally, the binding interface is augmented by a triad of A-minor interactions formed between the A9 oligonucleotide and the lower stem of the ENE (Mitton-Fry et al. 2010). Analogous RNA triple helices with stabilizing functions have been found in vertebrate lncRNAs: MALAT1, and MENβ, and have been bioinformatically predicted in HOTAIR (Kalwa et al. 2016).

PAN RNA secondary structure has been resolved by SHAPE-MaP RNA probing in three different biological contexts, i.e., in the nucleus, in the cytoplasm and within virions (Fig. 5d) (Sztuba-Solinska et al. 2017). It has been shown that PAN RNA folds into a branched secondary structure including three domains, with the domain I and III being the most structurally constrained and characterized by low SHAPE reactivity and low Shannon entropy. Domain I includes the MRE motif that has been predicted initially to fold into three adjacent hairpins (Massimelli et al. 2011; Sei and Conrad 2011). However, SHAPE-MaP has demonstrated that the MRE core adopts an unstructured, single-stranded conformation. This finding has been recently strengthened by crystallographic studies of the globular domain of ORF57 from a close homolog of KSHV, herpesvirus saimiri (HVS) in complex with a PAN MRE. It has been shown that ORF57 has an apparent binding specificity for uridine nucleotides in a bent, non-base paired conformation (Tunnicliffe et al. 2019). PAN SHAPE-MaP study has also revealed that domain II shows higher structural flexibility, likely to accommodate long-range tertiary interactions, like the formation of the ENE triple helix, or to support compact folding of adjacent regions and provide an accessible scaffold for protein interactions. The structural modularity noted for PAN is based on two-dimensional "connectivity" maps and does not shed light on long-range interactions that might be intimately related to PAN function. Also, we are currently lacking the insight into potential epitranscriptomic modifications and their role in modulating PAN RNA structure and function.

7.3 Human Cytomegalovirus LncRNAs

Human cytomegalovirus (HCMV) is a beta-herpesvirus that replicates in glandular epithelial tissue and ultimately develops a lifelong latent infection in the host. HCMV encodes at least four lncRNAs: RNA1.2, RNA2.7, RNA4.9, and RNA5.0 (Schwarz and Kulesza 2014). The function of RNA1.2 is currently unknown (Schwarz and Kulesza 2014), but it has been proposed to encode 30 kDa polyprotein (Gatherer et al. 2011). β-2.7 is an unspliced, polyadenylated lncRNA, found predominantly in cytoplasm (Gatherer et al. 2011). The *β-2.7* gene exhibits early gene kinetics showing maximal amplification between 8 and 14 h post-infection (p.i.) and remains transcriptionally active through the virus replication cycle. β-2.7 lncRNA binds directly to the GRIM19, a subunit of mitochondrial enzyme complex I, to protect virus-infected cells from apoptosis and results in continued ATP production. Interaction of β2.7 RNA with GRIM19 has been shown to inhibit rotenone stress-induced apoptosis in neuronal cells (Reeves et al. 2007), which represents a refined strategy by which HCMV can regulate the metabolic viability of the infected host cell (Reeves et al. 2007). A small ~800 nt domain of the transcript, referred to as p137, has been identified to protect cells stressed by rotenone with the same efficacy as the full-length lncRNA (Kuan et al. 2012), leading to its use in the development of a novel therapeutic for Parkinson's disease in rats (Kuan et al. 2012; Poole et al. 2016). RNA4.9 has been proposed to play a role in transcriptional repression of viral immediate-early (IE) gene expression during latency (Rossetto et al. 2013a; Noriega et al. 2014). It has been shown to associate with PRC2 to enhance silencing of the viral major immediate early promoter (MIEP) during latent infection by the recruitment of histone modifiers (Rossetto et al. 2013a). RNA5.0 is expressed at extraordinarily high levels during lytic infection (Kulesza and Shenk 2004; Gatherer et al.

2011), yet its deletion does not result in any replication defects of the virus (Kulesza and Shenk 2004). The structures of all HCMV-encoded lncRNAs remain to be identified.

7.4 *Human Immunodeficiency Virus-1 Antisense RNA (ASP)*

HIV-1, in addition to the full-length genomic RNA (gRNA) and variously spliced mRNAs, translates an antisense transcript ASP RNA. ASP is a 2.6-kb, non-polyadenylated nuclear lncRNA produced from the 3′ long terminal repeat (LTR) U3 region of proviral DNA (Saayman et al. 2014). ASP has been shown to recruit PRC2 to the HIV-1 5′ long terminal repeat (LTR), resulting in the accumulation of suppressive histone H3K27 trimethylation that facilitates the establishment of HIV latency (Zapata et al. 2017). Silencing of ASP has been shown to activate HIV gene expression (Saayman et al. 2014). The structure of ASP RNA has not been resolved, however, recently developed approach for detecting ASP RNA from patients will likely lead to its further structural and functional characterization (Mancarella et al. 2019).

The ASP RNA contains ORF that has been proposed to encode a putative protein of 189 amino acid residues referred to as AntiSense Protein (ASP). The existence of both, the lncRNA and the protein, has been controversial for many years, yet, new evidence argues in favor of its expression. It has been demonstrated that the full-length ASP protein can be expressed ex vivo from the HIV-1 3′ LTR (Torresilla et al. 2013). Also, ASP protein has been detected in freshly infected cells (Briquet and Vaquero 2002; Clerc et al. 2011; Laverdure et al. 2012), and two recent independent clinical studies have shown the in vivo expression of ASP by detecting a cell-mediated immune response against several *asp* epitopes within 30% of individuals infected with subtype B viruses (Bet et al. 2015; Berger et al. 2015). Cassan et al. (2016) have shown that ASP is specific to major (M) group of HIV-1 and that the correlation between its presence and the prevalence of HIV-1 groups and M subtypes is statistically significant and results from significant selection pressure. Furthermore, studies have suggested that ASP protein can form stable aggregates, be located partially at the plasma membrane, and is associated with autophagy.

7.5 *Subgenomic Flaviviral RNAs (sfRNAs)*

Flaviviruses (FV) encompass a large group of single-stranded, positive-sense RNA viruses including Yellow fever virus (YFV), Dengue viruses (DEN), West Nile virus (WNV), and Zika virus (ZIKV) (Payne 2017). These viruses produce 300–500 nt long subgenomic flaviviral RNA (sfRNA). sfRNA results from stalling of the cellular 5′-3′ exoribonuclease XRN1/Pacman on conserved RNA structure in the 3′ untranslated region (UTR) of the viral genome. This structure has been initially predicted to fold into a stem-loop (SL) and dumbbell (DB) RNA (Funk et al. 2010). The application of DMS, SHAPE probing (Chapman et al. 2014b), and crystallography has resulted in the 2.5 Å resolution of the RNA-XRN1 complex and identified an interwoven pseudoknot that causes steric hindrance stalling the XRN1. This region is often referred to as XRN1-resistant RNAs (xrRNAs). It has been shown that the xrRNA sequence directly contacts the surface of XRN1, positioned in a cleft, with its ring structure braced over the entry to the active site (Chapman et al. 2014a). These XRN1-xrRNA contacts further stabilize the xrRNA fold preventing conformational changes that are otherwise required for XRN1 helicase activity. RNA structure probing and mutagenesis studies have provided further evidence for pseudoknot formation and its function as a stalling signal (Kruijt et al. 2005).

Compared to their biogenesis, not much is known about the downstream function of sfRNAs. It has been suggested that sfRNA may act as a suppressor of RNA interference (RNAi) in both mammalian and arthropod cells (Schnettler et al. 2012, 2014). The highly abundant sfRNA most likely acts as a decoy for DICER to prevent it from cleaving other double-stranded RNAs. In Dengue virus serotype 2 (DENV-2), sfRNA has been shown to interact with the host proteins G3BP1, G3BP2, and CAPRIN1 and inhibit translation of the several interferon-stimulated genes (Bidet et al. 2014) SfRNA produced by a dominant Puerto Rico DEN-2 has shown sequence-dependent binding to and prevention of tripartite motif 25 (TRIM25) deubiquitylation, which is critical for sustained and amplified RIG-I-induced type I interferon expression (Manokaran et al. 2015). The production of sfRNA in WNV has been shown to increase viral replication and is essential for virus-induced cytopathicity and pathogenicity in mice. In Zika virus, sfRNAs have been shown to antagonize the innate interferon response and RNA interference (Göertz et al. 2018), as well as to bind to a fragile X mental retardation protein (FMRP) in cell culture and in mouse with significant implications in the pathogenesis of ZIKV infections (Soto-Acosta et al. 2018).

8 LncRNA and Therapy

Discovery of lncRNAs and their involvement in disease (Ayers 2013; Mayama et al. 2016; Wang et al. 2017a, c) has opened new perspectives in the understanding of the genetic roots for multiple diseases, but also the development of new therapeutic approaches. Currently, there are no approved lncRNA-targeting drugs; however, an increasing number of lncRNA-based approaches are in preclinical phases, bringing the prospect of personalized medicine within reach.

LncRNAs can be targeted by the following approaches: (1) Post-transcriptional RNA degradation pathways aiming at the knockdown of pathogenic RNAs, either using siRNAs that invoke a DICER- and ARGONAUTE (AGO)-dependent cleavage pathway, or antisense oligonucleotides (ASOs) that target the RNA of interest for degradation via an RNase H-dependent mechanism. (2) Modulation of lncRNA genes by steric blockade of the promoter or by using genome-editing techniques.

(3) Loss of function by creating steric inhibition of RNA:protein interactions, or disruption of its secondary structure by using RNA- binding small molecules, or ASOs.

Post-transcriptional targeting has been successfully applied to target specific mRNA in clinical trials. This includes fomivirsen, which is used to treat cytomegalovirus retinitis, and mipomersen used to treat homozygous familial hypercholesterolemia. Thus, there is no reason to assume that a comparable approach would not be effective against pathogenic lncRNAs. Because oligonucleotide-based compounds rely on direct complementarity between oligonucleotides and their target RNA, the screening and testing are relatively straightforward. ASOs, in particular, can be explored to target all classes of lncRNAs, because they can enter the nucleus and knockdown nuclear lncRNAs (Zong et al. 2015), in contrast to siRNA and short hairpin RNAs (shRNA) that mainly operate within the cytoplasm. For example, nuclear-localized MALAT1 has been successfully downregulated in skeletal muscle with the systemic administration of ASOs that might have therapeutic applications in cancer (Wheeler et al. 2012). Also, the subcutaneous administration of ASOs targeting MALAT1 effectively inhibits human lung cancer cell proliferation in a mouse xenograft model (Gutschner et al. 2013). Since oligonucleotide-based drugs are target-specific, the potential side effects are generally reduced. Though recently, there have been reports of liver toxicity associated with mipomersen usage (Geary et al. 2015). While unmodified oligonucleotides are available, certain chemical modifications (such as 2'-O methyl, morpholinos, and locked nucleic acids) increase their lifespan in the complex cellular and reduce degradation products, which may also have effects on the cells.

Some lncRNAs can be noncompliant to knockdown by either ASOs or RNAi, especially if their subcellular localization is not accessible to either RNase H or the RNAi machinery. It could also occur if the lncRNA is highly structured or blocked due to excessive protein binding or hybridizing to other cellular nucleic acids. In these cases, the targeting of lncRNA promoters provides an alternative strategy. For example, DTA-H19, a double-stranded DNA plasmid carrying the gene for the A subunit of diphtheria toxin under the regulation of the H19 gene promoter, has been administered to elicit an antitumor response in various solid tumors (Amit et al. 2011). This approach has been tested initially in a mouse bladder cancer model, however now the testing expanded to non-small cell lung carcinoma, colon, pancreatic, and ovarian cancers (Sidi et al. 2008; Gofrit et al. 2014). Moreover, the bacterial clustered regularly interspaced short palindromic repeats (CRISPR)/CRISPR-associated 9 (CAS9) system has emerged as a promising method to achieve transcriptional silencing of lncRNA-expressing loci. In this approach, dead-CAS9 is fused to transcriptional repressors, and this fusion protein is targeted to a specific gene promoter by guide RNAs to achieve transcriptional silencing (Thakore et al. 2015). CRISPR has been used to delete lncRNA-21A, UCA1, and AK023948 in human cell lines (Ho et al. 2015), and RIAN, a 23 kb maternally expressed lncRNA gene, in mice in vivo (Han et al. 2014). These approaches can be further enhanced with the use of the CRISPRi system which relies on the integration of inducible CAS9 construct into the *AAVS1* locus of induced pluripotent stem cells (iPSCs).

Here, lncRNAs can be targeted by the addition of RNA guides (Zhu et al. 2014). Recently the RNA-targeting CRISPR/CAS13 system was identified (Abudayyeh et al. 2017), providing another promising approach to knockdown lncRNAs.

Another major interest in targeting lncRNAs is to upregulate the expression of genes of therapeutic interest. Such upregulation is crucial for activating tumor suppressors in cancers, other crucial growth factors or transcription factors whose expression confers a growth disadvantage to proliferating cells. Several lncRNAs work as natural antisense transcripts (NATs); they overlap with protein-coding genes and affect the transcription of their overlapping genes *in cis* (Katayama et al. 2005). Thus, antisense targeting NATs, or "antagoNATs", provides a precise way for gene. For example, continuous in vivo infusion of AntagoNATs targeting mouse antisense lncRNA BDNF-AS produced a specific upregulation in Brain-Derived Neurotrophic Factor (BDNF) mRNA and protein. Mice treated with AntagoNATs displayed a marked increase in proliferating cells and neuronal survival in the brain as compared to controls (Modarresi et al. 2012). Also repeated administration of AntagoNATs targeting monkey APOA1-AS resulted in the upregulation of HDL levels (Halley et al. 2014).

As discussed, many lncRNAs manifest their function through structure-mediated interactions with effector molecules (Brockdorff 2013; Hendrickson et al. 2016). Thus, modified ASOs can be employed to bind specifically to and block the lncRNA interaction interface, resulting in loss of function. Also, small molecules recognizing unique RNA motifs and RNA structure-mediated interactions can serve as a useful tool for targeting lncRNAs. For example, ellipticine has been identified as a potent modulator of HOTAIR, brain-derived neurotrophic factor antisense (BDNF-AS) lncRNAs, and EZH2 histone methyltransferase. Also, recent studies identified small chemotypes that specifically recognize and bind to MALAT1 triple helix, a unique motif that regulates its stability (Donlic et al. 2018; Abulwerdi et al. 2019). Thus, small molecules emerge as novel RNA-specific therapeutics that hopefully soon will be included in a vademecum for the treatment of a variety of human disorders.

9 Conclusions

The expansion of the lncRNA field came with the groundbreaking FANTOM, ENCODE, and GENCODE projects (Derrien et al. 2012; Dunham et al. 2012). These projects for the first time seriously challenged the central dogma by demonstrating that while there are approximately 20,000 protein-coding genes and 2600 miRs, the number of known lncRNAs readily exceeds 50,000. Novel lncRNAs are continuously being discovered, emanating from the expansive DNA between protein-coding genes in animals, but also found in simple eukaryotes, in the bacterial and archaeal branches of the tree of life and even beyond. As discussed, lncRNAs are by no means a homogeneous group but comprise transcripts with very different biogenesis, fates, interactions, and biological functions, thus defining them as a whole is, to say the least, difficult endeavor. Attempts have been made to divide lncRNAs considering their genomic localization, subcellular localization, RNA features, and modes of action. Nonetheless, lncRNAs molecular and functional versatility continuously prove that more surprising discoveries are yet to come.

Currently, we are moving forward from efforts focusing merely on lncRNAs annotation, to the development of novel strategies that provide the insight into their exact mode of action, the identification of their effector molecules, and the cross-talk lncRNAs have with each other. The aspiring goal in the field is to understand how aberrancy in these communication pathways may be implicated in pathogenesis and dysregulation of cellular processes. The location and magnitude of these interactions for specific lncRNAs and their effectors must be precisely regulated to adapt the gene expression and cellular processes in response to various stimuli. Dissecting these pathways might provide the key to understanding many human disorders. For example, the comprehensive interrelationship of lncRNAs and WNT signaling pathway, an evolutionarily conserved pathway that controls critical biological processes including embryonic development, homeostasis, and cell fate, has appeared as a significant tumorigenic signaling network in various cancers (Yang et al. 2018). Likewise, the cross-talk between specific lncRNAs has been shown to regulate and remodel the extracellular matrix within the tumor microenvironment (D'Angelo and Agostini 2018).

Most current studies focus primarily on studying lncRNA expression in the total cell populations. With single-cell transcriptomics on the rise, expression of lncRNAs can be tracked at the single-cell level providing substantial information on their function. As Bumgarner et al. (2012) have shown in yeast, there is variation in the expression levels of the lncRNAs in individual cells. Also, single-cell RNA-seq has revealed dynamic changes in lncRNA expression during cell reprogramming (Kim et al. 2015). Thus, combining the single-cell transcriptomics with high-resolution quantification and spatial position of lncRNAs and single-molecule RNA fluorescence in situ hybridization (smFISH) could aid in functionally classifying lncRNA based on their subcellular localization (Soares et al. 2018). RNA single-molecule FISH (smFISH) has been widely used to detect and quantify individual lncRNA molecules (Raj et al. 2008; Dunagin et al. 2015; Cabili et al. 2015). smFISH can also be applied concomitantly with immunofluorescence labeling (IF-FISH) (Hinten et al. 2016).

Once we tackle lncRNAs functionality at the single-cell level and unravel their involvement in complex regulatory networks, the exciting aspect of the lncRNA-related field would be their application as biomarkers (Meseure et al. 2015). Since mutations or aberrant expressions of lncRNAs have been shown to result in cellular dysfunction leading to various disease state, comparative profiling of lncRNAs isolated from different body fluids such as a serum, plasma, urine, or sputum can serve as a potential method for early detection of various diseases. For example, BRAF-activated non-coding RNA (BANCR) lncRNA has been shown to be upregulated in human malignant melanoma and patients with high levels of expression have a lower survival rate (Li et al. 2014). Also, three lncRNA signatures have been identified recently in esophageal squamous cell carcinoma, including

ENST00000435885.1, XLOC_013014, and ENST00000547963.1. The expression of these lncRNAs classified the patients into two groups with significantly different overall survival rates (Chen et al. 2014). Recent studies have shown that the prostate-specific lncRNA PCA3 levels in urine can accurately predict tumor volume and pathological features, which may guide treatment (Roobol et al. 2010; Ploussard et al. 2011). Thus, PCA3 can be used as a noninvasive urine-based test for large-scale screening protocols and for predicting prostate carcinomas aggressiveness. This prominent example of rapid translation of lncRNA research into clinical practice offers a prototype for developing different lncRNAs as biomarkers.

Scientists have also been entertaining the idea of employing lncRNAs as therapeutic agents themselves. For example, XIST transgene has been used for transcriptional repression of chromosome 21, which potentially can become the basis of a therapeutic strategy against Down syndrome (Chiang et al. 2018). Also, a complex of small domain of HCMV β2.7 lncRNA coupled to rabies virus glycoprotein peptide has a potential application against Parkinson's disease and other neurological disorders, as it readily crosses the blood-brain barrier (Kuan et al. 2012) and the lncRNA itself inhibits rotenone stress-induced apoptosis in neuronal cells.

Recently, Miao et al. (2017) performed the estimate of development prospects in the field of lncRNA studies. According to the number of publications, they divided the stage of lncRNA research into two phases: the initial stage (2007–2011), with the number of publications increasing slowly, and the golden period (2012–2016), when there was a sharp growth of publications related to lncRNAs, with the prediction curve strongly indicating rapid progress in this field in the following years. We are at the verge of fascinating discoveries that indisputably will prompt researchers to build a comprehensive framework of regulatory circuits embedding lncRNAs, thereby deciphering a bit further the puzzle of life biodiversity and complexity.

Acknowledgements G.T., H.C., and J.S-S. are funded by Alabama Agricultural Experiment Station, Hatch Funding Program and start-up funds from the Department of Biological Sciences, College of Science and Mathematics, and Office of the Vice President for Research, Auburn University.

Authors contributions G.T., H.C., and J.S-S. contribute to the manuscript writing and figure preparation.

References

Abudayyeh OO, Gootenberg JS, Essletzbichler P et al (2017) RNA targeting with CRISPR–Cas13. Nature 550:280–284

Abulwerdi FA, Xu W, Ageeli AA et al (2019) Selective small-molecule targeting of a triple helix encoded by the long noncoding RNA, MALAT1. ACS Chem Biol 14:223–235

Al-Rugeebah A, Alanazi M, Parine NR (2019) MEG3: an oncogenic long non-coding RNA in different cancers. Pathol Oncol Res 25:859–874

Alam T, Medvedeva YA, Jia H et al (2014) Promoter analysis reveals globally differential regulation of human long non-coding RNA and protein-coding genes. PLoS One 9:e109443

Alcid EA, Tsukiyama T (2014) ATP-dependent chromatin remodeling shapes the long noncoding RNA landscape. Genes Dev 28:2348–2360

Amit D, Tamir S, Birman T et al (2011) Development of targeted therapy for bladder cancer mediated by a double promoter plasmid expressing diphtheria toxin under the control of IGF2-P3 and IGF2-P4 regulatory sequences. Int J Clin Exp Med 4:91–102

Amodio N, Stamato MA, Juli G et al (2018) Drugging the lncRNA MALAT1 via LNA gapmeR ASO inhibits gene expression of proteasome subunits and triggers anti-multiple myeloma activity. Leukemia 32:1948–1957

Amort T, Soulière MF, Wille A et al (2013) Long non-coding RNAs as targets for cytosine methylation. RNA Biol 10:1002–1008

Arias C, Weisburd B, Stern-Ginossar N et al (2014) KSHV 2.0: a comprehensive annotation of the Kaposi's sarcoma-associated herpesvirus genome using next-generation sequencing reveals novel genomic and functional features. PLoS Pathog 10:e1003847

Ariel F, Jegu T, Latrasse D et al (2014) Noncoding transcription by alternative RNA polymerases dynamically regulates an auxin-driven chromatin loop. Mol Cell 55:383–396

Ariel F, Romero-Barrios N, Jégu T et al (2015) Battles and hijacks: noncoding transcription in plants. Trends Plant Sci 20:362–371

Arieti F, Gabus C, Tambalo M et al (2014) The crystal structure of the split end protein SHARP adds a new layer of complexity to proteins containing RNA recognition motifs. Nucleic Acids Res 42:6742–6752

Askarian-Amiri ME, Crawford J, French JD et al (2011) SNORD-host RNA Zfas1 is a regulator of mammary development and a potential marker for breast cancer. RNA 17:878–891

Augui S, Nora EP, Heard E (2011) Regulation of X-chromosome inactivation by the X-inactivation centre. Nat Rev Genet 12:429–442

Ayers D (2013) Long non-coding RNAs: novel emergent biomarkers for cancer diagnostics. J Cancer Res Treat 1:31–35

Bánfai B, Jia H, Khatun J et al (2012) Long noncoding RNAs are rarely translated in two human cell lines. Genome Res 22:1646–1657

Bazin J, Bailey-Serres J (2015) Emerging roles of long non-coding RNA in root developmental plasticity and regulation of phosphate homeostasis. Front Plant Sci 6:400

Beaulieu YB, Kleinman CL, Landry-Voyer A-M et al (2012) Polyadenylation-dependent control of long noncoding RNA expression by the poly(A)-binding protein nuclear 1. PLoS Genet 8: e1003078

Ben Amor B, Wirth S, Merchan F et al (2008) Novel long non-protein coding RNAs involved in Arabidopsis differentiation and stress responses. Genome Res 19:57–69

Beniaminov A, Westhof E, Krol A (2008) Distinctive structures between chimpanzee and human in a brain noncoding RNA. RNA 14:1270–1275

Berger CT, Llano A, Carlson JM et al (2015) Immune screening identifies novel T cell targets encoded by antisense reading frames of HIV-1. J Virol 89:4015–4019

Bernard D, Prasanth KV, Tripathi V et al (2010) A long nuclear-retained non-coding RNA regulates synaptogenesis by modulating gene expression. EMBO J 18:3082–3093

Bernstein BE, Mikkelsen TS, Xie X et al (2006) A bivalent chromatin structure marks key developmental genes in embryonic stem cells. Cell 125:315–326

Bet A, Maze E, Bansal A et al (2015) The HIV-1 antisense protein (ASP) induces CD8 T cell responses during chronic infection. Retrovirology 12:15

Bhan A, Mandal SS (2015) LncRNA HOTAIR: a master regulator of chromatin dynamics and cancer. Biochim Biophys Acta 1856:151–164

Bhat RA, Thimmappaya B (1984) Adenovirus mutants with DNA sequence perturbations in the intragenic promoter of VAI RNA gene allow the enhanced transcription of VAII RNA gene in HeLa cells. Nucleic Acids Res 12:7377–7388

Bhatia G, Sharma S, Upadhyay SK et al (2019) Long non-coding RNAs coordinate developmental transitions and other key biological processes in grapevine. Sci Rep 9:3552

Bidet K, Dadlani D, Garcia-Blanco MA (2014) G3BP1, G3BP2 and CAPRIN1 are required for translation of interferon stimulated mRNAs and are targeted by a dengue virus non-coding RNA. PLoS Pathog 10:e1004242
Biswas S, Thomas AA, Chen S et al (2018) MALAT1: an epigenetic regulator of inflammation in diabetic retinopathy. Sci Rep 8:6526
Blanchard SC (2009) Single-molecule observations of ribosome function. Curr Opin Struct Biol 19:103–109
Block KF, Puerta-Fernandez E, Wallace JG et al (2011) Association of OLE RNA with bacterial membranes via an RNA-protein interaction. Mol Microbiol 79:21–34
Böhmdorfer G, Sethuraman S, Jordan Rowley M et al (2016) Long non-coding RNA produced by RNA polymerase V determines boundaries of heterochromatin. elife 5:e19092
Borah S, Darricarrère N, Darnell A et al (2011) A viral nuclear noncoding RNA binds re-localized poly(A) binding protein and is required for late KSHV gene expression. PLoS Pathog 7: e1002300
Briquet S, Vaquero C (2002) Immunolocalization studies of an antisense protein in HIV-1-infected cells and viral particles. Virology 292:177–184
Brockdorff N (2013) Noncoding RNA and polycomb recruitment. RNA 19:429–442
Brockdorff N (2018) Local tandem repeat expansion in Xist RNA as a model for the functionalisation of ncRNA. Non-Coding RNA 4:28
Brockdorff N, Ashworth A, Kay GF et al (1992) The product of the mouse Xist gene is a 15 kb inactive X-specific transcript containing no conserved ORF and located in the nucleus. Cell 71:515–526
Brown CJ, Hendrich BD, Rupert JL et al (1992) The human XIST gene: analysis of a 17 kb inactive X-specific RNA that contains conserved repeats and is highly localized within the nucleus. Cell 71:527–542
Brown JA, Valenstein ML, Yario TA et al (2012) Formation of triple-helical structures by the 3′-end sequences of MALAT1 and MEN noncoding RNAs. Proc Natl Acad Sci U S A 109:19202–19207
Brown JA, Bulkley D, Wang J et al (2014) Structural insights into the stabilization of MALAT1 noncoding RNA by a bipartite triple helix. Nat Struct Mol Biol 21:633–640
Bumgarner SL, Neuert G, Voight BF et al (2012) Single-cell analysis reveals that noncoding RNAs contribute to clonal heterogeneity by modulating transcription factor recruitment. Mol Cell 45:470–482
Busan S, Weeks KM (2017) Visualization of RNA structure models within the integrative genomics viewer. RNA 23:1012–1018
Cabili M, Trapnell C, Goff L et al (2011) Integrative annotation of human large intergenic noncoding RNAs reveals global properties and specific subclasses. Genes Dev 25:1915–1927
Cabili MN, Dunagin MC, McClanahan PD et al (2015) Localization and abundance analysis of human lncRNAs at single-cell and single-molecule resolution. Genome Biol 16:20
Campbell M, Kim KY, Chang P-C et al (2014) A lytic viral long noncoding RNA modulates the function of a latent protein. J Virol 88:1843–1848
Cao J (2014) The functional role of long non-coding RNAs and epigenetics. Biol Proced Online 16:42
Carlevaro-Fita J, Johnson R (2019) Global positioning system: understanding long noncoding RNAs through subcellular localization. Mol Cell 73:869–883
Carlevaro-Fita J, Polidori T, Das M et al (2019) Ancient exapted transposable elements promote nuclear enrichment of human long noncoding RNAs. Genome Res 29:208–222
Carrieri C, Cimatti L, Biagioli M et al (2012) Long non-coding antisense RNA controls Uchl1 translation through an embedded SINEB2 repeat. Nature 491:454–457
Cassan E, Arigon-Chifolleau A-M, Mesnard J-M et al (2016) Concomitant emergence of the antisense protein gene of HIV-1 and of the pandemic. Proc Natl Acad Sci U S A 113:11537–11542

Chapman EG, Costantino DA, Rabe JL et al (2014a) The structural basis of pathogenic subgenomic flavivirus RNA (sfRNA) production. Science 344:307–310

Chapman EG, Moon SL, Wilusz J et al (2014b) RNA structures that resist degradation by Xrn1 produce a pathogenic dengue virus RNA. elife 3:e01892

Chavez-Calvillo G, Martin S, Hamm C et al (2018) The structure-to-function relationships of gammaherpesvirus-encoded long non-coding RNAs and their contributions to viral pathogenesis. Non-Coding RNA 4:24

Chen Z, Li J, Tian L et al (2014) MiRNA expression profile reveals a prognostic signature for esophageal squamous cell carcinoma. Cancer Lett 350:34–42

Cheng J (2005) Transcriptional maps of 10 human chromosomes at 5-nucleotide resolution. Science 308:1149–1154

Chiang J-C, Jiang J, Newburger PE et al (2018) Trisomy silencing by XIST normalizes down syndrome cell pathogenesis demonstrated for hematopoietic defects in vitro. Nat Commun 9:5180

Chillón I, Pyle AM (2016) Inverted repeat Alu elements in the human lincRNA-p21 adopt a conserved secondary structure that regulates RNA function. Nucleic Acids Res 44:9462–9471

Chooniedass-Kothari S, Emberley E, Hamedani MK et al (2004) The steroid receptor RNA activator is the first functional RNA encoding a protein. FEBS Lett 566:43–47

Chu C, Zhang QC, Da Rocha ST et al (2015) Systematic discovery of Xist RNA binding proteins. Cell 161.404–416

Clamp M, Fry B, Kamal M et al (2007) Distinguishing protein-coding and noncoding genes in the human genome. Proc Natl Acad Sci U S A 104:19428–19433

Clark MB, Johnston RL, Inostroza-Ponta M et al (2012) Genome-wide analysis of long noncoding RNA stability. Genome Res 22:885–898

Clarke PA, Pe'eru T, Ma Y et al (1994) Structural features of adenovirus 2 virus-associated RNA required for binding to the protein kinase DAI. Nucleic Acids Res 22:4364–4374

Clerc I, Laverdure S, Torresilla C et al (2011) Polarized expression of the membrane ASP protein derived from HIV-1 antisense transcription in T cells. Retrovirology 8:74

Conrad NK (2016) New insights into the expression and functions of the Kaposi's sarcoma-associated herpesvirus long noncoding PAN RNA. Virus Res 212:53–63

Conrad NK, Steitz JA (2005) A Kaposi's sarcoma virus RNA element that increases the nuclear abundance of intronless transcripts. EMBO J 24:1831–1841

Csorba T, Questa JI, Sun Q et al (2014) Antisense COOLAIR mediates the coordinated switching of chromatin states at FLC during vernalization. Proc Natl Acad Sci U S A 111:16160–16165

D'Angelo E, Agostini M (2018) Long non-coding RNA and extracellular matrix: the hidden players in cancer-stroma cross-talk. Non-coding RNA Res 3:174–177

Derrien T, Johnson R, Bussotti G et al (2012) The GENCODE v7 catalog of human long noncoding RNAs: analysis of their gene structure, evolution, and expression. Genome Res 22:1775–1789

Derti A, Garrett-Engele P, MacIsaac KD et al (2012) A quantitative atlas of polyadenylation in five mammals. Genome Res 22:1173–1183

Di C, Yuan J, Wu Y et al (2014) Characterization of stress-responsive lncRNAs in Arabidopsis thaliana by integrating expression, epigenetic and structural features. Plant J 80:848–861

Diederichs S (2014) The four dimensions of noncoding RNA conservation. Trends Genet 30:121–123

Dinescu S, Ignat S, Lazar A et al (2019) Epitranscriptomic signatures in lncRNAs and their possible roles in cancer. Genes (Basel) 10:52

Ding Y, Tang Y, Kwok CK et al (2014) In vivo genome-wide profiling of RNA secondary structure reveals novel regulatory features. Nature 505:696–700

Dodd DW, Gagnon KT, Corey DR (2013) Digital quantitation of potential therapeutic target RNAs. Nucleic Acid Ther 23:188–194

Dominissini D, Moshitch-Moshkovitz S, Schwartz S et al (2012) Topology of the human and mouse m6A RNA methylomes revealed by m6A-seq. Nature 485:201–206

Dong Y, Yoshitomi T, Hu JF et al (2017) Long noncoding RNAs coordinate functions between mitochondria and the nucleus. Epigenetics Chromatin 10:41
Donlic A, Morgan BS, Xu JL et al (2018) Discovery of small molecule ligands for MALAT1 by tuning an RNA-binding scaffold. Angew Chem Int Ed Engl 57:13242–13247
Dragomir M, Chen B, Calin GA (2018) Exosomal lncRNAs as new players in cell-to-cell communication. Transl Cancer Res 7:S243–S252
Dunagin M, Cabili MN, Rinn J, Raj A (2015) Visualization of lncRNA by single-molecule fluorescence in situ hybridization. In: Methods in molecular biology. Humana Press, New York, NY, pp 3–19
Dunham I, Kundaje A, Aldred SF et al (2012) An integrated encyclopedia of DNA elements in the human genome. Nature 489:57–74
Engreitz JM, Sirokman K, McDonel P et al (2014) RNA-RNA interactions enable specific targeting of noncoding RNAs to nascent pre-mRNAs and chromatin sites. Cell 159:188–199
Engreitz J, Lander ES, Guttman M (2015) Nuclear bodies and noncoding RNAs. Springer, New York
Espejo RT, Plaza N (2018) Multiple ribosomal RNA operons in bacteria; their concerted evolution and potential consequences on the rate of evolution of their 16S rRNA. Front Microbiol 9:1232
Fang R, Moss WN, Rutenberg-Schoenberg M et al (2015) Probing Xist RNA structure in cells using targeted structure-seq. PLoS Genet 11:e1005668
Fang S, Zhang L, Guo J et al (2018) NONCODEV5: a comprehensive annotation database for long non-coding RNAs. Nucleic Acids Res 46:D308–D314
Fedoroff NV (2012) Transposable elements, epigenetics, and genome evolution. Science 338:758–767
Fort A, Hashimoto K, Yamada D et al (2014) Deep transcriptome profiling of mammalian stem cells supports a regulatory role for retrotransposons in pluripotency maintenance. Nat Genet 46:558–566
Frankish A, Diekhans M, Ferreira A-M, Johnson R, Jungreis I, Loveland J et al (2019) GENCODE reference annotation for the human and mouse genomes. Nucleic Acids Res 47(D1):D766–D773
Funk A, Truong K, Nagasaki T et al (2010) RNA structures required for production of subgenomic flavivirus RNA. J Virol 84:11407–11417
Gallo A, Vella S, Miele M et al (2017) Global profiling of viral and cellular non-coding RNAs in Epstein–Barr virus-induced lymphoblastoid cell lines and released exosome cargos. Cancer Lett 388:334–343
Ganem D (2006) KSHV infection and the pathogenesis of Kaposi's sarcoma. Annu Rev Pathol Mech Dis 1:273–296
Garalde DR, Snell EA, Jachimowicz D et al (2018) Highly parallel direct RNA sequencing on an array of nanopores. Nat Methods 15:201–206
Gascoigne DK, Cheetham SW, Cattenoz PB et al (2012) Pinstripe: a suite of programs for integrating transcriptomic and proteomic datasets identifies novel proteins and improves differentiation of protein-coding and non-coding genes. Bioinformatics 28:3042–3050
Gatherer D, Seirafian S, Cunningham C et al (2011) High-resolution human cytomegalovirus transcriptome. Proc Natl Acad Sci 108:19755–19760
Ge Y, Yan X, Jin Y et al (2015) MiRNA-192 and miRNA-204 directly suppress lncRNA HOTTIP and interrupt GLS1-mediated glutaminolysis in hepatocellular carcinoma. PLoS Genet 11:e1005726
Geary RS, Baker BF, Crooke ST (2015) Clinical and preclinical pharmacokinetics and pharmacodynamics of mipomersen (Kynamro®): a second-generation antisense oligonucleotide inhibitor of apolipoprotein B. Clin Pharmacokinet 54:133–146
Geisler S, Lojek L, Khalil AM et al (2012) Decapping of long noncoding RNAs regulates inducible genes. Mol Cell 45:279–291
Ghosh SK, Patton JR, Spanjaard RA (2012) A small RNA derived from RNA coactivator SRA blocks steroid receptor signaling via inhibition of Pus1p-mediated pseudouridylation of SRA:

evidence of a novel RNA binding domain in the N-terminus of steroid receptors. Biochemistry 51:8163–8172
Gibb EA, Vucic EA, Enfield KSS et al (2011) Human cancer long non-coding RNA transcriptomes. PLoS One 6:e25915
Ginno PA, Lott PL, Christensen HC et al (2012) R-loop formation is a distinctive characteristic of unmethylated human CpG island promoters. Mol Cell 45:814–825
Göertz GP, Abbo SR, Fros JJ et al (2018) Functional RNA during Zika virus infection. Virus Res 254:41–53
Gofrit ON, Benjamin S, Halachmi S et al (2014) DNA based therapy with diphtheria toxin-a BC-819: a phase 2b marker lesion trial in patients with intermediate risk nonmuscle invasive bladder cancer. J Urol 191:1697–1702
Gudenas BL, Wang L (2018) Prediction of LncRNA subcellular localization with deep learning from sequence features. Sci Rep 8:16385
Guil S, Soler M, Portela A et al (2012) Intronic RNAs mediate EZH2 regulation of epigenetic targets. Nat Struct Mol Biol 19:664–670
Gupta RA, Shah N, Wang KC et al (2010) Long non-coding RNA HOTAIR reprograms chromatin state to promote cancer metastasis. Nature 464:1071–1076
Gutschner T, Hämmerle M, Eißmann M et al (2013) The noncoding RNA MALAT1 is a critical regulator of the metastasis phenotype of lung cancer cells. Cancer Res 73:1180–1189
Guttman M, Amit I, Garber M et al (2009) Chromatin signature reveals over a thousand highly conserved large non-coding RNAs in mammals. Nature 458:223–227
Guttman M, Garber M, Levin JZ et al (2010) Ab initio reconstruction of cell type-specific transcriptomes in mouse reveals the conserved multi-exonic structure of lincRNAs. Nat Biotechnol 28:503–510
Hacisuleyman E, Goff LA, Trapnell C et al (2014) Topological organization of multichromosomal regions by the long intergenic noncoding RNA Firre. Nat Struct Mol Biol 21:198–206
Halley P, Kadakkuzha BM, Faghihi MA et al (2014) Regulation of the apolipoprotein gene cluster by a long noncoding RNA. Cell Rep 6:222–230
Hamazaki N, Nakashima K, Hayashi K, Imamura T (2017) Detection of bidirectional promoter-derived lncRNAs from small-scale samples using pre-amplification-free directional RNA-seq method. In: Methods in molecular biology. Humana Press, New York, NY, pp 83–103
Han J, Zhang J, Chen L et al (2014) Efficient in vivo deletion of a large imprinted lncRNA by CRISPR/Cas9. RNA Biol 11:829–835
Hardwick SA, Bassett SD, Kaczorowski D et al (2019) Targeted, high-resolution RNA sequencing of non-coding genomic regions associated with neuropsychiatric functions. Front Genet 10:309
Harris KA, Breaker RR (2018) Large noncoding RNAs in bacteria. Microbiol Spectr 6:6
Harris KA, Zhou Z, Peters ML et al (2018) A second RNA-binding protein is essential for ethanol tolerance provided by the bacterial OLE ribonucleoprotein complex. Proc Natl Acad Sci U S A 115:E6319–E6328
Hasegawa Y, Brockdorff N, Kawano S et al (2010) The matrix protein hnRNP U is required for chromosomal localization of Xist RNA. Dev Cell 19:469–476
Hautbergue GM (2017) RNA nuclear export: from neurological disorders to cancer. In: Advances in experimental medicine and biology. Humana Press, New York, NY, pp 89–109
Hawkes EJ, Hennelly SP, Novikova IV et al (2016) COOLAIR antisense RNAs form evolutionarily conserved elaborate secondary structures. Cell Rep 16:3087–3096
Hawkins PG, Morris KV (2010) Transcriptional regulation of Oct4 by a long non-coding RNA antisense to Oct4-pseudogene 5. Transcription 1:165–175
He S, Liu S, Zhu H (2011) The sequence, structure and evolutionary features of HOTAIR in mammals. BMC Evol Biol 11:102
He S, Guynet C, Siguier P et al (2013) IS 200/IS 605 family single-strand transposition: mechanism of IS 608 strand transfer. Nucleic Acids Res 41:3302–3313
Hendrickson DG, Kelley DR, Tenen D et al (2016) Widespread RNA binding by chromatin-associated proteins. Genome Biol 17:28

Heo JB, Sung S (2011) Vernalization-mediated epigenetic silencing by a long Intronic noncoding RNA. Science 331:76–79
Hezroni H, Koppstein D, Schwartz MG et al (2015) Principles of long noncoding RNA evolution derived from direct comparison of transcriptomes in 17 species. Cell Rep 11:1110–1122
Hinten M, Maclary E, Gayen S, Harris C, Kalantry S (2016) Visualizing long noncoding RNAs on chromatin. In: Methods in molecular biology. Humana Press, New York, NY, pp 147–164
Ho T-T, Zhou N, Huang J et al (2015) Targeting non-coding RNAs with the CRISPR/Cas9 system in human cell lines. Nucleic Acids Res 43:e17
Holoch D, Moazed D (2015) RNA-mediated epigenetic regulation of gene expression. Nat Publ Gr 16:71–84
Hon CC, Ramilowski JA, Harshbarger J et al (2017) An atlas of human long non-coding RNAs with accurate 5′ ends. Nature 543:199–204
Hoque M, Ji Z, Zheng D et al (2013) Analysis of alternative cleavage and polyadenylation by 3′ region extraction and deep sequencing. Nat Methods 10:133–139
Hu M, Polyak K (2006) Serial analysis of gene expression. Nat Protoc 1:1743–1760
Huang JZ, Chen M, Chen D et al (2017) A peptide encoded by a putative lncRNA HOXB-AS3 suppresses colon cancer growth. Mol Cell 68:171–184.e6
Hubé F, Velasco G, Rollin J et al (2011) Steroid receptor RNA activator protein binds to and counteracts SRA RNA-mediated activation of MyoD and muscle differentiation. Nucleic Acids Res 39:513–525
Hutchinson JN, Ensminger AW, Clemson CM et al (2007) A screen for nuclear transcripts identifies two linked noncoding RNAs associated with SC35 splicing domains. BMC Genomics 8:39
Ingolia NT, Lareau LF, Weissman JS (2011) Ribosome profiling of mouse embryonic stem cells reveals the complexity and dynamics of mammalian proteomes. Cell 147:789–802
Janssen BD, Hayes CS (2012) The tmRNA ribosome-rescue system. Adv Protein Chem Struct Biol 86:151–191
Jean Finnegan E, Kovac KA, Jaligot E et al (2005) The downregulation of FLOWERING LOCUS C (FLC) expression in plants with low levels of DNA methylation and by vernalization occurs by distinct mechanisms. Plant J 44:420–432
Ji P, Diederichs S, Wang W et al (2003) MALAT-1, a novel noncoding RNA, and thymosin β4 predict metastasis and survival in early-stage non-small cell lung cancer. Oncogene 22:8031–8041
Johnson R, Guigó R (2014) The RIDL hypothesis: transposable elements as functional domains of long noncoding RNAs. RNA 20:959–976
Johnson JM, Edwards S, Shoemaker D et al (2005) Dark matter in the genome: evidence of widespread transcription detected by microarray tiling experiments. Trends Genet 21:93–102
Julsing MK, Rijpkema M, Woerdenbag HJ et al (2007) Functional analysis of genes involved in the biosynthesis of isoprene in Bacillus subtilis. Appl Microbiol Biotechnol 75:1377–1384
Kallen AN, Zhou XB, Xu J et al (2013) The imprinted H19 LncRNA antagonizes Let-7 MicroRNAs. Mol Cell 52:101–112
Kalwa M, Hänzelmann S, Otto S et al (2016) The lncRNA HOTAIR impacts on mesenchymal stem cells via triple helix formation. Nucleic Acids Res 44:10631–10643
Kapusta A, Kronenberg Z, Lynch VJ et al (2013) Transposable elements are major contributors to the origin, diversification, and regulation of vertebrate long noncoding RNAs. PLoS Genet 9: e1003470
Katayama S, Tomaru Y, Kasukawa T et al (2005) Antisense transcription in the mammalian transcriptome. Science 309:1564–1566
Kelley D, Rinn J (2012) Transposable elements reveal a stem cell-specific class of long noncoding RNAs. Genome Biol 13:R107
Kertesz M, Wan Y, Mazor E et al (2010) Genome-wide measurement of RNA secondary structure in yeast. Nature 467:103–107
Khare SP, Habib F, Sharma R et al (2012) HIstome – a relational knowledgebase of human histone proteins and histone modifying enzymes. Nucleic Acids Res 40:D337–D342

Kim D-H, Sung S (2017) Vernalization-triggered intragenic chromatin loop formation by long noncoding RNAs. Dev Cell 40:302–312.e4
Kim DH, Marinov GK, Pepke S et al (2015) Single-cell transcriptome analysis reveals dynamic changes in lncRNA expression during reprogramming. Cell Stem Cell 16:88–101
Kim D-H, Xi Y, Sung S (2017) Modular function of long noncoding RNA, COLDAIR, in the vernalization response. PLoS Genet 13:e1006939
Kim HJ, Kim LK, Lee SH et al (2018) Expression levels of the long noncoding RNA steroid receptor activator promote cell proliferation and invasion and predict patient prognosis in human cervical cancer. Oncol Lett 16:5410–5418
König J, Zarnack K, Rot G et al (2010) ICLIP reveals the function of hnRNP particles in splicing at individual nucleotide resolution. Nat Struct Mol Biol 17:909–915
Korostowski L, Raval A, Breuer G et al (2011) Enhancer-driven chromatin interactions during development promote escape from silencing by a long non-coding RNA. Epigenetics Chromatin 4:21
Kruijt M, DE Kock MJD, de Wit PJGM (2005) Receptor-like proteins involved in plant disease resistance. Mol Plant Pathol 6:85–97
Kuan W-L, Poole E, Fletcher M et al (2012) A novel neuroprotective therapy for Parkinson's disease using a viral noncoding RNA that protects mitochondrial complex I activity. J Exp Med 209:1–10
Kulesza CA, Shenk T (2004) Human cytomegalovirus 5-kilobase immediate-early RNA is a stable intron. J Virol 78:13182–13189
Kurihara Y, Matsui A, Hanada K et al (2009) Genome-wide suppression of aberrant mRNA-like noncoding RNAs by NMD in Arabidopsis. Proc Natl Acad Sci U S A 106:2453–2458
Lachner M (2003) An epigenetic road map for histone lysine methylation. J Cell Sci 116:2117–2124
Lanz RB, McKenna NJ, Onate SA et al (1999) A steroid receptor coactivator, SRA, functions as an RNA and is present in an SRC-1 complex. Cell 97:17–27
Lanz RB, Razani B, Goldberg AD et al (2002) Distinct RNA motifs are important for coactivation of steroid hormone receptors by steroid receptor RNA activator (SRA). Proc Natl Acad Sci U S A 99:16081–16086
Lanz RB, Chua SS, Barron N et al (2003) Steroid receptor RNA activator stimulates proliferation as well as apoptosis in vivo. Mol Cell Biol 23:7163–7176
Launer-Felty K, Wong CJ, Cole JL (2015) Structural analysis of adenovirus VAI RNA defines the mechanism of inhibition of PKR. Biophys J 108:748–757
Laverdure S, Gross A, Arpin-Andre C et al (2012) HIV-1 antisense transcription is preferentially activated in primary monocyte-derived cells. J Virol 86:13785–13789
Lee YJ, Glaunsinger BA (2009) Aberrant herpesvirus-induced polyadenylation correlates with cellular messenger RNA destruction. PLoS Biol 7:e1000107
Lee HJ, Gopalappa R, Sunwoo H et al (2019) En bloc and segmental deletions of human XIST reveal X chromosome inactivation-involving RNA elements. Nucleic Acids Res 47:3875–3887
Leucci E, Patella F, Waage J et al (2013) microRNA-9 targets the long non-coding RNA MALAT1 for degradation in the nucleus. Sci Rep 3:2535
Li F, Huang C, Li Z et al (2014) Suppression of RNA silencing by a plant DNA virus satellite requires a host calmodulin-like protein to repress RDR6 expression. PLoS Pathog 10:e1003921
Li F, Xiao Y, Huang F et al (2015a) Spatiotemporal-specific lncRNAs in the brain, colon, liver and lung of macaque during development. Mol BioSyst 11:3253–3263
Li X, Zhu P, Ma S et al (2015b) Chemical pulldown reveals dynamic pseudouridylation of the mammalian transcriptome. Nat Chem Biol 11:592–597
Liang D, Wilusz JE (2014) Short intronic repeat sequences facilitate circular RNA production. Genes Dev 28(20):2233–2247
Lin R, Maeda S, Liu C et al (2007) A large noncoding RNA is a marker for murine hepatocellular carcinomas and a spectrum of human carcinomas. Oncogene 26:851–858

Liu N, Pan T (2016) N6-methyladenosine–encoded epitranscriptomics. Nat Struct Mol Biol 23:98–102
Liu N, Dai Q, Zheng G et al (2015) N6-methyladenosine-dependent RNA structural switches regulate RNA-protein interactions. Nature 518:560–564
Liu H, Begik O, Lucas MC et al (2019) Accurate detection of m6A RNA modifications in native RNA sequences. Nat Commun 10:4079
Losko M, Kotlinowski J, Jura J (2016) Long noncoding RNAs in metabolic syndrome related disorders. Mediat Inflamm 2016:1–12
Lu Z, Zhang QC, Lee B et al (2016) RNA duplex map in living cells reveals higher-order transcriptome structure. Cell 165:1267–1279
Lu Z, Carter AC, Chang HY (2017) Mechanistic insights in X-chromosome inactivation. Philos Trans R Soc B Biol Sci 372:20160356
Luo G, Wang M, Wu X, Tao D, Xiao X, Wang L et al (2015) Long non-coding RNA MEG3 inhibits cell proliferation and induces apoptosis in prostate cancer. Cell Physiol Biochem 37(6):2209–2220
Ma Y, Mathews MB (1996) Structure, function, and evolution of adenovirus-associated RNA: a phylogenetic approach. J Virol 70:5083–5099
Mancarella A, Procopio FA, Achsel T et al (2019) Detection of antisense protein (ASP) RNA transcripts in individuals infected with human immunodeficiency virus type 1 (HIV-1). J Gen Virol 100:863–876
Mann H, Ben-Asouli Y, Schein A et al (2003) Eukaryotic RNase P. Mol Cell 12:925–935
Manokaran G, Finol E, Wang C et al (2015) Dengue subgenomic RNA binds TRIM25 to inhibit interferon expression for epidemiological fitness. Science 350:217–221
Marques AC, Hughes J, Graham B et al (2013) Chromatin signatures at transcriptional start sites separate two equally populated yet distinct classes of intergenic long noncoding RNAs. Genome Biol 14:R131
Martin KC, Ephrussi A (2009) mRNA localization: gene expression in the spatial dimension. Cell 136:719–730
Massimelli MJ, Kang J-GG, Majerciak V et al (2011) Stability of a long noncoding viral RNA depends on a 9-nt core element at the RNA 5′ end to interact with viral ORF57 and cellular PABPC1. Int J Biol Sci 7:1145–1160
Massimelli MJ, Majerciak V, Kruhlak M et al (2013) Interplay between polyadenylate-binding protein 1 and Kaposi's sarcoma-associated herpesvirus ORF57 in accumulation of Polyadenylated nuclear RNA, a viral long noncoding RNA. J Virol 87:243–256
Matsumoto A, Pasut A, Matsumoto M et al (2017) MTORC1 and muscle regeneration are regulated by the LINC00961-encoded SPAR polypeptide. Nature 541:228–232
Mattick JS, Rinn JL (2015) Discovery and annotation of long noncoding RNAs. Nat Struct Mol Biol 22:5–7
Mayama T, Marr A, Kino T (2016) Differential expression of glucocorticoid receptor noncoding RNA repressor Gas5 in autoimmune and inflammatory diseases. Horm Metab Res 48:550–557
McHugh CA, Chen C-K, Chow A et al (2015) The Xist lncRNA interacts directly with SHARP to silence transcription through HDAC3. Nature 521:232–236
McKay DB, Xi L, Barthel KKB et al (2014) Structure and function of steroid receptor RNA activator protein, the proposed partner of SRA noncoding RNA. J Mol Biol 426:1766–1785
Meseure D, Drak Alsibai K, Nicolas A et al (2015) Long noncoding RNAs as new architects in cancer epigenetics, prognostic biomarkers, and potential therapeutic targets. Biomed Res Int 2015:1–14
Meyer KD, Saletore Y, Zumbo P et al (2012) Comprehensive analysis of mRNA methylation reveals enrichment in 3′ UTRs and near stop codons. Cell 149:1635–1646
Miao Y, Xu S-Y, Chen L-S et al (2017) Trends of long noncoding RNA research from 2007 to 2016: a bibliometric analysis. Oncotarget 8:83114–83127
Mitton-Fry RM, DeGregorio SJ, Wang J et al (2010) Poly(A) tail recognition by a viral RNA element through assembly of a triple helix. Science 330:1244–1247

Miyagawa R, Tano K, Mizuno RIE et al (2012) Identification of cis- and trans-acting factors involved in the localization of MALAT-1 noncoding RNA to nuclear speckles. RNA 18:738–751
Modarresi F, Faghihi MA, Lopez-Toledano MA et al (2012) Inhibition of natural antisense transcripts in vivo results in gene-specific transcriptional upregulation. Nat Biotechnol 30:453–459
Moindrot B, Cerase A, Coker H et al (2015) A pooled shRNA screen identifies Rbm15, Spen, and Wtap as factors required for Xist RNA-mediated silencing. Cell Rep 12:562–572
Mondal T, Subhash S, Vaid R et al (2015) MEG3 long noncoding RNA regulates the TGF-β pathway genes through formation of RNA–DNA triplex structures. Nat Commun 6:7743
Monfort A, Di Minin G, Postlmayr A et al (2015) Identification of Spen as a crucial factor for Xist function through forward genetic screening in haploid embryonic stem cells. Cell Rep 12:554–561
Munschauer M, Nguyen CT, Sirokman K et al (2018) The NORAD lncRNA assembles a topoisomerase complex critical for genome stability. Nature 561:132–136
Murphy LC, Simon SLR, Parkes A et al (2000) Altered expression of estrogen receptor coregulators during human breast tumorigenesis. Cancer Res 60:6266–6271
Mus E, Hof PR, Tiedge H (2007) Dendritic BC200 RNA in aging and in Alzheimer's disease. Proc Natl Acad Sci 104:10679–10684
Nakagawa S, Ip JY, Shioi G et al (2012) Malat1 is not an essential component of nuclear speckles in mice. RNA 18:1487–1499
Necsulea A, Soumillon M, Warnefors M et al (2014) The evolution of lncRNA repertoires and expression patterns in tetrapods. Nature 505:635–640
Nicot N, Hausman J-F, Hoffmann L et al (2005) Housekeeping gene selection for real-time RT-PCR normalization in potato during biotic and abiotic stress. J Exp Bot 56:2907–2914
Niemczyk M, Ito Y, Huddleston J et al (2013) Imprinted chromatin around DIRAS3 regulates alternative splicing of GNG12-AS1, a long noncoding RNA. Am J Hum Genet 93:224–235
Nishimoto Y, Nakagawa S, Hirose T et al (2013) The long non-coding RNA nuclear-enriched abundant transcript 1_2 induces paraspeckle formation in the motor neuron during the early phase of amyotrophic lateral sclerosis. Mol Brain 6:31
Noh JH, Kim KM, Abdelmohsen K et al (2016) HuR and GRSF1 modulate the nuclear export and mitochondrial localization of the lncRNA RMRP. Genes Dev 30:1224–1239
Noh JH, Kim KM, McClusky WG et al (2018) Cytoplasmic functions of long noncoding RNAs. Wiley Interdiscip Rev RNA 9:e1471
Noriega VM, Haye KK, Kraus TA et al (2014) Human cytomegalovirus modulates monocyte-mediated innate immune responses during short-term experimental latency in vitro. J Virol 88:9391–9405
Novikova IV, Hennelly SP, Sanbonmatsu KY (2012) Structural architecture of the human long non-coding RNA, steroid receptor RNA activator. Nucleic Acids Res 40:5034–5051
Novikova IV, Dharap A, Hennelly SP et al (2013) 3S: shotgun secondary structure determination of long non-coding RNAs. Methods 63:170–177
Obaid M, Udden SMN, Deb P et al (2018) LncRNA HOTAIR regulates lipopolysaccharide-induced cytokine expression and inflammatory response in macrophages. Sci Rep 8:15670
Parker MT, Knop K, Sherwood AV et al (2019) Nanopore direct RNA sequencing maps an Arabidopsis N6 methyladenosine epitranscriptome. bioRxiv. https://doi.org/10.1101/706002
Patil DP, Chen C-K, Pickering BF et al (2016) m6A RNA methylation promotes XIST-mediated transcriptional repression. Nature 537:369–373
Payne S (2017) Family flaviviridae. In: Viruses. Elsevier, Amsterdam, Netherlands, pp 129–139
Paytuví Gallart A, Hermoso Pulido A, Anzar Martínez de Lagrán I et al (2016) GREENC: a Wiki-based database of plant lncRNAs. Nucleic Acids Res 44:D1161–D1166
Pintacuda G, Wei G, Roustan C et al (2017) hnRNPK recruits PCGF3/5-PRC1 to the Xist RNA B-repeat to establish polycomb-mediated chromosomal silencing. Mol Cell 68:955–969.e10

Plessy C, Bertin N, Takahashi H et al (2010) Linking promoters to functional transcripts in small samples with nanoCAGE and CAGEscan. Nat Methods 7:528–534

Ploussard G, Durand X, Xylinas E et al (2011) Prostate cancer antigen 3 score accurately predicts tumour volume and might help in selecting prostate Cancer patients for active surveillance. Eur Urol 59:422–429

Ponjavic J, Ponting CP, Lunter G (2007) Functionality or transcriptional noise? Evidence for selection within long noncoding RNAs. Genome Res 17:556–565

Poole E, Kuan WL, Barker R et al (2016) The human cytomegalovirus non-coding Beta2.7 RNA as a novel therapeutic for Parkinson's disease – translational research with no translation. Virus Res 212:64–69

Puerta-Fernandez E, Barrick JE, Roth A et al (2006) Identification of a large noncoding RNA in extremophilic eubacteria. Proc Natl Acad Sci 103:19490–19495

Quinn JJ, Chang HY (2016) Unique features of long non-coding RNA biogenesis and function. Nat Rev Genet 17:47–62

Rackham O, Shearwood A-MJ, Mercer TR et al (2011) Long noncoding RNAs are generated from the mitochondrial genome and regulated by nuclear-encoded proteins. RNA 17:2085–2093

Raj A, van den Bogaard P, Rifkin SA et al (2008) Imaging individual mRNA molecules using multiple singly labeled probes. Nat Methods 5:877–879

Raya RR, H'bert EM (2009) Isolation of phage via induction of lysogens. In: Clokie MR, Kropinski AM (eds) Bacteriophages, Methods in molecular biology, vol 501. Humana Press, New York, NY

Reeves MB, Davies AA, McSharry BP et al (2007) Complex I binding by a virally encoded RNA regulates mitochondria-induced cell death. Science 316:1345–1348

Rinn JL, Kertesz M, Wang JK et al (2007) Functional demarcation of active and silent chromatin domains in human HOX loci by noncoding RNAs. Cell 129:1311–1323

Roobol MJ, Schröder FH, van Leenders GLJH et al (2010) Performance of prostate cancer antigen 3 (PCA3) and prostate-specific antigen in prescreened men: reproducibility and detection characteristics for prostate cancer patients with high PCA3 scores ($\geq$100). Eur Urol 58:893–899

Rossetto CC, Pari GS (2011) Kaposi's sarcoma-associated herpesvirus noncoding polyadenylated nuclear RNA interacts with virus- and host cell-encoded proteins and suppresses expression of genes involved in immune modulation. J Virol 85:13290–13297

Rossetto CC, Pari G (2012) KSHV PAN RNA associates with demethylases UTX and JMJD3 to activate lytic replication through a physical interaction with the virus genome. PLoS Pathog 8: e1002680

Rossetto CC, Tarrant-Elorza M, Pari GS (2013a) Cis and trans acting factors involved in human cytomegalovirus experimental and natural latent infection of CD14 (+) monocytes and CD34 (+) cells. PLoS Pathog 9:e1003366

Rossetto CC, Tarrant-Elorza M, Verma S et al (2013b) Regulation of viral and cellular gene expression by Kaposi's sarcoma-associated herpesvirus polyadenylated nuclear RNA. J Virol 87:5540–5553

Roundtree IA, Luo GZ, Zhang Z et al (2017) YTHDC1 mediates nuclear export of N6-methyladenosine methylated mRNAs. elife 6:e31311

Rouskin S, Zubradt M, Washietl S et al (2014) Genome-wide probing of RNA structure reveals active unfolding of mRNA structures in vivo. Nature 505:701–705

Saayman S, Ackley A, Turner A-MW et al (2014) An HIV-encoded antisense long noncoding RNA epigenetically regulates viral transcription. Mol Ther 22:1164–1175

Sahakyan A, Yang Y, Plath K (2018) The role of Xist in X-chromosome dosage compensation. Trends Cell Biol 28:999–1013

Schnettler E, Sterken MG, Leung JY et al (2012) Noncoding flavivirus RNA displays RNA interference suppressor activity in insect and mammalian cells. J Virol 86:13486–13500

Schnettler E, Tykalová H, Watson M et al (2014) Induction and suppression of tick cell antiviral RNAi responses by tick-borne flaviviruses. Nucleic Acids Res 42:9436–9446

Schwartz S, Bernstein DA, Mumbach MR et al (2014) Transcriptome-wide mapping reveals widespread dynamic-regulated pseudouridylation of ncRNA and mRNA. Cell 159:148–162

Schwarz TM, Kulesza CA (2014) Long ncRNAs expressed during human cytomegalovirus infections. Future Virol 9:587–594

Sei E, Conrad NK (2011) Delineation of a core RNA element required for Kaposi's sarcoma-associated herpesvirus ORF57 binding and activity. Virology 419:107–116

Seiler J, Breinig M, Caudron-Herger M et al (2017) The lncRNA VELUCT strongly regulates viability of lung cancer cells despite its extremely low abundance. Nucleic Acids Res 45:5458–5469

Shen X, Corey DR (2018) Chemistry, mechanism and clinical status of antisense oligonucleotides and duplex RNAs. Nucleic Acids Res 46:1584–1600

Sheng L, Ye L, Zhang D et al (2018) New insights into the long non-coding RNA SRA: physiological functions and mechanisms of action. Front Med 5:244

Sherpa C, Rausch JW, Le Grice SF (2018) Structural characterization of maternally expressed gene 3 RNA reveals conserved motifs and potential sites of interaction with polycomb repressive complex 2. Nucleic Acids Res 46:10432–10447

Shukla GC, Singh J, Barik S (2011) MicroRNAs: processing, maturation, target recognition and regulatory functions. Mol Cell Pharmacol 3:83–92

Sidi AA, Ohana P, Benjamin S et al (2008) Phase I/II marker lesion study of Intravesical BC-819 DNA plasmid in H19 over expressing superficial bladder cancer refractory to Bacillus Calmette-Guerin. J Urol 180:2379–2383

Siegfried NA, Busan S, Rice GM et al (2014) RNA motif discovery by SHAPE and mutational profiling (SHAPE-MaP). Nat Methods 11:959–965

Simon MD, Wang CI, Kharchenko PV et al (2011) The genomic binding sites of a noncoding RNA. Proc Natl Acad Sci U S A 108:20497–20502

Skourti-Stathaki K, Proudfoot NJ, Gromak N (2011) Human senataxin resolves RNA/DNA hybrids formed at transcriptional pause sites to promote Xrn2-dependent termination. Mol Cell 42:794–805

Sleutels F, Zwart R, Barlow DP (2002) The non-coding air RNA is required for silencing autosomal imprinted genes. Nature 415:810–813

Smith MA, Gesell T, Stadler PF et al (2013) Widespread purifying selection on RNA structure in mammals. Nucleic Acids Res 41:8220–8236

Smola MJ, Christy TW, Inoue K et al (2016) SHAPE reveals transcript-wide interactions, complex structural domains, and protein interactions across the Xist lncRNA in living cells. Proc Natl Acad Sci U S A 113:10322–10327

Soares RJ, Maglieri G, Gutschner T et al (2018) Evaluation of fluorescence in situ hybridization techniques to study long non-coding RNA expression in cultured cells. Nucleic Acids Res 46: e4–e4

Somarowthu S, Legiewicz M, Chillón I et al (2015) HOTAIR forms an intricate and modular secondary structure. Mol Cell 58:353–361

Song J, Angel A, Howard M et al (2012) Vernalization – a cold-induced epigenetic switch. J Cell Sci 125:3723–3731

Soto-Acosta R, Xie X, Shan C et al (2018) Fragile X mental retardation protein is a Zika virus restriction factor that is antagonized by subgenomic flaviviral RNA. elife 7:e39023

Squires JE, Patel HR, Nousch M et al (2012) Widespread occurrence of 5-methylcytosine in human coding and non-coding RNA. Nucleic Acids Res 40:5023–5033

Stoddard BL (2011) Homing endonucleases: from microbial genetic invaders to reagents for targeted DNA modification. Structure 19:7–15

Su ZD, Huang Y, Zhang ZY et al (2018) ILoc-lncRNA: predict the subcellular location of lncRNAs by incorporating octamer composition into general PseKNC. Bioinformatics 34:4196–4204

Sun Y, Ma L (2019) New insights into long non-coding RNA MALAT1 in cancer and metastasis. Cancers (Basel) 11:216

Sun Q, Csorba T, Skourti-Stathaki K et al (2013) R-loop stabilization represses antisense transcription at the Arabidopsis FLC locus. Science 340:619–621
Sunwoo H, Wu JY, Lee JT (2015) The Xist RNA-PRC2 complex at 20-nm resolution reveals a low Xist stoichiometry and suggests a hit-and-run mechanism in mouse cells. Proc Natl Acad Sci U S A 112:E4216–E4225
Sunwoo H, Colognori D, Froberg JE et al (2017) Repeat E anchors Xist RNA to the inactive X chromosomal compartment through CDKN1A-interacting protein (CIZ1). Proc Natl Acad Sci U S A 114:10654–10659
Sztuba-Solinska J, Rausch JW, Smith R et al (2017) Kaposi's sarcoma-associated herpesvirus polyadenylated nuclear RNA: a structural scaffold for nuclear, cytoplasmic and viral proteins. Nucleic Acids Res 45:6805–6821
Tani H, Torimura M, Akimitsu N (2013) The RNA degradation pathway regulates the function of GAS5 a non-coding RNA in mammalian cells. PLoS One 8:e55684
Thakore PI, D'Ippolito AM, Song L et al (2015) Highly specific epigenome editing by CRISPR-Cas9 repressors for silencing of distal regulatory elements. Nat Methods 12:1143–1149
Torresilla C, Larocque E, Landry S et al (2013) Detection of the HIV-1 minus-strand-encoded antisense protein and its association with autophagy. J Virol 87:5089–5105
Tripathi V, Ellis JD, Shen Z et al (2010) The nuclear-retained noncoding RNA MALAT1 regulates alternative splicing by modulating SR splicing factor phosphorylation. Mol Cell 39:925–938
Tsai MC, Manor O, Wan Y et al (2010) Long noncoding RNA as modular scaffold of histone modification complexes. Science 329:689–693
Tunnicliffe RB, Levy C, Ruiz Nivia HD et al (2019) Structural identification of conserved RNA binding sites in herpesvirus ORF57 homologs: implications for PAN RNA recognition. Nucleic Acids Res 47:1987–2001
Underwood JG, Uzilov AV, Katzman S et al (2010) FragSeq: transcriptome-wide RNA structure probing using high-throughput sequencing. Nat Methods 7:995–1001
Uroda T, Anastasakou E, Rossi A et al (2019) Conserved pseudoknots in lncRNA MEG3 are essential for stimulation of the p53 pathway. Mol Cell 75:982–995.e9
Vachon VK, Conn GL (2016) Adenovirus VA RNA: an essential pro-viral non-coding RNA. Virus Res 212:39–52
Vakoc CR, Mandat SA, Olenchock BA et al (2005) Histone H3 lysine 9 methylation and HP1γ are associated with transcription elongation through mammalian chromatin. Mol Cell 19:381–391
Vezzoli A, Bonadies N, Allen MD et al (2010) Molecular basis of histone H3K36me3 recognition by the PWWP domain of Brpf1. Nat Struct Mol Biol 17:617–619
Viphakone N, Sudbery I, Griffith L et al (2019) Co-transcriptional loading of RNA export factors shapes the human transcriptome. Mol Cell 75:310–323.e8
Voigt P, Tee WW, Reinberg D (2013) A double take on bivalent promoters. Genes Dev 27:1318–1338
Wahid AM, Coventry VK, Conn GL (2008) Systematic deletion of the adenovirus-associated RNA I terminal stem reveals a surprisingly active RNA inhibitor of double-stranded RNA-activated protein kinase. J Biol Chem 283:17485–17493
Wallace JG, Zhou Z, Breaker RR (2012) OLE RNA protects extremophilic bacteria from alcohol toxicity. Nucleic Acids Res 40:6898–6907
Wang W, Zhu Y, Li S et al (2016) Long noncoding RNA MALAT1 promotes malignant development of esophageal squamous cell carcinoma by targeting β-catenin via Ezh2. Oncotarget 7:25668–25682
Wang J, Du S, Wang J et al (2017a) The prognostic value of abnormally expressed lncRNAs in colorectal cancer: a meta-analysis. PLoS One 12:e0179670
Wang J, Meng X, Dobrovolskaya OB et al (2017b) Non-coding RNAs and their roles in stress response in plants. Genomics Proteomics Bioinformatics 15:301–312
Wang J, Ye C, Xiong H et al (2017c) Dysregulation of long non-coding RNA in breast cancer: an overview of mechanism and clinical implication. Oncotarget 8:5508–5522

Ward M, McEwan C, Mills JD et al (2015) Conservation and tissue-specific transcription patterns of long noncoding RNAs. J Hum Transcr 1:2–9
Watters KE, Yu AM, Strobel EJ et al (2016) Characterizing RNA structures in vitro and in vivo with selective 2′-hydroxyl acylation analyzed by primer extension sequencing (SHAPE-Seq). Methods 103:34–48
Weile C, Gardner PP, Hedegaard MM et al (2007) Use of tiling array data and RNA secondary structure predictions to identify noncoding RNA genes. BMC Genomics 8:244
Weinberg Z, Perreault J, Meyer MM et al (2009) Exceptional structured noncoding RNAs revealed by bacterial metagenome analysis. Nature 462:656–659
Wery M, Descrimes M, Vogt N et al (2016) Nonsense-mediated decay restricts LncRNA levels in yeast unless blocked by double-stranded RNA structure. Mol Cell 61:379–392
West JA, Davis CP, Sunwoo H et al (2014) The long noncoding RNAs NEAT1 and MALAT1 bind active chromatin sites. Mol Cell 55:791–802
Wheeler TM, Leger AJ, Pandey SK et al (2012) Targeting nuclear RNA for in vivo correction of myotonic dystrophy. Nature 488:111–115
Wierzbicki AT, Haag JR, Pikaard CS (2008) Noncoding transcription by RNA polymerase pol IVb/pol V mediates transcriptional silencing of overlapping and adjacent genes. Cell 135:635–648
Williams T, Ngo LH, Wickramasinghe VO (2018) Nuclear export of RNA: different sizes, shapes and functions. Semin Cell Dev Biol 75:70–77
Wilusz JE, Freier SM, Spector DL (2008) 3′ end processing of a long nuclear-retained noncoding RNA yields a tRNA-like cytoplasmic RNA. Cell 135:919–932
Wilusz JE, JnBaptiste CK, Lu LY et al (2012) A triple helix stabilizes the 3′ ends of long noncoding RNAs that lack poly(A) tails. Genes Dev 26:2392–2407
Wu JL, Meng FM, Li HJ (2018) High expression of lncRNA MEG3 participates in non-small cell lung cancer by regulating microRNA-7-5p. Eur Rev Med Pharmacol Sci 22:5938–5945
Wylie AA, Murphy SK, Orton TC et al (2000) Novel imprinted DLK1/GTL2 domain on human chromosome 14 contains motifs that mimic those implicated in IGF2/H19 regulation. Genome Res 10:1711–1718
Xu N, Segerman B, Zhou X et al (2007) Adenovirus virus-associated RNAII-derived small RNAs are efficiently incorporated into the RNA-induced silencing complex and associate with polyribosomes. J Virol 81:10540–10549
Yang X, Yang Y, Sun BF et al (2017) 5-methylcytosine promotes mRNA export-NSUN2 as the methyltransferase and ALYREF as an m 5 C reader. Cell Res 27:606–625
Yang G, Shen T, Yi X et al (2018) Crosstalk between long non-coding RNAs and Wnt/β-catenin signalling in cancer. J Cell Mol Med 22:2062–2070
Yin QF, Yang L, Zhang Y et al (2012) Long noncoding RNAs with snoRNA ends. Mol Cell 48:219–230
Yoon J-H, Abdelmohsen K, Kim J et al (2013) Scaffold function of long non-coding RNA HOTAIR in protein ubiquitination. Nat Commun 4:2939
Yoon J-H, Kim J, Gorospe M (2015) Long noncoding RNA turnover. Biochimie 117:15–21
Yu K, Chedin F, Hsieh C-L et al (2003) R-loops at immunoglobulin class switch regions in the chromosomes of stimulated B cells. Nat Immunol 4:442–451
Zapata JC, Campilongo F, Barclay RA et al (2017) The human immunodeficiency virus 1 ASP RNA promotes viral latency by recruiting the polycomb repressor complex 2 and promoting nucleosome assembly. Virology 506:34–44
Zeng C, Fukunaga T, Hamada M (2018) Identification and analysis of ribosome-associated lncRNAs using ribosome profiling data. BMC Genomics 19:414
Zhang X, Rice K, Wang Y et al (2010) Maternally expressed gene 3 (MEG3) noncoding ribonucleic acid: isoform structure, expression, and functions. Endocrinology 151:939–947
Zhang B, Gunawardane L, Niazi F et al (2014) A novel RNA motif mediates the strict nuclear localization of a long noncoding RNA. Mol Cell Biol 34:2318–2329

Zhang H, Zhu C, Zhao Y et al (2015) Long non-coding RNA expression profiles of hepatitis C virus-related dysplasia and hepatocellular carcinoma. Oncotarget 6:43770–43778
Zhang B, Mao YS, Diermeier SD, Novikova IV, Nawrocki EP, Jones TA et al (2017) Identification and characterization of a class of MALAT1-like genomic loci. Cell Rep 19(8):1723–1738
Zhao X, Patton JR, Ghosh SK et al (2007) Pus3p- and Pus1p-dependent pseudouridylation of steroid receptor RNA activator controls a functional switch that regulates nuclear receptor signaling. Mol Endocrinol 21:686–699
Zhao Y, Dunker W, Yu YT et al (2018) The role of noncoding RNA pseudouridylation in nuclear gene expression events. Front Bioeng Biotechnol 6:8
Zheng GXY, Do BT, Webster DE et al (2014) Dicer-microRNA-Myc circuit promotes transcription of hundreds of long noncoding RNAs. Nat Struct Mol Biol 21:585–590
Zhou Y, Zhong Y, Wang Y et al (2007) Activation of p53 by MEG3 non-coding RNA. J Biol Chem 282:24731–24742
Zhou KI, Parisien M, Dai Q et al (2016) N6-methyladenosine modification in a long noncoding RNA hairpin predisposes its conformation to protein binding. J Mol Biol 428:822–833
Zhu F, Xi D-H, Yuan S et al (2014) Salicylic acid and Jasmonic acid are essential for systemic resistance against tobacco mosaic virus in Nicotiana benthamiana. Mol Plant-Microbe Interact 27:567–577
Zhuo H, Tang J, Lin Z et al (2016) The aberrant expression of MEG3 regulated by UHRF1 predicts the prognosis of hepatocellular carcinoma. Mol Carcinog 55:209–219
Zong X, Huang L, Tripathi V, Peralta R, Freier SM, Guo S et al (2015) Knockdown of nuclear-retained long noncoding RNAs using modified DNA antisense oligonucleotides. In: Nakagawa S, Hirose T (eds) Nuclear bodies and noncoding RNAs, Methods in molecular biology, vol 1262. Humana Press, New York, NY

Evolving Roles of Long Noncoding RNAs

K. Lakshmi Narayanan, Xizi Wu, Haichao Wei, and Jia Qian Wu

Contents

Authors K. Lakshmi Narayanan and Xizi Wu contributed equally to this chapter and are listed as co-first authors.

K. Lakshmi Narayanan · X. Wu · H. Wei
The Vivian L. Smith Department of Neurosurgery, McGovern Medical School, The University of Texas Health Science Center at Houston, Houston, TX, USA

Center for Stem Cell and Regenerative Medicine, UT Brown Institution of Molecular Medicine, Houston, TX, USA
e-mail: Xizi.Wu@uth.tmc.edu; Haichao.Wei@uth.tmc.edu

J. Q. Wu (✉)
The Vivian L. Smith Department of Neurosurgery, McGovern Medical School, The University of Texas Health Science Center at Houston, Houston, TX, USA

Center for Stem Cell and Regenerative Medicine, UT Brown Institution of Molecular Medicine, Houston, TX, USA

MD Anderson Cancer Center UTHealth Graduate School of Biomedical Sciences, Houston, TX, USA
e-mail: jiaqian.wu@uth.tmc.edu

S. Jurga, J. Barciszewski (eds.), *The Chemical Biology of Long Noncoding RNAs*,
RNA Technologies 11, https://doi.org/10.1007/978-3-030-44743-4_2

Abstract The advancement of genomic research has allowed the large-scale discovery of long noncoding RNAs (lncRNAs). lncRNAs are defined as large and diverse class RNA transcripts longer than 200 bp with no protein-coding capacity. lncRNAs can be categorized based on their cellular location, structures, and mechanisms of function. lncRNAs are involved in many important biological processes such as gene transcription, epigenetic regulation and development. Although lncRNAs are not as well-conserved as protein-coding genes in sequence, it was reported that their structures are more conserved and are important for their functions. lncRNAs are highly cell type-specific and their expressions are dysregulated during disease pathogenesis. Characterization of lncRNAs may provide better solutions for diagnosis, prognosis, and targeted therapies. In this book chapter, we summarize the classification, localization, and structural studies of lncRNA. We will also review various mechanisms and physiological functions of lncRNAs, and biomarker potential of lncRNAs in neurological disorders. Finally, we will discuss the recent advancements in antisense oligonucleotide (ASO) chemistry for targeted delivery and other methods in targeting lncRNAs for potential disease treatment.

Keywords Long noncoding RNA · Function mechanism · Structure · Antisense oligonucleotide chemistry · Neurological disorders

1 Introduction

The advances in next-generation sequencing technologies have made it possible to uncover the complexity of human genome and transcriptome. 70–90% of the mammalian genome is transcribed at some point during development; however, only <2% of the genome is associated with protein-coding genes (Djebali et al. 2012). A large number of disease-causing mutations are in the noncoding regions of the genome according to the Encyclopedia of DNA Elements (ENCODE) project

(Hangauer et al. 2013). Noncoding RNAs (ncRNAs) have diverse biological roles. Briefly, ncRNAs can be divided into two categories based on their length: short noncoding RNAs (<200 nt), such as microRNAs (miRNAs), and long noncoding RNAs (lncRNAs, > = 200 nt) (Zampetaki et al. 2018). Recently, the human GENCODE project revealed that approximately 17,904 lncRNAs have been identified (according to release version 31) and the number is continuously growing. Even though lncRNAs are expressed in low levels, they are widely implicated in developmental processes and disease progression (Dong et al. 2016). Specifically, these lncRNAs have critical regulatory functions involving, for example, chromatin organization, gene transcription, post-transcriptional regulation, nuclear structural constitution, and protein metabolism (Cuevas-Diaz Duran et al. 2019). However, the mechanisms of many lncRNAs are still unknown and ongoing studies are in process to understand their roles in disease pathogenesis and their potential to serve as biomarkers and therapeutic targets. In the following sections, we will describe their intrinsic characteristics, functions, and disease-related implications.

2 The Classification, Localization, and Structures of lncRNAs

2.1 Definition and Classification of lncRNAs

2.1.1 lncRNAs

The lncRNA is a type of functional RNA molecule that can be transcribed from the genome but cannot be translated into proteins. Like protein-coding genes, most of lncRNAs, are transcribed by RNA polymerase II and are capped, polyadenylated, and spliced, while some are transcribed by RNA polymerase III (St Laurent et al. 2015). lncRNAs exist in a variety of species, including animals, plants, yeast, prokaryotes, and even viruses. However, unlike other well-known RNAs, such as mRNAs, lncRNAs are poorly conserved in sequence, whereas the secondary structure of lncRNAs has been reported to be conserved across different species (Ulitsky et al. 2011).

2.1.2 lncRNAs Classification

lncRNAs can be characterized into different categories based on, for example, their locations and transcription directions which contribute to their biological and pathological functions (St Laurent et al. 2015). Those transcribed from intergenic regions are named intergenic lncRNAs and transcripts overlapping with protein-coding genes are termed genic lncRNAs. Genic lncRNA can be divided into three categories: exonic (GE), intronic (GI), and overlapping (GO) according to their intersections with gene exons, introns, or entire genes. Another type of classification is based on lncRNA transcription directions in relation to protein-coding genes. Accordingly, there are two categories: sense lncRNAs and antisense lncRNAs. They may overlap with part of protein-coding genes or the entire sequence of a protein-coding gene. Sense lncRNAs are transcribed from the same strand of protein-coding genes. Antisense lncRNAs are transcribed from the opposite strand of protein-coding genes and may or may not overlap with portions of coding sequences and they are also termed as natural antisense transcripts (NATs) which include *cis*-NATs and *trans*-NATs (Cuevas-Diaz Duran et al. 2019; Jarroux et al. 2017). *Cis*-NATs are transcribed from the same genomic locus and they often regulate the expression of the corresponding sense transcripts, whereas *trans*-NATs are the transcripts derived from other genomic locations and can impact the expression of non-paired genes (Henz et al. 2007).

2.2 Subcellular Localization

The lncRNAs in the cells are transported to different subcellular locations to perform various regulatory functions in diverse biological processes (Carlevaro-Fita and Johnson 2019). lncRNAs exist both in the nucleus and cytoplasm (van Heesch et al. 2014). The heterogonous lncRNA localization patterns can give clues to lncRNA mechanisms. To visualize the subcellular localization of lncRNA, fluorescence in situ hybridization (FISH) is a widely used standard technique. By using single-molecule RNA FISH, 34 different lncRNAs have been examined and classified into five groups based on localization: large foci in the nucleus, large nuclear foci and single molecules scattered through the nucleus, predominantly nuclear without foci, cytoplasmic and nuclear, and predominantly cytoplasmic in different cell lines (Cabili et al. 2015).

lncRNAs localized in the nucleus often have important roles, including chromatin remodeling, transcriptional regulation, and acting as a part of the nuclear architecture (Chen and Carmichael 2010). lncRNAs can function in remodeling chromatin structure by interacting with chromatin modulating proteins and RNA-binding proteins (RBPs), and preventing integration of chromatin modulators like histone deacetylase to specific gene sites (Jain et al. 2016). Additionally, previous studies indicate some lncRNAs containing distinct motifs, such as *MALAT1* and *XIST*, can bind to nuclear protein to be retained in the nucleus (Gandhi et al. 2018). As transcriptional and post-transcriptional regulators, the nuclear-retained lncRNAs can accumulate and affect genes located in close proximity, in the same chromosome at a distance or in different chromosomes. Furthermore, lncRNAs can be integrated into chromosomal foci or be involved in several nuclear processes to constitute a chromosome. lncRNAs can function in nucleus formation (Quinodoz and Guttman 2014) and/or maintain specific nuclear domains, including nucleolus, nuclear speckles, and paraspeckles (Hacisuleyman et al. 2014). Even though the previous study showed lncRNA can translocate within the nucleus (Cabili et al. 2015), the related mechanism is still unclear.

lncRNAs can also be exported to the cytoplasm to execute their functions. It is commonly believed that cytoplasmic lncRNAs work through interacting with RBPs and function as regulatory elements in the cytoplasm (Noh et al. 2018). First of all, cytoplasmic lncRNA can interfere with protein post-translational modifications. For example, *lincRNA-p21* is induced by hypoxia-induced protein alpha (HIF-1α) during hypoxia. It conversely binds HIF-1α and von Hippel–Lindau (VHL) and reduces VHL-mediated HIF-1α ubiquitination, thus increasing HIF-1α accumulation and promoting glycolysis (Yang et al. 2014). Besides, lncRNAs serve as decoys for miRNA and proteins to regulate gene expression. For instance, *HULC* works as *miR-372* sponge and prevents its translational repression to *PRKACB* mRNA which thereby enhances *PRKACB* translation (Noh et al. 2018). It is also shown that lncRNA enables to influence mRNA translation. For example, mRNAs can be degraded via staufen 1 (STAU1)-mediated mRNA decay (SMD), when their 3′ untranslated region (3′ UTR) binds to STAU1. STAU1 is able to bind double-stranded RNA (dsRNA) which is formed by lncRNAs and Alu elements of 3′' UTR in target mRNA (Gong and Maquat 2011).

2.3 *Structural Studies*

2.3.1 Structural Determination of lncRNAs and Potential Structural Analysis Techniques

Intriguingly, compared with the poorly conserved nucleotide sequence among different species, lncRNAs are well-conserved in their secondary structures to execute their functions. In this section, we introduce biochemical assays and advanced techniques for the better understanding of lncRNA structures and functions.

2.3.2 lncRNA Secondary Structure Predication

lncRNA secondary structure can be determined by enzymatic and chemical probing methods. Enzymatic probing utilizes nucleases to digest paired or unpaired RNAs into fragments (Fig. 1a), whereas chemical probing uses chemicals to reconstruct nucleotides (Fig. 1b). After RNA is isolated and fragmented through cleavage or modification, cDNA products will be synthesized by reverse transcription and further processed to sequencing (Fig. 1). By using next-generation sequencing (Djebali et al.), cDNA analysis can determine the nucleotide location and modification frequency. These probes are widely applicable both in vitro and in vivo (Zampetaki et al. 2018).

The common enzymatic probing methods in practice are parallel analysis of RNA structure (PARS) and fragmentation sequencing (Frag-seq). In PARS, as shown in Fig. 1a, refolded single-stranded RNAs (ssRNAs) and dsRNAs in vitro are treated

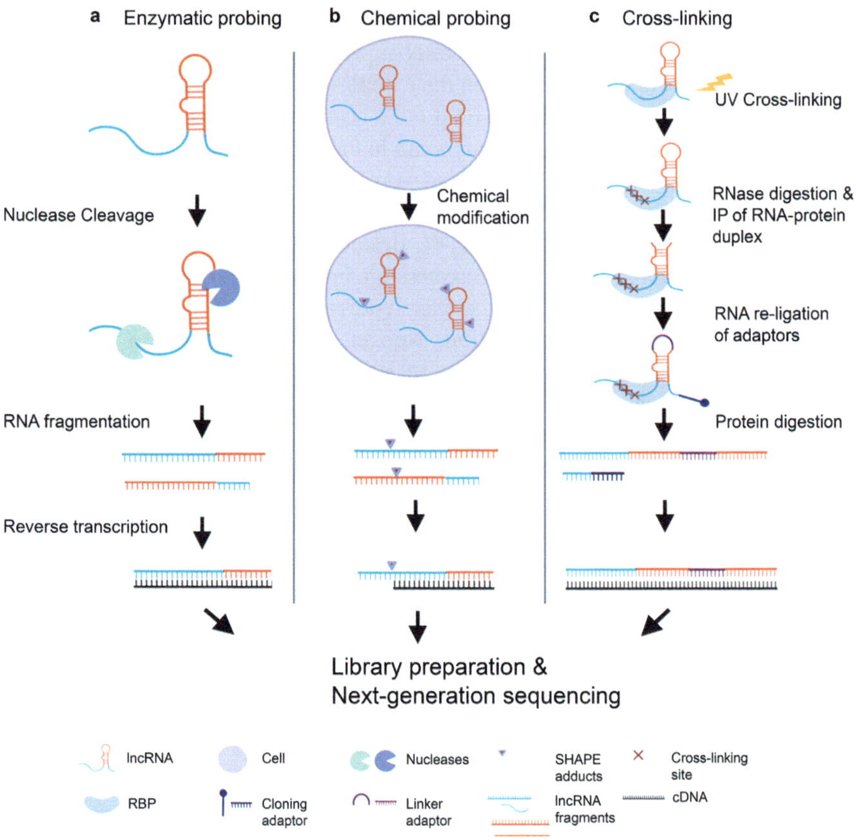

Fig. 1 Examples of the experimental procedure for lncRNA structural determination. (**a**) Enzymatic probing method (PARS). lncRNA is treated with nucleases and cleaved into RNA fragments in vitro, followed by library preparation for deep NGS. (**b**) Chemical probing method (SHAPE). lncRNA is digested and modified (acylated) by SHAPE chemicals. The SHAPE adducts can be detected by reverse transcriptase-mediated primer extension and block reverse transcription before the adduct position. Then it is combined with library preparation and NGS. (**c**) Cross-linking method (hiCLIP). lncRNA is cross-linked with RBP via UV, followed by RNase digestion, IP, RNA re-ligation of cloning and linker adaptors, protein digestion, library preparation and processed NGS. *UV* ultraviolet, *IP* immunoprecipitation, *RBP* RNA-binding protein. (This figure was created with BioRender.com)

with RNase S1 or V1, respectively. Both S1 and V1 cleave 3′-phosphodiester bonds of RNA and leave a 5′-phosphate (5′P). After random fragmentation, these V1 and S1 nuclease-cleaved fragments are ligated to the adaptors for deep sequencing. By transcriptome mapping, V1-generated RNAs have a large number of reads aligned to the same region because the region has been cut multiple times, which indicated the RNA is more likely to have double-stranded conformation. In contrast, the fragments cut by RNase S1 generate less reads aligned to the same base which implies single-stranded conformation (Kertesz et al. 2010). Frag-seq is also a high-throughput RNA structure probing method. Unlike PARS, in Frag-seq ssRNA is cleaved by nuclease P1 and yields 5′-monophosphate and 3′-OH products, and leaves the control without nuclease treatment. The control is used to estimate the occurrence of fragments with an endogenous 5′-phosphate, compared with fragments with a 5′-phosphate produced by nuclease cleavage. Additionally, an equal mass of input 20–100 nucleotide RNAs are treated with polynucleotide kinase and ATP, catalyzing 5-′-phosphorylation and 3′-cyclic phosphate removal, which is helpful to examine endogenous breaks that do not leave a 5′-phosphate and 3′- OH. Subsequently, uncleaved RNA lacking the modification and the fragments with wrong sizes were removed by filtration. The fragments with modification were adapters-ligated and then the final libraries were made and sequenced. Orientation information for each fragment is also maintained. By comparing with control, P1 generated fragments can be distinguished and the orientation is determined. This approach makes it possible to identify the secondary structures (Underwood et al. 2010).

The chemical methods widely used are dimethyl sulfate (DMS) probing and selective 2′-hydroxyl acylation by primer extension (SHAPE). DMS is a base-specific reagent that can easily penetrate cell membrane and methylate the base-pairing with A and C of RNA in loops, bulges, mismatches, and joining regions (Cordero et al. 2012). The DMS-modified A's and C's can indicate the methylation footprints by NGS (Zampetaki et al. 2018). Furthermore, based on the above method, the targeted structure-seq has been developed, which combines DMS probing and gene-specific primer for library preparation. It enables investigators to analyze target RNA conformation in cells by a small number of sequencing reads (Fang et al. 2015). With respect to SHAPE (Fig. 1b), by treating with different SHAPE reagents such as isatoic anhydride, N-methyl (IA), N-methyl isatoic anhydride (NMIA), 1-methyl-7-nitro-isatoic anhydride (1M7), or benzoyl cyanide (BzCN), 1-methyl-7-nitro-isatoic anhydride, or benzoyl cyanide, all four types of nucleotides are selectively 2-hydroxyl acylated in the unpaired RNA region. It gives rise to covalent SHAPE adducts which can be detected by reverse transcriptase-mediated primer extension. Reverse transcription is blocked at one nucleotide before the position of an addict. The truncated fragments can be used to predict RNA secondary structures and assess RNA dynamic modification (Weeks and Mauger 2011). Combined with NGS, SHAPE and mutational profiling (SHAPE-MaP) and SHAPE-seq have been developed. Based on that, in-cell SHAPE-Seq couples in-cell probing of RNA structure with SHAPE-seq to measure the gene expression and thereby study the structure–function relationship (Watters et al. 2016). Furthermore, in vivo click SHAPE (icSHAPE) utilizes chemical NAI-N3 to modify all four types of based within living cells followed by selective chemical enrichment of NAI-N3-modified RNA. The modified RNA is tagged, purified, and prepared for libraries. The subsequent NGS will determine the modified RNA structure at base-resolution (Flynn et al. 2016).

2.3.3 Predict lncRNA Tertiary Structure

All the above methods were used to identify whether the lncRNA is single-stranded or double-stranded. It is quite challenging to determine lncRNA tertiary structures by traditional 3D structural methods which are applied to other types of RNAs, since lncRNA possesses larger size and more flexibility (Qian et al. 2019). The prediction is currently based on the knowledge of secondary structures and features of other types of RNAs, further combining cross-linking method and computational analysis (Yan et al. 2016). Through detecting physical proximity between nucleotides, researchers utilized the cross-linking method to investigate RNA interacting with other compartments. This method uses ultraviolet to crosslink RNA duplexes with RBPs, followed by immunoprecipitation (CLIP) (Lin and Miles 2019). CLIP has multiple variants such as high-throughput sequencing combined with CLIP (HITS-CLIP), individual-nucleotide CLIP (iCLIP) to detect at residual amino acids level, hybrid and individual nucleotide CLIP (hiCLIP), and so on (Sugimoto et al. 2015). For example, hiCLIP utilizes UV to crosslink RNA duplexes with proteins, followed by RNase digestion, immunoprecipitation, RNA re-ligation of cloning and linker adaptors, protein digestion, and NGS (Fig. 1c) (Sugimoto et al. 2015).

Computational prediction methods have been developed on the basis of chemical (e.g., FARNA and NAST), mathematical (e.g., BARNACLE), molecular (e.g., CG Model), and physical features of RNA. However, it is still difficult to determine the structures of larger RNA molecules (Yan et al. 2016). Therefore, recent studies recommended to combine computational and experimental strategies to decode RNA structure (Quarrier et al. 2010). Intriguingly, the secondary structural map of RepA using chemical probing in vitro has been validated. Then, they combined UV crosslinking experiments with computational three-dimensional (3D) modeling to generate a 3D model of *RepA* functional domains. Their study indicated the presence of tertiary architecture in lncRNA molecules and it's link to specific functions (Liu et al. 2017).

2.3.4 Structure–Function Relationships

According to the multiple structural motifs that RNA possesses, RNA tertiary structures include stem-loop, pseudoknot, triplex, and G-quadruplex which assist lncRNA actions in *cis* (intramolecular, within the same RNA) or in *trans* (intermolecular, with RNA, DNA or protein) manner. Some lncRNAs like *Xist* have a double stem-loop structure to recruit polycomb complex for gene repression; *MALAT1* can form a triple helix structure during post-transcription to stabilize itself and secure functional performance (Qian et al. 2019). Some other high-order structures of lncRNAs facilitate the conformation of lncRNA complex with proteins, for instance, *HOTAIR*. Two of the four sub-domains from *HOTAIR* are predicted as protein-binding domains and highly conserved in the secondary structure elements surrounding protein-binding motifs (Blythe et al. 2016). Furthermore, lncRNAs such as circular RNAs (circRNAs) can form a circular structure by *trans*-splicing and exon skipping which couples individual 3′ exon splice with downstream 5′ exon splice site. Circular intronic lncRNAs (ciRNAs) can retain introns and form small nucleolar RNAs (snoRNA) motif to assist ciRNA accumulation (Lasda and Parker 2014).

3 Mechanisms of Action and Physiological Functions

lncRNAs are expressed in low levels and they are involved in many regulation processes of gene expression. The mechanisms of lncRNA can be described based on its structure, cellular localization, and interactions with other molecules (Zampetaki et al. 2018; Carlevaro-Fita and Johnson 2019). In this section, we will discuss lncRNA functions based on its interaction with proteins, DNAs, and RNAs.

3.1 lncRNA Interaction with Proteins

Many studies reveal that lncRNAs can interact with multiple proteins during gene transcription and translation. lncRNAs can recruit or act as a scaffold for chromatin-modifying complexes, such as SWI/SNF or PRC subunits, and thereby stabilize these proteins at specific gene loci to attain appropriate gene regulation (Marchese et al. 2017). For example, *lncTCF7* promotes *TCF7* transcription by recruiting the SWI/SNF complex to the *TCF7* promoter (Sun et al. 2018). Additionally, *HOTAIR* can target multiple chromatin-modifying complexes acting as a scaffold and bridge between PRC2 and the LSD1/CoREST/REST complex to repress gene expression (Wang and Chang 2011). Moreover, lncRNA can also work with epigenetic factors to remodel chromatin, thus regulates gene transcription. lncRNA is also known to facilitate methylation by histone H3K4 methyltransferases to induce gene expression, such as lncRNA *Fendrr* (Grote et al. 2013). In another example, *Kcnq1ot1* interacts with histone methyltransferases G9a and PRC2 and recruit polycomb complexes to form a repression complex for bidirectional silencing of genes in the *Kcnq1* domain (Wang and Chang 2011). Furthermore, lncRNAs are known to interact with SR splicing factors. For instance, lncRNA *MALAT1* that is located in nuclear speckles can act as a molecular sponge to recruit SR splicing factors, and affect the distribution of splicing factors in nuclear speckles and ultimately control alternative splicing in situ (Tripathi et al. 2013).

lncRNAs can also interfere with the interaction between proteins and protein-modifying enzymes, affecting posttranslational modifications and important signaling pathways. For instance, lncRNA *NKILA* can bind nuclear factor-κB (NF-κB)/inhibitor of κB (IκB) as ternary complex, inhibit IκB kinase (IKK)-mediated IκB phosphorylation by masking the phosphorylation sites, and lead to NF-κB inactivation. *lincRNA-p21* located in the upstream of *CDKN1A* gene can act as a transcriptional repressor in p53 pathway by interaction with heterogeneous nuclear ribonucleoprotein K (hnRNP-K) to inducing apoptosis (Rashid et al. 2016). Furthermore, lncRNAs can also interact with RNA/DNA binding proteins to regulate various functions. Several lncRNAs are known to bind with DNA-binding proteins (DBPs), and transcription factors to regulate gene expression. For example, the *linc-YY1* interacts with yin yang 1 (YY1) through its middle domain, to expel YY1/polycomb repressive complex (PRC2) from target promoters, and thereby activate the gene expression (Zhou et al. 2015).

As mentioned in the previous section, lncRNA can be translocated and have functions in specific subcellular locations. Importin and some RBPs are involved in this process. For example, lncRNA *NRON* was found to control the nuclear translocation of the nuclear factor of activated T cells (NFAT) by interacting with importin-β. *RMRP*, which was abundant in the nucleus, can mobilize into cytoplasm under the guide of chromosome region maintenance 1 (CRM1), and eventually locates in mitochondria (Noh et al. 2018).

3.2 lncRNA Interaction with DNA

lncRNAs can target specific DNA sequences and form RNA-DNA hybrid duplexes or RNA-DNA triplexes structure. RNA–DNA hybrid duplexes are formed by Watson–Crick base paring, whereas in triplex structure, the third strand is generated by weaker Hoogsteen hydrogen bond (Li et al. 2016). Triplexes structures are usually unstable and can only be produced in rare conditions. A lateral mesoderm-specific lncRNA *Fendrr* is involved in mammalian cardiogenesis and has been suggested to regulate neighboring and distant genes by recruiting either PRC2 or the trithorax group/mixed lineage leukemia (TrxG/Mll) protein complexes. Moreover, *Fendrr* can also form RNA–DNA triplexes in the promoter region of genes during the development process (e.g., *Foxf1* and *Pitx2*) (Grote et al. 2013). The RNA–DNA triplex can recruit PRC2 as an anchor, which in turn leaves the repressive histone marks in the above-mentioned gene promoters and triggers the differentiation process. Another example is that lncRNA *Khps1* interacts with a homo-purine stretch upstream of the transcription start site of gene *Sphk1*, forming an RNA–DNA triplex that anchors the lncRNA and associated effector proteins to the gene promoter (Postepska-Igielska et al. 2015). Additionally, AG-rich motifs also enable lncRNAs to form a triple helix with DNA. lncRNA *PARTICLE* is transcribed from the promoter of *MAT2A*. It was shown the AG motif of *PARTICLE* forms a triple helix with the *MAT2A* promoter to recruit PRC2, which in turn methylates the CpGs of the respective promoter and downregulate the *MAT2A* expression after the cells are exposed to radiation (O'Leary et al. 2015). It has been reported recently that lncRNA *HOTAIR* can also form a triplex with the AG-rich motifs in the promoter region of target genes in mesenchymal stem cells. Triplex structure formed by *HOTAIR* recruits PRC2 and LSD1 and ultimately silences the gene (Kalwa et al. 2016).

lncRNAs form triplexes with molecules and regulate gene expression in either a *cis*-acting manner (on neighboring genes) or a *trans*-acting manner (on distant genes) (Guttman and Rinn 2012). *Cis*-acting lncRNAs are usually transcribed from the promoter of target genes and form a triplex with promoter DNA sequences to regulate gene expression. The triplex-forming motifs are either fully or partially symmetrical and the transcribed lncRNA with symmetrical triplex-forming motifs must change orientation to form a triplex structure with the target DNA sequence. On the other hand, *trans*-acting lncRNAs specifically target the binding sites through the triplex formation with distant DNA sequences. If the triplex-forming motif is symmetrical, it will target dsDNA with the same sequence; if not, it will target dsDNA in the reverse direction (Li et al. 2016).

3.3 lncRNA Interaction with RNAs

We have mentioned earlier that alternative splicing is a vital regulatory process to generate a variety of gene transcripts. Recent studies showed that lncRNA can regulate alternative splicing through three main mechanisms. In addition to interacting with specific splicing factors or being involved in chromatin remodeling as stated previously, lncRNAs, particularly NATs, can form RNA–RNA duplexes with pre-mRNAs in splicing regulation (Romero-Barrios et al. 2018). For example, in epithelial cells, the 5′ UTR intron of *Zeb2* mRNA contains an internal ribosome entry site (IRES) proximal to the start codon. Under normal condition, 5′ UTR intron is eliminated upon binding of the spliceosome, which prevents *Zeb2* translation. During epithelial–mesenchymal transition, activated lncRNA *Zeb2-NATs* form an RNA-RNA duplex in 5′ UTR of *Zeb2* and prevents the binding of spliceosome. It generates the mRNA with conserved 5′ UTR containing IRES and allows ribosome access to IRES, thus facilitates *Zeb2* translation (Beltran et al. 2008). In addition, lncRNAs also play a vital role in regulating mRNA stability by forming a duplex with target RNAs, implying lncRNA can be involved in the process of mRNA synthesis and degradation (Qian et al. 2019). For instance, lncRNA *BACE1-AS* increases the stability of *BACE1* mRNA and upregulate the β-site APP cleavage enzyme 1 (BACE1) protein level by forming RNA duplex with *BACE1* mRNA. *BACE1-AS* competes with *miR-485-5p* and thereby increases the stability of *BACE1* mRNA (Faghihi et al. 2010). Additionally, it has been reported that lncRNAs are involved in protein translation by interaction with rRNA, ribosome, or eukaryotic initiation factors (Li et al. 2019). Recent studies showed that 5′-region of the nucleolar-specific lncRNA *LoNA* bound and sequestered nucleolin to suppress rRNA transcription, and its snoRNA like 3′ end recruited and diminished fibrillarin activity to reduce rRNA methylation (Li et al. 2018). Moreover, as mentioned above, lncRNA can also interplay with miRNAs, serving as competitive endogenous RNAs (ceRNAs) or miRNA sponge (Noh et al. 2018).

In addition to interacting with other RNAs, lncRNA is able to encode small RNAs, such as miRNAs, snoRNAs, PIWI-interacting RNAs (piRNAs), and also participates in their functional pathways. Some lncRNAs can be regarded as the precursor of miRNAs and contain miRNA sequences in intron or exon (Yamamura et al. 2018). It is widely known that some miRNAs function as tumor suppressors. During tumorigenesis, miRNA releasing from the lncRNA is inhibited, which ultimately leads to the downregulation of miRNA and tumor, such as *H19* to *miR-675* (Keniry et al. 2012). lncRNAs have also been found to encode snoRNAs, such as *SNHGs* which encode snoRNAs involved in multiple cancers. Moreover, lncRNAs can also be piRNA precursors and take part in piRNA-associated pathways (Yamamura et al. 2018).

4 lncRNAs as Potential Biomarkers and Therapeutic Targets in Neurological Disorders

Recent studies provided vital evidence that long noncoding RNA molecules have regulatory roles in normal CNS development and are highly involved in various neurological disorders. In the following section, we will cover some examples of lncRNAs as potential biomarkers and therapeutic targets in neurological disorders (as shown in Table 1).

4.1 Alzheimer's Disease (AD)

AD is the most prevalent form of dementia primarily affecting the elderly population and causes impairment in memory, decision-making, and behavior. AD is characterized by dysregulated processing of amyloid precursor protein (APP). Abnormal accumulation of toxic amyloid beta peptide (Aβ) in the hippocampus region of the brain is one of the known mechanisms involved in the AD pathogenesis (Salta and De Strooper 2012). BACE1 is involved in the cleavage of APP to toxic Aβ peptide. The expression of *BACE1-AS* is highly upregulated in the AD patient's brain, signifying that *BACE1-AS* is involved in the AD disease pathology (Modarresi et al. 2011). Recent studies have shown that *BACE1-AS* binds with the *BACE1* mRNA, which increases *BACE1* mRNA stability and promotes the synthesis of BACE1 protein. The increased BACE1 protein in turn increases the production of toxic Aβ in the cells (Faghihi et al. 2010). The elevated levels of Aβ can further promote the *BACE1-AS* overexpression. Therefore, *BACE1* and *BACE1-AS* have been identified as potential biomarkers and treatment targets of AD (Decourt and Sabbagh 2011).

Brain cytoplasmic 200 (BC200) is a translational regulator that targets eukaryotic translation initiation factor 4A. *BC200* is an lncRNA implicated in AD and brain aging. *BC200* can interact directly with RNA-binding proteins and it is involved in abnormal transport of mRNA leading to abnormal protein localization (Wang et al.

Table 1 Some examples of lncRNAs involved in various neurological disorders

lncRNA	Disease	Description	Mechanism of action	References
BDNF-AS	HD	Transcribed from the opposite strand of *BDNF* gene. *BDNF-AS* modulates synaptic function and depletion of *BDNF-AS* promotes adult neurogenesis	Upregulates the transcription of *BDNF* and has a neuroprotective property	Sunwoo et al. (2017)
HttAS-v1	HD	Transcribed from the opposite strand of huntingtin gene	*HttAS-v1* reduces endogenous normal HTT transcript levels	Estrada-Sanchez and Rebec (2013)
Abhd11os	HD	Loss of *Abhd11os* contributes to striatal vulnerability	Overexpression of *Abhd11os* is neuroprotective	Xie et al. (2010)
lncSCIR1	SCI	Regulates astrocyte proliferation and migration	Knockdown of *lncSCIR1* promotes proliferation and migration in SCI	Zhou et al. (2018)
MALAT1	SCI	Highly conserved and highly expressed in neuronal cells	Upregulated in SCI through effector miR199b	McColl et al. (2017)
BC200	AD	Involved in the regulation of synaptic excitability	Induces the translation of *APP* and lead to toxic accumulation of Aβ in patient brain	Liu et al. (2012)
BACE1-AS	AD	Transcribed from the opposite strand of *BACE1* gene	*BACE1-AS* binds to *BACE1* mRNA and increases the stability and the production of BACE1 protein	Salta and De Strooper (2012)
GDNFOS	AD	Transcribed from the opposite strand of *GDNF*	Negatively regulates the expression of GDNF protein	Airavaara et al. (2011)
UCHL1-AS1	PD	Transcribed from the opposite strand of *UCHL1*	Increases the production of UCHL1 protein by promoting the translational process	Carrieri et al. (2015)
NaPINK1	PD	Transcribed from the opposite strand of *PINK* locus	Binds to PINK protein and increases its stability as well as its neuroprotective function	Elkouris et al. (2019)
Gomafu	SZ	*Gomafu* is also called as *MIAT* (myocardial infarction-associated transcript)	Binds to serine/arginine-rich splicing factor 1 and causes misregulated splicing of the target genes	Barry et al. (2014)

2018). *BC200* is highly expressed in somata and its level decreases in dendrites. This aberrant expression severely affects microtubule-dependent transport and could lead to axonal-dendritic blockage during the early stages of AD. Previous study showed that high expression of *BC200* in Broadmann area 9 of AD affected brain region when compared to control normal brain and they also found that *BC200* is often clustered in the perikaryal location. These changes suggest that dendritic loss has occurred accompanied by overexpression in somata leading to abnormal localization and distribution of *BC200*. The abnormal distribution of *BC200* might be responsible for the AD mechanism. A study on AD postmortem brain tissue indicated that *BC200* RNA levels are significantly increased in the AD brains compared to the age-matched controls. This upregulation of *BC200* levels in these AD brains increased in parallel with the disease progression (Mus et al. 2007). A recent study using AD patient's plasma samples revealed that the *BC200* did not show any significant increase when compared to controls (Feng et al. 2018). The discrepancy between these two studies may be due to the different types of samples used in the studies. More comprehensive studies are needed to further validate *BC200* as a biomarker.

lncRNA *17A* is synthesized by RNA polymerase III during inflammatory stimulation. *17A* is involved in the regulation of G-protein-coupled receptor 51 (GPCR51) splicing to isoform variant -B R2 of the B-type receptor for the neurotransmitter gamma-amino butyric acid (GABA-B receptor). This spliced variant B severely affects the GABA-B biological function by activation of potassium channels and increases the accumulation of intracellular 3′-5-cyclic adenosine monophosphate. The dysregulation in the GABA-B receptor signaling leads to an increased Aβ42/Aβ40 ratio. Studies have shown that *17A* overexpression causes increased production of toxic Aβ42 (Massone et al. 2011), suggesting that *17A* could be directly involved in the AD pathogenesis.

Another lncRNA is *51A* that maps antisense to intron-1 of sortilin-related receptor gene (*SORL1*). *SORL1* has been shown to be involved in the AD pathogenesis (Rogaeva et al. 2007). *SORL1* interacts with APP in endosomes and it affects the *trans*-Golgi-mediated trafficking and proteolytic process of APP. *51A* synthesis may promote the alternative splicing of *SORL1* variant, thereby decreases *SORL1* expression. This might shift the APP from retro cyclic pathway to β secretase cleavage pathway and lead to increased toxic accumulation of Aβ (Ciarlo et al. 2013).

Brain-derived neurotrophic factor (BDNF) is one of the key neurotrophic factors that play a key role in the normal development of CNS and its expression has been downregulated in many neurological conditions including AD. *BDNF-AS* is known to decrease the expression of BDNF and studies have shown that silencing *BDNF-AS* prevented the Aβ-induced toxicity (Guo et al. 2018).

Sox2OT is another lncRNA expressed during the neuronal cell differentiation. *Sox2OT* is known to regulate the Sox2 expression which is one of the key pluripotency-associated transcription factors (Shahryari et al. 2015). A study using a mouse model of AD reported that *Sox2ot* and *1810014B01Rik* could potentially serve as biomarkers in the early and late stages of AD (Arisi et al. 2011). This still needs to be comprehensively validated in patient samples. Remarkable efforts have been made and are still in the process to identify novel and highly specific lncRNAs altered in AD as biomarkers. This effort will also help to understand potential signaling pathways widely affected during AD.

4.2 Spinal Muscular Atrophy (SMA)

SMA is a motor neuron disorder caused by the loss of motor neurons in the spinal cord leading to weakness of the voluntary muscles. The deficiencies in survival motor neuron proteins SMN1 and SMN2 due to abnormal splicing affect the normal survival of the motor neurons (Chaytow et al. 2018). In the United States, about 400 infants are born with SMA every year and they are suffering from muscle weakness and are delayed in reaching the developmental milestones. lncRNA *SMN-AS1* located in the *SMN2* locus normally affects the *SMN2* transcription and leads to decreased expression of SMN2 protein levels. At present, the best treatment option available for SMA patients is Nusinersen. Nusinersen is the oligonucleotide-based therapeutics to treat SMA. Nusinersen is also called as Spinraza. Nusinersen is an 18-mer phosphorothioate 2′-O-methoxy antisense oligonucleotide, which increases the SMN protein levels in a randomized clinical trial on affected infants. Nusinersen showed improvement in reaching motor development milestones. Nusinersen corrects the splicing defects in the SMN2 protein and increased the SMN2 protein levels. The combination of two antisense oligonucleotides anti-SMN-AS1 oligonucleotide together with Nusinersen showed maximum increased SMN proteins and studies in mouse models of SMA showed improved survival (d'Ydewalle et al. 2017). The main reported side-effects of Nusinersen were constipation and respiratory infections. Nusinersen is cost-prohibitive currently (~ $350,000 per year). The development of cost-effective drug targeting lncRNA *SMN-AS1* is still under investigation.

4.3 Spinal Cord Injury (SCI)

SCI is one of the major devastating disorders that affected about 500,000 people globally causing permanent disability in movement and chronic pain (McColl et al. 2017). There is currently no cure for the affected people. lncRNA *Malat1* is upregulated in SCI through effector *miR199b* and activates the production of pro-inflammatory cytokines in microglia. Targeted knocking down of spinal *Malat1* significantly reduced the pro-inflammatory cytokines in contusion epicenter including reduction in the expression of microglial marker IBA-1 and dramatic improvement in the locomotor function of the hind limb in the mouse model of SCI (Zhou et al. 2018). Another lncRNA *lncSCIR1* is significantly downregulated in the mouse contusive SCI model which is inversely correlated with the bone morphogenetic protein 7 (BMP7) and adrenomedullin (ADM) (Wang et al. 2015). Both of the proteins were involved in the induction of astrogliosis in the affected spinal cord. Knocking down *lncSCIR1* promoted the proliferation and migration of the cultured astrocytes. Further studies are needed to validate the *lncSCIR1* as a therapeutic target for SCI.

4.4 Huntington's Disease (HD)

HD is a rare autosomal dominant inherited neurodegenerative disorder caused by the pathogenic expansion of toxic CAG repeats in the exon1 of huntingtin gene (*HTT*) affecting neurons in the striatum and cortex region of the brain (Estrada-Sanchez and Rebec 2013). HD primarily affects movements, memory, and eventually leading to premature death due to toxic accumulation of polyglutamine. Studies are in progress to prevent the abnormal transcription of toxic *mHTT* (mutant *HTT*) and to reduce the accumulation of *mHTT*. *HttAS1_v1* is the antisense transcript of HTT, and is expressed at low levels in the frontal cortex of HD patients. The low *HttAS1_v1* expression leads to high toxic expression of *mHTT* mRNA in cortex and triggers the HD pathogenicity (Chung et al. 2011). Another lncRNA *NEAT1* is involved in the formation and maintenance of paraspeckles found in the mammalian cells (Sunwoo et al. 2017). A study using HD mouse model and HD patients found that the *NEAT1* levels are increased (Cheng et al. 2018). Further studies revealed *NEAT1* upregulation in cellular models and suggested its role in neuroprotective mechanism against neuronal damage (Sunwoo et al. 2017). Additionally, *Bdnf-AS* antisense transcript of *Bdnf* gene is known to block the expression of *Bdnf*. Oligonucleotides targeting *Bdnf-AS* increased the expression of BDNF protein in mouse HD models (Modarresi et al. 2012). A similar approach in HD patients showed a promising increase in the BDNF protein levels in the forebrain (Xie et al. 2010). Another lncRNA *Abhd11os* is selectively reduced in the multiple mouse models of HD and overexpression of *Abhd11os* in the mouse brain increased the neuroprotection property (Francelle et al. 2015). Moreover, many additional lncRNAs were found to be dysregulated in the brains of the HD patients including *NEAT1, DGCR5*, and *MEG3* (Mondal et al. 2015).

4.5 Angelman Syndrome (AS)

AS is a genetic disorder affecting the nervous system. In affected children, it causes severe neurodevelopmental delay, microcephaly, ataxia, and seizures. AS is caused by the deletion mutation in the ubiquitin protein ligase E3A gene (*UPE3A*) on chromosome 15 and thereby no functionally active UPE3A produced in the affected patients. The presence of antisense lncRNA *UBE3A-ATS* also inactivates the *UPE3A*. Mouse study using intracerebroventricular (ICV) administration of anti-Ube3a-ATS oligonucleotides showed partial restoration of the UBE3A protein levels and improvement in behavior and cognitive functions. This study showed great promise; there are no significant changes in the body weight and astrocyte marker (GFAP) expression after 1 month of the ICV injection in mice (Meng et al. 2015). Ionis Pharmaceuticals is developing the antisense oligonucleotide-based compounds to target *UBE3A-ATS* in AS patients.

4.6 Schizophrenia (SZ)

SZ is a chronic and severe mental disorder and visible symptoms appear between the ages of 16 and 30 (Herold et al. 2018). Symptoms include hallucination, distorted thoughts, and lack of co-ordination. At present, it is not certain about the genetic cause of SZ, but there are significant associations of lncRNAs with SZ. *GOMAFU* is one of the lncRNAs that is downregulated in the cortical grey matter of the SZ in the affected patients, whereas the deletion of *Gomafu* in mice resulted in increased levels of dopamine in the brain (Barry et al. 2014). *GOMAFU* is known to be associated with the myocardial infarction in human and it is also shown to be involved in the retinal cell fate determination in the mouse (Clark and Blackshaw 2017). Another lncRNA *DLG2-AS* associated with the SZ is antisense to the gene *DLG2*, an important protein involved in the assembly of the NMDA receptor (Reggiani et al. 2017). A study using SZ patients' peripheral blood mononuclear cells has identified about 125 lncRNA differentially expressed when compared to the healthy controls. However, it requires a lot of efforts and resources to validate each of the lncRNA and identify their functional roles before therapeutics can be developed.

4.7 Dravet Syndrome (DS)

DS is a genetic disorder caused by heterozygous loss of function of the voltage-gated sodium channel Nav1.1 gene *SCN1A*, which leads to febrile seizures resistant to anticonvulsants treatment, progressive psychomotor retardation, and unexpected sudden death (Escayg and Goldin 2010). This disease is caused by haplo-insufficiency of *SCN1A* gene function. The therapeutic strategy would be to increase the expression of normal *SCN1A* allele to compensate for the loss of function of the mutant *SCN1A* allele. The presence of antisense transcript *SCN1ANAT* lncRNA blocks the upregulation of the normal *SCN1A* allele and ASO to target *SCN1ANAT* leads to increased expression of the normal *SCN1A* allele and significant improvement in the seizure frequency in mouse models (Hsiao et al. 2016).

4.8 Parkinson's Disease (PD)

One of the main causes of PD is the selective loss of dopaminergic neuron in the substantia nigra and basal ganglia, which resulted in decreased production of the neurotransmitter dopamine controlling body movements. Phosphatase and tensin homolog-induced putative kinase 1 (PINK1) is dysregulated in PD. *NaPINK1*, a lncRNA transcribed from the antisense strand of *PINK1* locus, was shown to stabilize the *PINK1* expression. Further studies are needed to validate the therapeutic aspect of *NaPINK1* in PD (Elkouris et al. 2019). Another antisense transcript *AS*

Uchl1 regulates the expression of ubiquitin carboxy-terminal hydrolase L1 protein (UCHL1) involved in normal brain function (Wei et al. 2018). *AS Uchl1* increases the Uchl1 protein level and the manipulation of the UCHL1 expression by targeting *AS Uchl1* has been proposed as a promising therapeutic approach for PD.

5 Advances in Antisense Oligonucleotide (ASO) Chemistry for Targeted Delivery

Advancement in the next-generation sequencing led to the discovery of many lncRNA associated with human diseases. Manipulation of the disease-associated lncRNAs can be achieved by using oligonucleotide therapeutics. The oligonucleotide therapeutics field has undergone rapid advancements in the recent years for the precise targeted delivery to treat a particular disease with fewer side effects. In this section, we will cover some of the important oligonucleotide chemistries used to target the lncRNAs associated with diseases.

Antisense oligonucleotides (ASOs) are synthetic single-stranded nucleic acids with length ranging between 8 and 50 nucleotides that are complementary to target RNA through standard Watson–Crick base pairing resulting in RNA degradation or altered transcription/translation (Rinaldi and Wood 2018). Initially, the first generation of ASOs was used in the form of synthetic unmodified DNA, which is extremely susceptible to degradation by exo- and endo-nucleases activity (Dias and Stein 2002). Because of the fast turnover and insufficient intracellular concentrations of the first-generation ASOs, these ASOs are not widely applicable for therapies. New-generation ASOs have been developed since then. Second- and third-generation ASOs with backbone-modified antisense oligonucleotides improved in their nuclease resistance, affinity for the target RNA or both. The following are some examples of antisense oligonucleotide chemistry.

5.1 Peptide Nucleic Acids (PNA)

Peptide nucleic acids are generated by the entire replacement of the sugar phosphate backbone with polyamide linkages but still retain its hybridization ability with target RNA (PNA-RNA) or DNA (PNA-DNA) based on Watson–Crick base pairing (Egholm et al. 1993). PNAs are usually uncharged. They have high binding affinity to RNA or DNA and are highly resistant to nuclease and protease degradation (Schwarz et al. 1999). PNA works through steric inhibition of the target mRNA rather than RNase H-mediated cleave of the target RNA and thereby it prevents transcription/translation process (Pellestor and Paulasova 2004). The main disadvantages of PNA are its less cellular uptake and poor solubility in water. However, the recently modified PNAs with peptide conjugates improved water solubility and cellular uptake. The clinical uses of PNAs in neurodegenerative disorders are limited due to poor pharmacokinetic properties and they are also rapidly cleared from the systemic circulation leading to less efficacious effect.

5.2 *Locked Nucleic Acid (LNA)*

Locked nucleic acid (LNA) is an analog of a nucleic acid that is locked in a C3′-endo (β-D-LNA) or C2′-endo (α-L-LNA) conformation by linking the ribose 2′-O- and 4′-C-atoms via a methylene bridge (Obika et al. 1999). LNA can stably hybridize with DNA or RNA forming LNA-DNA or LNC-RNA duplex to increase the specificity and potency of ASOs. In addition to high biostability, LNA also has good mismatch discrimination, metabolic stability, and nontoxic effect at effective dosage in the model animal system compared to other nucleotide modifications (Soler-Bistue et al. 2019). LNAs are extremely potent both in vivo and in vitro, which makes it suitable for in vivo testing in animal models to study pharmacodynamics and pharmacokinetics.

5.3 *Phosphorodiamidate Morpholino Oligonucleotides (PMO)*

Phosphorodiamidate morpholino oligonucleotides (PMO) are short single-stranded DNA analogs with morpholine rings in the backbone connected by phosphorodiamidate linkages, resulting in a neutrally charged molecule with highly precise binding to nucleic acids (Moulton and Jiang 2009). PMO effectively bind target mRNA by Watson–Crick base pairing to blocks the mRNA translation or modulate pre-mRNA splicing through sequence-specific steric inhibition process (Nan and Zhang 2018). PMO are resistant to endo- and exonucleases, proteases, esterases, and other enzymes in the biological fluids and thereby it possesses good half-life suitable for the in vivo studies (Hudziak et al. 1996). The increasing evidence shows PMOs are of low toxicity and cost, good pharmacokinetics and bioavailability, which makes it potentially a safe and effective treatment for human disease (Nan and Zhang 2018).

6 Targeting lncRNAs for Potential Disease Treatment

lncRNAs play significant roles in many neurological disorders, cancer, autoimmune diseases, and cardiovascular diseases. In this section, we have highlighted some biomarker detection methods including RNA-seq, in situ hybridization, microarray,

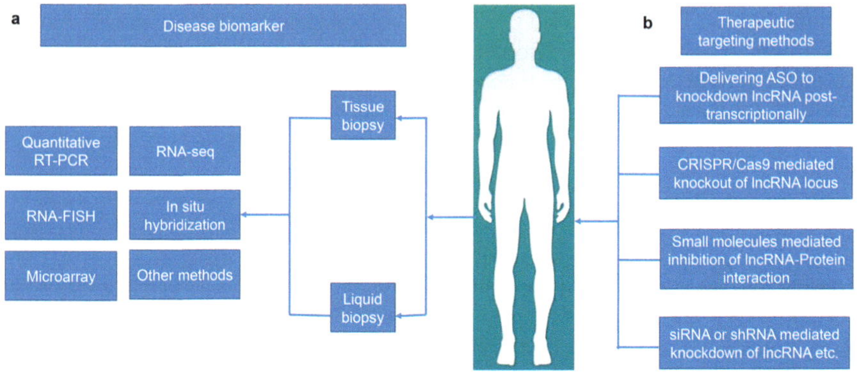

Fig. 2 Diagnostic and therapeutic potential of lncRNAs. (**a**) lncRNAs from patient tissue or liquid biopsy could be tested by various methods and could potentially serve as diagnostic/prognostic biomarkers. (**b**) The disease-associated lncRNAs could act as potential targets for treatment, for example, lncRNAs can be targeted using CRISPR/Cas9, ASOs, siRNA/shRNA, and small molecules

RNA-FISH, and quantitative PCR which were broadly utilized to identify and validate the novel disease-associated lncRNAs (Fig. 2a). The disease-associated lncRNAs have been targeted by using methods such as ASO, CRISPR/Cas9, small molecules, and small interference RNA (siRNA) as shown in Fig. 2b.

6.1 Delivering ASOs to Target lncRNAs

As mentioned above, ASOs are under clinical development for more than a decade and many advancements have been made to deliver the ASOs to the disease-causing targets more specifically with less side-effects (Verma 2018). The recent advancements in ASO chemical modifications including backbone and sugar moieties made it possible to target disease-associated lncRNAs to increase/decrease protein levels, modify the splicing, and inhibit the miRNAs (Chery 2016). Most of the ASOs targeting diseases-causing lncRNAs involved in neurological disorders mentioned earlier are still under the preclinical development in animal models to validate the targets, evaluate off-target side effects and pharmaco-dynamic/pharmaco-kinetic properties. At present, the studies are still in progress for targeting lncRNAs for human use.

6.2 CRISPR/Cas-Mediated Alteration of lncRNAs

In recent years, clustered regularly interspaced short palindromic repeats (CRISPR) technology made it possible to selectively activate (CRISPRa) or repress (CRISPRi)

the target of the preferred genomic locations to alter the disease pathogenicity (Gilbert et al. 2014). Using the second-generation CRISPR technology, it is now possible to base-edit both at DNA and RNA levels to correct the mutation for particular disease precisely. The involvement of lncRNAs in disease pathogenesis is more complex and altering the lncRNA expression using CRISPR should be validated to eliminate the off-target effects to completely prevent the side effects (Goyal et al. 2017). Even though CRISPR technology has potential to target the disease-causing lncRNA specifically, many preclinical studies failed due to potential off-target effects.

6.3 Small Molecules Mediated Alteration of lncRNAs

lncRNAs play many diverse biological functions based on tissue specificity, localization, structure, and interacting partners. In a recent study, lncRNA *GAS5* decreases significantly in type 2 diabetes mellitus (T2DM) patient's serum. *GAS5* binds to the promoter region of the insulin receptor to regulate its expression and depletion inhibits insulin signaling and glucose uptake. Using a high-throughput screening method, a small molecule NP-C86 was identified which binds to *GAS5* with high affinity and thereby it stabilizes and increases *GAS5* levels. The increased *GAS5* levels ultimately increase the glucose uptake in diabetic patient adipocytes (Shi et al. 2019). Development of a comprehensive screening platform to identify the small molecules based on structure–activity relationship (SAR) for disease-causing lncRNAs could lead to small molecule therapy for many disorders.

6.4 RNA Interference (RNAi)-Mediated Knockdown of lncRNA

RNAi is a most commonly used technique to knockdown the target mRNA mediated through RNAi-induced silencing complex, the multi-protein complex machinery to degrade the target mRNAs (Meister et al. 2004). It has been shown that the degradation of the target mRNAs occurs in the cytoplasm. A large proportion of lncRNAs are predominantly localized in the nucleus, thus RNA interference (siRNA and short-hairpin RNA (shRNA)) may not be effective approaches and would result in less knocking down of these nucleus lncRNAs. lncRNAs can interact with a variety of cellular proteins and nucleic acids. Depending on the lncRNA binding partners, lncRNA can also shuttle between the nucleus and cytoplasm. RNAi may be effective for the lncRNA localized in the cytoplasm.

7 Conclusions

This chapter has provided a broad overview about the lncRNA classification, structures, and its regulatory role in various physiological functions. Many lncRNAs are known to be involved in the disease pathogenesis. We addressed the biomarker and therapeutic potential of the lncRNAs in some of the neurological disorders. Oligonucleotide-based therapeutics are gaining increasing attention due to its high specificity and less side-effect. In this chapter, we also summarized some important antisense oligonucleotide chemistries developed in the recent years along with various molecular approaches for effective targeting of the disease-associated lncRNA.

Acknowledgments This work was supported by grants from the National Institutes of Health R01 NS088353; 1R21EY028647-01, 1R21AR071583-01; The Staman Ogilvie Fund-Memorial Hermann Foundation; and Mission Connect, a program of the TIRR Foundation.

References

Airavaara M, Pletnikova O, Doyle ME et al (2011) Identification of novel GDNF isoforms and *cis*-antisense *GDNFOS* gene and their regulation in human middle temporal gyrus of Alzheimer disease. J Biol Chem 286:45093–45102

Arisi I, D'Onofrio M, Brandi R et al (2011) Gene expression biomarkers in the brain of a mouse model for Alzheimer's disease: mining of microarray data by logic classification and feature selection. J Alzheimers Dis 24:721–738

Barry G, Briggs JA, Vanichkina DP et al (2014) The long non-coding RNA *Gomafu* is acutely regulated in response to neuronal activation and involved in schizophrenia-associated alternative splicing. Mol Psychiatry 19:486–494

Beltran M, Puig I, Peña C et al (2008) A natural antisense transcript regulates *Zeb2/Sip1* gene expression during Snail1-induced epithelial–mesenchymal transition. Genes Dev 22:756–769

Blythe AJ, Fox AH, Bond CS (2016) The ins and outs of lncRNA structure: how, why and what comes next? Biochim Biophys Acta 1859:46–58

Cabili MN, Dunagin MC, McClanahan PD et al (2015) Localization and abundance analysis of human lncRNAs at single-cell and single-molecule resolution. Genome Biol 16:20

Carlevaro-Fita J, Johnson R (2019) Global positioning system: understanding long noncoding RNAs through subcellular localization. Mol Cell 73:869–883

Carrieri C, Forrest AR, Santoro C et al (2015) Expression analysis of the long non-coding RNA antisense to Uchl1 (*AS Uchl1*) during dopaminergic cells' differentiation in vitro and in neurochemical models of Parkinson's disease. Front Cell Neurosci 9:114

Chaytow H, Huang YT, Gillingwater TH et al (2018) The role of survival motor neuron protein (SMN) in protein homeostasis. Cell Mol Life Sci 75:3877–3894

Chen LL, Carmichael GG (2010) Decoding the function of nuclear long non-coding RNAs. Curr Opin Cell Biol 22:357–364

Cheng C, Spengler RM, Keiser MS et al (2018) The long non-coding RNA *NEAT1* is elevated in polyglutamine repeat expansion diseases and protects from disease gene-dependent toxicities. Hum Mol Genet 27:4303–4314

Chery J (2016) RNA therapeutics: RNAi and antisense mechanisms and clinical applications. Postdoc J 4:35–50

Chung DW, Rudnicki DD, Yu L et al (2011) A natural antisense transcript at the Huntington's disease repeat locus regulates *HTT* expression. Hum Mol Genet 20:3467–3477
Ciarlo E, Massone S, Penna I et al (2013) An intronic ncRNA-dependent regulation of SORL1 expression affecting Abeta formation is upregulated in post-mortem Alzheimer's disease brain samples. Dis Model Mech 6:424–433
Clark BS, Blackshaw S (2017) Understanding the role of lncRNAs in nervous system development. Adv Exp Med Biol 1008:253–282
Cordero P, Kladwang W, VanLang CC et al (2012) Quantitative dimethyl sulfate mapping for automated RNA secondary structure inference. Biochemistry 51:7037–7039
Cuevas-Diaz Duran R, Wei H, Kim DH et al (2019) Long non-coding RNAs: important regulators in the development, function and disorders of the central nervous system. Neuropathol Appl Neurobiol 45:538–556
d'Ydewalle C, Ramos DM, Pyles NJ et al (2017) The antisense transcript *SMN-AS1* regulates SMN expression and is a novel therapeutic target for spinal muscular atrophy. Neuron 93:66–79
Decourt B, Sabbagh MN (2011) BACE1 as a potential biomarker for Alzheimer's disease. J Alzheimers Dis 24(Suppl 2):53–59
Dias N, Stein CA (2002) Antisense oligonucleotides: basic concepts and mechanisms. Mol Cancer Ther 1:347–355
Djebali S, Davis CA, Merkel A et al (2012) Landscape of transcription in human cells. Nature 489:101–108
Dong X, Muppani NR, Wu J (2016) Long noncoding RNAs: critical regulators for cell lineage commitment in the central nervous system. In: Transcriptomics and gene regulation. Springer, Dordrecht, pp 73–97
Egholm M, Buchardt O, Christensen L et al (1993) PNA hybridizes to complementary oligonucleotides obeying the Watson-Crick hydrogen-bonding rules. Nature 365:566–568
Elkouris M, Kouroupi G, Vourvoukelis A et al (2019) Long non-coding RNAs associated with neurodegeneration-linked genes are reduced in Parkinson's disease patients. Front Cell Neurosci 13:58
Escayg A, Goldin AL (2010) Sodium channel SCN1A and epilepsy: mutations and mechanisms. Epilepsia 51:1650–1658
Estrada-Sanchez AM, Rebec GV (2013) Role of cerebral cortex in the neuropathology of Huntington's disease. Front Neural Circuits 7:19
Faghihi MA, Zhang M, Huang J et al (2010) Evidence for natural antisense transcript-mediated inhibition of microRNA function. Genome Biol 11:R56
Fang R, Moss WN, Rutenberg-Schoenberg M et al (2015) Probing *Xist* RNA structure in cells using targeted structure-seq. PLoS Genet 11:e1005668
Feng L, Liao YT, He JC et al (2018) Plasma long non-coding RNA *BACE1* as a novel biomarker for diagnosis of Alzheimer disease. BMC Neurol 18:4
Flynn RA, Zhang QC, Spitale RC et al (2016) Transcriptome-wide interrogation of RNA secondary structure in living cells with icSHAPE. Nat Protoc 11:273–290
Francelle L, Galvan L, Gaillard MC et al (2015) Striatal long noncoding RNA *Abhd11os* is neuroprotective against an N-terminal fragment of mutant huntingtin in vivo. Neurobiol Aging 36(1601):e1607–e1616
Gandhi M, Caudron-Herger M, Diederichs S (2018) RNA motifs and combinatorial prediction of interactions, stability and localization of noncoding RNAs. Nat Struct Mol Biol 25:1070–1076
Gilbert LA, Horlbeck MA, Adamson B et al (2014) Genome-scale CRISPR-mediated control of gene repression and activation. Cell 159:647–661
Gong C, Maquat LE (2011) lncRNAs transactivate STAU1-mediated mRNA decay by duplexing with 3′ UTRs via Alu elements. Nature 470:284–288
Goyal A, Myacheva K, Gross M et al (2017) Challenges of CRISPR/Cas9 applications for long non-coding RNA genes. Nucleic Acids Res 45:e12
Grote P, Wittler L, Hendrix D et al (2013) The tissue-specific lncRNA *Fendrr* is an essential regulator of heart and body wall development in the mouse. Dev Cell 24:206–214

Guo CC, Jiao CH, Gao ZM (2018) Silencing of lncRNA *BDNF-AS* attenuates Abeta25-35-induced neurotoxicity in PC12 cells by suppressing cell apoptosis and oxidative stress. Neurol Res 40:795–804

Guttman M, Rinn JL (2012) Modular regulatory principles of large non-coding RNAs. Nature 482:339–346

Hacisuleyman E, Goff LA, Trapnell C et al (2014) Topological organization of multichromosomal regions by the long intergenic noncoding RNA *Firre*. Nat Struct Mol Biol 21:198–206

Hangauer MJ, Vaughn IW, McManus MT (2013) Pervasive transcription of the human genome produces thousands of previously unidentified long intergenic noncoding RNAs. PLoS Genet 9: e1003569

Henz SR, Cumbie JS, Kasschau KD et al (2007) Distinct expression patterns of natural antisense transcripts in Arabidopsis. Plant Physiol 144:1247–1255

Herold CJ, Lasser MM, Seidl UW et al (2018) Neurological soft signs and psychopathology in chronic schizophrenia: a cross-sectional study in three age groups. Front Psych 9:98

Hsiao J, Yuan TY, Tsai MS et al (2016) Upregulation of haploinsufficient gene expression in the brain by targeting a long non-coding RNA improves seizure phenotype in a model of Dravet syndrome. EBioMedicine 9:257–277

Hudziak RM, Barofsky E, Barofsky DF et al (1996) Resistance of morpholino phosphorodiamidate oligomers to enzymatic degradation. Antisense Nucleic Acid Drug Dev 6:267–272

Jain AK, Xi Y, McCarthy R et al (2016) *LncPRESS1* is a p53-regulated lncRNA that safeguards pluripotency by disrupting SIRT6-mediated de-acetylation of histone H3K56. Mol Cell 64:967–981

Jarroux J, Morillon A, Pinskaya M (2017) History, discovery, and classification of lncRNAs. Adv Exp Med Biol 1008:1–46

Kalwa M, Hanzelmann S, Otto S et al (2016) The lncRNA *HOTAIR* impacts on mesenchymal stem cells via triple helix formation. Nucleic Acids Res 44:10631–10643

Keniry A, Oxley D, Monnier P et al (2012) The *H19* lincRNA is a developmental reservoir of *miR-675* that suppresses growth and Igf1r. Nat Cell Biol 14:659–665

Kertesz M, Wan Y, Mazor E et al (2010) Genome-wide measurement of RNA secondary structure in yeast. Nature 467:103–107

Lasda E, Parker R (2014) Circular RNAs: diversity of form and function. RNA 20:1829–1842

Li Y, Syed J, Sugiyama H (2016) RNA-DNA triplex formation by long noncoding RNAs. Cell Chem Biol 23:1325–1333

Li D, Zhang J, Wang M et al (2018) Activity dependent *LoNA* regulates translation by coordinating rRNA transcription and methylation. Nat Commun 9:1726

Li Z, Zhao W, Wang M et al (2019) The role of long noncoding RNAs in gene expression regulation. In: Gene expression profiling in cancer. IntechOpen, London

Lin C, Miles W (2019) Beyond CLIP: advances and opportunities to measure RBP–RNA and RNA–RNA interactions. Nucleic Acids Res 47:5490–5501

Liu XA, Rizzo V, Puthanveettil SV (2012) Pathologies of axonal transport in neurodegenerative diseases. Transl Neurosci 3:355–372

Liu F, Somarowthu S, Pyle AM (2017) Visualizing the secondary and tertiary architectural domains of lncRNA *RepA*. Nat Chem Biol 13:282–289

Marchese FP, Raimondi I, Huarte M (2017) The multidimensional mechanisms of long noncoding RNA function. Genome Biol 18:206

Massone S, Vassallo I, Fiorino G et al (2011) *17A*, a novel non-coding RNA, regulates GABA B alternative splicing and signaling in response to inflammatory stimuli and in Alzheimer disease. Neurobiol Dis 41:308–317

McColl MA, Gupta S, Smith K et al (2017) Promoting long-term health among people with spinal cord injury: what's new? Int J Environ Res Public Health 14(12):E1520

Meister G, Landthaler M, Patkaniowska A et al (2004) Human Argonaute2 mediates RNA cleavage targeted by miRNAs and siRNAs. Mol Cell 15:185–197

Meng L, Ward AJ, Chun S et al (2015) Towards a therapy for Angelman syndrome by targeting a long non-coding RNA. Nature 518:409–412
Modarresi F, Faghihi MA, Patel NS et al (2011) Knockdown of *BACE1-AS* nonprotein-coding transcript modulates beta-amyloid-related hippocampal neurogenesis. Int J Alzheimers Dis 2011:929042
Modarresi F, Faghihi MA, Lopez-Toledano MA et al (2012) Inhibition of natural antisense transcripts in vivo results in gene-specific transcriptional upregulation. Nat Biotechnol 30:453–459
Mondal T, Subhash S, Vaid R et al (2015) *MEG3* long noncoding RNA regulates the TGF-beta pathway genes through formation of RNA-DNA triplex structures. Nat Commun 6:7743
Moulton J, Jiang S (2009) Gene knockdowns in adult animals: PPMOs and vivo-morpholinos. Molecules 14:1304–1323
Mus E, Hof PR, Tiedge H (2007) Dendritic *BC200* RNA in aging and in Alzheimer's disease. Proc Natl Acad Sci U S A 104:10679–10684
Nan Y, Zhang Y-J (2018) Antisense phosphorodiamidate morpholino oligomers as novel antiviral compounds. Front Microbiol 9:750
Noh JH, Kim KM, McClusky WG et al (2018) Cytoplasmic functions of long noncoding RNAs. Wiley Interdiscip Rev RNA 9:e1471
O'Leary VB, Ovsepian SV, Carrascosa LG et al (2015) *PARTICLE*, a triplex-forming long ncRNA, regulates locus-specific methylation in response to low-dose irradiation. Cell Rep 11:474–485
Obika S, Morio K, Hari Y et al (1999) Preparation and properties of 2′,5′-linked oligonucleotide analogues containing 3′-O,4′-C-methyleneribonucleosides. Bioorg Med Chem Lett 9:515–518
Pellestor F, Paulasova P (2004) The peptide nucleic acids (PNAs), powerful tools for molecular genetics and cytogenetics. Eur J Hum Genet 12:694–700
Postepska-Igielska A, Giwojna A, Gasri-Plotnitsky L et al (2015) LncRNA *Khps1* regulates expression of the proto-oncogene *SPHK1* via triplex-mediated changes in chromatin structure. Mol Cell 60:626–636
Qian X, Zhao J, Yeung PY et al (2019) Revealing lncRNA structures and interactions by sequencing-based approaches. Trends Biochem Sci 44:33–52
Quarrier S, Martin JS, Davis-Neulander L et al (2010) Evaluation of the information content of RNA structure mapping data for secondary structure prediction. RNA 16:1108–1117
Quinodoz S, Guttman M (2014) Long noncoding RNAs: an emerging link between gene regulation and nuclear organization. Trends Cell Biol 24:651–663
Rashid F, Shah A, Shan G (2016) Long non-coding RNAs in the cytoplasm. Genomics Proteomics Bioinformatics 14:73–80
Reggiani C, Coppens S, Sekhara T et al (2017) Novel promoters and coding first exons in *DLG2* linked to developmental disorders and intellectual disability. Genome Med 9:67
Rinaldi C, Wood MJ (2018) Antisense oligonucleotides: the next frontier for treatment of neurological disorders. Nat Rev Neurol 14:9–21
Rogaeva E, Meng Y, Lee JH et al (2007) The neuronal sortilin-related receptor SORL1 is genetically associated with Alzheimer disease. Nat Genet 39:168–177
Romero-Barrios N, Legascue MF, Benhamed M et al (2018) Splicing regulation by long noncoding RNAs. Nucleic Acids Res 46:2169–2184
Salta E, De Strooper B (2012) Non-coding RNAs with essential roles in neurodegenerative disorders. Lancet Neurol 11:189–200
Schwarz FP, Robinson S, Butler JM (1999) Thermodynamic comparison of PNA/DNA and DNA/DNA hybridization reactions at ambient temperature. Nucleic Acids Res 27:4792–4800
Shahryari A, Jazi MS, Samaei NM et al (2015) Long non-coding RNA *SOX2OT*: expression signature, splicing patterns, and emerging roles in pluripotency and tumorigenesis. Front Genet 6:196
Shi Y, Parag S, Patel R et al (2019) Stabilization of lncRNA *GAS5* by a small molecule and its implications in diabetic adipocytes. Cell Chem Biol 26(319–330):e316

Soler-Bistue A, Zorreguieta A, Tolmasky ME (2019) Bridged nucleic acids reloaded. Molecules 24: e2297

St Laurent G, Wahlestedt C, Kapranov P (2015) The landscape of long noncoding RNA classification. Trends Genet 31:239–251

Sugimoto Y, Vigilante A, Darbo E et al (2015) hiCLIP reveals the in vivo atlas of mRNA secondary structures recognized by Staufen 1. Nature 519:491–494

Sun Q, Hao Q, Prasanth KV (2018) Nuclear long noncoding RNAs: key regulators of gene expression. Trends Genet 34:142–157

Sunwoo JS, Lee ST, Im W et al (2017) Altered expression of the long noncoding RNA *NEAT1* in Huntington's disease. Mol Neurobiol 54:1577–1586

Tripathi V, Shen Z, Chakraborty A et al (2013) Long noncoding RNA *MALAT1* controls cell cycle progression by regulating the expression of oncogenic transcription factor B-MYB. PLoS Genet 9:e1003368

Ulitsky I, Shkumatava A, Jan CH et al (2011) Conserved function of lincRNAs in vertebrate embryonic development despite rapid sequence evolution. Cell 147:1537–1550

Underwood JG, Uzilov AV, Katzman S et al (2010) FragSeq: transcriptome-wide RNA structure probing using high-throughput sequencing. Nat Methods 7:995–1001

van Heesch S, van Iterson M, Jacobi J et al (2014) Extensive localization of long noncoding RNAs to the cytosol and mono- and polyribosomal complexes. Genome Biol 15:R6

Verma A (2018) Recent advances in antisense oligonucleotide therapy in genetic neuromuscular diseases. Ann Indian Acad Neurol 21:3–8

Wang KC, Chang HY (2011) Molecular mechanisms of long noncoding RNAs. Mol Cell 43:904–914

Wang J, Hu B, Cao F et al (2015) Down regulation of *lncSCIR1* after spinal cord contusion injury in rat. Brain Res 1624:314–320

Wang DQ, Fu P, Yao C et al (2018) Long non-coding RNAs, novel culprits, or bodyguards in neurodegenerative diseases. Mol Ther Nucleic Acids 10:269–276

Watters KE, Abbott TR, Lucks JB (2016) Simultaneous characterization of cellular RNA structure and function with in-cell SHAPE-seq. Nucleic Acids Res 44:e12

Weeks KM, Mauger DM (2011) Exploring RNA structural codes with SHAPE chemistry. Acc Chem Res 44:1280–1291

Wei CW, Luo T, Zou SS et al (2018) The role of long noncoding RNAs in central nervous system and neurodegenerative diseases. Front Behav Neurosci 12:175

Xie Y, Hayden MR, Xu B (2010) BDNF overexpression in the forebrain rescues Huntington's disease phenotypes in YAC128 mice. J Neurosci 30:14708–14718

Yamamura S, Imai-Sumida M, Tanaka Y et al (2018) Interaction and cross-talk between non-coding RNAs. Cell Mol Life Sci 75:467–484

Yan K, Arfat Y, Li D et al (2016) Structure prediction: new insights into decrypting long noncoding RNAs. Int J Mol Sci 17(1):E132

Yang F, Zhang H, Mei Y et al (2014) Reciprocal regulation of HIF-1α and *lincRNA-p21* modulates the Warburg effect. Mol Cell 53:88–100

Zampetaki A, Albrecht A, Steinhofel K (2018) Long non-coding RNA structure and function: is there a link? Front Physiol 9:1201

Zhou L, Sun K, Zhao Y et al (2015) *Linc-YY1* promotes myogenic differentiation and muscle regeneration through an interaction with the transcription factor YY1. Nat Commun 6:10026

Zhou HJ, Wang LQ, Wang DB et al (2018) Long noncoding RNA *MALAT1* contributes to inflammatory response of microglia following spinal cord injury via the modulation of a miR-199b/IKKbeta/NF-kappaB signaling pathway. Am J Physiol Cell Physiol 315:C52–C61

Biogenesis and Function of the Noncoding Isoform-Type LncRNAs

Yasuhiko Kato and Hajime Watanabe

Contents

Abstract In a single gene locus, multiple isoforms are generated in different circumstances owing to alternative promoter usage and alternative splicing. Although noncoding isoforms are also transcribed from the same locus, these long noncoding RNAs (lncRNAs) have been poorly defined and poorly characterized due to the difficulty of annotation by short-read RNA sequencing. In this review, we describe methods for the identification of noncoding isoform-type lncRNAs and introduce nine genes that produce noncoding isoforms. The molecular and the physiological functions of these noncoding isoforms have been discussed, including the regulation of the expression of overlapping protein-coding isoforms.

Keywords Noncoding isoform · Long noncoding RNA · Alternative promoter · Alternative splicing · Long-read RNA sequencing · RNA capture · Overlapping RNA

Y. Kato (✉) · H. Watanabe
Department of Biotechnology, Graduate School of Engineering, Osaka University, Osaka, Japan
e-mail: kato_yasuhiko@bio.eng.osaka-u.ac.jp; watanabe@bio.eng.osaka-u.ac.jp

S. Jurga, J. Barciszewski (eds.), *The Chemical Biology of Long Noncoding RNAs*, RNA Technologies 11, https://doi.org/10.1007/978-3-030-44743-4_3

1 Introduction

Long noncoding RNAs (lncRNAs) are pervasively transcribed in the eukaryotic genome (Mellor et al. 2016), and are important for the control of master regulatory genes involved in cell differentiation and cell development (Orom et al. 2010; Fatica and Bozzoni 2014). The discovery of lncRNAs has been mainly achieved through RNA sequencing (RNA-seq) analysis. Short-read RNA-seq has revealed that lncRNAs do not overlap with protein-coding RNAs in the sense orientation, including antisense, intronic, or intergenic lncRNAs (Derrien et al. 2012; Iyer et al. 2015; Hon et al. 2017; You et al. 2017) (Fig. 1). This is because the short-read RNA-seq is not suitable to distinguish between the coding genes and the noncoding genes when they overlap with each other (Housman and Ulitsky 2016). However, lncRNA is also transcribed as a noncoding isoform from the protein-coding gene and it overlaps with the sense orientation of the protein-coding isoform (Fig. 1). In this review, this type of lncRNA has been defined as "noncoding isoform-type lncRNA". These lncRNAs are difficult to annotate and their functions have been poorly analyzed. In this chapter, we first introduce recent advances in the methods for the identification of the noncoding isoform-type lncRNAs. We then introduce their biogenesis and functions with examples of several lncRNAs that have been studied till date.

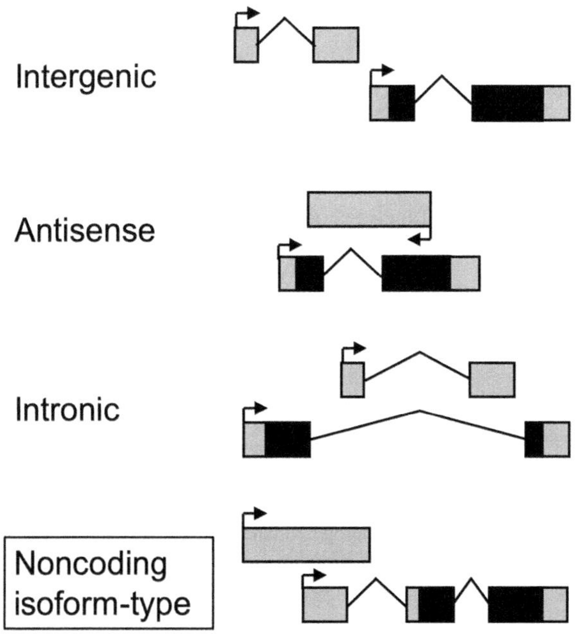

Fig. 1 Classification of lncRNAs based on their positional relationship with the protein-coding genes Grey and black bars indicate the noncoding regions and the protein-coding regions

2 Methods for Genome-Wide Identification of the Noncoding Isoform-Type LncRNAs

Long-read RNA-seq using PacBio can overcome the limitation of short-read RNA-Seq that cannot identify the noncoding isoforms (Sharon et al. 2013; Au et al. 2013; Abdel-Ghany et al. 2016). Using this technology, more than 10 kb of cDNAs can be sequenced in a single read, which leads to the determination of the full-length of transcripts. This technique enables the detection of multiple isoforms from a single gene locus and is referred to as Iso-Seq. Kuo et al. conducted chicken brain and embryo transcriptome analyses using Iso-Seq (Kuo et al. 2017). Of identified 60 k transcripts, more than 20 k were novel lncRNA transcripts. Interestingly, ~ 3 k transcripts were classified as noncoding isoform-type lncRNAs (also named as "sense exonic overlapping lncRNAs" in a study by Kuo et al). These authors also indicated that human and mouse have 20 k and 10 k noncoding isoform-type lncRNAs, respectively, as determined by their sense exonic overlap prediction tool on the Ensembl human and mouse processed_transcript models (Kuo et al. 2017).

PacBio Iso-Seq is hindered by low throughput (Sharon et al. 2013) but this can be compensated by the RNA Capture Sequencing in which the RNAs in the targeted genomic region are enriched using oligonucleotides and are sequenced (Mercer et al. 2012). To examine the transcriptional landscape of human chr21, RNA Capture was combined with short- and long-read RNA-seq (Deveson et al. 2018), which identified 7310 unique multi-exonic isoforms variants, including the noncoding isoforms at the protein-coding loci. Compared to the PacBio Iso-Seq, recent Nanopore-based sequencing technology has shown several advantages such as minimal preparation, a less-biased template length or GC content, and virtually unlimited read length (Byrne et al. 2017; Oikonomopoulos et al. 2016). These new approaches have been hypothesized to contribute to the formation of a comprehensive map of lncRNAs (Uszczynska-Ratajczak et al. 2018).

Despite difficulties for annotation and functional analyses, a lower number of the noncoding isoform-type lncRNAs were individually characterized. In the following sections, we have demonstrated the biogenesis and functions of the noncoding isoform-type lncRNAs from nine different genes that belong to the genomes of humans, the crustacean *Daphnia*, or the yeast (Table 1 and Figs. 2 and 3).

Table 1 List of genes that produce noncoding isoform-type lncRNAs

Locus (number of exons)	LncRNA	Species	Shared exons[a]	Biogenesis of lncRNA			Molecular archetype	References
				AP/AS	Cap	poly A		
Scaffold								
SRA1 (5 exons)	SRA	Human	Exons 1–5	AP/AS	nd	O	Scaffold	Lanz et al. (1999)
RUNX1 (10 exons)	RUNXOR	Human	Exons 1–10	AP	nd	nd	Scaffold	Wang et al. (2014)
Decoy								
DHFR (6 exons)	Interfering transcript	Human	Exon 1 & 2	AP	nd	nd	Decoy	Martianov et al. (2007)
CEBPA (single exon)	ecCEBPA	Human	Exon 1	AP	O	×	Decoy	Ruscio et al. (2013)
TGFB2 (8 exons)	TGFB2-OT1 (FLJ11812)	Human	Exon 8	AS[b]	O	O	Decoy	Ge et al. (2014) and Huang et al., (2015)
FBP1 + (yeast)	mlonRNAs	Fission yeast	Exon 1	AP	O	O	Decoy[c]	Hirota et al. (2008)
Unknown molecular mechanisms								
ASCC3 (42 exons)	ASCC3 short, ncRNA	Human	Exons 1–3	AS (ALE)	O	O	Unknown	Williamson et al. (2017)
DSX1 (4 exons)	DAPALR	Daphnia (crustacean)	Exon 3	AP	O	×	Unknown	Kato et al. (2018)
LCK (13 exons)	HULLK	Human	Exons 6–13	AP	nd	O	Unknown	Ta et al. (2019)

AP alternative promoter, *AS* alternative splicing, *ALE* alternative last exon, *cap* 5′-capping, *nd* not described

[a]Exons shared between the noncoding and the protein-coding isoforms

[b]By comprehensive analyses of human full-length cDNAs, FLJ11812 was originally identified as a capped RNA (Ota et al. 2004)

[c]In addition to functions as a decoy, mlonRNA transcription also has functions in chromatin remodeling (Hirota et al. 2008)

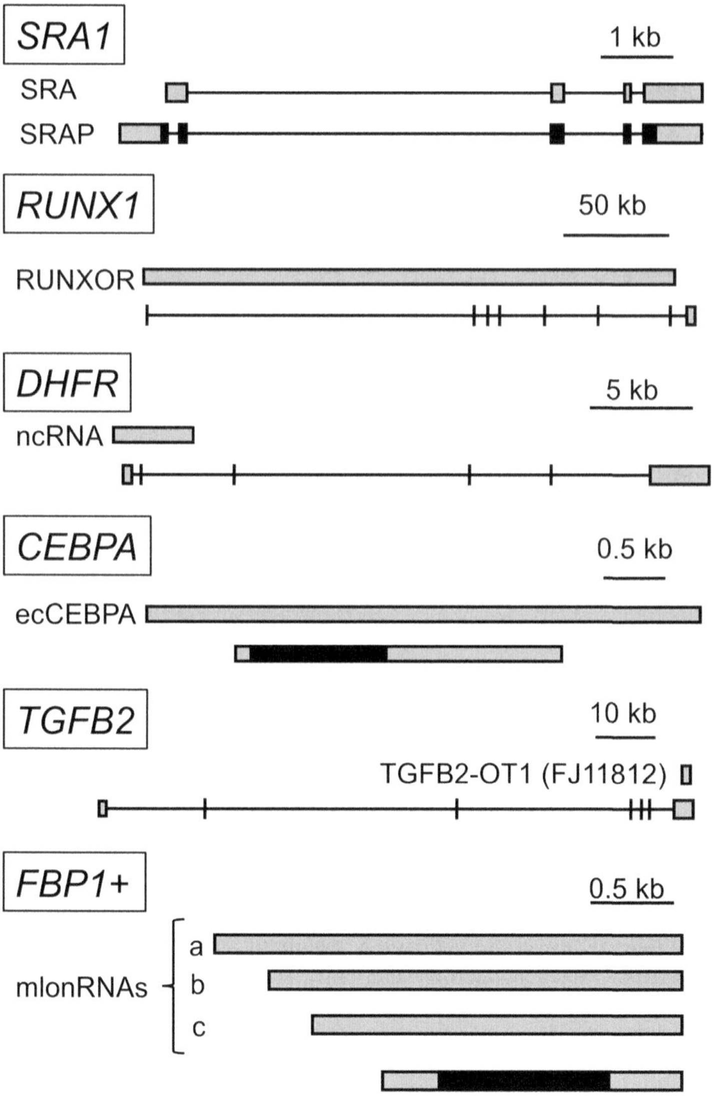

Fig. 2 Positional relationship of six noncoding-isoform-type lncRNAs with scaffold or decoy function
Position of the noncoding transcript is shown above of the protein-coding isoform. Grey and black bars indicate the noncoding regions and the protein-coding regions

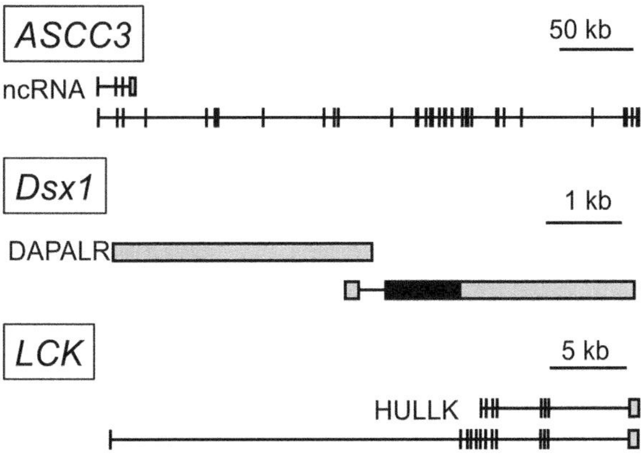

Fig. 3 Positional relationship of three noncoding-isoform-type lncRNAs with unknown molecular function Position of the noncoding transcript is shown above of the protein-coding isoform. Grey and black bars indicate the noncoding regions and the protein-coding regions

3 Biogenesis of the Noncoding Isoform-Type LncRNAs

3.1 Promoter Usage for the Expression of the Protein-Coding and the Noncoding Isoforms

3.1.1 Promoter-Sharing

Human *ASCC3* gene produces two isoforms, a long protein-coding and a short noncoding isoforms, both of which are transcribed from the same promoter (Table 1 and Fig. 3). These isoforms share the first three exons. However, UV irradiation leads to a slowdown of transcript elongation and stops transcription in the intron four of the long isoform, resulting in the generation of an alternative last exon. This mechanism leads to the production of 25 kb of the short isoform in contrast to the production of 375.5 kb of the long isoform harboring 42 exons (Williamson et al. 2017). In the human *TGFB2* gene, its noncoding isoform TGF2B-OT1 (originally named as FLJ11812) overlaps the exon 8 containing the 3′ UTR. Ge et al. reported that this lncRNA was generated by alternative splicing and TIA1 (TIA1 cytotoxic granule-associated RNA binding protein) controlled this processing (Ge et al. 2014) (Table 1 and Fig. 2).

3.1.2 Alternative Promoter Usage

The other seven genes described in this review rely on alternative promoter usage for the transcription of their noncoding isoforms (Table 1 and Figs. 2 and 3). In the human *SRA1* gene, the noncoding isoform SRA is transcribed in the downstream transcription starting site of the protein-coding isoform SRAP within the exon 1. Although the splicing pattern is similar between the noncoding and the protein-coding isoforms, intron 1 has been shown to be retained in some SRA transcript variants (Lanz et al. 1999; Cooper et al. 2011). In human *LCK* locus, transcription of the HULLK lncRNA starts at the exon 6 of the protein-coding isoform, continues until the terminal exon 13, and is spliced in a manner that is similar to the splicing of the LCK transcript (Ta et al. 2019). All of the other five lncRNAs are synthesized upstream of their protein-coding isoforms and partly or completely overlap with their transcribed regions. These lncRNAs are not subjected to splicing and are produced as intron-less transcripts (Table 1 and Figs. 2 and 3).

3.2 Post-transcriptional Processing

3.2.1 5′-Capping

Presence of 5′-capping has been analyzed in three lncRNAs: the human TGF2B-OT (FLJ11812) (Ota et al. 2004), the human ecCEBPA (Di Ruscio et al. 2013), and the *Daphnia* DAPALR (Kato et al. 2018) with RNA ligase-mediated (RLM) and oligo-capping RACE methods (Maruyama and Sugano 1994). In addition, a short non-coding isoform from the human *ASCC3* gene (Williamson et al. 2017) and the mlonRNAs transcripts (Hirota et al. 2008) are RNA polymerase II (RNAPII)-dependent, indicating that these lncRNAs also have 5′-cap structure.

3.2.2 Polyadenylation

Polyadenylation was detected in four human lncRNAs (SRA, TGFB2-OT1, ASCC3 short noncoding isoform, HULLK) and yeast mlonRNAs. In contrast, the human ecCEBPA and the *Daphnia* DAPALR are non-polyadenylated lncRNAs (Table 1). DAPALR 3′ processing occurs at an apical loop of the stem-loop structure (Kato et al. 2018), which has been unique among the lncRNAs studied so far. After this processing, a new hairpin stem loop may be generated at the 3′ end of the mature transcript, which may lead to protection from degradation by the 3′–5′ exonuclease activity. Previously, stabilization of the non-polyadenylated lncRNAs MALAT1 and NEAT1 by 3′ terminal RNA triplex structure has been reported (Wilusz et al. 2008).

4 Function of the Noncoding Isoform-Type LncRNAs

4.1 Scaffold

4.1.1 The Human *SRA1* Gene

The *SRA1* gene produces the noncoding isoform SRA and the protein-coding isoform SRAP. Lanz et al. discovered SRA by a yeast two-hybrid screening system to find cofactors that interact with the AF1 domain of the human progesterone receptor (PR) (Lanz et al. 1999). SRA was detected in the multiple tissues among which liver and skeletal muscle showed the strong expression. SRA overexpression enhanced PR transactivation in HeLa cells. Similarly, this lncRNA could enhance transactivation of the human glucocorticoid receptor (GR), androgen receptor (AR), estrogen receptor (ER), thyroid hormone receptor (TR), retinoic acid receptors (RAR and RXR), or peroxisome proliferator-activated receptor (PPAR). Truncation of the AF1 domain of the PR or GR abolished the effect of SRA. The knockdown of endogenous SRA inhibited PR-dependent gene activation. No SRA-encoded protein was detected and several SRA mutants harboring multiple stop codons showed ability to enhance transactivation of PR, demonstrating that SRA acts as an lncRNA. In addition, de novo protein synthesis inhibitor cycloheximide did not suppress the function of SRA for GR-mediated transactivation, providing clear evidence that the translation of SRA is not necessary for its function. Taken together, the noncoding isoform SRA functions as a coactivator for the steroid receptors by binding to their AF-1 domain (Lanz et al. 1999).

Lanz et al. investigated whether SRA functions as a component of ribonucleotide protein complexes (Lanz et al. 1999). They prepared whole-cell extract from human breast carcinoma cell line T-47D and fractioned protein complexes by gel filtration chromatography. The endogenous SRA was included in complexes of 600–700 kDa containing coactivator SRC-1. Association of SRA with SRC-1 was confirmed by coimmunoprecipitation using an antibody against SRC-1, indicating that SRA forms ribonucleoprotein complexes containing the SRC-1 (Lanz et al. 1999). SRA also interacts with the other transcription factors such as MyoD, Oct4, Nanog, and FOXO, which results in the formation of a complex with cofactors including SRC-1 and P68 (Sheng et al. 2018). Such features indicate a role of this lncRNA as a scaffold (Fig. 4). In addition, SRA binds to CTCF (Yao et al. 2010), TrxG (Wongtrakoongate et al. 2015) and PRC (Zhao et al. 2010) for chromatin modification and epigenetic regulation. Structural analyses of SRA using SHAPE (Mortimer and Weeks 2007), in-line (Regulski and Breaker 2008), DMS (Tijerina et al. 2007), and RNase V1 (Nichols and Yue 2008) probing suggested that stem loops and sub-structural domains may be important for its function (Novikova et al. 2012). Due to the interaction with diverse factors, SRA has many physiological functions such as steroidogenesis (Xu et al. 2009; Kelly et al. 2010), muscle differentiation (Caretti et al. 2006), and adipogenesis (Xu et al. 2010). Further, the SRA lncRNA is involved in human diseases including breast cancer (Yan et al. 2016) and obesity (Liu et al. 2014). SRAP may be associated with SRA and inhibit its action on MyoD during muscle differentiation, implying that alternative production of noncoding and protein-coding isoforms is important for proper MyoD regulation (Hubé et al. 2011).

4.1.2 The Human *RUNX1* Gene

The *RUNX1* (runt-related transcription factor 1), also known as *AML1* (Acute myeloid leukemia 1) and *CEBFA2* (Core-binding factor subunit alpha-2), is a master regulator of hematopoiesis and is located on Chr21. The *RUNX1* gene shows

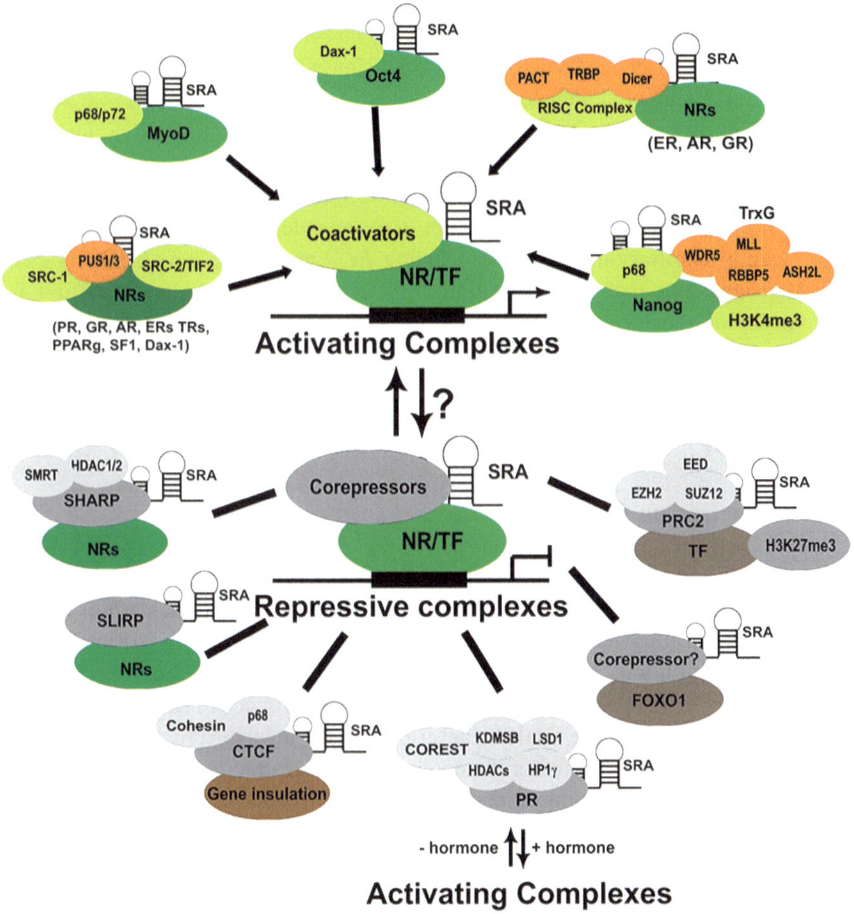

Fig. 4 Function of SRA as a scaffold
PUS pseudouridine synthase, *RISC* RNA-induced silencing complex, *Trxn* trithorax complex, *SHARP* the SMRT/HDAC1-associated repressor protein, *SLIRP* SRA stem-loop interacting RNA binding protein, CTCF, CCCTC-binding factor, HP1γ, heterochromatin protein 1γ, and *LSD1* lysine-specific demethylase 1, *PRC2* polycomb repressive complex 2. (This schematic representation was reproduced from Sheng et al. 2018)

chromosomal translocations in hematopoietic malignancies, especially acute myeloid leukemia. The t(8;21) translocation is a common phenomenon that leads to the fusion of Runx1 with the *ETO* gene. Its noncoding isoform RUNXOR (Runx1 overlapping RNA) was identified to interact with the RUNX1 promoter by using the R3C (RNA-guided Chromatin Conformation Capture) (Wang et al. 2014). This noncoding isoform is unspliced and includes all of 10 exons of the *RUNX1* gene in addition to its promoter region, resulting in the generation of approximately 260 kb of the transcript. The RUNXOR was expressed in leukemia-derived cell lines (KG-1, KG-1a, K562, and Kasumi-1) as well as the protein-coding isoform. The expression of this lncRNA was higher in AML patients' bone marrow compared to the control sample, suggesting that this lncRNA might be involved in chromosomal translocation of the *RUNX1* gene. To detect RNA/DNA interaction with the low noise/signal background, Wang et al. developed a "reverse transcription-associated trap" (RAT) assay. Using this assay, they found that the 3′ region of the RUNXOR interacted with *RUNX1* promoters P1 and P2. Interestingly, this lncRNA associated with the most common breakpoint for RUNX1 chromosomal translocation, which is located between exon 6 and exon 7. In addition, they revealed that RUNXOR associates with several chromosomal regions including *ETO* at chromosome 8 (Wang et al. 2014), suggesting that this noncoding isoform may function as a scaffold for chromosomal translocations.

4.2 Decoy

4.2.1 The Human *DHFR* Gene

The *DHFR* (dihydrofolate reductase) gene encodes an enzyme that converts dihydrofolate to tetrahydrofolate, which is a key molecule in one-carbon transfer reactions controlling various physiological processes including biosynthesis of purines and thymidine, amino acid homeostasis (glycine, serine, and methionine), epigenetic maintenance, and redox defense. The growing cells have been known to express higher levels of the DHFR transcript than the levels expressed by quiescent cells. This gene produces a minor noncoding isoform upstream of the protein-coding isoform. Among the total transcripts from the *DHFR* locus, the noncoding ones are considered to occupy only 1% in the HeLa cells (Masters and Attardi 1985).

In the quiescent state, the transcription of the noncoding RNA is activated, but the protein-coding mRNA expression is repressed in the U2OS cells (Martianov et al. 2007). To investigate a role of the minor noncoding isoform, Martianov et al. reconstituted transcription for the TATA-less *DHFR* gene in vitro. Addition of the synthetic noncoding transcript inhibited transcription of the protein-coding isoform, indicating that the noncoding isoform regulates transcription of the protein-coding isoform in trans. Of TATA-binding protein (TBP) and transcription factor IIB (TFIIB) that occupied the promoter of the protein-coding isoform, TFIIB interacted with the noncoding transcript in vitro, suggesting that the minor transcript was important for disrupting the formation of the pre-initiation complex (PIC). Using a combination of loss- and gain-of-function experiments, Martianov et al. revealed that the noncoding transcript functions as decoys for TFIIB (Fig. 5a) and dissociates the preinitiation complex from the promoter of the protein-coding isoform of the DHFR gene in vivo (Martianov et al. 2007), demonstrating tight control of the growth-dependent DHFR gene expression by the noncoding isoform-type lncRNA.

The promoter for the protein-coding isoform is GC-rich and contains several G-track sequences. These properties had been known to lead to the formation of stable purine-purine-pyrimidine triplex structures (H form) between DNA and RNA

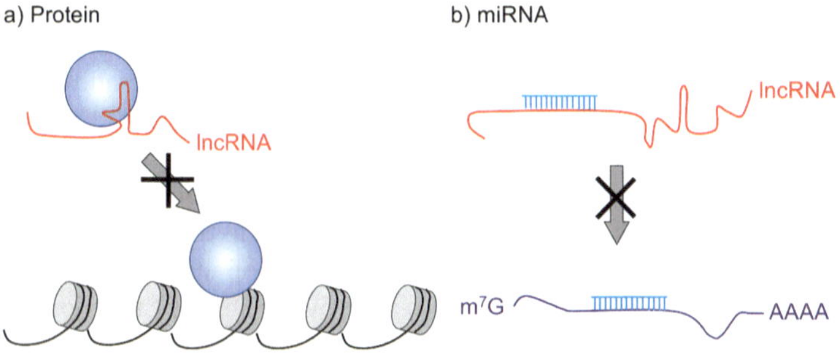

Fig. 5 Function of lncRNAs as a decoy. (**a**) Protein and (**b**) miRNA
LncRNAs bind proteins or regulatory RNAs such as miRNAs and can titrate away from their targets

(Letai et al. 1988). By a standard H-form band-shift assay, Martianov et al. found that the stable triplex was formed between the promoter and the noncoding transcript, suggesting that the triplex structures may allow the noncoding transcript to target and repress the promoter that drives expression of the protein-coding isoform (Martianov et al. 2007).

4.2.2 The Human *CEBPA* Gene

The transcription factor CEBPA (CCAAT/enhancer-binding protein alpha), also known as C/EBPα, induces cell growth arrest and functions as a tumor suppressor. This intronless *CEBPA* gene is methylated and silenced in lung cancer (Tada et al. 2006) and in acute myeloid leukemia (Hackanson et al. 2008). In order to examine the effect of transcription on the DNA methylation of the *CEBPA* gene, Di Ruscio et al. searched for RNAs that were transcribed in its neighboring regions (Di Ruscio et al. 2013). Strand-specific reverse transcriptase PCR, northern blot analysis, 5′ and 3′ RACE experiments found ~4.5 kb of a non-polyadenylated RNA that encompasses ~2.6 kb of the entire CEBPA protein-coding isoform in the same-sense orientation. This noncoding isoform named ecCEBPA (extra-coding CEBPA transcript) is transcribed from the upstream to downstream of the protein-coding mRNA in human myeloid leukemia cell lines HL60 and U937. Expression of ecCEBPA occurs in the S phase before its overlapping mRNA transcription begins.

To elucidate the role of ecCEBPA in the regulation of *CEBPA* expression, Di Ruscio et al. performed both loss- and gain-of-function experiments (Di Ruscio et al. 2013). The knockdown of ecCEBPA increased DNA methylation level of the distal promoter of the CEBPA gene, resulting in reduced expression of the protein-coding mRNA. In contrast, overexpression of a part of ecCEBPA that does not overlap with CEBPA mRNA led to decrease of the DNA methylation level, which increased the *CEBPA* expression. These data demonstrate that ecCEBPA inhibits DNA methylation on the promoter of the *CEBPA* gene.

RNA immunoprecipitation with anti-DNMT1 antibody revealed physiological interaction between ecCEBPA and DNMT1 (Di Ruscio et al. 2013). This finding was consistent with a report that DNMT1 expression and its enzymatic activity peak in S-phase. By RNA electrophoresis mobility shift assays, DNMT1 was found to bind stem-loop-structured RNAs. In addition, Di Ruscio et al. divided DNMT1 protein into five regions and fused each region to glutathione S-transferase and performed pull down assays, which found that the catalytic domain including the target recognition domain bound RNAs with stem-loop structures. Together, the noncoding isoform ecCEBPA functions as a decoy for DNMT1 that downregulates CRBPA expression (Fig. 5a). This relationship between the noncoding isoform and DNMT1 suggests a mechanism for site-specific modification of aberrant DNA methylation (Di Ruscio et al. 2013).

4.2.3 The Human *TGFB2* Gene

The TGFB2 (transforming growth factor beta-2) is one of three TGFβ paralogs in humans. The TGFβ family controls cellular behaviors such as the proliferation, differentiation, migration, and specification of synthetic repertoire. Its noncoding isoform FLJ11812 was discovered as a downstream target of the mTOR signaling (Ge et al. 2014). In vascular endothelial cells, an mTOR activator 3BDO (3-benzyl-5-((2-nitrophenoxy) methyl)-dihydrofuran-2(3H)-one) downregulated expression of this lncRNA. The sequence of FLJ11812 is present within the 3′ UTR of the *TGFB2* gene. Overexpression of FLJ11812 increased expression of an autophagy-related marker protein MAP 1LC3B and seemed to induce autophagic flux in vascular endothelial cells, suggesting a role of FLJ11812 for autophagy. The noncoding isoform did not change the expression of the protein-coding isoform (Ge et al. 2014).

In FLJ11812 sequence, Bioinformatics approach found a potential binding site of mir-4459, which was also present in the 3′ UTR of ATG13, an important gene for controlling autophagy. Luciferase reporter assay revealed that this miRNA was bound to both FLJ11812 and the 3′ UTR of ATG13. Overexpression of FLJ11812 led to increase of ATG13 protein level, demonstrating that this noncoding isoform functions as a decoy for miR-4459 (Ge et al. 2014) (Fig. 5b). Later, the same research group changed name of this lncRNA from FLJ11812 to TGFB2-OT1 (TGFB2 overlapping transcript 1) and found that this noncoding transcripts can also be decoys for miR-3916 and miR-4488. Titrating away of these miRNAs led to the upregulation of CERS (ceramide synthase 1), NAT8L (N-acetyltransferase 8-like [GCN-related, putative]), LARP1 (La ribonucleoprotein domain family, member 1), resulting in the regulation of autophagy and inflammation (Huang et al. 2015).

4.2.4 The Yeast *FBP1+* Gene

The *FBP1+* gene (fructose-1,6-bisphosphatase 1) encodes a key enzyme for gluconeogenesis in *Schizosaccharomyces pombe* (fission yeast). This gene is silent during the growth in glucose-containing medium but is significantly activated in response to glucose starvation. Hirota et al. found that three different lengths of noncoding isoforms are transcribed in the *FBP1*+ gene promoter during starvation-induced derepression (Hirota et al. 2008). Of the three, at 0–20 min after glucose starvation, the long isoform named “a” is transcribed from the most distant position to the *FBP1* + transcription start site (TSS). This transcript covers the UAS1, UAS2, and TATA for the *FBP1*+ induction. Later TSSs for the noncoding transcripts shift 5′–3′ in a stepwise manner toward the *FBP1*+ TSS, resulting in production of middle and short isoforms (b and c) at 10–60 and 20–60 min, respectively. These b and c isoforms overlap UAS2 and TATA sequences. ChIP assay revealed that regions for RNAPII binding are also translocated from 5′ to 3′ in upstream of the *FBP1*+ promoter after glucose starvation. The micrococcal nuclease (MNase) digestion assay found that, in a similar manner with shift of transcriptionally active region, chromatin remodeling shifts 5′–3′(Hirota et al. 2008).

Hirota et al. further investigated a correlation between chromatin remodeling and transcriptional initiation events (Hirota et al. 2008). Insertion of a transcriptional terminator between UAS2 and TATA led to inhibition of *FBP1*+ induction and chromatin remodeling as well as the terminator insertion between UAS1 and UAS2. Interestingly, in upstream of a transcription terminator, chromatin remodeling was not affected, indicating importance of RNAPII passage for the chromatin remodeling (Hirota et al. 2008). Galipon et al. named these noncoding isoform-type lncRNAs as mlonRNAs (metabolic stress-induced long noncoding RNAs (Galipon et al. 2013). In the follow-up studies, mlonRNAs was found to bind and inhibit the Tup-1 like corepressors, resulting in binding of the Atf1 transcription activator for the *FBP1*+ promoter (Takemata et al. 2016), suggesting that the lncRNAs have a decoy function for corepressors (Fig. 5a). In addition, Takemata et al. revealed that the Atf1 binding leads to histone acetylation through the recruitment of histone acetyltransferase Gcn5 (Takemata et al. 2016).

4.3 *Unknown Molecular Mechanism*

4.3.1 The Human *ASCC3* Gene

The *ASCC3* encodes a 3′–5′ DNA helicase and has been identified as one of three subunits comprising the activating signal co-integrator complex (ASCC). This ASCC complex functions as a transcriptional coactivator of nuclear receptors (Jung et al. 2002) and associates with ALKBH3 protein for the maintenance of genomic integrity (Dango et al. 2011). ASCC3 protein interacts with RNAPII and CSB (Cockayne syndrome B protein) which controls both TC-NER (transcription-coupled nucleotide excision repair) and global transcription response to DNA damage, demonstrating the direct effect of ASCC3 on transcription and/or DNA repair (Boeing et al. 2016). More recently, the ASCC3 protein was found to suppress global transcription after UV irradiation (Williamson et al. 2017).

The noncoding isoform of this gene was identified by transcriptome analyses after UV irradiation. First, DRB/GRO-seq (5,6-dichloro-1-β-D-ribofuranosylbenzimidazole/global run-on sequencing) found that UV irradiation reduced the rate of the *ASCC3* transcription elongation (Williamson et al. 2017). Second, mRNA splicing was analyzed using mRNAs from UV-treated cells by next-generation sequencing because mRNA processing is coupled to transcript elongation. These analyses revealed that the short noncoding isoform is generated by alternative last exon (ALE) splicing in response to UV. Knockdown of the short ASCC3 isoform interfered transcription recovery after UV irradiation and led to increased sensitivity to UV. In addition, when ALE was removed to abolish the short isoform expression, transcriptional recovery was inhibited. These indicate that the noncoding isoform functions in recovery from the repressed state after UV irradiation and plays an important role for UV resistance (Williamson et al. 2017).

Knocking out the short noncoding isoform resulted in 108 differentially expressed genes compared to control cells after UV irradiation. Of the 108, 79 genes were downregulated, the majority of which were upregulated by knockdown of the long protein-coding isoform. Five genes (IFIT2, IFIH1, HERC5, IFI44, DDX60) were markedly affected by the *ASCC3* locus. Their increased expressions in response to UV irradiation were significantly impaired in the short isoform-deficient cells. In contrast, all of those genes were "over-induced" by knockdown of the long isoform, demonstrating antagonistic regulation by the short and long isoforms. The short noncoding isoform seems to, at least in part, repress the ASCC complex which maintains transcriptional repression (Williamson et al. 2017).

4.3.2 The *Daphnia Dsx1* Gene

The *Dsx1* (*Doublesex1*) gene encodes a transcription factor containing DNA-binding domain named as the DM domain in the crustacean *Daphnia* that produces only male offspring in response to environmental stimuli (Kato et al. 2011). Dsx1 expression is maintained and upregulated only in the male population and controls male determination and differentiation. While investigating post-transcriptional regulation of this gene, Kato et al. developed chimeric reporter mRNAs in which *Dsx1* 5′ UTR was linked to *DsRed2* ORF with a *Dsx1* 3′ UTR. Interestingly, despite lacking a *Dsx1* ORF, this chimeric reporter RNA induced a male trait, the elongation of the first antennae in female embryos (Kato et al. 2018). In addition, the RNA coding for 205 nt of *Dsx1* 5′ UTR was found to induce the same phenotype. This short RNA also upregulated *Dsx1* expression, indicating that the *Dsx1* 5′ UTR is involved in *Dsx1* activation and in male trait development. This finding led to the identification of a noncoding isoform-type lncRNA DAPALR (doublesex1 alpha promoter-associated long RNA) that is transcribed upstream of the *Dsx1* transcription start site and overlaps with its 5′ UTR (Kato et al. 2018).

DAPALR expression is upregulated in males by the juvenile hormone and bZIP transcription factor Vrille, both of which are known to activate *Dsx1* expression

(Kato et al. 2011; Mohamad Ishak et al. 2017). Its expression is 10 times lower than that of the *Dsx1* and is maintained during male development. Silencing of DAPALR in males reduced *Dsx1* expression, resulting in the feminization of the first antenna, which is a major male trait in this species. In addition, the knockdown animals could produce offspring. In contrast, overexpression of DAPALR and its functional element overlapping with *Dsx1* 5′ UTR led to *Dsx1* activation, suggesting that DAPALR may transactivate *Dsx1* expression (Kato et al. 2018), although its molecular action remains unclear.

4.3.3 The Human *LCK* Gene

The LCK (lymphocyte-specific protein kinase) belongs to an src family of non-receptor protein tyrosine kinase. This kinase has long been known to be related to malignancies of both lymphoid and non-lymphoid tissues. By using an RNAi phenotypic screen with shRNAs that targeted the kinases, the LCK gene was identified as a potential positive regulator of growth of prostate cancer (PCa) cells. However, interestingly, the LCK proteins could not be detected in those cancer cell lines. This unexpected finding led to the identification of the noncoding isoform HULLK (hormone upregulated lncRNA within LCK) (Ta et al. 2019).

The expression of HULLK was increased by treatment of a synthetic androgen R1881 in the PCa cells but this hormone effect was inhibited in AR (androgen receptor)-null PCa cells, indicating that this lncRNA is regulated by androgen in an AR-dependent manner. Both p300 inhibitor A-145 and Brd4 inhibitor JQ1 could repress the androgen-dependent upregulation of HULLK expression, indicating that AR recruits these two coactivators for activating HULLK expression. AR expression was not changed by silencing and overexpression of HULLK, suggesting this lncRNA does not reciprocally control AR expression. HULLK expression was detected in seven PCa cells (LNCap, C4-2, CWR22Rv1, PC3, DU145, VCap, LAPC4) and correlates well with high-grade prostate cancer in multiple cohorts. Functional analyses of HULLK suggest that this lncRNA possibly drives PCa cell growth and may increase the sensitivity of the cells to androgen for proliferation. The HULLK-target genes and its molecular action remains unknown.

5 Concluding Remarks and Future Perspectives

Recent advances in long-read RNA-Seq technologies using the PacBio Iso-seq and Nanopore have led to increased identification of the noncoding isoform-type lncRNAs. These transcripts are currently known to occupy a large portion of lncRNAs. Characterization of even a limited number of noncoding isoform-type lncRNAs has revealed the diversity of mechanisms of their biogenesis and functions. To better understand lncRNAs, it is necessary to investigate unknown molecular mechanisms of the identified RNAs such as ASCC3 short noncoding isoform,

DAPALR, and HULLK. In addition, there is also a need to evaluate novel noncoding isoform transcripts identified through long-read RNA-Seq. Understanding noncoding isoform-type lncRNAs may not only contribute to the development of a complete lncRNA map, but also to a future understanding of an alternative gene expression and the physiological significance of such an expression.

References

Abdel-Ghany SE, Hamilton M, Jacobi JL et al (2016) A survey of the sorghum transcriptome using single-molecule long reads. Nat Commun 7:11706

Au KF, Sebastiano V, Afshar PT et al (2013) Characterization of the human ESC transcriptome by hybrid sequencing. Proc Natl Acad Sci U S A 110:E4821–E4830

Boeing S, Williamson L, Encheva V et al (2016) Multiomic analysis of the UV-induced DNA damage response. Cell Rep 15:1597–1610

Byrne A, Beaudin AE, Olsen HE et al (2017) Nanopore long-read RNAseq reveals widespread transcriptional variation among the surface receptors of individual B cells. Nat Commun 8:16027

Caretti G, Schiltz RL, Dilworth FJ et al (2006) The RNA helicases p68/p72 and the noncoding RNA SRA are coregulators of MyoD and skeletal muscle differentiation. Dev Cell 11:547–560

Cooper C, Vincett D, Yan Y et al (2011) Steroid receptor RNA activator bi-faceted genetic system: heads or tails? Biochimie 93:1973–1980

Dango S, Mosammaparast N, Sowa ME et al (2011) DNA unwinding by ASCC3 helicase is coupled to ALKBH3-dependent DNA alkylation repair and cancer cell proliferation. Mol Cell 44:373–384

Derrien T, Johnson R, Bussotti G et al (2012) The GENCODE v7 catalog of human long noncoding RNAs: analysis of their gene structure, evolution, and expression. Genome Res 22:1775–1789

Deveson IW, Brunck ME, Blackburn J et al (2018) Universal alternative splicing of noncoding exons. Cell Syst 6:245–255.e5

Di Ruscio A, Ebralidze AK, Benoukraf T et al (2013) DNMT1-interacting RNAs block gene-specific DNA methylation. Nature 503:371–376

Fatica A, Bozzoni I (2014) Long non-coding RNAs: new players in cell differentiation and development. Nat Rev Genet 15:7–21

Galipon J, Miki A, Oda A et al (2013) Stress-induced lncRNAs evade nuclear degradation and enter the translational machinery. Genes Cells 18:353–368

Ge D, Han L, Huang S et al (2014) Identification of a novel MTOR activator and discovery of a competing endogenous RNA regulating autophagy in vascular endothelial cells. Autophagy 10:957–971

Hackanson B, Bennett KL, Brena RM et al (2008) Epigenetic modification of CCAAT/enhancer binding protein expression in acute myeloid leukemia. Cancer Res 68:3142–3151

Hirota K, Miyoshi T, Kugou K et al (2008) Stepwise chromatin remodelling by a cascade of transcription initiation of non-coding RNAs. Nature 456:130–134

Hon C-C, Ramilowski JA, Harshbarger J et al (2017) An atlas of human long non-coding RNAs with accurate 5′ ends. Nature 543:199–204

Housman G, Ulitsky I (2016) Methods for distinguishing between protein-coding and long noncoding RNAs and the elusive biological purpose of translation of long noncoding RNAs. Biochim Biophys Acta 1859:31–40

Huang S, Lu W, Ge D et al (2015) A new microRNA signal pathway regulated by long noncoding RNA TGFB2-OT1 in autophagy and inflammation of vascular endothelial cells. Autophagy 11:2172–2183

Hubé F, Velasco G, Rollin J et al (2011) Steroid receptor RNA activator protein binds to and counteracts SRA RNA-mediated activation of MyoD and muscle differentiation. Nucleic Acids Res 39:513–525

Iyer MK, Niknafs YS, Malik R et al (2015) The landscape of long noncoding RNAs in the human transcriptome. Nat Genet 47:199–208

Jung D-J, Sung H-S, Goo Y-W et al (2002) Novel transcription coactivator complex containing activating signal cointegrator 1. Mol Cell Biol 22:5203–5211

Kato Y, Kobayashi K, Watanabe H, Iguchi T (2011) Environmental sex determination in the branchiopod crustacean Daphnia magna: deep conservation of a Doublesex gene in the sex-determining pathway. PLoS Genet 7:e1001345

Kato Y, Perez CAG, Mohamad Ishak NS et al (2018) A 5′ UTR-overlapping LncRNA activates the male-determining gene doublesex1 in the crustacean Daphnia magna. Curr Biol 28:1811–1817.e4

Kelly VR, Xu B, Kuick R et al (2010) Dax1 up-regulates Oct4 expression in mouse embryonic stem cells via LRH-1 and SRA. Mol Endocrinol 24:2281–2291

Kuo RI, Tseng E, Eory L et al (2017) Normalized long read RNA sequencing in chicken reveals transcriptome complexity similar to human. BMC Genomics 18:323

Lanz RB, McKenna NJ, Onate SA et al (1999) A steroid receptor coactivator, SRA, functions as an RNA and is present in an SRC-1 complex. Cell 97:17–27. https://doi.org/10.1016/S0092-8674(00)80711-4

Letai AG, Palladino MA, Fromm E et al (1988) Specificity in formation of triple-stranded nucleic acid helical complexes: studies with agarose-linked polyribonucleotide affinity columns. Biochemistry 27:9108–9112

Liu S, Sheng L, Miao H et al (2014) SRA gene knockout protects against diet-induced obesity and improves glucose tolerance. J Biol Chem 289:13000–13009

Martianov I, Ramadass A, Serra Barros A et al (2007) Repression of the human dihydrofolate reductase gene by a non-coding interfering transcript. Nature 445:666–670

Maruyama K, Sugano S (1994) Oligo-capping: a simple method to replace the cap structure of eukaryotic mRNAs with oligoribonucleotides. Gene 138:171–174

Masters JN, Attardi G (1985) Discrete human dihydrofolate reductase gene transcripts present in polysomal RNA map with their 5′ ends several hundred nucleotides upstream of the main mRNA start site. Mol Cell Biol 5:493–500

Mellor J, Woloszczuk R, Howe FS (2016) The interleaved genome. Trends Genet 32:57–71

Mercer TR, Gerhardt DJ, Dinger ME et al (2012) Targeted RNA sequencing reveals the deep complexity of the human transcriptome. Nat Biotechnol 30:99–104

Mohamad Ishak NS, Nong QD, Matsuura T et al (2017) Co-option of the bZIP transcription factor Vrille as the activator of Doublesex1 in environmental sex determination of the crustacean Daphnia magna. PLoS Genet 13:e1006953

Mortimer SA, Weeks KM (2007) A fast-acting reagent for accurate analysis of RNA secondary and tertiary structure by SHAPE chemistry. J Am Chem Soc 129:4144–4145

Nichols NM, Yue D (2008) Ribonucleases. Curr Protoc Mol Biol 84(1):3.13.1–3.13.8. Chapter 3, Unit

Novikova IV, Hennelly SP, Sanbonmatsu KY (2012) Structural architecture of the human long non-coding RNA, steroid receptor RNA activator. Nucleic Acids Res 40:5034–5051

Oikonomopoulos S, Wang YC, Djambazian H et al (2016) Benchmarking of the Oxford Nanopore MinION sequencing for quantitative and qualitative assessment of cDNA populations. Sci Rep 6:31602

Orom UA, Derrien T, Beringer M et al (2010) Long noncoding RNAs with enhancer-like function in human cells. Cell 143:46–58

Ota T, Suzuki Y, Nishikawa T et al (2004) Complete sequencing and characterization of 21,243 full-length human cDNAs. Nat Genet 36:40–45

Regulski EE, Breaker RR (2008) In-line probing analysis of riboswitches. Methods Mol Biol 419:53–67

Sharon D, Tilgner H, Grubert F, Snyder M (2013) A single-molecule long-read survey of the human transcriptome. Nat Biotechnol 31:1009–1014
Sheng L, Ye L, Zhang D et al (2018) New insights into the long non-coding RNA SRA: physiological functions and mechanisms of action. Front Med 5:244
Ta HQ, Whitworth H, Yin Y et al (2019) Discovery of a novel long noncoding RNA overlapping the LCK gene that regulates prostate cancer cell growth. Mol Cancer 18:113
Tada Y, Brena RM, Hackanson B et al (2006) Epigenetic modulation of tumor suppressor CCAAT/ enhancer binding protein α activity in lung cancer. J Natl Cancer Inst 98:396–406
Takemata N, Oda A, Yamada T et al (2016) Local potentiation of stress-responsive genes by upstream noncoding transcription. Nucleic Acids Res 44:5174–5189
Tijerina P, Mohr S, Russell R (2007) DMS footprinting of structured RNAs and RNA–protein complexes. Nat Protoc 2:2608–2623
Uszczynska-Ratajczak B, Lagarde J, Frankish A et al (2018) Towards a complete map of the human long non-coding RNA transcriptome. Nat Rev Genet 19:535–548
Wang H, Li W, Guo R et al (2014) An intragenic long noncoding RNA interacts epigenetically with the RUNX1 promoter and enhancer chromatin DNA in hematopoietic malignancies. Int J Cancer 135:2783–2794
Williamson L, Saponaro M, Boeing S et al (2017) UV irradiation induces a non-coding RNA that functionally opposes the protein encoded by the same gene. Cell 168:843–855.e13
Wilusz JE, Freier SM, Spector DL (2008) 3′ end processing of a long nuclear-retained noncoding RNA yields a tRNA-like cytoplasmic RNA. Cell 135:919–932
Wongtrakoongate P, Riddick G, Fucharoen S, Felsenfeld G (2015) Association of the long non-coding RNA steroid receptor RNA activator (SRA) with TrxG and PRC2 complexes. PLoS Genet 11:e1005615
Xu B, Yang W-H, Gerin I et al (2009) Dax-1 and steroid receptor RNA activator (SRA) function as transcriptional coactivators for steroidogenic factor 1 in steroidogenesis. Mol Cell Biol 29:1719–1734
Xu B, Gerin I, Miao H et al (2010) Multiple roles for the non-coding RNA SRA in regulation of Adipogenesis and insulin sensitivity. PLoS One 5:e14199
Yan R, Wang K, Peng R et al (2016) Genetic variants in lncRNA SRA and risk of breast cancer. Oncotarget 7:22486–22496
Yao H, Brick K, Evrard Y et al (2010) Mediation of CTCF transcriptional insulation by DEAD-box RNA-binding protein p68 and steroid receptor RNA activator SRA. Genes Dev 24:2543–2555
You B-H, Yoon S-H, Nam J-W (2017) High-confidence coding and noncoding transcriptome maps. Genome Res 27:1050–1062
Zhao J, Ohsumi TK, Kung JT et al (2010) Genome-wide identification of polycomb-associated RNAs by RIP-seq. Mol Cell 40:939–953

Long Non-coding RNAs in a Single-Cell Type: Function and Subcellular Localization

Raphael Severino Bonadio, Enrico Alessio, and Stefano Cagnin

Contents

Abstract After the conclusion of the human genome project, the annotated genes varied among 26,000 and 39,000. However, this gene count is significantly smaller than what was observed later in 2012 when it was evidenced that most of the human genome is transcribed. The vast majority of the transcribed genome does not code for translated RNAs that are involved in the protein synthesis, but comprises non-coding RNAs. This is true not only for the human genome but also for most of the mammalian genomes. Among non-coding RNAs, the most abundant are long non-coding RNAs (lncRNAs). They are emerging as important players in the regulation of several aspects of cellular biology. For instance, they are involved in

R. S. Bonadio · E. Alessio
Department of Biology, University of Padova, Padova, Italy

S. Cagnin (✉)
Department of Biology, University of Padova, Padova, Italy

CRIBI Biotechnology Center, University of Padova, Padova, Italy

CIR-Myo Myology Center, University of Padova, Padova, Italy
e-mail: stefano.cagnin@unipd.it

S. Jurga, J. Barciszewski (eds.), *The Chemical Biology of Long Noncoding RNAs*, RNA Technologies 11, https://doi.org/10.1007/978-3-030-44743-4_4

cell differentiation, normal functioning of differentiated cells, and also involved in the development of pathological conditions such as tumors. They present an expression that is more cell-specific than coding RNAs and for this reason, it is important to study them in single cells instead of using bulk tissues or a cell population. Recent improvements in techniques that allow genomic and transcriptomic analyses of single cells and those for fluorescence amplification have given a considerable boost in the comprehension of cell specificity, subcellular localization, and function of lncRNAs. The purpose of this chapter is to discuss different methods that are available to detect RNA expression using a single cell approach evidencing the advantages and disadvantages of each of them. We will consider how the analysis of single cells has contributed to the better comprehension of lncRNA functions and how it was involved in proposing new paradigms. To conclude the chapter, we will consider published databases to retrieve information on genomic and transcriptomic experiments on single cells.

Keywords Single-cell RNA sequencing · Single cell · Single-cell expression · Single-cell databases · Long non-coding RNAs

1 Introduction

The genome of multicellular eukaryotes is pervasively transcribed. Considering that only about 2% of the human genome codes for protein-coding genes, the most abundant part of transcribed RNA is non-coding. Non-coding RNAs (ncRNAs) include different classes that can be broadly divided into structural RNAs (rRNAs and tRNAs) and regulatory RNAs (Dahariya et al. 2019). Most known non-coding RNAs are rRNAs and tRNAs that are involved in mRNA translation. Regulatory RNAs are divided into three classes: small, medium, and long non-coding RNAs. Small non-coding RNAs are 20–50 nucleotide long and include microRNAs (miRNAs) involved in post-transcriptional regulation (Lu and Rothenberg 2018), small interfering RNAs (siRNAs) that are double-stranded RNAs used for gene silencing through RNA interfering pathway (Nikam and Gore 2018), piwi interacting RNAs (piRNAs) representing the largest class of small non-coding RNAs in animal cells to allow the regulation of genetic elements in germ cell lines such as transposons (Hirakata and Siomi 2016), cis-regulatory RNAs (cisRNAs) (Dahariya et al. 2019), centromere repeat-associated short interacting RNAs (crasiRNAs) proposed to be essential for cellular stability and chromosome segregation (Lindsay et al. 2012), and telomere-specific small RNAs (telsRNAs) that are heterochromatin-associated pi-like small RNAs (Cao et al. 2009). Medium non-coding RNAs are 50–200 nucleotides long and include small cytoplasmic RNAs (scRNAs) (Bird and Sells 1986), small nuclear RNAs (snRNAs) (Meier 2017) localized in splicing speckles and Cajal bodies of the cell nucleus in eukaryotic cells and are involved in pre mRNA processing, small nucleolar RNAs (snoRNAs) (Xing and Chen 2018) involved in processes of rRNA modifications, transcription initiation RNAs (tiRNAs) that modulate local epigenetic structure to regulate transcription factors localization (Taft et al. 2011), and promoter-upstream transcripts (prompts) transcribed from genomic regions located approximately 0.5 to 2.5 kilobases upstream of active transcription start sites (Preker et al. 2008). Long non-coding RNAs (lncRNAs) are defined as transcripts longer than 200 nucleotides, without protein-coding potential. For a classification of lncRNAs see (Alessio et al. 2020). In the human genome ~144,000 loci for lncRNAs have been identified while in the mouse genome ~126,000 (http://www.noncode.org/). Recently, 28,000 human lncRNA genes with their genuine 5′ transcriptional start site were described (Hon et al. 2017). However, this number can increase using expression studies at the higher definition as lncRNAs are more tissue-, cell-specific, and less expressed than messenger RNA (mRNAs) (Derrien et al. 2012; Xu et al. 2017). Since the mechanisms of action of most annotated lncRNAs are still unknown, understanding their subcellular localization and cell-specific expression will allow a better comprehension of their function. The new frontier of single-cell analysis will allow understanding the RNA cell-specific expression improving our knowledge of lncRNAs. A great advancement in the comprehension of lncRNA functional roles could be provided by associating their expression at a single-cell level with subcellular localization using a fluorescent in situ hybridization technique (FISH).

In this chapter, we will discuss improvements in single-cell analysis focusing on advantages and disadvantages of each approach and discovering their application to find new lncRNAs, to evaluate their function, and to propose new paradigms in the field of cell biology.

2 Methods for Transcriptional Analysis of Single Cells

In the early ‘00s, single-cell gene expression studies were mainly carried out with low-throughput techniques, such as qPCR and FISH (Levsky et al. 2002; Ginsberg et al. 2004, 2006). However, these approaches allow recovering information on specific targets and with the rapid advance of high-throughput platforms, the transcriptome analysis became more popular. The first reports of single-cell RNA sequencing were published in 2009 (Tang et al. 2009; Lao et al. 2009), and since then novel strategies have been developed to increase the sensitivity and accuracy of the methods. Also, the number of single cells investigated in each experimental approach increased exponentially in the last decade (from tens to ten thousand single cells). The reduced reagent volumes required on reactions and lower consumable costs contributed to developing new methods, but there are still challenges to overcome, as will be discussed in the following paragraphs. The standard workflow for the transcriptional analysis of single cells includes: single-cell isolation, sequencing library preparation followed by RNA sequencing, and bioinformatic analyses.

2.1 Single-Cell Isolation

In order to set-up an experiment with single cells, some methods require tissue pre-processing and/or cell labeling. The enzymatic digestion or mechanical disruption of tissues is usually performed to obtain a cell suspension. Alternatively, fixatives can be used to preserve tissue structure when cell dissociation is not required. As a consequence, pre-analytical processing can disturb native gene expression profiles of cells and this might interfere with the interpretation of biological phenomena. Reducing the processing time with minimal manipulation of samples is still a challenge to overcome. Depending on the biological question to be addressed and on the type of starting material, a pool of cells can be enriched based on its particular morphophysiological features. The enrichment techniques are classified according to the properties of the cells in two main groups. The first group is based on physical parameters (e.g., size, shape, density, electrical charge, and deformability) and methods include microfiltration, density gradient centrifugation, and microchip capture platforms. The second group is based on biological features (e.g., protein expression profile) and the techniques that can be used are immunoaffinity, fluorescent-activated cell sorting, and magnetic-activated cell sorting (Valihrach et al. 2018). Several methods have been established to isolate single cells, each with particular pros and cons. The performance of each method will depend on three parameters: throughput (number of cells analyzed), purity (fraction of target cells collected after the enrichment), and recovery (amount of target cells obtained as compared to initially available target cells).

Limiting Dilution Method The limiting dilution is a method to isolate single cells based on the preparation of a highly diluted cell suspension prior to seeding them (Chemello et al. 2019a, b; Alessio et al. 2019; Zhang et al. 2016) (Fig. 1a). This process was extensively used for decades on monoclonal cell culture production and relies on Poisson's distribution statistics. Therefore, to obtain such diluted concentrations, the calculations estimate approximately 0.5–0.9 cells per aliquot, but quite often are different cell numbers seeded on well plates or tubes (0, 1, 2, etc.). After seeding, wells must be checked under the microscope, making a laborious process. Actually, the efficiency of this method to isolate single cells is very low due to its statistical nature and for this reason this method is not very common nowadays.

Laser Capture Microdissection The laser capture microdissection (LCM) principle is based on the visualization of cells under an inverted microscope associated with an infrared or ultraviolet laser system which is able to capture a single cell from a solid tissue (Cheng et al. 2013; Nichterwitz et al. 2016) (Fig. 1b). After the identification of the cell of interest under the microscope, a focused laser pulse melts a thin transparent thermoplastic film on a cap above the cell, which attaches to the film. Then, the targeted cell can be recovered from the film into a microcentrifuge tube with specific buffers (Kummari et al. 2015). The LCM is the only technique available that does not require tissue dissociation. This can be advantageous because it can avoid sample contamination and consequently avoid issues on downstream

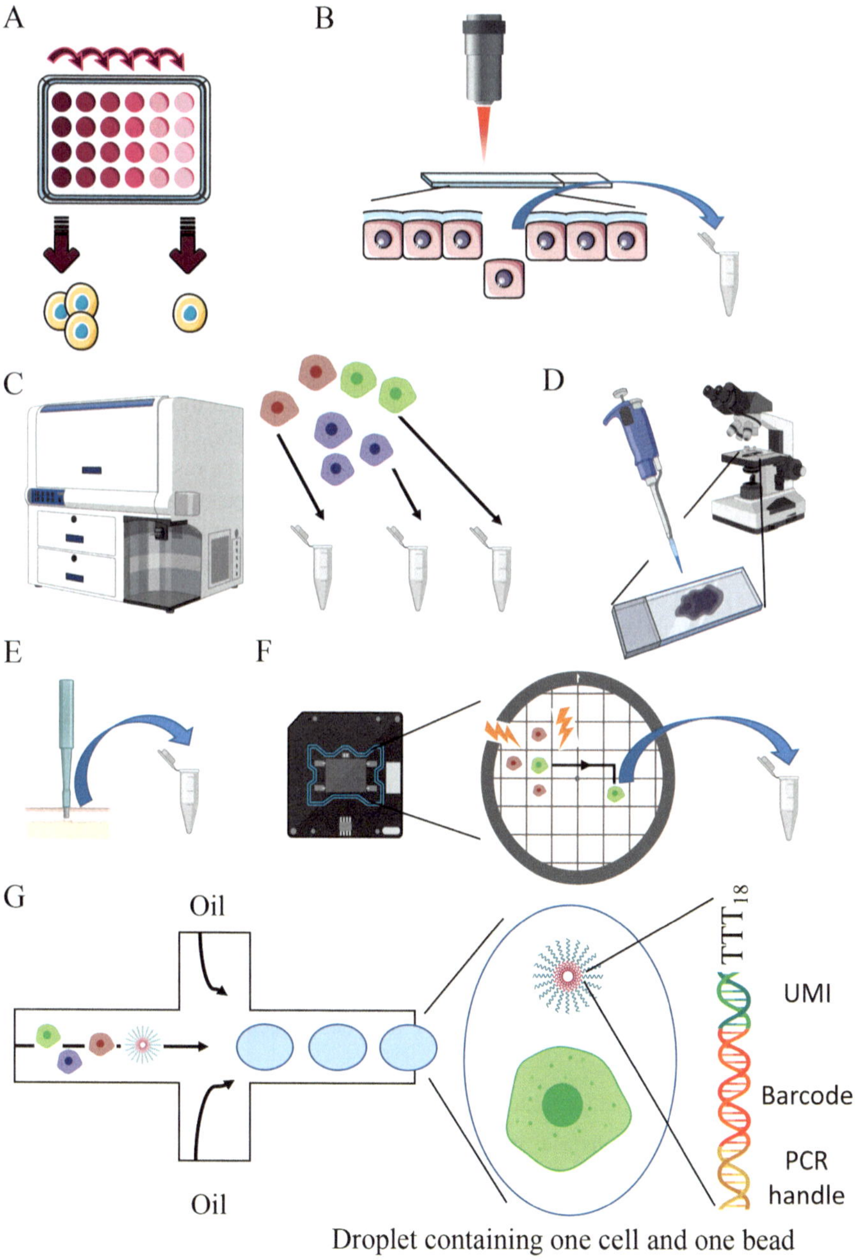

Fig. 1 Methods to isolate single cells. (**a**) Limiting dilution method consists of diluting the sample until obtaining a single cell on each well. (**b**) Laser capture microdissection utilizes the ability of dissecting specific cells from a tissue using a laser to cut the tissue in association with a microscope to inspect the sample. (**c**) Fluorescent-activated cell sorting can separate cells according to specific characteristics. It is fundamental to have tags to differently label different cells. (**d**) Micromanipulation is based on visual inspection of the tissue to manually isolate single cells after tissue disaggregation. (**e**) Punching technologies are based on the principle of capturing cells on

molecular analyses. Other advantages include visual inspection and spatial information of cells on the tissue and the possibility to store the remaining tissue. In this approach, there is the possibility of comparing the gene expression status of adjacent cells and investigate their particularities or how they interact with each other, for instance. However, the visual inspection might require specific training and some preparations need the use of fixatives, which can reduce RNA quality and lead to biased results. Moreover, during LCM, the laser cutting energy may introduce technical artifacts, such as slicing the cells or introducing DNA or RNA damage.

Fluorescent-Activated Cell Sorting The fluorescent-activated cell sorting (FACS) is a method that can be used both during cell population enrichment (prior to another single-cell isolation technique) or to collect single cells, depending on the platform used. The principle of this technique is based on tagging cells with a fluorescent monoclonal antibody specific to surface antigens or intracellular molecules, allowing the instrument to separate positive from negative cells and debris (Gross et al. 2015) (Fig. 1c). The cell suspension flows through the channels on the instrument and each cell is individually exposed to a laser, which allows the detection of cell features (e.g., fluorescence, size, and granularity). Then, the equipment applies a positive or negative charge to the droplet containing the individual cell and the electrostatic deflection system sort them into collection tubes. Also, negative selection can be performed on unmarked cells. Thus, this method allows the analysis of quantitative and qualitative parameters. One example of a platform able to isolate single cells is the BD FACSAria III Cell Sorter that can detect up to 18 markers simultaneously. Even though FACS has higher efficiency than the previous methods, it is still considered a low throughput technique because it requires a large input (more than 10,000 cells) to a reduced output (96 well plates with individual cells). Another drawback of this technique is the necessity of antibodies to target cells, which in some cases are not available.

Micromanipulation Technique The micromanipulation technique consists in aspirating single cells using a motor-controlled micropipette coupled to an inverted microscope (Fig. 1d). Cell suction is performed by an ultrathin glass pipette (tip opening of 2–3 μm) which applies a slight positive pressure into the culture media. This pressure avoids the influx of liquid and consequently capturing unwanted

Fig. 1 (continued) microwells and "punch" them on collection vials. (**f**) The dielectrophoresis system uses the electrokinetic properties of cells to move them on specific positions on a chip from where the cell is collected or physically separated from each other. (**g**) Microfluidic platforms can produce droplets of water in an oil phase each containing a single cell and a barcoded bead. Each bead contains several primers composed by PCR handles, cell barcodes, and unique molecular identifiers (UMIs) sequences. PCR handles are common in all beads and allows PCR amplification after cDNA synthesis. Cell barcode sequences, identical across all the primers of the same microparticle but different from those on different beads, allow recovery of the cells' origin. UMIs, different on each primer, allow mRNA transcripts to be digitally counted and to identify PCR duplicates. An oligod(T) sequence is present at the end of all primers for capturing polyadenylated RNAs and priming reverse transcription

biomaterial. After detecting the target cell by visual inspection, the tip touches the cell and the pressure is released, so the cell attaches to the tip. Then, the collected single cell can be dispensed in microtubes for further analysis. This method played an important role in assisted reproductive medicine and developmental biology research (Malter 2016). However, it requires well-trained operators because this process is manual and requires fine motor skills. Other issues can occur depending on the operator, for example, sample contamination and extended timing of the experiment. These problems can be overcome by using an automated capillary-based platform to collect single cells. For instance, the AVISO CellCelector (Baron et al. 2018) (http://www.als-jena.com/zellselektion.html?&L=1) is a technology that combines the precision of automated robotics with intelligent image processing software, making the single-cell collection easier and faster than traditional micro-manipulation (Szczerba et al. 2019). The system works with bright field, phase contrast, and fluorescence and also allows documentation before and after single-cell isolation to monitor efficiency. The major downside is that this platform is not suited to work with cells that have strong adhesive properties.

Punching Technology The punching technology is used to isolate single cells based on loading a cell suspension into a chip to capture individual cells into microwells, identify the interesting ones by visual examination and then "punch" them into vials for downstream analysis (Fig. 1e). The Puncher Platform (Vycap) fills the cell suspension in the microchip by negative pressure. Once a cell has landed on a well, the pore is blocked, and no other cell can be pulled inside that well. Then, a six-channel fluorescent microscope allows automatic detection and pre-selection of target cells. After the identification and image acquisition, a needle punches the bottom of the well containing the cell into a microtube. This device is especially suited to perform studies with rare cells (1 per 10,000) in large volumes (up to 50 mL) such as circulating tumor cells (CTC) and fetal trophoblasts (Stevens et al. 2018). The CellRaft AIR System (CellMicrosystems) is another punching platform, but the strategy to isolate single cells is different. First cells are seeded in the CytoSort Array which is available in two formats: 10,000 wells (200 × 200 μm) or 40,000 microwells (100 × 100 μm). After cells attach to the bottom of magnetic wells, they can be identified by brightfield, fluorescence, or confocal microscopy. The system has a motorized needle that punches the well in the culture media and a magnetic wand that collect the microwell and the single-cell attached to it. The CytoSort Array is compatible with coatings such as Matrigel or extracellular matrix components allowing broader applications and the capture process has minimal disturbance on cells (Wang et al. 2010).

Dielectrophoresis Technique The dielectrophoresis (DEP) is carried out by the DEPArray NxT (Menarini Silicon Biosystems) instrument. This technology explores the electrokinetic properties of cells to trap them in non-uniform electric fields, move them into electrode grids on a chip, and collect single cells in standard vials. Microelectrodes are ordered in a pattern that allows the application of electric fields in multiple directions. As the culture media has conductive properties, cells can be manipulated by negative dielectrophoresis moving away from the higher electric field. The system allows the visualization of cells during the capture process and the low voltage applied does not interfere with cell viability (Di Trapani et al. 2018) (Fig. 1f).

Droplet-Based Microfluidics The droplet-based microfluidics employs the encapsulation of single cells into aqueous droplets embedded in an immiscible oil phase (Utada et al. 2005). These devices consider several parameters (viscosity and interfacial tension of fluids, channel geometry, etc.) and control precisely the speed injection of both fluids to generate the monodisperse droplets. The encapsulation efficiency also relies on Poisson statistics, and due to the nature of the method is expected to find some small percentage of droplets without cells or with more than one cell, for example (Konry et al. 2016). In fact, the protocols have been optimized during the last years to increase capturing efficiency and enable massively parallelized analyses. However, due to its randomness, this process requires a large number of cells. Barcoded beads can be added on the droplets to identify the correspondent cell of origin. Each bead is covered by thousands of DNA molecules containing a poly(dT) expansion to capture poly-A RNAs, a short DNA sequence specific to each bead (barcode) and another sequence specific to each molecule, called unique molecular identifiers (UMI). The UMIs are used to identify duplications of reads due to different efficiency of cDNA synthesis and remove such errors (Macosko et al. 2015) (Fig. 1g). Details about reverse transcription particularities are provided in the next section. There are two types of droplet-based platforms currently available. The ones that contain reagents to perform reverse transcription and encapsulate them with the cells (e.g., Chromium System – 10× Genomics; InDrop System – 1CellBio; C1 System – Fluidigm; ddSEQ Single-Cell Isolator – Bio-Rad-Illumina) and those that require an additional step to break off the droplets and prepare the cDNA library (e.g., Nadia – Dolomite Bio). Details about reverse transcription particularities are provided in the next section. These instruments are designed to analyze up to 10,000 single cells (except for the C1 System that cannot process more than 800 cells) and generate high throughput data in a parallel manner, which avoids cross-contamination and increases reproducibility (Valihrach et al. 2018). However, the RNA capture efficiency of current protocols is not optimal (around 65% of maximum efficiency) (Valihrach et al. 2018) and the platforms require clean samples, without cell clumps or debris, in order to avoid microfluidic channel clogging. The current trend is to use instruments that provide end-to-end solutions, such as the C1 System and the Chromium System, that process from the single-cell isolation to data analysis in a single platform. To compare cited methods, see Table 1.

The methods to isolate single cells have become easier over the years. Nevertheless, some tissue samples are not easily dissociated into a suspension of single cells, especially the frozen or fixed samples. In those cases, an alternative is to use specific protocols to isolate single nuclei and perform the nuclei scRNA-seq. In fact, there is a good correlation of gene expression in the nucleus compared to gene expression in the cell (Bakken et al. 2018; Grindberg et al. 2013). Another limitation of the high-throughput techniques is the lack of information about the spatial context of the cells.

Table 1 Comparison of techniques used to isolate single cells

Techniques	Limiting dilution	LCM	FACS	Micro-manipulation	Punching technology	DEP	Capillary-based	High-throughput droplet-based	Low-throughput droplet-based
Main platforms	N/A	Several platforms	Several platforms	N/A	CellRaft AIR Puncher platform	DEPArray NxT	AVISO CellCelector	Chromium system Nadia InDrop system ddSEQ Single-cell isolator	C1 system
Throughput	Low	Low	High	Low	Low (<100 cells)	Low (<100 cells)	Low (<100 cells)	High (between 6k and 10k cells)	Low (<800 cells)
Visual control	Possible	Yes	No	Yes	Yes	Yes	Yes	No	Yes
Cell selection	Possible (morphological analysis)	Yes	Yes	Yes (morphological analysis)	Yes	Yes	Yes	No	Yes
Input of cells	Low	Low	High	Low	Medium	Medium	Medium	High	Medium
Advantages	Low cost	Spatial information, storage of tissue	Capture rare cells, fast analysis	Low cost	Active cell selection, high transfer efficiency	Active cell selection, cell-cell interaction analysis	Low price of consumables	Suitable for processing a high number of cells	Suitable for RNA-Seq, DNA-Seq, miRNA-Seq, epigenomics, qRT-PCR analysis
Disadvantages	Time-consuming, low efficiency	Fixatives can damage RNA and introduce bias	Require antibodies/molecular markers	Laborious, low efficiency	Bioinformatic analysis not provided	High price of consumables	Require skills to operate, bioinformatics not provided	High cost, profiles of only polyadenylated RNAs (need specific protocols for non-polyadenylated RNAs)	Size-based cell selection

Some methods based on single-molecule fluorescent in situ hybridization have been developed to map gene expression on subcellular compartments by using up to 1000 targets (Moffitt et al. 2016a, b; Shah et al. 2016). Scaling up these techniques will allow a better understanding of molecular events that drive cellular heterogeneity.

2.2 Sequencing Library Preparation

After successful single-cell isolation, all protocols share an initial step to proceed with the scRNA-seq. First, cells must be lysed and RNAs must undergo reverse transcription and amplification. This is required because the RNA mass present in each cell is around 10–50 pg, depending on the cell type, and the input for RNA-seq protocols can reach 1 μg of total RNA (Boon et al. 2011). The cell lysis is a critical step to obtain high-quality results because it is estimated that only 10–20% of transcripts are converted into cDNA (Islam et al. 2014). This is a consequence of a limitation of the current protocols that are optimized to amplify polyadenylated RNAs (such as mRNAs and some lncRNAs) by using oligo-dT primers to initiate the cDNA reaction. Capturing other classes of RNAs without specification of target sequences are a current challenge, but some studies provided protocols to the enrichment of other classes of non-coding RNAs (e.g., for circular RNAs (Fan et al. 2015), for total RNA (Sheng et al. 2017; Hayashi et al. 2018), or for miRNAs (Faridani et al. 2016; Chemello et al. 2019a, b)). The cDNA amplification can be performed by the template switching or the linear amplification principles (Grun and van Oudenaarden 2015). The template switching consists of exploring the ability of Moloney murine leukemia virus (MMLV) reverse transcriptase that synthesize a short sequence (mostly cytosines) at the 3′ end of the first-strand cDNA. Then, a poly-G primer pair with the C-tail on the mRNA and a second adaptor are used to extend the cDNA molecule. There are two approaches to carry out library preparation, both using tagmentation with Tn5 transposase: the STRT-seq or the SMART-seq. In the STRT-seq only the 5′ end fragment is captured and sequenced, but this method allows the integration of DNA barcodes and UMIs. In the SMART-seq approach, the full-length cDNAs are sequenced, but it does not allow integration of UMIs (Wellenreuther et al. 2004; Biscontin et al. 2010; Grun and van Oudenaarden 2015). Alternatively, the CEL-seq method is based on the linear amplification principle, which uses the T7 RNA polymerase to synthesize multiple copies of RNA, a process called in vitro transcription. Once RNAs were amplified, another round of reverse transcription is performed and the reads mostly coverage the 3′ ends of transcripts (Hashimshony et al. 2012). Each method allows for different types of studies. For example, to investigate gene expression, heterogeneity is an ideal approach that allows the sequencing of the 3′ untranslated region (UTR) of the polyadenylated RNAs (based on the integration of UMIs), while to study splicing patterns, a technology, such as SMART-seq, that provides a full transcript coverage should be considered. Moreover, all amplification steps were largely discussed because of their ability to maintain a linear amplification range to preserve gene expression proportions (Wang 2005; Day et al. 2007).

2.3 RNA Sequencing and Bioinformatic Analysis

Following library preparation, samples are sequenced and analyzed with bioinformatic tools. To reach statistically meaningful results and detect rare cells, high-throughput platforms are recommended because they can process a high number of cells simultaneously. Single-cell RNA sequencing (scRNA) is the gold standard approach since the other methods to measure gene expression have huge limitations. For example, with qRT-PCR it would be extremely laborious and inefficient to detect numerous targets in hundreds of cells, while microarrays require pre-established transcriptomic probes and can introduce some level of background noise which interferes on low abundant transcripts (Zhao et al. 2014; White et al. 2011). Raw data need to be processed and filtered to reduce technical noise. Thus, the first step is identifying the quality of reads and performing the trimming of low-quality bases, barcodes, and adaptors before to proceed with the sequence alignment to a reference database. However, a minimum of 35 base pairs (bp) read length is required to avoid false-positive hits (Grun and van Oudenaarden 2015). If UMIs were added during the library construction, a statistical correction needs to be applied to convert multiple identical reads into a transcript count estimate (Grun et al. 2014). Spike-in controls can be used to identify and filter out samples with low sequencing efficiency, for example, RNAs from apoptotic cells or incompletely lysed cells. Besides, spike-in controls can be used during normalization, since the number of spike-in RNAs should be equal for all samples and to estimate absolute levels of gene expression if the exact number of cells is known (Hebenstreit 2012). For experiments without these internal controls, alternative protocols were developed (Lun et al. 2016). Normalization is necessary especially to solve issues derived from zero-inflated counts in the data expression matrix, avoiding cell-specific bias. The power of single-cell sequencing will depend on the number of cells analyzed and the library complexity. Since the efficiency of scRNA-seq can vary as a consequence of technical factors, it is important to sequence every single cell with sufficient sequencing depth. This ensures that even low expressed genes can be detected and quantified properly. Even experiments with low input samples can cost-effectively provide good results when the library is constructed using protocols that enrich the 5′ or 3′ portion of RNAs, which requires less sequencing depth than standard protocols (Hwang et al. 2018). For instance, full-length transcript libraries are sequenced at 106 reads per cell or 15–25 × 106 reads per cell in studies that aim to investigate alternative splicing. On the other hand, by sequencing 3′ end libraries researchers only need 104–105 reads per cell (Haque et al. 2017). Due to its particular features, several tools have been developed to analyze scRNA-seq data. There are almost 500 tools available on the web (Ballarino et al. 2016), covering over 30 categories including quality control, read alignment, normalization, and gene networks analysis. For more details about scRNA-seq data processing see (Luecken and Theis 2019). One of the most interesting applications of scRNA-seq is the identification of specific cell types in a heterogeneous pool of cells. In this line of work, new molecular markers can be associated with already known cells, but also novel cell types can be discovered by this technology. Hierarchical clustering methods in combination with PCA analysis can be used to explore this heterogeneity. LncRNAs are promising molecules to be investigated since they are expressed in a cell-specific manner and can be useful molecular tools for future studies and applications.

3 Single-Cell Analysis of LncRNAs

The different roles of lncRNAs have been studied using different approaches, focusing on their localization, tissue, and cell specificity and timing of their expression. Many studies have described lncRNAs as regulatory elements, especially for the development and maintenance of tissue-specific characteristics in different organisms (e.g., human, mouse, fruit fly, etc.) (Kern et al. 2018; Tsoi et al. 2015; Schor et al. 2018). The study of lncRNAs at the single-cell level is a much more recent development that became more popular with the improvement of single-cell genome-wide techniques like single-cell RNA sequencing (scRNA seq) or fluorescent in situ hybridization (FISH).

There are many advantages in studying lncRNAs at the single-cell level due to the more cell type-specific expression of lncRNAs than coding genes and the function of lncRNA dependent on their subcellular localization. Studying their localization using FISH can allow uncovering sub-cytoplasmic and sub-nuclear patterns that would be lost when using differential extraction of RNA from the nucleus and cytoplasm. These techniques also allow measuring the differences in lncRNAs expression in different cells of the same tissue. Interestingly, some studies have been able to measure significant differences in lncRNAs expression between cells of early stages embryos, suggesting that these cells start to differentiate from each other earlier than previously thought.

3.1 LncRNA Expression in Embryonic Cells

One of the first works that measured the expression of a specific lncRNA at the single-cell level was published in 2003 by Hartshorn and colleagues (Hartshorn et al. 2003). They used laser zona drilling to separate the blastomeres from eight-cell stage embryos and then measured the expression of the lncRNA Xist (X-inactive–specific transcript) from each cell separately. They used a multiplex Xist/Sry real-time RT-PCR assay with a single-copy sensitivity to identify the sex of the samples (Sry gene is localized in the Y chromosome) and at the same time quantify the number of Xist RNA molecules present in individual cells (Xist gene is localized in the X chromosome and allows X chromosome inactivation to balance X chromosome number in human females). This analysis uncovered the differences in the accumulation of the Xist RNA between individual blastomeres of the same eight-cell embryo. In particular, they showed how most of the Xist molecules of an embryo are contained in two of the eight blastomeres supporting the conclusion that early mammalian embryo cells are not functionally equivalent to each other. They also proposed that the Xist expression differences could be the consequence of differences in DNA methylation or that it could be dependent on the order in which cells divide.

In 2005, the same group introduced a new method (Hartshorn et al. 2005) to analyze RNA and/or DNA in small samples. The procedure they developed allows avoiding chemical extraction of nucleic acids or their binding to a matrix by carrying out all steps by sequentially diluting the sample in a single tube. They used this method to quantify the genomic regions and RNA expression in single mouse embryos and isolated blastomeres of three protein-coding genes (Sry, hsp70.1, hsp70.3) and one lncRNA (Xist). They showed how this method allows eliminating any RNA loss and can be accurate enough to detect single molecules.

Later on, the same group published a manuscript describing the first example of multiplex RT LATE PCR (Hartshorn et al. 2007). This novel assay was constructed to obtain a real-time measurement of Oct4 and Xist templates; they also used already existing methods to evaluate the specificity and quantitative accuracy of RT LATE PCR. They developed a new method that allows the accurate measurement of the copy number of different RNAs independently from their relative abundance in a cell. They analyzed sets of blastomeres isolated from eight-cell stage embryos, all of which are considered pluripotent and morphologically equivalent. However, the results obtained from this novel method show that the RNA levels of both Oct4 and Xist vary significantly between the cells obtained from the same embryo. By analyzing multiple embryos, they were able to show that the expression levels of Xist and Oct4 do not correlate even though both genes are upregulated at this stage of development. By comparing RNA extracted from male and female embryos they also confirmed that the assay they developed is not affected by gene expression differences related to the sex of the embryo (Hartshorn et al. 2007).

Also they laid the foundation for better study differences in cells in early embryonic stages when cells are retained all similar. They showed how coding and non-coding transcripts are expressed at different levels in the different cells of an eight cell embryo and highlighted the importance of getting the most detailed picture possible regarding the expression of RNAs within each cell to better understand the various roles of each cell in a tissue (Hartshorn et al. 2007).

Xist RNA was one of the first lncRNAs to be discovered in the early 1990s when it was evidenced its involvement in the X chromosome inactivation for a dose compensation (Brown 1991; Brown et al. 1992; Borsani et al. 1991; Brockdorff et al. 1991, 1992). Brigs and colleagues set out to measure at the single cell level the expression of the lncRNA Xist starting from single cells of human embryos during the oocyte to embryo transition (Briggs et al. 2015). They compared the expression of Xist in embryos cells with the expression in induced pluripotent stem cells (iPSCs) that had been newly reprogrammed using mRNAs. They were able to pinpoint the start of Xist expression at the four-cells embryonal stage at concurrently to the activation of the embryonic genome. By using a combination of FISH, qRT-PCR, and FACS, Brigs and colleagues uncovered that the iPSCs do initially express Xist but the expression of this lncRNA quickly diminishes within a few passages of in-vitro culture. Another observation was that the expression patterns of genes from the X chromosomes change in cells at later passages as a result of the localization at the inactive X chromosome of both lncRNA Xist and the protein H3K27me3.

Using techniques like PCR, FISH, and FACS requires targeting specific transcripts. This is a limitation whenever there is no information about which gene might be involved in the processes in analysis. This means that the development of genome-wide (or transcriptome-wide) single-cell techniques was a huge step forward to help boost the relevance and range of single-cell studies. One of these techniques is single-cell RNA sequencing (scRNA-seq) which was first applied to just coding but then quickly expanded to the study of non-coding RNAs as well. Thanks to the technical advancements in this field many studies were made possible, one of the interesting results from these studies is the discovery that several genes, including lncRNAs, are expressed regardless of the X chromosome inactivation.

scRNA-seq has been used to analyze 124 individual cells from human embryonic stem cells (hESCs) and human preimplantation embryos (Yan et al. 2013). Thanks to the use of an Illumina HiSeq2000 sequencing platform, they generated a dataset of 22,687 maternally expressed genes that includes 8701 lncRNAs. This is a considerable step forward considering that the previous transcriptome-wide techniques based on cDNA microarrays could detect at most 9735 maternally expressed genes (Wells and Patrizio 2008). Of the 8701 maternally expressed lncRNAs found with scRNA-seq, 2733 were previously unknown, many of these genes also resulted specifically expressed during certain developmental stages. It has been revealed that lncRNAs are differentially expressed during the seven embryonic stages considered supporting their function during development. They also identified as functionally relevant lncRNAs that maintain expression in all the embryos analyzed.

Next, Petropoulos and collaborators sequenced the transcriptomes of 1529 individual cells obtained from 88 human preimplantation embryos (Petropoulos et al. 2016). This work improved the number of analyzed embryos to reduce at a minimum the effects that are related to the individual embryo-specific differences. With this analysis, they uncovered that the establishment of the trophectoderm, epiblast, and primitive endoderm lineages is preceded by an intermediate state during which lineage-specific genes are co-expressed in the individual cells. They also showed that, before the embryo implantation, all cells belonging to the three lineages will have achieved dosage compensation of genes transcribed from the X chromosome. They also compared the transcription patterns in human embryos to the murine ones and found that in human embryos, the lncRNA XIST is transcribed from both the maternal and the paternal alleles allowing both the X chromosome dosage compensation in females and the transcription of coding genes. This would suggest that, in humans, the X chromosome inactivation could be compatible with an incomplete but reciprocal inactivation of both X chromosomes in females.

As previously discussed, there are two options to analyze RNA expression from single cells using the RNA sequencing: a) to evaluate the RNA full length or b) to analyze a tag for each RNA (5′ or 3′ end). All these methods are based on the capture of polyadenylated RNAs using an oligoc(T) probe to retrotranscribe the RNA but losing information on non-poly-adenylated RNAs. In the field of ESCs, a new method was developed to sequence full-length RNAs from total RNA extracted from single cells (Hayashi et al. 2018). This novel technique, that they called "random displacement amplification sequencing" (RamDA seq), was developed to provide high sensitivity to non-polyadenylated RNA and achieve near-complete coverage of transcripts in their full length. To study the variations in the expression of non-polyadenylated RNAs, Hayashi and colleagues used their newly designed technique to analyze mouse embryonic stem cells at different stages of development. The resulting time-course analysis revealed hundreds of dynamically regulated transcripts that include lncRNAs, genes coding for histone proteins, and enhancer RNAs. Using this technique on cells at different stages of development also allowed them to unveil for the first time the cell type-specific expression at the single-cell level of 7580 transcripts 458 of which were not polyadenylated.

3.2 LncRNAs in Stem Cells: A Single-Cell Approach

A deep-sequencing approach was used for the development of a digital transcriptome analysis protocol that focuses on isolated mouse blastomeres and oocytes (Tang et al. 2010, 2011). This protocol includes the isolation and lysis of single cells (oocytes and blastomeres) followed by the reverse transcription of the whole cell lysate. Single-cell cDNA is then amplified for 20/29 cycles of PCR and sequenced using the SOLiD platform. They were able to obtain the expression of at least 75% more genes compared to previous microarray protocols and reveal significant differences in the transcriptomes at the single-cell level when comparing blastomeres from different embryos. By analyzing the expression of the different alleles of the same gene, Tang and colleagues demonstrated that the allele-specific gene expression (ASE) is similar when comparing blastomeres from the same embryo while only 6% of cells coming from the same embryo have a more stochastic expression.

With the increasing number of papers published on single-cell transcriptomics, the amount of available data has been growing and growing. This huge set of data is the perfect starting point for in silico studies that aim to develop new approaches to the analysis of RNA sequencing information.

The data published by Tang and colleagues were subsequently analyzed to provide a catalog of lncRNAs expressed in mouse embryos (Zhang et al. 2014). With this new analysis, they were able to identify 5563 novel lncRNAs expressed from 3492 loci during the mouse cleavage embryonic stage. The lncRNAs identified resulted having a specific expression dependent on the stage of development (Qiu et al. 2016), this is probably because these transcripts are involved in the regulation of specific mechanisms that are required during the stage of development during which they are overexpressed.

A new approach for the study of single-cell sequencing data was developed allowing the development of a novel algorithm authors called "single-cell topological data analysis" (scTDA) (Rizvi et al. 2017). This algorithm aims to study the temporal, unbiased transcriptional regulation using topology-based computational analyses. Compared to other methods previously available, scTDA is described as a nonlinear statistical framework that is model-independent and unsupervised that can be used to characterize transient states of cells. They used their newly developed algorithm to analyze the in vitro differentiation of mouse embryonic stem cells (mESCs) after induction of differentiation into motor neurons. They were able to identify four distinct states the cells go through during the differentiation process: pluripotent cells, precursor cells, progenitor cells, and fully differentiated cells. This distinction was possible thanks to the use of scTDA to identify stage-dependent changes in transcription factors, RNA binding proteins, and lncRNAs supporting the notion of their cell stage-specific expression.

To study the landscape of lncRNA expression in single isolated hematopoietic stem cells (HSCs) and their precursors (pre-HSCs), bioinformatics and functional screenings were combined. Analyzing the levels of H19, they uncovered the importance of this lncRNA for the emergence of HSCs from pre-HSCs in the aorta-gonads-mesonephros region (Zhou et al. 2019). Specifically, an initial deficiency of H19 was shown to stop the cells transitioning from endothelial to hematopoietic that however resulted being independent from the expression of the miR-675 which derives from H19. The lower expression of H19 in pre-HSCs was also linked to an increased methylation of the promoters of Runx1, Spi1, and various other master hematopoietic transcription factors that were, as a result, downregulated. One of the effects that they linked to a lower expression of H19 was the increase in the activity of S-adenosylhomocysteine hydrolase which can explain the changes in methylation state of the promoters since this protein is a regulator of DNA methylation.

The reprogramming of fibroblasts is a technique that allows the production of different cell types from a donor. Reprogramming process is performed using specific transcription factors (c-Myc, Oct3/4, Sox2 and Klf4 (Takahashi and Yamanaka 2006)) and is widely used for therapeutic purposes or drug screening (Elitt et al. 2018; Muller et al. 2018; Castro-Vinuelas et al. 2018). Recently, also microRNAs have been proposed as vehicles to induce cells to pluripotency (Bao et al. 2013) demonstrating that not only coding but also non-coding RNAs are involved in this process.

To allow the query of single-cell transcriptome data in a more intuitive way, self-organizing maps (SOMs) were used (Zhou et al. 2019). They used this structure to organize the single-cell expression patterns of over 16,000 genes, 437 of which were lncRNAs, at different stages of reprogramming toward pluripotency. They also performed loss-of-function studies and highlighted the role of lncRNAs in the repression of lineage-specific genes or in the regulation of metabolic gene expression. These findings proved that the cells undergoing reprogramming will overexpress sets of lncRNAs that are functionally relevant for the reprogramming itself. Pseudogenes were also highlighted as important factors in the process of cellular reprogramming, opening up this field of study to the world of non-coding RNAs which are still an underrepresented class in the study of cellular reprogramming, cellular state plasticity, and cellular transformation.

3.3 LncRNAs in Single Differentiated Cells

Several studies have focused on specialized/differentiated cells rather than on pluripotent/embryonic cells allowing the analysis of functions of ncRNAs in more specialized systems. Alterations in differentiated cells can be informative on how pathologies can affect the normal cell behavior.

A catalog of >8000 human lncRNAs by unifying RNA seq data from more than four billion reads across 24 cell and tissue types was developed. Each entry in the developed database contains the description of more than 30 properties for each feature (e.g., sequence, structural, transcriptional, and orthology descriptions). This work highlights the tissue specificity of lncRNAs, which is much stronger compared to coding RNAs (Cabili et al. 2011). They showed the existence of a correlation between the expression and genes located in the same genomic region, probably influenced by a common transcriptional regulation. Scientists did not limit their analysis to strictly non-coding and coding RNAs but described a third class of RNAs having high evolutionary conservation and including short ORFs. They proposed that these transcripts might serve as both lncRNAs and as mRNAs required for the translation of small peptides (Cabili et al. 2011).

The same group used a single-molecule RNA FISH to detect the hybridization of custom fluorescent-labeled probes that recognize the lncRNA sequence; with this tool they studied the expression and subcellular localization of lncRNAs in three different types of cells: hFFs, hLFs, and HeLa (Cabili et al. 2015). They aimed to provide a stringent validation of the number of molecules and their spatial localization of analyzed lncRNAs as well as to obtain a high-resolution quantification of their expression. They were able to observe several distinct patterns of subcellular and subnuclear localization: some lncRNAs were found in bright sub-nuclear foci while others were spatially restricted to the cytoplasm. Some of the analyzed lncRNAs were known to have an extremely low expression when their expression was analyzed extracting RNA from cell cultures, between these transcripts there were some that at the single level resulted having a very high expression in a small number of "jackpot" cells, suggesting that their expression might be necessary but only on a small number of cells in the population. The lncRNAs that were observed forming bright foci in the nucleus were also shown to lose their localization during mitosis suggesting that their function is probably not that of mitotic bookmarking factors. Finally, they highlight that the correlation in expression between lncRNAs and neighboring genes is not always observed when the lncRNA has a divergent position compared to the coding RNA. A new system was developed to culture cells from human brain tissue removed during neurosurgery based on a papain-aided procedure (Spaethling et al. 2017). This group produced single-cell analyses of the transcriptomes from over 300 cells and used deep sequencing to detect over 12,000 expressed genes. They used expressed genes to classify different cells finding that hundreds of mRNAs, lncRNAs, and pri-miRNAs were enriched in a specific cell type. In particular, 935 lncRNAs were identified and 113 were found to be differentially expressed throughout the various cell types. For example, the lncRNA LINC01314 was found to be specifically expressed in astrocytes while LINC00152 was found to be specifically expressed by endothelial cells, MEG3 in neurons, LINC00520 in microglia, LINC01105 in oligodendrocytes and finally NEAT1 was found to be highly expressed in both astrocytes and endothelial cells while SNHG6 was found to be highly expressed from all analyzed cell types (Spaethling et al. 2017). This work again sustains the importance of the cell-specific expression of lncRNAs not only for the correct development but also to maintain the cell specificity. In this regard, we evaluated the importance of lncRNAs in maintaining different metabolic properties of skeletal muscle myofibers. Myofibers are the smallest complete contractile system of skeletal muscle influencing its contraction velocity and metabolism. Myofibers can be divided into fast and slow according to their contraction velocity and metabolism. Fast myofibers prevalently use glucose as an energy source, while slow myofibers use oxidative respiration. After manually recovering of single muscle fiber (Chemello et al. 2019a, b), we used microarray technology to compare the expression of lncRNAs in single fast and slow contracting myofibers. Moreover, we classified lncRNAs according to their subcellular localization. We revealed that cells with different metabolic properties express different lncRNAs supporting the idea that lncRNAs influence cell metabolism. We revealed that some differentially expressed lncRNAs localize differently in cellular compartments according to cell metabolism and lncRNA isoform (Alessio et al. 2019). This last result suggests that the same lncRNA could have different roles in different cells and that the same might be true for different isoforms of the same lncRNA. We validated our results by using both qRT-PCR and RNA-FISH, on both C2C12 myoblasts and myofibers. We functionally discussed the involvement of the lncR6NA Pvt1 in the modulation of muscle atrophy and mitochondrial network. Our findings suggested that Pvt1 can modulate the expression of c-Myc, which in turn regulates Bcl-2, having an effect on autophagy and apoptosis that are strictly linked to the confirmation of the mitochondrial network (Alessio et al. 2019).

3.4 *Single-Cell Expression of LncRNA in Tumors*

The function of lncRNAs in cervical carcinoma formation by analyzing the single-cell RNA sequencing data from 37 Hela-S3 cells was explored. They found that an average of 511 lncRNAs is expressed in each cell and that these transcripts display a much greater cell specificity compared to protein-coding RNAs. By performing a co-expression network analysis, they were able to identify seven different modules one of which was enriched in pathways of mitotic, packaging of telomere ends, and chromosome maintenance (Wang and Roy 2017).

The increased cell specificity of lncRNAs respect to coding genes was confirmed (Wu et al. 2019). They analyzed the expression of lncRNAs in 979 human bone marrow-derived CD34+ single cells combining scRNA-seq with de novo transcriptome reconstruction. They were able to identify a total of 3173 lncRNAs, 2365 of which were not known at the time of the analysis. From their analysis it was clear how lncRNAs expression had high variability at the single-cell level, an observation that is only possible thanks to the single-cell approach the group adopted. They showed how these transcripts followed a highly dynamic lineage-specific pattern of expression during hematopoiesis. Interestingly, during hematopoiesis, the expression of lncRNAs resulted highly correlated to the expression of neighboring coding genes. They characterized the lncRNAs that resulted preferentially upregulated in hematopoietic stem cells and their differentiated lineages. Wu and colleagues also studied to the role of lncRNAs in single cells obtained from patients suffering from myelodysplastic syndromes with a particular focus on aneuploidy cells. They described a highly ordered network of lncRNAs that participate in the regulation of the early hematopoiesis and coordinate the aberrant mRNAs and lncRNAs transcribed in dysplastic hematopoiesis.

In glioblastoma (GBM), different subpopulations of cells have specific functional status revealing the existence of highly invasive cells within the heterogeneous GBM tumors (Pang et al. 2019). They used a scRNA-seq technique based on SMART-seq to reconstruct the pseudotemporal branched trajectory that single tumor cells take during their proliferation. The initial stages showed features and phenotypes similar to those of GSC, however with proliferation the invasive activity has been observed to increase to high levels. This allowed Pang and colleagues to determine the path along which GSCs gradually transform to acquire their final highly invasive phenotype, they called this the "stem-to-invasion-path." Cells following this path are characterized by an incremental expression of GBM invasion-associated genes while at the same time a downregulation of GSCs-related genes.

4 Databases of Gene Expression Data of Single Cells

The development of different techniques to physically separate single cells allowed to accumulate genomic and transcriptomic data on single cells. Such as sequencing from bulk tissues, also single-cell DNA or RNA sequencing experiments are deposited, usually as raw data, in the Sequence Read Archive (SRA) (https://www.ncbi.nlm.nih.gov/sra) or in the Gene Expression Omnibus (GEO) (https://www.ncbi.nlm.nih.gov/geo/) databases (Fortriede et al. 2020). Unfortunately, data on these repositories are not easy to use and need bioinformatics skills to make readable information. For this purpose, several databases were developed with user-friendly interfaces. The European Bioinformatics Institute (EMBL-EBI) developed a single-cell expression atlas (https://www.ebi.ac.uk/gxa/sc/home) containing

Table 2 Databases storing single-cell data

Database name	Organisms	Web link
PanglaoDB	Mouse, human	https://panglaodb.se/
Single-cell database	Mouse	http://betsholtzlab.org/VascularSingleCells/database.html
JingleBells	Different organisms (mouse, human, zebrafish, brown rat)	http://jinglebells.bgu.ac.il/nonimmune/
Brain atlas	Mouse, human	https://portal.brain-map.org/atlases-and-data/rnaseq
Single-cell RNA sequencing	Human	https://bioinfo.uth.edu/scrnaseqdb/
Single-cell data with Nadia (DolomiteBio)		https://www.dolomite-bio.com/product/scrna-seq-data-sets/?utm_medium=affiliate&utm_source=biostar&utm_campaign=data&utm_content=dolomitebio
Sanger institute experiments	Mouse, human	https://hemberg-lab.github.io/scRNA.seq.datasets/
BioTuring		https://bioturing.com/bbrowser
Cancer single-cell atlas	Human	http://biocc.hrbmu.edu.cn/CancerSEA/

52 experiments for cells from H. sapiens, 51 for M. musculus, 7 for D. rerio, 1 for R. norvegicus, 1 for S. mansoni, 1 for A. gambiae, 1 for D. melanogaster, and 1 for C. elegans. This database does not only contain data from animals but also from plants (5 for A. thaliana), fungi (1 for S. cerevisiae), and protists (2 for P. berghei). The Institute of Physical and Chemical Research in Japan (RIKEN) also developed a single-cell centric database (http://single-cell.clst.riken.jp/). The RIKEN database contains experiments performed on H. sapiens and M. musculus single cells. Broad institute in the United States developed its database for single-cell datasets (https://singlecell.broadinstitute.org/single_cell). It contains 117 studies and data for about 3,780,000 cells. Several other databases were developed and published. For a comprehensive list of databases presenting single-cell experiment results not described in the text see Table 2. All databases give the possibility to view cell clusters and to search for information starting from a specific experiment. Differently not all databases give the possibility to recruit information querying a specific gene. This last restriction may be a strong limitation for the database usage. However, studying a gene or a group of genes in specific tissues trying to understand which cells in that tissue contribute to the activity of studied genes without performing single-cell experiments and therefore exploiting the information already available may be important to formulate new experimental hypothesis.

5 Discussion and Perspectives

LncRNAs are involved in several aspects of cell biology: the regulation of cell cycle (Kitagawa et al. 2013), tumorigenesis (Martini et al. 2017), development (Brazao et al. 2016; Delas and Hannon 2017), muscle differentiation and disease (Neguembor et al. 2014; Ballarino et al. 2016), and metabolism (Sun and Wong 2016). The first non-coding transcripts studied were rRNA and tRNA because of their involvement in the RNA transduction process. Nowadays, it would be interesting to evaluate the involvement of lncRNAs in reprogramming cells. It was already demonstrated that miRNAs can be used in cell reprogramming (Martinez and Gregory 2010) but knowing that lncRNAs can modulate chromatin status and that also lncRNAs are differentially expressed during this cell reprogramming can open the possibility of using them to reprogram cells probably in a more efficient manner and avoiding using oncogenes such as c-Myc.

A deeper mechanistic exploration of lncRNA expression, subcellular localization, and interacting partners will allow a better interpretation of their function in cell reprogramming and several other biological contexts. We and others demonstrated the importance of using a single-cell approach to evidence the function of specific ncRNAs. This result is correlated to the cell-specific expression of ncRNAs not detectable analyzing RNA expression using bulk samples because each cell contributes differently to the expression of each gene. For these reasons, recently developed techniques for the analysis of single cells (the first single-cell RNA-seq study was published in 2009 (Tang et al. 2009) and in 2011 the first single-cell DNA-seq (Navin et al. 2011)) may improve the knowledge of pathways that allow complex interaction among same and different cell types within tissues. Moreover, these techniques may be useful for understanding altered pathways that trigger the development of different pathologies, such as tumors.

Single-cell RNA-seq techniques evidenced the existence of many more cell types than previously estimated (Chiu et al. 2014), but a functional explanation of this phenomenon has not been provided yet. The two alternative hypotheses proposed were: a) the "functionally neutrality" of alterations, and b) an "adaptive"-based mechanism. In the first case, it is hypothesized the existence of multiple equi-phenotypic molecular states (different combinations of gene expression may lead to the same phenotype) (Kim and Eberwine 2010). The second hypothesis develops an evolutionary concept having cell-to-cell variability a better response to environmental changes (Kim et al. 2015).

Unfortunately, using singe-cell sequencing it is not possible to detect ten thousand genes such as when bulk tissues are used. In fact, from a single-cell analysis, only about 3000–4000 transcripts are detected. The analysis of proteins in a single-cell approach is more and more complicated due to the difficulty of protein amplification and results in about 2000 analyzed proteins (Marx 2019). It would be interesting to understand if this discrepancy between results obtained with single cells and bulk samples is due to a detection limit or if effectively single cells express such a limited number of genes and the union of the molecules expressed from different cells composing the tissue recognizes the 10,000 activated genes that are found using bulk samples.

Another limit of most single-cell expression protocols is that RNA retrotranscription (RT) is performed using oligod(T) as primers for the RT enzyme. This means that only RNAs having poly(A) have been retrotranscribed losing information on RNAs without poly(A), such as, for example, mature miRNAs. Adaptor ligations to both 5′ and 3′ ends may be an alternative method to evaluate the expression of non-polyadenylated RNAs in single cells. This method was used by Faridani and colleagues (Faridani et al. 2016) and allowed them to identify about 3800 miRNAs from HEK293, naïve embryonic stem cells, and primed embryonic stem cells. Also in our laboratory, it was developed a SMART-seq-based method to identify mature miRNAs expressed in skeletal muscle single fibers evidencing 137 miRNAs. We showed an average correlation of almost 0.8 among single fibers SMART-seq and qRT-PCR methods for 55 tested miRNAs (Alessio et al. 2019). As we profiled miRNAs and mRNAs from the same myofibers (Chemello et al. 2019a, b), also others proposed the Holo-seq method to profile mRNAs and miRNAs from the same sample obtaining the expression of about 250 miRNAs from single embryonic stem cells (Xiao et al. 2018), a comparable number to that we obtained from single muscle fibers.

In conclusion, single-cell analysis is based on the dissociation of each cell, process that leads to losing the spatial information. Spatial information is particularly important for cell function because the ambient that cells experiment influences their function (Dupont 2016; Walker et al. 2018). Recently, to overcome this problem it was developed several approaches that can give information on gene expression without losing spatial map (Thrane et al. 2018; Cottrell et al. 2018; Eng et al. 2019). Probably, the integration of single-cell and spatial transcriptomic analyses will allow a better comprehension of tissues' functions and cell–cell interactions.

References

Alessio E, Buson L, Chemello F et al (2019) Single cell analysis reveals the involvement of the long non-coding RNA Pvt1 in the modulation of muscle atrophy and mitochondrial network. Nucleic Acids Res 47:1653–1670

Alessio E, Bonadio RS, Buson L et al (2020) A single cell but many different transcripts: a journey into the world of long non-coding RNAs. Int J Mol Sci 21(1):302

Bakken TE, Hodge RD, Miller JA et al (2018) Single-nucleus and single-cell transcriptomes compared in matched cortical cell types. PLoS One 13:e0209648

Ballarino M, Morlando M, Fatica A et al (2016) Non-coding RNAs in muscle differentiation and musculoskeletal disease. J Clin Invest 126:2021–2030

Bao X, Zhu X, Liao B et al (2013) MicroRNAs in somatic cell reprogramming. Curr Opin Cell Biol 25:208–214

Baron CS, Kester L, Klaus A et al (2018) Single-cell transcriptomics reveal the dynamic of haematopoietic stem cell production in the aorta. Nat Commun 9:2517–2532

Bird RC, Sells BH (1986) Cytoskeleton involvement in the distribution of mRNP complexes and small cytoplasmic RNAs. Biochim Biophys Acta 868:215–225

Biscontin A, Casara S, Cagnin S et al (2010) New miRNA labeling method for bead-based quantification. BMC Mol Biol 11:44

Boon WC, Petkovic-Duran K, Zhu Y et al (2011) Increasing cDNA yields from single-cell quantities of mRNA in standard laboratory reverse transcriptase reactions using acoustic microstreaming. J Vis Exp 53:e3144

Borsani G, Tonlorenzi R, Simmler MC et al (1991) Characterization of a murine gene expressed from the inactive X chromosome. Nature 351:325–329

Brazao TF, Johnson JS, Muller J et al (2016) Long noncoding RNAs in B-cell development and activation. Blood 128:e10–e19

Briggs SF, Dominguez AA, Chavez SL et al (2015) Single-cell XIST expression in human preimplantation embryos and newly reprogrammed female induced pluripotent stem cells. Stem Cells 33:1771–1781

Brockdorff N, Ashworth A, Kay GF et al (1991) Conservation of position and exclusive expression of mouse Xist from the inactive X chromosome. Nature 351:329–331

Brockdorff N, Ashworth A, Kay GF et al (1992) The product of the mouse Xist gene is a 15 kb inactive X-specific transcript containing no conserved ORF and located in the nucleus. Cell 71:515–526

Brown SD (1991) XIST and the mapping of the X chromosome inactivation centre. BioEssays 13:607–612

Brown CJ, Hendrich BD, Rupert JL et al (1992) The human XIST gene: analysis of a 17 kb inactive X-specific RNA that contains conserved repeats and is highly localized within the nucleus. Cell 71:527–542

Cabili MN, Trapnell C, Goff L et al (2011) Integrative annotation of human large intergenic noncoding RNAs reveals global properties and specific subclasses. Genes Dev 25:1915–1927

Cabili MN, Dunagin MC, McClanahan PD et al (2015) Localization and abundance analysis of human lncRNAs at single-cell and single-molecule resolution. Genome Biol 16:20

Cao F, Li X, Hiew S et al (2009) Dicer independent small RNAs associate with telomeric heterochromatin. RNA 15:1274–1281

Castro-Vinuelas R, Sanjurjo-Rodriguez C, Pineiro-Ramil M et al (2018) Induced pluripotent stem cells for cartilage repair: current status and future perspectives. Eur Cell Mater 36:96–109

Chemello F, Alessio E, Buson L et al (2019a) Isolation and transcriptomic profiling of single myofibers from mice. Bio-protocol 9:e3378

Chemello F, Grespi F, Zulian A et al (2019b) Transcriptomic analysis of single isolated myofibers identifies miR-27a-3p and miR-142-3p as regulators of metabolism in skeletal muscle. Cell Rep 26:3784–3797 e8

Cheng L, Zhang S, MacLennan GT et al (2013) Laser-assisted microdissection in translational research: theory, technical considerations, and future applications. Appl Immunohistochem Mol Morphol 21:31–47

Chiu IM, Barrett LB, Williams EK et al (2014) Transcriptional profiling at whole population and single cell levels reveals somatosensory neuron molecular diversity. elife 3:e04660

Cottrell TR, Thompson ED, Forde PM et al (2018) Pathologic features of response to neoadjuvant anti-PD-1 in resected non-small-cell lung carcinoma: a proposal for quantitative immune-related pathologic response criteria (irPRC). Ann Oncol 29:1853–1860

Dahariya S, Paddibhatla I, Kumar S et al (2019) Long non-coding RNA: classification, biogenesis and functions in blood cells. Mol Immunol 112:82–92

Day RC, McNoe L, Macknight RC (2007) Evaluation of global RNA amplification and its use for high-throughput transcript analysis of laser-microdissected endosperm. Int J Plant Genomics 2007:61028

Delas MJ, Hannon GJ (2017) lncRNAs in development and disease: from functions to mechanisms. Open Biol 7:170121–170131

Derrien T, Johnson R, Bussotti G et al (2012) The GENCODE v7 catalog of human long noncoding RNAs: analysis of their gene structure, evolution, and expression. Genome Res 22:1775–1789

Preker P, Nielsen J, Kammler S et al (2008) RNA exosome depletion reveals transcription upstream of active human promoters. Science 322:1851–1854
Qiu JJ, Ren ZR, Yan JB (2016) Identification and functional analysis of long non-coding RNAs in human and mouse early embryos based on single-cell transcriptome data. Oncotarget 7:61215–61228
Rizvi AH, Camara PG, Kandror EK et al (2017) Single-cell topological RNA-seq analysis reveals insights into cellular differentiation and development. Nat Biotechnol 35:551–560
Schor IE, Bussotti G, Males M et al (2018) Non-coding RNA expression, function, and variation during Drosophila embryogenesis. Curr Biol 28:3547–3561 e9
Shah S, Lubeck E, Zhou W et al (2016) In situ transcription profiling of single cells reveals spatial organization of cells in the mouse hippocampus. Neuron 92:342–357
Sheng K, Cao W, Niu Y et al (2017) Effective detection of variation in single-cell transcriptomes using MATQ-seq. Nat Methods 14:267–270
Spaethling JM, Na YJ, Lee J et al (2017) Primary cell culture of live neurosurgically resected aged adult human brain cells and single cell transcriptomics. Cell Rep 18:791–803
Stevens M, Oomens L, Broekmaat J et al (2018) VyCAP's puncher technology for single cell identification, isolation, and analysis. Cytometry A 93:1255–1259
Sun X, Wong D (2016) Long non-coding RNA-mediated regulation of glucose homeostasis and diabetes. Am J Cardiovasc Dis 6:17–25
Szczerba BM, Castro-Giner F, Vetter M et al (2019) Neutrophils escort circulating tumour cells to enable cell cycle progression. Nature 566:553–557
Taft RJ, Hawkins PG, Mattick JS et al (2011) The relationship between transcription initiation RNAs and CCCTC-binding factor (CTCF) localization. Epigenetics Chromatin 4:13–26
Takahashi K, Yamanaka S (2006) Induction of pluripotent stem cells from mouse embryonic and adult fibroblast cultures by defined factors. Cell 126:663–676
Tang F, Barbacioru C, Wang Y et al (2009) mRNA-seq whole-transcriptome analysis of a single cell. Nat Methods 6:377–382
Tang F, Barbacioru C, Nordman E et al (2010) RNA-seq analysis to capture the transcriptome landscape of a single cell. Nat Protoc 5:516–535
Tang F, Barbacioru C, Nordman E et al (2011) Deterministic and stochastic allele specific gene expression in single mouse blastomeres. PLoS One 6:e21208
Thrane K, Eriksson H, Maaskola J et al (2018) Spatially resolved transcriptomics enables dissection of genetic heterogeneity in stage III cutaneous malignant melanoma. Cancer Res 78:5970–5979
Tsoi LC, Iyer MK, Stuart PE et al (2015) Analysis of long non-coding RNAs highlights tissue-specific expression patterns and epigenetic profiles in normal and psoriatic skin. Genome Biol 16:24–39
Utada AS, Lorenceau E, Link DR et al (2005) Monodisperse double emulsions generated from a microcapillary device. Science 308:537–541
Valihrach L, Androvic P, Kubista M (2018) Platforms for single-cell collection and analysis. Int J Mol Sci 19:807–827
Walker C, Mojares E, Del Rio Hernandez A (2018) Role of extracellular matrix in development and cancer progression. Int J Mol Sci 19:3028–3059
Wang E (2005) RNA amplification for successful gene profiling analysis. J Transl Med 3:28–39
Wang J, Roy B (2017) Single-cell RNA-seq reveals lincRNA expression differences in Hela-S3 cells. Biotechnol Lett 39:359–366
Wang Y, Phillips C, Xu W et al (2010) Micromolded arrays for separation of adherent cells. Lab Chip 10:2917–2924
Wellenreuther R, Schupp I, Poustka A et al (2004) SMART amplification combined with cDNA size fractionation in order to obtain large full-length clones. BMC Genomics 5:36–44
Wells D, Patrizio P (2008) Gene expression profiling of human oocytes at different maturational stages and after in vitro maturation. Am J Obstet Gynecol 198:455 e1–9; discussion 455 e9–11
White AK, VanInsberghe M, Petriv OI et al (2011) High-throughput microfluidic single-cell RT-qPCR. Proc Natl Acad Sci U S A 108:13999–14004

Wu Z, Gao S, Zhao X et al (2019) Long noncoding RNAs of single hematopoietic stem and progenitor cells in healthy and dysplastic human bone marrow. Haematologica 104:894–906

Xiao Z, Cheng G, Jiao Y et al (2018) Holo-seq: single-cell sequencing of holo-transcriptome. Genome Biol 19:163–182

Xing YH, Chen LL (2018) Processing and roles of snoRNA-ended long noncoding RNAs. Crit Rev Biochem Mol Biol 53:596–606

Xu Q, Song Z, Zhu C et al (2017) Systematic comparison of lncRNAs with protein coding mRNAs in population expression and their response to environmental change. BMC Plant Biol 17:42–57

Yan L, Yang M, Guo H et al (2013) Single-cell RNA-seq profiling of human preimplantation embryos and embryonic stem cells. Nat Struct Mol Biol 20:1131–1139

Zhang K, Huang K, Luo Y et al (2014) Identification and functional analysis of long non-coding RNAs in mouse cleavage stage embryonic development based on single cell transcriptome data. BMC Genomics 15:845–857

Zhang K, Gao M, Chong Z et al (2016) Single-cell isolation by a modular single-cell pipette for RNA-sequencing. Lab Chip 16:4742–4748

Zhao S, Fung-Leung WP, Bittner A et al (2014) Comparison of RNA-seq and microarray in transcriptome profiling of activated T cells. PLoS One 9:e78644

Zhou J, Xu J, Zhang L et al (2019) Combined single-cell profiling of lncRNAs and functional screening reveals that H19 is pivotal for embryonic hematopoietic stem cell development. Cell Stem Cell 24:285–298 e5

Long Noncoding RNAs as Scaffolds for Multiprotein Signaling Complexes

Sonam Dhamija and Manoj B. Menon

Contents

Abstract Long noncoding RNAs (lncRNAs) are gaining prominence as more evidence emerges for crucial roles of these molecules in diverse biological processes. Differential expression analyses have contributed the most in the discovery of functional lncRNAs. Among the generally deciphered functions, the sequence-complementarity-based mechanisms including miRNA sponge activity have predominated. However, some of the functionally characterized lncRNAs perform interesting scaffolding functions by nucleating multiprotein complexes. Multidomain lncRNAs can act as flexible linkers for assembling dynamic signaling complexes, much like protein adaptors. Even though these are among the most difficult lncRNA functions with respect to predictability and characterization, recent

S. Dhamija
Division of Cancer Research, Department of Thoracic Surgery, Medical Center – University of Freiburg, Freiburg, Germany

M. B. Menon (✉)
Kusuma School of Biological Sciences, Indian Institute of Technology Delhi, New Delhi, India
e-mail: menon@bioschool.iitd.ac.in

S. Jurga, J. Barciszewski (eds.), *The Chemical Biology of Long Noncoding RNAs*, RNA Technologies 11, https://doi.org/10.1007/978-3-030-44743-4_5

large-scale interactome studies have identified protein-association or scaffolding as a prominent regulatory role of lncRNAs. This chapter focuses on this category of lncRNAs with a scaffolding function. Here we describe some examples of nuclear and cytoplasmic noncoding RNAs with protein scaffolding functions, discuss their mechanisms of action, and their relevance in the respective cellular context.

Keywords RNA scaffold · lncRNA · Nucleoprotein complexes · RNA structure · RNA binding proteins

1 Introduction

The central dogma of molecular biology established a hierarchy wherein DNA acts as the genetic storage mechanism, RNA as messengers, and proteins as functional/structural components of living systems. There were obvious deviations from the general rule evident by the presence of viruses with RNA genomes and more particularly the discovery of reverse transcribing retroviruses. Ribosomal RNAs (rRNAs) and transfer RNAs (tRNAs), the two classes of most abundant noncoding RNAs (ncRNAs), were discovered as early as the 1950s. However, the messenger function dominated the role of RNAs well into the second half of twentieth century (Jarroux et al. 2017). The functional relevance of noncoding RNAs was revealed when the first class of enzymatic RNA or ribozymes were discovered in *Tetrahymena thermophila* (Cech 1990). Further discovery of RNAs with regulatory roles – eukaryotic microRNAs and bacterial small RNAs, discovered in the pre-genomic era, strengthened the functional importance of RNAs independent of coding potential (Lee et al. 1993). Walter Gilbert first proposed the term "RNA world" in his 1986 commentary which hypothesized a primordial evolutionary stage that relied entirely on self-replicating RNAs with enzymatic and regulatory roles (Gilbert 1986). In early 2000s, sequencing of the human genome provided evidence that only 1.2% of the genome encodes for proteins (Lander et al. 2001). According to recent studies, the number of protein-coding genes in the human genome is estimated to be well below 19,000 (Ezkurdia et al. 2014). Subsequent advent of the high-throughput sequencing established that the genome is pervasively transcribed and this led to the discovery of a large number of noncoding RNAs (ncRNAs) (Kapranov et al. 2007; Lander et al. 2001). Spatial and temporal regulation of the proteome function, mediated by proteins as well as the noncoding transcriptome, could well be the factors which discriminate the simple biology of a worm or fly from the complexity of human anatomy and physiology.

Long noncoding RNAs (LncRNAs) are transcripts of >200 nucleotides (nt) in length that do not encode for proteins. LncRNAs are derived from intergenic, intronic, and/or exonic regions of mRNAs in sense or antisense direction (Diederichs 2014). Majority of the annotated lncRNAs are transcribed by RNA Polymerase

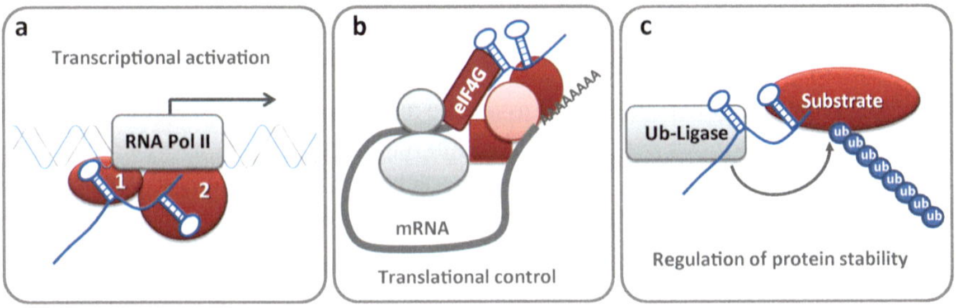

Fig. 1 *LncRNAs modulate cellular processes by forming complexes with proteins*. Three of the major functions known for lncRNAs are transcriptional control, regulation of mRNA translation, and control of protein stability. All three of these usually involve lncRNA–protein interactions. (**a**) Several lncRNAs regulate transcription by binding and scaffolding transcription factor complexes (protein 1 and 2 are indicated as possible transcriptional regulators), enabling RNA polymerase recruitment and transcription at gene promoters. (**b**) LncRNAs participate in mRNA translation control by several mechanisms including binding to eIF4G (translation initiation factor) and other mRNA/3′-UTR binding proteins. (**c**) LncRNAs also regulate protein stability with general mechanisms involving the recruitment of ubiquitin ligases to their cognate substrate proteins, facilitating ubiquitination and proteasomal degradation

II. Most of them are capped, polyadenylated, and spliced similar to mRNAs. Due to the limited homology among different species, functional characterization of lncRNAs is extremely challenging and only sequence information is available for many of them. Competing endogenous RNA (ceRNA) or miRNA-sponge function predominates the reported roles of lncRNAs, largely owing to the predictability of such activities by sequence analysis (Dhamija and Menon 2018). Most studies have relied on differential expression in a particular context as indication of lncRNA function in the process. Determination of cell-type specificity of expression and subcellular distribution is often performed to get hints on their functionality. LncRNAs have been implicated in structural and regulatory roles in a number of cellular processes thus coordinating higher-order physiology. One of their major functions involves bringing different proteins together in space and time by serving as tethering factors or scaffolds. In the perceived theory of RNA world, the system is hypothesized to have evolved into a world of ribonucleoproteins ("RNP world"). Several proteins bind RNA, form ribonucleoproteins, and regulate RNA metabolism. mRNA 3′UTR binding factors, which regulate RNA stability and translation, are examples of such complexes. In these cases proteins are still the functional determinants and RNA is usually the target of protein activity. However, the RNA components in the RNA-dominant ribonucleoprotein complexes are expected to be much more than targets or bystanders. In line with this, several lncRNAs have already been shown to function by forming protein–RNA and protein–RNA–DNA complexes (Fig. 1).

2 Scaffolding in Signal Transduction

Scaffolding proteins are proteins that assemble other molecules together in space and time to coordinate processes like localization, complex formation, signaling, cell–cell communication, etc. Scaffolding proteins achieve this by simply tethering to the proteins involved, and in some cases also cause allosteric regulation of the bound proteins, enhancing the signal efficiency. One of the classical examples in signaling is the scaffold Ste5 which simultaneously binds to Ste11, Ste7 and Fus3, the 3-kinases of the Mitogen-activated protein kinase cascade required for mating in *Saccharomyces cerevisiae* (Fig. 2a) (Good et al. 2009). In a broader sense, proteins which nucleate multiprotein complexes are considered as scaffolding proteins, while those tether two different proteins together are called "adaptors." Scaffolds and adaptor proteins could themselves be targets of regulation which often help in fine-tuning signaling responses. MAPK signaling pathways in general are typical examples for scaffold-coordinated orchestration of signaling specificity (reviewed in Witzel et al. 2012). There are numerous MAPKs and upstream kinases and the specificity in their signaling scheme is often achieved by the scaffolding proteins which bring a specific set of kinases together. Similar to the example above, the mammalian protein KSR (kinase suppressor of Ras) physically assembles the kinases Raf, MEK, and ERK, insulating them from other competing pathways and facilitating efficient and sequential phosphorylation events (Fig. 2a). KSR in turn is phosphorylated by the activated ERK. ERK-mediated phosphorylation reduces Raf binding to the KSR, and thus attenuates MEK activation (McKay et al. 2009). This type of feedback regulation is not unique to this signaling module. Different themes

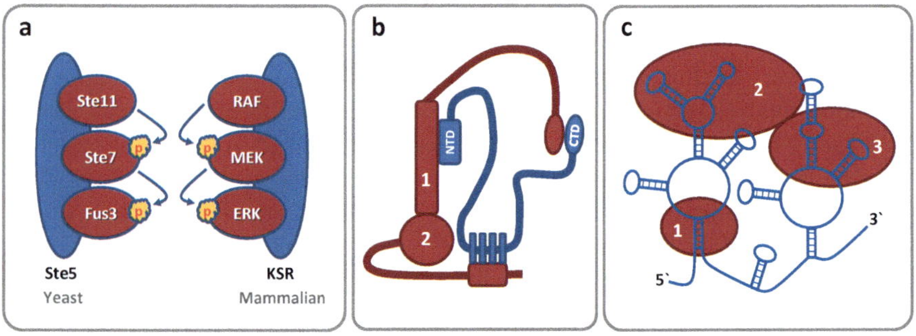

Fig. 2 *RNAs can act as scaffolds akin to multidomain proteins*. (**a**) Well-known scaffolding proteins (in blue) in yeast and mammalian system. Ste5 scaffold in yeast and KSR in the mammalian system scaffold the MAP kinase-signaling modules and promote activation of signaling pathways by holding them in a complex. While Ste5 bind Ste11, Ste7, and Fus3 to mediate mating signals in yeast, KSR parallelly binds the RAS/MEK/ERK pathway components of the mammalian MAPK pathway. (**b**) In blue, a scaffolding protein with three different domains is depicted. Protein 1 and 2 (in red) are brought together by the scaffolding protein, facilitating possible enzymatic and non-enzymatic interaction between these molecules. (**c**) Similarly, proteins 1, 2, and 3 (in red) are held together by the RNA scaffold (in blue) helping in protein–protein signaling

of the same regulatory mechanisms exist in other MAPK pathways providing much needed flexibility with required specificity.

3 Noncoding RNA Scaffolds for Proteins – rRNAs Guide the Way

The prerequisite for a protein to act as a scaffold is the presence of multiple domains which can independently or cooperatively bind two or more interaction partners at the same time. The modular organization of scaffolding proteins with several interaction domains like SH3 (SRC Homology 3), SH2 (SRC Homology 2), WW (40 amino acid domain with two invariant tryptophan residues), etc. makes it possible to assemble such multiprotein complexes (Fig. 2b). It is evident that long RNA molecules with their unique and dynamic secondary structures could be attractive candidates for scaffolding function (Fig. 2c). RNAs fold into highly dynamic or stable structures capable of generating specific interaction domains for proteins. The most relevant example of a large three-dimensional RNA scaffold-assembly is the ribosome itself. A mammalian ribosome consists of two major subunits designated as 40S (smaller subunit) and 60S (larger subunit) based on their sedimentation rates. The larger and smaller subunits further associate to give the 80S translation-competent ribosome. At the ultrastructural level, ribosomes are intricate ribonucleoprotein complexes wherein four rRNA species scaffold 80 ribosomal proteins (Bastide and David 2018). It can be considered as a three-dimensional structure with RNA skeletal framework, associated with proteins to form a translating molecular machine. Ribosome assembly is a multistep process involving the sequential generation and maturation of ribonucleoproteins. It is tightly regulated and coordinated with the energy and metabolic needs of the cell. Eukaryotic ribosomal subunits originally assemble in the nucleolus, an organelle which specializes in ribosome biogenesis. Further maturation proceeds when the ribosomal subunits translocate into the cytoplasm. There are more than 200 assembly factors involved in the process and many of the proteins associate with rRNAs co-transcriptionally to generate intermediates with the necessary folding and structure. Recent evidences indicate that lncRNAs also participate in steps of ribosomal biogenesis (Verheyden et al. 2018). In addition to the basic RNA-protein scaffold, the ribosomal subunits and the translation competent 80S ribosome could further accommodate additional regulatory proteins, making them master scaffolds for protein–protein interaction and translational control in the cytosol/endoplasmic reticulum membrane.

4 LncRNAs as Protein Scaffolds – Examples and Attempts to Classify Them

Two major mechanisms by which lncRNAs function in biological processes are (i) enzymatic activity (e.g., ribozymes) and (ii) sequence-dependent target binding (e.g., ceRNA). Some long RNAs clearly participate in structure/protein-binding-dependent scaffolding function, rather than those based on enzymatic activity or sequence-dependent gene regulatory function. However, it should be noted that there are some RNAs whose functional categorization is rather unclear. For example, the telomerase RNA-component (TERC) is a classic example of a trans-acting RNA scaffold which assembles a functional telomerase complex. However, it also provides the sequence template for telomere extension and also is indispensable for enzymatic activity. As described above, the possibility to generate multiple protein binding structures, separated by inherently flexible loops and disordered regions, make lncRNAs perfect candidates for scaffolding functions. Some of those lncRNAs have been well characterized. While majority of them function at the level of chromatin and transcriptional regulation, some of them have already been described to control specific protein modification events in the cytosol (Table 1).

4.1 HOTAIR – A Multidomain LncRNA with Diverse Scaffolding Functions

One of the classic functions of lncRNAs is scaffolding chromatin-modifying complexes through different domains. Hox transcript antisense intergenic RNA (HOTAIR), for instance, binds Polycomb repressor complex 2 (PRC2) by its 5′ domain and Lysine-specific histone demethylase 1 (LSD1) complex at the 3′ domain, recruiting them to genes located *in-trans* and silencing their expression (Tsai et al. 2010). The scaffolding function of HOTAIR also plays a role in post-translational regulation of protein stability or ubiquitination. HOTAIR interacts specifically with two different RNA-binding E3 ubiquitin ligases, Dzip3 and Mex3b. It can also interact with the protein targets of these ligases, leading to substrate ubiquitination and subsequent degradation. The levels of Ataxin-1 and Snurportin-1, substrates of Dzip3 and Mex3b respectively, are thus regulated by HOTAIR which keeps them in complex with their specific ligases. Interestingly, this process is further regulated by the RNA-binding protein (RBP), HuR which reduces the stability of HOTAIR by binding to it and promoting Ago2/let7-mediated degradation of the lncRNA (Yoon et al. 2013).

Table 1 List of lncRNAs with potential scaffold function, their interactors, and functional classification

LncRNA	Major protein/signaling partners	Potential class	Main references
ANRIL	PRC1 (Cbx7) and PRC2 (Suz12)	Chromatin scaffold	Yap et al. (2010) and Kotake et al. (2011)
BCAR4	SNIP1 & PNUTS	Chromatin/transcription scaffold	Xing et al. (2014)
CamK-A	PNCK and IκBα	Kinase scaffold	Sang et al. (2018)
CCAT1	PRC2, SUV39H1	Chromatin/transcription scaffold	Zhang et al. (2017)
Evf2	DLX, BRG1, BAF170, BAF155	Chromatin/transcription scaffold	Cajigas et al. (2018)
Gadd7	TDP-43	DNA repair	Liu et al. (2012)
GUARDIN	BRCA1, BARD1	Others	Hu et al. (2018)
HOTAIR	PRC2, LSD1, DNMT1, Dzip3/Ataxin-1, Mex3b/Snurportin-1	Chromatin scaffold, ubiquitin ligase scaffold	Yoon et al. (2013) and Tsai et al. (2010)
HOXA11-AS	EZH2, LSD1	Chromatin scaffold	Sun et al. (2016)
Kcnq1ot1	PRC2, G9a	Chromatin scaffold	Pandey et al. (2008)
LINC01198	PTEN, NEDD4	Ubiquitin ligase scaffold	Chen et al. (2019)
LINK-A	BRK and LRRK2	Kinase scaffold	Lin et al. (2016)
LINP1	Ku80 and DNA-PKcs	Kinase scaffold/DNA repair	Zhang et al. (2016)
Lnc-DC	STAT3	Transcription scaffold	(Wang et al. 2014)
Lnczc3h7a	RIG-I & TRIM25	Ubiquitin ligase scaffold	Lin et al. (2019)
MALAT1	PRC2, E2F, splicing factors, and other coactivators	Chromatin/transcription scaffold	Yang et al. (2011), Hirata et al. (2015) and Fan et al. (2014)
NEAT1	PRC2, inflammasome, SFPQ, NONO-PSF	Chromatin scaffold (paraspeckles)	Chen et al. (2018), Zhang et al. (2019a) and An et al. (2018)
NKILA	p65, IκBα	Others	Liu et al. (2015)
NRON	NFAT1, NFAT1-kinases, and IQGAP1	Kinase scaffold	Sharma et al. (2011)
PCAT1	FKBP51	Others	Shang et al. (2019)
ROCKI	APEX1	Transcription scaffold	Zhang et al. (2019b)
SAMMSON	p32, CARF, and XRN2	Others	Leucci et al. (2016) and Vendramin et al. (2018)
Tre-RNA	hnRNPK, FXR1, FXR2, PUF60, SF3B3, eIF4G1	Others	Gumireddy et al. (2013)
TUG1	PRC2, PRC1, YY1	Chromatin/transcription scaffold	Yang et al. (2011)

(continued)

Table 1 (continued)

LncRNA	Major protein/signaling partners	Potential class	Main references
TUNA	PTBP1, hnRNPK, and Nucleolin	Chromatin/transcription scaffold	Lin et al. (2014)
XIST	SAF-A, RMB15, hnRNPK, Spen, etc.	Chromatin scaffold	Pintacuda et al. (2017) and Splinter et al. (2011)

4.2 NEAT1 as a Scaffold for Paraspeckle Formation

Nuclear Paraspeckle Assembly Transcript 1 or Nuclear Enriched Abundant Transcript 1 (NEAT1), as the names suggest, is an abundantly expressed long noncoding RNA essential for the organization of nuclear RNA granules called paraspeckles (reviewed in An et al. 2018). It is ubiquitously expressed and forms the basic scaffold for >60 nuclear proteins and assembles the paraspeckles. Majority of the paraspeckle proteins are RNA-binding proteins and many of them harbor prion-like domains enabling phase separation, facilitating speckle formation. The paraspeckles also anchor other RNA species which mostly define their function. Sequestration or nuclear retention of hyper-edited RNA, sequestration of transcription factors, and modulation of miRNA biogenesis are some of the described roles of this class of NEAT1-primed nuclear speckles. Variations in the expression of NEAT1 and its splice variants seem to modulate paraspeckle size and morphology. Moreover, stress-responsive nature of paraspeckles and their association with neurodegenerative diseases indicates dynamic nature and potential clinical relevance. Interestingly, a recent study has shown a cytosolic function for NEAT1. NEAT1 was shown to participate in cytosolic inflammasome activation probably by scaffolding and stabilizing the multiprotein inflammasome complex (Zhang et al. 2019a). The versatility of this lncRNA is clear from the fact that it also acts as a scaffold for the PRC2 complex much like HOTAIR, while also displaying some ceRNA activity in the context of tumor progression (Klec et al. 2019).

4.3 LncRNA Scaffolds in Chromatin Remodeling and Transcriptional Control

A large number of the characterized lncRNAs are nuclear and are strongly associated with the chromatin. There is a line of thought which believe that lncRNAs are just transcriptional noise and many of them are artefacts and products of immature transcription and post-transcriptional splicing. It is true that that there are many incompletely spliced transcripts and products of unsuccessful transcription initiation which add to the nuclear RNA repertoire. They are often not exported out and are retained in the nucleus as part of the quality control mechanism. However, at least some of them could be performing some additional functions in the nucleus, much like genuine lncRNAs. Despite all skepticism, one of the best characterized functions of lncRNAs is their role in global and specific chromatin remodeling and transcriptional control. Like NEAT1 and HOTAIR, many lncRNAs function as scaffolds for chromatin-remodeling complexes like the polycomb repressor complexes 1 or 2 (PRC1/2). The classical example is the XIST (X-inactive specific transcript), which scaffolds and recruits a variety of proteins to the X-chromosome to establish and maintain X-chromosome inactivation (Splinter et al. 2011). Experimental evidence clearly indicates a role for the three-dimensional structure of this RNA in recruiting and scaffolding a panel of proteins at the inactive X-chromosome (reviewed in Pintacuda et al. 2017). The expression of the tumor suppressor PTEN is regulated by its antisense transcript PTENpg1 which suppresses PTEN expression by recruiting EZH2 and DNMT3a to its promoter (Johnsson et al. 2013). It is interesting to note that a large number of lncRNAs interact with EZH2, a PRC2 component, and alter the localization and functions of this polycomb complex, especially in the context of cancer (Su et al. 2018). Moreover some of them like the lncRNA Firre (Functional Intergenic Repeating RNA Element) interact with chromosomes as well as chromatin-associated and nuclear matrix proteins to influence higher order nuclear architecture (Hacisuleyman et al. 2014). BCAR4 (Breast Cancer Anti-Estrogen Resistance 4), a lncRNA with high expression in several tumor-types, regulates non-canonical hedgehog signaling by acting as a transcriptional scaffold for the GLI2 transcription factor (Xing et al. 2014). BCAR4 nucleates a multiprotein complex consisting of GLI2 by directly interacting with SNIP1, a transcriptional coactivator and PNUTS, a regulatory subunit of protein phosphatase-1 (PP1). In a sequence of events consisting of p300 histone acetyltransferase activation and PP1-mediated RNA polymerase dephosphorylation, the lncRNA-scaffolded protein complex leads to GLI2-dependent transcription and enhances cancer cell migration. MALAT1 (Metastasis Associated Lung Adenocarcinoma Transcript 1) is another highly investigated and reviewed nuclear lncRNA which performs multiple functions in the vicinity of the chromosomes (reviewed in McKay et al. 2009). Splicing is another chromatin-proximal RNA modifying process, which is also regulated by lncRNAs like MALAT1. Recent studies have revealed the presence of super enhancers, which are regulatory sites which can impart control of gene transcription over large genomic distances. Super enhancer–associated lncRNAs do exist and they act as master regulators of tissue specificity and tumorigenesis (Khan and Zhang 2016; Orom et al. 2010; Cajigas et al. 2018). In many of these cases, the extended protein interactome of these RNA molecules are not known so far.

4.4 *LncRNA Scaffolds in DNA Repair*

DNA repair is a complex process mediated by multiprotein complexes and provides graduated responses to various levels of genotoxic challenges faced by the cells ranging from minor base-dimerization to double stranded breaks and large lesions.

Growth-arrested DNA damage-inducible gene 7 (gadd7), a lncRNA identified in Chinese hamster cells is induced in response to DNA damage and growth halting signals and plays role in the induction of G1/S arrest upon DNA damage. RNA pull-down analysis revealed interaction of gadd7 with TAR DNA-binding protein (TDP-43) via GU/UG repeats. Upon UV irradiation, the interaction between lncRNA and TDP-43 is enhanced which prevents TDP-43 binding to its target mRNAs including the 3′UTR of CDK6. This leads to rapid decay of CDK6 mRNA and inhibition of cell cycle progression, thus helping cells maintain genome integrity (Liu et al. 2012). Neither the other interaction partners of this lncRNA nor its human homologs are known until now. Another lncRNA LINP1 (lncRNA in non-homologous end joining pathway 1) identified as a transcript overexpressed in human triple negative breast cancer (TNBC) makes tumors resistant to radiotherapy. Mechanistically, LINP1 serves as a scaffold for Ku80 (XRCC5) and DNA-PKcs (DNA-dependent protein kinase, catalytic subunit/XRCC7) proteins involved in Non-Homologous End-Joining (NHEJ) repair pathway. Post-irradiation treatment, LINP1 is recruited to the chromatin where it enhances the interaction between Ku80 and DNA-PKcs, facilitating repair of DNA double-strand breaks (DSBs). This increased efficiency of repair mediated by LINP1 scaffold makes cancer cells resistant to radiotherapy (Zhang et al. 2016). A more general role has been proposed recently for lncRNAs in the general DNA-damage response. In response to DNA damage, a specific RNA Polymerase II complex recruited to the damaged DNA synthesizes damage-induced long noncoding RNAs (dilncRNAs) at the DNA ends. These dilncRNAs and associated small RNAs promote DNA damage response foci formation by binding 53BP1 and other repair proteins (Michelini et al. 2017). It is not clear whether the dilncRNAs directly participate as scaffolds in this process or will act as precursors for the small RNAs required for the process of foci formation.

4.5 NRON – The Noncoding Repressor of NFAT

NRON is probably one of the top lncRNA candidates currently investigated for their therapeutic relevance. As the name suggests, it was identified as a negative regulator of Nuclear Factor of Activated T cells (NFAT), a family of calcium-regulated transcription factors with role in immune-response and myogenesis (Willingham et al. 2005). Nuclear translocation of NFAT1 is regulated by controlled phosphorylation–dephosphorylation events. NFAT1 is constitutively phosphorylated and locked in an inactive conformation in the cytoplasm. Cytosolic calcium-dependent phosphatase calcineurin is responsible for the dephosphorylation of NFATs. Upon calcium stimulation, activated calcineurin dephosphorylates NFATs leading to their nuclear translocation. Multiple pathways modulating calcineurin regulation have been investigated, however there is very little information regarding the regulation of the NFAT kinases and how they achieve and maintain quantitative basal phosphorylation of NFATs. NRON nucleates a multiprotein complex consisting of NFAT1, IQGAP1, and a number of NFAT1 kinases, facilitating constitutive NFAT phosphorylation in resting cells. NFAT1 activating stimuli also induce dissociation of this complex, in addition to calcineurin activation. This is the perfect example for the regulation of signaling or protein phosphorylation by an RNA-based scaffold. Calcium signaling has been a target of therapeutic intervention in the context of chronic inflammation as well as organ transplantation. The calcineurin inhibitors cyclosporins and their derivatives are actively being used in clinics to induce transient immunosuppression and to prevent organ rejection. A better understanding of this pathway, with potential relevance also in skeletal/cardiac muscle will make NRON a potential candidate for targeting by RNA-based therapeutics. Interestingly, a recent study has reported another NFAT-independent function of NRON, wherein it was shown to support HIV1 latency by inducing the degradation of the viral Tat protein (Li et al. 2016). Once again, NRON forms a multiprotein complex-linking Tat protein to the ubiquitin ligase CUL4B and the proteasomal subunit PSMD11, facilitating its ubiquitination and proteasomal degradation. This could be considered as a strategy by the pathogen, wherein a lncRNA abundantly expressed in the host CD4+ T cells is being utilized by the virus to maintain latency.

4.6 *LncRNAs as Protein Kinase Scaffolds: LINK-A and CamK-A*

Interactome-analysis studies using mass spectrometry (MS) approaches have revealed novel functional aspects of lncRNA-mediated regulation in cell signaling, particularly in the context of cancer. For example, RNA pulldown assay followed by MS established a role for the cytoplasmic lncRNA – LINK-A (Long-Intergenic Noncoding RNA for Kinase Activation) in signal transduction (Lin et al. 2016). LINK-A plays a crucial role in HB-EGF (Heparin-binding EGF-like growth factor) – induced HIF1α signaling in triple-negative breast cancer (TNBC). Upon ligand-binding induced EGF-receptor – GPNMB (glycoprotein nmb) heterodimerization, the recruitment and activation of downstream protein–tyrosine kinase BRK is dependent on the scaffolding function of LINK-A. LINK-A mediates recruitment of BRK to GPNMB. LINK-A binding induces conformational change in BRK which ceases the auto-inhibition of BRK, thus mediating its auto-phosphorylation and activation. BRK and another LINK-A bound kinase LRRK2 (leucine-rich repeat kinase 2) phosphorylate HIF1α at two distinct sites, promoting its stabilization and transcriptional activation respectively. The strong correlation between expression of LINK-A and its associated phosphorylated signaling intermediates could potentially serve as biomarkers and therapeutic targets in TNBC (Lin et al. 2016).

Another study identified a tumorigenic lncRNA CamK-A (lncRNA for calcium-dependent kinase activation) that is highly expressed in multiple human cancers and activates calcium (Ca^{2+})-dependent NF-κB (nuclear factor κB) signaling via mechanisms similar to that of LINK-A (Sang et al. 2018). CamK-A was found to interact with both PNCK (pregnancy upregulated non-ubiquitously-expressed CaM kinase) and IκBα (NF-κB inhibitor alpha). CamK-A binding with PNCK brings about a conformational change in PNCK, which makes the calmodulin binding and the autophosphorylation sites accessible, resulting in phosphorylation and activation of PNCK. Under conditions of high cytosolic Ca^{2+}, for instance, during hypoxia in solid tumors, the activated PNCK phosphorylates IκBα resulting in its degradation and activation of NF-κB signaling. This enhances cytokine production, glycolytic rates, and infiltration of macrophages, thus modulating the composition of the tumor microenvironment (Sang et al. 2018). LINK-A and CAMK-A provide important insights into allosteric regulation mediated by lncRNA scaffolds. These studies also call upon the importance of identification of protein interaction partners and characterization of their binding regions in facilitating the functional analysis of lncRNAs. In both cases, lncRNA binding within or near the autoinhibitory regions of kinases releases the auto-inhibitory domains, leading to kinase activation. Considering the conservation of activation mechanisms and domain architecture among protein kinases, it could be speculated that many more lncRNAs could be doing similar functions in the regulation of other kinases. Availability of more and more lncRNA-interactome data may help in identifying a family of "kinase regulatory lncRNA scaffolds."

4.7 The Translational Regulatory LncRNA (tre-RNA) and Others

Tre-RNA was originally found to be upregulated in metastatic breast cancer and was shown to regulate the expression of epithelial marker E-cadherin at the level of protein translation, hence the name (Gumireddy et al. 2013). Tre-RNA assembles a multiprotein complex on the 3′UTR regions of E-cadherin mRNA and suppresses its translation. Interestingly, such scaffolding activity by RNAs at the 3′UTRs could present another general mechanism of lncRNA scaffold function. LncRNA molecules are often associated with the translation machinery and regulate target mRNA specific as well as general translation (Vucicevic et al. 2016). Originally, lncRNAs were originally predicted based on the absence of significantly long open-reading frames (ORFs). The identification of micro-peptide encoding short ORFs in many of the transcripts previously characterized as lncRNAs has brought some ambiguity to this definition. In this context, it is difficult to interpret the experimental result when a lncRNA is shown to associate with the translation machinery. They could be encoding micro-peptides by acting as templates for translation or could well be genuine regulators of protein translation. LncRNAs with scaffolding activity also regulate ribosome biogenesis and thus participate in general translation control. SAMMSON a pro-oncogenic lncRNA was recently shown to simultaneously modulate rRNA biogenesis in the nucleus as well as mitochondria, increasing the fitness levels and energy efficiency of cancer cells. SAMMSON was shown to sequester

RNA metabolizing enzymes in different cellular compartments by forming ribonucleoprotein complexes, thus regulating ribosome biogenesis in the mitochondria as well as nucleoli (Vendramin et al. 2018; Verheyden et al. 2018).

4.8 LncRNAs as Ubiquitin Ligase Scaffolds

As described in the case of HOTAIR, there are a number of lncRNAs reported to scaffold ubiquitin ligase–substrate pairs, facilitating protein ubiquitination. LncRNAs have been shown to link E3 ligases and their substrate proteins mediating both K48-linked ubiquitination that directs proteasomal degradation and K63-dependent ubiquitination which modulates target protein function. LINC01198, a lncRNA upregulated in glioma, was shown to promote the degradation of the tumor suppressor phosphatase PTEN (phosphatase and tensin homolog). LINC01198 simultaneously binds to PTEN and its E3 ligase, NEDD4–1 (neural precursor cell expressed, developmentally downregulated 4), promoting PTEN ubiquitination and proteasomal degradation. Consequently, LINC01198 positively regulates the PTEN target, oncogenic PI3K/Akt-signaling pathway promoting glioma progression (Chen et al. 2019). Another ubiquitin ligase accessory lncRNA Lnczc3h7a seems to function in innate immune response signaling. This lncRNA facilitates the K63-linked ubiquitination of the RNA sensor RIG-I (retinoic acid-inducible gene-I) during early antiviral response. Lnczc3h7a binds to both RIG-I and the ubiquitin ligase TRIM25, holding them in a complex during viral infection. Interestingly, RIG-I binds the lncRNA only when it is in complex with TRIM25. Binding of lncRNA strengthens RIG-1-TRIM25 interaction, induces RIG-1 ubiquitination, and consequently the host response by enhancing the production of Type I Interferons (IFNs) and pro-inflammatory cytokines (Lin et al. 2019).

5 Future Perspectives

A recent genome-wide lncRNA interactome study predicted protein scaffolding as one the most prevalent functions of lncRNAs (Ribeiro et al. 2018). The report also could establish that most of the disease-relevant lncRNAs are involved in scaffolding protein complexes. This finding is more significant if one considers the fact that a large majority of known lncRNAs are actually shown to function as ceRNAs or miRNA sponges. More interestingly, according to this study lncRNA adaptors and scaffolds seem to participate in more than half of all known protein complexes and signaling networks. Based on interactome studies coupled to predictive analysis, they identified 847 lncRNAs with potential scaffolding function (Ribeiro et al. 2018). It is evident that the numbers of scaffold-forming lncRNAs are far more than the few of them which have already been characterized. The list of candidate scaffold lncRNAs generated by this study will be a good way to start investigations into this class of noncoding transcripts.

Three-dimensional-structure-driven functions of nucleic acids are difficult to predict or characterize compared to the mechanisms involving sequence specific base pairing. This is obviously due to the ease with which sequence-based features can be analyzed in silico, leading to the identification of target molecules which hybridize to the candidate lncRNA. The capability to accurately predict the secondary structure of long RNA molecules is a challenge toward better prediction of function and interactors. Even though tools like M-fold are available from a long time now, accurate prediction of RNA secondary structure is still difficult. Long RNA molecules have too many degrees of freedom to allow accurate prediction of a single structure with any level of confidence. In addition, lncRNA folding in nature is guided by several interaction partners. It is important to understand that many or most of the RNAs bind proteins co-transcriptionally. The algorithms usually calculate folds by energy minimization and often provide more than a couple of probable structures for an RNA species of even moderate size. Could circular RNAs be efficient scaffolds especially due to their high half-lives? Could it be possible that the constraints imposed by lack of free ends make them capable of forming more stable structures? In this context, are there efficient and reliable structure prediction algorithms available for circular RNAs? These are still major bottlenecks in the predictive analysis of lncRNA function. Till date the scaffold functions of lncRNAs have been identified mostly based on lncRNA interactome studies using MS analyses. The sensitivity and reliability of such techniques have been constantly evolving (Barra and Leucci 2017). However, will it ever become a predictable function much like sponge or ceRNA activity? We do not completely understand all permutation–combinations of protein–RNA binding specificities to be able to make accurate predictions of potential protein interaction partners of a given lncRNA. Databases with experimentally validated protein–lncRNA interactions will be a good start point. Fortunately, investigations have already been initiated in this direction and such servers have started to appear (An et al. 2019). As more and more candidates with distinct signaling functions are revealed, we may be able to look for commonalities in their mode of actions, classify them and make it possible to predict new functions from their sequence or secondary structures. As discussed earlier, chromatin scaffolds, kinase regulators and ubiquitin ligase scaffolds are possible groups where we already have more than a couple of lncRNAs to categorize.

References

An H, Williams NG, Shelkovnikova TA (2018) NEAT1 and paraspeckles in neurodegenerative diseases: a missing lnc found? Noncoding RNA Res 3:243–252

An G, Sun J, Ren C et al (2019) LIVE: a manually curated encyclopedia of experimentally validated interactions of lncRNAs. Database (Oxford) 2019:baz011

Barra J, Leucci E (2017) Probing long non-coding RNA-protein interactions. Front Mol Biosci 4:45
Bastide A, David A (2018) The ribosome, (slow) beating heart of cancer (stem) cell. Oncogenesis 7:34
Cajigas I, Chakraborty A, Swyter KR et al (2018) The Evf2 Ultraconserved enhancer lncRNA functionally and spatially organizes Megabase distant genes in the developing forebrain. Mol Cell 71:956–972 e959
Cech TR (1990) Nobel lecture. Self-splicing and enzymatic activity of an intervening sequence RNA from Tetrahymena. Biosci Rep 10:239–261
Chen Q, Cai J, Wang Q et al (2018) Long noncoding RNA NEAT1, regulated by the EGFR pathway, contributes to glioblastoma progression through the WNT/beta-catenin pathway by scaffolding EZH2. Clin Cancer Res 24:684–695
Chen WL, Chen HJ, Hou GQ et al (2019) LINC01198 promotes proliferation and temozolomide resistance in a NEDD4-1-dependent manner, repressing PTEN expression in glioma. Aging (Albany, NY) 11:6053–6068
Dhamija S, Menon MB (2018) Non-coding transcript variants of protein-coding genes – what are they good for? RNA Biol 15:1025–1031
Diederichs S (2014) The four dimensions of noncoding RNA conservation. Trends Genet 30:121–123
Ezkurdia I, Juan D, Rodriguez JM et al (2014) Multiple evidence strands suggest that there may be as few as 19,000 human protein-coding genes. Hum Mol Genet 23:5866–5878
Fan Y, Shen B, Tan M et al (2014) TGF-beta-induced upregulation of malat1 promotes bladder cancer metastasis by associating with suz12. Clin Cancer Res 20:1531–1541
Gilbert W (1986) Origin of life: the RNA world. Nature 319:618–618
Good M, Tang G, Singleton J et al (2009) The Ste5 scaffold directs mating signaling by catalytically unlocking the Fus3 MAP kinase for activation. Cell 136:1085–1097
Gumireddy K, Li A, Yan J et al (2013) Identification of a long non-coding RNA-associated RNP complex regulating metastasis at the translational step. EMBO J 32:2672–2684
Hacisuleyman E, Goff LA, Trapnell C et al (2014) Topological organization of multichromosomal regions by the long intergenic noncoding RNA firre. Nat Struct Mol Biol 21:198–206
Hirata H, Hinoda Y, Shahryari V et al (2015) Long noncoding RNA MALAT1 promotes aggressive renal cell carcinoma through Ezh2 and interacts with miR-205. Cancer Res 75:1322–1331
Hu WL, Jin L, Xu A et al (2018) GUARDIN is a p53-responsive long non-coding RNA that is essential for genomic stability. Nat Cell Biol 20:492–502
Jarroux J, Morillon A, Pinskaya M (2017) History, discovery, and classification of lncRNAs. Adv Exp Med Biol 1008:1–46
Johnsson P, Ackley A, Vidarsdottir L et al (2013) A pseudogene long-noncoding-RNA network regulates PTEN transcription and translation in human cells. Nat Struct Mol Biol 20:440–446
Kapranov P, Cheng J, Dike S et al (2007) RNA maps reveal new RNA classes and a possible function for pervasive transcription. Science 316:1484–1488
Khan A, Zhang X (2016) dbSUPER: a database of super-enhancers in mouse and human genome. Nucleic Acids Res 44:D164–D171
Klec C, Prinz F, Pichler M (2019) Involvement of the long noncoding RNA NEAT1 in carcinogenesis. Mol Oncol 13:46–60
Kotake Y, Nakagawa T, Kitagawa K et al (2011) Long non-coding RNA ANRIL is required for the PRC2 recruitment to and silencing of p15(INK4B) tumor suppressor gene. Oncogene 30:1956–1962
Lander ES, Linton LM, Birren B et al (2001) Initial sequencing and analysis of the human genome. Nature 409:860–921
Lee RC, Feinbaum RL, Ambros V (1993) The C. elegans heterochronic gene lin-4 encodes small RNAs with antisense complementarity to lin-14. Cell 75:843–854
Leucci E, Vendramin R, Spinazzi M et al (2016) Melanoma addiction to the long non-coding RNA SAMMSON. Nature 531:518–522

Li J, Chen C, Ma X et al (2016) Long noncoding RNA NRON contributes to HIV-1 latency by specifically inducing tat protein degradation. Nat Commun 7:11730
Lin N, Chang KY, Li Z et al (2014) An evolutionarily conserved long noncoding RNA TUNA controls pluripotency and neural lineage commitment. Mol Cell 53:1005–1019
Lin A, Li C, Xing Z et al (2016) The LINK-A lncRNA activates normoxic HIF1alpha signalling in triple-negative breast cancer. Nat Cell Biol 18:213–224
Lin H, Jiang M, Liu L et al (2019) The long noncoding RNA Lnczc3h7a promotes a TRIM25-mediated RIG-I antiviral innate immune response. Nat Immunol 20:812–823
Liu X, Li D, Zhang W et al (2012) Long non-coding RNA gadd7 interacts with TDP-43 and regulates Cdk6 mRNA decay. EMBO J 31:4415–4427
Liu B, Sun L, Liu Q et al (2015) A cytoplasmic NF-kappaB interacting long noncoding RNA blocks IkappaB phosphorylation and suppresses breast cancer metastasis. Cancer Cell 27:370–381
McKay MM, Ritt DA, Morrison DK (2009) Signaling dynamics of the KSR1 scaffold complex. Proc Natl Acad Sci U S A 106:11022–11027
Michelini F, Pitchiaya S, Vitelli V et al (2017) Damage-induced lncRNAs control the DNA damage response through interaction with DDRNAs at individual double-strand breaks. Nat Cell Biol 19:1400–1411
Orom UA, Derrien T, Beringer M et al (2010) Long noncoding RNAs with enhancer-like function in human cells. Cell 143:46–58
Pandey RR, Mondal T, Mohammad F et al (2008) Kcnq1ot1 antisense noncoding RNA mediates lineage-specific transcriptional silencing through chromatin-level regulation. Mol Cell 32:232–246
Pintacuda G, Young AN, Cerase A (2017) Function by structure: spotlights on Xist long non-coding RNA. Front Mol Biosci 4:90
Ribeiro DM, Zanzoni A, Cipriano A et al (2018) Protein complex scaffolding predicted as a prevalent function of long non-coding RNAs. Nucleic Acids Res 46:917–928
Sang LJ, Ju HQ, Liu GP et al (2018) LncRNA CamK-A regulates ca(2+)-signaling-mediated tumor microenvironment remodeling. Mol Cell 72:71–83 e77
Shang Z, Yu J, Sun L et al (2019) LncRNA PCAT1 activates AKT and NF-kappaB signaling in castration-resistant prostate cancer by regulating the PHLPP/FKBP51/IKKalpha complex. Nucleic Acids Res 47:4211–4225
Sharma S, Findlay GM, Bandukwala HS et al (2011) Dephosphorylation of the nuclear factor of activated T cells (NFAT) transcription factor is regulated by an RNA-protein scaffold complex. Proc Natl Acad Sci U S A 108:11381–11386
Splinter E, de Wit E, Nora EP et al (2011) The inactive X chromosome adopts a unique three-dimensional conformation that is dependent on Xist RNA. Genes Dev 25:1371–1383
Su M, Xiao Y, Tang J et al (2018) Role of lncRNA and EZH2 interaction/regulatory network in lung cancer. J Cancer 9:4156–4165
Sun M, Nie F, Wang Y et al (2016) LncRNA HOXA11-AS promotes proliferation and invasion of gastric cancer by scaffolding the chromatin modification factors PRC2, LSD1, and DNMT1. Cancer Res 76:6299–6310
Tsai MC, Manor O, Wan Y et al (2010) Long noncoding RNA as modular scaffold of histone modification complexes. Science 329:689–693
Vendramin R, Verheyden Y, Ishikawa H et al (2018) SAMMSON fosters cancer cell fitness by concertedly enhancing mitochondrial and cytosolic translation. Nat Struct Mol Biol 25:1035–1046
Verheyden Y, Goedert L, Leucci E (2018) Control of nucleolar stress and translational reprogramming by lncRNAs. Cell Stress 3:19–26
Vucicevic D, Gehre M, Dhamija S et al (2016) The long non-coding RNA PARROT is an upstream regulator of c-Myc and affects proliferation and translation. Oncotarget 7:33934–33947
Wang P, Xue Y, Han Y et al (2014) The STAT3-binding long noncoding RNA lnc-DC controls human dendritic cell differentiation. Science 344:310–313

Willingham AT, Orth AP, Batalov S et al (2005) A strategy for probing the function of noncoding RNAs finds a repressor of NFAT. Science 309:1570–1573

Witzel F, Maddison L, Bluthgen N (2012) How scaffolds shape MAPK signaling: what we know and opportunities for systems approaches. Front Physiol 3:475

Xing Z, Lin A, Li C et al (2014) lncRNA directs cooperative epigenetic regulation downstream of chemokine signals. Cell 159:1110–1125

Yang L, Lin C, Liu W et al (2011) ncRNA- and Pc2 methylation-dependent gene relocation between nuclear structures mediates gene activation programs. Cell 147:773–788

Yap KL, Li S, Munoz-Cabello AM et al (2010) Molecular interplay of the noncoding RNA ANRIL and methylated histone H3 lysine 27 by polycomb CBX7 in transcriptional silencing of INK4a. Mol Cell 38:662–674

Yoon JH, Abdelmohsen K, Kim J et al (2013) Scaffold function of long non-coding RNA HOTAIR in protein ubiquitination. Nat Commun 4:2939

Zhang Y, He Q, Hu Z et al (2016) Long noncoding RNA LINP1 regulates repair of DNA double-strand breaks in triple-negative breast cancer. Nat Struct Mol Biol 23:522–530

Zhang E, Han L, Yin D et al (2017) H3K27 acetylation activated-long non-coding RNA CCAT1 affects cell proliferation and migration by regulating SPRY4 and HOXB13 expression in esophageal squamous cell carcinoma. Nucleic Acids Res 45:3086–3101

Zhang P, Cao L, Zhou R et al (2019a) The lncRNA Neat1 promotes activation of inflammasomes in macrophages. Nat Commun 10:1495

Zhang Q, Chao TC, Patil VS et al (2019b) The long noncoding RNA ROCKI regulates inflammatory gene expression. EMBO J 38

Landscape of Long Noncoding RNA Genes, Pseudogenes, and Protein Genes in Segmental Duplications in the Critical Human Chromosomal Region 22q11.2

Nicholas Delihas

Contents

Abstract Human chromosomal region 22q11.2 is prone to deletions that result in genetic disorders termed 22q11.2 deletion syndrome. Approximately 1 in 4000 infants are born with this deletion. The 22q11.2 region has a large number of DNA duplicated sequences termed segmental duplications or low-copy repeats (LCR22) that are known to take part in the deletion process. Here we analyze sequences that neighbor LCR22 segmental duplications to help determine their origins. Sequences immediately outside of and within the LCR22 boundaries are found highly conserved relative to analogous sequences in chimpanzee chr22. We propose that neighbor sequences form a framework for LCR22 sequence expansion, and that this framework originated in a common primate ancestor. The second part of this chapter deals with genes contained in LCR22 segmental duplications. Most LCR22 genes have not been analyzed, but they may have a bearing on the molecular basis for developmental abnormalities that result from a loss of 22q11.2. We find the largest LCR22s, LCR22A and LCR22D, primarily contain long noncoding RNA genes and pseudogenes. An analysis of linear and circular RNA transcript expression

N. Delihas (✉)
Department of Molecular Genetics and Microbiology, Renaissance School of Medicine, Stony Brook University, Stony Brook, NY, USA
e-mail: Nicholas.Delihas@stonybrook.edu

S. Jurga, J. Barciszewski (eds.), *The Chemical Biology of Long Noncoding RNAs*, RNA Technologies 11, https://doi.org/10.1007/978-3-030-44743-4_6

in developing fetal tissues from these LCR22 genes is also presented; this may be useful in future analyses of ncRNA gene functions in fetal development.

Keywords Chromosomal segmental duplications · Long noncoding RNA genes · Pseudogenes · Circular RNAs · 22q11.2 · Evolution

1 Introduction

Segmental duplications, also termed low-copy repeats (LCR), are large DNA repeat sequences found in primate chromosomes (Dennis et al. 2017). They are approximately 10–900 Kbp in length, generally share greater than 90% nucleotide sequence identity and are considered to be an important part of primate evolution (Dennis et al. 2017). Human chromosome 22 (chr22) contains the largest number of segmental duplications per chromosomal length (Redaelli et al. 2019) and thus has the highest density of duplications compared with other human chromosomes. The chr22 region, termed 22q11.2, contains eight DNA sequence duplications, which are partial duplications of each other, LCR22A-H (Guo et al. 2011; Demaerel et al. 2018) (Fig. 1). Two of these, LCR22A and LCR22D, are known to be involved in meiotic non-allelic homologous recombination, which leads to deletion of the region that is found in the majority of human 22q11.2 deletions and results in the onset of genetic diseases (Guo et al. 2018). A chromosomal misalignment can occur during meiosis when there is recombination involving an LCR22 instead of the normal site of recombination. This is due to high sequence similarities. On the other hand, an important biological and evolutionary function of segmental duplications is to provide an abundance of duplicated sequences that can develop into new genes (Dennis et al. 2017; Samonte and Eichler 2002; Delihas 2020). Thus, segmental duplications are a double-edged sword: They have evolved in primates as a source of evolution, but they can also lead to genomic disorders due to similarities in sequence with resultant misalignment during genetic recombination (Stankiewicz and Lupski 2002).

Here we analyze the two largest LCR22s, LCR22A and LCR22D, with respect to chromosomal boundary sequences and genes present in the LCR22 duplications, and compare chromosomal positions of these genes and gene positional order with those

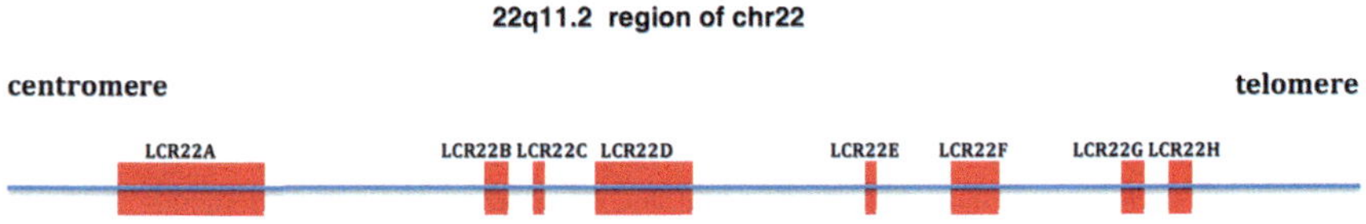

Fig. 1 Representation of eight LCR22 chromosomal duplications, shown in red highlight, LCR22A-H in chr22 region 22q11.2. LCR22s are partial DNA repeat sequences of each other

found in the chimpanzee genome. This has enabled a prediction of how segmental duplications arose evolutionarily. The study described here suggests that the framework for chromosomal expansion by low-copy repeats was present in a primate ancestor and that at least one repeat, the LCR22A is likely to be human specific. In a pioneer study, Babcock and coworkers (Babcock et al. 2007) proposed that the origins of LCR22 segmental duplications occurred in primates and experimentally showed that the gorilla genome displays a large expansion of LCR22s.

Noncoding RNA (ncRNA) genes, which are present in segmental duplications and are discussed in this chapter, are the pseudogenes and long intergenic noncoding RNA (lincRNA) genes; however, also included are the circular RNAs (circRNA)s that stem from both protein and RNA genes. ncRNAs are very diverse; their origins and structures greatly differ (Dykes and Emanueli 2017); however the three classes that are discussed here, lincRNAs, pseudogene transcripts, and circRNAs, are all characterized as long noncoding RNAs (lncRNAs) and have a length greater than 200 nt.

lincRNAs are transcripts from RNA genes that are situated between other genes in chromosomes and do not overlap genes. These RNAs contain exons, have a $5'$ cap, and are polyadenylated at the $3'$ end, but they are not translated into proteins. There are thousands of lincRNAs that have been detected but their functions for the most part are unknown (Hung and Chang 2010). However, one of the first lincRNAs discovered and functionally characterized is HOTAIR. HOTAIR interacts with the Polycomb Repressive Complex 2 (PRC2) and regulates transcription by altering histone methylation (Rinn et al. 2007; Hung and Chang 2010). The Polycomb Repressive Complex 2 contains regulatory ncRNA genes and the proteins that are involved in the methylation of histones and the silencing of genes.

Pseudogenes were first found and characterized by Jacq et al. in 1977 (Jacq et al. 1977), who described a tandem DNA repeat of 5S RNA genes where the 5S genes have a copy of the 5S DNA sequence that contains mutations. They were termed pseudogenes, and they arise from copies of the parent gene that has mutated but still produces transcripts, or as we know now, also from a protein gene that has mutated and no longer can produce a protein but continues to produce transcripts; pseudogenes have a sequence that is altered from the parent gene and thus does not produce a protein product or have the function of the RNA from the parent RNA gene (Zheng et al. 2007; Tutar 2012). Thousands of pseudogenes are known that stem from both protein genes and RNA genes.

There are several categories of pseudogenes:

1. Non-processed pseudogenes are gene loci that represent a copy of a protein gene or an RNA gene that has sustained mutational changes. Some also contain additional sequences that originated from other genomic regions. Non-processed pseudogenes that are copies of protein genes contain mutations that render the gene incapable of forming a protein. Non-processed pseudogenes that originate from a copy of an RNA gene have different functions due to mutations and can contain additional sequences from other genomic regions.

2. Processed pseudogenes are generated by retrotransposition into the genome of a processed messenger RNA from a protein gene that is reverse transcribed into DNA. Processed pseudogenes do not have introns but have a poly A sequence at the 3′ end.
3. Unitary pseudogenes are formed by mutations in a protein gene that can render it inactive to produce a viable protein due to disabling mutations, which results in an inactive protein gene. It remains as a unitary pseudogene in the same chromosomal location, has exons and introns, but produces only a transcript (Zhang et al. 2010).

Of the thousands of pseudogenes known, determination of the functions of their transcripts is daunting, however several have been investigated (Tutar 2012). There has been extensive interest and work on the non-processed pseudogene *PTENP1*, which has a very high identity with its parent gene, phosphatase and tensin homolog (*PTEN*), a tumor suppressor gene. Pseudogene *PTENP1* can regulate levels of PTEN (Poliseno et al. 2010), possibly via the pathway involving the binding (sponging) of miRNAs that may suppress PTEN mRNA (Zhang et al. 2017).

There is little information on functional roles of processed pseudogenes, but there are examples of insertion of a DNA sequence of mRNAs into the genome by retrotransposition that creates a different phenotype. This was found with various breeds of domestic dogs that have short legs. Retrotransposition of the fibroblast growth factor 4 (*fgf4*) mRNA sequence is associated with the phenotype produced by a heritable skeletal disorder of the legs (Brown et al. 2017). With human cancers, cancer cell genomes were found to have processed pseudogene insertions that are not present in their normal cell counterparts (Ewing et al. 2013).

circRNAs are generated from linear RNAs via covalent linkage of 3′ and 5′ ends of the linear RNAs that are present in cells, but they can also arise from back-splicing and the circularization of exons during RNA splicing (Lasda and Parker 2014). Structurally, they represent covalently closed circles; in addition, they have no poly-A tail and are resistant to exoribonucleases, thus making them likely to be more stable than linear RNAs. Functionally, they can bind (sponge) microRNAs and regulate at the post-transcriptional level, or bind other molecules such as transcriptional regulators thereby regulating transcription (Lasda and Parker 2014). Other functions have also been described (Yu and Kuo 2019), but there is a paucity of knowledge of possible broad functions of thousands of circRNAs.

In this chapter, we document and analyze the lincRNA genes and pseudogenes found in chr22 LCR22s. We also present published data on linear RNA and circular RNA (circRNA) transcripts and their levels of expression from lincRNA genes and pseudogenes, as well as from protein genes during human development; these transcripts may have a bearing on gene functions during fetal developmental.

A gene map and list of genes in the chromosomal 22q11.2 region was previously formed (Guna et al. 2015). However current progress in ncRNA genomic annotations reveals a much larger number of ncRNA genes that are present in segmental duplications, and these are included here. We find that a sequence expansion of human chr22 with LCR22A and D appears to have involved the development of new genes that are primarily ncRNA genes. These add to the list of total genes that are lost in a 22q11.2 deletion, which occurs during fetal formation and results in genetic diseases such as DiGeorge Syndrome (McDonald-McGinn et al. 2015).

2 Analysis of LCR22 Segmental Duplications

2.1 LCR22A

To define the sequences and genes that boarder the LCR22s and to ascertain their possible origins, guidepost gene locations in human chr22 analogous to similar locations in chr22 of chimpanzee were ascertained. Several genes which show evolutionarily conservation in vertebrates were found to form reliable guideposts that help confirm the terminal ends of the human LCR22s (e.g., *TUBA8*, *USP18*, *DGCR6*, and *DGCR2* in the case of LCR22A). These genes are present in both chimpanzee and human genomes and they provide a view of both the LCR22A segmental duplication boundaries and the expansion of the human sequence (Fig. 2). For example, the evolutionarily conserved *TUBA8* gene sequence at the chromosomal centromere side is unique to and found only in one copy in chr22, whereas parts of the *USP18* sequence are found duplicated several times in chr22 within LCR22A and LCR22D. Thus we consider *USP18* part of LCR22A and *TUBA8* as the first gene outside of the segmental duplication (Fig. 2b). At the telomere side, we propose that LCR22A ends with carbonic anhydrase 15 pseudogene *CA15P1* and before *DGCR2* (Fig. 2b). The *CA15P1* sequence is found duplicated in chr22; the *DGCR2* sequence is not, thus the *CA15P1* gene appears to form the endpoint of LCR22A at the telomere side. The total length of the human LCR22A segmental

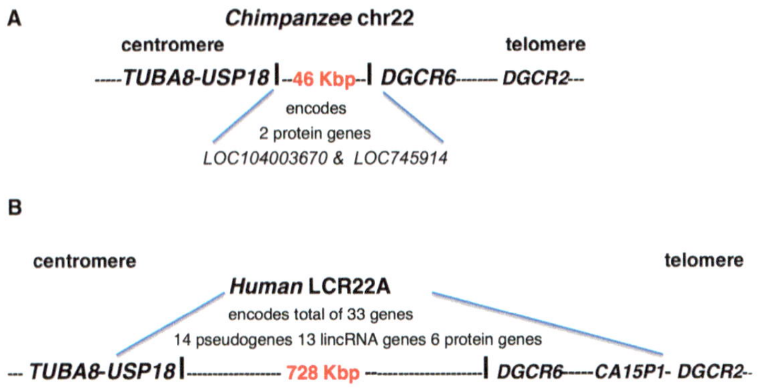

Fig. 2 (**a**). Schematic of the chimpanzee region of chr22 with guidepost genes: *TUBA8, USP18, DGCR6, and DGCR2*. The 46 Kbp sequence shown in chimpanzee chr22 are also present in human LCR22A and linked to human *USP18* (not shown). (**b**). Human LCR22A region in chr22 that shows the DNA sequence expansion region (728 Kbp) situated between *USP18* and *DGCR6*. The human LCR22A end boundaries, genes *USP18* and *CA15P1* are also shown

duplication, which includes the *CA15P1* gene, is ~885 Kbp, which is the DNA length from the 5′ end of the *USP18* gene to the 3′ end of the *CA15P1* gene.

The 46 Kbp sequence that is present in chimpanzee, encodes two predicted proteins, and is found linked to the *USP18* gene but is also present and linked to *USP18* in humans; however in humans the sequence encodes the lincRNA gene *FAM230D*. This is one of several examples of an ancestral sequence that can evolve into a protein or RNA gene when duplicated (Delihas 2020).

The length between *USP18* and *DGCR6* is 46 Kbp in chimpanzee and 728 Kbp in humans (Fig. 2). This gives an idea of the expansion of the DNA sequence in the human chromosomal region where LCR22A resides compared with the analogous region in chimpanzee. However there is no significant sequence expansion between *DGCR6* and *DGCR2* (which is part of the LCR22A) in chimpanzee and humans, that is, the chimpanzee sequence between *DGCR6* and *DGCR2* is 117 Kbp and in human the sequence is 122 Kbp with a 98% identity between the two sequences. Thus, the LCR22A length is greater than the actual expansion of sequence.

The similarities of the following genes and their relative positions between the chimpanzee and human chr22 genomes:

TUBA8–USP18 |-46 Kbp conserved sequence–variable sequence expansion-|-*DGCR6–DGCR2* (Fig. 2) suggests that this gene arrangement may have originated in a common primate ancestor and serves as a proto-LCR22 forming unit. It perhaps represents the starting point for sequence expansion by segmental duplication and contains the proper recognition sites, for example, *Alu* transposable elements that take part in chromosomal rearrangements (Babcock et al. 2003). As the chimpanzee region has not expanded in sequence between *USP18* and *DGCR6* (Fig. 2), there may be a lower probability of chromosomal rearrangements at this site in the chimpanzee chr22. It should be noted that the gene arrangement: *TUBA8-USP18* |-2 genes ~46 Kbp -|-*DGCR6* without a sequence expansion is also present in other primates: *Macaca mulatta* (rhesus monkey), *Pongo abelii* (Sumatran orangutan)], *Macaca nemestrina* (pig-tailed macaque).

2.2 LCR22D

LCR22D is the second largest segmental duplication in the human 22q11.2 region. A comparison with the analogous region in chimpanzee chr22 indicates that the region is complex and difficult to analyze. Figure 3 shows a simplified framework of vertebrate-conserved guidepost genes used, *LRRC74B* and *RIMBP3C–UBE2L3–YDJC* highlight the distance between guideposts in the chimpanzee and human regions, which are 1.2 Mbp in the chimpanzee genome and 502 Kbp in humans (Fig. 3). The centromere LCR22D boundary genes (not shown in Fig. 3) are discussed below.

A more detailed view of guidepost genes that define the LCR22D boundary ends is in Fig. 4. In humans, there are two genes between *LRRC74B* and the end border of LCR22D on the centromere side, one is *LOC112268300*, which is only partially

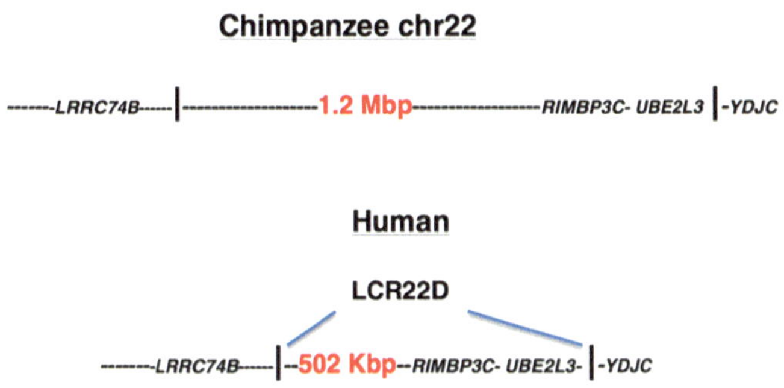

Fig. 3 Schematic representation of key guidepost genes used to compare the chimpanzee and human chromosomes in the region of LCR22D

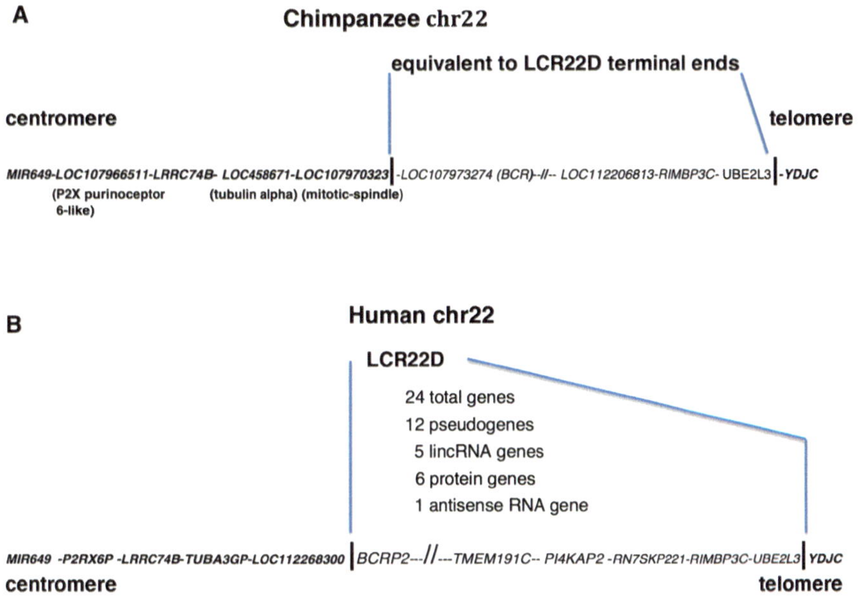

Fig. 4 Schematic drawing of representative guidepost genes positioned outside of the LCR22D and genes that are within the human LCR22D segmental duplication. (**a**). Gene arrangements in the chimpanzee genome. (**b**). Gene arrangements outside of and within LC22D boundaries. A list of gene types that are within human LCR22D is also shown

annotated and termed an uncharacterized pseudogene with a sequence similar to 1 Protein. The *LOC112268300* sequence is not found repeated in chr22 and thus appears to be outside of LCR22D. This helps define the LCR22D end at the centromere side. The analogous gene in chimpanzee, *LOC107970323* (Fig. 4a) is annotated as a mitotic-spindle organizing protein 2B-like pseudogene but it has a nt sequence highly similar (98% identity) to that of human *LOC112268300* over the entire length of ~21 kbp. The other gene, human *TUBA3GP*, tubulin alpha 3 g pseudogene is related to chimpanzee *LOC458671,* the tubulin alpha-3 chain-like pseudogene. Thus, the five genes that are on the centromere side of LCR22D (Fig. 4), which are used as guideposts between humans and chimpanzee, appear to be related and have very similar sequences. On the telomere side, *RIMBP3C, UBE2L3,* and *YDJC* are orthologous genes in humans and chimpanzee. DNA sequences from *RIMBP3C* and *UBE2L3* are also found repeated in segmental duplication LCR22A, therefore we consider these genes to be within the LCR22D boundary. *YDJC*, termed chitooligosaccharide deacetylase homolog is found only in one copy in chr 22 and is thus considered the first gene outside of LCR22D on the telomere side.

The large 1.2 Mbp sequence that separates the centromere and telomere guidepost genes in the chimpanzee genome (Fig. 3a) indicates the complexity in this chromosomal region and that major rearrangements may have occurred in the chimpanzee. For example, in humans*TMEM191C* is situated within LCR22D, very close to human *RN7SKP221* (Fig. 4b). In chimpanzee, *TMEM191C* is in a very distal location and found ~3.7 Mbp removed from the chimpanzee *RN7SKP221* gene in chr22, thus suggesting a chromosomal rearrangement. This may partly explain the large difference between the chromosomal locations of guidepost genes, on the centromere and telomere sides, between the human and chimpanzee genomes . We do not know if the 1.2 Mbp sequence that separates the centromere and telomere guidepost genes in the chimpanzee genome is part of a chimpanzee specific sequence expansion or not.

The apparent genomic rearrangement hampers a further analysis of sequence expansion and LCR22D, which was more straightforward with LCR22A. On the other hand, part of the human sequence that is within the LCR22D on the centromere side, a 28 Kbp region that includes the *BCRP2* gene (Fig. 4b), has a high identity with the equivalent region of the in chimpanzee where *LOC107973274 (BCR)* resides (Fig. 4a). Thus, this sequence forms a *BCR* protein gene in the chimpanzee but the related pseudogene sequence, *BCRP2* in humans. The entire chimpanzee *BCR* sequence is present in the human *BCRP2* sequence with a 98% identity. On the telomere side, the annotated genes, chimpanzee *LOC112206813* and human *RN7SKP221* are different, but ~55 Kbp sequence that resides to the left of *RIMBC3*, which is within LCR22D boundary and includes the *RN7SKP221* gene sequence (Fig. 4b), is highly similar with a 98% identity to the sequence to the left of *RIMBC3* in chimpanzee (Fig. 4a). Therefore the sequences at both the centromere and telomere ends that are within the LCR22D boundaries are also present in the chimpanzee genome next to guidepost genes. As these sequences within the human LCR22D are highly similar to the chimpanzee chr22 sequences, we propose that the structural element containing guidepost genes and the short sequences within the LCR22D were in place in a common primate ancestor to humans and chimpanzee, as a proto-LCR22D forming unit. This is analogous to the proposed proto-LCR22A.

3 Genes Present in LCR22A and LCR22D

LCR22A contains 33 annotated genes (Fig. 2; Table 1). The majority are ncRNA genes with nearly an equal number of lincRNA genes and pseudogenes; there are also 6 protein genes. Table 2 and Fig. 4 show that LCR22D carries a total of 24 genes: 12 pseudogenes, 5 lincRNA genes, 1 antisense RNA gene, and 6 protein genes. As mentioned previously, pseudogenes are defined as processed, unprocessed, and unitary pseudogenes (Vanin 1985; Zheng et al. 2007; Zhang et al. 2010). More than half of the pseudogenes in LCR22A are unprocessed pseudogenes. RNA genes in Tables 1 and 2 that are annotated starting with *LOC* are only partially characterized and are predicted genes; however, known RNA transcripts stem from these gene loci. The data indicate that the LCR22 segmental duplications primarily form new ncRNA genes.

Tables 1 and 2 also show the RNA transcripts expressed in different tissues by these genes. Transcripts produced by RNA and protein genes consist of both linear RNAs and the circular RNAs (circRNAs). circRNAs form from linear RNAs and arise via the covalent joining of the 5′ and 3′ ends of the RNA strands to form a closed loop. Thousands of circRNAs have been identified that are produced by protein and RNA genes in human cells, and these RNAs display functions in cellular and molecular processes, as well show an involvement in disease processes such as cancer (Guo et al. 2014; Lasda and Parker 2014; Qu et al. 2015).

Data that specify the expression levels of circRNAs in different tissues during fetal development from both RNA and protein genes in LCR22A and LCR22D are available from the Salzman and co-workers (Szabo et al. 2015) and are displayed on the NCBI website (https://www.ncbi.nlm.nih.gov/gene). In addition, expression of linear RNAs in developing brain tissues from LCR22 genes are available from Atlas (https://www.ebi.ac.uk/gxa/licence.html]) (Petryszak et al. (2016) and Ensembi (http://useast.ensembl.org/Homo_sapiens/Info/Index). The data presented in Tables 1 and 2 were assembled from these websites and show RNA expression from genes in six developing human fetal tissues for circRNAs, marked with one asterisk (∗). The expressed linear RNAs are marked with two asterisks (∗∗), where RNA expression shown is only from developing brain tissues and the spinal cord.

3.1 Analysis of ncRNA Transcript Expression During Human Development and Somatic Tissues

Knowledge of ncRNA genes, RNA transcript, and functions from ncRNA genes in LCR22 segmental duplications is important to better understand how human developmental abnormalities occur with a deletion of the 22q11.2 chromosomal region. Unfortunately data on lincRNA and pseudogene functions during fetal development are currently in their infancy; however, ncRNA gene transcript expression values, which are available, may pinpoint genes that have specific functions during fetal

Table 1 Genes present in LCR22A

Gene	Type	Expression in fetal tissues[a, b]
USP18	ubiquitin specific peptidase 18	All tissues, major in lung, 17 weeks∗; major, multiple sections of developing brain at different stages of development, spinal cord∗∗
FAM230D	lincRNA	Adrenal, intestine, lung, stomach∗; developing brain 10 weeks, choroid plexus∗∗
LOC105372858	lincRNA	Selective expression in fetal heart 10 weeks, minor in kidney∗; not determined∗∗
GGTLC5P	gamma-glutamyltransferase light chain 5 pseudogene	Not determined∗; developing brain 10 weeks, choroid plexus∗∗
LOC105372942	lincRNA	All tissues, major in lung, 17 weeks∗; not determined∗∗
FAM230J	lincRNA	Adrenal, intestine, lung, stomach∗; major in developing brain 10 weeks, choroid plexus, minor spinal cord Carnegie Stage 16∗∗
PPP1R26P3	protein phosphatase 1 regulatory subunit 26 pseudogene 3	Not determined∗; developing brain choroid plexus 10 weeks∗∗
LOC107987325	lincRNA	Major heart 10 weeks, minor kidney, 16 weeks∗; not determined∗∗
FAM230A	NMD RNA Nonsense-mediated decay transcript	Stomach 10 weeks, lung 20 weeks, intestine 10 weeks, minor in heart 11 weeks∗; developing brain, choroid plexus, 10 weeks∗∗
GGTLC3	gamma-glutamyltransferase light chain family member 3	Not determined∗; developing brain, choroid plexus 10 weeks∗∗
TMEM191B	transmembrane protein 191B	All tissues, major in adrenal, intestine∗; developing brain in multiple anatomical regions∗∗
PI4KAP1	phosphatidylinositol 4-kinase alpha pseudogene 1	All tissues, major in adrenal, 10 weeks, stomach, 20 weeks, intestine, 15 weeks∗; Major in developing brain in multiple anatomical regions, spinal cord∗∗
RN7SKP131	RN7SK pseudogene 131	Not determined∗; developing brain, choroid plexus, 10 weeks∗∗
RIMBP3	RIMS binding protein 3	Not determined∗; developing brain, basal ganglion, 12 weeks, other stages and regions of brain development, spinal cord∗∗
SUSD2P2	sushi domain containing 2 pseudogene 2	Not determined∗; developing brain, choroid plexus, 10 weeks∗∗
LOC105372871	lincRNA	Kidney, 10 weeks, intestine, 11 weeks, adrenal, 18 weeks, stomach, 16 weeks, heart, 18 weeks∗; not determined∗∗

(continued)

Table 1 (continued)

Gene	Type	Expression in fetal tissues[a, b]
CA15P2	CA15 pseudogene 2	Not determined*; developing brain, choroid plexus, 10 weeks*
PPP1R26P2	protein phosphatase 1 regulatory subunit 26 pseudogene 2	Not determined*; developing brain, choroid plexus, 10 weeks**
PPP1R26P4	protein phosphatase 1 regulatory subunit 26 pseudogene 4	Not determined*; developing brain, choroid plexus, 10 weeks**
FAM230E	lincRNA	Adrenal, 10 weeks heart, 10 weeks, intestine,10 weeks, lung 17 weeks, stomach, 10 weeks*; developing brain, choroid plexus, 10 weeks, spinal cord, Carnegie Stage 16**
LOC105377182	lincRNA	All tissues, major in lung, 17 weeks*; not determined**
GGT3P	gamma-glutamyltransferase 3 pseudogene	Major in intestine, 15 weeks, lung 17 weeks minor in other tissues*; developing brain in multiple anatomical developing brain in multiple anatomical regions, spinal cord**
E2F6P1	E2F transcription factor 6 pseudogene 1	Not determined*; developing brain, choroid plexus, 10 weeks**
LOC101929738	putative POM121-like protein 1, pseudogene	Not determined*; developing brain choroid plexus, 10 weeks**
LOC102725072	Putative uncharacterized protein DKFZp434K191 pseudogene	Adrenal 18 weeks, stomach 18 weeks, heart 10 weeks, lung 11 weeks, kidney 10 weeks, intestine 11 weeks*; not determined**
BCRP7	BCR pseudogene 7	Not determined* developing brain in multiple anatomical regions, spinal cord**
FAM230F	lincRNA	Stomach 10 weeks, lung 17 weeks, intestine 17 weeks, heart 11 weeks*; developing brain, choroid plexus, 10 weeks**
DGCR6	DiGeorge syndrome critical region protein gene 6	All tissues, major in adrenal, 16 weeks, heart 11–18 weeks, low in kidney*; developing brain in multiple anatomical regions, spinal cord**
PRODH	proline dehydrogenase 1	Major in intestine, 15 weeks, lung 17 weeks minor in other tissues*; major in different stages developing brain and spinal cord**
DGCR5	DiGeorge syndrome critical region gene 5 lincRNA,	All tissues, major in lung, 17 weeks, kidney 10 weeks, intestine 11 weeks, adrenal, 16 weeks*; major in different stages of developing brain and spinal cord**
DGCR9	DiGeorge syndrome critical region gene 9 lncRNA	Not determined*; major in different stages developing brain**

(continued)

Table 1 (continued)

Gene	Type	Expression in fetal tissues[a, b]
DGCR10	DiGeorge syndrome critical region gene 10 lncRNA	Not determined∗; major in different stages developing brain∗∗
CA15P1	CA15 pseudogene 1	Not determined∗; minor in different stages developing brain∗∗

[a]circRNA expression (∗) in six different tissues during fetal development. Data from: www.ncbi.nlm.nih.gov/gene/ (Szabo et al. 2015)
[b]linear RNA expression data (∗∗) from Ensembl and Expression Atlas [http://useast.ensembl.org/, (Petryszak et al. 2016)] representing linear RNA transcript expression from different stages of developing brain tissues

development. For example, one of the LCR22 genes that stands out due to its specific expression in fetal tissue is the lincRNA gene *LOC105372858* (Table 1). This gene displays circRNA expression specifically in heart tissue during fetal development at 10 weeks (Table 1; Fig. 5), albeit it displays a weak signal, RPKM = 0.006 (Fig. 5). There is also a very minor circRNA expression in kidney at 16 weeks' developmental stage. The *LOC105372858* sequence consists entirely of a segment of the 5′ half of lincRNA *FAM230C* gene and displays a high identity with this segment. The *LOC105372858* sequence appears to be a part of a common conserved remnant sequence found upstream of the *FAM230* gene family genes (Delihas 2018), which in this case, developed into the ncRNA gene *LOC105372858*. The linear RNA transcript from this gene has two exons. Functions of *LOC105372858* are unknown, but the circRNA stemming from transcription of this gene may be of special interest as it is almost selectively expressed in developing heart tissue (with minor expression in kidney), and abnormal heart development is a consequence of a 22q11.2 deletion (McDonald-McGinn et al. 2015). Another lincRNA gene, *LOC107987325* found in LCR22A (Table 1), also shows circRNA expression primarily in heart tissue at developmental stage 10 weeks, RPKM = 0.034 but it is also has a significant expression in kidney, RPKM = 0.011 at 16 weeks.

The levels of circRNA expression from RNA genes in LCR22A and LCR22D, for the most part, are weak compared to that of protein genes (determined from NCBI, https://www.ncbi.nlm.nih.gov/genome); Salzman and co-workers (Szabo et al. 2015). Genes such as *PRODH, DGCR6,* and *UBR2L3,* which are phylogenetically conserved protein genes and found in many vertebrates, show ~10 times or more the levels of circRNA expression than lincRNA genes. For example, lincRNA *FAM230D* shows a circRNA expression of RPKM = 0.033 in adrenal tissues; *UBE2L3*, the protein gene present in most vertebrates shows a value of RPKM = 10.2 in adrenal tissue. On the other hand, several pseudogenes are highly expressed; pseudogene *PI4KAP2* circRNA is expressed in developing adrenal tissues at RPKM = 10.3. There is also a high expression of linear RNAs in midbrain developing tissue from this pseudogene, thus making the function of *PI4KAP2* (a phosphatidylinositol 4-kinase alpha pseudogene) of special interest. Another major difference in expression of circRNAs is with those circRNAs stemming from protein genes; they are expressed ubiquitously with expression found in most tissues. This is not the case with many lincRNA genes shown in Tables 1 and 2, such

Table 2 Genes in LCR22D

Gene	Type	Expression in fetal tissues[a, b]
BCRP2	BCR pseudogene 2 unprocessed pseudogene	Significant all tissues, major in adrenal, heart, intestine*; major, multiple sections of developing brain at different stages of development**
POM121L7P	POM121 transmembrane nucleoporin like 7 pseudogene processed pseudogene	Not determined*; developing brain, only in choroid plexus**
LOC112268299	lincRNA	Not determined*; not determined**
E2F6P2	E2F transcription factor 6 pseudogene 2 processed pseudogene	Not determined*; developing brain, only in choroid plexus**
FAM230B	lincRNA	Minor, intestine, adrenal, stomach, lung* developing brain, only in choroid plexus**
LOC105372935	lincRNA	Significant all tissues, major in lung 17 weeks*; not determined**
GGT2	gamma-glutamyltransferase 2	Major, all tissues, major in intestine, lung, heart*; developing brain, only in forebrain, basal ganglion, cerebellum; spinal cord**
E2F6P3	E2F transcription factor 6 pseudogene 3 processed pseudogene	Not determined* developing brain, only in choroid plexus**
POM121L8P	POM121 transmembrane nucleoporin like 8, pseudogene rocessed pseudogene	All tissues, major in adrenal, 18 weeks, lung 10 weeks, stomach 18 weeks*; minor developing brain, only in choroid plexus**
BCRP6	BCR pseudogene 6 unprocessed pseudogene	Not determined*; developing brain, only in choroid plexus**
FAM230H	lincRNA	Strongest in stomach, intestine, weaker in adrenal, heart, lung*; developing brain, only in choroid plexus**
PPP1R26P5	protein phosphatase 1 regulatory subunit 26 pseudogene 5 processed pseudogene	Not determined*; developing brain, only in choroid plexus
CA15P3	CA15 pseudogene 3	Not determined* Not determined**
LOC105377190	lincRNA	Minor in kidney, 10 weeks, intestine 11 weeks, lesser in stomach, adrenal, heart*; not determined**
SUSD2P1	sushi domain containing 2 pseudogene 1	Not determined*; not determined**
RIMBP3B	RIMS binding protein 3B	Not determined*; developing brain, only in choroid plexus**
RN7SKP63	RN7SK pseudogene 63 uncharacterized pseudogene	Not determined*; developing brain, only in choroid plexus**

(continued)

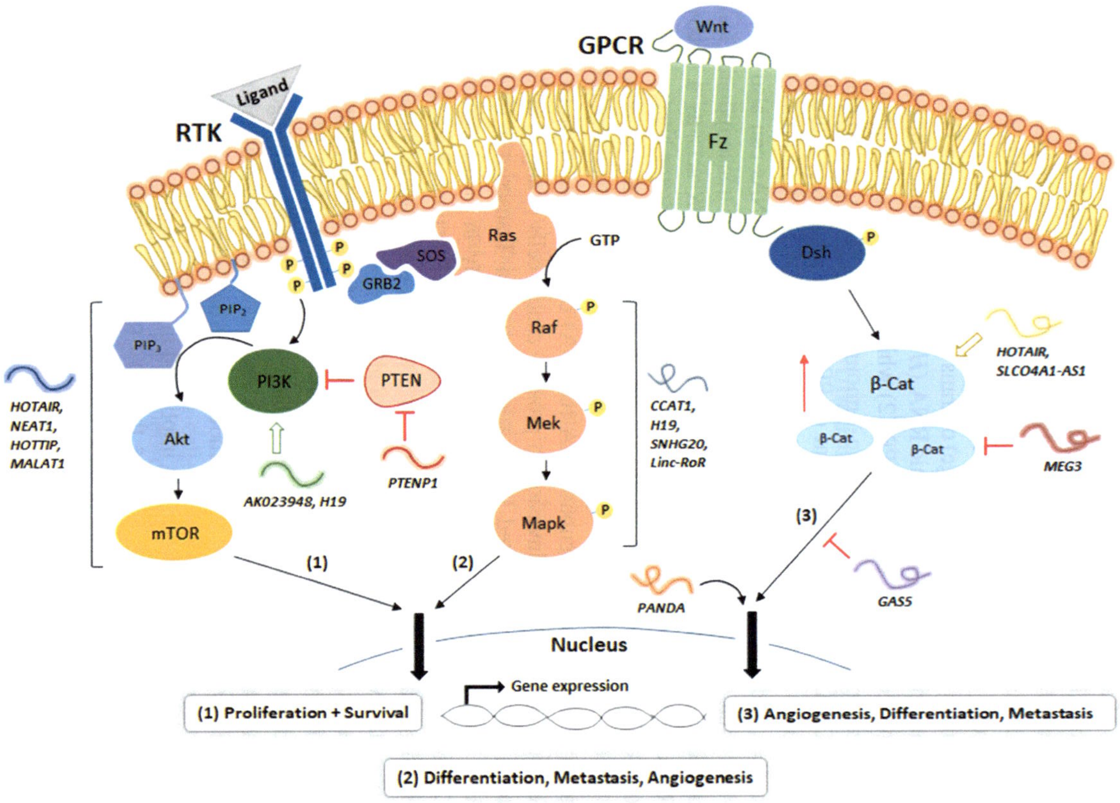

Fig. 1 *Long non-coding RNAs promote hallmarks of cancer through dysregulation of intrinsic cellular signalling pathways*. Here examples of lncRNAs which affect oncogenesis are shown. This list is non-exhaustive and highlights how oncogenic and tumour suppressive lncRNAs target essential signalling cascades to modulate

cancer programmes. The PI3K/Akt/mTOR pathway (1) primarily upregulates proliferation and survival. The lncRNAs *HOTAIR*, *NEAT1*, *HOTTIP*, and *MALAT1* have unknown targets within the PI3K/Akt cascade. *AK023948* and *H19* either target PI3K subunit p85 or act on the protein complex directly. *PTENP1* inhibits the repressor of PI3K (PTEN) directly. In the MAPK/ERK pathway, the result is differentiation, metastasis, and angiogenesis. *PANDA* directly targets the p53 response element in the nucleus to modulate downstream MAPK/ERK effects, whereas *CCAT1*, *H19*, and *Linc-RoR* have diverse targets which act on central players within the cytoplasmic portion of the cascade. Lastly, lncRNA-driven perturbations to the Wnt/β-catenin signalling pathway induce angiogenesis and stem cell differentiation which are precursors to metastatic dissemination. Here, *HOTAIR* increases β-catenin signalling to drive cancer progression versus *GAS5* and *MEG3* which actively inhibit metastasis and angiogenesis. *SLCO4A1-AS1* stabilizes β-catenin by blocking inhibitory phosphorylation. Therefore, numerous lncRNAs act on different regions of critical signalling networks which underlie cancer progression and tumourigenesis. Highlighting a few examples here provides evidence to support the role of the non-coding transcriptome in cancer

feedback loop of uninhibited cellular proliferation. Another lncRNA, Plasmacytoma Variant Translocation 1 (*PVT1*), is co-amplified with Myc in a myriad of cancers and prevents proteolytic degradation of Myc, leading to unrepressed cellular growth (Schmitt and Chang 2016; Iaccarino 2017).

2. *Metastasis*: Typically, epithelial cells are confined within the basement membrane, preventing movement between different tissues. Cancer cells can cross the basement membrane and disseminate to sites far from the original tumour. At the chromatin level, *HOTAIR*-mediated remodelling leads to increased metastatic potential (Schmitt and Chang 2016). Ectopic overexpression of *HOTAIR* in breast cells induces PRC2 activity and leads to genome-wide reprogramming of chromatin states that enable cell motility and matrix invasion (Schmitt and Chang 2016). *MALAT1* is a regularly occurring lncRNA implicated in metastatic cancers. However, its identity as an oncogenic lncRNA is controversial (Eastlack et al. 2018; Toraih et al. 2018). Some studies have shown that *MALAT1* can suppress metastasis in breast cancer (Kim et al. 2018) while others show *MALAT1* promotes EMT. Specifically, in oesophageal and bladder cancer cell lines, *MALAT1* triggers EMT through Notch1 signalling and Wnt signalling respectively (Ying et al. 2012; Chen et al. 2018). However, *MALAT1* knockout in breast cancer cell lines and *MALAT1* inactivation in mice models promote metastasis, while *MALAT1* rescue suppresses metastasis (Kim et al. 2018). These results show that some lncRNAs can have opposite functions in different tissues and malignancies.
3. *Controlled cellular death*: Apoptosis modulates cellular growth and causes cells to undergo programmed death in physiological conditions (e.g. during development). Several lncRNAs have been implicated in this process. For example, P21-associated non-coding RNA DNA damage-activated (*PANDA*) is a lncRNA that may deregulate p53-mediated apoptosis. In the right biological context, *PANDA* is important for growth and development, but its overexpression can lead to oncogenesis. It has been reported that *PANDA* acts as a decoy to inhibit apoptosis by inactivating the transcription factor NF-YA to prevent expression of apoptotic genes in osteosarcoma (Kotake et al. 2017). On the other hand, in diffuse large B-cell lymphoma, *PANDA* suppresses cell proliferation and apoptosis by inactivating the MAP/ERK signalling pathway (Wang et al. 2017), thus serving as a tumour suppressor. Additionally, overexpression of *PANDA* has been found to deregulate cellular senescence and promotes HCC through suppression of IL-8 (Peng et al. 2017) and interactions with PRC and scaffold-attachment-factor A (*SAFA*) (Puvvula et al. 2014).
4. *Angiogenesis*: Under normal physiological conditions, angiogenesis is necessary for growth, development, and wound healing. Solid malignancies rapidly consume surrounding nutrients, triggering localized hypoxia, as well as the release of angiogenic factors that stimulate the production of new blood vessels. As a result, these neoplasms create opportunity for greater access to nutrients and create an access route to distant organs. Unsurprisingly, lncRNAs have been found implicated in angiogenesis (Kumar and Goyal 2017; Tang and Hann 2018). One of the transcripts associated with this function is LncRNA associated with

microvascular invasion in HCC (lncRNA *MVIH*), which is upregulated in HCC and inhibits the secretion of phosphoglycerate kinase 1, an inhibitor of angiogenesis (Yuan et al. 2012). Overexpression of maternally expressed gene 3 (*MEG3*) lncRNA prevents breast epithelium transformation by downregulating the Akt pathway and subsequent vascular endothelial growth factor A, and thus inhibits angiogenesis (Al-Rugeebah et al. 2019). Interestingly, deregulation or loss of *MEG3* expression is common in many cancers (Al-Rugeebah et al. 2019). *HOTAIR* has also been shown to play a significant role in angiogenesis and tumour microenvironment (Tang and Hann 2018). Its influence on tumour vascular redistribution and other mechanisms of cancer promotion (hypoxia pathways, immune dysregulation, cancer stem cells) supports the notion that angiogenic changes are essential to sufficient tumour progression.

In conclusion, lncRNAs are implicated in many hallmarks of cancer, and can play both oncogenic and oncosuppressive functions. Table 1 shows several examples of lncRNAs implicated in cancer initiation and progression.

2 Therapies for Cancer Treatment

The most successful treatment for malignant neoplasms is surgical excision. Inoperable cancers can be treated by a variety of modalities, which include chemotherapy, radiotherapy, and biological therapies. Some of these treatments can also be administered before surgery to reduce the tumour burden (neo-adjuvant therapy), or after surgery to prevent cancer recurrence (adjuvant therapy). In this section, we will briefly outline the mechanism of action and clinical utility of all these therapies. This will be an essential step towards understanding the molecular mechanisms of cancer treatment resistance and the role of lncRNAs in these phenomena.

Chemotherapeutic agents trigger cancer cell death by inducing DNA damage directly (e.g. platinum agents) or by a variety of indirect mechanisms, which include inhibiting DNA and RNA synthesis (antimetabolites), altering microtubule dynamics (e.g. taxanes), and blocking the activity of enzymes required for correct DNA replication (e.g. topoisomerase inhibitors) (Mustacchi and Laurentiis 2015; Luengo et al. 2017; Wang and Tse-Ding 2019). All these mechanisms trigger cell death programmes in proliferating cells. Since cancer cells are characterised by persistent proliferation, conventional chemotherapy affects malignant tissues more than most normal tissues. Chemotherapy can be the only option available when the cancer is too widespread to use radiotherapy approaches, if there is no rational base for the use of biological therapies or when cancer cells become resistant to the other treatments. Chemotherapy can lead to relevant side effects since it targets also healthy proliferating cells. Common side effects of chemotherapy include myelosuppression and gastrointestinal toxicity (nausea, vomiting, diarrhoea) (Kayl and Meyers 2006).

Radiotherapy is one of the most common and effective strategies to treat cancer patients. This treatment modality delivers high-energy ionising radiations into the

et al. 2014). In the next section we will discuss the main mechanisms by which cancer cells resit to therapy.

3 Mechanisms of Cancer Cells' Resistance to Conventional Therapies

Unsuccessful cancer treatment can be caused by "intrinsic" or "acquired" resistance. Intrinsic resistance is caused by cancer cells which are not susceptible to the treatment from the beginning (e.g. some pancreatic cancers are intrinsically resistant to most chemotherapy regimens (Coppola et al. 2017)); acquired resistance is observed in malignancies that become treatment-resistant after an initial response to the therapy (Shachar et al. 2018). Intrinsic resistance can be ascribed to the multiclonal nature of cancer: At any given moment, malignancies are composed of different cellular subpopulations, each with a defined set of genetic and epigenetic alterations. It is therefore conceivable that one of these alterations can enable the survival of a pre-existing subpopulation, when the malignancy is exposed to a specific treatment. The molecular mechanisms underpinning acquired cancer resistance are associated with the microevolutionary nature of cancer. Treatments that eliminate most cancer cells can favour the proliferation of clones that have acquired an advantageous mutation (or an epigenetic alteration) in that specific environment, resulting in emergence of resistant clones. Mechanisms of cancer cell's resistance to therapy are treatment- and cancer type-specific; however, many traits are shared across different types of malignancies (Bhosle and Hall 2009). The main mechanisms of resistance include alterations in membrane-protein transporters (Fletcher et al. 2016), increased DNA Damage Repair (DDR) rate (Shah et al. 2016), recurrence of cancer stem-like cells (CSCs) (Zhao 2016), altered levels of reactive oxygen species (Galadari et al. 2017) and EMT (Du and Shim 2016), autophagy and inhibition of cell death (Qin et al. 2016). We will briefly discuss these mechanisms in the sections below.

3.1 Alterations in Membrane-Protein Transporters

The activity of membrane proteins that prevent the drug from accumulating in the cytosol is a main cause of drug resistance in cancer cells (Nikolaou et al. 2018). These proteins are often called multidrug-resistant (MDR) transporters as their actions can result in the development of resistance to multiple types of drugs. There are two major families of MDR-associated proteins: the adenosine triphosphate-binding cassette (ABC) transporter superfamily and the solute carrier transporters family. The latter mediates the uptake of anti-cancer drugs; therefore, it has been shown that the activity of these transporters is reduced in different types of cancers (Sherlach and Roepe 2014).

ABC proteins excrete potentially cytotoxic molecules, including most anticancer agents. There are around 50 different ABC transporters that have been linked to cancer drug resistance. However, some of the most commonly studied are P-glycoprotein (P-gp, aka MDR1 and ABCB1), MDR-associated protein (MRP1, aka ABCC1) and breast cancer resistance protein (BCRP) (Gottesman 2001). A meta-analysis covering 31 clinical studies concluded that MDR1-expressing breast tumours were three times more likely to fail response to chemotherapy than patients whose tumours were MDR1-negative (Trock et al. 1997). MRP1 levels are detectable in different tumours including neuroblastoma, non-small-cell lung cancer (NSCLC) or ovarian carcinoma (Haber et al. 2006). MRP1 has been proposed to be a predictive biomarker for disease-free survival in acute myeloid leukaemia (AML) in patients treated with double induction therapy (i.e. mitoxantrone, cytosine arabinoside, etoposide and m-amsacrine or daunorubicin and cytosine arabinoside, according to age stratification) (Schaich et al. 2005). Overexpression of BCRP in patients with AML, with normal karyotype, did not influence the achievement of remission but these patients showed an increased time of relapse and disease-free survival (Damiani et al. 2006). All these proteins pump out different hydrophobic compounds, which results in a wide and overlapping substrate specificity in transporting most of the conventional drugs used in the clinics (Sarkadi et al. 2006).

3.2 DNA Damage Repair

Response to drugs and radiotherapy is also determined by the efficiency of DDR mechanisms in cancer cells (Nikolaou et al. 2018). DDR is mediated by different protein complexes. Here we will discuss the main mechanisms of DDR implicated in therapy resistance: nucleotide excision repair (NER), base excision repair (BER), mismatch repair (MMR), and repair of double strand breaks (DSB) via homologous recombination (HR) and non-homologous end-joining (NHEJ).

NER is involved in repairing the damage of lesions caused by UV light and commonly used drugs such as platinum agents (Wang and Lippard 2005). Excision repair cross-complementation group 1 (ERCC1) is an important protein of the NER pathway, which is overexpressed upon platinum agent-based treatments in patients with NSCLC (Ceppi et al. 2006). ERCC1-deficient lung adenocarcinomas display enhanced cisplatin sensitivity (Jokic et al. 2016).

Another important pathway of DNA repair is BER, which targets specific DNA glycosylases within base sequences to identify and repair segments of damaged DNA. Numerous factors involved in this process have been implicated in multiple cancer types, including AP endonuclease 1 (APE1), DNA Polymerase β (Pol β) and X-ray cross complementing 1/Ligase IIIα (Wallace et al. 2012). Although combination therapies with molecules targeting APE1 and Pol β have been shown to potentiate cytotoxic effects in vivo, the clinical applicability of these therapies is limited by inefficient delivery methods and target-specificity.

MMR detects sequence abnormalities on newly synthesized DNA (insertions, deletions, mis-incorporated bases), and actively stabilizes and corrects mismatched bases. Genomic defects in MMR genes such as *MLH1* and *MSH2* have been implicated in cancer treatment response in numerous malignancies. In ovarian cancer, it has been demonstrated that MMR genes like *MLH1* are downregulated by platinum-based chemotherapies. This leads to an accumulation of unrepaired DNA without initiation of apoptosis and therefore to chemotherapy resistance. Other studies have shown that repeated cisplatin treatment in human colon carcinoma cell lines will select for resistance in the form of coordinated loss of p53 and MMR activity (Martin et al. 2008).

Lastly, DSBs are lethal derangements repaired via HR or NHEJ. HR occurs during S and G2 phases of the cell cycle when homologous sequences (sister chromatids) can be used as templates, whereas NHEJ utilizes overhanging strands of DNA to isolate and complete sequences breaks (Mao et al. 2008). Dysregulation of DSB repair mechanisms has been implicated in carcinogenesis, and new evidence suggests that cancer cells can self-induce DSBs to promote cancerogenesis and cancer cell stemness (Liu et al. 2017b). Thus, active clinical trials targeting DSB repair proteins like Poly(ADP-ribose) polymerase 1 are underway in *BRCA*-mutated ovarian cancer, HER2 negative breast cancer, NSCLC, and gastric cancer (ClinicalTrials.gov identifiers NCT01844986, NCT02032823, NCT01924533, and NCT01788332 respectively) (Srivastava and Raghavan 2015).

In conclusion, critical drivers within the DDR network may be used as promising targets for mitigating chemotherapy resistance in the future.

3.3 Cancer Stem Cells

There is strong evidence showing that some solid tumours and leukaemia contain cancer stem cells (CSCs), which are defined as the cancer cells with the ability to self-renew indefinitely and to differentiate into other cancer cell types, thereby giving rise to primary and metastatic cancers (Eun et al. 2017; Thomas and Majeti 2017). A growing body of evidence indicates that the number of type of CSCs in a tumour determines its propensity to invade other organs and to become resistant to a wide range of treatments. Several explanations have been proposed to explain the treatment resistance of CSCs. Since CSCs are less proliferative than "differentiated" cancer cells, they are intrinsically resistant to radiotherapy and chemotherapy (Crea et al. 2009; Al-Dimassi et al. 2014). In addition, CSCs express higher levels of drug efflux pumps and are surrounded by a protective microenvironment that reduces the efficacy of anticancer treatments (Bao et al. 2006; Carnero and Lleonart 2016). Finally, CSCs are extremely plastic, that is, they can rapidly change gene expression patterns and adapt to new challenges in their environment. This plasticity is often mediated by the activation of specific epigenetic repressors (e.g. EZH2) (Crea et al.

2011). Targeting these epigenetic pathways results in CSC elimination in vitro and in vivo.

Unfortunately, most of the current treatments fail to eliminate all the cancer cell subpopulations and leave CSCs intact. The surviving CSCs can therefore proliferate and lead to cancer recurrence and metastatic progression (Ghisolfi et al. 2012). Based on this evidence, it is not surprising that the treatment resistance of some malignancies is directly correlated with the proportion of CSCs in the tumour (Lopez-Ayllon et al. 2014). These observations have prompted us and others to propose new types of combination therapies for cancers; in each therapeutic regime, there should be at least one treatment (e.g. radiotherapy, chemotherapy) that targets the bulk of differentiated or semi-differentiated cancer cells, in combination with at least one drug that selectively targets CSCs (e.g. EZH2 inhibitors). The first type of treatment would reduce the tumour mass and produce an objective response, while the second would prevent the generation of metastasis and the development of drug-resistant clones.

3.4 *Oxidative Stress and EMT*

Treatment with radiotherapy and with some anticancer drugs leads to oxidative stress in cancer cells, a mechanisms that can promote cell death by damaging DNA proteins and other macromolecules (Jaiyesimi et al. 1995). Treatment of breast cancer cells with high doses of tamoxifen activates oxidative stress-dependent apoptosis independently of oestrogen receptor (ER). However, exposure to tamoxifen-induces oxidative stress and increases the expression of multi-drug resistance transporters and antioxidant genes in the recipient breast cancer cells. Therefore, the surviving cancer cells are prone to develop more aggressive phenotypes and respond poorly to further tamoxifen treatment (Bekele et al. 2016). Furthermore, altered production of reactive oxygen or nitrogen species has been shown to activate the epithelial-to-mesenchymal transition (EMT), which is another mechanism of therapy resistance (Bao et al. 2013). During EMT, cancer cells originating from epithelial tissues lose their polarity and become mesenchymal stem cells. In addition to reactive oxygen species, several other phenomena can activate EMT in cancer cells. The most common triggers of EMT in cancer cells include hypoxia, immune response, and signals produced by the tumour-associated microenvironment (e.g. TGF-b). Cancer cell EMT is often reversible and is mediated by epigenetic modulators (e.g. EZH2), which often cooperate with specific transcription factors (e.g. TWIST, ZEB, SNAIL). Cancer cells undergoing EMT develop CSC-like properties and metastasize (Lyakhovich and Lleonart 2016). Hence, it is not surprising that EMT has been widely linked to cancer progression and increased treatment resistance. However, the relationship between EMT and drug resistance is not completely elucidated. Proposed mechanisms of EMT-dependent drug resistance include epigenetic changes that favour the concurrent activation of mesenchymal and anti-apoptotic genes, as well as more efficient DNA damage repair in mesenchymal cells.

3.5 Impaired Autophagy and Apoptosis Inhibition

Autophagy, meaning "self-eating", is a cellular process that starts with the formation of an autophagosome, a double-membrane vesicle that incorporates cytoplasmic components, protein aggregates and organelles, which then are degraded by fusion with a lysosome (Lyakhovich and Lleonart 2016). Under normal physiological conditions, basal levels of autophagy are critically important for cellular homeostasis, tumour suppression, stem cell differentiation, and self-renewal (Lyakhovich and Lleonart 2016). Several studies have shown that the loss of various proteins involved in autophagy (e.g. Beclin 1, autophagy-related proteins) leads to impaired autophagy and subsequent oncogenesis (Lyakhovich and Lleonart 2016). On the other hand, autophagy also promotes carcinogenesis by enhancing stress tolerance and cell survival, especially under toxic stress conditions (Lyakhovich and Lleonart 2016).

Emerging evidence suggests that autophagy plays an important role in the plastic response of cancer cells to therapy. Under hypoxic conditions, increased levels of autophagy have been reported to improve CSCs' survival to therapy and in some cases mediate conversion of non-CSCs to a CSC phenotype (Zhu et al. 2013). CSCs can utilize autophagy to enhance EMT transition and confer chemo-resistance and immune-resistance. Abolition of autophagy using pharmacological agents (ATG5/7 inhibitors) resensitizes EC109/CDDP cells to therapy (Yu et al. 2014). However, treatment of colorectal cancer with oxaliplatin induces autophagy and favours colorectal CSCs proliferation, thereby increasing the survival of resistant cells. Therefore, altered autophagy pathways play a role in the development of cancer cell resistance as well as in the CSCs phenotype (Yang et al. 2015).

Another common mechanism of resistance to both chemotherapy and radiotherapy is inhibition of apoptosis (Al-Dimassi et al. 2014). Apoptosis is controlled cellular death, initiated in response to homeostatic and regulatory mechanisms that can be propagated through either the extrinsic or intrinsic pathway. The activation of these pathways results in cell death, a class of proteolytic enzymes called caspases. The extrinsic pathway involves cellular signalling through cell death receptors (e.g. TFNR1, FAS, TRAIL), leading to caspase activation; the intrinsic pathway involves proapoptotic factors that permeabilize the outer mitochondrial membrane, leading to release of mitochondrial apoptogenic proteins that promote caspase activation or caspase-independent cell death. Incidentally,

Considering this evidence, it is not surprising that apoptotic pathways can be modulated by cancer cells in different ways, to achieve treatment resistance. For example, exposure of cancer cells to radiotherapy can increase the levels of the protein Survivin, a member of the inhibitors of apoptosis (IAP) family, which inhibits both caspase-3 and -7 to block apoptosis (Ambrosini et al. 1997). Furthermore, NSCL cancer cells treated with cisplatin have been found to up-regulate Annexin A2. Silencing of annexin A2 resensitizes NSCL cancer cells to cisplatin treatment by activating p53-dependent cell apoptosis (Feng et al. 2017). Thus, repressed apoptosis is a mechanism of therapy resistance and enhancing apoptotic signals can improve treatment success.

3.6 *Intrinsic and Acquired Alterations That Cause Resistance to Biological Therapies*

In addition to the mechanisms described in the precedent paragraphs, genetic and epigenetic alterations at specific *loci* have been implicated in the acquisition of resistance to targeted biological therapies. For example, patients with NSCLC are often treated with biological therapy, based on their "personalized" screening for mutations or epigenetic alteration of the target gene/protein. The epidermal growth factor receptor (EGFR) is the most common target of this type and EGFR inhibitors are widely used in the clinics. However, all NSCLC patients exposed to these targeted treatments will develop resistance to the drug at some point (Asao et al. 2019). The mechanisms by which NSCLC cells acquire resistance to EGFR inhibitors include (1) intrinsic activating mutations in the *EGFR locus*, which are found in 10–15% of NSCLC patients in Western countries (Manuscript 2014); these patients never respond to the biological therapy; (2) secondary EGFR alterations, which include acquired mutations and epigenetic alterations selected by the treatment itself (e.g. T790M mutation) (Pao et al. 2005); and (3) activation of alternative pathways to compensate for the loss of EGFR function (e.g. Ubiquitin pathway activation) (Takezawa et al. 2012). The last mechanism can be ascribed to both genetic mutations and epigenetic alterations.

4 Mechanisms of LncRNA-Mediated Drug Resistance

As mentioned in the previous section, increasing improvements in cancer treatment and the development of new therapeutic approaches have been in some cases counteracted by treatment failures and by reduced patients' response rates, which are caused by intrinsic or (more frequently) acquired treatment resistance. So far, we have introduced the overall importance of lncRNAs in cancer and illustrated the main mechanisms of cancer cells' resistance to treatments. Here we will discuss the emerging role of lncRNAs in orchestrating cancer cell resistance to therapy.

4.1 *The Role of LncRNAs in Chemotherapy Resistance*

Emerging evidence indicates that numerous lncRNAs are implicated in chemotherapy resistance. To exemplify the main interactions between chemotherapeutics and lncRNAs, we will discuss in depth some chemotherapeutics and the role of one specific transcript, which has been implicated in the response to the described types of drugs. For this purpose, we will consider *H19* since it is one of the most studied lncRNAs and since it has been shown to act via both nuclear and cytoplasmic mechanisms (Fig. 2).

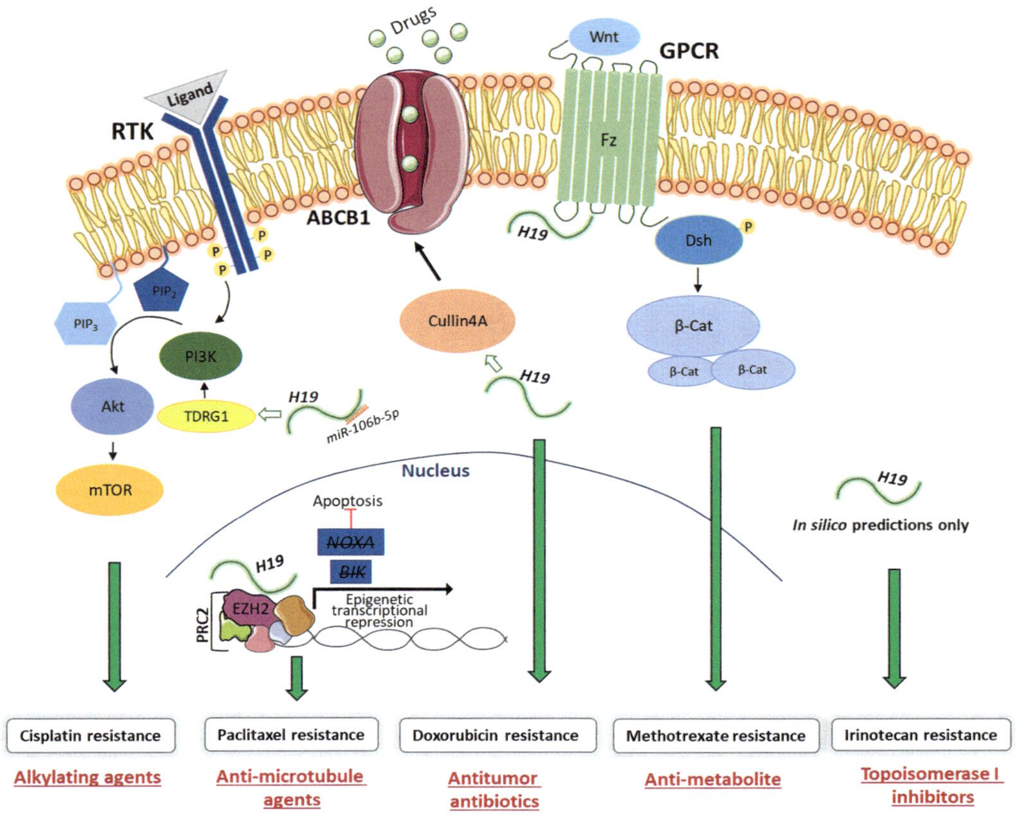

Fig. 2 *LncRNAs play key roles in chemoresistance*. Long non-coding RNAs regulate chemoresistance, mostly by activating drug resistance pathways. In this figure are represented some mechanisms that drive chemoresistance to different classes of chemotherapeutics (in red) using a specific lncRNA (*H19*, in green) as an example. LncRNAs can sequester miRNAs, recruit epigenetic effectors or stabilize proteins and act via other possible functions, thereby regulating crucial pathways that can promote chemotherapy resistance. From the left: *H19* sequesters miR-106b-5p, thereby releasing the block on RDRG1 expression that promotes cisplatin resistance via stimulation of the PI3K/Akt/mTOR pathway; *H19* can also recruit EZH2 at the promoter of proapoptotic genes to repress their transcription and inhibit apoptosis, resulting in paclitaxel resistance; this lncRNA can also promote doxorubicin resistance via Cullin4A increase that determines increase of the drug efflux transporter ABCB1; via

Platinum agents are among the most commonly used anticancer drugs. Platinum agents form crosslinks between complementary DNA strands. This results in multiple DNA replication errors and accumulation of double-strand breaks, with consequent activation of programmed cell death. Cisplatin is one of the most used platinum agents and although it is highly effective in most cases, patients can experience treatment failure and tumour relapse (Wei et al. 2018). This can happen in testicular cancer (i.e. seminoma), where lncRNA *H19* has been shown to promote cisplatin resistance via sequestering miR-106b-5p and thereby activating *TDRG1* (testis development–related gene 1) (Wei et al. 2018). *TDRG1* is an oncogenic nc-RNA specifically expressed in testis that stimulates the PI3K/Akt/mTOR signalling to promote cancer progression (Wei et al. 2018). This evidence has been confirmed both in vitro and in vivo (Wei et al. 2018).

LncRNAs have been also implicated in many cases of resistance to anti-microtubule agents, plant-derived compounds that alter microtubule dynamics and block cell cycle progression, thereby inducing cell death. Taxanes are the best characterized and clinically effective drugs of this type. Taxanes are so called since they derive from the yew tree *Taxus baccata*, from which the first taxane (paclitaxel) was extracted. *H19* activates an anti-apoptotic response in paclitaxel-resistant breast cancer cells, which is mediated by the epigenetic silencing of the pro-apoptotic factors BIK and NOXA (Si et al. 2016). This epigenetic alteration is caused by the nuclear activity of *H19*, which binds and recruits the PRC2 protein EZH2 to the promoter of the pro-apoptotic genes, thereby promoting the resistance phenotype (Si et al. 2016).

Cytotoxic antitumour antibiotics are a class of chemotherapeutics that stop cells from dividing, via several mechanisms of action. Anthracyclines are an important family of cytotoxic antibiotics, isolated from *Streptomyces* bacterium. One of the most used in the clinics is doxorubicin. It has been shown that doxorubicin treatment induces *H19* expression in breast cancer cells (Zhu et al. 2017). In addition, *H19* silencing promotes cells sensitivity to doxorubicin (Zhu et al. 2017). In this case *H19* enhances Cullin4A expression (a ubiquitin ligase involved in several cellular key processes), which in turn increases the expression of the ABCB1 transporter. ABCB1 expression results in the reduction of intracellular paclitaxel concentrations. As already mentioned, ABC transporters are indeed largely known to efflux drugs, reducing their intracellular levels and increasing drug resistance (Zhu et al. 2017).

H19 is also upregulated in colorectal cancer cells that are resistant to the antimetabolite methotrexate. Antimetabolites have a chemical structure that is similar to the structure of some nucleobases or nucleosides. This feature gives them the ability to stop the synthesis of both DNA and RNA (Wu et al. 2017). In the mentioned study, modulation of *H19* affects cells response to methotrexate via stimulation of the Wnt/β-catenin pathway. This pathway regulates cells signal transduction and promotes epithelial to mesenchymal transition (EMT) (Liang et al. 2015), metastasis and drug resistance (Yu et al. 2012).

Camptothecins (i.e. irinotecan, topotecan) are a class of chemotherapeutics that inhibit topoisomerase I, an enzyme that rearranges the tri-dimensional structure of the DNA by causing temporary single strand breaks. These drugs block the ligation step of topoisomerases, leading to DNA strand breaks and cell death (165). *H19* has been highlighted as a potential gene involved in irinotecan resistance based on in silico analysis in colorectal cancer; further experimental validation is needed in order to understand *H19* role in resistance to this class of chemotherapeutics (Sun et al. 2019).

H19 as other lncRNAs shows great potential in chemotherapy resistance (Fig. 2) but further studies are needed to reveal the full potential of these molecules, since these studies are all very recent and this field of research is still largely unexplored. As discussed for *H19,* many other lncRNAs have been reported to affect chemotherapy resistance and most of them have been shown to drive resistance to difference drugs (Liu et al. 2015, 2016b, 2017a; Yue et al. 2016; Dong et al. 2018; Pucci et al. 2018; Zhang et al. 2019). This aspect is very important in the context of lncRNAs inhibition to increase drug response, since the targeting of a single lncRNA could restore drug sensitivity in different cancers and/or to many treatments. This will be discussed in detail in the last section of this chapter. In the following paragraphs we will focus on the roles of lncRNAs in the other two large families of cancer treatments: biological therapies and radiotherapy.

4.2 Interactions of LncRNAs with Biological Therapies

We have already underlined the importance of biological therapies in cancer treatment and briefly introduced some mechanisms of resistance to this type of anticancer drugs.

Despite the successful outcome of biological therapies in cancer treatment, cancer cells can develop mutation in target genes or activate alternative pathways to survive and proliferate, thereby becoming resistant. As we will discuss in the next paragraphs, lncRNAs play a prominent role in these phenomena. Some lncRNAs have been shown to promote sensitivity to biological treatments, while others have been shown to drive resistance.

For example, lncRNA *BCAR4* increases breast cancer sensitivity to the TKI lapatinib (Godinho et al. 2012), while silencing the *TUC338* lncRNA inhibits HCC tumour growth and sensitizes the cells to the TKI sorafenib. The latter effect is caused by an increased expression of the oncosuppressor GTPase-activating protein RASAL1 (Jin et al. 2017). *TUC338* effect has been observed both in vitro and in vivo, suggesting that its inhibition could be considered in the clinical setting to develop treatments that restore sorafenib sensitivity (Jin et al. 2017).

Besides lapatinib and sorafenib, lncRNAs emerge also in gefitinib response, such as *CASC9* that is upregulated in lung adenocarcinoma cells resistant to gefitinib compared to sensitive cells. Its silencing increases cells sensitivity; conversely *linc00227* is downregulated in the resistant cells (Ma et al. 2017). This shows that lncRNAs can be considered as targets to increase drug response via their inhibition or also cellular stimulation when they act as oncosuppressors, although most lncRNAs have been found upregulated in cancer.

In this regard, a study on HCC cells reveals that around 1500 lncRNAs are differentially expressed between gefitinib sensitive and resistant cells, meaning that some of them are upregulated, while others downregulated (Wu et al. 2016).

In lung adenocarcinoma lncRNA *GAS5* is downregulated and associated with oncosuppressor phenotypes like reduction in tumour size and pathological stages and upregulated in gefitinib-sensitive cells where it seems to stimulate apoptosis (Dong et al. 2015). Moreover, when overexpressed in nude mice it reduces tumour growth and correlate with EGFR and IGF-1R proteins (Dong et al. 2015).

Trastuzumab is effectively used in HER2-positive breast cancers as it selectively targets this transmembrane receptor tyrosine kinase on the cancer cells. Despite this activity, ~30% of patients develop resistance to trastuzumab and tumour recurrence (Merry et al. 2016).

GAS5 is decreased in breast cancer cells resistant to trastuzumab and reduces cancer proliferation via inhibition of *miR-21* (Li et al. 2016a). This miRNA in turn causes the reactivation of the oncosuppressor PTEN. Notably this lncRNA is also downregulated in breast cancer tissues from patients treated with trastuzumab (Li et al. 2016a).

Growth factor signalling through other receptors may be also responsible for trastuzumab resistance. Since lncRNAs can be involved in a large variety of pathways, especially in promoting cancer aggressive phenotypes, it is conceivable that several of them can participate in this resistance (Merry et al. 2016). For example, the lncRNA urothelial cancer associated 1 (*UCA1*) is involved in trastuzumab resistance by sequestering miR-18a and releasing the repression of Yes-associated protein 1 (YAP1), which promotes the survival of breast cancer cells exposed to trastuzumab (Zhu et al. 2018). *UCA1* is also expressed in colorectal cancer (CRC), especially in cetuximab-resistant cells and in the exosomes produced by these cells. The latter observation suggests that this lncRNA could be used as a biomarker of drug treatment response (Yang et al. 2018d). This lncRNA can indeed be also stably detected in CRC patients' serum and its expression is lower in patients with complete response to the biological therapy (Yang et al. 2018d). Additionally, the presence of *UCA1* in exosomes could promote the dissemination of the resistant phenotype to cells that are initially sensitive to the drug (Yang et al. 2018d). Therefore, *UCA1* inhibition could have broad effects in overcoming the resistance to biological therapies.

As mentioned above, hormonal therapy can be considered a biological therapy approach and some already mentioned lncRNAs are implicated in hormonal treatment response. We already mentioned that *BCAR4* drives lapatinib sensitivity, but this transcript seems to have an opposite effect when breast cancer cells are exposed to the oestrogen receptor inhibitor tamoxifen (Godinho et al. 2012). A clinical study shows that *BCAR4*-positive breast cancer patients treated with tamoxifen have shorter progression free survival (Godinho et al. 2012) and shorter metastasis-free and overall survival than *BCAR4*-negative patients (Godinho et al. 2010).

Among all the lncRNAs recently involved in hormonal treatment resistance, a strong role seems to be played by *UCA1* that was studied in relation to tamoxifen resistance in several studies (Li et al. 2016b; Liu et al. 2016a; Wu and Luo 2016; Xu et al. 2016). *UCA1* is found upregulated in both tamoxifen resistant human breast cancer cells LCC2 and in the exosomes released by these cells (Xu et al. 2016). Another analysis has confirmed and further investigated the role of this lncRNA and the mechanism by which it mediates breast cancer cell's resistance to tamoxifen. This study shows that *UCA1* has anti-apoptotic functions in cells treated with tamoxifen by activating mTOR signalling (Wu and Luo 2016). The AKT/mTOR pathway has already been shown hyperactivated in tamoxifen-resistant, ER^+ breast cancer cells (Block et al. 2012) and to be activated by *UCA1* in lung (Cheng et al. 2015) and bladder cancers (Li et al. 2014). Another study has highlighted the miRNA-sponge role of this lncRNA in tamoxifen resistant breast cancer cells (Li et al. 2016b). *UCA1* is induced upon tamoxifen treatment and sequesters miR-128a, thereby rescuing HIF1α inhibition; the activation of this axis favours drug resistance (Li et al. 2016b). Additionally, *UCA1* promotes breast cancer migration and tamoxifen resistance by driving β-Catenin nuclear translocation, thereby activating the Wnt pathway (Liu et al. 2016a).

These studies are other examples of the multifaceted roles and mechanisms that lncRNAs can have, not just in different conditions and cells but also in the same cells and context when they act simultaneously on several factors and pathways, to drive synergistically a specific cell response.

Other lncRNAs have been shown to interact with hormonal treatment-resistance pathways in response to tamoxifen (Basak et al. 2018; Ma et al. 2019) but also to other hormonal treatments, such as dexamethasone (Wu and Wang 2018) and glucocorticoids (Zhao et al. 2019). In this context, we have recently characterized *HORAS5* (aka *linc0161*) as a driver of prostate cancer resistance to androgen-deprivation therapy, via the stabilization of *AR* mRNA. This lncRNA enhances the AR-dependent transcriptional programme and promotes prostate cancer cell survival (Parolia et al. 2019).

Overall, *UCA1* seems to be particularly relevant in the context of biological therapies since it probably regulates several pathways in different contexts. Other lncRNAs could emerge in the future with similar potentiality in the drug resistance landscape.

4.3 LncRNAs in Radiotherapy Resistance

As described in the Sects. 2 and 3 of this chapter, cancer cells can respond to radiotherapy-induced stress by altering some molecular pathways that promote cell survival. In this way they become resistant to the radiotherapy (Grdina et al. 2013), a phenomenon that can cause cancer recurrence and metastasis (Lee et al. 2017). Therefore, the study of molecular interactors that could take part in this process is increasingly relevant and lncRNAs recently emerged among the key players in this field.

As mentioned in the previous paragraphs, LncRNAs can regulate many cancer-related pathways and, besides others, they can also regulate the expression of genes involved in DNA damage response and repair that are crucial in radiotherapy efficacy (Bartel and Ulitsky 2012).

LncRNAs have been linked to radiotherapy response. Examples are *GACAT3* that enhances lung carcinoma cells sensitivity to radiotherapy (Yang et al. 2018c) or *POU6F2-AS2* that increases oesophageal squamous cell carcinoma radio-resistance, by decreasing radiotherapy-induced DNA damage via multiple mechanisms (Liu et al. 2016a). Interestingly *POU6F2-AS2* carries out this function by driving YBX1 (Y-box binding protein) chromatin localization and its downstream pathways (Liu et al. 2016a).

Specific lncRNAs can influence the outcome of radiotherapy by acting in different cancers. Higher plasma levels of *GAS5* have been associated with poorer response to chemo-radiotherapy in head and neck cancer patients (Fayda et al. 2016). Interestingly *GAS5* reduces DNA damage in glioblastoma multiforme cells response to treatment (chemotherapy in this case) (Huang et al. 2019). Curiously *GAS5* is downregulated in prostate cancer cells and its upregulation mediates by α-Solanine (an alkaloid steroid found in potatoes) promotes radiosensitivity (Yang et al. 2019a). We have mentioned above *GAS5's* role in promoting biological therapies sensitivity. This multifaceted action, according to tissue and cancer type expression pattern, is typical of lncRNAs and is due to their ability to fold in complex 3D structures, thereby interacting with a large variety of different molecules and subcellular components (see Sect. 1 of this chapter). Since each cell type will express a different set of proteins, lncRNAs action is highly tissue- and cell-specific.

One of the most well-studied lncRNAs in cancer is *HOTAIR*, which has been also found to regulate cancer response to radiotherapy. *HOTAIR* is upregulated in pancreatic cancer cells upon radiation and its silencing increases the efficacy of radiotherapy, by upregulating cell autophagy (Wu et al. 2018). *HOTAIR*'s role in response to irradiation has been also highlighted in breast cancer, where it is upregulated in cancer tissues compared to normal and in cancer cells where its expression also increases after ionizing radiation (Hu et al. 2019). In this study *HOTAIR* silencing sensitizes cancer cells to radiation via cell survival inhibition, DNA damage increase and cell cycle arrest (Hu et al. 2019). *HOTAIR* is therefore a radiotherapy resistance-associated lncRNA that sequesters oncosuppressive miRNAs and is a good candidate as therapeutic target to radiosensitize breast cancer cells (Hu et al. 2019).

Overall several lncRNAs have been discussed in this paragraph involved in radiotherapy resistance but many others could be discovered in the next few years. It seems that lncRNAs potential in drug resistance is open to continuous discoveries.

5 Targeting LncRNAs to Overcome Drug Resistance

In the previous section, we have shown that lncRNAs orchestrate drug resistance via numerous mechanisms. LncRNAs are characterized by tissue- and disease-specific expression profiles (Kong et al. 2019). Hence, the production of lncRNA-targeting drugs could pave the way for cancer-specific treatments that cause minimal side effects.

By definition, lncRNAs do not encode proteins with enzymatic activity. Hence, classical pharmacological tools (enzyme inhibitors, antibodies) are ineffective against these macromolecules. Nevertheless, laboratories investigating the functional roles of lncRNAs have produced a range of molecular tools to modulate the expression of these transcripts. Some of these tools could be tested in future clinical trials to inhibit the expression of oncogenic lncRNAs and, in the future, to reactivate oncosuppressive lncRNAs.

Stable modulation of lncRNA expression has been achieved both in vitro and in vivo using lentiviral transduction (Panda et al. 2018; Zhang et al. 2018) and CRISPR-Cas9-dependent genome editing (Lavalou et al. 2019). While the former technique has been extensively used for both lncRNA silencing and overexpression, the latter is mainly used to eliminate the expression of functional lncRNAs by partial deletion of the target gene. This approach has been shown to trigger the compensatory upregulation of partially deleted lncRNAs sequences and therefore requires extensive in vivo validation (Lavalou et al. 2019). As for lentivirus-dependent gene modulation, so far, no vector has been shown to direct sufficient quantities of modified DNA to the cancer tissue. For these reasons, the clinical application of both lentiviral and genome-editing techniques is promising but still in its infancy (Table 2). Notably, a phase I clinical trial using lentiviral vectors to anti-cancer immunity has been recently reported (Somaiah et al. 2019). When this and similar trials will be completed, it will be interesting to assess the safety and efficacy of this technique, as well as its applicability to lncRNA targeting. So far, CRISPR-Cas9 technology has been mainly employed for genetic reprogramming of immune cells (e.g. via PD1 deletion) ex vivo, followed by re-injection of the modified cells into the host (Zhan et al. 2019). These pioneering studies are still enrolling patients, hence

Table 2 Main methods that are currently employed to modulate the expression of lncRNAs in experimental conditions. The "clinical development" column refers to clinical trials using the same type of technology on any target (coding or non-coding) in solid tumours

Methodology	Effect on lncRNA expression	Clinical development	Reference
ASO	Silencing	Phase III/clinical approval	Chi et al. (2017)
siRNA	Silencing	Phase I/II	Zuckerman and Davis (2015)
CRISPR-Cas	Overexpression/silencing	Phase I	Somaiah et al. (2019)
Lentiviral vector	Overexpression/silencing	Phase I	Zhan et al. (2019)

there are no definitive data on the feasibility and safety of these techniques, nor on their effectiveness of targeting cancer-specific lncRNAs.

While stable modulation of lncRNA expression is a long-term goal of translational research, transient lncRNA silencing techniques are closer to clinical tests. Current methods to study lncRNA functions in vivo and in vitro include transient silencing via small-interfering RNAs (siRNAs) or antisense oligonucleotides (ASOs) (Chery 2016). Synthetic siRNAs are double-stranded RNAs, analogues to endogenous small non-coding RNAs. When transfected into mammalian cells, these siRNAs are processed by the RNA-induced silencing complex (RISC) (Pratt and MacRae 2009). The RISC unwinds the siRNA to a single strand component and matches it to its complementary RNA target. The newly formed siRNA-target RNA complex is degraded by endonucleases. This technology has been successfully employed to silence both coding and non-coding RNAs and to study their function in cancer initiation and progression (Mahmoodi Chalbatani et al. 2019). However, since the RISC complex is primarily located in the cytoplasm (Pratt and MacRae 2009) siRNAs are usually ineffective in silencing nuclear lncRNAs.

Another common methodology of transient lncRNA silencing is via ASOs, that is single-stranded DNA polymers (13–200 nt), which target complementary RNA sequences (Fig. 3). Upon cellular internalization, ASOs bind the target RNA and trigger the degradation of the RNA/DNA complex by the RNase-H enzyme (Chery 2016). Since RNase-H is expressed in the nucleus and in the cytoplasm (Liang et al. 2017), ASOs are effective in silencing nuclear and cytoplasmic lncRNAs. Notably, ASOs have been shown to modulate gene expression via additional mechanisms, including splicing inhibition. Since some lncRNAs are actively spliced (Crea et al. 2016b), it will be worth investigating whether splicing-inhibiting ASOs could be useful inhibitors of specific lncRNA variants.

Among the methodologies discussed here, siRNA- and ASO-dependent silencing are the most promising for clinical application. For example, we have shown that in vivo silencing of the lncRNA *HORAS5* significantly reduces the growth of CRPC (Parolia et al. 2019). This result was obtained upon intra-tumoural injection of siRNAs conjugated with a lipidic transfection reagent, a technique that is hardly translatable to the clinical setting. Currently, the clinical application of siRNA therapies is hampered by poor stability, off-target effects, and insufficient tissue-specificity. Several clinical and industrial investigators have shown that both lipidic and non-lipidic nanoparticles can overcome these issues by improving both the stability and tissue selectivity of siRNAs (Mahmoodi Chalbatani et al. 2019). However, none of these delivery technologies has emerged as clearly superior for clinical applications. Some siRNAs are currently in Phase I/II clinical trials for different diseases, including cancer (Zuckerman and Davis 2015). However, no clinical study on this type of drugs has shown clear anticancer activity.

The most advanced technique for RNA targeting in oncology is represented by ASOs. We and others have shown that non-coding RNA-targeting ASOs can efficiently arrest the proliferation and/or the metastatic potential of cancer cells (Crea et al. 2016a; Gordon et al. 2019). More recently, we have shown that the lncRNA *HORAS5* favours taxane resistance in prostate cancer by activating an anti-

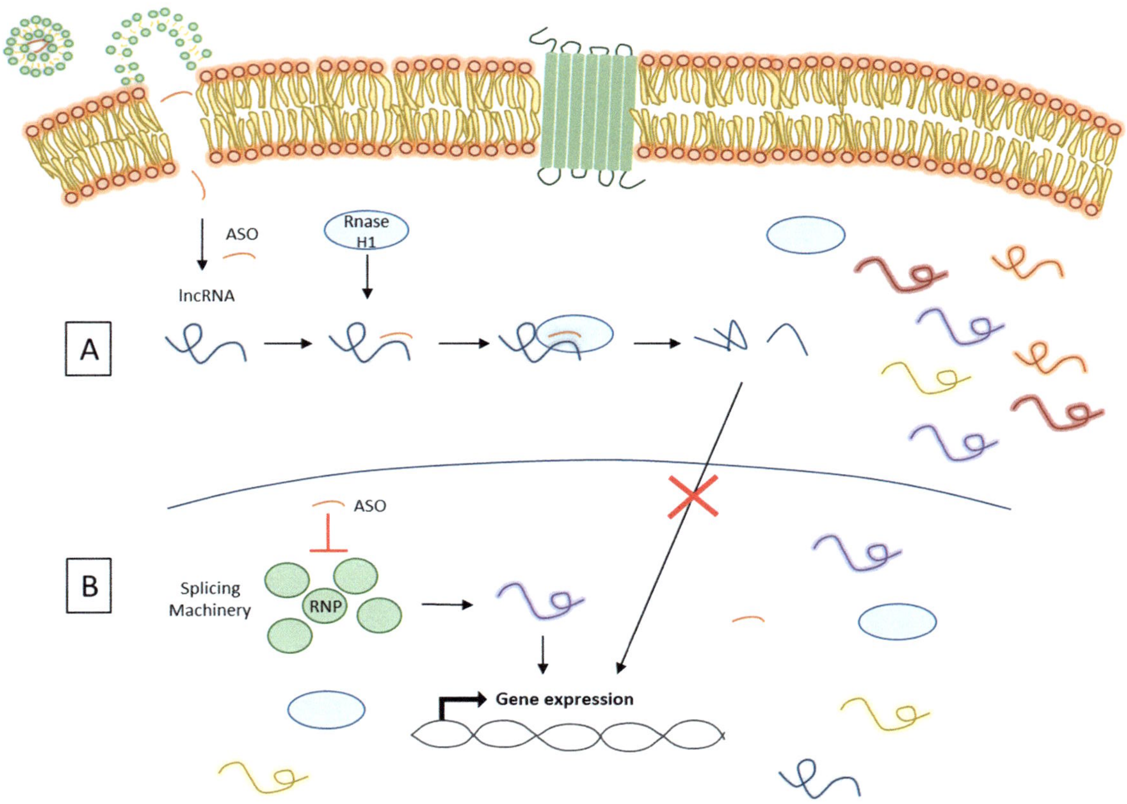

Fig. 3 *Antisense oligonucleotides (ASO) silence lncRNA-mediated signalling through increased degradation or inhibited splicing*. Here, two examples of how ASO modulate lncRNA function are shown. (**a**) ASOs introduced via lipidic or non-lipidic transfection infiltrate the cell cytoplasm and nucleus and complementarily bind to target lncRNAs. Cytoplasmic or nuclear RNAse H are recruited to degrade lncRNA thereby decreasing the load of lncRNA and silencing lncRNA-mediated signalling (Chery 2016). (**b**) ASOs are also known to influence splicing mechanisms, preserving exons or skipping exons. Here, the ASO targets the ribonuclease proteins, preventing formation of the splicing machinery. It is postulated that ASOs can similarly modulate lncRNA activity

apoptotic pathway (Pucci et al. 2019). We have designed and effectively tested ASOs targeting *HORAS5* in combination with taxanes and showed that the two treatments are synergistic. Although there is no clinical trial on lncRNA-targeting ASOs, Phase II/III trials have shown that these molecules can have anticancer effect and silence the expression of protein-coding genes (Duffy et al. 2016) (Chi et al. 2017). Notably, ASOs structure has been successfully modified to increase both the stability in biological fluids and the intracellular penetration. For all these reasons, we believe that ASOs are currently the most promising lncRNA-targeting drugs available.

Future preclinical and clinical studies will need to address numerous challenges before the approval of lncRNA-targeting ASOs that prevent the progression of chemo-resistant malignancies. Among those challenges, identifying the optimal drug combination and treatment schedule will be of paramount importance. In addition, lncRNA expression shows a notable degree of inter-individual variability (Crea et al. 2014). For this reason, it will be important to develop tests that predict the expression of a specific lncRNA in the patient's tumour. Only patients with sufficiently elevated expression of the lncRNA will be treated with the specific ASO. Non-invasive diagnostic tests could be used to monitor the efficacy of the lncRNA-targeting drugs. Recent studies indicate that the expression of some lncRNAs in biological fluids (e.g. serum) correlates with the intratumoural levels of these transcripts (Crea et al. 2014; Sun et al. 2018). We think that the combined use of non-invasive diagnostic techniques and lncRNA-targeting drugs will pave the way for a new type of personalised cancer therapy, which could overcome the emergence of drug resistance.

References

Al-Dimassi S, Abou-Antoun T, El-Sibai M (2014) Cancer cell resistance mechanisms: a mini review. Clin Transl Oncol 16:511–516

Al-Rugeebah A, Alanazi M, Parine NR (2019) MEG3: an oncogenic long non-coding RNA in different cancers. Pathol Oncol Res 25:859–874

Ambrosini G, Adida C, Altieri DC (1997) A novel anti-apptosis gene, survivin, expressed in cancer and lymphoma. Nat Med 3:917–921

Arora A, Scholar EM (2005) Role of tyrosine kinase inhibitors in cancer therapy. J Pharmacol Exp Ther 315:971–979

Asao T, Takahashi F, Takahashi K (2019) Resistance to molecularly targeted therapy in non-small-cell lung cancer. Respir Investig 57:20–26

Baker S, Dahele M, Lagerwaard FJ et al (2016) A critical review of recent developments in radiotherapy for non-small cell lung cancer. Radiat Oncol 11:115

Bao S, Wu Q, McLendon RE et al (2006) Glioma stem cells promote radioresistance by preferential activation of the DNA damage response. Nature 444:756–760

Bao B, Azmi A, Li Y et al (2013) Targeting CSCs in tumor microenvironment: the potential role of ROS-associated miRNAs in tumor aggressiveness. Curr Stem Cell Res Ther 9:22–35

Bartel DP, Ulitsky I (2012) lincRNAs: genomics, evolution, and mechanisms. Changes 29:997–1003

Basak P, Chatterjee S, Bhat V et al (2018) Long non-coding RNA H19 acts as an estrogen receptor modulator that is required for endocrine therapy resistance in ER + breast cancer cells. Cell Physiol Biochem 51:1518–1532
Bekele RT, Venkatraman G, Liu RZ et al (2016) Oxidative stress contributes to the tamoxifen-induced killing of breast cancer cells: implications for tamoxifen therapy and resistance. Sci Rep 6:1–17
Bhosle J, Hall G (2009) Principles of cancer treatment by chemotherapy. Surgery 27:173–177
Block M, Gründker C, Fister S et al (2012) Inhibition of the AKT/mTOR and erbB pathways by gefitinib, perifosine and analogs of gonadotropin-releasing hormone I and II to overcome tamoxifen resistance in breast cancer cells. Int J Oncol 41:1845–1854
Carnero A, Lleonart M (2016) The hypoxic microenvironment: a determinant of cancer stem cell evolution. BioEssays 38:65–74
Caudell JJ, Torres-Roca JF, Gillies RJ et al (2017) The future of personalised radiotherapy for head and neck cancer. Lancet Oncol 18:e266–e273
Ceppi P, Volante M, Novello S et al (2006) ERCC1 and RRM1 gene expressions but not EGFR are predictive of shorter survival in advanced non-small-cell lung cancer treated with cisplatin and gemcitabine. Ann Oncol 17:1818–1825
Chakravarty D, Sboner A, Nair SS et al (2014) The oestrogen receptor alpha-regulated lncRNA NEAT1 is a critical modulator of prostate cancer. Nat Commun 5:1–16
Chen M, Xia Z, Chen C et al (2018) LncRNA MALAT1 promotes epithelial-to-mesenchymal transition of esophageal cancer through Ezh2-Notch1 signaling pathway. Anti-Cancer Drugs 29:767–773
Cheng N, Cai W, Ren S et al (2015) Long non-coding RNA UCA1 induces non-T790M acquired resistance to EGFR-TKIs by activating the AKT/mTOR pathway in EGFR-mutant non-small cell lung cancer. Oncotarget 6:23582–23593
Chery J (2016) RNA therapeutics: RNAi and antisense mechanisms and clinical applications. Postdoc J 4:35–50
Chi KN, Higano CS, Blumenstein B et al (2017) Custirsen in combination with docetaxel and prednisone for patients with metastatic castration-resistant prostate cancer (SYNERGY trial): a phase 3, multicentre, open-label, randomised trial. Lancet Oncol 18:473–485
Citrin DE (2017) Recent developments in radiotherapy. N Engl J Med 377:1065–1075
Coppola S, Carnevale I, Danen EHJ et al (2017) A mechanopharmacology approach to overcome chemoresistance in pancreatic cancer. Drug Resist Updat 31:43–51
Crea F, Danesi R, Farrar WL (2009) Cancer stem cell epigenetics and chemoresistance. Epigenomics 1:63–79
Crea F, Hurt EM, Mathews LA et al (2011) Pharmacologic disruption of polycomb repressive complex 2 inhibits tumorigenicity and tumor progression in prostate cancer. Mol Cancer 10:1–10
Crea F, Watahiki A, Quagliata L et al (2014) Identification of a long non-coding RNA as a novel biomarker and potential therapeutic target for metastatic prostate cancer. Oncotarget 5:764–774
Crea F, Quagliata L, Michael A et al (2016a) Integrated analysis of the prostate cancer small-nucleolar transcriptome reveals SNORA55 as a driver of prostate cancer progression. Mol Oncol 10:693–703
Crea F, Venalainen E, Ci X et al (2016b) The role of epigenetics and long noncoding RNA MIAT in neuroendocrine prostate cancer. Epigenomics 8:721–731
Damiani D, Tiribelli M, Calistri E et al (2006) The prognostic value of P-glycoprotein (ABCB) and breast cancer resistance protein (ABCG2) in adults with de novo acute myeloid leukemia with normal karyotype. Haematologica 91:825–828
Delaney G, Jacob S, Featherstone C et al (2005) The role of radiotherapy in cancer treatment: estimating optimal utilization from a review of evidence-based clinical guidelines. Cancer 104:1129–1137
Diamantopoulos MA, Tsiakanikas P, Scorilas A (2018) Non-coding RNAs: the riddle of the transcriptome and their perspectives in cancer. Ann Transl Med 6:241–241

Dong S, Qu X, Li W et al (2015) The long non-coding RNA, GAS5, enhances gefitinib-induced cell death in innate EGFR tyrosine kinase inhibitor-resistant lung adenocarcinoma cells with wide-type EGFR via downregulation of the IGF-1R expression. J Hematol Oncol 8:1–13
Dong X, Fang Z, Yu M et al (2018) Knockdown of long noncoding RNA HOXA-AS2 suppresses chemoresistance of acute myeloid leukemia via the miR-520c-3p/S100A4 Axis. Cell Physiol Biochem 51:886–896
Du B, Shim JS (2016) Targeting epithelial-mesenchymal transition (EMT) to overcome drug resistance in cancer. Molecules 21:1–15
Duffy AG, Ulahannan SV, Rahma OE et al (2016) Modulation of tumor eIF4E by antisense inhibition: a phase I/II translational clinical trial of ISIS 183750—an antisense oligonucleotide against eIF4E-in combination with irinotecan in solid tumors and irinotecan-refractory colorectal cancer. Int J Cancer 139:1648–1657
Eastlack SC, Dong S, Mo YY et al (2018) Expression of long noncoding RNA MALAT1 correlates with increased levels of Nischarin and inhibits oncogenic cell functions in breast cancer. PLoS One 13:1–13
Eun K, Ham SW, Kim H (2017) Cancer stem cell heterogeneity: origin and new perspectives on CSC targeting. BMB Rep 50:117–125
Fayda M, Isin M, Tambas M et al (2016) Do circulating long non-coding RNAs (lncRNAs) (LincRNA-p21, GAS 5, HOTAIR) predict the treatment response in patients with head and neck cancer treated with chemoradiotherapy? Tumor Biol 37:3969–3978
Feng X, Liu H, Zhang Z et al (2017) Annexin A2 contributes to cisplatin resistance by activation of JNK-p53 pathway in non-small cell lung cancer cells. J Exp Clin Cancer Res 36:1–14
Fletcher JI, Williams RT, Henderson MJ et al (2016) ABC transporters as mediators of drug resistance and contributors to cancer cell biology. Drug Resist Updat 26:1–9
Galadari S, Rahman A, Pallichankandy S et al (2017) Reactive oxygen species and cancer paradox: to promote or to suppress? Free Radic Biol Med 104:144–164
Gao S, Wang P, Hua Y et al (2016) ROR functions as a ceRNA to regulate Nanog expression by sponging miR-145 and predicts poor prognosis in pancreatic cancer. Oncotarget 7:1608–1618
Gene Therapy of Cancer (Third Edition)
Ghisolfi L, Keates AC, Hu X et al (2012) Ionizing radiation induces stemness in cancer cells. PLoS One 7:1–11
Godinho MFE, Sieuwerts AM, Look MP et al (2010) Relevance of BCAR4 in tamoxifen resistance and tumour aggressiveness of human breast cancer. Br J Cancer 103:1284–1291
Godinho MFE, Wulfkuhle JD, Look MP et al (2012) BCAR4 induces antioestrogen resistance but sensitises breast cancer to lapatinib. Br J Cancer 107:947–955
Gordon MA, Babbs B, Cochrane DR et al (2019) The long non-coding RNA MALAT1 promotes ovarian cancer progression by regulating RBFOX2-mediated alternative splicing. Mol Carcinog 58:196–205
Gottesman MMAS (2001) Overview: ABC transporters and human disease. J Bioenerg Biomembr 33:453–458
Grdina DJ, Murley JS, Miller RC et al (2013) Survivin-associated adaptive response in radiation therapy. Cancer Res 73:4418–4428
Haber M, Smith J, Bordow SB et al (2006) Association of high-level MRP1 expression with poor clinical outcome in a large prospective study of primary neuroblastoma. J Clin Oncol 24:1546–1553
He X, Tan X, Wang X et al (2014) C-Myc-activated long noncoding RNA CCAT1 promotes colon cancer cell proliferation and invasion. Tumour Biol 35:12181–12188
Housman G, Byler S, Heerboth S et al (2014) Drug resistance in cancer: an overview. Cancers (Basel) 6:1769–1792
Hu X, Ding D, Zhang J et al (2019) Knockdown of lncRNA HOTAIR sensitizes breast cancer cells to ionizing radiation through activating miR-218. Biosci Rep 29:1–9
Huang H, Jiang R, Lian Z et al (2019) miR-222/GAS5 is involved in DNA damage and cytotoxic effects induced by temozolomide in T98G cell line. J Appl Toxicol 39:726–734

Iaccarino I (2017) LncRNAs and MYC: an intricate relationship. Int J Mol Sci 18:1–14

Iyer MK, Niknafs YS, Malik R et al (2015) The landscape of long noncoding RNAs in the human transcriptome. Nat Genet 47:199–208

Jaiyesimi IA, Buzdar AU, Decker DA et al (1995) Use of tamoxifen for breast cancer: twenty-eight years later. J Clin Oncol 13:513–529

Jin W, Chen L, Cai X et al (2017) Long non-coding RNA TUC338 is functionally involved in sorafenib-sensitized hepatocarcinoma cells by targeting RASAL1. Oncol Rep 37:273–280

Jokic M, Vlasic I, Rinneburger M et al (2016) Ercc1 deficiency promotes tumorigenesis and increases cisplatin sensitivity in a Tp53 context-specific manner. Mol Cancer Res 14:1110–1123

Kayl AE, Meyers CA (2006) Side-effects of chemotherapy and quality of life in ovarian and breast cancer patients. Curr Opin Obstet Gynecol 18:24–28

Kim J, Piao H, Kim B et al (2018) Long non-coding RNA MALAT1 suppresses breast cancer metastasis. Nat Genet 50:1705–1715

Koirala P, Huang J, Ho TT et al (2017) LncRNA AK023948 is a positive regulator of AKT. Nat Commun 8:1–10

Kong Y, Lu Z, Liu P et al (2019) Long noncoding RNA: genomics and relevance to physiology. Compr Physiol 9:933–946

Kotake Y, Goto T, Naemura M et al (2017) Long noncoding RNA PANDA positively regulates proliferation of osteosarcoma cells. Anticancer Res 37:81–86

Krishnamurthy N, Kurzrock R (2018) Targeting the Wnt/beta-catenin pathway in cancer: update on effectors and inhibitors. Cancer Treat Rev 62:50–60

Kumar MM, Goyal R (2017) LncRNA as a therapeutic target for angiogenesis. Curr Top Med Chem 17:1750–1757

Lavalou P, Eckert H, Damy L et al (2019) Strategies for genetic inactivation of long noncoding RNAs in zebrafish. RNA 25:897–904

Lee SY, Jeong EK, Ju MK et al (2017) Induction of metastasis, cancer stem cell phenotype, and oncogenic metabolism in cancer cells by ionizing radiation. Mol Cancer 16:1–25

Li Z, Li X, Wu S et al (2014) Long non-coding RNA UCA1 promotes glycolysis by upregulating hexokinase 2 through the mTOR-STAT3/microRNA143 pathway. Cancer Sci 105:951–955

Li W, Zhai L, Wang H et al (2016a) Downregulation of LncRNA GAS5 causes trastuzumab resistance in breast cancer. Oncotarget 7:27778–27786

Li X, Wu Y, Liu A, Tang X (2016b) Long non-coding RNA UCA1 enhances tamoxifen resistance in breast cancer cells through a miR-18a-HIF1α feedback regulatory loop. Tumor Biol 37:14733–14743

Liang WC, Fu WM, Wong CW et al (2015) The LncRNA H19 promotes epithelial to mesenchymal transition by functioning as MiRNA sponges in colorectal cancer. Oncotarget 6:22513–22525

Liang XH, Sun H, Nichols JG et al (2017) RNase H1-dependent antisense oligonucleotides are robustly active in directing RNA cleavage in both the cytoplasm and the nucleus. Mol Ther 25:2075–2092

Liu J, Wan L, Lu K et al (2015) The long noncoding RNA MEG3 contributes to cisplatin resistance of human lung adenocarcinoma. PLoS One 10:1–16

Liu J, Sun X, Zhu H et al (2016a) Long noncoding RNA POU6F2-AS2 is associated with esophageal squamous cell carcinoma. J Biochem 160:195–204

Liu Y, Xu N, Liu B et al (2016b) Long noncoding RNA RP11-838N2.4 enhances the cytotoxic effects of temozolomide by inhibiting the functions of miR-10a in glioblastoma cell lines. Oncotarget 7:43835–43851

Liu R, Zeng Y, Zhou CF et al (2017a) Long noncoding RNA expression signature to predict platinum-based chemotherapeutic sensitivity of ovarian cancer patients. Sci Rep 7:1–10

Liu X, Li F, Huang Q et al (2017b) Self-inflicted DNA double-strand breaks sustain tumorigenicity and stemness of cancer cells. Cell Res 27:764–783

Lopez-Ayllon BD, Moncho-Amor V, Abarrategi A et al (2014) Cancer stem cells and cisplatin-resistant cells isolated from non-small-lung cancer cell lines constitute related cell populations. Cancer Med 3:1099–1111

Luengo A, Gui DY, Vander Heiden MG (2017) Targeting metabolism for cancer therapy. Cell Chem Biol 24:1161–1180
Lyakhovich A, Lleonart ME (2016) Bypassing mechanisms of mitochondria-mediated Cancer stem cells resistance to chemo- and radiotherapy. Oxidative Med Cell Longev 2016:1–10
Ma P, Zhang M, Nie F et al (2017) Transcriptome analysis of EGFR tyrosine kinase inhibitors resistance associated long noncoding RNA in non-small cell lung cancer. Biomed Pharmacother 87:20–26
Ma Y, Bu D, Long J et al (2019) LncRNA DSCAM-AS1 acts as a sponge of miR-137 to enhance Tamoxifen resistance in breast cancer. J Cell Physiol 234:2880–2894
Mahmoodi Chalbatani G, Dana H, Gharagouzloo E et al (2019) Small interfering RNAs (siRNAs) in cancer therapy: a nano-based approach. Int J Nanomedicine 14:3111–3128
Mao Z, Bozzella M, Seluanov A et al (2008) DNA repair by nonhomologous end joining and homologous recombination during cell cycle in human cells. Cell Cycle 7:2902–2906
Martin LP, Hamilton TC, Shilder RJ (2008) Platinum resistance: the role of DNA repair pathways. Clin Cancer Res 14:1291–1295
Mercer TR, Dinger ME, Mattick JS (2009) Long non-coding RNAs: insights into functions. Nat Rev Genet 10:155–159
Merry CR, McMahon S, Forrest ME et al (2016) Transcriptome-wide identification of mRNAs and lincRNAs associated with trastuzumab-resistance in HER2-positive breast cancer. Oncotarget 7:53230–53244
Molina J, Adjei A (2006) The Ras/Raf/MAPK pathway. J Thorac Oncol 1:7–9
Mustacchi G, De Laurentiis M (2015) The role of taxanes in triple-negative breast cancer: literature review. Drug Des Devel Ther 9:4303–4318
National Cancer Institute. In: Target Cancer Ther. https://www.cancer.gov/about-cancer/treatment/types/targeted-therapies/targeted-therapies-fact-sheet
Nikolaou M, Pavlopoulou A, Georgakilas AG et al (2018) The challenge of drug resistance in cancer treatment: a current overview. Clin Exp Metastasis 35:309–318
Ning L, Li Z, Wei D et al (2017) LncRNA, NEAT1 is a prognosis biomarker and regulates cancer progression via epithelial-mesenchymal transition in clear cell renal cell carcinoma. Cancer Biomark 19:75–83
Panda S, Setia M, Kaur N et al (2018) Noncoding RNA Ginir functions as an oncogene by associating with centrosomal proteins. PLoS Biol 16:1–50
Pao W, Miller VA, Politi KA et al (2005) Acquired resistance of lung adenocarcinomas to gefitinib or erlotinib is associated with a second mutation in the EGFR kinase domain. PLoS Med 2:0225–0235
Parolia A, Venalainen E, Xue H et al (2019) The long noncoding RNA HORAS5 mediates castration-resistant prostate cancer survival by activating the androgen receptor transcriptional program. Mol Oncol 13:1121–1136
Peng C, Hu W, Weng X et al (2017) Over expression of long non-coding RNA PANDA promotes hepatocellular carcinoma by inhibiting senescence associated inflammatory factor IL8. Sci Rep 7:1–11
Pérez-Herrero E, Fernández-Medarde A (2015) Advanced targeted therapies in cancer: drug nanocarriers, the future of chemotherapy. Eur J Pharm Biopharm 93:52–79
Ponjavic J, Ponting CP, Lunter G (2007) Functionality or transcriptional noise? Evidence for selection within long noncoding RNAs. Genome Res 17:556–565
Pratt AJ, MacRae IJ (2009) The RNA-induced silencing complex: a versatile gene-silencing machine. J Biol Chem 284:17897–17901
Pucci P, Rescigno P, Sumanasuriya S et al (2018) Hypoxia and noncoding RNAs in Taxane resistance. Trends Pharmacol Sci 39:695–709
Pucci P, Venalainen E, Alborelli I et al (2019) 1986PD HORAS5 promotes cabazitaxel resistance in castration resistant prostate cancer via a BCL2A1-dependent survival mechanism. Ann Oncol 30:v798

Puvvula PK, Desetty RD, Pineau P et al (2014) Long noncoding RNA PANDA and scaffold-attachment-factor SAFA control senescence entry and exit. Nat Commun 5:1–16
Qin S, Zhang B, Xiao G et al (2016) Fibronectin protects lung cancer cells against docetaxel-induced apoptosis by promoting Src and caspase-8 phosphorylation. Tumor Biol 37:13509–13520
Qu Z, Adelson DL (2012) Evolutionary conservation and functional roles of ncRNA. Front Genet 3:1–11
Sanchez Calle A, Kawamura Y, Yamamoto Y et al (2018) Emerging roles of long non-coding RNA in cancer. Cancer Sci 109:2093–2100
Sarkadi B, Homolya L, Szakács G, Váradi A (2006) Human multidrug resistance ABCB and ABCG transporters: participation in a chemoimmunity defense system. Physiol Rev 86:1179–1236
Schaich M, Soucek S, Thiede C et al (2005) MDR1 and MRP1 gene expression are independent predictors for treatment outcome in adult acute myeloid leukaemia. Br J Haematol 128:324–332
Schmitt AM, Chang HY (2016) Long noncoding RNAs in cancer pathways. Cancer Cell 29:452–463
Shachar S, Jolly T, Jones E et al (2018) Management of triple-negative breast cancer in older patients: how is it different? Oncology 32:58–63
Shah MY, Martinez-Garcia E, Phillip JM et al (2016) MMSET/WHSC1 enhances DNA damage repair leading to an increase in resistance to chemotherapeutic agents. Oncogene 35:5905–5915
Sherlach KS, Roepe PD (2014) Drug resistance associated membrane proteins. Front Physiol 5:1–4
Si X, Zang R, Zhang E et al (2016) LncRNA H19 confers chemoresistance in ERα-positive breast cancer through epigenetic silencing of the pro-apoptotic gene BIK. Oncotarget 7:81452–81462
Somaiah N, Block MS, Kim JW et al (2019) First-in-class, first-in-human study evaluating LV305, a dendritic-cell tropic lentiviral vector, in sarcoma and other solid tumors expressing NY-ESO-1. Clin Cancer Res:1–11
Srivastava M, Raghavan SC (2015) DNA double-strand break repair inhibitors as cancer therapeutics. Chem Biol 22:17–29
Su Y, Lu J, Chen X et al (2019) Long non-coding RNA HOTTIP affects renal cell carcinoma progression by regulating autophagy via the PI3K/Akt/Atg13 signaling pathway. J Cancer Res Clin Oncol 145:573–588
Sun L, Su Y, Liu X et al (2018) Serum and exosome long non coding RNAs as potential biomarkers for hepatocellular carcinoma. J Cancer 9:2631–2639
Sun F, Liang W, Qian J (2019) The identification of CRNDE, H19, UCA1 and HOTAIR as the key lncRNAs involved in oxaliplatin or irinotecan resistance in the chemotherapy of colorectal cancer based on integrative bioinformatics analysis. Mol Med Rep 20:3583–3596
Takahashi K, Yan IK, Haga H et al (2014a) Modulation of hypoxia-signaling pathways by extracellular linc-RoR. J Cell Sci 127:1585–1594
Takahashi K, Yan IK, Kogure T et al (2014b) Extracellular vesicle-mediated transfer of long non-coding RNA ROR modulates chemosensitivity in human hepatocellular cancer. FEBS Open Bio 4:458–467
Takezawa K, Pirazzoli V, Arcila ME et al (2012) HER2 amplification: a potential mechanism of acquired resistance to egfr inhibition in EGFR -mutant lung cancers that lack the second-site EGFR T790M mutation. Cancer Discov 2:922–933
Tang Q, Hann SS (2018) HOTAIR: an oncogenic long non-coding RNA in human cancer. Cell Physiol Biochem 47:893–913
Thomas D, Majeti R (2017) Biology and relevance of human acute myeloid leukemia stem cells. Blood 129:1577–1585
Toraih EA, Ellawindy A, Fala SY et al (2018) Oncogenic long noncoding RNA MALAT1 and HCV-related hepatocellular carcinoma. Biomed Pharmacother 102:653–669
Trock BJ, Leonessa F, Clarke R (1997) Multidrug resistance in breast cancer: a meta-analysis of MDR1/gp170 expression and its possible functional significance. J Natl Cancer Inst 89:917–931
Vivanco I, Sawyers CL (2002) The phosphatidylinositol 3-kinase-AKT pathway in humancancer. Nat Rev Cancer 2:489–501

Wallace SS, Murphy DL, Sweasy JB (2012) Base excision repair and cancer Susan. Cancer Lett 327:73–89
Wang KC, Chang HY (2011) Molecular mechanisms of long noncoding RNAs. Mol Cell 43:904–914
Wang D, Lippard SJ (2005) Cellular processing of platinum anticancer drugs. Nat Rev Drug Discov 4:307–320
Wang W, Tse-Ding Y (2019) Recent advances in use of topoisomerase inhibitors in combination cancer therapy. Curr Top Med Chem 19:730–740
Wang Y, Zhang M, Xu H et al (2017) Discovery and validation of the tumor-suppressive function of long noncoding RNA PANDA in human diffuse large B-cell lymphoma through the inactivation of MAPK/ERK signaling pathway. Oncotarget 8:72182–72196
Wang N, Yu Y, Xu B et al (2019) Pivotal prognostic and diagnostic role of the long non-coding RNA colon cancer-associated transcript 1 expression in human cancer (Review). Mol Med Rep 19:771–782
Wei J, Gan Y, Peng D et al (2018) Long non-coding RNA H19 promotes TDRG1 expression and cisplatin resistance by sequestering miRNA-106b-5p in seminoma. Cancer Med 7:6247–6257
Wu C, Luo J (2016) Long non-coding RNA (lncRNA) urothelial carcinoma-associated 1 (UCA1) enhances tamoxifen resistance in breast cancer cells via inhibiting mTOR signaling pathway. Med Sci Monit 22:3860–3867
Wu Y, Wang H (2018) LncRNA NEAT1 promotes dexamethasone resistance in multiple myeloma by targeting miR-193a/MCL1 pathway. J Biochem Mol Toxicol 32:1–6
Wu Y, Yu DD, Hu Y et al (2016) Genome-wide profiling of long non-coding RNA expression patterns in the EGFR-TKI resistance of lung adenocarcinoma by microarray. Oncol Rep 35:3371–3386
Wu KF, Liang WC, Feng L et al (2017) H19 mediates methotrexate resistance in colorectal cancer through activating Wnt/β-catenin pathway. Exp Cell Res 350:312–317
Wu C, Yang L, Qi X et al (2018) Inhibition of long non-coding RNA HOTAIR enhances radiosensitivity via regulating autophagy in pancreatic cancer. Cancer Manag Res 10:5261–5271
Xiong T, Li J, Chen F et al (2019) PCAT-1: a novel oncogenic long non-coding RNA in human cancers. Int J Biol Sci 15:847–856
Xu CG, Yang MF, Ren YQ et al (2016) Exosomes mediated transfer of lncRNA UCA1 results in increased tamoxifen resistance in breast cancer cells. Eur Rev Med Pharmacol Sci 20:4362–4368
Yang HZ, Ma Y, Zhou Y et al (2015) Autophagy contributes to the enrichment and survival of colorectal cancer stem cells under oxaliplatin treatment. Cancer Lett 361:128–136
Yang B, Gao G, Wang Z et al (2018a) Long non-coding RNA HOTTIP promotes prostate cancer cells proliferation and migration by sponging miR-216a-5p. Biosci Rep 38:1–9
Yang W, Redpath RE, Zhang C et al (2018b) Long non-coding RNA h19 promotes the migration and invasion of colon cancer cells via MAPK signaling pathway. Oncol Lett 16:3365–3372
Yang X, Zhang W, Cheng S-Q et al (2018c) High expression of lncRNA GACAT3 inhibits invasion and metastasis of non-small cell lung cancer to enhance the effect of radiotherapy. Eur Rev Med Pharmacol Sci 22:1315–1322
Yang YN, Zhang R, Du JW et al (2018d) Predictive role of UCA1-containing exosomes in cetuximab-resistant colorectal cancer 11 medical and health sciences 1112 oncology and carcinogenesis. Cancer Cell Int 18:1–11
Yang J, Hao T, Sun J et al (2019a) Long noncoding RNA GAS5 modulates α-Solanine-induced radiosensitivity by negatively regulating miR-18a in human prostate cancer cells. Biomed Pharmacother 112:1–8
Yang Y, Wang F, Huang H et al (2019b) LncRNA SLCO4A1-AS1 promotes growth and invasion of bladder cancer through sponging miR-335-5p to upregulate OCT4. Onco Targets Ther 12:1351–1358

Ying L, Chen Q, Wang Y et al (2012) Upregulated MALAT-1 contributes to bladder cancer cell migration by inducing epithelial-to-mesenchymal transition. Mol BioSyst 8:2289–2294

Yu M, Ting DT, Stott SL et al (2012) RNA sequencing of pancreatic circulating tumour cells implicates WNT signaling in metastasis. Nature 487:510–513

Yu L, Gu C, Zhong D et al (2014) Induction of autophagy counteracts the anticancer effect of cisplatin in human esophageal cancer cells with acquired drug resistance. Cancer Lett 355:34–45

Yu J, Han Z, Sun Z et al (2018) LncRNA SLCO4A1-AS1 facilitates growth and metastasis of colorectal cancer through β-catenin-dependent Wnt pathway. J Exp Clin Cancer Res 37:1–12

Yuan SX, Yang F, Yang Y et al (2012) Long noncoding RNA associated with microvascular invasion in hepatocellular carcinoma promotes angiogenesis and serves as a predictor for hepatocellular carcinoma patients' poor recurrence-free survival after hepatectomy. Hepatology 56:2231–2241

Yue B, Cai D, Liu C et al (2016) Linc00152 functions as a competing endogenous RNA to confer oxaliplatin resistance and holds prognostic values in colon cancer. Mol Ther 24:2064–2077

Zhai W, Zhu R, Ma J et al (2019) A positive feed-forward loop between LncRNA-URRCC and EGFL7/P-AKT/FOXO3 signaling promotes proliferation and metastasis of clear cell renal cell carcinoma. Mol Cancer 18:1–15

Zhan T, Rindtorff N, Betge J et al (2019) CRISPR/Cas9 for cancer research and therapy. Semin Cancer Biol 55:106–119

Zhang B, Lu HY, Xia YH et al (2018) Long non-coding RNA EPIC1 promotes human lung cancer cell growth. Biochem Biophys Res Commun 503:1342–1348

Zhang PF, Wu J, Wu Y et al (2019) The lncRNA SCARNA2 mediates colorectal cancer chemoresistance through a conserved microRNA-342-3p target sequence. J Cell Physiol 234:10157–10165

Zhao J (2016) Cancer stem cells and chemoresistance: the smartest survives the raid. Pharmacol Ther 160:145–158

Zhao Q, Zhao S, Li J et al (2019) TCF7L2 activated HOXA-AS2 decreased the glucocorticoid sensitivity in acute lymphoblastic leukemia through regulating HOXA3/EGFR/Ras/Raf/MEK/ERK pathway. Biomed Pharmacother 109:1640–1649

Zhu H, Wang D, Liu Y et al (2013) Role of the Hypoxia-inducible factor-1 alpha induced autophagy in the conversion of non-stem pancreatic cancer cells into CD133 + pancreatic cancer stem-like cells. Cancer Cell Int 13:1–8

Zhu QN, Wang G, Guo Y et al (2017) LncRNA H19 is a major mediator of doxorubicin chemoresistance in breast cancer cells through a cullin4A-MDR1 pathway. Oncotarget 8:91990–92003

Zhu HY, Bai WD, Ye XM et al (2018) Long non-coding RNA UCA1 desensitizes breast cancer cells to trastuzumab by impeding miR-18a repression of Yes-associated protein 1. Biochem Biophys Res Commun 496:1308–1313

Zuckerman JE, Davis ME (2015) Clinical experiences with systemically administered siRNA-based therapeutics in cancer. Nat Rev Drug Discov 14:843–856

Long Noncoding RNAs as Drivers of Acquired Chemoresistance in Hepatocellular Carcinoma

Johanna K. DiStefano and Caecilia Sukowati

Contents

J. K. DiStefano (✉)
Diabetes and Fibrotic Disease Unit, Translational Genomics Research Institute, Phoenix, AZ, USA
e-mail: jdistefano@tgen.org

C. Sukowati
Department of Medicine, University of Udine, Udine, Italy

Italian Liver Foundation, AREA Science Park Basovizza, Trieste, Italy
e-mail: caecilia.sukowati@fegato.it

S. Jurga, J. Barciszewski (eds.), *The Chemical Biology of Long Noncoding RNAs*, RNA Technologies 11, https://doi.org/10.1007/978-3-030-44743-4_8

Abstract Hepatocellular carcinoma (HCC) is a leading cause of cancer-related death worldwide. While recent improvements in the clinical management of HCC have yielded more favorable outcomes for patients, five-year survival rates for these individuals remain poor. Effective treatment of HCC is complicated by a complex disease etiology, significant intertumoral and intratumoral heterogeneity, the frequent absence of early-stage symptoms, and a high potential for metastasis. HCC is highly refractory to chemotherapy, largely due to the acquisition of multidrug resistance, which limits effective pharmacological treatments for HCC patients. While the mechanisms underlying the development of chemoresistance in HCC are diverse, a number of recent studies have positioned long noncoding RNAs (lncRNAs) as key participants in the process. Here, we provide an overview of HCC and the obstacles posed by acquired chemoresistance and summarize the empirical evidence supporting a role for lncRNAs in the development of acquired resistance, focusing on the specific molecular mechanisms by which these molecules attenuate chemosensitivity. We also discuss the contribution of extracellular vesicles as mediators of lncRNA-induced chemoresistance. Although newly emerging, the literature has already yielded important insights into the complex mechanisms by which lncRNAs modulate chemoresistance in HCC.

Keywords Acquired chemoresistance · Liver cancer · Extracellular vesicles · Long noncoding RNAs · Sorafenib · Doxorubicin · Hepatocellular carcinoma

1 Hepatocellular Carcinoma

1.1 Epidemiology and Risk Factors

Liver cancer refers to any cancer that originates and spreads from cells in the liver. As of this writing, liver cancer is the sixth most commonly diagnosed cancer and the fourth leading cause of cancer-related mortality worldwide (Bray et al. 2018). In men, incidence and mortality are approximately twice that in women, thus ranking liver cancer fifth in terms of global cases and second with respect to deaths (Bray et al. 2018). Hepatocellular carcinoma (HCC) is the most common form of liver cancer and accounts for about 90% of liver cancer cases (Balogh et al. 2016). The incidence and mortality of HCC have been increasing in North America and many European regions and declining in traditionally high-risk countries, such as Japan and some parts of China (Kulik and El-Serag 2019). In the United States, HCC has an annual incidence of at least 6 per 100,000 individuals (El-Serag and Kanwal 2014) and represents the fastest rising cause of cancer-related death (Mittal and El-Serag 2013). Cases of HCC are also predicted to rise in response to population aging and growth (Akinyemiju et al. 2017).

The geographic distribution of HCC is paralleled by its etiology and major risk factors. Approximately 90% of HCC cases are associated with a known underlying etiology (Akinyemiju et al. 2017). HCC predominantly arises in the context of liver cirrhosis, which is estimated to affect 1–2% of the global population and result in more than one million deaths/year worldwide (Forouzanfar et al. 2015). Cirrhosis is the terminal stage of progressive fibrotic liver diseases, and is mainly caused by chronic viral hepatitis (HBV and HCV), excess alcohol consumption, and nonalcoholic fatty liver disease (NAFLD)/nonalcoholic steatohepatitis (NASH) (Tsochatzis et al. 2014). Less than 20% of HCC cases develop in the non-cirrhotic liver (Alkofer et al. 2011). The incidence of NAFLD/NASH-related HCC is anticipated to increase in concert with the emerging global epidemic of metabolic disease and related comorbidities (Dhamija et al. 2019).

1.2 Treatment

Treatment strategies for HCC depend on the profile of the tumor, degree of liver dysfunction, and the performance status of the patient (Kulik and El-Serag 2019). The most widely used HCC staging is based on the consensus of the Barcelona Clinic for Liver Cancer (BCLC) staging system, which pairs these parameters with a recommended guided therapy (Llovet et al. 1999). It is the main clinical algorithm for the stratification of patients according to prognosis and treatment allocation (Llovet et al. 2016).

Based on the consensus for the clinical management of HCC (Llovet et al. 2012), surgical intervention is the main curative treatment for HCC and leads to the best outcomes (five-year survival of 60–80%) in well-selected candidates (Galle et al. 2018). Liver resection and liver transplantation represent the first option in patients with early-stage HCC on an intention-to-treat. In clinical practice, surgery can often be extended to other stages of HCC upon down-staging by any non-surgical means (Galle et al. 2018). Image-guided radiofrequency ablation (RFA) is the treatment of choice for patients with early-stage HCC when liver transplantation or hepatectomy is precluded. For patients in intermediate HCC stage, palliative treatment using transarterial chemoembolization (TACE) is recommended (Llovet et al. 1999, 2012). For advanced HCC, sorafenib (discussed below) is the standard treatment, although it only extends overall survival by approximately 3 months (Llovet et al. 2008). Survival benefit from sorafenib in all subgroups of patients with advanced HCC appears to be highest in patients without extrahepatic manifestation (Bruix et al. 2017a). At this time, other tyrosine kinase inhibitors are under investigation in the first-line setting, either alone or in combination with sorafenib (Sun and Cabrera 2018). In early- and intermediate-stage HCCs, combination therapies with RFA and/or TACE with tyrosine kinase inhibitors are also used in some patients.

1.3 Chemotherapy Agents

In systemic therapies for advanced HCC, sorafenib, an oral multi-kinase inhibitor, was the first approved therapy that demonstrated an improvement in median overall survival (Bruix et al. 2017a; Llovet et al. 2008). Until now, five systemic therapies (sorafenib and lenvatinib in the first-line setting and regorafenib, cabozantinib, and ramucirumab in the second-line setting) are already under phase III trial and are showing modest survival benefits (Abou-Alfa et al. 2018; Bruix et al. 2017b; Kudo et al. 2018; Llovet et al. 2008). However, many patients are highly refractory to systemic therapy. For instance, resistance to sorafenib is quite common (Zhu et al. 2017) and in some patients, HCC heterogeneity appears to be a significant determinant (Niu et al. 2017). Evidence supporting the impact of HCC heterogeneity on patient response to systemic therapies is emerging. In an analysis of primary cell cultures from 55 different tumoral regions from ten HBV-associated HCCs, there was considerable intratumor heterogeneity and branched evolution in all ten tumors (Gao et al. 2017). Genetic alterations that can be targeted by systemic therapy such as sorafenib were identified in intratumor sub-regions from four of these HCCs and were associated with drug sensitivity. In contrast, cells from the remaining sub-regions, which did not have these alterations, were not sensitive to therapy (Gao et al. 2017).

In an exploratory study of patient response to regorafenib (seven responders versus ten non-responders), next-generation sequencing revealed 49 variants in 27 oncogenes or tumor suppressor genes, together with nine plasma miRNAs, which contributed to drug response (Teufel et al. 2019). A recent study of 126 proteins in 34 liver cancer cell lines reported the association between genetic alterations and gene expression patterns with treatment response against a number of anticancer agents, including sorafenib, trametinib, and FGFR4 inhibitors (Caruso et al. 2019). The use of genetic information may be helpful in selecting patients for clinical trials.

For patients receiving chemotherapies (mainly doxorubicin, cisplatin, and 5-fluorouracil [5-FU]), many antineoplastic agents have been tested, but response rates have been low, ranging from only 10% to 15% (Abou-Alfa et al. 2008). Significant toxicity and low effectiveness are also severe limitations of currently available chemotherapeutic HCC drugs (Sukowati et al. 2010).

In TACE patients, studies highlighted the use of drug-eluted-beads TACE (DEB-TACE) with comparable or better results compared to conventional TACE (Lee et al. 2017; Nam et al. 2016; Woo and Heo 2015). A recent network meta-analysis study using 17 randomized controlled trials showed that the combination of doxorubicin and mitomycin plus gemcitabine may be the most optimal regimen in TACE-treated HCC (Guo et al. 2018). A retrospective imaging feature study of 32 patients who underwent DEB-TACE showed that HCC intertumoral heterogeneity was significantly related to the complete response of the patients (Reis et al. 2017). Interestingly, cellular response to doxorubicin from one individual HCC also varied. From a single HCC case (an ES grade 1, HCV+), different cellular clones showed different sensitivity to doxorubicin (Parent et al. 2014). The authors suggested that disparate sensitivities may be related to the cancer stem cell (CSC) features of the cells. CSCs are responsible for tumor initiation, relapse, and metastasis. In addition, these cells are resistant against conventional therapies. Hepatic CSC exhibits high expression of the ATP-binding cassette sub-family G member 2 (ABCG2) protein, which is a multidrug transporter that removes drugs from the cytoplasm, thereby maintaining a non-toxic intracellular environment (Holczbauer et al. 2013; Hu et al. 2008; Jia et al. 2013; Shi et al. 2008). ABCG2 plays a role in chemoresistance, the main defensive mechanism of cancer cells against chemotherapy (Dean 2009; Doyle and Ross 2003). This protein is also considered as the molecular determinant of the side population (SP) of progenitor/stem cells, mainly based on its functional activity in pumping Hoeschst 33,342 stain from cells (Goodell et al. 1997; Zhou et al. 2001). SP cells are more resistant to anticancer drugs, including doxorubicin, 5-FU, and gemcitabine, than non-SP cells (Haraguchi et al. 2006a; Haraguchi et al. 2006b). However, phenotypically separated CSC shows a similar tendency. CSC CD133+ cells had a lower sensitivity to both chemotherapy (doxorubicin) and radiotherapy compared to CD133- cells (Lan et al. 2013; Li et al. 2016a; Piao et al. 2012), mediated by preferential activation of Akt/PKB and Bcl-2 cell survival response (Chiba et al. 2012). Other CSCs have also been reported to convey drug resistance (Chiba et al. 2012; Haraguchi et al. 2010; Ho et al. 2012).

1.4 HCC: Chemoresistance and Heterogeneity

Despite the rapid emergence of targeted therapy development, intermediate and advanced-stage HCC remain largely incurable due to low response rate and therapeutic resistance (Chen et al. 2019). HCC heterogeneity represents a challenge for the establishment of a robust HCC classification and may contribute to treatment failure and drug resistance in many cases of HCC (Friemel et al. 2015; Li et al. 2019b). For example, intermediate (BCLC stage B) and advanced HCC (BCLC stage C) comprise a wide range of heterogeneous patients with different tumor profiles and liver functions. However, recommended treatments for BCLC stage B and stage C are standardized to TACE and sorafenib, respectively. Thus, the response of treatment might be different between patients. Thus, several groups have proposed an urgent sub-classification of BCLC B and C to improve the effectiveness of therapeutic options (Golfieri et al. 2019; Sun et al. 2019). Such heterogeneity may cause an imbalance among the enrolled patient population, premature withdrawal from clinical trials, and variable response to treatment (Liu et al. 2019).

Human genetic heterogeneity contributes to HCC development and chemosensitivity. Acquired resistance to anticancer drugs is often associated with increased expression of the ATP-binding cassette subfamily B member 1 (*ABCB1*) gene, also known as the multidrug resistance 1 (*MDR1*) gene, which is a member of the ATP-binding cassette (ABC) transporter superfamily (Ye et al. 2013). The

ABCB1 protein, an ATP-dependent drug efflux pump, functions to remove xenobiotic compounds from cells (Roninson et al. 1986; Ye et al. 2013). The upregulation of *ABCB1* expression promotes greater efflux of xenobiotics, which leads to reduced chemosensitivity. Several studies have shown that genetic variants in the *ABCB1* gene increased susceptibility to HCC. A meta-analysis of eight studies, comprised of eleven cohorts (4407 HCC cases and 4436 controls), calculated the probability of *ABCB1* mutant/wild type alleles for HCC development. The presence of the *ABCB1* variant allele significantly increased the risk for HCC development under all genetic models analyzed (allelic, dominant, or recessive models; $P < 0.001$ for each) (Wang et al. 2015). An investigation of *ABCB1* and *ABCG2* (ATP-binding cassette sub-family G member 2) genetic variants in 47 sorafenib-treated patients reported statistically significant evidence for the association between *ABCB1* 3435 C > T, *ABCG2* 34 G > A and *ABCG2* 1143 C > T and sorafenib exposure (Tandia et al. 2017). Interestingly, heterozygous patients showed the lowest sorafenib plasma levels and had a tendency toward a better clinical response.

1.5 HCC: Mechanisms of Acquired Chemoresistance

Molecular and cellular mechanisms to resist the introduction of xenobiotics are complex processes (Fig. 1). Following exposure to antineoplastic agents and multi-kinase inhibitors, alone or in combination regimens, HCC might activate an escape pathway to mitigate cell death. Several known mechanisms include the induction of drug transporters and changes in apoptosis/survival pathways. Acquired sorafenib resistance has been associated with the dysregulation of various signaling pathways (as reviewed in Nishida et al. 2015), modulation of autophagy and hypoxia (Long et al. 2019; Mendez-Blanco et al. 2018; Prieto-Dominguez et al. 2016; Song et al. 2019), and alterations in cell metabolism (Kim et al. 2017; Lee et al. 2018). However, cell response to chemotherapy also depends on the integrative crosstalk of various intracellular signals and the microenvironment (Chen et al. 2015). Further, cancer cell plasticity and genomic instability are, and will remain, major barriers to the efficiency and effectiveness of drug delivery (Saunders et al. 2012).

In addition to the complexity of the molecular pathways dysregulated in drug resistance, epigenetic regulation via alterations of noncoding RNAs (ncRNAs), including microRNAs (miRNA: <~28 nucleotides) and long noncoding RNAs (lncRNAs: > 200 nucleotides), are emerging as critical points of regulation. These functional RNA molecules have been shown to be important regulatory molecules in many, if not all, biological processes across organisms. Whereas, the involvement of miRNAs in acquired chemoresistance in HCC, mainly through the modulation of drug transporters and signaling pathways, has been well characterized (Tricoli et al. 2019; Xu et al. 2018b), mounting evidence is now revealing that lncRNAs are also mechanistically involved in HCC chemoresistance.

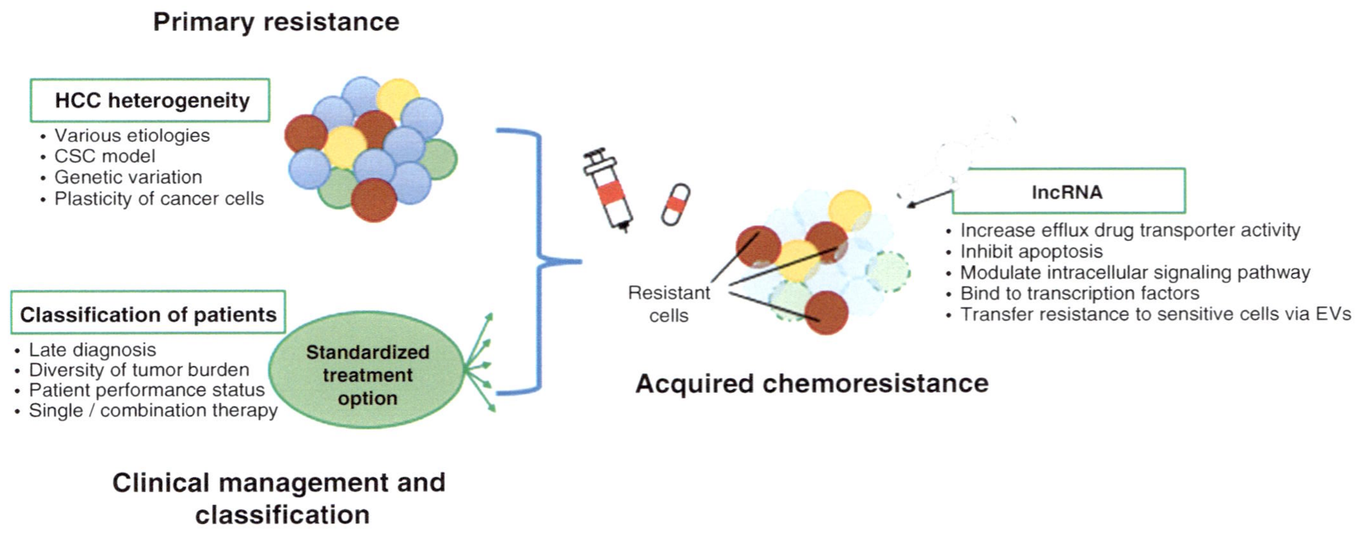

Fig. 1 Factors involved in acquired chemoresistance in HCC. Chemoresistance is attributed to various factors, including the nature of the HCC and the host genetic background, as well as the treatment administered. Upon exposure to chemo- or systemic therapy against HCC, lncRNAs participate in diverse processes to avoid cell death and transfer drug resistance capacity to chemosensitive cells

2 LncRNAs Associated with Acquired Chemoresistance in HCC

While the mechanisms underlying the development of chemoresistance in HCC are diverse, a number of recent studies have positioned lncRNAs as key participants in the process. These molecules are known to participate in a wide range of cellular activities including differentiation, proliferation, migration, invasion, and apoptosis (DiStefano 2018). lncRNAs also regulate gene expression, sequester miRNAs, and affect genomic imprinting, alternative splicing, and other molecular processes (Moran et al. 2012; Nagano and Fraser 2011; Wang and Chang 2011; Yan and Wang 2012). An increasing number of experimental studies have shown that lncRNAs promote the development of multidrug resistance in cancer cells. In the following paragraphs, we provide an overview of the lncRNAs that have been linked to acquired resistance to the most common chemotherapeutic agents used in the treatment of HCC, including doxorubicin, cisplatin, sorafenib, 5-fluorouracil, and oxaliplatin (Table 1).

2.1 *H19*

The *H19* gene encodes a 2.3 kb maternally expressed lncRNA (Rachmilewitz et al. 1992) that has been associated with cell proliferation and regulation of gene expression (Gabory et al. 2010). The *H19* gene is located on chromosome 11p15.5, approximately 100 kb distal of insulin-like growth factor 2 (*IGF2*), and these two genes are transcribed from a conserved imprinted gene cluster (Gabory et al. 2010). *H19* expression has been linked with most cancers and appears to play a role in all stages of tumorigenesis, although both tumor-promoting and tumor-suppressing actions of the lncRNA have been reported (Raveh et al. 2015). Likewise, in studies of HCC development and progression, measurement of *H19* expression has also yielded conflicting results (He et al. 2014; Matouk et al. 2007; Schultheiss et al. 2017; Yoshimizu et al. 2008), suggesting that the role of *H19* in HCC and other cancers is complicated and factors such as cell type, tumor type, or tumor may contribute to differences in expression.

With respect to chemoresistance, in a comparison of doxorubicin-resistant human hepatoma HepG2 cells and chemosensitive parent cells, expression of *H19* was significantly upregulated (Tsang and Kwok 2007). Knockdown of *H19* expression resulted in a significant increase in doxorubicin potency (as measured by the half maximal inhibitory concentration [IC_{50}]), while overexpression of the lncRNA promoted chemoresistance. *ABCB1* was also upregulated In HepG2 cells with acquired doxorubicin resistance, relative to the parent cell line, suggesting a mechanism by which increased efflux may underlie the development of chemoresistance. Interestingly, *H19* knockdown corresponded with suppressed ABCB1 transcript and protein levels in both doxorubicin-resistant HepG2 cells and parent HepG2 cells,

Table 1 lncRNAs involved in acquired chemoresistance in HCC

lncRNA	Drug	Model	Dir[a]	Mechanism	Ref
H19	Dox[b]	HepG2	Up	Regulation of ABCB1 promoter methylation	Tsang and Kwok (2007)
	Dox Sora[c]	HepG2 PLC/PRF/5 Huh7	Down	Increasing cytotoxic action (doxorubicin) or decreasing cell proliferation (sorafenib)	Schultheiss et al. (2017)
HOTAIR	Dox, CDDP[d]	HepG2	Up	Not known	Yang et al. (2011)
	CDDP	Huh7	NA	Modulation of pSTAT3-ABCB1	Zhou et al. (2017)
MALAT1	Dox, CDDP	HepG2	Up	Not known	Lai et al. (2012)
TUG1	CDDP	HCC patients HepG2 SMMC-7221	Up	Regulation of ABCB1 expression	Yang et al. (2016)
HULC	Oxa[e], 5-FU[f], pirarubicin	HepG2 Xenograft model	Up	Participation in Sirt1-USP22-autophagy pathway	Xiong et al. (2017)
lncARSR	Dox	HepG2 SMMC-7221	Up	Modulation of PTEN-PI3k/Akt pathway	Li et al. (2017b)
HANR	Dox	Hep3B Huh7 Xenograft model	Up	Modulation of GSKIP/GSK3β axis	Xiao et al. (2017)
NR2F1-AS1	Oxa	HepG2 Huh7	Up	Competitive sponging of miR-363, leading to increased expression of ABCC1	Huang et al. (2018)
SNHG16	5-FU	Hep3B Huh7	Down	Competitive sponging of miR-93	Xu et al. (2018a)
KRAL	5-FU	HepG2 SMMC-7221	Down	Regulation of KEAP1 through competitive sponging of miR-141	Wu et al. (2018)
NRAL	CDDP	HepG2 SMMC-7221	Up	Regulation of NRF2 through competitive sponging of miR-340-5p	Wu et al. (2019)
NEAT1	Sora Dox	HepG2 PLC/PRF/5 Huh7	Up	Not known	Kessler et al. (2019)
PDZD7	5-FU Sora	HepG2 Bel-7402	Up	Lnc-PDZD7-miR-101-EZH2-ATOH8 axis	Zhang et al. (2019)

[a]Direction of expression change in chemoresistance, [b]doxorubicin, [c]sorafenib, [d]cisplatin, [e]oxaliplatin, [f]5-fluorouracil

resulting in increased cellular doxorubicin levels. Promoter methylation, known to be critical for regulating *ABCB1* expression, was lower in doxorubicin-resistant-HepG2 cells (~50%) compared to HepG2 cells (88%), and *H19* knockdown increased promoter methylation in both cell lines, consistent with downregulation of gene expression. However, in an independent study, *H19* expression was found to be downregulated, not upregulated in chemoresistant cells (Schultheiss et al. 2017). Three different hepatoma cell lines, HepG2, Huh7, and PLC/PRF/5, were rendered chemoresistant to doxorubicin or sorafenib and then used to assess *H19* expression. In all chemoresistant cell lines, *H19* expression was reduced. In functional studies, overexpression of *H19* corresponded with suppressed tumor cell survival and proliferation following doxorubicin or sorafenib treatment. In human HCC samples drawn from four different cohorts, *H19* expression was decreased relative to normal tissues, consistent with a protective role for *H19* in HCC. To our knowledge, these are the only investigations of *H19* in HCC chemoresistance, and more work is needed to gain a clearer understanding of this lncRNA in modulating sensitivity to anticancer drugs. With regard to the discrepant results, it is worth noting that additional studies are needed to unravel the mechanisms by which *H19* contributes to the etiology, progression, metastasis, and chemoresistance of HCC.

2.2 *HOTAIR*

The *HOTAIR* (Homeobox [HOX] transcript antisense RNA) gene is located on chromosome 12 and encodes a 2.2 kb transcript. HOTAIR was initially identified as part of an investigation of transcriptional and epigenetic features of HOX loci in primary human fibroblasts (Rinn et al. 2007). Upregulated expression of *HOTAIR* has been associated with many different kinds of cancer (Bhan and Mandal 2015; Gupta et al. 2010; Loewen et al. 2014; Luo et al. 2016). In HCC, *HOTAIR* expression was significantly upregulated in tumors of >60% of HCC patients and in four different liver cancer cell lines (Bel-7402, HCCLM3, HepG2, and SMMC-7721) (Yang et al. 2011). *HOTAIR* expression was found to independently predict HCC recurrence in liver transplantation patients, and high *HOTAIR* levels were associated with a shorter recurrence-free survival. Knockdown of *HOTAIR* expression in HepG2 cells corresponded with a dose-dependent increase in sensitivity to cisplatin (2 μg/ml and 4 μg/ml) and doxorubicin (1 μg/ml and 2 μg/ml) (Yang et al. 2011). These results were replicated in Huh7 hepatoma cells, where *HOTAIR* knockdown led to decreased cell viability in response to cisplatin (Zhou et al. 2017). *HOTAIR* depletion also inhibited Huh7 proliferation and migration, downregulated expression of transcript and protein levels of ABCB1, and reduced levels of phosphorylated STAT3 (pSTAT3). Inhibition of STAT3 with the Janus kinase 2 (JAK2) inhibitor, AG490, reduced levels of pSTAT3 and ABCB1, which were decreased even further with *HOTAIR* knockdown. Combined, the results indicated that *HOTAIR* mediates cisplatin sensitivity through a STAT3-ABCB1 mechanism, although the effects of cisplatin sensitivity in Huh7 cells in the presence of AG490 and *HOTAIR* siRNA were not investigated. While these two studies provide preliminary evidence that *HOTAIR* may play a role in chemoresistance in HCC cells, it is worth noting that the magnitude of effects on chemosensitivity was modest. Further, these results did not determine whether *HOTAIR* directly promotes chemoresistance or instead, elicits decreased chemosensitivity as an indirect result of other cellular changes induced by *HOTAIR*, and it remains to be seen whether *HOTAIR* expression is altered in cisplatin-resistant HCC patients compared to those who are sensitive to cisplatin. Despite these limitations, numerous studies have established a convincing role for *HOTAIR* in multiple pathways associated with HCC tumor progression and metastasis (DiStefano 2017).

2.3 *MALAT1*

MALAT1 (metastasis-associated lung adenocarcinoma transcript 1), also known as NEAT2 (nuclear paraspeckle assembly transcript 2), is an 8.8 kb gene with a plus-strand orientation located on human chromosome 11q13.1. *MALAT1* is strongly conserved among mammalian species and may function, in part, to regulate pre-mRNA splicing and gene expression (Sun and Ma 2019). Although initially associated with poor clinical outcomes and overall survival in patients with early-stage non-small cell lung cancer or NSCLC (Ji et al. 2003), aberrant *MALAT1* expression has since been observed in a variety of other cancers, including leukemia, pancreatic cancer, breast cancer, and others (Sun and Ma 2019). In HCC, *MALAT1* was significantly upregulated in neoplastic samples relative to expression in noncancerous tissue adjacent to HCC tumors and normal tissue (Luo et al. 2006). Like NSCLC patients (Ji et al. 2003), HCC patients with high *MALAT1* expression experienced worse outcomes compared to those with low *MALAT1* levels (Lai et al. 2012). MALAT1 expression has also been preliminarily associated with chemoresistance. In that study, treatment of HepG2 cells with doxorubicin (0.5 μg/ml) or cisplatin (4 μg/ml) increased the number of apoptotic cells in the presence of siRNA against *MALAT1* compared to cells treated with control siRNA, which also corresponded with decreased cell proliferation and inhibition of cell migration and invasion (Lai et al. 2012). As with *HOTAIR*, it is not known whether *MALAT1* directly affects chemosensitivity in liver cancer cells, or whether the induced cytotoxicity of doxorubicin and cisplatin is reflective of the biological impact exerted by the lncRNA. Further, to our knowledge, this is the only study linking aberrant *MALAT1* expression in resistance to chemotherapies relative to HCC, and greater exploration of this relationship is necessary.

2.4 *TUG1*

TUG1 (Taurine upregulated gene 1) is a 6.7 kb lncRNA located on chromosome 22q12 that was first identified in a screen for genes regulated by the cysteine derivative, taurine, in mouse retina cells and found to contribute to photoreceptor formation (Young et al. 2005). Since then, aberrant expression of *TUG1* has been detected in several types of cancer and in models of these tumors (Li et al. 2016b), knockdown of *TUG1* expression suppressed cell proliferation, invasion, and colony formation (Xu et al. 2015; Zhang et al. 2014). *TUG1* also appears to be upregulated in HCC and promotes cell growth and apoptosis through a mechanism involving Krüppel-like factor 2 (KLF2) silencing (Huang et al. 2015). In HCC tissues, *TUG1* levels were approximately two-fold higher in cisplatin-resistant patients compared to cisplatin-sensitive patients (Yang et al. 2016). Up to three-fold higher expression of *TUG1* was observed in cisplatin-resistant hepatoma cells (SMMC-7721/CDDP and HepG2/CDDP) compared to parental cells. Knockdown of *TUG1* in CDDP-resistant SMMC-7721 and HepG2 cells resulted in lower cell viability, lower cell survival rate, and an increase in apoptotic cells compared to cells treated with control siRNA. In contrast, *TUG1* overexpression in SMMC-7221 and HepG2 cells exhibited higher cell viability in response to cisplatin treatment compared to cells treated with control plasmid. Treatment of cisplatin-resistant cells with *TUG1* siRNA reduced transcript and protein levels of ABCB1, while overexpression of *TUG1* in parent cells corresponded with elevated ABCB1 levels, suggesting that the mechanism by which this lncRNA mediates cisplatin resistance depends, at least in part, on the regulation of this gene. However, the specific manner in which *TUG1* exerts effects on *ABCB1* expression remains to be determined.

A major strength of this study was the demonstration of *TUG1* upregulation in HCC tissues from cisplatin-resistant patients compared to cisplatin-sensitive patients prior to confirmation using in vitro model systems. The identification of similar expression patterns in human tissue and cell models provides strong support for this lncRNA as a modulator of HCC chemoresistance. Because the mechanism by which *TUG1* may affect cisplatin resistance is via a protein with broad substrate specificity, it would also be interesting to see if *TUG1* mediates chemoresistance to other drugs used in the pharmacological treatment of HCC.

2.5 *HULC*

HULC (highly upregulated in liver cancer) was identified using a two-step process of (1) generating HCC-specific cDNA libraries by subtractive suppressive hybridization and creating an HCC-specific cDNA microarray comprised of ~7000 cDNA clones from these libraries, and (2) screening microarrays with samples from patients with HCC, focal nodular hyperplasia, cirrhosis, and non-neoplastic liver (Panzitt et al. 2007). *HULC* expression was 33-fold higher in 76% of HCCs compared to the non-neoplastic liver pool. *HULC* levels were most abundant in cytoplasm, where the lncRNA is associated with ribosomes (Panzitt et al. 2007). Characterization of this lncRNA identified a chromosomal location on 6p24.3 and a transcript size of ~0.5 kb. *HULC* was also primarily found in the cytoplasm and was co-purified with ribosomes of carcinoma cells (Panzitt et al. 2007). *HULC* was found to be evolutionarily conserved in primates, but neither mouse nor rat genomes appeared to have a HULC homolog (Panzitt et al. 2007). A cAMP response element binding (CREB) protein binding site in the proximal promoter region was found to be critical for HULC promoter activity in liver cancer (Wang et al. 2010).

Since this initial study, numerous reports of the association between *HULC* and multiple properties of HCC have followed (DiStefano 2017), and a pathway by which *HULC* was found to stabilize silent information regulator 1 (Sirt1) protein and induce autophagy in HCC cells was characterized (Xiong et al. 2017). Treatment of HepG2 cells with oxaliplatin, 5-fluorouracil, or pirarubicin upregulated *HULC* expression in a dose-dependent manner (Xiong et al. 2017). Knockdown of *HULC* expression corresponded with upregulated levels of cleaved-poly (ADP-ribose) polymerase (PARP), increased inhibition of cell survival in response to oxaliplatin, and enhanced inhibition of cell growth mediated by treatment with 5-fluoruouracil and pirarubicin. Interestingly, silencing of *HULC*, *SIRT1*, and *USP22*, a deubiquitinase, or inhibition of autophagy augmented apoptosis in response to oxaliplatin treatment, collectively suggesting that the *HULC-USP22-SIRT1*-autophagy pathway may contribute to the development of oxaliplatin resistance in liver cancer cells. When this pathway was explored in xenograft tumor models in nude mice, knockdown of *HULC* or *SIRT1* attenuated tumor growth and oxaliplatin enhanced this effect, suggesting that targeting *HULC* or *SIRT1* expression may sensitize HCCs to this antitumor compound. These results laid reasonable groundwork implicating *HULC* in the development of chemoresistance and indicate that targeting the *HULC/SIRT1*/autophagy pathway may improve sensitivity in HCC chemotherapy.

2.6 *lncARSR*

LncARSR (activated in RCC with sunitinib resistance) was initially found to promote resistance to sunitinib, a multi-targeted receptor tyrosine kinase (RTK) inhibitor, in renal cell carcinoma (RCC) (Qu et al. 2016a). In that study, lncARSR was identified through the screening of parental and sunitinib-resistant RCC cells and validation in patient-derived xenograft models (PDXs). Of the lncRNAs upregulated in PDXs showing poor sunitinib response, but not those with good response, only one lncRNA suppressed resistance in subsequent loss-of-function studies. Knockdown of this lncRNA corresponded with decreased chemoresistance compared to the other lncRNAs and was therefore named *lncARSR* and selected for molecular characterization. *LncARSR* (Ensembl ID: *ENST00000424980*) was determined to be located on chromosome 9 with a full-length of 591 nucleotides and comprised of four exons.

Subsequent work found that *lncARSR* overexpression in SMMC-7721 and HepG2 cells attenuated the doxorubicin-mediated loss of cell viability, while *lncARSR* knockdown increased sensitivity to doxorubicin in these cells, demonstrating that the lncRNA modulates resistance to multiple anti-cancer compounds in liver cancer cells (Li et al. 2017b). Injection of *lncARSR*-depleted cells into mice increased sensitivity of xenografts to doxorubicin, thereby validating the in vitro findings (Li et al. 2017b). *lncARSR* was found to bind with the transcript of *PTEN*, a tumor suppressor, and promote its degradation, and was negatively correlated with *PTEN* expression in HCC tissues. PTEN inhibits tumor progression through the PI3K/Akt pathway (Georgescu 2010). Treatment of cells with LY294002, a PI3K/Akt pathway inhibitor, or *PTEN* depletion attenuated the effects of *lncARSR* on doxorubicin resistance in HCC cells. In HCC tissues, *lncARSR* was significantly upregulated compared to adjacent noncancerous liver tissues, and HCC patients with high *lncARSR* expression presented with poor clinicopathological features, such as liver cirrhosis, large tumor size, advanced BCLC stage, and poor recurrence-free and overall survival (Li et al. 2017b).

2.7 *HANR*

Transcriptomic profiling of HCC and adjacent non-cancerous peri-tumor tissues revealed significant upregulation of a new lncRNA that was named *HANR* (HCC-associated long noncoding RNA) (Xiao et al. 2017). Elevated *HANR* expression was also observed in Hep3B and Huh7 hepatoma cell lines compared to normal immortalized hepatocytes. In HCC patients, *HANR* expression was positively correlated with advanced tumor stage, distant metastasis, and worse overall survival compared to a patient with low *HANR* expression. Knockdown of *HANR* expression resulted in decreased cell viability, increased cell apoptosis and increased caspase-3/7 activity compared to control conditions; opposite findings were observed with *HANR* overexpression. In Hep3B and Huh7 cells treated with doxorubicin, *HANR* knockdown enhanced cytotoxic effects, while *HANR* overexpression attenuated them. In nude mouse xenograft models, *HANR* augmented the cytotoxic effects of doxorubicin and the combination of *HANR* knockdown and doxorubicin yielded the strongest tumor suppression. *HANR* was visualized predominantly in the cytoplasm and was found to interact with glycogen synthase kinase beta (GSK3β) interaction protein (GSKIP) in RNA pulldown assays. While neither *HANR* nor doxorubicin exerted effects on GSKIP expression, knockdown or overexpression of *HANR* increased or decreased phosphorylation levels of GSK3β. Although preliminary, these results indicate that *HANR* modulate sensitivity to doxorubicin in liver cancer cells through participation in the GSKIP/GSK3β axis.

A subsequent study confirmed the upregulation of *HANR* expression in HCCs and the association between *HANR* knockdown and attenuated cell proliferation, migration, and invasion of HCC cells (Shi et al. 2018). MiR-214 was identified to be a potential target of *HANR* using bioinformatic algorithms, and this relationship was functionally confirmed in vitro. Treatment of HepG2 cells with knocked-down *HANR* expression (shHANR) with miR-214 inhibitor led to increased proliferation, migration, and invasion, indicating rescue of the shHANR-induced tumor phenotype. The *HANR*-miR-214 interaction was found to regulate EZH2 and TGFB, leading to a restoration of the tumor phenotype. Although preliminary, these two studies provide evidence suggesting that the newly discovered *HANR* may play roles in both HCC pathogenesis and HCC chemoresistance. At this time, features of HANR, such as its chromosomal location, size, and tissue expression, remain undetermined.

2.8 *NR2F1-AS1*

In an analysis of lncRNAs showing differential expression between oxaliplatin-resistant Huh7 cells and parental cells, *NR2F1-AS1* (nuclear factor subfamily 2, group F, member 1, antisense 1) was identified as a novel lncRNA showing at least two-fold higher expression in chemoresistant cells (Huang et al. 2018). This lncRNA (ENSG00000237187) localizes to chromosome 5q15 and is 176 kb long. *NR2F1-AS1* expression was also upregulated in HCC tissue from 47 oxaliplatin-resistant patients compared to matched oxaliplatin-sensitive patients. Knockdown of *NR2F1-AS1* expression improved oxaliplatin sensitivity in HCC cells and correlated with reduced expression of several drug-resistance genes, including *ABCB1*, *ABCC5* (ATP binding cassette subfamily C member 5), and *LRP1* (LDL receptor-related protein 1) compared to control cells. In addition, *NR2F1-AS1* knockdown in oxaliplatin-resistant Huh7 and HepG2 cells reduced both cell migration and invasion compared to control cells, while in a xenograft nude mouse model, *NR2F1-AS1* depletion corresponded with suppressed tumor growth. Using miRNA-target binding algorithms, miR-363 was predicted to bind with the *NR2F1-AS1* 3′ untranslated region (3'UTR) and this interaction was verified using luciferase assay. Interestingly, miR-363 expression was reduced in oxaliplatin-resistant cells compared with parental cell lines but increased in the same cells following N*R2F1-AS1* knockdown. In addition to *NR2F1-AS1*, miR-363 was predicted to target the 3'UTR of *ABCC1* (ATP binding cassette subfamily C member 1), which, like *ABCB1*, is known to contribute to the development of drug resistance. *ABCC1* levels decreased in response to treatment with *NR2F1-AS1* siRNA and increased in the presence of miR-363 siRNA. Treatment with siRNAs against both miR-363 siRNA and *NR2F1-AS1* resulted in expression levels intermediary between the two treatments. The authors hypothesized that *NR2F1-AS1* may modulate oxaliplatin resistance via a mechanism involving endogenous sponging of miR-363, leading to higher *ABCC1* expression. Correspondingly, increased levels of *ABCC1* would result in greater removal of intracellular oxaliplatin, leading to reduced cytotoxicity of the antineoplastic agent.

2.9 *SNHG16*

SNHG16 (small nucleolar RNA host gene 16) is a ubiquitously expressed 7.6 kb lncRNA located on human chromosome 17q25.1 that has been associated with a variety of cancers, including esophageal (Zhang et al. 2017), bladder (Cao et al. 2018), gastric (Lian et al. 2017), breast (Cai et al. 2017), and neuroblastoma (Yu et al. 2009). *SNHG16* has also been shown to promote cancer cell proliferation and migration and predict poor prognosis (Cao et al. 2018; Yu et al. 2009). In HCC, *SNHG16* expression was downregulated in tumors compared to pair-matched noncancerous tissues in HCC patients and in human HCC cell lines relative to normal liver cells (Xu et al. 2018a). In Hep3B and Huh7 cells, *SNHG16* overexpression inhibited cell proliferation, consistent with earlier findings in other cell lines, and corresponded with decreased cell viability in response to treatment with 5-fluorouracil. A potential interaction between *SNHG16* and miR-93 was predicted algorithmically and verified experimentally with a luciferase assay. Overexpression of *SNHG16* reduced miR-93 levels, suggesting an miRNA sponging mechanism, while overexpression of miR-93 corresponded with enhanced HCC cell proliferation and increased 5-fluorouracil resistance. The findings indicate that decreased *SNHG16* expression in HCC may result in de-repression of miR-93 and subsequently lead to the development of chemoresistance.

2.10 *KRAL and NRAL*

In a comparison of lncRNA expression patterns between 5-fluorouracil-resistant HepG2 cells and parental cells (HepG2/5-FU vs HepG2), > 3000 differentially regulated transcripts were identified (Wu et al. 2018). Of these, the authors focused on the most downregulated lncRNA (ENST000004977918), which showed a 20-fold change in the expression of 5-fluorouracil-resistant cells compared to parent cells. Because of its proximity to the gene encoding Kelch-like ECH-associated protein 1 (*KEAP1*) on 19q10.14, this lncRNA was named KEAP1 regulation-associated lncRNA (*KRAL*). Because *KRAL* expression was negatively associated with 5-FU resistance, the researchers hypothesized that 5-fluorouracil may downregulate *KRAL* expression. To address this possibility, HepG2 and SMMC-721 cells were treated with different non-cytotoxic doses of 5-fluorouracil and then KRAL expression was measured. In both cell lines, 5-fluorouracil decreased *KRAL* expression in a dose-dependent way. The drug produced similar effects on *KEAP1* expression. To determine whether *KRAL* regulated *KEAP1* expression, the lncRNA was stably overexpressed in HepG2/5-FU and SMMC-7221/5-FU cells. Overexpressed *KRAL* markedly enhanced mRNA and protein levels of KEAP1. Conversely, the silencing of *KRAL* significantly suppressed *KEAP1* expression. *KRAL* was mainly located in the cytoplasm of HCC cells, suggesting a role in post-transcriptional regulation. In addition, *KRAL* directly interacted with miR-141, which is known to target *KEAP1* and promote 5-fluorouracil resistance in HCC cells (Shi et al. 2015). Sequestration by *KRAL* led to less miR-141 being available to target *KEAP1* for degradation. Overexpression of *KRAL* in HepG2/5-FU and SMMC-7721/5-FU cells improved sensitivity to 5-fluorouracil and increased the percentage of apoptotic cells; these results were partially attenuated by silencing *KEAP1* or overexpressing miR-141. Mechanistically, these findings suggest that *KRAL* competes with *KEAP1* for miR-141 binding; when *KRAL* expression is downregulated, *KEAP1* levels decrease due to increased targeting by miR-141.

KEAP1 negatively regulates the expression of the transcription factor, NRF2 (also known as NFE2L2: nuclear factor, erythroid 2 like 2) through activation of ubiquitin-mediated proteasomal degradation (Furukawa and Xiong 2005). NRF2 provides protection against oxidative stress by regulating target genes involved in glutathione synthesis, reduction of reactive oxygen species, drug transport, and detoxification (Taguchi et al. 2011). Despite the protective effect of activated NRF2, constitutive NRF2 expression is linked with cancer cell growth and proliferation, as well as the development of resistance to chemotherapeutic agents (Kensler and Wakabayashi 2010; Taguchi et al. 2011). In an investigation of lncRNAs associated with resistance to cisplatin in hepatoma cells, a novel transcript, termed *NRAL* (*NRF2* regulation-associated lncRNA) due to its proximity to *NRF2*, was found to be ~30-fold higher in cisplatin-resistant HepG2 and SMMC-7721 cells compared to parent cells (Wu et al. 2019). *NRAL* (ENST00000412153) is an annotated 0.5 kb lncRNA located on 2q13.1, although little is known about its expression or function. Overexpression of *NRAL* corresponded to enhanced sensitivity to cisplatin, while knockdown of *NRAL* expression resulted in greater cell viability in the presence of cisplatin compared to control cells. *NRAL* knockdown led to a greater number of apoptotic cells and cleaved caspase 3 activity, suggesting a mechanism whereby *NRAL* contributes to acquired cisplatin resistance by blocking apoptosis. Putative binding sites for miR-340-5p were identified in the *NRAL* primary sequence, suggesting that, like *KRAL*, this lncRNA may also function as a competitive endogenous RNA. Of note, miR-340-5p is associated with cisplatin-resistance in HCC (Shi et al. 2014) and is expressed at lower levels in cisplatin-resistant HepG2 and SMMC-7721 cells compared to parent cells (Wu et al. 2019). Transfection of cells with miR-340-5p mimic or siRNA resulted in decreased or increased *NRAL* expression, respectively. Conversely, *NRAL* overexpression reduced miR-340-5p levels, while knockdown of *NRAL* increased miR-340-5p expression. These results established a functional interaction between the two noncoding RNAs and a mutual inhibition between *NRAL* and miR-340-5p. As with the earlier study by this research group (Wu et al. 2018), the miRNA known to regulate the lncRNA also targeted the nearby coding gene (*NRF2*), suggesting a mechanism of endogenous sponging. To determine whether miR-340-5p participates in *NRAL*-mediated HCC resistance to cisplatin, miR-340-5p was ectopically expressed under conditions of *NRAL* overexpression (which corresponded with higher cisplatin resistance), leading to greater inhibition of cell viability. Opposite results were observed with a combination of *NRAL* knockdown and miR-340-5p siRNA, leading to a reduction of cisplatin sensitivity. Knockdown of miR-340-5p led to increased mRNA and protein levels of NRF2, which was attenuated by *NRAL* knockdown. Mechanistically, *NRAL* and *NRF2* compete for miR-340-5p. Upregulation of *NRAL* in HCC would theoretically lead to sponging of miR-340-5p, which would de-repress NRF2 expression and contribute to the development of cisplatin resistance.

2.11 NEAT1

NEAT1 (nuclear paraspeckle assembly transcript 1) is a relatively well-characterized 22 kb lncRNA that is transcribed from the multiple endocrine neoplasia type 1 locus on 11q13.1. *NEAT1* is retained in the nucleus where it provides a scaffold for protein binding in the assembly of RNA-protein nuclear bodies (i.e., paraspeckles) involved in the regulation of gene expression (Fox et al. 2018). *NEAT1* has been shown to be induced by p53 and contributes to the suppression of transformation and initiation of cancer in response to oncogenic factors (Mello et al. 2017) and downregulated in acute promyelocytic leukemia (Zeng et al. 2014). However, overexpression of this lncRNA has been observed in solid tumors, including HCC, lung cancer, and colorectal cancer, and high *NEAT1* levels have been associated with worse prognosis (Yu et al. 2017). *NEAT1* was also found to be upregulated in ovarian cancer (Chen et al. 2016) and in ovarian cancer tissues and cells with paclitaxel (PTX) resistance (An et al. 2017). In that study, *NEAT1* was shown to functionally interact with miR-194, leading to the upregulation of ZEB in PTX-resistant ovarian cancer cells (An et al. 2017). In an independent study, knockdown of *NEAT1* corresponded with enhanced sensitivity of human cancer cells to chemotherapeutics such as bleomycin and doxorubicin, as well as poly(ADP-ribose) polymerase (PARP) inhibitors, and Nutlin-3a, which inhibits the tumor-suppressing activity of p53 (Adriaens et al. 2016).

Recently, the role of *NEAT1* in HCC chemoresistance was investigated using three different human hepatoma cell lines (HepG2, PLC/PRF/5, and Huh7 cells) resistant to sorafenib or doxorubicin (Kessler et al. 2019). In sorafenib- and doxorubicin-resistant cell lines, *NEAT1* expression was upregulated and paraspeckle formation increased; in contrast, chemosensitive cells did not form paraspeckles (Kessler et al. 2019). Interestingly, knockdown of *NEAT1* in HCC cells was associated with increased apoptosis and decreased viability and proliferation (Fang et al. 2017; Liu et al. 2017), indicating that *NEAT1* might affect chemoresistance independent of paraspeckles (Li et al. 2017a). Collectively, the results indicate that *NEAT1* plays a role in the pathogenesis of a broad number of tumors and the development of acquired resistance to number of chemotherapies.

2.12 *Lnc-PDZD7*

Zhang et al. (2019) performed a lncRNA microarray analysis of 18 HCC tissues and corresponding para-tumorous tissues with the goal of identifying transcripts associated with stemness features. The *lnc-PDZD7* (lnc-PDZ domain-containing 7) was selected from the 26 differentially expressed transcripts for further characterization. This relatively uncharacterized lncRNA (ENSG0000023662) is known to reside on human chromosome 10 and have a transcript size of 342 bp. In situ hybridization demonstrated a largely cytoplasmic localization. In assessments of different clinicopathological attributes, *lnc-PDZD7* expression was correlated with tumor stage, tumor size, tumor differentiation, and vascular invasion. The use of TACE treatment was more common in patients with high *lnc-PDZD7* compared to those with low expression levels. In patients receiving adjuvant TACE, disease-free survival rate was lower in patients with a high *lnc-PDZD7* expression compared to those with low expression, suggesting that high *lnc-PDZD7* expression may predict poor response to adjuvant TACE. Because of the association between *lnc-PDZD7* expression and poor response to TACE, the authors speculated that the lncRNA might also contribute to HCC chemoresistance. HepG2 cells with knocked-down *lnc-PDZD7* expression showed decreased cell viability, colony formation, and in vivo tumorigenicity in response to treatment with 5-fluorouracil. Conversely, *lnc-PDZD7* overexpression in Bel-7402 cells resulted in reduced sensitivity to sorafenib, but only at a maximum concentration of 80 μg/ml.

Mechanistically, *lnc-PDZD7* was shown to upregulate *EZH2* (enhancer of zeste 2 polycomb repressive complex 2 subunit) by competitively binding miR-101 (Zhang et al. 2019). Overexpression of *EZH2* in *lnc-PDZD7*-depleted HepG2 cells reduced cell viability in response to 5-fluorouracil treatment, while *EZH2* knockdown in *lnc-PDZD7*-overexpressing Bel-7402 cells significantly reduced cell viability in response to sorafenib treatment. In addition, the *lnc-PDZD7-EZH2* axis mediated chemoresistance through the downregulation of atonal basic helix-loop-helix (bHLH) transcription factor 8 (*ATOH8*) in HCC cells. Higher levels of *lnc-PDZD7* and *EZH2* and a lower level of *ATOH8* were observed in a patient showing poor response to TACE, while lower levels of *lnc-PDZD7* and *EZH2* and a higher level of *ATOH8* were observed in a patient showing a good response to TACE. It is tempting to speculate that the same patterns of expression might be found in patients showing different sensitivities to 5-fluorouracil or sorafenib.

3 Extracellular Vesicles, LncRNAs, and HCC Chemoresistance

Extracellular vesicles (EVs) are a heterogeneous group of cell-derived spherical structures that play a role in a variety of physiological and pathological processes (van Niel et al. 2018). EV cargo carries specific molecular messages conveyed through the transmission of proteins, nucleic acids, lipids, and sugars to recipient cells (Yanez-Mo et al. 2015). Although EVs vary according to size, morphology, biogenic pathway, and cell type of origin, they can be broadly classified into two major classes: exosomes and microvesicles (van Niel et al. 2018). Exosomes are 50–150 nm endosomal vesicles, while microvesicles (50–500 nm), which include apoptotic bodies, are generated from the outward budding of the plasma membrane. Cancer cells, including HCC cells (Kogure et al. 2011), are known to secrete EVs in abundance and these EVs exert a range of effects from promoting metastasis following chemotherapy (Keklikoglou et al. 2019) to enhancing tumor angiogenesis (Yamada 2017). In HCC, EV cargo has been linked with epithelial-mesenchymal transition, tumor vascularization, metastasis, and immune response (Abudoureyimu et al. 2019; Yang et al. 2017). Recent work revealed that EVs released from HCC cells contained oncomiRs (miRNAs associated with cancer) capable of activating hepatic stellate cells to a cancer-associated hepatic stellate cell phenotype, which reciprocally released EVs that promoted oncogenic activity in HCC cells (Li et al. 2019a). In addition to microenvironment effects, circulating EVs carrying tumor-specific transcripts have been identified in HCC patients (Zheng et al. 2019).

There are a number of mechanisms by which EVs mediate the development of chemoresistance, as summarized in Mashouri et al. (2019), including the transfer of nucleic acids to recipient cells and augmented drug efflux (Shedden et al. 2003). For example, cisplatin-resistant cancer cells were found to release EVs containing 2.6-fold more platinum (a cisplatin metabolite) than EVs released from chemosensitive cells (Safaei et al. 2005). In EVs derived from HCC cells, sorafenib resistance was induced through activation of the hepatocyte growth factor/c-Met/Akt signaling pathway and inhibition of sorafenib-mediated apoptosis (Qu et al. 2016b).

Although the contribution of EV-derived lncRNAs to acquired drug resistance in cancer is emerging, relatively little is known of this phenomenon in HCC. However, given the importance of EVs in mediating numerous processes related to carcinogenesis, progression, and metastasis, mention of these studies is warranted. HepG2 cells cultured in EV-depleted medium and treated with HepG2-derived EVs showed enhanced cell viability in response to doxorubicin or sorafenib (Takahashi et al. 2014b). EV release was also potentiated by TGFB treatment, which is known to affect chemosensitivity (Biswas et al. 2007). Two lncRNAs found to be abundantly expressed in malignant hepatocytes, *linc-ROR* (regulator of reprogramming) and *linc-VLDLR*, were also upregulated in EVs. Knockdown of *linc-ROR*, a stress-responsive lncRNA (Takahashi et al. 2014a), resulted in a dose-dependent reduction in cell viability in response to treatment with sorafenib, camptothecin, or doxorubicin. Depleted *linc-ROR* also corresponded with a 1.8-fold increase in caspase-3/7 activity and in the number of apoptotic cells in the presence of sorafenib, compared with control cells. Treatment of HepG2 cells with sorafenib increased *linc-ROR* expression intracellularly and within EVs, and in recipient HepG2 cells following incubation with HepG2-derived EVs, *linc-ROR* expression increased in a dose-dependent manner. Incubation with EVs derived from HepG2 cells in which *linc-ROR* had been knocked down resulted in decreased cell viability in response to sorafenib treatment, suggesting that EVs may transfer *linc-ROR* to recipient cells via

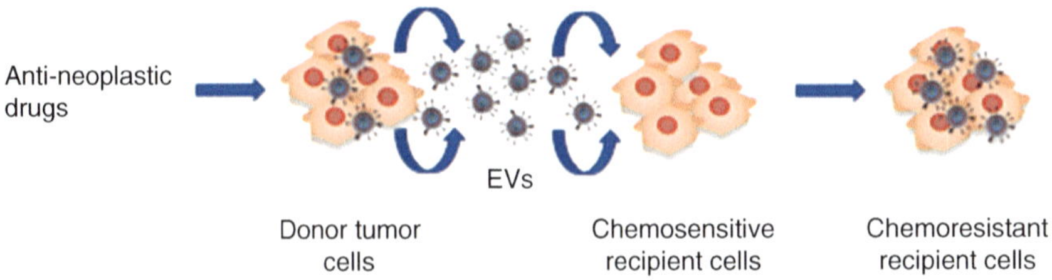

Fig. 2 Acquired chemoresistance through the EV-mediated transfer of lncRNAs. In the presence of anti-cancer drugs, chemoresistant tumor cells can transfer lncRNAs via EVs to modulate pharmacological response in recipient chemosensitive tumor cells. Identification of EV-derived lncRNAs, such as *linc-ROR* and *linc-VLDLR*, that mediate chemoresistance in recipient tumor cells provides an enhanced understanding of the mechanisms by which tumor cells develop resistance to anticancer agents

EVs. More recently, *linc-ROR* was found to promote resistance to radiotherapy in HepG2 and SMMC-7221 cells through a mechanism involving miR-145 sponging and subsequent upregulation of *RAD18* (RAD18 E3 ubiquitin protein ligase), a DNA repair protein (Chen et al. 2018). Whether *linc-ROR* promotes resistance to chemo- and radiotherapy in HCC via a single mechanism and whether *linc-ROR* delivered via EVs has a different effect on HCC cells than intracellular upregulation of the lncRNA remains to be seen.

In contrast, knockdown of the second lncRNA, *linc-VLDLR*, promoted HepG2 cell cycle progression and cell proliferation (Takahashi et al. 2014c). Like *linc-ROR*, *linc-VLDLR* expression increased in response to sorafenib, camptothecin, and doxorubicin, while knockdown of the lncRNA reduced cell viability in response to treatment with these agents. Levels of *linc-VLDLR* in EVs derived from tumor cells were also elevated in response to these drugs. *Linc-VLDLR* expression in recipient cells increased in a dose-dependent fashion following treatment with HepG2-derived EVs and decreased in response to incubation with EVs derived from HepG2 cells treated with *linc-VLDLR* siRNA. In cells exposed to HepG2-derived EVs, recipient cell viability increased in response to treatment with sorafenib or doxorubicin, but this effect was attenuated in cells transfected with *linc-VLDLR*. Mechanistically, *linc-VLDLR* knockdown corresponded with decreased expression of the ABCG2 mRNA and protein, while *ABCG2* overexpression attenuated the effect of *linc-VLDLR* knockdown on HepG2 cell death in response to sorafenib. Taken together, these two studies provide preliminary evidence that EV-derived lncRNAs modulate recipient cell response to chemotherapy (Fig. 2).

4 Summary and Conclusions

Acquired resistance to chemotherapeutic agents is a common occurrence in HCC. Because pharmacological approaches are generally implemented in patients with advanced HCC who are ineligible for surgical interventions, it is important to develop more effective clinical approaches to enhance chemosensitivity in these patients. Identification of mechanisms by which chemoresistance develops in HCC cells are expected to lead to the improved chemotherapeutic treatment of HCC. Emerging studies show that lncRNAs modulate cellular responses to cytotoxic drugs through specific mechanistic pathways; these lncRNAs could therefore provide new opportunities for approaches to enhance therapeutic efficacy.

Although a number of studies have reported the association between aberrant lncRNA expression and acquired chemoresistance, most of these have been based solely on in vitro investigation. Only two lncRNAs, *TUG1* and *NR2F1-AS1*, have been shown to have expression levels that differed significantly between chemoresistant and chemosensitive HCC patients. For the remaining lncRNAs, as well as lncRNAs identified in the future, it will be critical to demonstrate that expression levels vary between resistant and sensitive HCC samples. It will also be important to indicate clinically significant lncRNAs that can be applied in heterogeneous HCC patients with diverse underlying etiologies. Along the same lines, data from mouse models will be useful in establishing the in vivo consequences of aberrant lncRNA expression.

Many research articles published thus far have focused on a specific lncRNA in a single agent treatment (e.g., 5-fluorouracil, sorafenib, and doxorubicin). Even though HCC treatment depends on the standardized consensus of HCC staging, combination therapy and/or first- and second-line treatment is a common practice in the clinical setting. A critical question to consider is whether HCC cells develop different ways of resisting treatment depending on the pharmacology of the drug. Should each patient receive a personalized treatment based on HCC status, specific biomarkers, and clinical performance? Although these are issues that should be addressed for the optimized delivery of HCC care, prevention of HCC development remains the best strategy, while early HCC screening and detection, particularly in high-risk patients, will provide greater opportunity for use of curative approaches (i.e., surgery), rather than palliative treatment (i.e., chemotherapy).

Although the role of lncRNAs in the acquisition of chemoresistance in HCC is a robust, emerging field of study, substantial gaps in our understanding remain. However, given the novel functional role of these molecules in modulating cellular responses to cytotoxic agents, alone or via EV-mediated intercellular communication, lncRNAs have a high potential to improve the efficacy of anti-cancer treatments in HCC.

References

Abou-Alfa GK, Huitzil-Melendez FD, O'Reilly EM et al (2008) Current management of advanced hepatocellular carcinoma. Gastrointest Cancer Res 2:64–70

Abou-Alfa GK, Meyer T, Cheng AL et al (2018) Cabozantinib in patients with advanced and progressing hepatocellular carcinoma. N Engl J Med 379:54–63

Abudoureyimu M, Zhou H, Zhi Y et al (2019) Recent progress in the emerging role of exosome in hepatocellular carcinoma. Cell Prolif 52:e12541

Adriaens C, Standaert L, Barra J et al (2016) p53 induces formation of NEAT1 lncRNA-containing paraspeckles that modulate replication stress response and chemosensitivity. Nat Med 22:861–868

Akinyemiju T, Abera S, Ahmed M et al (2017) The burden of primary liver Cancer and underlying etiologies from 1990 to 2015 at the global, regional, and National Level: results from the global burden of disease study 2015. JAMA Oncol 3:1683–1691

Alkofer B, Lepennec V, Chiche L (2011) Hepatocellular cancer in the non-cirrhotic liver. J Visc Surg 148:3–11

An J, Lv W, Zhang Y (2017) LncRNA NEAT1 contributes to paclitaxel resistance of ovarian cancer cells by regulating ZEB1 expression via miR-194. Onco Targets Ther 10:5377–5390

Balogh J, Victor D III, Asham EH et al (2016) Hepatocellular carcinoma: a review. J Hepatocell Carcinoma 3:41–53

Bhan A, Mandal SS (2015) LncRNA HOTAIR: a master regulator of chromatin dynamics and cancer. Biochim Biophys Acta 1856:151–164

Biswas S, Guix M, Rinehart C et al (2007) Inhibition of TGF-beta with neutralizing antibodies prevents radiation-induced acceleration of metastatic cancer progression. J Clin Invest 117:1305–1313

Bray F, Ferlay J, Soerjomataram I et al (2018) Global cancer statistics 2018: GLOBOCAN estimates of incidence and mortality worldwide for 36 cancers in 185 countries. CA Cancer J Clin 68:394–424

Bruix J, Cheng AL, Meinhardt G et al (2017a) Prognostic factors and predictors of sorafenib benefit in patients with hepatocellular carcinoma: analysis of two phase III studies. J Hepatol 67:999–1008

Bruix J, Qin S, Merle P et al (2017b) Regorafenib for patients with hepatocellular carcinoma who progressed on sorafenib treatment (RESORCE): a randomised, double-blind, placebo-controlled, phase 3 trial. Lancet 389:56–66

Cai C, Huo Q, Wang X et al (2017) SNHG16 contributes to breast cancer cell migration by competitively binding miR-98 with E2F5. Biochem Biophys Res Commun 485:272–278

Cao X, Xu J, Yue D (2018) LncRNA-SNHG16 predicts poor prognosis and promotes tumor proliferation through epigenetically silencing p21 in bladder cancer. Cancer Gene Ther 25:10–17

Caruso S, Calatayud AL, Pilet J et al (2019) Analysis of liver Cancer cell lines identifies agents with likely efficacy against hepatocellular carcinoma and markers of response. Gastroenterology 157:760–776

Chen J, Jin R, Zhao J et al (2015) Potential molecular, cellular and microenvironmental mechanism of sorafenib resistance in hepatocellular carcinoma. Cancer Lett 367:1–11

Chen ZJ, Zhang Z, Xie BB et al (2016) Clinical significance of up-regulated lncRNA NEAT1 in prognosis of ovarian cancer. Eur Rev Med Pharmacol Sci 20:3373–3377

Chen Y, Shen Z, Zhi Y et al (2018) Long non-coding RNA ROR promotes radioresistance in hepatocelluar carcinoma cells by acting as a ceRNA for microRNA-145 to regulate RAD18 expression. Arch Biochem Biophys 645:117–125

Chen S, Cao Q, Wen W et al (2019) Targeted therapy for hepatocellular carcinoma: challenges and opportunities. Cancer Lett 460:1–9

Chiba T, Suzuki E, Negishi M et al (2012) 3-Deazaneplanocin a is a promising therapeutic agent for the eradication of tumor-initiating hepatocellular carcinoma cells. Int J Cancer 130:2557–2567

Dean M (2009) ABC transporters, drug resistance, and cancer stem cells. J Mammary Gland Biol Neoplasia 14:3–9

Dhamija E, Paul SB, Kedia S (2019) Non-alcoholic fatty liver disease associated with hepatocellular carcinoma: An increasing concern. Indian J Med Res 149:9–17

DiStefano JK (2017) Long noncoding RNAs in the initiation, progression, and metastasis of hepatocellular carcinoma. Non-coding RNA Res 2:129–136

DiStefano JK (2018) The emerging role of Long noncoding RNAs in human disease. Methods Mol Biol 1706:91–110

Doyle L, Ross DD (2003) Multidrug resistance mediated by the breast cancer resistance protein BCRP (ABCG2). Oncogene 22:7340–7358
El-Serag HB, Kanwal F (2014) Epidemiology of hepatocellular carcinoma in the United States: where are we? Where do we go? Hepatology 60:1767–1775
Fang L, Sun J, Pan Z et al (2017) Long non-coding RNA NEAT1 promotes hepatocellular carcinoma cell proliferation through the regulation of miR-129-5p-VCP-IkappaB. Am J Physiol Gastrointest Liver Physiol 313:G150–G156
Forouzanfar MH, Alexander L, Anderson HR et al (2015) Global, regional, and national comparative risk assessment of 79 behavioural, environmental and occupational, and metabolic risks or clusters of risks in 188 countries, 1990-2013: a systematic analysis for the global burden of disease study 2013. Lancet 386:2287–2323
Fox AH, Nakagawa S, Hirose T et al (2018) Paraspeckles: where Long noncoding RNA meets phase separation. Trends Biochem Sci 43:124–135
Friemel J, Rechsteiner M, Frick L et al (2015) Intratumor heterogeneity in hepatocellular carcinoma. Clin Cancer Res 21:1951–1961
Furukawa M, Xiong Y (2005) BTB protein Keap1 targets antioxidant transcription factor Nrf2 for ubiquitination by the Cullin 3-Roc1 ligase. Mol Cell Biol 25:162–171
Gabory A, Jammes H, Dandolo L (2010) The H19 locus: role of an imprinted non-coding RNA in growth and development. BioEssays 32:473–480
Galle PR, Forner A, Llovet JM et al (2018) EASL clinical practice guidelines: management of hepatocellular carcinoma. J Hepatol 69:182–236
Gao Q, Wang ZC, Duan M et al (2017) Cell culture system for analysis of genetic heterogeneity within hepatocellular carcinomas and response to pharmacologic agents. Gastroenterology 152 (232–242):e234
Georgescu MM (2010) PTEN tumor suppressor network in PI3K-Akt pathway control. Genes Cancer 1:1170–1177
Golfieri R, Bargellini I, Spreafico C et al (2019) Patients with Barcelona clinic liver Cancer stages B and C hepatocellular carcinoma: time for a subclassification. Liver Cancer 8:78–91
Goodell MA, Rosenzweig M, Kim H et al (1997) Dye efflux studies suggest that hematopoietic stem cells expressing low or undetectable levels of CD34 antigen exist in multiple species. Nat Med 3:1337–1345
Guo T, Wu P, Liu P et al (2018) Identifying the best anticancer agent combination in TACE for HCC patients: a network meta-analysis. J Cancer 9:2640–2649
Gupta RA, Shah N, Wang KC et al (2010) Long non-coding RNA HOTAIR reprograms chromatin state to promote cancer metastasis. Nature 464:1071–1076
Haraguchi N, Inoue H, Tanaka F et al (2006a) Cancer stem cells in human gastrointestinal cancers. Hum Cell 19:24–29
Haraguchi N, Utsunomiya T, Inoue H et al (2006b) Characterization of a side population of cancer cells from human gastrointestinal system. Stem Cells 24:506–513
Haraguchi N, Ishii H, Mimori K et al (2010) CD13 is a therapeutic target in human liver cancer stem cells. J Clin Invest 120:3326–3339
He Y, Meng XM, Huang C et al (2014) Long noncoding RNAs: novel insights into hepatocelluar carcinoma. Cancer Lett 344:20–27
Ho DW, Yang ZF, Yi K et al (2012) Gene expression profiling of liver cancer stem cells by RNA-sequencing. PLoS One 7:e37159
Holczbauer A, Factor VM, Andersen JB et al (2013) Modeling pathogenesis of primary liver cancer in lineage-specific mouse cell types. Gastroenterology 145:221–231
Hu C, Li H, Li J et al (2008) Analysis of ABCG2 expression and side population identifies intrinsic drug efflux in the HCC cell line MHCC-97L and its modulation by Akt signaling. Carcinogenesis 29:2289–2297
Huang MD, Chen WM, Qi FZ et al (2015) Long non-coding RNA TUG1 is up-regulated in hepatocellular carcinoma and promotes cell growth and apoptosis by epigenetically silencing of KLF2. Mol Cancer 14:165

Huang H, Chen J, Ding CM et al (2018) LncRNA NR2F1-AS1 regulates hepatocellular carcinoma oxaliplatin resistance by targeting ABCC1 via miR-363. J Cell Mol Med 22:3238–3245
Ji P, Diederichs S, Wang W et al (2003) MALAT-1, a novel noncoding RNA, and thymosin beta4 predict metastasis and survival in early-stage non-small cell lung cancer. Oncogene 22:8031–8041
Jia Q, Zhang X, Deng T et al (2013) Positive correlation of Oct4 and ABCG2 to chemotherapeutic resistance in CD90(+)CD133(+) liver cancer stem cells. Cell Reprogram 15:143–150
Keklikoglou I, Cianciaruso C, Guc E et al (2019) Chemotherapy elicits pro-metastatic extracellular vesicles in breast cancer models. Nat Cell Biol 21:190–202
Kensler TW, Wakabayashi N (2010) Nrf2: friend or foe for chemoprevention? Carcinogenesis 31:90–99
Kessler SM, Hosseini K, Hussein UK et al (2019) Hepatocellular carcinoma and nuclear Paraspeckles: induction in Chemoresistance and prediction for poor survival. Cell Physiol Biochem 52:787–801
Kim MJ, Choi YK, Park SY et al (2017) PPARdelta Reprograms Glutamine Metabolism in Sorafenib-Resistant HCC. Mol Cancer Res 15:1230–1242
Kogure T, Lin WL, Yan IK et al (2011) Intercellular nanovesicle-mediated microRNA transfer: a mechanism of environmental modulation of hepatocellular cancer cell growth. Hepatology 54:1237–1248
Kudo M, Finn RS, Qin S et al (2018) Lenvatinib versus sorafenib in first-line treatment of patients with unresectable hepatocellular carcinoma: a randomised phase 3 non-inferiority trial. Lancet 391:1163–1173
Kulik L, El-Serag HB (2019) Epidemiology and Management of Hepatocellular Carcinoma. Gastroenterology 156(477–491):e471
Lai MC, Yang Z, Zhou L et al (2012) Long non-coding RNA MALAT-1 overexpression predicts tumor recurrence of hepatocellular carcinoma after liver transplantation. Med Oncol 29:1810–1816
Lan X, Wu YZ, Wang Y et al (2013) CD133 silencing inhibits stemness properties and enhances chemoradiosensitivity in CD133-positive liver cancer stem cells. Int J Mol Med 31:315–324
Lee S, Kim KM, Lee SJ et al (2017) Hepatic arterial damage after transarterial chemoembolization for the treatment of hepatocellular carcinoma: comparison of drug-eluting bead and conventional chemoembolization in a retrospective controlled study. Acta Radiol 58:131–139
Lee M, Ko H, Yun M (2018) Cancer metabolism as a mechanism of treatment resistance and potential therapeutic target in hepatocellular carcinoma. Yonsei Med J 59:1143–1149
Li J, Chen JN, Zeng TT et al (2016a) CD133+ liver cancer stem cells resist interferon-gamma-induced autophagy. BMC Cancer 16:15
Li Z, Shen J, Chan MT et al (2016b) TUG1: a pivotal oncogenic long non-coding RNA of human cancers. Cell Prolif 49:471–475
Li R, Harvey AR, Hodgetts SI et al (2017a) Functional dissection of NEAT1 using genome editing reveals substantial localization of the NEAT1_1 isoform outside paraspeckles. RNA 23:872–881
Li Y, Ye Y, Feng B et al (2017b) Long noncoding RNA lncARSR promotes doxorubicin resistance in hepatocellular carcinoma via modulating PTEN-PI3K/Akt pathway. J Cell Biochem 118:4498–4507
Li J, Yan Y, Ang L et al (2019a) Extracellular vesicles-derived OncomiRs mediate communication between Cancer cells and Cancer-associated hepatic stellate cells in hepatocellular carcinoma microenvironment. Carcinogenesis May 29, pii: bgz096
Li L, Knutsdottir H, Hui K et al (2019b) Human primary liver cancer organoids reveal intratumor and interpatient drug response heterogeneity. JCI Insight 4(2):pii: 121490
Lian D, Amin B, Du D et al (2017) Enhanced expression of the long non-coding RNA SNHG16 contributes to gastric cancer progression and metastasis. Cancer Biomark 21:151–160

Liu Z, Chang Q, Yang F et al (2017) Long non-coding RNA NEAT1 overexpression is associated with unfavorable prognosis in patients with hepatocellular carcinoma after hepatectomy: a Chinese population-based study. Eur J Surg Oncol 43:1697–1703

Liu TH, Shao YY, Lu LC et al (2019) Considerations of heterogeneity in clinical trials for hepatocellular carcinoma. Expert Rev Gastroenterol Hepatol 13:615–621

Llovet JM, Bru C, Bruix J (1999) Prognosis of hepatocellular carcinoma: the BCLC staging classification. Semin Liver Dis 19:329–338

Llovet JM, Ricci S, Mazzzaferro V et al (2008) Sorafenib in advanced hepatocellular carcinoma. N Engl J Med 359:378–390

Llovet JM, Ducreux M, Lencioni R et al (2012) EASL-EORTC clinical practice guidelines: management of hepatocellular carcinoma. J Hepatol 56:908–943

Llovet JM, Zucman-Rossi J, Pikarsky E et al (2016) Hepatocellular carcinoma. Nat Rev Dis Primers 2:16018

Loewen G, Jayawickramarajah J, Zhuo Y et al (2014) Functions of lncRNA HOTAIR in lung cancer. J Hematol Oncol 7:90

Long Q, Zou X, Song Y et al (2019) PFKFB3/HIF-1alpha feedback loop modulates sorafenib resistance in hepatocellular carcinoma cells. Biochem Biophys Res Commun 513:642–650

Luo JH, Ren B, Keryanov S et al (2006) Transcriptomic and genomic analysis of human hepatocellular carcinomas and hepatoblastomas. Hepatology 44:1012–1024

Luo ZF, Zhao D, Li XQ et al (2016) Clinical significance of HOTAIR expression in colon cancer. World J Gastroenterol 22:5254–5259

Mashouri L, Yousefi H, Aref AR et al (2019) Exosomes: composition, biogenesis, and mechanisms in cancer metastasis and drug resistance. Mol Cancer 18:75

Matouk IJ, DeGroot N, Mezan S et al (2007) The H19 non-coding RNA is essential for human tumor growth. PLoS One 2:e845

Mello SS, Sinow C, Raj N et al (2017) Neat1 is a p53-inducible lincRNA essential for transformation suppression. Genes Dev 31:1095–1108

Mendez-Blanco C, Fondevila F, Garcia-Palomo A et al (2018) Sorafenib resistance in hepatocarcinoma: role of hypoxia-inducible factors. Exp Mol Med 50:134

Mittal S, El-Serag HB (2013) Epidemiology of hepatocellular carcinoma: consider the population. J Clin Gastroenterol 47(Suppl):S2–S6

Moran VA, Perera RJ, Khalil AM (2012) Emerging functional and mechanistic paradigms of mammalian long non-coding RNAs. Nucleic Acids Res 40:6391–6400

Nagano T, Fraser P (2011) No-nonsense functions for long noncoding RNAs. Cell 145:178–181

Nam HC, Jang B, Song MJ (2016) Transarterial chemoembolization with drug-eluting beads in hepatocellular carcinoma. World J Gastroenterol 22:8853–8861

Nishida N, Kitano M, Sakurai T et al (2015) Molecular mechanism and prediction of Sorafenib Chemoresistance in human hepatocellular carcinoma. Dig Dis 33:771–779

Niu L, Liu L, Yang S et al (2017) New insights into sorafenib resistance in hepatocellular carcinoma: responsible mechanisms and promising strategies. Biochim Biophys Acta Rev Cancer 1868:564–570

Panzitt K, Tschernatsch MM, Guelly C et al (2007) Characterization of HULC, a novel gene with striking up-regulation in hepatocellular carcinoma, as noncoding RNA. Gastroenterology 132:330–342

Parent R, Plissonnier ML, Bancel B et al (2014) Diversity of hepatocellular carcinoma clones bearing hematopoietic malignancies-related chromosomal translocation. J Cell Biochem 115:666–677

Piao LS, Hur W, Kim TK et al (2012) CD133+ liver cancer stem cells modulate radioresistance in human hepatocellular carcinoma. Cancer Lett 315:129–137

Prieto-Dominguez N, Ordonez R, Fernandez A et al (2016) Modulation of autophagy by Sorafenib: effects on treatment response. Front Pharmacol 7:151

Qu L, Ding J, Chen C et al (2016a) Exosome-transmitted lncARSR promotes Sunitinib resistance in renal Cancer by acting as a competing endogenous RNA. Cancer Cell 29:653–668

Qu Z, Wu J, Wu J et al (2016b) Exosomes derived from HCC cells induce sorafenib resistance in hepatocellular carcinoma both in vivo and in vitro. J Exp Clin Cancer Res 35:159
Rachmilewitz J, Goshen R, Ariel I et al (1992) Parental imprinting of the human H19 gene. FEBS Lett 309:25–28
Raveh E, Matouk IJ, Gilon M et al (2015) The H19 Long non-coding RNA in cancer initiation, progression and metastasis - a proposed unifying theory. Mol Cancer 14:184
Reis SP, Sutphin PD, Singal AG et al (2017) Tumor enhancement and heterogeneity are associated with treatment response to drug-eluting bead chemoembolization for hepatocellular carcinoma. J Comput Assist Tomogr 41:289–293
Rinn JL, Kertesz M, Wang JK et al (2007) Functional demarcation of active and silent chromatin domains in human HOX loci by noncoding RNAs. Cell 129:1311–1323
Roninson IB, Chin JE, Choi KG et al (1986) Isolation of human mdr DNA sequences amplified in multidrug-resistant KB carcinoma cells. Proc Natl Acad Sci U S A 83:4538–4542
Safaei R, Larson BJ, Cheng TC et al (2005) Abnormal lysosomal trafficking and enhanced exosomal export of cisplatin in drug-resistant human ovarian carcinoma cells. Mol Cancer Ther 4:1595–1604
Saunders NA, Simpson F, Thompson EW et al (2012) Role of intratumoural heterogeneity in cancer drug resistance: molecular and clinical perspectives. EMBO Mol Med 4:675–684
Schultheiss CS, Laggai S, Czepukojc B et al (2017) The long non-coding RNA H19 suppresses carcinogenesis and chemoresistance in hepatocellular carcinoma. Cell Stress 1:37–54
Shedden K, Xie XT, Chandaroy P et al (2003) Expulsion of small molecules in vesicles shed by cancer cells: association with gene expression and chemosensitivity profiles. Cancer Res 63:4331–4337
Shi GM, Xu Y, Fan J et al (2008) Identification of side population cells in human hepatocellular carcinoma cell lines with stepwise metastatic potentials. J Cancer Res Clin Oncol 134:1155–1163
Shi L, Chen ZG, Wu LL et al (2014) miR-340 reverses cisplatin resistance of hepatocellular carcinoma cell lines by targeting Nrf2-dependent antioxidant pathway. Asian Pac J Cancer Prev 15:10439–10444
Shi L, Wu L, Chen Z et al (2015) MiR-141 activates Nrf2-dependent antioxidant pathway via Down-regulating the expression of Keap1 conferring the resistance of hepatocellular carcinoma cells to 5-fluorouracil. Cell Physiol Biochem 35:2333–2348
Shi Y, Yang X, Xue X et al (2018) HANR promotes hepatocellular carcinoma progression via miR-214/EZH2/TGF-beta axis. Biochem Biophys Res Commun 506:189–193
Song Z, Liu T, Chen J et al (2019) HIF-1alpha-induced RIT1 promotes liver cancer growth and metastasis and its deficiency increases sensitivity to sorafenib. Cancer Lett 460:96–107
Sukowati CH, Rosso N, Croce LS et al (2010) Hepatic cancer stem cells and drug resistance: relevance in targeted therapies for hepatocellular carcinoma. World J Hepatol 2:114–126
Sun W, Cabrera R (2018) Systemic treatment of patients with advanced, Unresectable hepatocellular carcinoma: emergence of therapies. J Gastrointest Cancer 49:107–115
Sun Y, Ma L (2019) New insights into Long non-coding RNA MALAT1 in Cancer and metastasis. Cancers (Basel) 11:216
Sun JY, Yin T, Zhang XY et al (2019) Therapeutic advances for patients with intermediate hepatocellular carcinoma. J Cell Physiol 234:12116–12121
Taguchi K, Motohashi H, Yamamoto M (2011) Molecular mechanisms of the Keap1-Nrf2 pathway in stress response and cancer evolution. Genes Cells 16:123–140
Takahashi K, Yan IK, Haga H et al (2014a) Modulation of hypoxia-signaling pathways by extracellular linc-RoR. J Cell Sci 127:1585–1594
Takahashi K, Yan IK, Kogure T et al (2014b) Extracellular vesicle-mediated transfer of long non-coding RNA ROR modulates chemosensitivity in human hepatocellular cancer. FEBS Open Bio 4:458–467
Takahashi K, Yan IK, Wood J et al (2014c) Involvement of extracellular vesicle long noncoding RNA (linc-VLDLR) in tumor cell responses to chemotherapy. Mol Cancer Res 12:1377–1387

Tandia M, Mhiri A, Paule B et al (2017) Correlation between clinical response to sorafenib in hepatocellular carcinoma treatment and polymorphisms of P-glycoprotein (ABCB1) and of breast cancer resistance protein (ABCG2): monocentric study. Cancer Chemother Pharmacol 79:759–766

Teufel M, Seidel H, Kochert K et al (2019) Biomarkers associated with response to Regorafenib in patients with hepatocellular carcinoma. Gastroenterology 156:1731–1741

Tricoli L, Niture S, Chimeh U et al (2019) Role of microRNAs in the development of hepatocellular carcinoma and drug resistance. Front Biosci (Landmark Ed) 24:382–391

Tsang WP, Kwok TT (2007) Riboregulator H19 induction of MDR1-associated drug resistance in human hepatocellular carcinoma cells. Oncogene 26:4877–4881

Tsochatzis EA, Bosch J, Burroughs AK et al (2014) Liver cirrhosis. Lancet 383:1749–1761

van Niel G et al (2018) Shedding light on the cell biology of extracellular vesicles. Nat Rev Mol Cell Biol 19:213–228

Wang KC, Chang HY (2011) Molecular mechanisms of long noncoding RNAs. Mol Cell 43:904–914

Wang J, Liu X, Wu H et al (2010) CREB up-regulates long non-coding RNA, HULC expression through interaction with microRNA-372 in liver cancer. Nucleic Acids Res 38:5366–5383

Wang ZC, Liu LZ, Liu XY et al (2015) Genetic polymorphisms of the multidrug resistance 1 gene MDR1 and the risk of hepatocellular carcinoma. Tumour Biol 36:7007–7015

Woo HY, Heo J (2015) Transarterial chemoembolization using drug eluting beads for the treatment of hepatocellular carcinoma: now and future. Clin Mol Hepatol 21:344–348

Wu L, Pan C, Wei X et al (2018) lncRNA KRAL reverses 5-fluorouracil resistance in hepatocellular carcinoma cells by acting as a ceRNA against miR-141. Cell Commun Signal 16:47

Wu LL, Cai WP, Lei X et al (2019) NRAL mediates cisplatin resistance in hepatocellular carcinoma via miR-340-5p/Nrf2 axis. J Cell Commun Signal 13:99–112

Xiao J, Lv Y, Jin F et al (2017) LncRNA HANR promotes tumorigenesis and increase of Chemoresistance in hepatocellular carcinoma. Cell Physiol Biochem 43:1926–1938

Xiong H, Ni Z, He J et al (2017) LncRNA HULC triggers autophagy via stabilizing Sirt1 and attenuates the chemosensitivity of HCC cells. Oncogene 36:3528–3540

Xu Y, Wang J, Qiu M et al (2015) Upregulation of the long noncoding RNA TUG1 promotes proliferation and migration of esophageal squamous cell carcinoma. Tumour Biol 36:1643–1651

Xu F, Zha G, Wu Y et al (2018a) Overexpressing lncRNA SNHG16 inhibited HCC proliferation and chemoresistance by functionally sponging hsa-miR-93. Onco Targets Ther 11:8855–8863

Xu X, Tao Y, Shan L et al (2018b) The role of MicroRNAs in hepatocellular carcinoma. J Cancer 9:3557–3569

Yamada NO (2017) Extracellular vesicles: emerging mediators of intercellular communication and tumor angiogenesis. Ann Transl Med 5:59

Yan B, Wang Z (2012) Long noncoding RNA: its physiological and pathological roles. DNA Cell Biol 31(Suppl 1):S34–S41

Yanez-Mo M, Siljander PR, Andreu Z et al (2015) Biological properties of extracellular vesicles and their physiological functions. J Extracell Vesicles 4:27066

Yang Z, Zhou L, Wu LM et al (2011) Overexpression of long non-coding RNA HOTAIR predicts tumor recurrence in hepatocellular carcinoma patients following liver transplantation. Ann Surg Oncol 18:1243–1250

Yang L, Du Y, Yu P et al (2016) Long non-coding RNA TUG1 regulates the development of multidrug resistance in hepatocellular carcinoma via P-gp and MDR1. Int J Clin Exp Med 9:21388–21396

Yang N, Li S, Li G et al (2017) The role of extracellular vesicles in mediating progression, metastasis and potential treatment of hepatocellular carcinoma. Oncotarget 8:3683–3695

Ye CG, Yeung JH, Huang GL et al (2013) Increased glutathione and mitogen-activated protein kinase phosphorylation are involved in the induction of doxorubicin resistance in hepatocellular carcinoma cells. Hepatol Res 43:289–299

Yoshimizu T, Miroglio A, Ripoche MA et al (2008) The H19 locus acts in vivo as a tumor suppressor. Proc Natl Acad Sci U S A 105:12417–12422

Young TL, Matsuda T, Cepko CL (2005) The noncoding RNA taurine upregulated gene 1 is required for differentiation of the murine retina. Curr Biol 15:501–512

Yu M, Ohira M, Li Y et al (2009) High expression of ncRAN, a novel non-coding RNA mapped to chromosome 17q25.1, is associated with poor prognosis in neuroblastoma. Int J Oncol 34:931–938

Yu X, Li Z, Zheng H et al (2017) NEAT1: a novel cancer-related long non-coding RNA. Cell Prolif 50:e12329

Zeng C, Xu Y, Xu L et al (2014) Inhibition of long non-coding RNA NEAT1 impairs myeloid differentiation in acute promyelocytic leukemia cells. BMC Cancer 14:693

Zhang Y, Tang B, Song J et al (2019) Lnc-PDZD7 contributes to stemness properties and chemosensitivity in hepatocellular carcinoma through EZH2-mediated ATOH8 transcriptional repression. J Exp Clin Cancer Res 38:92

Zhang EB, Yin DD, Sun M et al (2014) P53-regulated long non-coding RNA TUG1 affects cell proliferation in human non-small cell lung cancer, partly through epigenetically regulating HOXB7 expression. Cell Death Dis 5:e1243

Zhang K, Chen J, Song H et al (2017) SNHG16/miR-140-5p axis promotes esophagus cancer cell proliferation, migration and EMT formation through regulating ZEB1. Oncotarget 9:1028–1040

Zheng Q, Zhao J, Yu H et al (2019) Tumor-specific transcripts are frequently expressed in hepatocellular carcinoma with clinical implication and potential function. Hepatology 71:259. https://doi.org/10.1002/hep.30805

Zhou S, Schuetz JD, Bunting KD et al (2001) The ABC transporter Bcrp1/ABCG2 is expressed in a wide variety of stem cells and is a molecular determinant of the side-population phenotype. Nat Med 7:1028–1034

Zhou JJ, Cheng D, He XY et al (2017) Knockdown of long non-coding RNA HOTAIR sensitizes hepatocellular carcinoma cell to cisplatin by suppressing the STAT3/ABCB1 signaling pathway. Oncol Lett 14:7986–7992

Zhu YJ, Zheng B, Wang HY et al (2017) New knowledge of the mechanisms of sorafenib resistance in liver cancer. Acta Pharmacol Sin 38:614–622

Single-Cell Analysis May Shed New Lights on the Role of LncRNAs in Chemoresistance in Gastrointestinal Cancers

Bernadette Neve, Nicolas Jonckheere, Audrey Vincent, and Isabelle Van Seuningen

Contents

Abstract A major challenge in the treatment of cancer is dealing with intrinsic and/or acquired chemoresistance and with the development of metastatic lesions. The underlying mechanisms of chemoresistance are complex as a consequence of cancer heterogeneity, such as different tumour tissue origin, inter-tumour heterogeneity between patients and intra-tumour heterogeneity within cell populations of tumours. Classifying tumours according to gene expression-based molecular subtypes predicts the response to therapy and has been proposed to innovate towards personalized therapy. We here highlight the molecular subtypes observed in gastrointestinal cancers. In addition, we discuss the implication of long non-coding RNAs (lncRNAs), a class of RNAs without protein-coding sequence and over 200 nucleotides long, in chemoresistance of gastrointestinal cancers.

Furthermore, with the development of single-cell RNA sequencing technologies 10 years ago, it has become clear that intercellular transcriptome heterogeneity of similar cell types may contribute to intra-tumour heterogeneity. Single-cell transcriptome profiling may identify specific cell phenotypes prone to develop chemoresistance that previously remained undistinguished by global gene expression or cell morphology. CRISPR/Cas9 methods may permit to elucidate the role of

B. Neve · N. Jonckheere · A. Vincent · I. Van Seuningen (✉)
Université de Lille, CNRS, Inserm, CHU Lille, UMR9020-U1277 – CANTHER – Cancer Heterogeneity, Plasticity and Resistance to Therapies, Lille cedex, France
e-mail: bernadette.neve@inserm.fr; nicolas.jonckheere@inserm.fr; audrey.vincent@inserm.fr; isabelle.vanseuningen@inserm.fr

S. Jurga, J. Barciszewski (eds.), *The Chemical Biology of Long Noncoding RNAs*, RNA Technologies 11, https://doi.org/10.1007/978-3-030-44743-4_9

lncRNAs in chemoresistance at a single-cell level. The single-cell approach may take cancer treatment a step closer towards personalized, or even cell-specific, therapy.

Keywords Long non-coding (lnc) RNAs · Chemotherapy · Resistance · Single cell · Gastrointestinal cancers

1 Gastrointestinal Cancers

Cancer is a heterogeneous disease, not only due to the origin of the tumour tissue but also because heterogeneity exists between similar cancers from different patients (inter-tumour heterogeneity) and within a single tumour (intra-tumour heterogeneity). We here focus on the implication of lncRNAs in chemoresistance of gastrointestinal cancers. We highlighted studies of oesophageal, gastric, pancreatic and colorectal cancers and of the chemotherapies used for these types of cancers.

Oesophageal Cancers Globally oesophageal cancer is the seventh most frequently diagnosed cancer and the sixth leading cause of cancer-related deaths (Ajani et al. 2019). Two types of cancers can be distinguished: 1) oesophageal squamous cell carcinoma (OSCC/ESCC), arising mainly in the upper and middle parts, and 2) oesophageal adenocarcinomas (OAC/EAC) of the lower part of the oesophagus. OSCC has earlier lymph node metastasis and is associated with a poorer prognosis. Tobacco, alcohol consumption and obesity are the major risk factors for OSCC. Pre- and postoperative chemotherapy with mitosis-inhibitor paclitaxel and DNA cross-linker carboplatin significantly improved overall survival (OS) of adenocarcinomas to an average of 43.2 months and improved the disease-free survival (DFS). Therapy with FOLFOX (anti-DNA metabolytes 5-fluorouracil (5-FU)/leucovorin and DNA cross-linker oxaliplatin) was also shown beneficial for these patients. Overexpression or amplification of the ErbB2 receptor tyrosine kinase 2 gene (ERBB2/HER2) has been implicated in the development of oesophageal and oesophageal-gastric junction cancers. For patients with such metastatic adenocarcinomas, ERBB2/HER2-directed antibody-drug conjugates (e.g. trastuzumab) is recommended as first-line therapy. For the treatment of unresectable metastatic tumours with high microsatellite instability (MSI) status or deficient mismatch repair (dMMR) states, vascular endothelial growth factor (VEGR) antibody (such as ramucirumab) and antibodies binding programmed death receptor 1 (PD-1) (pembrolizumab) are included as options for second-line or follow-up therapy for patients with metastatic disease.

Gastric Cancers Gastric cancer is the fifth most frequently diagnosed cancer and the third leading cause of death from cancer in the world. Most gastric cancers are malignant adenocarcinomas developing from the lining of the stomach; other gastric cancers include lymphomas, gastrointestinal stromal tumours (GIST) and carcinoid tumours (near the gastric-oesophageal junction) (Hayakawa et al. 2016). Environmental risk factors include *Helicobacter pylori* infection, smoking, high salt intake, other dietary factors and chronic gastric inflammation. The diagnosis of gastric cancer is often late at an advanced tumour stage with 70 to 80% of patients having lymph node infiltrations. Palliative chemotherapy is given in patients with advanced and metastatic disease, which include different combinations of anti-DNA metabolites (5-fluorouracil, methotrexate and leucovorin), DNA cross-linkers (cisplatin, oxaliplatin) and antibiotics (mitomycin, epirubicin, doxorubicin). ERBB2/HER2 status is relevant for the management of advanced or metastatic gastric adenocarcinoma tumours. The NCCN guidelines point out several biomarkers that should be assessed and considered in treatment strategies for gastric cancer (dMMR, MSI, programmed death ligand 1 (PD-L1) and tumour Epstein-Barr virus (EBV) status).

Pancreatic Cancers Pancreatic cancer is the seventh most common cause of cancer-related death in the world. However, only few patients present a resectable tumour (10–20%), and up to 80% of these patients receiving chemotherapy with curative intent do relapse, resulting on average in mortality within 2 years (Gbolahan et al. 2019). The risk factors for pancreatic cancer include inherited genetic predisposition, tobacco and alcohol exposure, diet and obesity, diabetes mellitus as well as chronic pancreatitis, abdominal surgeries and infections. About 95% of the pancreatic cancers originate from the enzyme secreting exocrine cells, with the most common type pancreatic ductal adenocarcinomas (PDAC), and 5% from the sugar-regulating hormone secreting endocrine cells (Tempero et al. 2019). For patients who present resectable tumours (10–20%), adjuvant therapy with the anti-DNA metabolites 5-fluorouracil/capecitabine and gemcitabine improved the 5-year survival to 30% (Gbolahan et al. 2019). In addition, the preferred intervention includes administration of a modified FOLFIRINOX cocktail (anti-DNA metabolites 5-FU, DNA cross-linker oxaliplatin and topoisomerase inhibitor irinotecan). Other regimens are gemcitabine with or without anti-EGFR antibodies (erlotinib) and gemcitabine in combination with an albumin-conjugated paclitaxel (abraxane). Although only a minority of patients with pancreatic cancer will be eligible for potentially curative resection, results of many trials have shown that for them adjuvant therapy is beneficial. To improve pancreatic cancer survival rates, identifying early diagnostic markers is crucial.

Colorectal Cancers Colorectal cancer (CRC, including both colon and rectal cancers) is the fourth most frequently diagnosed cancer, and it has become the second leading cause of cancer deaths both in the United States and in Europe. The risk of developing CRC increases by an unhealthy lifestyle, such as tobacco, alcohol and red meat consumption, low-fibre diets, sedentariness and obesity. Stool-based CRC screening has decreased disease mortality due to an early diagnosis (Rank and Shaukat 2017). The recommended treatment of non-metastatic CRC is surgical colectomy with curative intent in combination with adjuvant chemotherapy, such as FOLFOX (anti-DNA metabolites folinic acid, 5-fluorouracil and DNA cross-linker oxaliplatin) or CapeOX (capecitabine (pre 5-FU) and oxaliplatin). For all patients with metastatic CRC, the tumour should be analysed for RAS (KRAS and NRAS) or B-RAF mutations (increased of cellular proliferation by Mitogen-Activated Protein Kinase (MAPK) pathway activation) for a microsatellite instability (MSI) or mismatch repair (MMR) phenotype to define an optimized treatment with antibody-drug conjugates (anti-VEGF, bevacizumab; anti-BRAF, vemurafenib; PDL-PD1 inhibitors, pembrolizumab, nivolumab) (Benson et al. 2018).

2 Molecular Subtypes in Gastrointestinal Cancers

Heterogeneity of cancer comes from different origin of tumour tissue, from similar tumours of different patients and even from different cells within a single tumour. However, this heterogeneity is not taken into account by the decision-making process for therapy management, which mostly relies on a few critical clinical, histopathological or genetically defined classes (Ajani et al. 2016, 2019; Benson et al. 2018; Phelip et al. 2019; Tempero et al. 2019). Surgery to resect the tumour tissue in early stages in combination with chemotherapy targeting DNA stability and synthesis is the most curative treatment for cancer. The MSI or MMR phenotype may diverse the therapy. Especially at more advanced stages, analysing tumours for RAS (KRAS and NRAS), B-RAF mutations (increased of cellular proliferation by MAPK pathway activation), MSI and MMR phenotype is employed to define an optimized treatment with, e.g. antibody-drug conjugates (described in the "Chemotherapy Used in Gastrointestinal Cancers" section). Classifying tumours according to gene expression-based molecular subtypes predicts the response to therapy and has been proposed to innovate towards personalized therapy.

2.1 Coding mRNA Transcripts Profiles

Despite tumour heterogeneity, each tumour type presents similar fundamental cancer-related mechanisms. Actually, comparing molecular subtypes in gastrointestinal cancers may generalize five molecular subtypes (Bijlsma et al. 2017):

Immunogenic Subtypes These subtypes include tumours presenting microsatellite instability (MSI) in all gastrointestinal cancers. MSI is associated with mismatch repair (MMR) alterations associated with mutation in cancer driver genes and generation of cancer neoantigens that trigger an immune response (Guinney et al. 2015; Baretti and Le 2018). In addition, for this subtype, hypermethylation of CpG islands (CIMP) is a phenotype observed in the oesophageal MSI subtype (Cancer-Genome-Atlas-Research-Network 2017), gastric MSI subtype (Cristescu et al. 2015) and colorectal cancer CMS1 subtype (Guinney et al. 2015). The CIMP phenotype is also observed in pancreatic cancer but was not classified in the pancreatic immunogenic subtype (Bailey et al. 2016). The tumour tissue of the immunogenic subtype has infiltration of immune cells and presence of inflammatory response gene expression, and the patient may clinically benefit from immune checkpoint blockade therapy (PD-1 and PD-L1 pathway).

Epithelial Subtypes The epithelial subtype includes tumours that are microsatellite stable (MSS). This subtype has not been described in PDAC. The epithelial subtypes of oesophageal and gastric cancers have an increased WNT, p63; and E2F, Myc, RAS signalling pathway, respectively (Guinney et al. 2015; Cancer-Genome-Atlas-Research-Network 2017). Both WNT/β-catenin and p63 are critical regulators of epidermal stem cells (Kretzschmar and Clevers 2017; Soares and Zhou 2017). The oesophageal epithelial subtype tissue presents several mutations/amplifications that implicate cell cycle dysregulation. The gastric epithelial subtype (MSS/$TP53^+$) has frequent mutations in the TP53 and shows a high proliferation rate (Cristescu et al. 2015). The epithelial subtype in colorectal cancers, CMS2, has increased TP53, WNT, Myc signalling and gene amplifications of HNF4A, also regulating proliferation (Guinney et al. 2015).

Metabolic Subtypes A metabolic subtype is described in gastric, pancreatic and colorectal cancers. In the gastric metabolic subtype, the genes characteristically highly expressed include those that are also expressed in normal stomach mucosa, and the gene expression resembles that of spasmolytic-polypeptide-expressing metaplasia (SPEM), which has been proposed as an intermediate step in the development of gastric adenocarcinoma (Lei et al. 2013). This subtype was a high responder to 5-FU treatment. In pancreatic tumours, the metabolic subtype is either called "pancreatic progenitors" or "classical" and presents a slow proliferating, glycolytic and lipogenic phenotype (Collisson et al. 2011; Bailey et al. 2016). The KRAS mutations contribute to the metabolic adaptation. In colorectal cancer cell, the metabolic profile CMS3 is described with high expression of genes involved in sugar, amino acid, nucleotide and fatty acid metabolisms (Guinney et al. 2015). Both the gastric and colorectal cancer tumours of this subtype present chromosome instability.

Mesenchymal Subtypes The expression pattern in this subtype is indicative of the epithelial-mesenchymal transition (EMT) and of stromal infiltrates. Epithelial to mesenchymal transition in oesophageal OSCC is observed (Lv et al. 2017) but not associated with a specific subtype. In the gastric mesenchymal subtype (with high N-cadherin and low E-cadherin expression), the activation of several signalling pathways, such as p53, TGFß, VEGF and NFκB, is observed (Lei et al. 2013). In addition, this subtype has high activities of cancer stem cell (CSC) pathways. In pancreatic cancer, the mesenchymal subtype associates with squamous cell pancreatic carcinomas and is associated with a CIMP-positive phenotype (Bailey et al. 2016). Moreover, it shows increased expression of mesenchymal and CSC markers (Zhou et al. 2017). The colorectal mesenchymal subtype CMS4 also showed activation of TGFß and VEGF and a stromal infiltration expression profile and was enriched in CSC (Guinney et al. 2015). The mesenchymal subtypes display a resistance to anti-EGFR agents and are associated with a poor disease outcome (Bijlsma et al. 2017).

Organ-Specific Subtypes Some tumours originate from organ-specific cells and cannot be classified in a general subtype. In the pancreas, the "aberrantly differentiated endocrine exocrine" (ADEX) tumours display expression of progenitors, exocrine and endocrine differentiation markers (Bailey et al. 2016). ADEX is associated with a CIMP-positive phenotype. Amongst gastric cancers, the Epstein-Barr virus (EBV)-associated cancer is a separate class associated with mutations in the SWI/SNF factor ARID1A, the catalytic kinase subunit PIK3CA and BCL6-corepressor BCOR genes (Lim et al. 2016).

Only if analysing molecular subtypes contributes to decision-making for therapy, it will be incorporated as regular practice. This largely depends on the availability of innovative drugs but could also depend on reanalysing drugs previously discarded due to low responses in patients. For example, in colorectal cancers, low response rates to HSP90 inhibitors were observed in un-stratified patient populations. Recently, a strong in vivo anti-tumour activity of combination therapy with HSP90 inhibitors and 5-FU was observed in a chemoresistant PDX model from the mesenchymal subtype (Sveen et al. 2018).

2.2 Expression Profiles of NcRNAs

Tumour classification and non-coding RNAs expression, in particular miRNAs, have also been analysed. Molecular subtypes analysis for associated miRNAs in gastric cancers showed five specific miRNAs hypermethylated in the EBV-positive subtype (miR-148a, miR-196b, miR-212, miR-219–1 and miR-219-2), another four hypermethylated in both the EBV and MSI/Immunogenic subtypes (miR-9-1, miR-9-3, miR-34c and miR-137) and five in all subtypes (miR-10b, miR-124a-1, miR-124a-2, miR-124a-3 and miR-129-2) (Lim et al. 2016). Eight miRNAs are proposed as diagnostic markers (miR-21, miR-31, miR-32, miR-106a, miR-143, miR-182, miR-218 and miR-223) and 35 as prognostic indicators of gastric cancer (Alessandrini et al. 2018). In PDAC, 19 miRNAs were defined as predictive markers for overall survival, and several of them targeted different signalling pathways, such as p53, Cox2 and TGFβ signalling (Namkung et al. 2016). The miR-1244, miR-574-5p and miR-4474-5p seem to be overexpressed in the high-risk molecular subtype suggesting using anti-miRNA drugs may be risk lowering. In CRCs, miRNA expression can distinguish MSI/MSS subtypes, with a high expression of miR-196b and miR-106a in MSS and of miR-31 in MSI samples. The low expression of 16 miRNAs defined the mesenchymal subtype (Cantini et al. 2015). The miRNA profile in the subtypes associated with their predicted target gene expression. In particular, miR-194, miR-200b, miR-429 and miR-203 had a low expression in the mesenchymal subtype (stem/serrated/mesenchymal) and were shown to inhibit EMT and stemness properties.

Besides analysing the ncRNAs associations with molecular subtypes, clustering to ncRNA expression profiles has been performed. The miRNAs expression profiles have not yielded subtypes that were as strongly associated with a distinct biology as mRNA classification subtypes (Tang et al. 2016; Bijlsma et al. 2017). For CRCs, miRNAs expression analysis resulted in five clusters, including a proliferative and stem cell subtype (Cantini et al. 2015). These analyses also sorted PVT1 and HOTAIR as deregulated lncRNAs (Flippot et al. 2016; Wang et al. 2016). Since functional classification of lncRNAs is not yet established, the importance of other lncRNA differentially expressed remains to be elucidated. The importance of lncRNAs in the management of anticancer treatment is supported by their participation in drug resistance in cancer, such as discussed in the section below.

3 Chemotherapy Used in Gastrointestinal Cancers

Direct targets of classic adjuvant chemotherapy are mainly implicated in DNA stability (inflicting DNA breaks, inhibiting DNA repair or preventing DNA synthesis) (Fig. 1a), which result in inhibition of the cell cycle, alterations of the cancer transcriptome, proteome and cancer signalling pathways (signalome) in relation to, e.g. increased apoptosis, inhibiting autophagy, increased differentiation and inhibition of a stem-like phenotype. Recent therapy innovations include antibody-drug conjugates that target the signalome of cancer cells, angiogenesis and, in the case of PDL-PD1 inhibitors, the immune escape mechanisms of cancer cells that repress T-regulatory (TReg) immune cells. The chemotherapy treatments in gastrointestinal cancers include DNA cross-linkers, anti-DNA metabolites, antracyclin antibiotics, taxanes and antibody-drug conjugates.

DNA Cross-Linkers The currently used DNA cross-linkers are based on platinum compounds that form adducts within the DNA helix (Gustavsson et al. 2015). These are *cis*-diamine platin (cisplatin, carboplatin) or *trans*-cyclohexane-diamino adducts (oxaliplatin) to which mismatch repair proteins and damage-recognition proteins will bind. These effects block DNA synthesis and consequently inhibit the cell cycle and cell proliferation. The cytostatic antibiotic mitomycin has also alkylating properties causing DNA cross-linking.

Antimetabolites Chemotherapy drugs classified as anti-(DNA) metabolites interfere with the metabolism of nucleic acids (Gustavsson et al. 2015). For example, 5-fluorouracil (5-FU) is metabolized to 5-fluorodeoxyuridine monophosphate (FdUMP) that binds to the thymidylate synthase enzyme preventing the methylation of deoxyuridine monophosphate (dUMP) to deoxythymidine monophosphate (dTMP), a precursor for DNA synthesis (thymine nucleic acids). 5-FU-treated cells are inhibited in the S-phase of the cell cycle. Gemcitabine is metabolized to difluorodeoxycytidine 5′-triphosphate (dFdCTP), inhibiting the synthesis of dCTP, and incorporated in DNA leads to strand termination. Methotrexate is a folate antagonist that inhibits the conversion of dihydrotetrafolic acid (DHF) to tetrafolic

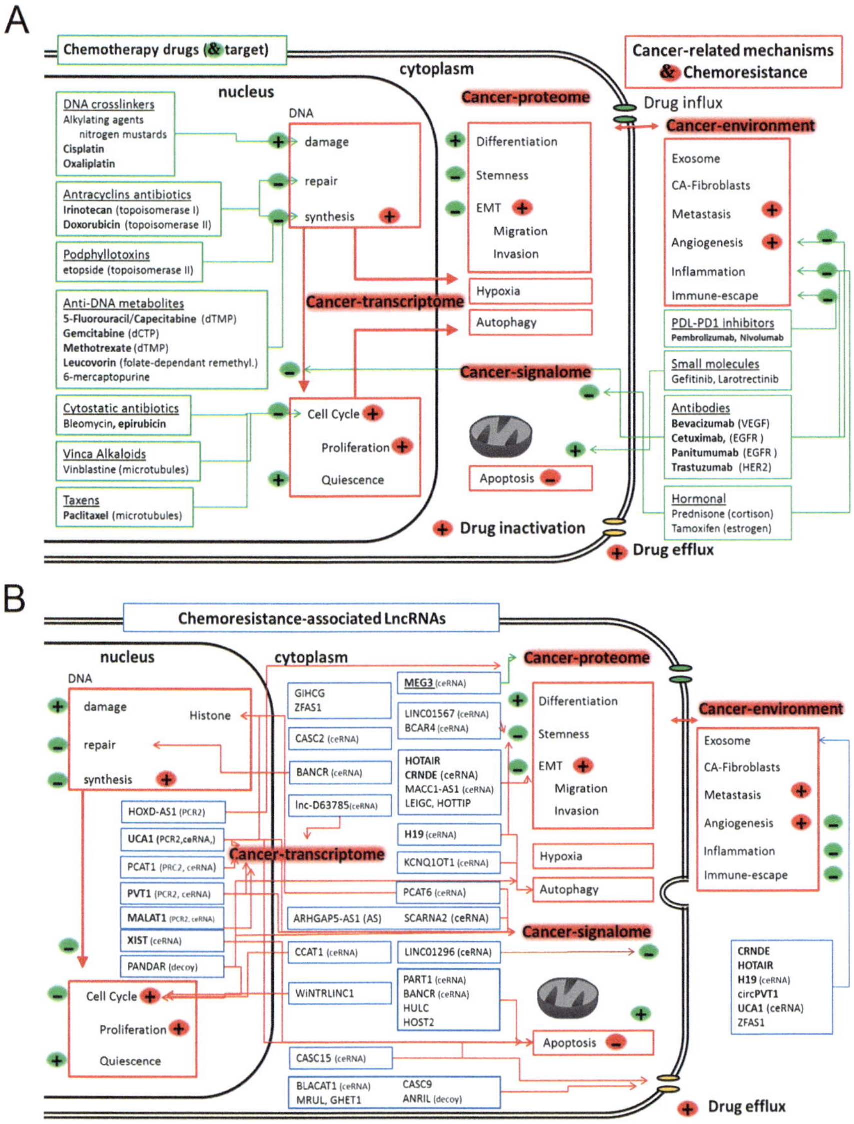

Fig. 1 Cellular cancer-related mechanisms. (**a**) Depicted are cancer-related mechanisms (in red) and where chemotherapy drugs intervene (in green). In bold, drugs used in gastrointestinal cancers. Indicated with green circles are mechanisms affected by chemotherapy, and with red circles, mechanisms related to chemoresistance. (**b**) Summary of LncRNAs (blue squares) that are reported to affect chemoresistance in gastrointestinal cancers. The affected cancer-related mechanisms are connected by red/green lines

acid (THF) that is needed by the thymidylate synthase enzyme to synthesize dTMP. Leucovorin is a tetrahydrofolate used to reduce the side effects of anti-DNA metabolites.

Anthracycline Antibiotics These antibiotics are inhibitors of the topoisomerase enzymes needed for DNA synthesis (Bjornsti and Kaufmann 2019). Topoisomerase I cleaves one single strand of the DNA helix allowing to relax the helix and then reseals the DNA. Topoisomerase I binding to irinotecan results in stagnation of the topoisomerase 1 and single-strand DNA (ssDNA) breaks. Upon collision with the replication fork in dividing cells, this stagnation results in double-strand breaks of the DNA, leading to inhibition of the cell cycle and apoptosis. Topoisomerase II cuts both strands of the DNA (dsDNA breaks) before religation and needs ATP for its supercoil-modulating activity. Doxorubicin and epirubicin (a cytostatic antibiotic) intercalate the DNA and induce topoisomerase II enzyme stagnation; collision with the replication causes DNA damage beyond repair and eventually leads to cell death.

Taxanes The taxane paclitaxel (abraxane), originally derived from the Pacific yew tree, acts as a mitotic inhibitor by binding to microtubules and preventing cells to sort the M-phase of the cell cycle and leading to apoptosis (Florian and Mitchison 2016).

Antibody-Drug Conjugates Antibody-drug conjugates are human(ized) monoclonal antibodies chemically conjugated with cytotoxic small molecules. They target signalling pathways through receptor interaction blockades, such as anti-ERBB2/HER2 (trastuzumab), anti-VEGR (ramucirumab), anti-epidermal growth factor receptor (EGFR) (cetuximab, panitumumab, erlotinib) and anti-PD-1 (pembrolizumab). The interaction of PD-1 (on regulatory T (TReg) immune cells) with PDL (on the cancer cells) negative regulates the anti-tumour response of T-cells and evokes immune-escape (Kalbasi and Ribas 2019).

Upon chemotherapy treatment, tumour cells show an innate, or develop an acquired, resistance to chemotherapeutic drugs. Chemoresistance does not necessarily implicate the drug-targeted mechanisms but involves diverse cancer-related mechanisms (depicted by red circles in Fig. 1a). Chemoresistance may, for example, be the result of limiting drug intracellular entrance and/or the accessibility of the plasma membrane and its tumour cell epitopes against immune recognition or antibody-based therapy by mucin expression (Jonckheere et al. 2014). Other mechanisms of chemoresistance are increased drug efflux by increased expression of ATP-binding cassette transporters or vesicle exocytosis. Moreover, alteration of apoptotic regulators or effectors (caspases, Bcl2-family proteins), enhanced DNA repair, altered autophagy and epigenic alterations resulting in more stem cell-like phenotypes have been reported in chemoresistance (Assaraf et al. 2019).

4 LncRNAs Implicated in Chemoresistance of Gastrointestinal Cancers

Transcripts that do not encode proteins and are larger than 200 nucleotides are nominated long non-coding RNAs (lncRNAs). There are over 15,000 of such transcripts defined in the current human GENCODE release (v28), so it comes as no surprise that they are also implicated in cancer-related mechanisms. In gastrointestinal cancers, most of the associated lncRNAs contribute to chemoresistance (Fig. 1b, red flashes). Although it may be a research bias effect that tumour suppressor lncRNAs are not readily detected, only Meg3 has been shown to play a chemo-sensitising role in all gastrointestinal cancers (Ghafouri-Fard and Taheri 2019) (Fig. 1b, green flash). Meg3 has been shown to act as an endogenous RNA competing for miRNA binding (ceRNA) and affecting the miRNA-mediated decay of the miRNA-target genes. In the cell nucleus, the lncRNAs HOXD-AS1, UCA1, PCAT1 PVT1 and MALAT1 have been shown to interact with the polycomb repressive complex-2 (PRC2) that stimulates epigenetic silencing by trimethylation of lysine 27 of histone H3 (H3K27me3) (Neve et al. 2018; Derderian et al. 2019; Li et al. 2019b; Sun and Ma 2019; Ye et al. 2019). These lncRNAs bind to subunits of the PRC2 complex, such as subunit SUZ12, embryonic ectoderm development (EED) and the enhancer of zeste homolog 2 (EZH2). XIST has also been shown to regulate PRC2 epigenetic silencing and function as a ceRNA in the nucleus (Bartolomei et al. 2019). In addition, the lncRNA PANDAR acts as a decoy in the nucleus through binding to tumour suppressor gene p53 protein, thereby inhibiting cyclin-dependent kinase inhibitor 1A (CDKN1A) in gastric cancer (Liu et al. 2019b).

Lots of lncRNAs have been shown to affect chemoresistance-related mechanism in gastrointestinal cancer by acting as ceRNA. Different genes of the cancer transcriptome are affected by GIHCG, ZFAS1, CASC2, BANCR and lnc-D63785 (Pan et al. 2017; Li et al. 2018b; Ma et al. 2018; Zhou et al. 2018; Liu et al. 2019a). LncRNAs CCAT1, GHET1 and WiNTRLINC1 accelerate the cell cycle (Xia et al. 2018; Hu et al. 2019; Yokota et al. 2019). Moreover, LINC01567, BCAR4 and H19 stimulate a stemness phenotype (Yu et al. 2017b; Lecerf et al. 2019; Ouyang et al. 2019), and HOTAIR, CRNDE, MACC1-AS, LEIGC and HOTTIP stimulate a mesenchymal phenotype (Han et al. 2014; Li et al. 2015; Zhang et al. 2018a; He et al. 2019; Qu et al. 2019). PANDAR, H19 and KCNQ1OT1 contribute to autophagy stimulating metabolism and survival in cancer cells through recycling of intracellular components (Liu et al. 2018b; Lecerf et al. 2019; Li et al. 2019d). The lncRNAs LINC01296, ARHGAP5-AS1, SCARNA2 and PCAT6 affect different cancerous signalling pathways (Liu et al. 2018a; Wu et al. 2019b; Zhang et al. 2019b; Zhu et al. 2019). Cell death by apoptosis is inhibited in gastrointestinal cancers by PART1, BANCR, HULC, HOST2 and CASC15 (Zhang et al. 2016; An and Cheng 2018; Kang et al. 2018; Ma et al. 2018; Gao et al. 2019). Chemoresistance by increased expression of ATP-binding cassette transporters is stimulated by CASC15, BLACAT1, MRUL, GHET1, CASC9 and ANRIL (Wang et al. 2014a, b; Shang et al. 2017; Zhang et al. 2017a, 2018b; Wu et al. 2018a, b; Gao et al. 2019). Mesenchymal cancer-associated fibroblasts (CAF) play a dynamic part in tumour cell proliferation and invasion (reviewed in pancreatic and colorectal cancer by (Koliaraki et al. 2017; von Ahrens et al. 2017)). In addition to autocrine and paracrine secretion by CAFs, translocation of metabolic substrates has been shown to transfer from CAFs to tumour cells via exosomes. Exosome secreted by CAFs or tumour cells may include CRNDE, HOTAIR, H19, PVT1, UCA1 and ZFAS1, which could impact distant cell signalling or promote a niche that sustains the tumour microenvironment.

Several lncRNAs seem to play a general role in cancer-associated cellular mechanisms and have been extensively described (CRNDE, H19, HOTAIR, MALAT1, MEG3, PVT1, UCA1 and XIST depicted in bold in Fig. 1b). We here discuss the role of these lncRNAs in chemoresistance and their mechanism of action in gastrointestinal cancers.

The Colorectal Neoplasia Differentially Expressed (CRNDE) The colorectal neoplasia differentially expressed gene is located on Chr16q12.2 and encodes at least 5 transcript variants ranging from 0.9 to 1.1 kb and including 6 different exons (NCBI RefSeq). Transcript NM_001308963.1 contains an ORF that may be translated to an 84-amino-acid nuclear protein (CRNDEP) with an unknown function (Lafrenie et al. 2015). The promoter of the lncRNA CRNDE may share regulatory elements with the neighbouring IRX5 gene on the opposite strand (1.1 kb apart). Indeed, in hepatocellular carcinoma, CRNDE is co-activated with IRX5, and both genes seem to be regulated by miR-136-5p (Zhu et al. 2018a, b). In CRC cells, this miRNA also targets E2F transcription factor 1 (E2F1)) and is implicated in CRNDE-stimulated metastasis and resistance to oxaliplatin (Gao et al. 2017). In addition, CRNDE contributes to chemoresistance of CRC through modulation of miR-181a-5p expression levels (Han et al. 2017). The implication of Wnt/beta-catenin signalling suggests that CRNDE may also affect EMT in CRC. CRNDE plays a role in 5-FU, oxaliplatin or irinotecan resistance in the chemotherapy of CRC (Sun et al. 2019). Furthermore, its presence was also shown in exosomes, where it may modulate miR-217 expression (Yu et al. 2017a). In gastric cancer, CRNDE was shown to affect miR-145, thereby regulating proliferation through E2F transcription factor 3 (E2F3) (Hu et al. 2017). In addition to its function as a competing endogenous RNA (ceRNA), CRNDE interacts with EZH2 in the PRC2 complex epigenetically silencing dual specificity phosphatase 5 (DUSP5; regulating MAPK/ERK signalling) and cyclin-dependent kinase inhibitor 1A (CDKN1A; inhibiting the cell cycle) expression (Ding et al. 2017). Through its interaction with heterogeneous nuclear ribonucleoprotein UL2 (hnRNP-UL2), CRNDE may activate the Ras/MAPK signalling pathway in CRC (Jiang et al. 2017).

The Imprinted Maternally Expressed Gene H19 The imprinted maternally expressed gene H19 is located on Chr11p15.5 close to the telomeric region. A 1.3 kb transcript with an alternative exon 1 and a 2.3 kb transcript with 5 exons are annotated in the UCSC genome browser (RefSeq genes). Onyango et al. reported an antisense, and paternally expressed, nucleolar protein overlapping the last exon of

H19 (Onyango and Feinberg 2011). The IGF2 gene, which is also expressed from the paternally inherited chromosome, is located 100 kb downstream of H19. Studies in mice showed that both Igf2 and H19 are imprinted in a reciprocal manner in most somatic cells, and they are regulated by the same downstream enhancer (Nordin et al. 2014). The first exon of the 2.3 kb H19 gene is host to the miR-675 gene, which was shown to promote gastric and colorectal cancers (Lecerf et al. 2019). Additional bioinformatic analyses also suggest that H19 plays a role in oxaliplatin and irinotecan resistance in CRC (Sun et al. 2019). Through sponging miR-194-5p and thereby regulating sirtuin 1 (SIRT1) transcript expression, H19 was associated with increased 5-FU resistance in CRC cells (Wang et al. 2018a, b). Low expression of miR-194 is associated with the mesenchymal subtype (Cantini et al. 2015). Furthermore, H19 is implicated in diverse cancer-related mechanisms that may interfere with chemoresistance. For example, H19 promotes cell proliferation and inhibited cell apoptosis in oesophageal OAC, gastric cancer and PDAC cells (Yang et al. 2014; Ma et al. 2016; Yan et al. 2017). In addition, H19 is implicated in EMT by acting as a ceRNA for miR-22-3p in gastric cancer cells, thereby affecting the downregulation of Snail1 (Gan et al. 2019), and by acting as a ceRNA for miR-138 and miR-200a in CRC, thereby inhibiting the mesenchymal marker genes ZEB1 and ZEB2 (Liang et al. 2015). Although evidenced in other cancer types, H19 can bind to EZH2 from the PRC2 complex modulating repressive histone methylation (H3K27me3).

Homeobox (HOX) Transcript Antisense RNA (HOTAIR) Homeobox (HOX) transcript antisense RNA gene is located on Chr12q13.13 within the homeobox gene C cluster (HOXC) and encodes three different transcripts with 4 to 6 exons. All transcripts are approximately 2.3 kb in size. The alternative exon 1 of transcript variant 1 (NR_047517.1) is located within the single intron of the opposite strand neighbouring gene HOXC11 (3.8 kb downstream of the two other transcripts). This HOTAIR transcript may regulate HOXC11 (Qu et al. 2019), which in fact increases in response to 5-FU in gastric cancer cell lines (Park et al. 2004). HOTAIR is also known to *trans*-repress the cluster region genes HOXD4 and HOXD8 through the interaction with Suz12 and EZH2 of the PRC2 complex that maintains the epigenetic repression mark H3K27Me3. HOXD4 promotes tumour cell proliferation in gastric adenocarcinoma (Liu et al. 2019b), whereas HOXD8 significantly reduces cell proliferation in CRC cells (Mansour and Senga 2017). HOTAIR is required to repress transcript expression of several imprinted *loci* such as Dlk1-Meg3 and Igf2-H19 (Li et al. 2013). Moreover, the epigenetic regulation of HOTAIR with EZH2/PRC2 plays a role in chemoresistance and allows EMT of CRC cells (Qu et al. 2019). In addition, HOTAIR/EZH2 silences tumour suppressor miR-34a in gastric cancer cells and in PDAC cells. This miRNA inhibits the cancer signalling pathways PI3K/Akt and Wnt/beta-catenin, implicated in cisplatin resistance of gastric cancer cells. MiR-34a also prevents oxaliplatin and 5-FU drug resistance in human colon cancer (Akao et al. 2011; Sun et al. 2017). In parallel, HOTAIR has been shown to function as a ceRNA in oesophageal cancer (binding to miR-125, miR-143, miR-148a and miR-204 (Ma et al. 2017; Xu and Zhang 2017; Wang et al. 2019)), in gastrointestinal stromal tumours (binding to miR-196a (Niinuma et al. 2012)), in gastric cancer (binding to miR-17-5p, miR-126, miR-217 and miR-454-3p (Dong et al. 2019; Jia et al. 2019; Jiang et al. 2019; Xiao et al. 2019)) and in human colon cancer (binding to miR-203a-3p, miR-218, miR-326 (Li et al. 2017c; Xiao et al. 2018; Pan et al. 2019)). As discussed in Section 2.2, miR-148a, miR-196b miR-143, miR-218 and miR-203 are markers of molecular subtypes.

Metastasis-Associated Lung Adenocarcinoma Transcript 1 (MALAT1) Metastasis-associated lung adenocarcinoma transcript 1 is a well-known lncRNA of approximately 6.7 kb in length (Sun and Ma 2019). The MALAT1 precursor gene has only one exon that is located on Chr11q13.1. Cleavage of this precursor transcript by RNase P results in a cytosolic tRNA-like small ncRNA (MALAT1-associated small cytoplasmic RNA, mascRNA of ~60 nt) and the nuclear MALAT1, which is stabilized by its 3′ triple helical structure. An antisense transcript starting 0.9 kb downstream of MALAT1 and affecting MALAT1 expression has recently been described (Zong et al. 2016; Gomes et al. 2019). MALAT1 plays a unique role in the organization of nuclear speckle bodies (Sun and Ma 2019). MALAT1 is post-transcriptionally modified by RNA methyltransferase-like protein 16 (METTL16) (Brown et al. 2016; Warda et al. 2017)), adding N^6-methyladenosine and mediating MALAT1 binding to heterogeneous nuclear ribonucleoprotein G (hnRNP-G) (Liu et al. 2017b). The hnRNP-G is part of the supraspliceosome involved in nuclear pre-mRNA processing. Indeed, MALAT1 regulates alternative splicing and interacts with SR splicing factors, modulating their distribution to nuclear speckles (Sun and Ma 2019). MALAT1's regulation of SF2/ASF splicing factors is permissive for cell cycle progression and proliferation in gastric cancer cells (Wang et al. 2014a, b). Additionally, the binding of MALAT1 to splicing factor SFPQ was implicated in tumour growth and metastasis in CRC (Ji et al. 2014). MALAT1 has also been shown to interact with the PRC2 complex contributing to oxaliplatin chemoresistance in CRC and to gastric cellular migration and invasion. Furthermore, MALAT1 is implicated in autophagy-associated chemoresistance via miR-23b-3p, miR-30b and miR-183 binding in gastric cancer (YiRen et al. 2017; Li et al. 2019a; Xi et al. 2019) and via miR-101 binding in CRC (Si et al. 2019). Many other miRNAs bind to MALAT1 including miR-106b-5p, miR-202, miR-218 and miR-1297 in gastrointestinal cancers (Li et al. 2017a, b; Zhang et al. 2017b; Zhuang et al. 2019).

The Maternally Expressed Gene 3 (MEG3) The maternally expressed gene 3 is located at Chr14q23.1 and encodes at least 15 transcripts with 6 to 10 exons ranging from 1.5 to 9.7 kb in size (NCBI RefSeq, (Ghafouri-Fard and Taheri 2019)). MEG3 gene is embedded in the DLK1-DIO3 imprinted region that contains paternally expressed protein-coding genes (DLK1, RTL1, DIO3) and maternally expressed ncRNA genes (MEG3, MEG8, RTL1-AS, miRNAs, piRNAs and snoRNAs) (Sellers et al. 2018; Li et al. 2019c). LncRNA HOTAIR was implicated in the repression of the expressed genes in this cluster. MEG3 expression can prevent Dlk1 gene activation (inhibiting Notch1) in embryonic stem cell signalling (Sanli et al. 2018). MEG3 is the only described lncRNA that plays a chemo-sensitising role in digestive cancer (Ghafouri-Fard and Taheri 2019). MEG3 functions as a tumour repressor and is guiding the EZH2/PRC2 repressive complex to its target genes, e.g. HOTAIR, a process mediated by the formation of triplex RNA-DNA helixes at GA-rich regions (Mondal et al. 2015). MEG3 inhibits the TGFß-stimulated cell proliferation and induces apoptosis in gastric carcinoma cells and in CRC cells (Ghafouri-Fard and Taheri 2019). In addition, MEG3 suppresses EMT of gastric cancer cells, also as ceRNA, through binding to miR-21 (Xu et al. 2018). miR-21 is a diagnostic marker for gastric cancer. Increasing p53 expression plays a central role in the action of MEG3, including in gastrointestinal cancer cells. Moreover, MEG3 increases oxaliplatin sensitivity in CRC cells (ceRNA of miR-141 Wang et al. 2018a, b). Additional miRNAs that are associated with MEG3 in gastrointestinal cancers are miR-181, miR-770, MiR-208a, miR-4261 and miR-181a-5p (Ding et al. 2019; Ghafouri-Fard and Taheri 2019; Ma et al. 2019).

The Plasmacytoma Variant Translocation 1 (PVT1) The plasmacytoma variant translocation 1 is one of the lncRNAs present at the cMyc locus at Chr8q24.21 (Derderian et al. 2019). The PVT1 gene encodes at least 6 alternative transcripts ranging from 0.9 kb to 2 kb and contains 9 different exons (NCBI-nucleotides). Intron 3 of PVT1 gene contains another lncRNA TMEM75 that is also associated with CRC (Jin et al. 2018). Moreover, copy-number amplification of PVT1 was shown in CRC (Takahashi et al. 2014). Actually, PVT1 is frequently co-increased with the Myc gene, through a gain of copy number, and is shown to stimulate high MYC protein levels (Tseng et al. 2014). Myc is the major driving force of tumourigenesis and implicated in metabolic reprogramming to favour glycolysis (Warburg effect) (Dejure and Eilers 2017). In gastric cancer cells, PVT1 is associated with EZH2/PRC2 and increases the cell cycling by repressing p15 and p16 (Kong et al. 2015). Gemcitabine resistance in pancreatic PDAC is mediated by EZH2/PRC2 and PVT1/Myc regulation (Yoshida et al. 2017; You et al. 2018). PVT1 is a host gene for 6 different miRNAs: miR-1204, miR-1205, miR-1206, miR-1207-5p, miR-1207-3p and miR-1208, which were recently shown to be upregulated in gastric cancer by copy-number variations (Anauate et al. 2019). The expression of miR-1205 is implicated in gastric cancer and that of miR-1207-5p in gastric, colorectal and pancreatic cancers ((Anauate et al. 2019), (Chen et al. 2014; Lu et al. 2016), and (Wang and Wu 2017; You et al. 2018), respectively)). In particular, gemcitabine inhibits cancer progression by stimulating miR-1207 expression. PVT1 can sponge several miRNAs promoting cell proliferation and invasion (e.g. miR-152, miR-186 and miR-1266 in gastric cancers (Chen et al. 2014; Derderian et al. 2019) and miR-26b, miR-128, miR-214-3p, miR-216a-5p and miR-455 in CRC (Chai et al. 2018; Derderian et al. 2019; Zeng et al. 2019).

The Urothelial Cancer-Associated 1 (UCA1) The urothelial cancer-associated 1 gene is located on Chr19p13.12 and encodes three transcripts of 1.4 kb, 2.2 kb, and 2.7 kb, all with a small variation of the three major exons (Xue et al. 2016). LINC01764 is an antisense lncRNA in the same gene locus that was recently shown to have an increased expression in cetuximab-resistant CRC cells (Jing et al. 2019). The expression of UCA1 is upregulated upon chemoresistance to cisplatin, doxorubicin and 5-fluorouracil in gastric cancer cells (Fang et al. 2016) and upon 5-fluorouracil chemoresistance in CRC cells (Bian et al. 2016). UCA1 can interact with EZH2/PRC2, which has also been shown in gastric cancer cells (Wang et al. 2017a, b). Moreover, UCA1 has been shown to stimulate cell proliferation, and silencing UCA1 expression in cancer cells has been shown to arrest the cell cycle in the G0/G1 phase (reviewed in (Neve et al. 2018)). Furthermore, UCA1 can bind to cytosolic proteins implicated in tumorigenic mechanisms, such as G protein-coupled receptor kinase 2 (GRK2) in gastric cancers, which leads to the activation of the ERK signalling pathway. In addition, UCA1 binds to mediators of the Hippo pathway (MOB1, Lats1 and YAP) activating YAP/TEAD-regulated transcription in pancreatic cancer cells. UCA1 also functions as a ceRNA for miR-204 in oesophageal cancer; for miR-7-p5, miR-27b, diagnostic marker miR-182 and miR-590 in gastric cancer; for miR-96 and miR-135a in pancreatic cancer; and for miR-143 and miR-204 in CRC. The miR-143 is also a diagnostic marker for gastric cancer.

The X-Inactive Specific Transcript (XIST) The X-inactive-specific transcript is located on chrXq13.2 and encodes a 19 kb-long lncRNA with 6 exons. To regulate the transcript expression level of X-linked genes in females (compared to the male XY chromosomes), the paternally inherited X chromosome is transcriptionally silenced through an epigenetic process, of which XIST is the master regulator (Bartolomei et al. 2019). Indeed, Xist transcript is uniquely observed at the inactivated X chromosome, except during mitosis. Transcription of a 37 kb antisense lncRNA TSIX (one exon) covers the promoter region of Xist, suggesting that Tsix is required for Xist repression on the maternally X chromosome that remains active. The XIST transcript is highly methylated with at least 78 N^6-methyladenosine (m6A) residues that can recruit the m6A reader, YTHDC1, important for XIST-mediated transcriptional repression. Epigenetic regulation by Xist is mediated through the recruitment of PRC2 to the X-inactivation centre, targeting H3K27 trimethylation (H3K27me3) to gene-rich islands, after which this epigenetic silencing spreads to the full 150 megabases of the chromosome (Simon et al. 2013). Only 78 human genes are not silenced by the X-inactivation (Balaton and Brown 2016), of which DDX3 has an oncogenic role in CRC and cetuximab resistance (Wu et al. 2017). The XIST transcript is significantly upregulated in cancers, including oesophageal, gastric and pancreatic cancer (Yang et al. 2018). In CRC, XIST plays a role in 5-FU and doxorubicin resistance (Xiao et al. 2017; Zhu et al. 2018a, b). Several studies report the involvement of XIST in protecting cancer-related genes from miRNA-mediated decay (ceRNA) (Yang et al. 2018). In oesophageal OSCC, these miRNAs are miR-101 and miR-494 (Chen et al. 2019); in gastric cancer miR-101, miR-497 and miR-185; in pancreatic cancer miR-34a-5p and miR-133a; and in CRC miR-124, miR-132-3p and miR-200b-3p. However, physiological parameters on miRNA gene target abundance and miRNA affinity have previously made Denzler et al. argue that the feasibility of gene regulation through changes in ceRNA levels is unlikely to occur in differentiated cells under physiological or disease circumstances (Denzler et al. 2016). The sex-specific expression of XIST provided the unique opportunity to study the biological feasibility of proposed XIST-miRNA interactions. The results of this study show that Xist-ceRNA targeting of mRNA transcript may be mediated through multiple miRNAs rather than one specific miRNA-coding-gene interaction and that only a subset of miRNAs are preferentially localized to the nucleus, where XIST may bind them (Palazzo et al. 2019).

5 Single-Cell Expression Analysis

The first single-cell RNA sequencing (scRNA-seq) method was described by Tang et al., using a polyT primer with an anchor sequence to reverse transcribe and identify the mRNAs from a single cell (Tang et al. 2009). Several high-throughput scRNA-seq technologies have recently been developed, including "Cell Expression by Linear amplification and Sequencing" (CEL)-seq2, microfluidic droplet-based (Drop)-seq, "Massively Parallel Single-Cell RNA" (MARS)-seq, and "Switching Mechanism at 5' End of RNA Template" (Smart)-seq2. It has become clear that similar cell types have an intercellular transcriptome heterogeneity with different mRNA abundances and transcript splicing patterns, which may contribute to intra-tumour heterogeneity (Wang et al. 2017a, b). Intra-tumour heterogeneity is averaged in the classic bulk RNA sequencing. Indeed, single-cell sequencing in oesophageal OSCC cells revealed two subpopulations of paclitaxel resistance cancer cells. These studies revealed KRT19 as a novel specific marker for paclitaxel resistance (Wu et al. 2018a, b). Moreover, it was revealed that the stemness phenotype of oesophageal OAC cells was related to the overexpression of cell cycle-associated genes(e.g. the cell cycle-regulated Aurora kinase A (AURKA) and cell division cycle 20 (CDC20) genes) (Wu et al. 2019a). This phenotype in oesophageal OSCC was also correlated with the signalling pathways of DNA replication and DNA damage repair. Moreover, the DNA damage repair-associated factor PARP4 was identified as a novel potential cancer stemness marker, suggesting PARP-1 inhibitors may serve as an effective supplementary therapy for OSCC.

Studies of scRNA expression in gastric cancer organoids defined heterogeneous epithelial and fibroblast cell subpopulations and identified R-spondin 3 as the endogenous factor, secreted by fibroblasts in the gastric stroma, that maintains the self-renewal capacity of the gastric epithelial stem cells (Chen et al. 2019). Furthermore, Drop-seq analysis of gastric cancer biopsies identified a cluster of metaplastic stemlike cells from which gastric cancer may arise. A panel of specific early gastric cancer marker genes (not expressed in other cell types) was identified, including two novel genes: the iron transporter SLC11A2 and the kallikrein serine protease KLK7 (Zhang et al. 2019a).

Analysis of PDAC cells with scRNA-seq indicated two distinct pancreatic cell populations, epithelial-like and mesenchymal-like, and distinct molecular subtypes of macrophages and CAF (Hosein et al. 2019). Apparently the mesenchymal-like cells appeared probably later in the disease process (Bernard et al. 2019). These single-cell studies revealed micro-environmental heterogeneities that accompany early cancer pathogenesis. For example, conflicting findings on pancreatic CAF may be reconciled by the identification of the three cellular subtypes (myofibroblastic (proximally to PDAC cells), inflammatory and antigen-presenting CAFs (distally and immune response related)) (Belle and DeNardo 2019; Elyada et al. 2019).

Single-cell analysis in rectal cancer cells suggested that tumours of the same molecular classification possess their own architecture, which may result in different diagnosis, prognosis and drug responses (Liu et al. 2017a). Moreover, it revealed that copy-number alterations are inherited and some not previously detected in bulk sequencing. Methods that combine scRNA-seq with CRISPR/Cas9-based gene perturbations have been recently developed (Dixit et al. 2016; Jaitin et al. 2016; Datlinger et al. 2017). The CRISPR/guide RNA, invalidating target gene expression, introduced in a single cell can now be assessed in scRNA-seq analysis. Analysis of guide RNA-related transcriptomes and loss-of-function screens revealed regulatory checkpoints of the EMT (McFaline-Figueroa et al. 2019). The results of this study showed that intermediate transition states may arise from a disabled signalling pathway component at a checkpoint causing staggering the normal continuum of a changing phenotype (EMT).

Transcript expression studies with tissue samples have suggested that lncRNAs are expressed, on average, at lower levels than mRNAs (Cabili et al. 2011; Cabili et al. 2015). However, scRNA-seq revealed that many lncRNAs are abundantly expressed in individual cells (including, e.g. MALAT1 and MEG3) and are cell type-specific (Liu et al. 2016; See et al. 2017). Moreover, 30% of the nuclear intergenic lncRNAs were novel lncRNAs only detectable by scRNA-seq and not by bulk RNA-seq (Liu et al. 2016). An example showing that profiling lncRNA expressions in scRNA-seq is feasible and informative was recently provided in renal cell carcinoma (Li et al. 2018a). In fact, hierarchical clustering of scRNA-seq data of metastatic cancer cell expression resulted in 4 clusters with each having at least 15 specific lncRNAs. Both the lncRNA CCAT1 and PVT1 from the cMyc locus at Chr8q24.21 were present in a cluster enriched in ribosomal proteins and with a GO-functional annotation of signal-recognition-particle-dependent protein targeting to membranes, which may reveal novel functions of these lncRNAs.

Recently, different strategies were employed to study lncRNAs by CRISPR/Cas9, such as deleting promoters, genes or exons with Cas9/guide RNA, and inactivation (CRISPRi) or activation (CRISPRa) transcription. These strategies can be combined with scRNA-seq (Dixit et al. 2016) to reveal lncRNA-regulated transcriptomes. This will render a wealth of new knowledge on lncRNA function. Moreover, unravelling lncRNAs at a single-cell level may elucidate novel lncRNAs implicated in chemoresistance and will render information on the implication of lncRNAs in checkpoints of the cancer-related mechanisms. Studies using single-cell analysis may take cancer treatment a step closer towards personalized, or even cell-specific, therapy.

Acknowledgements We would like to apologize to authors not cited in this chapter, due to limitations on references. This research was funded by "Institut National de la Santé et de la Recherche Médicale" (Inserm), "Centre National de la Recherche Scientifique" (CNRS), and grants from the "Comité du Nord de la Ligue Nationale contre le Cancer" and "Cancéropôle Nord-Ouest". The authors declare no conflict of interest. The funders had no role in the writing or in the decision to publish this chapter.

References

Ajani JA, D'Amico TA, Almhanna K et al (2016) Gastric cancer, V3.2016, NCCN Clin Prac Guidelines Oncol. J Natl Compr Cancer Netw 14:1286–1312

Ajani JA, D'Amico TA, Bentrem DJ et al (2019) Esophageal and Esophagogastric junction cancers, V2.2019, NCCN. J Natl Compr Cancer Netw 17:855–883

Akao Y, Noguchi S, Iio A et al (2011) Dysregulation of miR-34a expression causes drug-resistance to 5-FU in human colon cancer DLD-1 cells. Cancer Lett 300:197–204

Alessandrini L, Manchi M, De Re V et al (2018) Proposed molecular and miRNA classification of gastric cancer. I J Mol Sci 19:1683

An N, Cheng D (2018) The LncRNA HOST2 promotes gemcitabine resistance in human pancreatic Cancer cells. Pathol Oncol Res 7:018–0486

Anauate AC, Leal MF, Wisnieski F et al (2019) Analysis of 8q24.21 miRNA cluster expression and copy number variation in gastric cancer. Future Med Chem 11:947–958

Assaraf YG, Brozovic A, Gonçalves AC et al (2019) The multi-factorial nature of clinical multidrug resistance in cancer. Drug Resist Updat 46:100645

Bailey P, Chang DK, Nones K et al (2016) Genomic analyses identify molecular subtypes of pancreatic cancer. Nature 531:47–52

Balaton BP, Brown CJ (2016) Escape artists of the X chromosome. Trends Genet 32:348–359

Baretti M, Le DT (2018) DNA mismatch repair in cancer. Pharmacol Ther 189:45–62

Bartolomei MS, Loda A, Heard E (2019) Xist RNA in action: past, present, and future. PLOS Gen 15:e1008333

Belle JI, DeNardo DG (2019) A single-cell window into pancreas cancer fibroblast heterogeneity. Cancer Discov 9:1001–1002

Benson AB, Venook AP, Al-Hawary MM et al (2018) NCCN guidelines insights: Colon Cancer, V2.2018. J Natl Compr Cancer Netw 16:359–369

Bernard V, Semaan A, Huang J et al (2019) Single-cell transcriptomics of pancreatic Cancer precursors demonstrates epithelial and microenvironmental heterogeneity as an early event in neoplastic progression. Clin Cancer Res 25:2194–2205

Bian Z, Jin L, Zhang J et al (2016) LncRNA-UCA1 enhances cell proliferation and 5-fluorouracil resistance in colorectal cancer by inhibiting miR-204-5p. Sci Rep 6

Bijlsma MF, Sadanandam A, Tan P et al (2017) Molecular subtypes in cancers of the gastrointestinal tract. Nat Rev Gastroenterol Hepatol 14:333–342

Bjornsti M-A, Kaufmann SH (2019) Topoisomerases and cancer chemotherapy: recent advances and unanswered questions. F1000Research 8:1704

Brown JA, Kinzig CG, DeGregorio SJ et al (2016) Methyltransferase-like protein 16 binds the 3′-terminal triple helix of MALAT1 lncRNA. Proc Natl Acad Sci U S A 113:14013–14018

Cabili MN, Trapnell C, Goff L et al (2011) Integrative annotation of human large intergenic ncRNAs reveals global properties and specific subclasses. Genes Dev 25:1915–1927

Cabili MN, Dunagin MC, McClanahan PD et al (2015) Localization and abundance analysis of human lncRNAs at single-cell and single-molecule resolution. Genome Biol 16:20

Cancer-Genome-Atlas-Research-Network (2017) Integrated genomic characterization of oesophageal carcinoma. Nature 541:169–175

Cantini L, Isella C, Petti C et al (2015) MicroRNA-mRNA interactions underlying colorectal cancer molecular subtypes. Nat Commun 6:8878
Chai J, Guo D, Ma W et al (2018) A feedback loop consisting of RUNX2/LncRNA-PVT1/miR-455 is involved in the progression of colorectal cancer. Am J Cancer Res 8:538–550
Chen L, Lu MH, Zhang D et al (2014) miR-1207-5p and miR-1266 suppress gastric cancer growth and invasion by targeting telomerase reverse transcriptase. Cell Death Dis 30:553
Chen J, Lau BT, Andor N et al (2019a) Single-cell transcriptome analysis identifies distinct cell types and niche signaling in a primary gastric organoid model. Sci Rep 9
Chen Z, Hu X, Wu Y et al (2019b) LncRNA XIST promotes the development of esophageal cancer by sponging miR-494 to regulate CDK6 expression. Biomed Pharmacother 109:2228–2236
Collisson EA, Sadanandam A, Olson P et al (2011) Subtypes of pancreatic ductal adenocarcinoma and their differing responses to therapy. Nat Med 17:500–503
Cristescu R, Lee J, Nebozhyn M et al (2015) Molecular analysis of gastric cancer identifies subtypes associated with distinct clinical outcomes. Nat Med 21:449–456
Datlinger P, Rendeiro AF, Schmidl C et al (2017) Pooled CRISPR screening with single-cell transcriptome readout. Nat Methods 14:297–301
Dejure FR, Eilers M (2017) MYC and tumor metabolism: chicken and egg. EMBO 36:3409–3420
Denzler R, McGeary Sean E, Title Alexandra C et al (2016) Impact of MicroRNA levels, target-site complementarity, and cooperativity on competing endogenous RNA-regulated gene expression. Mol Cell 64:565–579
Derderian C, Orunmuyi AT, Olapade-Olaopa EO et al (2019) PVT1 signaling is a mediator of Cancer progression. Front Oncol 9:502
Ding J, Li J, Wang H et al (2017) LncRNA CRNDE promotes colorectal cancer cell proliferation via epigenetically silencing DUSP5/CDKN1A expression. Cell Death Dis 8:e2997–e2997
Ding L, Tian Y, Wang L et al (2019) Hypermethylated lncRNA MEG3 promotes the progression of gastric cancer. Aging 4:102309
Dixit A, Parnas O, Li B et al (2016) Perturb-Seq: dissecting molecular circuits with scalable single-cell RNA profiling of pooled genetic screens. Cell 167:1853-1866.e1817
Dong X, He X, Guan A et al (2019) Long non-coding RNA Hotair promotes gastric cancer progression via miR-217-GPC5 axis. Life Sci 217:271–282
Elyada E, Bolisetty M, Laise P et al (2019) Cross-species single-cell analysis of pancreatic ductal adenocarcinoma reveals antigen-presenting Cancer-associated fibroblasts. Cancer Discov 9:1102–1123
Fang Q, Chen X, Zhi X (2016) Long non-coding RNA (LncRNA) urothelial carcinoma associated 1 (UCA1) increases multi-drug resistance of gastric cancer via downregulating miR-27b. Med Sci Monit 22:3506–3513
Flippot R, Malouf G, Su X et al (2016) Cancer subtypes classification using long non-coding RNA. Oncotarget 7:54082–54093
Florian S, Mitchison TJ (2016) Anti-microtubule drugs. Methods Mol Biol:3542-3540_3525
Gan L, Lv L, Liao S (2019) LncRNA H19 regulates cell growth and metastasis via the miR223p/Snail1 axis in gastric cancer. Int J Oncol 54:2157–2168
Gao H, Song X, Kang T et al (2017) LncRNA CRNDE functions as a competing endogenous RNA to promote metastasis and oxaliplatin resistance by sponging miR-136 in colorectal cancer. Onco Targets Ther 10:205–216
Gao R, Fang C, Xu J et al (2019) LncRNA CACS15 contributes to oxaliplatin resistance in colorectal cancer by positively regulating ABCC1 through sponging miR-145. Arch Biochem Biophys 663:183–191
Gbolahan OB, Tong Y, Sehdev A et al (2019) Overall survival of patients with recurrent pancreatic cancer treated with systemic therapy: a retrospective study. BMC Cancer 19:468
Ghafouri-Fard S, Taheri M (2019) Maternally expressed gene 3 (MEG3): a tumor suppressor long non coding RNA. Biomed Pharmacother 118:109129
Gomes CP, Nobrega-Pereira S, Domingues-Silva B et al (2019) An antisense transcript mediates MALAT1 response in human breast cancer. BMC Cancer 19:771

Guinney J, Dienstmann R, Wang X et al (2015) The consensus molecular subtypes of colorectal cancer. Nat Med 21:1350–1356

Gustavsson B, Carlsson G, Machover D et al (2015) A review of the evolution of systemic chemotherapy in the management of colorectal cancer. Clin Colorectal Cancer 14:1–10

Han Y, Ye J, Wu D et al (2014) LEIGC long non-coding RNA acts as a tumor suppressor in gastric carcinoma by inhibiting the epithelial-to-mesenchymal transition. BMC Cancer 14:1471–2407

Han P, Li JW, Zhang BM et al (2017) The lncRNA CRNDE promotes colorectal cancer cell proliferation and chemoresistance via miR-181a-5p-mediated regulation of Wnt/beta-catenin signaling. Mol Cancer 16:017–0583

Hayakawa Y, Sethi N, Sepulveda AR et al (2016) Oesophageal adenocarcinoma and gastric cancer: should we mind the gap? Nat Rev Cancer 16:305–318

He W, Liang B, Wang C et al (2019) MSC-regulated lncRNA MACC1-AS1 promotes stemness and chemoresistance through fatty acid oxidation in gastric cancer. Oncogene 38:4637–4654

Hosein AN, Huang H, Wang Z et al (2019) Cellular heterogeneity during mouse pancreatic ductal adenocarcinoma progression at single-cell resolution. JCI Insight 4: pii: 129212

Hu Cheng E, Du Pei Z, Zhang Hui D et al (2017) LncRNA CRNDE promotes proliferation of gastric cancer cells by targeting miR-145. Cell Physiol Biochem 42:13–21

Hu M, Zhang Q, Tian XH et al (2019) lncRNA CCAT1 is a biomarker for the proliferation and drug resistance of esophageal cancer via the miR-143/PLK1/BUBR1 axis. Mol Carcinog 22:23109

Jaitin DA, Weiner A, Yofe I et al (2016) Dissecting immune circuits by linking CRISPR-pooled screens with single-cell RNA-Seq. Cell 167:1883-1896.e1815

Ji Q, Zhang L, Liu X et al (2014) Long non-coding RNA MALAT1 promotes tumour growth and metastasis in colorectal cancer through binding to SFPQ and releasing oncogene PTBP2 from SFPQ/PTBP2 complex. Br J Cancer 111:736–748

Jia J, Zhan D, Li J et al (2019) The contrary functions of lncRNA HOTAIR/miR-17-5p/PTEN axis and Shenqifuzheng injection on chemosensitivity of gastric cancer cells. J Cell Mol Med 23:656–669

Jiang H, Wang Y, Ai M et al (2017) LncRNA CRNDE stabilized by hnRNPUL2 accelerates cell proliferation and migration in colorectal carcinoma via activating Ras/MAPK signaling pathways. Cell Death Dis 8:e2862–e2862

Jiang D, Li H, Xiang H et al (2019) LncRNA HOTAIR knockdown increases miR-454-3p to suppress gastric cancer growth by targeting STAT3/cyclin D1. Med Sci Monit 25:1537–1548

Jin X, Liu G, Zhang X et al (2018) LncRNA TMEM75 promotes colorectal cancer progression by activation of SIM2. Gene 675:80–87

Jing C, Ma R, Cao H et al (2019) Long noncoding RNA and mRNA profiling in cetuximab-resistant colorectal cancer cells by RNA sequencing analysis. Cancer Med 8:1641–1651

Jonckheere N, Skrypek N, Van Seuningen I (2014) Mucins and tumor resistance to chemotherapeutic drugs. Biochim Biophys Acta Rev Cancer 1846:142–151

Kalbasi A, Ribas A (2019) Tumour-intrinsic resistance to immune checkpoint blockade. Nat Rev Immunol: 1038/s41577-41019-40218-41574

Kang M, Ren M, Li Y et al (2018) Exosome-mediated transfer of lncRNA PART1 induces gefitinib resistance in esophageal squamous cell carcinoma via functioning as a competing endogenous RNA. J Exp Clin Cancer Res 37:018–0845

Koliaraki V, Pallangyo CK, Greten FR et al (2017) Mesenchymal cells in Colon Cancer. Gastroenterol 152:964–979

Kong R, E-b Z, D-d Y et al (2015) LncRNA PVT1 indicates a poor prognosis of gastric cancer and promotes cell proliferation through epigenetically regulating p15 and p16. Mol Cancer 14:82

Kretzschmar K, Clevers H (2017) Wnt/β-catenin signaling in adult mammalian epithelial stem cells. Dev Biol 428:273–282

Lafrenie RM, Szafron LM, Balcerak A et al (2015) The novel gene CRNDE encodes a nuclear peptide (CRNDEP) which is overexpressed in highly proliferating tissues. PLoS One 10: e0127475

Lecerf C, Le Bourhis X, Adriaenssens E (2019) The long non-coding RNA H19: an active player with multiple facets to sustain the hallmarks of cancer. Cell Mol Life Sci 76:4673. https://doi.org/10.1007/s00018-019-03240-z

Lei Z, Tan IB, Das K et al (2013) Identification of molecular subtypes of gastric cancer with different responses to PI3K inhibitors and 5-fluorouracil. Gastroenterol 145:554–565

Li L, Liu B, Wapinski Orly L et al (2013) Targeted disruption of Hotair leads to homeotic transformation and gene derepression. Cell Rep 5:3–12

Li Z, Zhao X, Zhou Y et al (2015) The lncRNA HOTTIP promotes progression and gemcitabine resistance by regulating HOXA13 in pancreatic cancer. J Transl Med 13:84

Li J, Gao J, Tian W et al (2017a) Long non-coding RNA MALAT1 drives gastric cancer progression by regulating HMGB2 modulating the miR-1297. Cancer Cell Int 17:44

Li P, Zhang X, Wang H et al (2017b) MALAT1 is associated with poor response to oxaliplatin-based chemotherapy in colorectal cancer patients and promotes chemoresistance through EZH2. Mol Cancer Ther 16:739–751

Li P, Zhang X, Wang L et al (2017c) lncRNA HOTAIR contributes to 5FU resistance through suppressing miR-218 and activating NF-kappaB/TS signaling in colorectal cancer. Mol Ther Nucleic Acids 8:356–369

Li X, Meng X, Wei C et al (2018a) Dissecting LncRNA roles in renal cell carcinoma metastasis and characterizing genomic heterogeneity by single-cell RNA-seq. Mol Cancer Res 16:1879–1888

Li Y, Lv S, Ning H et al (2018b) Down-regulation of CASC2 contributes to cisplatin resistance in gastric cancer by sponging miR-19a. Biomed Pharmacother 108:1775–1782

Li H, He C, Wang X et al (2019a) MicroRNA-183 affects the development of gastric cancer by regulating autophagy via MALAT1-miR-183-SIRT1 axis and PI3K/AKT/mTOR signals. Artif Cells Nanomed Biotechnol 47:3163–3171

Li H, Ma X, Yang D et al (2019b) PCAT-1 contributes to cisplatin resistance in gastric cancer through epigenetically silencing PTEN via recruiting EZH2. J Cell Biochem 3:29370

Li J, Shen H, Xie H et al (2019c) Dysregulation of ncRNAs located at the DLK1DIO3 imprinted domain: involvement in urological cancers. Cancer Manag Res 11:777–787

Li Y, Li C, Li D et al (2019d) lncRNA KCNQ1OT1 enhances the chemoresistance of oxaliplatin in colon cancer by targeting the miR-34a/ATG4B pathway. Onco Targets Ther 12:2649–2660

Liang WC, Fu WM, Wong CW et al (2015) The lncRNA H19 promotes epithelial to mesenchymal transition by functioning as miRNA sponges in colorectal cancer. Oncotarget 6:22513–22525

Lim B, Kim JH, Kim M et al (2016) Genomic and epigenomic heterogeneity in molecular subtypes of gastric cancer. World J Gastroenterol 22:1190–1201

Liu SJ, Nowakowski TJ, Pollen AA et al (2016) Single-cell analysis of long non-coding RNAs in the developing human neocortex. Genome Biol 17:67

Liu M, Liu Y, Di J et al (2017a) Multi-region and single-cell sequencing reveal variable genomic heterogeneity in rectal cancer. BMC Cancer 17:787

Liu N, Zhou KI, Parisien M et al (2017b) N6-methyladenosine alters RNA structure to regulate binding of a low-complexity protein. Nucleic Acids Res 45:051–6063

Liu B, Pan S, Xiao Y et al (2018a) LINC01296/miR-26a/GALNT3 axis contributes to colorectal cancer progression by regulating O-glycosylated MUC1 via PI3K/AKT pathway. J Exp Clin Cancer Res 37:316

Liu J, Ben Q, Lu E et al (2018b) LncRNA PANDAR blocks CDKN1A gene transcription by competitive interaction with p53 protein in gastric cancer. Cell Death Dis 9:168

Liu G, Jiang Z, Qiao M et al (2019a) Lnc-GIHCG promotes cell proliferation and migration in gastric cancer through miR- 1281 adsorption. Mol Genet Genomic Med 7:e711

Liu H, Tian H, Zhao J et al (2019b) High HOXD4 protein expression in gastric adenocarcinoma tissues indicates unfavorable clinical outcomes. Saudi J Gastroenterol 25:46–54

Lu MH, Tang B, Zeng S et al (2016) LncRNA BC032469, a novel competing endogenous RNA, upregulates hTERT expression by sponging miR-1207-5p and promotes proliferation in gastric cancer. Oncogene 35:3524–3534

Lv Y-J, Wang W, Ji C-S et al (2017) Association between periostin and epithelial-mesenchymal transition in esophageal squamous cell carcinoma and its clinical significance. Oncol Lett 14:376–382

Ma L, Tian X, Wang F et al (2016) The lncRNA H19 promotes cell proliferation via E2F-1 in pancreatic ductal adenocarcinoma. Cancer Biol Ther 17:1051–1061

Ma J, Fan Y, Feng T et al (2017) HOTAIR regulates HK2 expression by binding endogenous miR-125 and miR-143 in oesophageal squamous cell carcinoma progression. Oncotarget 8:86410–86422

Ma S, Yang D, Liu Y et al (2018) LncRNA BANCR promotes tumorigenesis and enhances adriamycin resistance in colorectal cancer. Aging 10:2062–2078

Ma J, Li TF, Han XW et al (2019) Downregulated MEG3 contributes to tumour progression and poor prognosis in oesophageal squamous cell carcinoma by interacting with miR-4261, downregulating DKK2 and activating the Wnt/beta-catenin signalling. Artif Cells Nanomed Biotechnol 47:1513–1523

Mansour MA, Senga T (2017) HOXD8 exerts a tumor-suppressing role in colorectal cancer as an apoptotic inducer. Int J Biochem Cell Biol 88:1–13

McFaline-Figueroa JL, Hill AJ, Qiu X et al (2019) A pooled single-cell genetic screen identifies regulatory checkpoints in the continuum of the epithelial-to-mesenchymal transition. Nat Genet 51:1389–1398

Mondal T, Subhash S, Vaid R et al (2015) MEG3 lncRNA regulates the TGF-β pathway genes through formation of RNA–DNA triplex structures. Nat Commun 6:7743

Namkung J, Kwon W, Choi Y et al (2016) Molecular subtypes of pancreatic cancer based on miRNA expression profiles have independent prognostic value. J Gastroenterol Hepatol 31:1160–1167

Neve B, Jonckheere N, Vincent A et al (2018) Epigenetic regulation by lncRNAs: An overview focused on UCA1 in colorectal cancer. Cancers (Basel) 10: pii: E440

Niinuma T, Suzuki H, Nojima M et al (2012) Upregulation of miR-196a and HOTAIR drive malignant character in gastrointestinal stromal tumors. Cancer Res 72:1126–1136

Nordin M, Bergman D, Halje M et al (2014) Epigenetic regulation of the Igf2/H19 gene cluster. Cell Prolif 47:189–199

Onyango P, Feinberg AP (2011) A nucleolar protein, H19 opposite tumor suppressor (HOTS), is a tumor growth inhibitor encoded by a human imprinted H19 antisense transcript. Proc Natl Acad Sci U S A 108:16759–16764

Ouyang S, Zhou X, Chen Z et al (2019) LncRNA BCAR4, targeting to miR-665/STAT3 signaling, maintains cancer stem cells stemness and promotes tumorigenicity in colorectal cancer. Cancer Cell Int 19:72

Palazzo AF, Marshall EA, Stewart GL et al (2019) Beyond sequence homology: cellular biology limits the potential of XIST to act as a miRNA sponge. PLoS One 14:e0221371

Pan L, Liang W, Fu M et al (2017) Exosomes-mediated transfer of lncRNA ZFAS1 promotes gastric cancer progression. J Cancer Res Clin Oncol 143:991–1004

Pan S, Liu Y, Liu Q et al (2019) HOTAIR/miR-326/FUT6 axis facilitates colorectal cancer progression through regulating fucosylation of CD44 via PI3K/AKT/mTOR pathway. Biochim Biophys Acta, Mol Cell Res 1866:750–760

Park JS, Young Yoon S, Kim JM et al (2004) Identification of novel genes associated with the response to 5-FU treatment in gastric cancer cell lines using a cDNA microarray. Cancer Lett 214:19–33

Phelip JM, Tougeron D, Léonard D et al (2019) Metastatic colorectal cancer: French intergroup clinical practice guidelines for diagnosis, treatments and follow-up. Dig Liver Dis 51:1357–1363

Qu X, Alsager S, Zhuo Y et al (2019) HOX transcript antisense RNA (HOTAIR) in cancer. Cancer Lett 454:90–97

Rank KM, Shaukat A (2017) Stool based testing for colorectal Cancer: an overview of available evidence. Curr Gastroenterol Rep 19:39

Sanli I, Lalevee S, Cammisa M et al (2018) Meg3 non-coding RNA expression controls imprinting by preventing transcriptional upregulation in cis. Cell Rep 23:337–348
See K, Tan WLW, Lim EH et al (2017) Single cardiomyocyte nuclear transcriptomes reveal a lincRNA-regulated de-differentiation and cell cycle stress-response in vivo. Nat Commun 8:225
Sellers ZP, Schneider G, Maj M et al (2018) Analysis of the paternally-imprinted DLK1–MEG3 and IGF2–H19 tandem gene loci in NT2 embryonal carcinoma cells identifies DLK1 as a potential therapeutic target. Stem Cell Rev 14:823–836
Shang C, Sun L, Zhang J et al (2017) Silence of cancer susceptibility candidate 9 inhibits gastric cancer and reverses chemoresistance. Oncotarget 8:15393–15398
Si Y, Yang Z, Ge Q et al (2019) LncRNA Malat1 activated autophagy, hence promoting cell proliferation and inhibiting apoptosis by sponging miR-101 in colorectal cancer. Cell Mol Biol Lett 24:50
Simon MD, Pinter SF, Fang R et al (2013) High-resolution Xist binding maps reveal two-step spreading during X-chromosome inactivation. Nature 504:465–469
Soares E, Zhou H (2017) Master regulatory role of p63 in epidermal development and disease. Cell Mol Life Sci 75:1179–1190
Sun Y, Ma L (2019) New insights into long non-coding RNA MALAT1 in cancer and metastasis. Cancers 11:216
Sun C, Wang FJ, Zhang HG et al (2017) miR-34a mediates oxaliplatin resistance of colorectal cancer cells by inhibiting macroautophagy via transforming growth factor-beta/Smad4 pathway. World J Gastroenterol 23:1816–1827
Sun F, Liang W, Qian J (2019) The identification of CRNDE, H19, UCA1 and HOTAIR as the key lncRNAs involved in oxaliplatin or irinotecan resistance in the chemotherapy of colorectal cancer based on integrative bioinformatics analysis. Mol Med Rep 20:3583–3596
Sveen A, Bruun J, Eide PW et al (2018) Colorectal Cancer consensus molecular subtypes translated to preclinical models uncover potentially targetable cancer cell dependencies. Clin Cancer Res 24:794–806
Takahashi Y, Sawada G, Kurashige J et al (2014) Amplification of PVT-1 is involved in poor prognosis via apoptosis inhibition in colorectal cancers. Br J Cancer 110:164–171
Tang F, Barbacioru C, Wang Y et al (2009) mRNA-Seq whole-transcriptome analysis of a single cell. Nat Methods 6:377–382
Tang S, Wu WKK, Li X et al (2016) Stratification of digestive cancers with different pathological features and survival outcomes by MicroRNA expression. Sci Rep 6
Tempero MA, Malafa MP, Chiorean EG et al (2019) Guidelines insights: pancreatic adenocarcinoma, V1.2019. J Natl Compr Cancer Netw 17:202–210
Tseng Y-Y, Moriarity BS, Gong W et al (2014) PVT1 dependence in cancer with MYC copy-number increase. Nat 512:82–86
von Ahrens D, Bhagat TD, Nagrath D et al (2017) The role of stromal cancer-associated fibroblasts in pancreatic cancer. J Hematol Oncol 10:76
Wang X, Wu X (2017) The role of MiR-1207-5p in colorectal Cancer. Clin Lab 63:1875–1882
Wang J, Su L, Chen X et al (2014a) MALAT1 promotes cell proliferation in gastric cancer by recruiting SF2/ASF. Biomed Pharmacother 68:557–564
Wang Y, Zhang D, Wu K et al (2014b) LncRNA MRUL promotes ABCB1 expression in multidrug-resistant gastric cancer cell sublines. Mol Cell Biol 34:3182–3193
Wang ZL, Li B, Piccolo SR et al (2016) Integrative analysis reveals clinical phenotypes and oncogenic potentials of lncRNAs across 15 cancer types. Oncotarget 7:35044–35055
Wang W, Gao D, Wang X (2017a) Can single-cell RNA sequencing crack the mystery of cells? Cell Biol Toxicol 34:1–6
Wang Z-Q, Cai Q, Hu L et al (2017b) LncRNA UCA1 induced by SP1 promotes cell proliferation via recruiting EZH2 and activating AKT pathway in gastric cancer. Cell Death Dis 8:e2839
Wang H, Li H, Zhang L et al (2018a) Overexpression of MEG3 sensitizes colorectal cancer cells to oxaliplatin through regulation of miR-141/PDCD4 axis. Biomed Pharmacother 106:1607–1615

Wang M, Han D, Yuan Z et al (2018b) LncRNA H19 confers 5-Fu resistance in colorectal cancer by promoting SIRT1-mediated autophagy. Cell Death Dis 9:1149

Wang AH, Tan P, Zhuang Y et al (2019) High-resolution Xist binding maps reveal two-step spreading during X-chromosome inactivation. J Cell Mol Med

Warda AS, Kretschmer J, Hackert P et al (2017) Human METTL16 is a N 6 -methyladenosine (m 6 a) methyltransferase that targets pre-mRNAs and various non-coding RNAs. EMBO Rep 18:2004–2014

Wu DW, Lin PL, Wang L et al (2017) The YAP1/SIX2 axis is required for DDX3-mediated tumor aggressiveness and cetuximab resistance in KRAS-wild-type colorectal cancer. Theranostics 7:1114–1132

Wu H, Chen S, Yu J et al (2018a) Single-cell transcriptome analyses reveal molecular signals to intrinsic and acquired paclitaxel resistance in esophageal squamous cancer cells. Cancer Lett 420:156–167

Wu X, Zheng Y, Han B et al (2018b) LncRNA BLACAT1 modulates ABCB1 to promote oxaliplatin resistance of gastric cancer via sponging miR-361. Biomed Pharmacother 99:832–838

Wu H, Li Y, Hou Q et al (2019a) Single-cell intratumoral stemness analysis reveals the involvement of cell cycle and DNA damage repair in two different types of esophageal cancer. Oncol Rep 41:3201–3208

Wu H, Zou Q, He H et al (2019b) LncRNA PCAT6 targets miR-204 to modulate the chemoresistance of colorectal cancer cells to 5-fluorouracil-based treatment through HMGA2 signaling. Cancer Med 8:2484–2495

Xi Z, Si J, Nan J (2019) LncRNA MALAT1 potentiates autophagy associated cisplatin resistance by regulating the microRNA30b/autophagy related gene 5 axis in gastric cancer. Int J Oncol 54:239–248

Xia Y, Yan Z, Wan Y et al (2018) Knockdown of lncRNA GHET1 inhibits cell cycle progression and invasion of gastric cancer cells. Mol Med Rep 18:375–3381

Xiao Y, Yurievich UA, Yosypovych SV (2017) LncRNA XIST is a prognostic factor in colorectal cancer and inhibits 5-fluorouracil-induced cell cytotoxicity through promoting thymidylate synthase expression. Oncotarget 8:83171–83182

Xiao Z, Qu Z, Chen Z et al (2018) LncRNA HOTAIR is a prognostic biomarker for the proliferation and chemoresistance of colorectal cancer via MiR-203a-3p-mediated Wnt/ß-catenin signaling pathway. Cell Physiol Biochem 46:275–1285

Xiao J, Lai H, Wei SH et al (2019) lncRNA HOTAIR promotes gastric cancer proliferation and metastasis via targeting miR-126 to active CXCR4 and RhoA signaling pathway. Cancer Med 8:6768–6779

Xu F, Zhang J (2017) LncRNA HOTAIR functions as miRNA sponge to promote the epithelial to mesenchymal transition in esophageal cancer. Biomed Pharmacother 90:888–896

Xu G, Meng L, Yuan D et al (2018) MEG3/miR21 axis affects cell mobility by suppressing epithelial mesenchymal transition in gastric cancer. Oncol Rep 40:39–48

Xue M, Chen W, Li X (2016) UCA1: a lncRNA with a crucial role in cancer. J Cancer Res Clin Oncol 142:1407–1419

Yan J, Zhang Y, She Q et al (2017) LncRNA H19/miR-675 Axis promotes gastric Cancer via FADD/caspase 8/caspase 3 signaling pathway. Cell Physiol Biochem 42:2364–2376

Yang X, Song JH, Cheng Y et al (2014) LncRNA HNF1A-AS1 regulates proliferation and migration in oesophageal adenocarcinoma cells. Gut 63:881–890

Yang Z, Jiang X, Jiang X et al (2018) X-inactive-specific transcript: a long ncRNA with complex roles in human cancers. Gene 679:28–35

Ye Y, Yang S, Han Y et al (2019) HOXD-AS1 confers cisplatin resistance in gastric cancer through epigenetically silencing PDCD4 via recruiting EZH2. Open Biol 9:25

YiRen H, YingCong Y, Sunwu Y et al (2017) LncRNA MALAT1 regulates autophagy associated chemoresistance via miR-23b-3p sequestration in gastric cancer. Mol Cancer 16:017–0743

Yokota K, Tanaka Y, Harada H et al (2019) WiNTRLINC1/ASCL2/c-Myc Axis characteristics of colon cancer with differentiated histology at young onset and essential for cell viability. Ann Surg Oncol 26:4826. https://doi.org/10.1245/s10434-019-07780-3

Yoshida K, Toden S, Ravindranathan P et al (2017) Curcumin sensitizes pancreatic cancer cells to gemcitabine by attenuating PRC2 subunit EZH2, and the lncRNA PVT1 expression. Carcinogenesis 38:1036–1046

You L, Wang H, Yang G et al (2018) Gemcitabine exhibits a suppressive effect on pancreatic cancer cell growth by regulating processing of PVT1 to miR1207. Mol Oncol 12:2147–2164

Yu B, Du Q, Li H et al (2017a) Diagnostic potential of serum exosomal colorectal neoplasia differentially expressed long non-coding RNA (CRNDE-p) and microRNA-217 expression in colorectal carcinoma. Oncotarget 8:83745–83753

Yu X, Mi L, Dong J et al (2017b) Long intergenic non-protein-coding RNA 1567 (LINC01567) acts as a "sponge" against microRNA-93 in regulating the proliferation and tumorigenesis of human colon cancer stem cells. BMC Cancer 17:716

Zeng X, Liu Y, Zhu H et al (2019) Downregulation of miR-216a-5p by lncRNA PVT1 suppresses colorectal cancer progression via modulation of YBX1 expression. Cancer Manag Res 11:6981–6993

Zhang Y, Song X, Wang X et al (2016) Silencing of LncRNA HULC enhances chemotherapy induced apoptosis in human gastric cancer. J Med Biochem 35:137–143

Zhang X, Bo P, Liu L et al (2017a) Overexpression of long non-coding RNA GHET1 promotes the development of multidrug resistance in gastric cancer cells. Biomed Pharmacother 92:580–585

Zhang Y, Chen Z, Li MJ et al (2017b) Long non-coding RNA metastasis-associated lung adenocarcinoma transcript 1 regulates the expression of Gli2 by miR-202 to strengthen gastric cancer progression. Biomed Pharmacother 85:264–271

Zhang J, Yin M, Peng G et al (2018a) CRNDE: An important oncogenic long non-coding RNA in human cancers. Cell Prolif 51:e12440

Zhang Z, Feng L, Liu P et al (2018b) ANRIL promotes chemoresistance via disturbing expression of ABCC1 by regulating the expression of Let-7a in colorectal cancer. Biosci Rep 38:21

Zhang P, Yang M, Zhang Y et al (2019a) Dissecting the single-cell transcriptome network underlying gastric premalignant lesions and early gastric cancer. Cell rep 27:1934-1947.e1935

Zhang PF, Wu J, Wu Y et al (2019b) The lncRNA SCARNA2 mediates colorectal cancer chemoresistance through a conserved microRNA-342-3p target sequence. J Cell Physiol 234:10157–10165

Zhou P, Li B, Liu F et al (2017) The epithelial to mesenchymal transition (EMT) and cancer stem cells: implication for treatment resistance in pancreatic cancer. Mol Cancer 16:52

Zhou Z, Lin Z, He Y et al (2018) The LncRNA D63785 regulates chemotherapy sensitivity in human gastric cancer by targeting miR-422a. Mol Ther Nucleic Acids 12:405–419

Zhu J, Zhang R, Yang D et al (2018a) Knockdown of long non-coding RNA XIST inhibited doxorubicin resistance in colorectal cancer by upregulation of miR-124 and downregulation of SGK1. Cell Physiol Biochem 51:113–128

Zhu L, Liu Y, Chen Q et al (2018b) LncRNA colorectal neoplasia differentially expressed gene as a potential target to upregulate the expression of IRX5 by miR-136-5P to promote oncogenic properties in hepatocellular carcinoma. Cell Physiol Biochem 50:2229–2248

Zhu L, Zhu Y, Han S et al (2019) Impaired autophagic degradation of lncRNA ARHGAP5-AS1 promotes chemoresistance in gastric cancer. Cell Death Dis 10:383

Zhuang M, Zhao S, Jiang Z et al (2019) MALAT1 sponges miR-106b-5p to promote the invasion and metastasis of colorectal cancer via SLAIN2 enhanced microtubules mobility. EBioMedicine 41:286–298

Zong X, Nakagawa S, Freier SM et al (2016) Natural antisense RNA promotes 3′ end processing and maturation of MALAT1 lncRNA. Nucleic Acids Res 44:2898–2908

LncRNAs in the Development, Progression, and Therapy Resistance of Hormone-Dependent Cancer

Yuichi Mitobe, Kazuhiro Ikeda, Kuniko Horie-Inoue, and Satoshi Inoue

Contents

Abstract Sex hormones play crucial roles in the development and maintenance of sexual characteristics as well as the tumorigenesis of hormone-dependent cancers. While endocrine therapies targeting hormone receptors or hormone production are primarily effective against hormone-dependent cancers such as breast and prostate cancers, the emergence of therapy-resistant cancers after long-term treatment is a serious clinical issue. Therefore, the elucidation of mechanisms underlying endocrine therapy resistance is the first step toward the development of alternative clinical management for the advanced diseases. Recent advances in RNA research dissected critical long noncoding RNAs (lncRNAs) that play roles in the pathophysiology of hormone-dependent cancers through their unique molecular mechanisms. The

Y. Mitobe · K. Ikeda · K. Horie-Inoue
Division of Gene Regulation and Signal Transduction, Research Center for Genomic Medicine, Saitama Medical University, Saitama, Japan
e-mail: ymitobe@saitama-med.ac.jp; ikeda@saitama-med.ac.jp; khorie07@saitama-med.ac.jp

S. Inoue (✉)
Division of Gene Regulation and Signal Transduction, Research Center for Genomic Medicine, Saitama Medical University, Saitama, Japan

Department of Systems Aging Science and Medicine, Tokyo Metropolitan Institute of Gerontology, Tokyo, Japan
e-mail: sinoue@tmig.or.jp

S. Jurga, J. Barciszewski (eds.), *The Chemical Biology of Long Noncoding RNAs*, RNA Technologies 11, https://doi.org/10.1007/978-3-030-44743-4_10

mutual regulation of these lncRNAs and hormone signaling pathways has been elaborately studied in breast and prostate cancers. This chapter focuses on the history and future perspective of lncRNAs and hormone-dependent cancers.

Keywords Sex hormone · Estrogen · Androgen · Breast cancer · Prostate cancer

1 Introduction of LncRNA

Recent advances in high-throughput sequencing technologies for whole transcriptome analysis have revealed that 2% of the genomic regions encode protein-coding transcripts (Consortium 2012). The remaining regions also generate noncoding RNA molecules that are not translated into proteins. Long noncoding RNA (lncRNA) is defined as an RNA molecule that does not encode protein with a length longer than 200 nucleotides (Wang and Chang 2011) including ribosomal RNAs. The rest of noncoding RNAs are short noncoding RNAs, which are represented by transfer RNA (tRNA) and microRNA (miRNA). MiRNAs are involved in various biological processes including cancer progression (Lee and Dutta 2009). The GENCODE v7 catalog of human lncRNAs produced by GENCODE consortium, within the framework of the encyclopedia of DNA element (ENCODE) project, showed that 14,880 transcripts transcribed from 9277 manually annotated genes (Derrien et al. 2012). LncRNAs are generated also through the transcriptional pathways similar to those of protein-coding genes, whereas they exhibit a bias toward two-exon transcripts and predominantly localized in the chromatin and nucleus (Derrien et al. 2012). Multiple types of lncRNA functions have been characterized in terms of biological phenomena (Mitobe et al. 2018). For example, lncRNAs interact with protein, DNA, and RNA molecules and regulate their functions in both positive or negative ways.

2 Sex Hormone and Hormone-Dependent Cancer

Sex hormones including estrogens and androgens are essential factors in the development and maintenance of female and male organs and sexual characteristics. Besides their physiological functions, dysregulation of sex hormone signaling plays a critical role in the development and progression of hormone-dependent cancers, especially breast and prostate cancers.

2.1 Sex Hormone Signaling

Sex hormones estrogens and androgens are classified as steroid hormones and are synthesized from cholesterol through elaborate enzymatic reactions (Ansar Ahmed et al. 1985). The major type of estrogen 17β-estradiol is produced predominantly in the ovaries. Androgens including testosterone and its metabolite dihydrotestosterone (DHT) are mainly produced in Leydig cells of the testes. Testosterone is also produced in female cells and converted to estradiol through the enzymatic function of aromatase. Aromatase expression is tightly regulated in both female and male cells. The produced estrogens and androgens function as ligands for estrogen receptor (ER) and androgen receptor (AR), respectively, both of which belong to nuclear hormone receptors (McEwen and Milner 2017).

ERs, which consist of ERα and ERβ, are ligand-dependent transcription factors. ERα is abundantly expressed in female reproductive organs such as the uterus and mammary glands. The protein structure of ERα and ERβ consists of N-terminal A/B domain including activation function region [A/B(AF-1)], DNA-binding domain (DBD), hinge domain (H), and ligand-binding domain (LBD)/C-terminal AF2 (Fig. 1) (Muramatsu and Inoue 2000). Estrogen binds to the LBD of ER and the ligand-bound ER, which forms homodimers, and then bind to a specific DNA enhancer/promoter region of its target gene, defined as estrogen response element (ERE). Various transcriptional cofactors such as CBP/p300 and steroid receptor

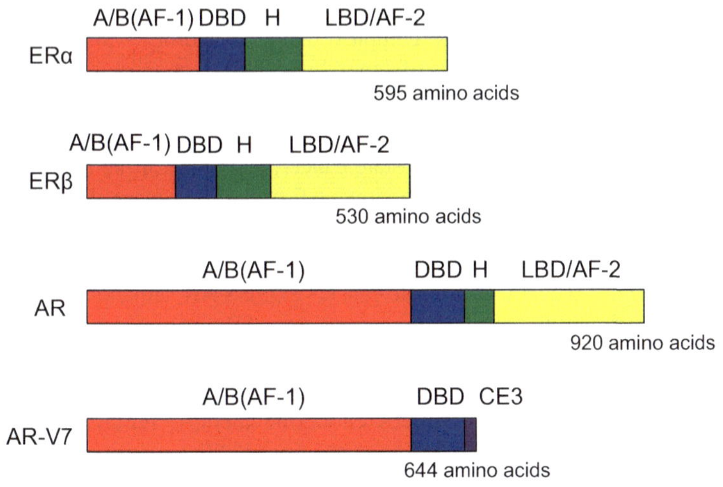

Fig. 1 Structures of estrogen receptors α and β (ERα and ERβ), androgen receptor (AR), and splice variant 7 (AR-V7). ER and AR are prototypic nuclear hormone receptors whose structure is characterized by following conserved domains: A/B(AF-1), N-terminal A/B domain including activation function-1; DBD, DNA-binding domain; H, hinge region; and LBD, ligand-binding domain including activation function-2 (AF-2). AR-V7 contains the cryptic exon 3 (CE3) in spite of the original AR C-terminal region with LBD and enables to exhibit ligand-independent activation of AR signaling

coactivator-1 (SRC-1)/p160 family members are also recruited to EREs and form complexes with ligand-bound ERs (Fig. 1) (Muramatsu and Inoue 2000).

Estrogen signaling is essential for female characterization such as mammary gland development, and estrogen dysregulation is critical for the development of breast cancer (Ghoncheh et al. 2016). Breast cancer is the most common cancer in women, and the number of breast cancer patients continues to increase worldwide (Waks and Winer 2019). Breast cancer is categorized into subtypes based on the expression of ER, progesterone receptor (PR), and epidermal growth factor receptor 2 (ERBB2/HER2): ER and/or PR-positive luminal subtype, HER2-positive subtype, and triple-negative breast cancer (TNBC) subtype that is ER/PR/HER2-negative. ER-positive breast cancers comprise 70% of all breast cancer cases (Waks and Winer 2019). Thus, endocrine therapy that targets hormone signaling is the first-line treatment for ER-positive breast cancer (Pondé et al. 2019). A selective ER modulator tamoxifen has been used for endocrine therapy for decades, although the drug has a partial agonistic action for ER, which increases the risk of endometrial cancer. Instead, full ER antagonist fulvestrant has recently been applied to clinical management. Inhibition of estrogen synthesis by aromatase inhibitors is another therapeutic option for ER-positive breast cancer patients (Pondé et al. 2019). Long-term endocrine therapy, however, often leads to the acquisition of therapy resistance, whose molecular mechanisms remain to be elucidated to conquer the serious clinical problem.

The androgen-AR interaction exhibits similarities with the estrogen-ER interaction. AR is also a ligand-dependent transcription factor and also consists of A/B (AF-1), DBD, H, and LBD/AF-2 domains (Fig. 1) (Tan et al. 2015). Upon binding androgens to AR, the ligand-AR complex binds to androgen response element (ARE) and recruits AR-associated cofactors. In developed countries, prostate cancer is the third most common cause of death from cancer among men (Litwin and Tan 2017). Similar to the estrogen signaling in breast cancer, androgen signaling is important for the development of prostate cancer, and some of the androgen target genes are applied to diagnosis and treatment for prostate cancer (Litwin and Tan 2017). For example, a typical androgen target prostate-specific antigen (PSA) is the broadly used biomarker for prostate cancer diagnosis and monitoring (Lilja et al. 2008). Because the majority of prostate cancers express AR, antiandrogen therapy is the first-line treatment for prostate cancer patients. AR antagonist bicalutamide has been used for years in endocrine therapy for prostate cancer patients. Nowadays, more potent antiandrogen reagents such as enzalutamide and abiraterone have emerged (Scher et al. 2010). Besides endocrine therapy, removal of testis, or castration, is originally effective for prostate cancer. After castration or long-term antiandrogen therapy, however, patients often suffered from castration-resistant prostate cancer (CRPC) (Kirby et al. 2011). Molecular mechanisms such as the amplification of AR signaling or hormone-independent AR activation have been shown as origins for CRPC. The amplification of AR signaling is mainly caused by the overexpression of AR and transcriptional coactivators that leads to hypersensitivity to androgens, and by abnormal AR splicing variants such as AR-V7 (Davies et al. 2018) (Fig. 1). AR-V7 expression is often observed in CRPC, and this AR variant is assumed to contribute to the pathophysiology of CRPC by exhibiting constitutive activation of AR signaling in a ligand-independent manner.

2.2 LncRNAs in Sex Hormone Signaling

2.2.1 SRA

Roles of lncRNAs in hormone signaling have been characterized prior to the era of high-throughput sequencing technologies (Fig. 2). An interactome analysis with the amino-terminal domain of the human PR showed *steroid receptor RNA activator* (*SRA*) (Lanz et al. 1999). *SRA* interacts with various hormone receptors, including PR, AR, ERα, and glucocorticoid receptor (GR), and has been reported to be

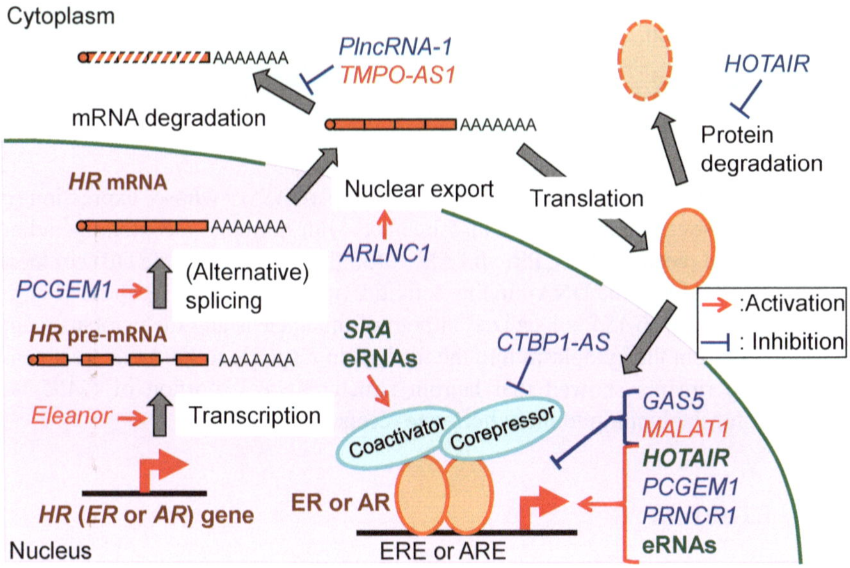

Fig. 2 Roles of lncRNAs in sex hormone signaling. Hormone receptors (HRs) ER and AR are ligand-dependent transcription factors that bind to estrogen response element (ERE) and androgen response element (ARE), respectively, in the transcriptional regulatory region of their target genes. HR-related transcriptional coactivators and corepressors also play roles in HR-mediated transcription. Recent studies identified critical HR-related lncRNAs, which modulate the HR expression at both transcriptional and posttranscriptional levels and contribute to HR-mediated gene regulation process. LncRNA involved in AR, ER, and AR/ER signaling are shown in blue, red, and green with bold, respectively. *ARlnc1* androgen-regulated long noncoding RNA 1, *CTBP1-AS* antisense transcript for c-terminal binding protein 1, *Eleanor* ESR1 locus enhancing and activating noncoding RNAs, *eRNAs* enhancer RNAs, *GAS5* growth arrest-specific transcript 5, *HOTAIR* HOX transcript antisense intergenic RNA, *MALAT1* metastasis associated in lung adenocarcinoma, *PCGEM* prostate-specific transcript 1, *PRNCR1* prostate cancer noncoding RNA1, *SRA* steroid receptor RNA activator, *TMPO-AS1* thymopoietin antisense RNA 1

involved in steroid receptor coactivator-1 (SRC-1) complexes to promote transcriptional activity of hormone receptors (Fig. 2) (Lanz et al. 1999).

2.2.2 HOTAIR

HOX transcript antisense intergenic RNA (*HOTAIR*), which is transcribed from the *HOX* loci, was identified as a metastasis-related lncRNA in breast cancer (Gupta et al. 2010). This lncRNA has been reported to be involved in the progression of various cancers by modulating diverse signaling pathways (Hajjari and Salavaty 2015) including hormone signaling. In ER-positive breast cancer, *HOTAIR* interacts with ERα protein and promotes ERα association to enhancer/promoter regions of estrogen target genes, leading to the development of hormone therapy resistance (Fig. 2) (Xue et al. 2016). *HOTAIR* also interacts with AR protein and inhibits the interaction between AR and ubiquitin ligase mouse double minute 2 homolog (MDM2), leading to the protection of AR from proteasome-mediated protein degradation (Fig. 2) (Zhang et al. 2015).

2.2.3 GAS5

Conversely, *growth arrest-specific transcript 5* (*GAS5*), whose expression is decreased in several cancer types, also interacts with hormone receptors including glucocorticoid receptor (GR), PR, and AR. *GAS5* directly binds to the DBD of these receptors and inhibits the DNA-binding activities of these hormone receptors (Fig. 2) (Kino et al. 2010). *GAS5* is localized in both of the nucleus and the cytoplasm and translocates from the cytoplasm into the nucleus in response to hormone treatment. Further experiments showed that hairpin structures in 3′ portion of *GAS5* are important for the binding to these hormone receptors.

2.2.4 CTBP1-AS

Hormone-related lncRNAs in breast and prostate cancers are listed in Table 1 (Knoll et al. 2015). Our group previously performed a screening of androgen-regulated lncRNAs based on an integrated analysis of chromatin immunoprecipitation (ChIP) sequencing and/or ChIP-chip with AR antibody and cap analysis of gene expression (CAGE) (Takayama et al. 2013). This analysis identified several androgen-inducible lncRNAs including *CTBP1-AS,* a natural antisense (AS) transcript for transcriptional repressor *c-terminal binding protein 1* (*CTBP1*) gene. *CTBP1-AS* lncRNA promotes androgen signaling activity by downregulating its sense gene *CTBP1*, which is shown as a corepressor of AR (Fig. 2). In addition to this *cis*-acting mechanism, it is demonstrated that *CTBP1-AS* interacts with RNA-binding protein splicing factor proline-glutamine rich (SFPQ/PSF) and histone deacetylase (HDAC) complex, and the latter supports transcriptional repressor activity of PSF. Notably, *CTBP1-AS* and

Table 1 Hormone-regulated lncRNAs

LncRNA	Androgen	Estrogen	Target
HOTAIR	–	↑↓	AR, ER, EZH2
NEAT1	–	↑	FOXN3
CTBP1-AS	↑	–	PSF
SOCS2-AS	↑	–	–
POTEF-AS	↑	–	–
PCA3	↑	–	PRUNE2
PCGEM1	↑	–	AR, *AR* RNA
ARLNC1	↑	–	*AR* mRNA
PlncRNA-1	↑	–	*AR* mRNA
PCAT1	↑	–	EZH2, BRCA2 mRNA
			Myc mRNA
MALAT1	↑	–	ER
TMPO-AS1	↑	↑	*ESR1* mRNA
Eleanor	–	↑	*ESR1* transcription, etc.
H19	–	↑	–
eRNAs	↑	↑	AR, ER
DANCR	↓	–	EZH2, RXRA

– No report

PSF downregulate the expression of *p53* and *SMAD3* genes, both of which function as cell cycle negative regulators in prostate cancer. These findings indicate that *CTBP1-AS* is a functional lncRNA in *cis* and *trans*.

2.2.5 SOCS2-AS1 and POTEF-AS1

In addition, *suppressor of cytokine signaling 2 antisense transcript 1* (*SOCS2-AS1*) (Misawa et al. 2016) and *protein family member F antisense transcript* (*POTEF-AS1*) (Misawa et al. 2017) are also demonstrated as androgen-inducible lncRNAs. *SOCS2-AS1* knockdown impairs prostate cancer cells viability and migration. Microarray-based global expression analysis showed that *SOCS2-AS1* is associated with apoptotic and AR signaling pathway. Among these apoptosis and AR-associated genes, an apoptosis inducer, tumor necrosis factor superfamily member 10 (TNFSF10) is induced by knockdown of *SOCS2-AS1* strongly. *SOCS2-AS1* interacts with AR and modulates AR and apoptosis pathways. *POTEF-AS1* also regulates prostate cancer cell viability. Notably, *SOCS2-AS1* and *POTEF-AS1* expression are elevated in a hormone therapy-resistant model cells, long-term androgen-deprived (LTAD) cells compared with parental cells, suggesting that these lncRNAs regulate in the acquisition of hormone therapy resistance.

2.2.6 PCA3

Prostate cancer antigen 3 (*PCA3*) *DD3* is one of the most differentially expressed lncRNAs in prostate cancer and is inducible by androgen treatment (Bussemakers et al. 1999). *PCA3* knockdown suppresses the proliferation and metastasis of prostate cancer cells. *PCA3* forms double-stranded RNA with the tumor suppressor gene, *a human homolog of the Drosophila prune gene* (*PRUNE2*) (Salameh et al. 2015). Intriguingly, *PCA3-PRUNE2* hybridization promotes RNA editing of *PRUNE2* and downregulation of *PRUNE2*. *PCA3* is a useful biomarker of prostate cancer and has been applied to a urine-based assay as a diagnostic test for prostate cancer (Groskopf et al. 2006). This is the first lncRNA approved for clinical use in prostate cancer diagnosis by the United States Food and Drug Administration (FDA). This diagnostic test with *PCA3* could improve specificity as well as both positive and negative predictive values compared to conventional serum PSA testing.

2.2.7 PCGEM and PRNCR1

Prostate-specific transcript 1 (*PCGEM1*) is an androgen-inducible and prostate cancer-upregulated lncRNA (Srikantan et al. 2000; Parolia et al. 2015). *Prostate cancer noncoding RNA1* (*PRNCR1*) was identified in a gene desert region of chromosome 8q24, a known prostate cancer susceptibility loci (Chung et al. 2011). Both *PCGEM1* and *PRNCR1* interact with AR and elicit both ligand-dependent and ligand-independent AR-mediated gene activation through the formation of a three-dimensional structural loop (Fig. 2) (Yang et al. 2013). In addition, an interactome analysis with *PCGEM1* showed that the lncRNA functionally interact with splicing factors including heterogeneous nuclear ribonucleoprotein A1 (hnRNPA1) and U2AF65 (Zhang et al. 2016). While the interaction of *PCGEM1* with hnRNPA1 suppresses the binding of the splicing factor to *AR* pre-mRNA, the interaction of *PCGEM1* with U2AF65 enhances the binding of the splicing factor to *AR* pre-mRNA and upregulates AR splicing variant AR-V7/AR3 (Fig. 2) (Zhang et al. 2016). Others also reported, however, that *PCGEM1* and *PRNCR1* do not interact with AR and are not involved in AR signaling pathways (Prensner et al. 2014c). Thus, the contribution of *PCGEM1* and *PRNCR1* to the AR signaling paradigm remains controversial.

2.2.8 ARLNC1

Androgen-regulated long noncoding RNA 1 (*ARLNC1*) is another androgen-inducible lncRNA, which is closely involved in the androgen signaling pathway. Knockdown of this lncRNA downregulates AR expression. Intriguingly, this lncRNA interacts with the 3′-untranslated region (UTR) of *AR* mRNA through

RNA-RNA hybridization and promotes *AR* mRNA nuclear export, leading to androgen signaling activation (Fig. 2) (Zhang et al. 2018). Importantly, knockdown of *ARLNC1* only suppresses proliferation of AR-positive prostate cancer cells, but not AR-negative cells, suggesting that *ARLNC1* plays an important role in androgen signaling positive feedback regulation.

2.2.9 PlncRNA-1

PlncRNA-1, a highly expressed lncRNA in prostate cancer, is induced by the androgen signaling pathway. This lncRNA is localized in the cytoplasm mainly, and knockdown of *PlncRNA-1* impairs cell proliferation through *AR* mRNA downregulation. This lncRNA binds to the *AR* mRNA 3'UTR and competes with miRNAs miR-34c and miR-297 to protect against the miRNA-mediated degradation of *AR* mRNA (Cui et al. 2013; Fang et al. 2016). Expression analysis using 48 prostate cancer tissue samples showed that *AR* mRNA and *PlncRNA-1* expression is significantly correlated.

2.2.10 PCAT1

Integrative analyses with 102 prostate cancer samples and cell lines identified 121 unannotated prostate cancer-associated noncoding transcripts (PCAT) that are differently expressed in prostate cancer (Prensner et al. 2011). Among these PCATs, *PCAT1* is an androgen-inducible lncRNA transcribed from a region adjacent to *PRNCR1* in prostate cancer susceptibility 8q24 loci (Chung et al. 2011). *PCAT1* expression is repressed by enhancer of zeste homolog 2 (EZH2), a component of polycomb repressive complex 2 (PRC2), and forms a complex with PRC2 to regulate its activity (Prensner et al. 2011). In addition to the epigenetic regulation, *PCAT1* downregulates tumor suppressor gene *breast cancer susceptibility gene 2* (*BRCA2*) by promoting *BRCA2* mRNA degradation (Prensner et al. 2014b). On the other hand, *PCAT1* protects Myc mRNA from microRNA-mediated degradation (Prensner et al. 2014a).

2.2.11 NEAT1 and MALAT1

Nuclear enriched abundant transcript 1 (*NEAT1*) is an lncRNA that is a known component of nuclear paraspeckle, and *NEAT1* associates with the proliferation of various cancer cells (Li et al. 2018). In prostate cancer cells, *NEAT1* expression is induced by estrogen treatment through ER activation, which uniquely contributes to the regulation of CRPC cell proliferation (Chakravarty et al. 2014). While *HOTAIR* was reported as a target of ER-mediated transcriptional repression in breast cancer (Xue et al. 2016), others showed that *HOTAIR* and *metastasis associate in lung adenocarcinoma* (*MALAT1*) are estrogen-inducible lncRNAs in prostate cancer cells

(Aiello et al. 2016). *MALAT1*, also known as *NEAT2*, was shown as a metastasis-promoting lncRNA in various cancer types (Sun and Ma 2019). Interestingly, *MALAT1* binds to EREs in the absence of estrogen and contributes to the inhibition of ERE-mediated transcription. On the response to estrogen treatment, *HOTAIR* interacts with EREs instead of *MALAT1* and activates ERE transcriptional activity, indicating that *HOTAIR* and *MALAT1* play opposite roles in estrogen signaling in prostate cancer (Fig. 2).

2.2.12 TMPO-AS1

TMPO-AS1 is an antisense lncRNA for thymopoietin (*TMPO*) that is originally identified as an E2F-regulated lncRNA (Parise et al. 2006). Our group recently identified *TMPO-AS1* as a cell proliferation-associated lncRNA that is inducible by estrogen treatment in ER-positive breast cancer cells (Mitobe et al. 2019). *TMPO-AS1* directly interacts with 3'-UTR of *estrogen receptor 1* (*ESR1*) mRNA and stabilizes the mRNA. In addition, *TMPO-AS1* also exhibits growth-stimulating effect in hormone therapy-resistant breast cancer cells. Clinically, *TMPO-AS1* high expression is significantly associated with poorer prognosis of breast cancer patients. Interestingly, reanalysis of microarray dataset from breast cancer patients treated by tamoxifen as adjuvant therapy (GSE9195) (Loi et al. 2008) showed that *TMPO-AS1* is one of 21 differentially expressed genes in tamoxifen-relapsing patients (Notas et al. 2015). In prostate cancer, *TMPO-AS1* is reported as an androgen-inducible oncogenic lncRNA (Huang et al. 2018).

2.2.13 Eleanor and H19

ESR1 locus enhancing and activating noncoding RNA (*Eleanor*) is a cluster of lncRNAs located in the neighboring of *ESR1* gene locus (Tomita et al. 2015). *Eleanor* expression is upregulated in hormone therapy-resistant MCF-7 cells. Interestingly, *Eleanor* lncRNAs *upregulate ESR1* mRNA expression by physical association with the *ESR1* gene locus. Chromosomal conformation capture study combined with high-throughput sequencing (4C-Seq) revealed that *Eleanor* plays a role in activating gene expression via long-range chromatin interaction (Fig. 2) (Abdalla et al. 2019). *H19* is also an estrogen-inducible lncRNA (Sun et al. 2015) and related to breast cancer cell proliferation and endocrine therapy resistance through regulating ER expression (Basak et al. 2018).

2.3 Enhancer RNA

Enhancer RNAs (eRNAs), a class of noncoding RNAs transcribed from enhancer/promoter regions, were originally identified in neuronal cells (Kim et al. 2010) and are focused on as a new paradigm of transcriptional mechanisms (Mikhaylichenko et al. 2018). In general, eRNAs are of lengths 50–2000 nucleotides without polyadenylation tail and transcribed from the DNA regions with H3K4me3 active histone modification mark (Kim et al. 2010). High-throughput transcriptome analyses (Li et al. 2013) independently showed that >1000 estrogen-regulated eRNAs are transcribed from genomic regions proximal to ER-binding sites. It has been shown that these eRNAs contribute to ER-mediated transcription in *cis* and *trans* (Fig. 2) (Li et al. 2013). Recent studies suggest that transcriptional complexes form biomolecular condensates, which are membraneless intracellular organelles similar to nucleoli or nuclear paraspeckles (Alberti et al. 2019). These biomolecular condensates are particularly formed by proteins with low complexity domain (LCD), which has an excess of one or a few types of amino acid residues compared to an average composition (Michelitsch and Weissman 2000). Since LCDs exhibit an intrinsically disordered conformation not amenable to conventional structural determination, they had been considered as nonfunctional domains. Nevertheless, pathogenic roles of LCDs have been revealed as mutations in prion-like domains that are prone to aggregation and have been found in familial amyotrophic lateral sclerosis (ALS)-related RNA-binding proteins that are fused in sarcoma (FUS/TLS) and heterogeneous nuclear ribonucleoprotein A1 (hnRNPA1) (Ito et al. 2017). It has been shown that a high concentration of LCD oligomerizes and forms a hydrogel that is similar to a biomolecular condensate defined as phase transition (Kato et al. 2012). The other groups also showed that LCDs of hnRNPA1 form protein-rich droplets at low temperatures and undergo the physical process called liquid-liquid phase separation (LLPS) (Molliex et al. 2015). LLPS could be a useful model of membraneless organelles and has been reported to be involved in various biological phenomena such as neuronal dysfunction (Murakami et al. 2015) and cancer development (Bouchard et al. 2018). Recent study has shown that acute estrogen-dependent activation of functional enhancers (by 5 minutes) concentrates complex of eRNAs and its associated proteins including ERα, RNA polymerase II, and other ERα-related transcriptional factors such as GATA3, FOXA1, AP2γ, MED1, and P300, leading to the occurrence of LLPS or phase transition (Nair et al. 2019).

In androgen treatment, a number of eRNAs transcribed from AREs are also annotated (Hsieh et al. 2014). For example, androgen-induced eRNA *KLK3e* is located in the AREs of *kallikrein-related peptidase 3 (KLK3)* gene, which encodes PSA, and is required for AR-mediated transcription in *cis*. Intriguingly, Hsieh et al. showed that this eRNA also modulates AR-mediated transcription in *trans* (Fig. 2) (Hsieh et al. 2014). It has been shown that *KLK3e* interacts with cyclin T1 protein, activates positive transcription elongation factor b (P-TEFb), and increases RNA polymerase II-mediated transcription in *trans* (Zhao et al. 2016). The altered *KLK3e* expression and P-TEFb activation mechanism could contribute to abnormal AR function and CRPC cell proliferation.

3 Tumor Progression-Associated LncRNAs in Breast and Prostate Cancers

Besides hormone-regulatory lncRNAs including eRNAs, there are also lncRNAs associated with breast and prostate cancers in a hormone-independent manner. Those lncRNAs interest with their binding partners to regulate their functions in either positive or negative ways (Fig. 3). In this section, we introduce lncRNAs involved in breast and prostate cancer development and progression.

3.1 Activation of LncRNA-Binding Partners

Cancer metastasis is a serious burden for cancer patients (Seyfried and Huysentruyt 2013). Several lncRNAs play important roles in cancer metastasis via the activation of their binding partners. While we focus on the aspect of *HOTAIR* as a hormone-regulatory lncRNA and its interaction with hormone receptors ER and AR (Xue et al.

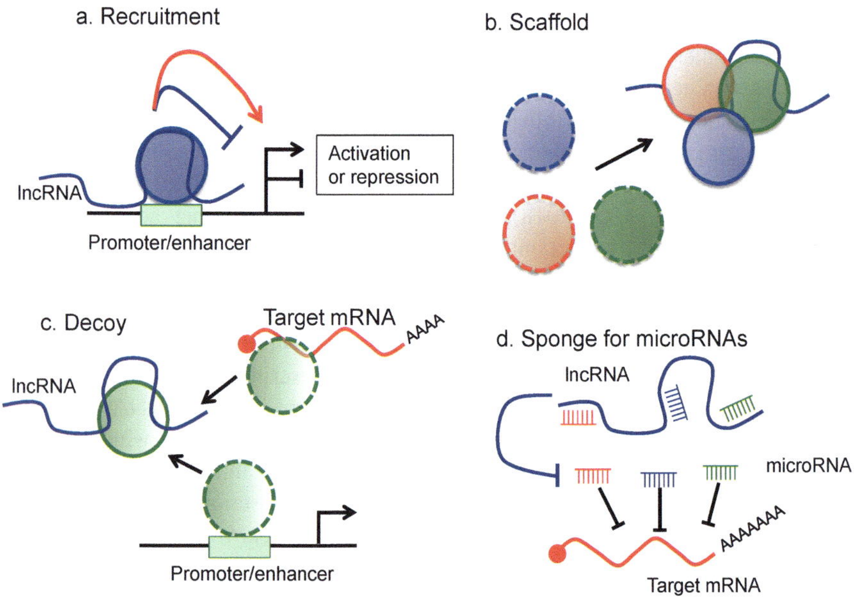

Fig. 3 Models of lncRNA actions in cancer development and progression. LncRNAs mainly exhibit their function through four different mechanisms. (**a**) Recruitment: LncRNA recruits transcription factors to promoter/enhancer regions. (**b**) Scaffold: LncRNA acts as a molecular scaffold for maintaining formation and stability of protein complex. (**c**) Decoy: LncRNA acts as a decoy RNA by binding to proteins leading to the inhibition of interaction with specific RNA or DNA region. (**d**) Sponge: LncRNA binds miRNA and prevents miRNA recruitment to its target mRNA

2016; Zhang et al. 2015), *HOTAIR* was originally identified in breast cancer in the context of metastasis rather than hormone signaling pathways (Gupta et al. 2010). *HOTAIR* is overexpressed in metastatic breast cancer, and the overexpression promotes breast cancer cell migration and invasion. *HOTAIR* interacts with PRC2 and is recruited to the transcription regulatory regions of metastatic suppressor genes, leading to the downregulation of their expression (Fig. 3a). *MALAT1* was identified as an invasion-related lncRNA and is also known as an oncogenic lncRNA in prostate cancer (Wang et al. 2015). *MALAT1* interacts with EZH2 and promotes cell migration and invasion of CRPC (Fig. 3a). Wang et al. identified *lncRNA associated with breast cancer brain metastasis* (*lnc-BM*) from a comparison study of lncRNA expression between parental and brain-metastatic breast cancer cells (Wang et al. 2017). *Lnc-BM* knockdown impaired the adhesion of cancer cells to brain endothelium and brain metastasis. Interactome analysis showed that *lnc-BM* interacts with janus kinase 2 (JAK2) protein and elevates the expression of metastatic genes such as *intercellular adhesion molecule-1* (*ICAM1*) and *C-C motif chemokine 2* (*CCL2*) based on the signaling pathway dependent on JAK and signal transducer and activator of transcription 3 (STAT3) (Fig. 3b).

Receptor tyrosine kinase-like orphan receptor 1 (ROR1) is overexpressed in bone metastatic breast cancer, and ROR1-dependent HER3 phosphorylation promotes metastasis via the activation of Hippo-yes-associated protein (YAP) pathway (Li et al. 2017b). A siRNA-based screening targeting lncRNAs revealed that an lncRNA *MST1/2-antagonizing for YAP activation* (*MAYA*) is involved in the ROR1-HER3 axis. Further analyses showed that *MAYA* interacts with NOP2/Sun methyltransferase 6 (NSUN6) and scribble cell polarity complex component (LLGL2) protein and inhibits the activity of macrophage stimulating 1 (MST1), an inhibitor of the Hippo pathway via methylation of MST1 (Fig. 3b). *NEAT1* is a component of nuclear paraspeckle (Fox et al. 2018) and is associated with cancer progression. In breast cancer, *NEAT1* forms a transcriptional repressor complex with forkhead box protein N3 (FOXN3) and SIN3 transcription regulator family member A (SIN3A) on the transcriptional regulatory regions of *GATA binding protein 3* (*GATA3*) and *tight junction protein 1* (*TJP1*) genes leading to the promotion of ER-positive breast cancer cell migration and invasion (Li et al. 2017c). EZH2 is a methyltransferase protein and a component of the transcriptional repressor complex and functions as a key player in cell invasion and migration in various cancers (Fig. 3a) (Kim and Roberts 2016). EZH2 has been reported to interact with lncRNAs such as *HOTAIR* as described above (Gupta et al. 2010). In prostate cancer, cytoplasmic lncRNA *differentiation antagonizing non-protein coding RNA* (*DANCR*, also known as *ANCR*) promotes cell invasion through the downregulation of TIMP metallopeptidase inhibitor 2/3 (TIMP2/3) by EZH2 recruitment on the *TIMP2/3* promoter. *DANCR* is overexpressed in prostate cancer (Fig. 3a) (Jia et al. 2016). Interestingly, *DANCR* expression is decreased by androgen treatment and could be associated with cell invasion after endocrine therapy. The function of *DANCR* in breast cancer remains controversial. It has been shown that *DANCR* is a downregulated lncRNA in breast cancer (Li et al. 2017a). *DANCR* binds to EZH2 protein and promotes its degradation. ShRNA-mediated knockdown of *DANCR* promotes normal mammary gland MCF-10 cell invasion and migration. On the other hand, the other group showed that *DANCR* promotes the phosphorylation of retinoid X receptor α (RXRA) via glycogen synthase kinase β (GSK3β) and impairs the transcriptional activity of RXRA in breast cancer (Tang et al. 2018b). Since RXRA represses the expression of phosphatidylinositol 4,5-biphosphate 3-kinase catalytic subunit α (PIK3CA), a mediator of PI3K/AKT pathway, *DANCR* activates PI3K/AKT pathway through the upregulation of PIK3CA, which leads to breast cancer cell proliferation.

HOXA transcript at the distal tip (*HOTTIP*) promotes a metastatic phenotype of prostate cancer via interacting and recruiting twist-related protein 1 (TWIST), a metastasis-related transcription factor that binds to the *HOXA9* gene promoter (Fig. 3a) (Malek et al. 2017). In breast cancer, increased expression of *HOTTIP* is significantly associated with poorer prognosis and related to breast cancer cell proliferation (Yang et al. 2017).

Long intergenic noncoding RNA for kinase activation (*LINK-A*) was identified in TNBC samples (Lin et al. 2016) and is involved in breast cancer development and metastasis. Originally, *LINK-A* was reported to physically associate with tyrosine protein kinase 6 (also known as breast tumor kinase, BRK) and to promote BRK-dependent phosphorylation of hypoxia-inducible factor 1α (HIF1α) at Tyr 565 (Fig. 3b). *LINK-A* stabilizes HIF1α through its protein phosphorylation under normoxic conditions and promotes breast cancer glycolysis (Lin et al. 2016). *LINK-A* also interacts with PtdIns(3,4,5)P3 and activates the AKT pathway (Lin et al. 2017). In addition, a recent study showed that *LINK-A* activates G protein-coupled receptor (GPCR) pathways, attenuates the expression of the antigen peptide-loading complex (PLC) components, and promotes the evasion from immune system attack (Hu et al. 2019). This study used *LINK-A* transgenic mice that exhibit the overexpression of *LINK-A* in mammary glands. Notably, *LINK-A* transgenic mice developed TNBC-like breast cancer, suggesting that *LINK-A* is an essential lncRNA for breast cancer and could be a promising therapeutic target. Indeed, it is confirmed that a GPCR antagonist effectively reduced the development of *LINK-A*-driven tumors where the GPCR pathway is activated.

A number of lncRNAs promote cell proliferation in cancers in a unique manner. Integrative methylation landscape analyses in 6475 tumors and 455 cancer cell lines revealed that 1006 lncRNA genes are hypomethylated, implying that they are overexpressed in cancers (Wang et al. 2018). Among them, epigenetically induced lncRNA 1 (*EPIC1*) overexpression is associated with poorer prognosis in breast cancer, and knockdown of *EPIC1* suppresses cell cycle transition from G1 to S phase. Global gene expression analysis showed that *EPIC1* is strongly associated with the Myc pathway, and, in fact, *EPIC1* physically interacts with Myc protein. *EPIC1* promotes Myc occupancy on its target promoter regions (Fig. 3a). They showed that *EPIC1* expression is also upregulated in prostate cancer cell line PC3, suggesting that *EPIC1* could play roles in prostate cancer progression. Androgen-inducible lncRNA *PCGEM1* also interacts with Myc protein and promotes Myc binding to tumor metabolism-related genes (Hung et al. 2014). *Plasmacytoma variant translocation 1* (*PVT1*) is an oncogenic lncRNA in various cancer types

(Cory et al. 1985) and upregulated in breast cancer samples. *PVT1* is located in the downstream of *MYC* loci. In breast cancer mouse models, an increase in *PVT1* and *MYC* copy-numbers (co-gain) shortens the survival of mice compared with mice having only increased *MYC* copy-number (Cory et al. 1985). *PVT1* suppression decreased Myc protein expression but not *MYC* mRNA. Further analysis showed that *PVT1* binds to and could stabilize the Myc protein (Tseng et al. 2014). Another group showed that *PVT1* binds to and stabilizes Kruppel-like factor 5 (KLF5) protein through regulating the expression of BCRA1-associated protein 1 (BAP1), a deubiquitinase protein (Tang et al. 2018a). Since KLF5 is involved in β-catenin pathway, *PVT1* promotes cell proliferation through regulating this pathway. In prostate cancer, *PVT1* also promotes cell proliferation through downregulating antiproliferative miRNA-146a (Liu et al. 2016). *PDCD4-AS1,* a natural antisense transcript of *programmed cell death protein 4* (*PDCD4*) gene, is downregulated in metastatic breast cancer (Jadaliha et al. 2018). *PDCD4-AS1* knockdown destabilizes tumor suppressor *PDCD4,* leading to the promotion of breast cancer cell migration. *PDCD4-AS1* could form RNA-RNA hybrid with *PDCD4* mRNA and protect it from the RNA degradation mediated by RNA-binding protein HuR.

The tumor suppressor protein p53 regulates the transcription of its target lncRNAs, and these p53-depedent lncRNA-mediated signaling pathway substantially contributes to cancer pathophysiology (Huarte et al. 2010; Schmitt et al. 2016). *GUARDIN* is a p53-inducible lncRNA and plays an important role for protecting genomic integrity through the upregulation of telomeric repeat binding factor 2 (TRF2) and the association with BCRA1 (Hu et al. 2018). Another p53-inducible lncRNA *lncRNA-p21* is upregulated by the treatment with enzalutamide, a potent AR inhibitor, in prostate cancer cells and in prostate cancer tissues from enzalutamide-treated patients. Increased *lncRNA-p21* (Huarte et al. 2010) interacts with methyltransferase protein EZH2. Interestingly, *lncRNA-p21* disrupts the PRC2 complex and releases EZH2 to methylate STAT3 protein, leading to CRPC progression (Luo et al. 2019).

Antisense noncoding RNA in the INK4/ARF locus (*ANRIL*) is transcribed from the AS locus of *p15/CDKN2B/INK4b-p16/CDKN2A/INK4a-p14/ARF* gene cluster, which encodes the members of inhibitors of cyclin-dependent kinase 4 (INK4) family (Pasmant et al. 2007). *ANRIL* expression is elevated in prostate cancer tissues compared with benign prostate tissues and is inversely correlated with the expression of a tumor suppressor *INK4a*. *ANRIL* downregulates *INK4a* via recruiting chromobox 7 (CBX7), a component of the PRC1, to its promoter region (Fig. 3a). In breast cancer, *ANRIL* expression is also increased, whereas CBX7 expression is decreased compared with that in normal breast tissues. In addition, underexpression of CBX7 is associated with poorer prognosis of breast cancer (Meseure et al. 2016), suggesting that *ANRIL* may function in a CBX7-independent manner in breast cancer. *ZEB1-AS1* is an AS transcript of *zinc finger E-box-binding homeobox 1* (*ZEB1*) gene that is a metastasis-promoting transcription factor. In prostate cancer, *ZEB1-AS1* knockdown suppresses ZEB1 expression and cell migration (Su et al. 2017). *ZEB1-AS1* recruits a histone methyltransferase myeloid/lymphoid or mixed-lineage leukemia 1 (MLL1) to the promoter region of *ZEB1* gene (Fig. 3a).

3.2 Inhibition of LncRNA-Binding Partners

Second chromosome locus associated with prostate 1 (*SChLAP1*) is expressed in aggressive prostate tumors and 25% of all prostate cancer samples (Prensner et al. 2013). High expression of *SChLAP1* is associated with increased mortality and recurrence of prostate cancer. Knockdown of *SChLAP1* impairs prostate cancer cell proliferation and invasion, and overexpression of *SChLAP1* promotes cell invasion. Pathway analysis for differentially expressed genes in prostate cancer cells with *SChLAP1* knockdown showed that *SChLAP1* is significantly associated with the switch/sucrose non-fermentable (SWI/SNF) complex. *SChLAP1* interacts strongly with SNF5 (also called SMARCB1, INI1, and BAF47), a component of the SWI/SNF complex (Wilson and Roberts 2011), and antagonizes SWI/SNF complex regulation (Fig. 3c).

Pseudogenes are defined as genomic loci that resemble bona fide genes. However, pseudogenes lack functional protein-coding ability as a result of mutations. Poliseno et al. reported that tumor suppressor PTEN and *PTEN pseudogene* (*PTENP1*) compete with one another for binding to miRNAs including *miRNA-20a*, *miRNA-19b*, *miRNA-21*, *miRNA-26a*, and *miRNA-214* (Poliseno et al. 2010). *PTEN* expression is well correlated with *PTENP1* in prostate cancer samples. *PTENP1* 3'-UTR overexpression attenuates the proliferation of prostate cancer cells through PTEN upregulation. *KRAPS1P,* a pseudogene of *KRAS,* also protects KRAS from the degradation through sharing miRNAs in prostate cancer samples. This type of RNA is now defined as a competing endogenous RNA (ceRNA) or miRNA sponge (Fig. 3d). While a vast number of ceRNAs have been reported (Tay et al. 2014), the effect of ceRNAs remains controversial, because quantitative analysis of miRNAs, lncRNAs, and target mRNAs showed that ceRNAs do not significantly influence gene expression through regulating its target miRNA abundance (Denzler et al. 2014).

LincRNA-RoR is a DNA damage-induced lncRNA and suppresses the p53 pathway (Zhang et al. 2013). *LincRNA-RoR* impairs p53 translation after DNA damage through complex formation with heterogeneous ribonucleoprotein I (hnNRP I). The same group found that *urothelial carcinoma associated 1* (*UCA1*) also binds to hnRNP I in breast cancer cells. *UCA1* was originally identified as an oncogenic lncRNA in bladder transitional cell carcinoma (Nortier et al. 2000). A breast cancer study showed that hnRNP I interacts with tumor suppressive p27 mRNA at the 5'-UTR and promotes p27 translation (Huang et al. 2014). *UCA1* inhibits hnRNP I activity leading to the suppression of p27 translation (Fig. 3c). In prostate cancer, *UCA1* has been also reported as an oncogenic lncRNA and might function as a ceRNA and protect *B-cell lymphoma 2* (*BCL2*) mRNA from miR-184-mediated degradation (Fig. 3d) (Zhang et al. 2017).

Nuclear factor (NF)-kappa B interacting lncRNA (*NKILA*) is an inflammatory cytokine-activated lncRNA in breast cancer cells and is located in cytoplasm. *NKILA* expression is lower in breast cancer samples compared with normal breast samples. *NKILA* overexpression suppresses the NF-κB pathway. *NKILA* interacts with inhibitor of kappa B (IκB) and p65 complexes and inhibits IκB kinase (IKK)-mediated phosphorylation of IκB (Fig. 3c) (Liu et al. 2015). Similar to *NKILA*, *PCAT1* is also involved in NF-κB and AKT signaling pathways (Shang et al. 2019). *PCAT1* knockdown suppresses the phosphorylation of p65 and AKT. *PCAT1* interacts with FK506 binding protein 51 (FKBP51), an interacting partner of IKK, and perturbs the binding of the PH domain and leucine-rich repeat protein phosphatase (PHLPP) to FKBP51 (Fig. 3c). *PCAT1* activates AKT and IKK by suppressing the function of PHLPP, which inhibits AKT and IKK functions.

MALAT1 is broadly known as an oncogenic lncRNA in many cancer types including prostate cancer as described above (Wang et al. 2015). However, recent study used the breast cancer mice model MMTV (mouse mammary tumor virus)-PyMT (polyomavirus middle T antigen) mice with *MALAT1* knockout as well as restoration and revealed that *MALAT1* rather suppresses lung metastasis of breast cancer cells (Kim et al. 2018). *MALAT1* binds to and inactivates TEA family domain member (TEAD) family protein that is involved in the YAP pathway and targets metastasis-related gene promoters such as *vascular endothelial growth factor a* (*VEGFA*) and *integrin beta 4* (*ITGB4*) (Fig. 3c). These findings would lead us to reconsider the ongoing *MALAT1*-targeting strategy against metastasis.

4 Conclusion

In this chapter, we described lncRNAs specifically involved in hormone signaling pathways. Sex hormone receptors are elaborately regulated by many lncRNAs in various ways such as through RNA export, RNA/protein stabilization, and recruitment of cofactors. Moreover, recent studies revealed that a vast number of eRNAs are transcribed from EREs and AREs and function in *cis* and *trans*. As ER and AR are ligand-dependent transcription factors, the mechanism of transcriptional switch between the on and off states will provide feasible models to analyze the function of lncRNA and eRNA.

An increasing number of reports show that various lncRNAs are associated with cancer development and progression and suggest that lncRNAs are promising diagnostic and therapeutic targets. In fact, several lncRNAs are currently being applied in diagnostic or therapeutic strategies such as *PCA3*. We here focused on functionally characterized lncRNAs, although many cancer-associated lncRNAs still remain to be characterized. Comprehensive analysis of novel lncRNAs as well as formerly identified lncRNAs may further uncover mechanisms that contribute to hormone-dependent cancer progression.

Acknowledgments The authors thank Dr. Kenichi Takayama for valuable discussion. This work was supported by the Project for Cancer Research and Therapeutic Evolution (P-CREATE, JP18cm0106144 to SI) from Japan Agency for Medical Research and Development, AMED, grants from the Japan Society for the Promotion of Science (17 K18061 to YM and 18 J00252 to YM).

References

Abdalla MOA, Yamamoto T, Maehara K et al (2019) The Eleanor ncRNAs activate the topological domain of the ESR1 locus to balance against apoptosis. Nat Commun 10:3778

Aiello A, Bacci L, Re A et al (2016) MALAT1 and HOTAIR long non-coding RNAs play opposite role in Estrogen-mediated transcriptional regulation in prostate Cancer cells. Sci Rep 6:38414

Alberti S, Gladfelter A, Mittag T (2019) Considerations and challenges in studying liquid-liquid phase separation and biomolecular condensates. Cell 176:419–434

Ansar Ahmed S, Penhale WJ, Talal N (1985) Sex hormones, immune responses, and autoimmune diseases. Mechanisms of sex hormone action. Am J Pathol 121:531–551

Basak P, Chatterjee S, Bhat V et al (2018) Long non-coding RNA H19 acts as an Estrogen receptor modulator that is required for endocrine therapy resistance in ER+ breast Cancer cells. Cell Physiol Biochem 51:1518–1532

Bouchard JJ, Otero JH, Scott DC et al (2018) Cancer mutations of the tumor suppressor SPOP disrupt the formation of active, phase-separated compartments. Mol Cell 72:19–36

Bussemakers MJ, van Bokhoven A, Verhaegh GW et al (1999) DD3: a new prostate-specific gene, highly overexpressed in prostate cancer. Cancer Res 59:5975–5979

Chakravarty D, Sboner A, Nair SS et al (2014) The oestrogen receptor alpha-regulated lncRNA NEAT1 is a critical modulator of prostate cancer. Nat Commun 5:5383

Chung S, Nakagawa H, Uemura M et al (2011) Association of a novel long non-coding RNA in 8q24 with prostate cancer susceptibility. Cancer Sci 102:245–252

Consortium EP (2012) An integrated encyclopedia of DNA elements in the human genome. Nature 489:57–74

Cory S, Graham M, Webb E et al (1985) Variant (6;15) translocations in murine plasmacytomas involve a chromosome 15 locus at least 72 kb from the c-myc oncogene. EMBO J 4:675–681

Cui Z, Ren S, Lu J et al (2013) The prostate cancer-up-regulated long noncoding RNA PlncRNA-1 modulates apoptosis and proliferation through reciprocal regulation of androgen receptor. Urol Oncol 31:1117–1123

Davies AH, Beltran H, Zoubeidi A (2018) Cellular plasticity and the neuroendocrine phenotype in prostate cancer. Nat Rev Urol 15:271–286

Denzler R, Agarwal V, Stefano J et al (2014) Assessing the ceRNA hypothesis with quantitative measurements of miRNA and target abundance. Mol Cell 54:766–776

Derrien T, Johnson R, Bussotti G et al (2012) The GENCODE v7 catalog of human long noncoding RNAs: analysis of their gene structure, evolution, and expression. Genome Res 22:1775–1789

Fang Z, Xu C, Li Y et al (2016) A feed-forward regulatory loop between androgen receptor and PlncRNA-1 promotes prostate cancer progression. Cancer Lett 374:62–74

Fox AH, Nakagawa S, Hirose T et al (2018) Paraspeckles: where long noncoding RNA meets phase separation. Trends Biochem Sci 43(2):124–135

Ghoncheh M, Pournamdar Z, Salehiniya H (2016) Incidence and mortality and epidemiology of breast Cancer in the world. Asian Pac J Cancer Prev 17:43–46

Groskopf J, Aubin SM, Deras IL et al (2006) APTIMA PCA3 molecular urine test: development of a method to aid in the diagnosis of prostate cancer. Clin Chem 52:1089–1095

Gupta RA, Shah N, Wang KC et al (2010) Long non-coding RNA HOTAIR reprograms chromatin state to promote cancer metastasis. Nature 464:1071–1076

Hajjari M, Salavaty A (2015) HOTAIR: an oncogenic long non-coding RNA in different cancers. Cancer Biol Med 12:1–9

Hsieh CL, Fei T, Chen Y et al (2014) Enhancer RNAs participate in androgen receptor-driven looping that selectively enhances gene activation. Proc Natl Acad Sci U S A 111:7319–7324

Hu WL, Jin L, Xu A et al (2018) GUARDIN is a p53-responsive long non-coding RNA that is essential for genomic stability. Nat Cell Biol 20:492–502

Hu Q, Ye Y, Chan LC et al (2019) Oncogenic lncRNA downregulates cancer cell antigen presentation and intrinsic tumor suppression. Nat Immunol 20:835–851

Huang J, Zhou N, Watabe K et al (2014) Long non-coding RNA UCA1 promotes breast tumor growth by suppression of p27 (Kip1). Cell Death Dis 5:e1008
Huang W, Su X, Yan W et al (2018) Overexpression of AR-regulated lncRNA TMPO-AS1 correlates with tumor progression and poor prognosis in prostate cancer. Prostate 78:1248–1261
Huarte M, Guttman M, Feldser D et al (2010) A large intergenic noncoding RNA induced by p53 mediates global gene repression in the p53 response. Cell 142:409–419
Hung CL, Wang LY, Yu YL et al (2014) A long noncoding RNA connects c-Myc to tumor metabolism. Proc Natl Acad Sci U S A 111:18697–18702
Ito D, Hatano M, Suzuki N (2017) RNA binding proteins and the pathological cascade in ALS/FTD neurodegeneration. Sci Transl med 9 pii: eaah5436
Jadaliha M, Gholamalamdari O, Tang W et al (2018) A natural antisense lncRNA controls breast cancer progression by promoting tumor suppressor gene mRNA stability. PLoS Genet 14: e1007802
Jia J, Li F, Tang XS et al (2016) Long noncoding RNA DANCR promotes invasion of prostate cancer through epigenetically silencing expression of TIMP2/3. Oncotarget 7:37868–37881
Kato M, Han TW, Xie S et al (2012) Cell-free formation of RNA granules: low complexity sequence domains form dynamic fibers within hydrogels. Cell 149:753–767
Kim KH, Roberts CW (2016) Targeting EZH2 in cancer. Nat Med 22:128–134
Kim TK, Hemberg M, Gray JM et al (2010) Widespread transcription at neuronal activity-regulated enhancers. Nature 465:182–187
Kim J, Piao HL, Kim BJ et al (2018) Long noncoding RNA MALAT1 suppresses breast cancer metastasis. Nat Genet 50:1705–1715
Kino T, Hurt DE, Ichijo T et al (2010) Noncoding RNA gas5 is a growth arrest- and starvation-associated repressor of the glucocorticoid receptor. Sci signal 3:ra8
Kirby M, Hirst C, Crawford ED (2011) Characterising the castration-resistant prostate cancer population: a systematic review. Int J Clin Pract 65:1180–1192
Knoll M, Lodish HF, Sun L (2015) Long non-coding RNAs as regulators of the endocrine system. Nat Rev Endocrinol 11:151–160
Lanz RB, McKenna NJ, Onate SA et al (1999) A steroid receptor coactivator, SRA, functions as an RNA and is present in an SRC-1 complex. Cell 97:17–27
Lee YS, Dutta A (2009) MicroRNAs in cancer. Annu Rev Pathol 4:199–227
Li W, Notani D, Ma Q et al (2013) Functional roles of enhancer RNAs for oestrogen-dependent transcriptional activation. Nature 498:516–520
Li Z, Hou P, Fan D et al (2017a) The degradation of EZH2 mediated by lncRNA ANCR attenuated the invasion and metastasis of breast cancer. Cell Death Differ 24:59–71
Li C, Wang S, Xing Z et al (2017b) A ROR1-HER3-lncRNA signalling axis modulates the hippo-YAP pathway to regulate bone metastasis. Nat Cell Biol 19:106–119
Li W, Zhang Z, Liu X et al (2017c) The FOXN3-NEAT1-SIN3A repressor complex promotes progression of hormonally responsive breast cancer. J Clin Invest 127:3421–3440
Li S, Li J, Chen C et al (2018) Pan-cancer analysis of long non-coding RNA NEAT1 in various cancers. Genes Dis 5:27–35
Lilja H, Ulmert D, Vickers AJ (2008) Prostate-specific antigen and prostate cancer: prediction, detection and monitoring. Nat Rev Cancer 8:268–278
Lin A, Li C, Xing Z et al (2016) The LINK-A lncRNA activates normoxic HIF1α signalling in triple-negative breast cancer. Nat Cell Biol 18:213–224
Lin A, Hu Q, Li C et al (2017) The LINK-A lncRNA interacts with PtdIns(3,4,5)P. Nat Cell Biol 19:238–251
Litwin MS, Tan HJ (2017) The diagnosis and treatment of prostate Cancer: a review. JAMA 317:2532–2542
Liu B, Sun L, Liu Q et al (2015) A cytoplasmic NF-κB interacting long noncoding RNA blocks IκB phosphorylation and suppresses breast cancer metastasis. Cancer Cell 27:370–381
Liu HT, Fang L, Cheng YX et al (2016) LncRNA PVT1 regulates prostate cancer cell growth by inducing the methylation of miR-146a. Cancer Med 5:3512–3519

Loi S, Haibe-Kains B, Desmedt C et al (2008) Predicting prognosis using molecular profiling in estrogen receptor-positive breast cancer treated with tamoxifen. BMC Genomics 9:239

Luo J, Wang K, Yeh S et al (2019) LncRNA-p21 alters the antiandrogen enzalutamide-induced prostate cancer neuroendocrine differentiation via modulating the EZH2/STAT3 signaling. Nat Commun 10:2571

Malek R, Gajula RP, Williams RD et al (2017) TWIST1-WDR5-Hottip regulates hoxa9 chromatin to facilitate prostate cancer metastasis. Cancer Res 77:3181–3193

McEwen BS, Milner TA (2017) Understanding the broad influence of sex hormones and sex differences in the brain. J Neurosci Res 95:24–39

Meseure D, Vacher S, Alsibai KD et al (2016) Expression of ANRIL-Polycomb complexes-CDKN2A/B/ARF genes in breast Tumors: identification of a two-gene (EZH2/CBX7) signature with independent prognostic value. Mol Cancer Res 14:623–633

Michelitsch MD, Weissman JS (2000) A census of glutamine/asparagine-rich regions: implications for their conserved function and the prediction of novel prions. Proc Natl Acad Sci U S A 97:11910–111915

Mikhaylichenko O, Bondarenko V, Harnett D et al (2018) The degree of enhancer or promoter activity is reflected by the levels and directionality of eRNA transcription. Genes Dev 32:42–57

Misawa A, Takayama K, Urano T, Inoue S (2016) Androgen-induced long noncoding RNA (lncRNA) SOCS2-AS1 promotes cell growth and inhibits apoptosis in prostate Cancer cells. J Biol Chem 291:17861–17880

Misawa A, Takayama KI, Fujimura T et al (2017) Androgen-induced lncRNA POTEF-AS1 regulates apoptosis-related pathway to facilitate cell survival in prostate cancer cells. Cancer Sci 108:373–379

Mitobe Y, Takayama KI, Horie-Inoue K et al (2018) Prostate cancer-associated lncRNAs. Cancer Lett 418:159–166

Mitobe Y, Ikeda K, Suzuki T et al (2019) ESR1-stabilizing lncRNA TMPO-AS1 promotes hormone-refractory breast cancer progression. Mol Cell Biol 39. https://doi.org/10.1128/MCB.00261-19

Molliex A, Temirov J, Lee J et al (2015) Phase separation by low complexity domains promotes stress granule assembly and drives pathological fibrillization. Cell 163:123–133

Murakami T, Qamar S, Lin JQ et al (2015) ALS/FTD mutation-induced phase transition of FUS liquid droplets and reversible hydrogels into irreversible hydrogels impairs RNP granule function. Neuron 88:678–690

Muramatsu M, Inoue S (2000) Estrogen receptors: how do they control reproductive and nonreproductive functions? Biochem Biophys Res Commun 270:1–10

Nair SJ, Yang L, Meluzzi D et al (2019) Phase separation of ligand-activated enhancers licenses cooperative chromosomal enhancer assembly. Nat Struct Mol Biol 26:193–203

Nortier JL, Martinez MC, Schmeiser HH et al (2000) Urothelial carcinoma associated with the use of a Chinese herb (Aristolochia fangchi). N Engl J Med 342:1686–1692

Notas G, Pelekanou V, Kampa M et al (2015) Tamoxifen induces a pluripotency signature in breast cancer cells and human tumors. Mol Oncol 9:1744–1759

Parise P, Finocchiaro G, Masciadri B et al (2006) Lap2alpha expression is controlled by E2F and deregulated in various human tumors. Cell Cycle 5:1331–1341

Parolia A, Crea F, Xue H et al (2015) The long non-coding RNA PCGEM1 is regulated by androgen receptor activity in vivo. Mol Cancer 14:46

Pasmant E, Laurendeau I, Héron D et al (2007) Characterization of a germ-line deletion, including the entire INK4/ARF locus, in a melanoma-neural system tumor family: identification of ANRIL, an antisense noncoding RNA whose expression coclusters with ARF. Cancer Res 67:3963–3969

Poliseno L, Salmena L, Zhang J et al (2010) A coding-independent function of gene and pseudogene mRNAs regulates tumour biology. Nature 465:1033–1038

Pondé NF, Zardavas D, Piccart M (2019) Progress in adjuvant systemic therapy for breast cancer. Nat Rev Clin Oncol 16:27–44

Prensner JR, Iyer MK, Balbin OA et al (2011) Transcriptome sequencing across a prostate cancer cohort identifies PCAT-1, an unannotated lincRNA implicated in disease progression. Nat Biotechnol 29:742–749

Prensner JR, Iyer MK, Sahu A et al (2013) The long noncoding RNA SChLAP1 promotes aggressive prostate cancer and antagonizes the SWI/SNF complex. Nat Genet 45:1392–1398

Prensner JR, Chen W, Han S et al (2014a) The long non-coding RNA PCAT-1 promotes prostate cancer cell proliferation through cMyc. Neoplasia 16:900–908

Prensner JR, Chen W, Iyer MK et al (2014b) PCAT-1, a long noncoding RNA, regulates BRCA2 and controls homologous recombination in cancer. Cancer Res 74:1651–1660

Prensner JR, Sahu A, Iyer MK et al (2014c) The IncRNAs PCGEM1 and PRNCR1 are not implicated in castration resistant prostate cancer. Oncotarget 5:1434–1438

Salameh A, Lee AK, Cardó-Vila M et al (2015) PRUNE2 is a human prostate cancer suppressor regulated by the intronic long noncoding RNA PCA3. Proc Natl Acad Sci U S A 112:8403–8408

Scher HI, Beer TM, Higano CS et al (2010) Antitumour activity of MDV3100 in castration-resistant prostate cancer: a phase 1-2 study. Lancet 375:1437–1446

Schmitt AM, Garcia JT, Hung T et al (2016) An inducible long noncoding RNA amplifies DNA damage signaling. Nat Genet 48:1370–1376

Seyfried TN, Huysentruyt LC (2013) On the origin of cancer metastasis. Crit Rev Oncog 18 (1–2):43–73

Shang Z, Yu J, Sun L et al (2019) LncRNA PCAT1 activates AKT and NF-κB signaling in castration-resistant prostate cancer by regulating the PHLPP/FKBP51/IKKα complex. Nucleic Acids Res 47:4211–4225

Srikantan V, Zou Z, Petrovics G, Xu L et al (2000) PCGEM1, a prostate-specific gene, is overexpressed in prostate cancer. Proc Natl Acad Sci U S A 97:12216–12221

Su W, Xu M, Chen X et al (2017) Long noncoding RNA ZEB1-AS1 epigenetically regulates the expressions of ZEB1 and downstream molecules in prostate cancer. Mol Cancer 16:142

Sun Y, Ma L (2019) New insights into long non-coding RNA. Cancers (Basel) 11

Sun H, Wang G, Peng Y et al (2015) H19 lncRNA mediates 17β-estradiol-induced cell proliferation in MCF-7 breast cancer cells. Oncol Rep 33:3045–3052

Takayama K, Horie-Inoue K, Katayama S et al (2013) Androgen-responsive long noncoding RNA CTBP1-AS promotes prostate cancer. EMBO J 32:1665–1680

Tan MH, Li J, Xu HE et al (2015) Androgen receptor: structure, role in prostate cancer and drug discovery. Acta Pharmacol Sin 36:3–23

Tang J, Li Y, Sang Y et al (2018a) LncRNA PVT1 regulates triple-negative breast cancer through KLF5/beta-catenin signaling. Oncogene 37:4723–4734

Tang J, Zhong G, Zhang H et al (2018b) LncRNA DANCR upregulates PI3K/AKT signaling through activating serine phosphorylation of RXRA. Cell Death Dis 9:1167

Tay Y, Rinn J, Pandolfi PP (2014) The multilayered complexity of ceRNA crosstalk and competition. Nature 505:344–252

Tomita S, Abdalla MOA, Fujiwara S et al (2015) A cluster of noncoding RNAs activates the ESR1 locus during breast cancer adaptation. Nat Commun 6:6966

Tseng YY, Moriarity BS, Gong W et al (2014) PVT1 dependence in cancer with MYC copy-number increase. Nature 512:82–86

Waks AG, Winer EP (2019) Breast Cancer treatment: a review. JAMA 321:288–300

Wang KC, Chang HY (2011) Molecular mechanisms of long noncoding RNAs. Mol Cell 43:904–914

Wang D, Ding L, Wang L et al (2015) LncRNA MALAT1 enhances oncogenic activities of EZH2 in castration-resistant prostate cancer. Oncotarget 6:41045–41055

Wang S, Liang K, Hu Q et al (2017) JAK2-binding long noncoding RNA promotes breast cancer brain metastasis. J Clin Invest 127:4498–4515

Wang Z, Yang B, Zhang M et al (2018) lncRNA epigenetic landscape analysis identifies EPIC1 as an oncogenic lncRNA that interacts with MYC and promotes cell-cycle progression in Cancer. Cancer Cell 33:706–720

Wilson BG, Roberts CW (2011) SWI/SNF nucleosome remodellers and cancer. Nat Rev Cancer 11:481–492

Xue X, Yang YA, Zhang A et al (2016) LncRNA HOTAIR enhances ER signaling and confers tamoxifen resistance in breast cancer. Oncogene 35:2746–2755

Yang L, Lin C, Jin C et al (2013) lncRNA-dependent mechanisms of androgen-receptor-regulated gene activation programs. Nature 500:598–602

Yang Y, Qian J, Xiang Y et al (2017) The prognostic value of long noncoding RNA HOTTIP on clinical outcomes in breast cancer. Oncotarget 8:6833–6844

Zhang A, Zhou N, Huang J et al (2013) The human long non-coding RNA-RoR is a p53 repressor in response to DNA damage. Cell Res 23:340–350

Zhang A, Zhao JC, Kim J et al (2015) LncRNA HOTAIR enhances the androgen-receptor-mediated transcriptional program and drives castration-resistant prostate Cancer. Cell Rep 13:209–221

Zhang Z, Zhou N, Huang J et al (2016) Regulation of androgen receptor splice variant AR3 by PCGEM1. Oncotarget 7:15481–15391

Zhang S, Dong X, Ji T et al (2017) Long non-coding RNA UCA1 promotes cell progression by acting as a competing endogenous RNA of ATF2 in prostate cancer. Am J Transl Res 9:366–375

Zhang Y, Pitchiaya S, Cieślik M et al (2018) Analysis of the androgen receptor-regulated lncRNA landscape identifies a role for ARLNC1 in prostate cancer progression. Nat Genet 50:814–824

Zhao Y, Wang L, Ren S et al (2016) Activation of P-TEFb by androgen receptor-regulated enhancer RNAs in castration-resistant prostate Cancer. Cell Rep 15:599–610

Tumorigenesis-Related Long Noncoding RNAs and Their Targeting as Therapeutic Approach in Cancer

Marianna Aprile, George Calin, Amelia Cimmino, and Valerio Costa

Contents

Abstract Long noncoding RNAs (lncRNA) are emerging as players in physiological processes and as crucial contributors to pathologic states, especially cancer. The direct contribution of lncRNAs to tumorigenesis has been demonstrated by loss- or gain-of-function experiments. In this chapter, we describe the most convincing evidences about their contribution to cancer hallmarks, with a focus on lncRNAs able to promote or suppress tumor formation either by their intrinsic properties or by their capacity to modulate known oncogenes/suppressors. Herein, we also describe the main lncRNAs-targeting approaches, discussing the main pros and cons of these methodologies. We further describe preclinical models widely used to address the

M. Aprile · A. Cimmino · V. Costa (✉)
Institute of Genetics and Biophysics "Adriano Buzzati-Traverso", CNR, Naples, Italy
e-mail: marianna.aprile@igb.cnr.it; amelia.cimmino@igb.cnr.it; valerio.costa@igb.cnr.it

G. Calin
Department of Experimental Therapeutics, The University of Texas MD Anderson Cancer Center, Houston, USA
e-mail: gcalin@mdanderson.org

S. Jurga, J. Barciszewski (eds.), *The Chemical Biology of Long Noncoding RNAs*,
RNA Technologies 11, https://doi.org/10.1007/978-3-030-44743-4_11

potential role of lncRNAs as both prognostic/diagnostic markers and therapeutic targets in the field of clinical oncology.

Keywords LncRNAs in cancer · Oncogenic lncRNAs · Tumor-suppressor lncRNAs · Therapeutic targeting

1 Introduction

Large-scale genomic studies reported that the fraction of protein-coding genes in the human genome is less than 2%, expanding our knowledge about the complexity of human transcriptome. Indeed, most of the genome is transcribed—frequently on both DNA strands—and tightly regulated. The human genome is characterized by thousands of transcriptionally active regions that generate transcripts with little-to-none protein-coding potential (Taft et al. 2010). However, challenging the traditional view of RNA as a bridge between DNA and proteins, they exert distinct functional roles (Bergmann and Spector 2014). Noncoding RNAs (ncRNAs) can be generally divided in two main classes according to size: small RNAs (about 20–30 nt), including microRNA (miRNAs), small nucleolar RNAs (snoRNAs), and Piwi-interacting RNAs (piRNAs), and highly heterogeneous family consisting of long noncoding RNAs (lncRNAs, >200 nt). The latter class, indicated by the generic "umbrella term" lncRNAs, was discovered by Rinn in the late 1990s as long interspersed transcribed regions lacking an open reading frame (ORF) as well as other characteristics crucial for their translation into protein products (Rinn et al. 2003). To date, large-scale *ab initio* reconstruction studies revealed that this class of ncRNAs comprises more than 91,000 transcripts (*www.mitranscriptome.org*). These transcripts are mainly generated by the RNA polymerase II and arise from intergenic regions, and they are generally 5′ capped, spliced, and 3′ polyadenylated (Consortium et al. 2007; Guttman et al. 2009; Iyer et al. 2015).

Recent high-throughput projects in the field of cancer genomics, and especially from the Cancer Genome Atlas (TCGA) Consortium, showed that several tumor-associated somatic alterations (i.e., mutations, gene rearrangements, and copy-number variations) occur within noncoding DNA portions, such as in regulatory regions (e.g., DNA enhancers) and in actively transcribed noncoding genes (Khurana et al. 2013). Simultaneously, a growing number of studies unequivocally demonstrated a role of lncRNAs in specific cancer hallmarks, including uncontrolled proliferation and metastasis, evasion of cell death, metabolic rewiring and immune escape (Geng et al. 2011; Niland et al. 2012, Huarte 2015; Table 1). Many of them have been reported as lncRNAs with oncogene- or tumor-suppressor-like functions or as indirect mediators of the activity of known oncogenes/suppressors (Fig. 1), including Myc and p53, both at transcriptional and posttranscriptional level (Gupta et al. 2010; Adriaens et al. 2016).

Table 1 Most relevant lncRNAs associated with cancer hallmarks

	Gene symbol	Cancer Hallmarks									
		SUSTAINED PROLIFERATIVE SIGNALING	INSENSITIVITY TO GROWTH SUPPRESSORS	EVASION OF APOPTOSIS	REPLICATIVE IMMORTALITY	INDUCED ANGIOGENESIS	TISSUE INVASION AND METASTASIS	ABNORMAL METABOLIC PATHWAYS	IMMUNE EVASION	GENOMIC INSTABILITY	INFLAMMATION
Oncogenes	*aHIF*					X					
	ANRIL	X	X	X			X			X	
	BCAR4						X				
	CCAT1	X	X				X				
	CCAT2	X					X			X	
	COMETT	X		X			X				
	HOTAIR						X				
	HOTTIP	X					X				
	HULC	X				X		X			
	Lilam	X									
	Linc-ROR						X				
	lncRNA-ATB						X				
	lncRNA-MIAT					X	X				
	LUNAR1	X									
	MALAT1						X				
	MINCR	X									
	MVIH	X					X				
	ncRAN	X					X				
	Orilnc	X	X								
	PCAT-1	X					X				
	PCGEM1		X	X				X			
	PVT1	X		X			X				
	SAMMSON							X			
	SCAT7										
	SChLAP1						X				
	THOR	X						X			
	TUG1	X					X				
	u-Eleanor	X									
	UCA1		X								
	VELUCT	X									
Tumor suppressors	*ANCR*						X				
	FILNC1							X			
	GAS5		X								
	Linc-PINT						X				
	LncAB	X	X								
	lncRNA-LET						X				
	MEG3	X		X							
	NBAT1	X					X				
	NBR2							X			
	NORAD									X	
	PCAT-29	X					X				
	PTENP1		X								
	TERRA				X						
	lincRNA-p21			X		X	X	X		X	
Dual role	*BANCR*						X				
	H19	X				X	X				
	NEAT1			X							
	PANDA				X					X	

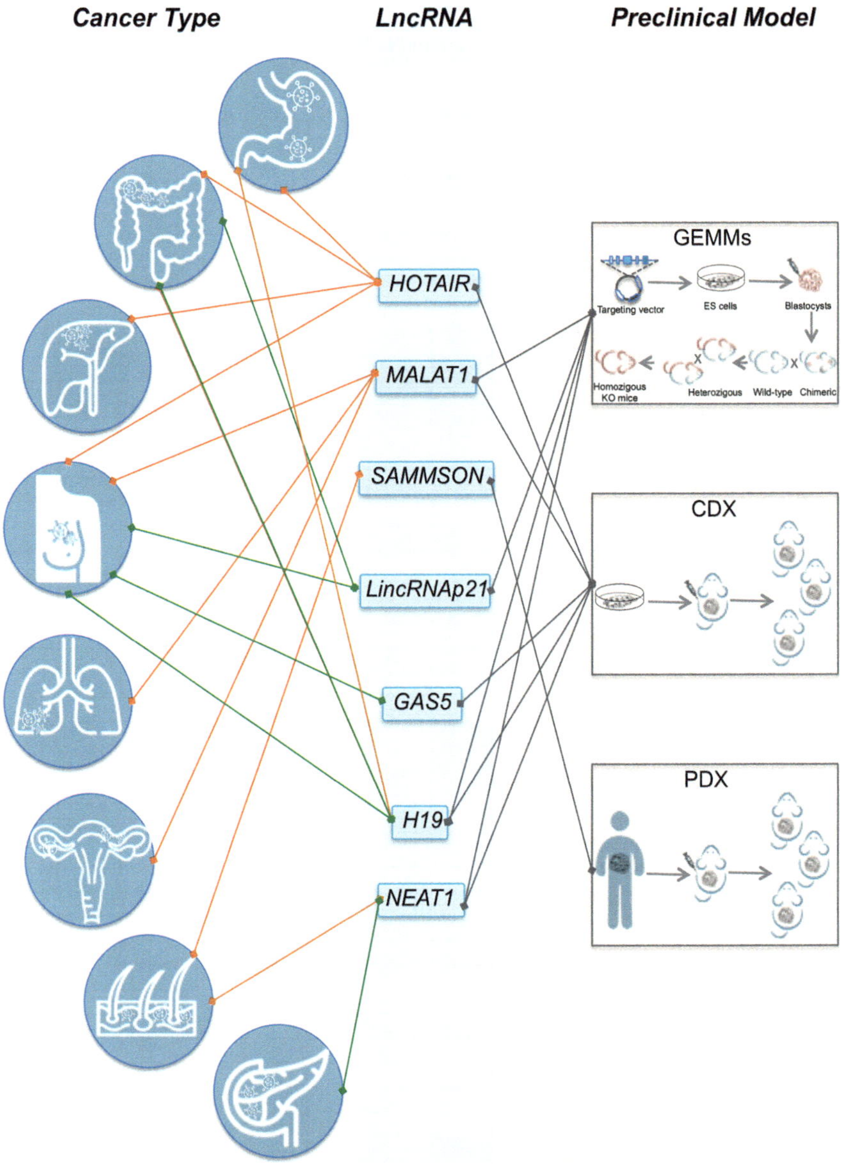

Fig. 1 Oncogenic and tumor-suppressor lncRNAs
Graphical summary of lncRNAs whose strong experimental evidences confirm a pro- or anti-tumorigenic role. Involvement of each lncRNA (middle panel) in each tumor (left panel) is indicated by colored lines. Orange and green lines indicate the oncogene- or tumor-suppressor-like function, respectively, of a given lncRNA. On the right panel, the experimental models (GEMM, CDX, or PDX) in which these lncRNAs have been studied

In this chapter, we provide a comprehensive description of the most relevant lncRNAs having clinical significance in cancer (Fig. 1), then we describe the main strategies to target them in preclinical studies, also discussing about the limitations of currently available targeting strategies.

2 LncRNAs in Tumorigenesis

The discovery of long noncoding RNA (lncRNA) has improved our understanding of many physiological processes governing cell function (Rinn et al. 2003), revealing also their potential implication in human diseases, including cancer. The comprehensive analysis of lncRNAs expression and alteration in more than 5,000 human tumor samples (across 13 cancer types TCGA) (Yan et al. 2015) has represented a milestone in the field. Indeed, transcriptional, genomic, and epigenetic investigation has definitely shown the deregulation of lncRNAs in many cancer types, highlighting that such alterations are highly tumor-specific and often associate with genomic rearrangements, such as SNPs and somatic copy-number alterations (CNAs), and epigenetic changes including promoter hypermethylation.

As schematized in Table 1, lncRNAs can contribute cancer hallmarks through direct oncogene- or tumor-suppressor-like activities or by modulating the activity of oncogenes/suppressors. Despite such modulation can occur through different mechanisms, these are mainly transcriptional and posttranscriptional.

2.1 LncRNAs with Oncogenic Properties

Here, we describe the most relevant examples of well-established lncRNAs having oncogenic properties (Table 1), then we report more recent examples of lncRNAs induced by oncogenes in different tumor types.

2.1.1 HOTAIR

The lncRNA HOX transcript antisense intergenic RNA (*HOTAIR*) resides within the *HOXC locus*, one of 39 genes encoding HOX transcription factors, located on human chr12q13.13. Three distinct transcript variants of *HOTAIR* lncRNA are annotated on NCBI RefSeq database (NR_047517.1, NR_047518.1 and NR_003716.3, consisting of six, four, and six exons, respectively; accessed on November 2019). It was the first lncRNA described as interacting with the PRC2 complex members EZH2 and SUZ12 but exerting its function in *trans* (Khalil et al. 2009). Indeed, *HOTAIR* can also recruit the lysine-specific demethylase (LSD1), part of the histone modification complex called CoREST/REST and involved in gene repression by chromatin remodeling. By adopting a secondary structure,

HOTAIR acts as a scaffold between the two complexes, resulting in the transcriptional repression of *HOXD locus* (Rinn et al. 2007; Tsai et al. 2010; Somarowthu et al. 2013; Sharma et al. 2015). *HOTAIR* has been reported as oncogene in several tumors, and its activity has been associated with tumor-promoting phenotypes, such as increased invasiveness and metastatic capacity. In line with its oncogenic role, high levels of *HOTAIR* in primary breast and colon tumors positively correlate with metastasis and poor outcome (Kogo et al. 2011; Gutschner et al. 2013). It also represents an independent prognostic marker for recurrence and shorter survival in hepatocarcinoma (Yang et al. 2011; Geng et al. 2011), suggesting *HOTAIR* as a potential therapeutic target in different cancer types (Fig. 1).

2.1.2 MALAT1

The lncRNA Metastasis-Associated Lung Adenocarcinoma Transcript 1 (*MALAT1*) maps on human chr11q13.1 and is located between *LOC101927789* and *SCYL1* genes. Three transcript variants of *MALAT1* are annotated on NCBI RefSeq database (NR_002819.4, NR_144567.1, and NR_144568.1, consisting of one, two, and three exons, respectively; accessed on November 2019). *MALAT1* lncRNA was identified in highly metastatic lung cancer where it was reported to be dramatically overexpressed (Ji et al. 2003). In a mouse model of lung cancer (xenograft), its loss caused a strong reduction in metastases formation, pointing to an oncogene-like role for this lncRNA (Gutschner et al. 2013). Initially defined as an oncogenic lncRNA in lung, *MALAT1* deregulation has been reported in other tumors (Fig. 1), indicating it as a key player in tumor progression. Silencing of *MALAT1* is likely to represent a very promising option to reduce the formation of metastases (Arun et al. 2016). Indeed, *MALAT1* is frequently mutated in luminal-type breast tumors (Nik-Zainal et al. 2016), and its depletion in ER^+ breast luminal cells dramatically impairs proliferation (Jadaliha et al. 2016). Conversely, in vitro and in vivo studies in ovarian cancer definitely demonstrated *MALAT1* role in uncontrolled proliferation, invasion, and metastasis of tumor cells, as well as in apoptosis (Jin et al. 2017). *Malat1* knockout (KO) mice are viable and have no evident developmental defects (Eissmann et al. 2012; Nakagawa et al. 2012; Zhang et al. 2012). However, KO or knockdown (KD) in breast cancer resulted in impaired tumor growth and metastasis with concomitant increased differentiation in cystic tumors (Arun et al. 2016). These reports strongly suggest that the lncRNA *MALAT1* has cell-/context-dependent functions in different cancers. Its key contribution to breast tumor progression and metastasis makes its KD a promising therapeutic option in breast cancer.

2.1.3 SAMMSON

The *Survival-Associated Mitochondrial Melanoma-Specific Oncogenic Noncoding RNA (SAMMSON)* maps on human chr3p13, and it is located between *MITF* and *MDFIC2*. One variant is currently annotated on NCBI RefSeq database

(NR_110000.1 consisting of four exons), despite at least nine transcripts are annotated in GENCODE v32 release.

It is a melanoma-specific lncRNA expressed in more than 90% of human melanomas. Its oncogenic activity in melanoma has been demonstrated by its Gapmer-mediated silencing which results in a marked reduction of tumor cells' clonogenic ability, massive cell death, and increased sensitivity of melanoma cells to therapies targeting MAPK in vitro and in patient-derived xenograft (PDX) (Leucci et al. 2016). In a more recent report, the same group revealed that, in the *nucleus, SAMMSON* is able to interfere with the binding of XRN2 to CARF (i.e., an RNA-binding protein), leading to cytoplasmic accumulation of aberrant CARF-p32 complexes that enhance mitochondrial and cytosolic translation fostering cancer cell fitness (Vendramin et al. 2018). Therefore, targeting *SAMMSON* represents a new effective, tissue-specific, and promising therapeutic option in melanoma.

2.1.4 VELUCT

Viability Enhancing LUng Cancer Transcript (*VELUCT*) maps on human chr2q37.3, but it is not currently annotated in NCBI Refseq nor in Ensemble and GENCODE databases. However, it is a striking example of how an extremely low abundant lncRNA can contribute to tumor phenotypes. Indeed, despite its extremely low abundance, *VELUCT* lncRNA KD causes a significant reduction of cell viability in lung cancer cell lines, suggesting a potential oncogenic function (Seiler et al. 2017). This report strongly indicates that even lncRNAs with very low expression can exert relevant roles in the cell, especially in the context of cancer cells.

2.2 *LncRNAs Acting Downstream Known Oncogenes*

Other lncRNAs having oncogene-like properties have been reported as induced downstream the activation—by point mutations or rearrangements—of known oncogenes. Among them, we focus on *Orilnc* (induced downstream *RAS;* Zhang et al. 2017), *BANCR,* and *COMET* (*BRAF*- and *RET*-induced lncRNAs; Flockhart et al. 2012; Esposito et al. 2019a).

2.2.1 *ORILNC1*

Orilnc1 lncRNA was recently identified in *BRAF*-mutant melanoma as regulated downstream the activation of RAS–RAF–MEK–ERK signaling pathway through AP-1 complex (Zhang et al. 2017). Silencing of Orilnc1 reduces Cyclin E1 expression inducing cell-cycle arrest in tumor cells; its KD also blocks tumor cell proliferation and growth, both in vitro and in vivo.

2.2.2 *COMETT*

The lncRNA Cytosolic Oncogenic Antisense To MET Transcript (*COMETT*, previously known as COrrelated-to-MET or COMET) was recently identified by the group of Dr. Costa in papillary thyroid carcinomas (Esposito et al. 2019a). It maps on human chr7q31.2 and is transcribed antisense to the oncogene *MET*. *COMETT* lncRNA has at least three known variants (annotated in GenBank as LN812953, LN812954, and LS991956, consisting of transcripts with four, five, and seven exons, respectively) and another shorter transcript annotated in NCBI RefSeq as *LINC01510*. We recently reported that depletion of *COMETT* lncRNA significantly impairs the expression levels of different MAPK pathway effectors, including *MET* oncogene (Esposito et al. 2019a). Repression of *COMETT* inhibits the viability and proliferation of tumor cells that harbor *BRAF* mutation or *RET* oncogene rearrangement. It causes a marked reduction in the motility and invasiveness of tumor cells. More interestingly, *COMETT* KD increases tumor cells' sensitivity to vemurafenib, a common inhibitor of mutated B-raf (Esposito et al. 2019a).

2.3 *LncRNAs-Modulating Myc Oncogene*

One of the most recurrent oncogenic CNAs across a variety of tumors occurs at *MYC* locus (Beroukhim et al. 2010). Targeting this oncogenic transcription factor in tumors has provided often unsuccessful and controversial results. Thus, the identification of new lncRNAs able to modulate *MYC* has paved the way to new therapeutic strategies in many cancers. *PVT1*, *CCAT1*, *CCAT2*, *PCAT1*, and *MINCR* are among the most promising lncRNAs regulating Myc activity in tumor cells (Shtivelman and Bishop 1989; Ling et al. 2013; Prensner et al. 2014; Kim et al. 2014; Doose et al. 2015).

2.3.1 PVT1

The lncRNA Plasmacytoma Variant Translocation 1 (*PVT1*) is located at human chr8q24.21 close to *MYC* locus and is co-amplified with it in almost all cancer types (Shtivelman and Bishop 1989). *PVT1* lncRNA has one known variant annotated in NCBI RefSeq (NR_003367.3 with nine exons) even though several transcripts are annotated on GENCODE (v32). In tumors where *PVT1* and *MYC* are co-amplified, *PVT1* lncRNA stabilizes Myc protein resulting in increased protein levels and enhanced proliferation (Tseng et al. 2014). *PVT1* fusion transcripts (with *NBEA* and WWOX in multiple myeloma) have been reported as contributing to tumorigenesis (Nagoshi et al. 2012).

2.3.2 PCAT1

A similar posttranscriptional regulation of *MYC* has been reported for *PCAT1*. Even this lncRNA maps on chr8q24.21 nearby *MYC* locus. Despite different transcripts are annotated on GENCODE v32, only one variant is annotated on NCBI RefSeq (NR_045262.2, consisting of two exons). This lncRNA, acting as competitive endogenous RNA (ceRNA), abrogates miR34–1 binding to *MYC* mRNA, resulting in increased translation and higher Myc protein levels in human prostate cancer (Prensner et al. 2014).

2.3.3 *CCAT1* and *CCAT2*

A different mechanism for Myc regulation has been also reported for the colon cancer-associated transcripts 1 and 2 (*CCAT1* and *CCAT2*, respectively). As *PVT1* and *PCAT1*, these lncRNAs map on chr8q24.21 nearby *MYC*. Their expression has been correlated with a cancer-associated single nucleotide polymorphism, the rs6983267, located in the super-enhancer region surrounding *MYC* gene at 8q24. *CCAT1* knockdown decreases colon cancer cell proliferation through CDKN1A/p21-mediated G1 cell-cycle arrest, and injection of *CCAT1* KD tumor cells prolongs tumor-free survival in xenografts (Kim et al. 2014). Conversely, *CCAT2* overexpression increases the expression of *MYC* by enhancing Wnt signaling through TCF7L2, promoting invasive tumor growth in xenograft models and higher metastatic capacity in liver (Ling et al. 2013).

2.3.4 MINCR

Finally, the *MYC*-induced lncRNA (*MINCR*), mapping on human chr8q24.3 and consisting of two exons, was identified in Burkitt lymphoma cells (Doose et al. 2015). It has been reported as an lncRNA with a key role in cell-cycle progression due to its contribution in *MYC*-mediated regulation of cell-cycle genes (Doose et al. 2015). Its KD is associated with the downregulation of genes encoding Aurora kinase A and B (i.e., *AURKA* and *AURKB*, respectively) and chromatin licensing and DNA replication factor 1 (*CDT1*), possibly explaining the impairment in cell proliferation observed in tumor cells upon knockdown of *MINCR*.

2.4 *LncRNAs with a Role of Tumor Suppressors*

An increasing number of studies are reporting lncRNAs acting as tumor suppressors (Table 1, Fig. 1), whereby their inactivation/deletion in tumors has a role in the onset and/or progression toward metastatic advanced cancer stages.

2.4.1 GAS5

The growth arrest-specific transcript 5 (*GAS5*) is one of the most highly expressed lncRNAs with a quite ubiquitous expression, as reported by serial analysis of gene expression (SAGE) performed in a broad spectrum of normal human tissues and cancers (Gibb et al. 2016). It maps on human chr1q25.1 and consists of eight noncoding exons. It is involved in embryogenesis (Hudson et al. 2014) as well as in many important cellular and physiological processes, including p53 signaling (Huarte et al. 2010), growth arrest (Schneider et al. 1988), and apoptosis (Mourtada-Maarabouni et al. 2009). *GAS5* is a decoy for glucocorticoid receptor (Kino et al. 2010) and is also able to bind to the androgen and progesterone receptors, having a role in hormone-dependent tumors (Hudson et al. 2014). *GAS5* is downregulated in breast, bladder, colon, pancreas, and prostate cancer, where its expression levels show an inverse correlation with tumor size, staging, and metastasis (Pickard and Williams 2015). Accordingly, *GAS5* overexpression in breast cancer xenograft models has a dramatic impact in vivo on tumor growth due to cell-cycle arrest and apoptotic induction (Mourtada-Maarabouni et al. 2009).

2.4.2 MEG3

The lncRNA Maternally Expressed Gene 3 (*MEG3*), originally identified as the human ortholog of the mouse gene trap *locus* 2 (*Gtl2*) (Schuster-Gossler et al. 1998), is one of the imprinted genes located within 14q32.3 (Miyoshi et al. 2000). At least 15 different transcript variants have been officially reported on NCBI RefSeq for this gene, even though further alternatively spliced isoforms are annotated on GENCODE v32 release. Hence, *MEG3* is one of the lncRNAs having the highest number of known transcripts. Its expression is completely lost in many tumors and cell lines (Zhou et al. 2007) due to different mechanisms. Despite the most frequent is gene deletion, epigenetics plays a key role in *MEG3* silencing in tumors. Indeed, this lncRNA falls within a *locus* tightly controlled by differentially methylated regions (DMRs), and the alteration of the methylation *status* at its promoter (hypermethylation) and intergenic region (hypomethylation) dramatically affects its expression (Zhou et al. 2007). Its role of tumor-suppressor gene is related to its ability to inhibit tumor cell proliferation in various cancer cell lines following in vitro reexpression (Zhang et al. 2003, 2010; Zhou et al. 2007; Braconi et al. 2011; Wang et al. 2012; Zhou et al. 2012).

2.4.3 LncAB

The lncRNA AB074169 (*lncAB*) is a single-exon lncRNA mapping on human chrXq23 within the intron 5 of the longer *PAK3* gene transcript. It was first identified in a comparative microarray analysis as significantly downregulated in papillary thyroid carcinoma samples *vs.* their adjacent counterparts, due to CpG island hypermethylation within the promoter (Gou et al. 2018). Accordingly, knockdown in vitro promoted tumor cell proliferation. Conversely, its forced overexpression resulted in cell-cycle arrest and tumor growth inhibition in vitro and in vivo. Mechanistically, *lncAB* binds to—and reduces the expression of—KHSRP, increasing p21 levels and simultaneously decreasing CDK2 expression to repress cell proliferation (Gou et al. 2018).

2.4.4 NORAD

The lncRNA Noncoding RNA Activated by DNA damage (*NORAD*), previously annotated as *LINC00657*, is a single-exon transcript mapping on human chr20q11.23 between *CNBD2* and *EPB41L1* genes. Despite it was first identified in a doxorubicin treatment screen in different colon cancer cell lines, it is ubiquitously and highly expressed almost in all human tissues. However, the highest expression has been measured in brain, especially in the frontal cortex (data from GTEx database at https://www.gtexportal.org/home/gene/NORAD).

Upon DNA damage, this lncRNA is induced by a p53-mediated mechanism (Lee et al. 2016). In colorectal carcinoma cells, it was reported to act in the cytosol as a potent decoy molecule for the RNA-binding proteins of the PUMILIO family (members 1 and 2) (Tichon et al. 2016). PUMILIO proteins stimulate mRNAs deadenylation and decapping causing posttranscriptional repression. Chromosomal instability is triggered in the absence of *NORAD* lncRNA by the repression of their targets, including several genes involved in DNA replication and DNA repair, ultimately affecting cell cycle and mitosis (Lee et al. 2016).

2.4.5 lincRNA-p21

The *lincRNA-p21*, also known as Tumor Protein P53 Pathway Corepressor 1 (*TP53COR1*), is a single-exon noncoding transcript mapping on human chr6p21.2. It was identified as an lncRNA induced in a p53-dependent manner upon DNA damage (Dimitrova et al. 2014). This lncRNA is transcribed in an antisense direction of the tumor-suppressor gene p21 (CDKN1A) and acts as transcriptional repressor in p53-dependent transcriptional responses (Huarte et al. 2010; Dimitrova et al. 2014). Targeting *lincRNA-p21* affects the expression of the adjacent gene *CDKN1A* and acts *in cis* on *CDKN1A* causing its activation, with the consequence to disrupt G1/S cell-cycle checkpoint (Dimitrova et al. 2014). In colorectal cancer, *lincRNA-p21* expression levels correlate with tumor staging and invasive phenotype (Zhai et al. 2013); in breast cancer cells, it interacts with HuR repressing *CTNNB1* and *JUNB* at transcriptional and posttranscriptional level (Yoon et al. 2012). Noteworthy, lincRNA-p21 is a unique example of lncRNA involved in the rewiring of metabolism of tumor cells, as it is induced by hypoxia and increases hypoxia-induced glycolysis (Yang et al. 2014).

2.5 LncRNAs with Dual Activity

Cancer-associated lncRNAs have been reported—in few cases—to promote or suppress tumor formation in a context-specific manner (Table 1, Fig. 1). For these lncRNAs, the contrasting or even divergent outcomes can often be explained, at least in part, by the wide genetic and phenotypic heterogeneity of tumors, by the use of distinct experimental systems (i.e., in vivo models *vs.* in vitro studies), or alternatively, they may reflect the presence of other tumor- or microenvironment-specific determinants. Some clear and well-known examples of lncRNAs with apparently divergent roles in tumors are provided here.

2.5.1 *H19*

H19 is a paternally imprinted gene mapping on human chr11p15.5. Three distinct transcript variants (consisting of five exons and differing for the splicing at the 5′) are annotated on NCBI RefSeq for this lncRNA. It was identified as highly expressed during mouse embryonic development and repressed at birth in most of the tissues. It was one of the first lncRNAs reported as reexpressed due to a loss of imprinting in tumor cells (Doucrasy et al. 1993; Berteaux et al. 2005) in many tumors, including breast, colorectal, and liver carcinomas (Davis et al. 1987; Bartolomei et al. 1991). *H19* is transcriptionally regulated by the tumor-suppressor p53 and by Myc and hypoxia-induced factor 1a (HIF-1a) oncogenes—frequently deleted (the former) or upregulated (the latter)—in various cancers types (Dugimont et al. 1998; Barsyte-Lovejoy et al. 2006; Matouk et al. 2010). Despite TCGA expression data reveal *H19* overexpression in colorectal and stomach (Weidle et al. 2017), a tumor-suppressive role has been reported in colorectal cancer (Apc$^{-/-}$ mice) of human rhabdoid tumor cell lines (Hao et al. 1993; Yoshimizu et al. 2008; Gabory et al. 2009). Despite the exact nature of such discrepancies has not been explored, such conflicting results are likely to be explained by the inter-species differences and/or by genetic heterogeneity of tumors.

H19 can impact tumor progression acting through different pathways. Indeed, it has been reported as a potential ceRNA for several different miRNAs and simultaneously as a precursor of miRNAs. The *H19* "sponge" activity—in line with its prevalent cytosolic localization—has been mainly reported for let-7 and miR-200 family members (Kallen et al. 2013).

As anticipated, *H19* lncRNA is an miR-675 precursor (Yoshimizu et al. 2008). In colon carcinomas, this miRNA is known to directly bind and inhibit the tumor-suppressor retinoblastoma protein (RB). Specifically, in colorectal cancer cell lines, high *H19* levels are paralleled by the concomitant increase of miR-675 and the reduction of RB that in turn promotes cell proliferation (Tsang et al. 2009). Therefore, targeting *H19* lncRNA in tumors where it has oncogene-like properties is likely to represent a promising therapeutic option.

2.5.2 *NEAT1*

The lncRNA Nuclear-Enriched Abundant Transcript 1 (*NEAT1*) maps on human chr11q13.1. Two different transcript variants are annotated for *NEAT1* on NCBI RefSeq, despite GENCODE v32 reports more than ten distinct transcripts in *NEAT1* locus. It has been reported as necessary for the formation of nuclear paraspeckles and is described as a direct target of p53 promoting skin tumorigenesis by increasing survival of tumor cells expressing mutated *KRAS* in genetically engineered mouse models (GEMM) (Adriaens et al. 2016). However, a recent report indicates a tumor-suppressing role for *NEAT1* in $p53^{-/-}$-mutant *Kras* mouse model of pancreatic ductal adenocarcinoma (Mello et al. 2017), indicating that—likewise *H19* and *BANCR* (described below)—this lncRNA can promote or suppress tumor formation in a context-specific manner.

2.5.3 *BANCR*

The *BRAF*-activated lncRNA (*BANCR*) is a noncoding transcript consisting of four exons mapping on chr9q21.11 nearby the gene encoding the amyloid beta precursor protein binding family A member 1 (*ABPA1*). It has been described either for its oncogenic or oncosuppressive roles in tumors. Indeed, *BANCR* has been identified in melanoma as a lncRNA with oncogene-like function in tumor cell migration (Flockhart et al. 2012; Li et al. 2014) and proliferation through MAPK pathway activation, but also reported to suppress tumorigenesis. Indeed, *BANCR* is a key regulator of epithelial-mesenchymal transition (EMT) in NSCLC, whose downregulation associates with metastatic disease and poor prognosis (Sun et al. 2014). Similarly, its reduced expression in papillary thyroid and in clear cell renal cell carcinomas (PTC and ccRCC, respectively) has been positively associated with tumor progression (Liao et al. 2017; Xue et al. 2018), suggesting a tumor-suppressing role, at least in these cancers.

2.6 Transcribed Ultraconserved Regions (T-UCRs)

Interestingly, another class of lncRNAs, the transcribed ultraconserved regions (T-UCRs)—also known as ultraconserved elements (UCEs or uc.) due to the highly conserved sequences (100% identity among human, mouse, and rat genomes (Bejerano et al. 2004))—have been reported to possess oncogenic properties (reviewed in Terracciano et al. 2017). Among these T-UCRs, uc.233, uc.270, and uc.190 have been described as having oncogene-like functions in pancreatic adenocarcinoma, pancreatic cancer cell lines, and transgenic mouse models (Jiang et al. 2016). More recently, for the uc.339, oncogene-like functions were reported (Vannini et al. 2017). Indeed, by acting as a sponge for miRNAs with a tumor-suppressive role (i.e., miR-95-5p, miR-339-3p, and miR-663b-3p), it in turn activates Cyclin E2 in patients with non-small cell lung cancer (NSCLC) (Vannini et al. 2017). A similar activity on miRNAs was demonstrated for the uc.63, whose overexpression is associated with the progression of castration-resistant prostate cancer via regulation of miR-130b mediated by MMP2 (Sekino et al. 2017). Moreover, the expression of uc.63 in breast cancer patients with luminal A subtype has been associated with a tumor-promoting effect (increased cancer cells' survival), and its expression has been correlated with a poor outcome. In line with this finding, uc.63 KD increases tumor cells apoptosis in vitro (Marini et al. 2017).

In a recent work, we described that uc.8+ —by targeting miR-596—stabilizes MMP9 in bladder cancer cells and associates with tumor progression (Olivieri et al. 2016). Moreover, we also found the interaction between the uc.8 and miR-596 is mediated by polycomb protein Yin Yang 1 (YY1), providing an additional layer of regulation in bladder cancer cells (Terreri et al. 2016).

To date, T-UCRs have been mainly described as having oncogenic properties. However, the tumor-suppressive activity of uc.160 has been reported in SGC-7901 and AGS gastric cancer cell lines, in which the uc.160 positively regulates the tumor-suppressor protein PTEN by targeting the oncogenic miR-155, a known PTEN-inhibiting molecule (Pang et al. 2018).

Despite the lack of a concrete strategy to target these oncosuppressor lncRNAs, addressing their role in the regulation of specific cell pathways may provide useful insights in the oncogenic processes. Noteworthy, the lack—or the dramatic reduction—of their expression may also have prognostic value in some cancers. Finally, innovative approaches based on the use of "dead" Cas9 (CRISPR/dCas9) fused to activating proteins (Konermann et al. 2015) may be hypothesized as reliable therapeutic options to reactivate these tumor-suppressor lncRNAs in specific cancer types.

3 Strategies to Target LncRNAs

Experimental proof of a direct contribution of lncRNAs to tumor cells' hallmarks is a primary endpoint for molecular oncology. Hence, loss- and/or gain-of-function experiments, both in vitro and in vivo, have been crucial to gain mechanistic insights in lncRNA activity, providing also the opportunity to develop new therapeutic approaches.

Currently available approaches to modulate lncRNAs can target their i) transcription (genome-level modulation), ii) stability/processing (posttranscriptional RNA-level modulation), and iii) binding capacity (steric inhibition of lncRNA-protein interactions).

3.1 Genomic Modulation of LncRNAs

The recent advances in genome editing approaches, and among them the wide diffusion of CRISPR-Cas9-based methods, allow reliable, reproducible, and specific transcriptional modulation of lncRNAs (reviewed in Esposito et al. 2019b). However, standard CRISPR-Cas9 was originally developed for editing protein-coding genes, for which a single Cas9 targeting the ORF introduces insertions/deletions causing frameshift with loss-of-function (Shalem et al. 2014). Hence, applying this approach to lncRNAs (lacking an ORF) has prompted researchers to develop new CRISPR-Cas9 tools for efficient knockout. The Double Excision CRISPR Knockout (DECKO) is a recently developed tool that applies a two-step cloning strategy to generate lentiviral vectors expressing simultaneously two-guide RNAs (gRNAs) (Aparicio-Prat et al. 2015). Such a pioneer strategy can overcome the limitation of standard Cas9-based systems developed for protein-coding regions. New advances in experimental approaches for the genome editing, such as the CRISPR interference (CRISPRi), allow to significantly reduce the transcription of lncRNA *loci* by adopting a modified version of Cas9 (i.e., the "dead"-Cas9, or dCas9) fused to various transcriptional repressors (Koch 2017; Gilbert et al. 2014). Among them, dCas9 can be fused with several protein cargoes with the capacity to inhibit (e.g., KRAB domain) or promote (e.g., VP64 effector) the transcription of targeted genes. A recent genome-wide study based on CRISPRi has targeted promoters of more than 16,000 lncRNAs, leading to the identification of about 500 of them essential for cancer cell growth and also showing the feasibility of targeting these molecules at transcriptional level as a possible therapeutic option in cancer (Zhu et al. 2016; Koch 2017).

Despite preclinical studies based on CRISPR systems revealed to be feasible (Gilbert et al. 2014; Koch 2017), how they will translate into cancer therapies is still a matter of debate.

3.2 Posttranscriptional Targeting of LncRNAs

The posttranscriptional targeting of lncRNAs occurs by oligonucleotide-based approaches relying on the formation of RNA-RNA or RNA-DNA duplex (Khvorova and Watts 2017; Stein and Castanotto 2017). The most common oligo-based system is double-stranded small-interfering RNAs (siRNAs) that are based on a degradation pathway in which Dicer and a multi-protein complex, i.e., the RNA-induced silencing complex (RISC), act with Argonaute 2 (Ago2) protein (Hannon and Rossi 2004). The adoption of siRNAs has proven successfully to knock down cytosolic lncRNAs in many cancer cell lines, but they failed to provide optimal silencing of nuclear lncRNAs, and in vivo experiments using siRNAs have been even more challenging. Indeed, their limited bioavailability in vivo and susceptibility to nucleases have largely restricted their use for clinical purposes. However, to prevent siRNAs from enzymatic degradation, some chemical modifications (e.g., 2'-O methyl sugar residues and phosphorothioate bonds at the 3′-end) have been adopted, improving their pharmacokinetics properties (Morrissey et al. 2005).

Alternatively, antisense oligonucleotides (ASOs), developed 40 years ago (Zamecnik and Stephenson 1978) as short 13-mer ssDNA designed to target the Rous sarcoma virus RNA, have been widely adopted to modulate lncRNA expression. Unlike siRNAs, ASOs are single-stranded RNA oligonucleotides (about 13–15 bp) binding to the target RNA via standard Watson-Crick pairing. Despite the formation of a duplex inhibits or modifies expression via steric hindrance and alteration of the splicing, the most common mechanism is the degradation mediated by RNAse H. Indeed, in line with the enrichment of RNase H in the nucleus, one of the peculiar characteristics of ASOs is their targeting of nuclear RNAs (Lima et al. 2007). Thus, it is a powerful tool to achieve significant lncRNA knockdown, as a large number of them is enriched in the nucleus (Bergmann and Spector 2014). Similar to siRNAs, ASOs chemistry has also evolved toward a new generation consisting of longer (15–20 nt) oligonucleotides with phosphorothioate modifications that make ASOs more resistant to endonucleases-mediated degradation (Stein et al. 1988; Allerson et al. 2005; Lima et al. 2007). To achieve optimal RNAse H-mediated knockdown of target RNAs combined with increased stability and resistance to degradation by endonucleases, other chemical modifications, such as 2'-O-methoxyethyl, have been introduced, reducing also off-target effects and conferring drug-like properties to ASOs (Yoo et al. 2004). These chimeric ASOs—commercially known as Gapmers ASOs—consist of RNA-DNA hybrids with 2'-O-methoxyethyl-modified sugar backbone. Further advances in ASO chemistry have led to the development of molecules with improved potency. Among them, the S-constrained ethyl (cEt) and the locked nucleic acids (LNAs) have demonstrated promising pharmacokinetics properties, which are fundament to translate in vitro and in vivo findings into the clinic.

Extending the chemical modifications to the whole sequence, the RNAse H-dependent activity is lost and knockdown of target RNAs cannot be achieved. However, such ASOs have been reported to efficiently work as splice-switching oligos (SSOs) to modulate pre-mRNA splicing through the binding to splicing regulatory elements (e.g., enhancers or silencer) (Hua et al. 2011). To date, FDA has approved the clinical use of splice-switching oligos only for spinal muscular atrophy (Geary et al. 2015).

3.3 *Steric Inhibition of LncRNA-Protein Interactions*

LncRNAs have been reported to act as scaffold molecules binding single proteins or entire complexes through their ability to form stable secondary and tertiary structures. Hence, mediating interactions among DNA, RNA, and proteins, they offer another therapeutic possibility, i.e., to interfere with their binding. One reliable approach is to target unique structural elements by using uniformly modified

ASOs (discussed above), which lack the capacity to trigger RNAse H-dependent degradation of RNAs, morpholinos, as well as small molecules able to cause a steric block of the potential (if known) contact inferface/s between lncRNA and proteins, finally resulting in a loss-of-function (Kloosterman et al. 2007).

The morpholinos—originally designed for developmental studies in zebrafish (Nasevicius and Ekker 2000; Moulton 2017)—were designed to block mRNA translation or to promote splice switching, similar to LNAs. Indeed, the first clinical approval for the use of morpholinos in humans was as splicing modulator of the dystrophin mRNA in patients with Duchenne muscular dystrophy (Aartsma-Rus et al. 2003; Alter et al. 2006; Goemans et al. 2011). Morpholinos are nonionic DNA analogs of about 25 nt in length capable to bind target mRNAs/pre-mRNAs (Moulton 2017; Staton and Giraldez, 2011). The main chemical modification is the replacement of sugars with methylenemorpholine rings and the presence of nonionic phosphorodiamidate bonds instead of anionic phosphates of DNA/RNAs (Blum et al. 2015). Hence, morpholinos can provide steric interference avoiding, or significantly suppressing, the binding of lncRNAs to RNA-binding proteins or to DNA targets (gene promoters or enhancers).

Another approach to target lncRNAs depends on their ability to form stable secondary/tertiary structures (Pegueroles and Gabaldón, 2016). In this regard, the development of experimental assays to map the secondary and/or tertiary structure of RNAs, such as selective 2′-hydroxylacylation by primer extension (SHAPE) (Wilkinson et al. 2006) and psoralen analysis of RNA interactions and structures (PARIS) (Lu et al. 2016), has been instrumental even for lncRNAs. Some clinical trials targeting bacterial/viral riboswitches with small-molecule inhibitors are underway (Howe et al. 2015), and it is plausible that the presence of similar structured elements in cancer-associated lncRNAs allows the use of small-molecule inhibitors also for this class of transcripts.

However, despite the use of oligonucleotides targeting lncRNAs in cancer has provided promising results, several caveats need to be considered. The introduction of chemical modifications has partially overcome the endonucleases-mediated degradation, improving overall the stability of these oligonucleotides, but one of the most relevant limits still depends on the activation of innate immune response. Indeed, it impedes the exogenous RNAs—including all the above-mentioned oligonucleotide types—to reach the cytosol, limiting their bioavailability (Crooke et al. 2017). Innate immunity has evolved to protect cells from potentially harmful viral RNAs, and Toll-like receptors (TLR) are the first cellular barrier against them. Upon target recognition, TLRs activate an intracellular cascade causing the massive production of type 1 interferon and other pro-inflammatory cytokines (Senn et al. 2005). Standard oligonucleotides, including siRNAs, containing a "CpG" motif can trigger such pro-inflammatory response activating TLR9, whereas ASOs have been specifically designed to minimize such responses by avoiding CpG motifs and using peculiar chemical modifications. Nonetheless, modified ASOs can still elicit—albeit at lower level—a pro-inflammatory response at high doses due to TLR-independent mechanisms (Senn et al. 2005). Indeed, although TLRs trigger a strong response to viral RNAs in the extracellular compartment or in the endosomes, they cannot sense viruses in the cytoplasm. The pathway of retinoic acid-inducible gene I-like (RIG-I) RNA helicases (RLHs) was identified as a TLR-independent mechanisms to detect intracellular viral infection (Yoneyama et al. 2004). The activation of these two pathways, other than initiating a pro-inflammatory harmful response, causes the entrapment of ASOs in the endosomal compartment, significantly reducing their bioavailability and their capacity to knockdown lncRNAs (Dowdy, 2017). Therefore, large-scale screenings to identify well-tolerated sequences (Kandimalla et al. 2013; Vollmer et al. 2004), as well as new strategies to conjugate oligonucleotides avoiding TLR activation and obtaining tissue-specific targeting, are under intensive investigation. Further studies to cope with the so-called "off-target" effects of these oligonucleotide-based approaches, as well as to evaluate their safety in humans, need to be considered before adopting such strategies in human therapies of cancers.

4 Preclinical Models: Old and New Tools to Explore LncRNA Targeting In Vivo

Preclinical models have been instrumental for investigating the role of lncRNAs in vivo and for addressing their potential as therapeutic targets and diagnostic/prognostic predictors. If, on the one hand, evolutionary differences between mice and humans represent obvious limitations to translating the research on mice in humans, on the other, genetic similarities between the two species render mice the most acceptable and feasible in vitro and in vivo model in biomedical research to investigate the mechanistic causes of many human diseases. Among the preclinical models useful in the context of lncRNA field, genetically engineered mouse models (GEMMs) recapitulate many aspects of human diseases, including cancers (Sharpless and Depinho 2006; Van Dyke and Jacks 2002). Thanks to the availability of mouse genome sequence and to advancements in genetic-engineering techniques, these models largely contributed to extend our knowledge of molecular, cellular, and anatomical changes occurring in tumorigenesis, also driving toward the improvement of therapeutic and diagnostic strategies (Walrath et al. 2010). Indeed, GEMMs have been useful to determine the effects of mutations, up- and downregulation of known or potential oncogene/tumor suppressor on tumor initiation/progression, and metastasis (Walrath et al. 2010; Singh et al. 2012). The technological improvements in conditional mutagenesis, in embryonic stem cells' manipulation, in gene knockdown (by RNAi), or in the use of programmable endonucleases and targeted transgenesis in zygotes have generated more precise mouse models leading to increased target specificity, experimental reproducibility and efficiency (Gurumurthy and Lloyd 2019). Beyond the large use of GEMMs for addressing the role of protein-coding genes in tumorigenesis—both by individual research efforts and systematic multiple-target approaches (Adams et al. 2004; Skarnes et al. 2004; To et al. 2004; Schnutgen et al. 2005; Hansen et al. 2008)—genetic KO of evolutionarily conserved lncRNAs (e.g., *H19, MALAT1, NEAT1*, Fig. 1) have been combined to oncogene-targeted GEMMs (Hao et al. 1993; Endo et al. 2013; Adriaens et al. 2016; Arun et al. 2016; Arun et al. 2018; Mello et al. 2017). For instance, *H19* knockout into Apc mouse model of colorectal cancer promotes polyps development and tumor onset, revealing a tumor-suppressive role (Yoshimizu et al. 2008; Gabory et al. 2009). Moreover, the overgrowth phenotype was observed in $H19^{\Delta 3}$ animals, also demonstrating that transgenic overexpression of *H19* rescued such phenotype through the modulation of several imprinted genes (Doucrasy et al. 1993). However, as previously discussed, *H19* has a potential dual role in tumorigenesis (Arun et al. 2018).

The use of GEMMs also revealed the dual role of *NEAT1* lncRNA in tumor formation, given that *Neat1* can modulate cancer formation—by dampening the activation of p53 mediated by oncogenes—and its KO in mice causes resistance to chemically induced skin cancer formation (Adriaens et al. 2016), whereas its loss increases tumorigenicity in the p53 null/mutant *Kras* mouse model of pancreatic ductal adenocarcinoma (Mello et al. 2017). Furthermore, genetic KO and ASO-mediated silencing in vivo for *Malat1* in the MMTV-PyMT mouse model of breast cancer has definitely clarified the contribution of *Malat1* lncRNA to metastasis formation (Arun et al. 2016).

For lncRNAs lacking evolutionary conservation, in vivo evaluation of their potential oncogenic/tumor-suppressive role has been mainly addressed by the use of cell line- or patient-derived xenograft models (CDX and PDX, respectively, Fig. 1). These represent humanized mouse models with a considerable value for preclinical evaluation and anticancer drug discovery. Specifically, human in vitro-cultured tumor cell lines (for CDXs) or fresh patient-derived tumors (for PDXs) are transplanted into nude mice, enabling to evaluate tumor growth and the effects of gene-specific targeting on tumorigenesis and/or the response to specific therapeutic treatments. However, CDXs rarely reflect original tissue architecture and require long-term in vitro culture of established cell lines, possibly introducing artifacts as increased homogeneity of tumor cells' population (Singh et al. 2012; Dong et al. 2010; Garber 2009). Theoretically, PDXs recapitulate tissue architecture more adequately, although they i) do not allow analyzing tumors at early stages of progression, ii) rarely metastasize, iii) require long times for establishment, and iv) have generally a low success rate (Kim et al. 2019; Garber 2009). Additionally, the intrinsic differences between mice and humans in terms of pharmacokinetics, combined to the incomplete tumor microenvironment in xenografts, can lead to results that do not fully reflect therapeutic responses to anticancer drugs used in clinical settings (Singh et al. 2012). One of the first CDX models used for studying in vivo the role of lncRNAs was generated for *H19* gene. Silencing of *H19* in human tumor cell lines and their xenograft transplantation into nude mice led to discover a role for *H19* lncRNA in cell proliferation and tumor growth. Indeed, its knockdown was reported to cause a marked abrogation of tumorigenicity capacity of transplanted cancer cells (Hao et al. 1993).

Other well-documented CDXs models for *HOTAIR*, *LUNAR* and *MALAT1* lncRNAs have been generated. Tail vein xenograft of *HOTAIR*-expressing gastric cancer cells (i.e., MKN74) in NOD/SCID/IL2Rγnull mice (Endo et al. 2013), and mammary fat pad and tail vein xenograft of *HOTAIR*-transduced breast cancer cells (MDA-MB-231 and SK-BR3) in SCID Beige mice (Gupta et al. 2010), demonstrated the role of *HOTAIR* in human tumorigenesis. Similarly, tail vein xenograft of human T-cell lymphoma cell line (i.e., CUTLL1) knockdown for *LUNAR* into sublethally irradiated immunodeficient Rag2null IL2Rγnull mice revealed the role of *LUNAR* in regulating tumor cells proliferation in T-cell acute lymphoblastic leukemia (Trimarchi et al. 2014). In addition, CDX models were also used to assess the role of conserved lncRNA in tumorigenesis. Among them, the xenograft of a squamous cell carcinoma cell line (EBC-1) knockdown for *MALAT1* in BALB/c nude mice was useful to study metastasis formation in lung cancer. Indeed, silencing of *MALAT1* by ASOs prevents metastasis formation after tumor implantation in mice, revealing the contribution of *MALAT1* lncRNA in metastatic process (Gutschner et al. 2013).

To further translate basic research on lncRNAs into preclinical settings, avoiding time-consuming cell culture propagation required for CDXs, PDXs models knockdown for specific lncRNAs, such as *SAMMSON and SCAT7,* have been recently developed, targeting cells by siRNAs or ASOs (Ali et al. 2018; Leucci et al. 2016). For instance, intravenous infusion with *SAMMSON*-targeting Gapmers significantly decreased cell proliferation and increased apoptosis in a PDX model of melanoma (Mel006, NMRI mice), and intra-tumor injection of Gapmers in PDX Mel010 suppressed tumor growth (Leucci et al. 2016). Similarly, injections of *SCAT7*-targeting LNAs in lung metastatic PDX mice models (NSG mice) significantly decreased tumor growth and volume (Ali et al. 2018). More recently, the development of patient-derived tumor organoids (PDTO) offers the potential to tailor the therapeutic approach on each patient's tumor, paving the way to personalized cancer therapies. Specifically, tissue-specific stem cells derived from adult human organs (including colon, stomach, liver, pancreas, and lung) are cultured in 3D conditions (e.g., on hydrogel with collagen or Matrigel) favoring cell proliferation and differentiation and originate self-organized progeny reflecting tissue architecture in vivo (Kim et al. 2019). However, although PTDO generation requires complex culture conditions—not yet efficiently established for all cancer types—they overcome some issues of mouse models, including the reduced time for their establishment, the presence of a tissue architecture and microenvironment, as well as maintaining of tumor heterogeneity (Arun et al. 2018, Granat et al. 2019). Moreover, PTDO offer both the possibility to analyze the effects of ex vivo knockdown and to validate in vivo the therapeutic response. PTDO are already revealing their great potential also for the manipulation of lncRNAs. Indeed, ASO-mediated *Malat1* silencing in MMTV-PyMT and Her2/neu-amplified mammary tumor 3D organoids inhibited branching ex vivo morphogenesis (Arun et al. 2016). Similarly, the knockdown of selected human Mammary Tumor-Associated RNAs (hMaTARs) in mammary organoids from MMTV-PyMT and MMTV-Neu-NDL mice markedly decreased cell proliferation, invasion, and organoid branching in cancer-specific context (Diermeier et al. 2016).

5 Concluding Remarks

The identification of lncRNAs' deregulation in tumors—especially by deep-sequencing approaches—and the subsequent understanding of their molecular functions using a combination of in vitro and in vivo models allowed a significant step forward in the understanding of lncRNA biology. Additionally, the rapid advancements in technologies and in experimental protocols have provided powerful and efficient tools to target lncRNAs in the context of human diseases, including cancers. In this regard, 2018 was nominated the year of RNA-based treatments, and since then, targeting lncRNAs in preclinical models of human cancers has shown very encouraging results.

Nonetheless, methodological and technological advancements are still needed to generate more realistic, cost- and time-effective, models to better address drug design on oncogenic lncRNAs and to boost the development of personalized therapeutic approaches for cancers (Day et al. 2015). Indeed, even though lncRNAs associated with cancer hallmarks have been reported and effectively targeted both in vitro cell and in vivo using mouse models, lncRNA-based approaches have not yet reached the clinical stage. Noteworthy, many lncRNAs are under evaluation as biomarkers, and others play relevant roles in the modulation of therapy responses in tumor cells (Pecero et al. 2019). At the time of writing, ten clinical trials (active, recruiting, or completed; https://clinicaltrials.gov) are evaluating lncRNAs as prognostic/diagnostic biomarkers in various cancer types (e.g., ovarian, breast, thyroid, lung, stomach cancer) or as predictors of chemotherapy efficacy (e.g., doxorubicin, cyclophosphamide, gemcitabine, and cisplatin in breast cancer).

To date, no clinical trials directly targeting lncRNAs in cancer are underway. However, we envision in the next few years a rapid expansion of this field, with patient-specific lncRNA profiling and targeting (by ASO or other small molecules) becoming routine in the clinical practice for cancer therapies.

References

Aartsma-Rus A, Janson AA, Kaman WE et al (2003) Therapeutic antisense-induced exon skipping in cultured muscle cells from six different DMD patients. Hum Mol Genet 12:907–914

Adams DJ, Biggs PJ, Cox T et al (2004) Mutagenic insertion and chromosome engineering resource (MICER). Nat Genet 36:867–871

Adriaens C, Standaert L, Barra J et al (2016) p53 induces formation of NEAT1 lncRNA-containing paraspeckles that modulate replication stress response and chemosensitivity. Nat Med 22:861–868

Ali MM, Akhade VS, Kosalai ST et al (2018) PAN-cancer analysis of S-phase enriched lncRNAs identifies oncogenic drivers and biomarkers. Nat Commun 9:883

Allerson CR, Sioufi N, Jarres R et al (2005) Fully 20-modified oligonucleotide duplexes with improved in vitro potency and stability compared to unmodified small interfering RNA. J Med Chem 48:901–904

Alter J, Lou F, Rabinowitz A et al (2006) Systemic delivery of morpholino oligonucleotide restores dystrophin expression bodywide and improves dystrophic pathology. Nat Med 12:175–177

Aparicio-Prat E, Arnan C, Sala I et al (2015) DECKO: single-oligo, dual-CRISPR deletion of genomic elements including long non-coding RNAs. BMC Genomics 16:846

Arun G, Diermeier S, Akerman M et al (2016) Differentiation of mammary tumors and reduction in metastasis upon Malat1 lncRNA loss. Genes Dev 30:34–51

Arun G, Diermeier SD, Spector DL (2018) Therapeutic targeting of long non-coding RNAs in Cancer. Trends Mol Med 24:257–277

Barsyte-Lovejoy D, Lau SK, Boutros PC et al (2006) The c-Myc oncogene directly induces the H19 noncoding RNA by allele-specific binding to potentiate tumorigenesis. Cancer Res 66:5330–5337

Bartolomei MS, Zemel S, Tilghman SM (1991) Parental imprinting of the mouse H19 gene. Nature 351:153–155

Bejerano G, Pheasant M, Makunin I et al (2004) Ultraconserved elements in the human genome. Science 304:1321–1325

Bergmann JH, Spector DL (2014) Long non-coding RNAs: modulators of nuclear structure and function. Curr Opin Cell Biol 26:10–18

Beroukhim R, Mermel CH, Porter D et al (2010) The landscape of somatic copy number alteration across human cancers. Nature 463:899–905

Berteaux N, Lottin S, Monté D et al (2005) H19mRNA-like noncoding RNA pro- motes breast cancer cell proliferation through positive control by E2F1. J Biol Chem 280:29625–29636

Blum M, De Robertis EM, Wallingford JB et al (2015) Morpholinos: antisense and sensibility. Dev Cell 35:145–149

Braconi C, Kogure T, Valeri N et al (2011) microRNA-29 can regulate expression of the long non-coding RNA gene MEG3 in hepatocellular cancer. Oncogene 30:4750–4756

Consortium EP, Birney E, Stamatoyannopoulos JA et al (2007) Identification and analysis of functional elements in 1% of the human genome by the ENCODE pilot project. Nature 447:799–816

Crooke ST, Wang S, Vickers TA et al (2017) Cellular uptake and trafficking of antisense oligonucleotides. Nat Biotechnol 35:230–237

Davis RL, Weintraub H, Lassar AB (1987) Expression of a single transfected cDNA converts fibroblasts to myoblasts. Cell 51:987–1000

Day CP, Merlino G, Van Dyke T (2015) Preclinical mouse cancer models: a maze of opportunities and challenges. Cell 163:39–53

Diermeier SD, Chang KC, Freier SM et al (2016) Mammary tumor-associated RNAs impact tumor cell proliferation, invasion, and migration. Cell Rep 17:261–274

Dimitrova N, Zamudio JR, Jong RM et al (2014) LincRNA-p21 activates p21 in *cis* to promote Polycomb target gene expression and to enforce the G1/S checkpoint. Mol Cell 54:777–790

Dong X, Guan J, English JC (2010) Patient-derived first generation xenografts of non-small cell lung cancers: promising tools for predicting drug responses for personalized chemotherapy. Clin Cancer Res 16:1442–1451

Doose G, Haake A, Bernhart SH et al (2015) MINCR is a MYC-induced lncRNA able to modulate MYC's transcriptional network in Burkitt lymphoma cells. Proc Natl Acad Sci U S A 112: E5261–E5270

Doucrasy S, Coll J, Barrois M et al (1993) Expression of the human fetal bac h19 gene in invasive cancers. Int J Oncol 2:753–758

Dowdy SF (2017) Overcoming cellular barriers for RNA therapeutics. Nat Biotechnol 35:222–229

Dugimont T, Montpellier C, Adriaenssens E et al (1998) The H19 TATA-less promoter is efficiently repressed by wild-type tumor suppressor gene product p53. Oncogene 16:2395–2401

Eissmann M, Gutschner T, Hämmerle M et al (2012) Loss of the abundant nuclear non- coding RNA MALAT1 is compatible with life and development. RNA Biol 9:1076–1087

Endo H, Shiroki T, Nakagawa T (2013) Enhanced expression of long non-coding RNA HOTAIR is associated with the development of gastric cancer. PLoS One 8:e77070

Esposito R, Esposito D, Pallante P et al (2019a) Oncogenic properties of the antisense lncRNA COMET in BRAF- and RET-driven papillary thyroid carcinomas. Cancer Res 79:2124–2135
Esposito R, Bosch N, Lanzós A et al (2019b) Hacking the Cancer genome: profiling therapeutically actionable long non-coding RNAs using CRISPR-Cas9 screening. Cancer Cell 35:545–557
Flockhart RJ, Webster DE, Qu K et al (2012) BRAFV600E remodels the melanocyte transcriptome and induces BANCR to regulate melanoma cell migration. Genome Res 22:1006–1014
Gabory A, Ripoche MA, Le Digarcher A et al (2009) H19 acts as a trans regulator of the imprinted gene network controlling growth in mice. Development 136:3413–3421
Garber K (2009) From human to mouse and back: 'tumorgraft' models surge in popularity. J Natl Cancer Inst 101:6–8
Geary RS, Baker BF, Crooke ST (2015) Clinical and preclinical pharmacokinetics and pharmacodynamics of mipomersen (Kynamro1): a second-generation antisense oligonucleotide inhibitor of apolipoprotein B. Clin Pharmacokinet 54:133–146
Geng YJ, Xie SL, Li Q et al (2011) Large intervening non-coding RNA HOTAIR is associated with hepatocellular carcinoma progression. J Int Med Res 39:2119–2128
Gibb EA, Vucic EA, Enfield KS et al (2016) Human cancer long non-coding RNA transcriptomes. PLoS One 6:e25915
Gilbert LA, Horlbeck MA, Adamson B et al (2014) Genome-scale CRISPR-mediated control of gene repression and activation. Cell 159:647–661
Goemans NM, Tulinius M, Van Den Akker JT et al (2011) Systemic administration of PRO051 in Duchenne's muscular dystrophy. N Engl J Med 364:1513–1522
Gou Q, Gao L, Nie X et al (2018) Long noncoding RNA AB074169 inhibits cell proliferation via modulation of KHSRP-mediated CDKN1a expression in papillary thyroid carcinoma. Cancer Res 78:4163–4174
Granat LM, Kambhampati O, Klosek S et al (2019) The promises and challenges of patient-derived tumor organoids in drug development and precision oncology. Animal Model Exp Med 2:150–161
Gupta RA, Shah N, Wang KC et al (2010) Long non-coding RNA HOTAIR reprograms chromatin state to promote cancer metastasis. Nature 464:1071–1076
Gurumurthy CB, Lloyd KCK (2019) Generating mouse models for biomedical research: technological advances. Dis Model Mech 12:dmm029462
Gutschner T, Hämmerle M, Eissmann M et al (2013) The noncoding RNA MALAT1 is a critical regulator of the metastasis phenotype of lung cancer cells. Cancer Res 73:1180–1189
Guttman M, Amit I, Garber M et al (2009) Chromatin signature reveals over a thousand highly conserved large non-coding RNAs in mammals. Nature 458:223–227
Hannon GJ, Rossi JJ (2004) Unlocking the potential of the human genome with RNA interference. Nature 431:371–378
Hansen GM, Markesich DC, Burnett MB et al (2008) Large-scale gene trapping in C57BL/6N mouse embryonic stem cells. Genome Res 18:1670–1679
Hao Y, Crenshaw T, Moulton T et al (1993) Tumour-suppressor activity of H19 RNA. Nature 365:764–767
Howe JA, Wang H, Fischmann TO et al (2015) Selective small-molecule inhibition of an RNA structural element. Nature 526:672–677
Hua Y, Sahashi K, Rigo F et al (2011) Peripheral SMN restoration is essential for long-term rescue of a severe spinal muscular atrophy mouse model. Nature 478:123–126
Huarte M (2015) The emerging role of lncRNAs in cancer. Nat Med 21:1253–1261
Huarte M, Guttman M, Feldser D et al (2010) A large intergenic noncoding RNA induced by p53 mediates global gene repression in the p53 response. Cell 142:409–419
Hudson WH, Pickard MR, De Vera IM et al (2014) Conserved sequence-specific lincRNA-steroid receptor interactions drive transcriptional repression and direct cell fate. Nat Commun 5:5395
Iyer MK, Niknafs YS, Malik R et al (2015) The landscape of long noncoding RNAs in the human transcriptome. Nat Genet 47:199–208

Jadaliha M, Zong X, Malakar P et al (2016) Functional and prognostic significance of long non-coding RNA MALAT1 as a metastasis driver in ER negative lymph node negative breast cancer. Oncotarget 7:40418–40436

Ji P, Diederichs S, Wang W et al (2003) MALAT-1, a novel noncoding RNA, and thymosin β4 predict metastasis and survival in early-stage non-small cell lung cancer. Oncogene 22:8031–8041

Jiang J, Azevedo-Pouly AC, Redis RS et al (2016) Globally increased ultraconserved noncoding RNA expression in pancreatic adenocarcinoma. Oncotarget 7:53165–53177

Jin Y, Feng SJ, Qiu S et al (2017) LncRNA MALAT1 promotes proliferation and metastasis in epithelial ovarian cancer via the PI3K-AKT pathway. Eur Rev Med Pharmacol Sci 21:3176–3184

Kallen AN, Zhou XB, Xu J et al (2013) The imprinted H19 lncRNA antagonizes Let-7 microRNAs. Mol Cell 52:101–112

Kandimalla ER, Bhagat L, Wang D et al (2013) Design, synthesis and biological evaluation of novel antagonist compounds of toll-like receptors 7, 8 and 9. Nucleic Acids Res 41:3947–3961

Khalil AM, Guttman M, Huarte M et al (2009) Many human large intergenic noncoding RNAs associate with chromatin-modifying complexes and affect gene expression. Proc Natl Acad Sci U S A 106:11667–11672

Khurana E, Fu Y, Colonna V et al (2013) Integrative annotation of variants from 1092 humans: application to cancer genomics. Science 342:1235587

Khvorova A, Watts JK (2017) The chemical evolution of oligonucleotide therapies of clinical utility. Nat Biotechnol 35:238–248

Kim T, Cui R, Jeon YJ et al (2014) Long-range interaction and correlation between MYC enhancer and oncogenic long noncoding RNA CARLo-5. Proc Natl Acad Sci U S A 111:4173–4178

Kim M, Mun H, Sung CO et al (2019) Patient-derived lung cancer organoids as in vitro cancer models for therapeutic screening. Nat Commun 10:3991

Kino T, Hurt DE, Ichijo T et al (2010) Noncoding RNA gas5 is a growth arrestand starvation-associated repressor of the glucocorticoid receptor. Sci Signal 3:ra8

Kloosterman WP, Lagendijk AK, Ketting RF et al (2007) Targeted inhibition of miRNA maturation with morpholinos reveals a role for miR-375 in pancreatic islet development. PLoS Biol 5:e203

Koch L (2017) Functional genomics: screening for lncRNA function. Nat Rev Genet 18:70–70

Kogo R, Shimamura T, Mimori K et al (2011) Long noncoding RNA HOTAIR regulates polycomb-dependent chromatin modification and is associated with poor prognosis in colorectal cancers. Cancer Res 71:6320–6226

Konermann S, Brigham MD, Trevino AE et al (2015) Genome-scale transcriptional activation by an engineered CRISPR-Cas9 complex. Nature 517:583–588

Lee S, Kopp F, Chang TC et al (2016) Noncoding RNA NORAD regulates genomic stability by sequestering PUMILIO proteins. Cell 164:69–80

Leucci E, Vendramin R, Spinazzi M et al (2016) Melanoma addiction to the long non-coding RNA SAMMSON. Nature 531:518–522

Li R, Zhang L, Jia L et al (2014) Long non-coding RNA BANCR promotes proliferation in malignant melanoma by regulating MAPK pathway activation. PLoS One 9:e100893

Liao T, Qu N, Shi RL et al (2017) BRAF-activated LncRNA functions as a tumor suppressor in papillary thyroid cancer. Oncotarget 8:238–247

Lima W et al (2007) The RNase H mechanism. In: Crooke ST (ed) Antisense drug technology. Principles, strategies, and applications, 2nd edn. CRC Press, New York, pp 47–74

Ling H, Spizzo R, Atlasi Y et al (2013) CCAT2, a novel noncoding RNA mapping to 8q24, underlies metastatic progression and chromosomal instability in colon cancer. Genome Res 23:1446–1461

Lu Z, Zhang QC, Lee B et al (2016) RNA duplex map in living cells reveals higher-order transcriptome structure. Cell 165:1267–1279

Marini A, Lena AM, Panatta E et al (2017) Ultraconserved long non-coding RNA uc.63 in breast cancer. Oncotarget 8:35669–35680

Matouk IJ, Mezan S, Mizrahi A et al (2010) The oncofetal H19 RNA connection: hypoxia, p53 and cancer. Biochim Biophys Acta 1803:443–451

Mello SS, Sinow C, Raj N et al (2017) Neat1 is a p53-inducible lincRNA essential for transformation suppression. Genes Dev 31:1095–1108

Miyoshi N, Wagatsuma H, Wakana S et al (2000) Identification of an imprinted gene, *Meg3/Gtl2* and its human homologue *MEG3*, first mapped on mouse distal chromosome 12 and human chromosome 14q. Genes Cells 5:211–220

Morrissey DV, Lockridge JA, Shaw L et al (2005) Potent and persistent in vivo anti- HBV activity of chemically modified siRNAs. Nat Biotechnol 23:1002–1007

Moulton JD (2017) Using morpholinos to control gene expression. Curr. Protoc. Nucleic Acid Chem 68:1–4

Mourtada-Maarabouni M, Pickard MR, Hedge VL et al (2009) GAS5, a non-protein-coding RNA, controls apoptosis and is downregulated in breast cancer. Oncogene 28:195–208

Nagoshi H, Taki T, Hanamura I et al (2012) Frequent PVT1 rearrangement and novel chimeric genes PVT1–NBEA and PVT1–WWOX occur in multiple myeloma with 8q24 abnormality. Cancer Res 72:4954–4962

Nakagawa S et al (2012) Malat1 is not an essential component of nuclear speckles in mice. RNA 18:1487–1499

Nasevicius A, Ekker SC (2000) Effective targeted gene 'knockdown' in zebrafish. Nat Genet 26:216–220

Nik-Zainal S, Davies H, Staaf J et al (2016) Landscape of somatic mutations in 560 breast cancer whole-genome sequences. Nature 534:47–54

Niland CN, Merry CR, Khalil AM (2012) Emerging roles for long non-coding RNAs in Cancer and neurological disorders. Front Genet 3:25

Olivieri M, Ferro M, Terreri S et al (2016) Long non-coding RNA containing ultraconserved genomic region 8 promotes bladder cancer tumorigenesis. Oncotarget 7:20636–20654

Pang L, Li Q, Zhang Y et al (2018) Transcribed ultraconserved noncoding RNA uc.160 acts as a negative regulator in gastric cancer. Am J Transl Res 10:2822–2833

Pecero ML, Salvador-Bofill J, Molina-Pinelo S (2019) Long non-coding RNAs as monitoring tools and therapeutic targets in breast cancer. Cell Oncol 42:1–12

Pegueroles C, Gabaldón T (2016) Secondary structure impacts patterns of selection in human lncRNAs. BMC Biol 14:60

Pickard M, Williams G (2015) Molecular and cellular mechanisms of action of tumour suppressor GAS5 LncRNA. Genes 6:484–499

Prensner JR, Chen W, Han S et al (2014) The long non-coding RNA PCAT-1 promotes prostate cancer cell proliferation through cMyc. Neoplasia 16:900–908

Rinn JL, Euskirchen G, Bertone P et al (2003) The transcriptional activity of human chromosome 22. Genes Dev 17:529–540

Rinn JL, Kertesz M, Wang et al (2007) Functional demarcation of active and silent chromatin domains in human HOX loci by noncoding RNAs. Cell 129:1311–1323

Schneider C, King RM, Philipson L et al (1988) Genes specifically expressed at growth arrest of mammalian cells. Cell 54:787–793

Schnutgen F, De-Zolt S, Van Sloun P et al (2005) Genome wide production of multipurpose alleles for the functional analysis of the mouse genome. Proc Natl Acad. Sci USA 102:7221–7226

Schuster-Gossler K, Bilinski P, Sado T et al (1998) The mouse *Gtl2* gene is differentially expressed during embryonic development, encodes multiple alternatively spliced transcripts, and may act as an RNA. Dev Dyn 1212:214–228

Seiler J, Breinig M, Caudron-Herger M et al (2017) The lncRNA VELUCT strongly regulates viability of lung cancer cells despite its extremely low abundance. Nucleic Acids Res 45:5458–5469

Sekino Y, Sakamoto N, Goto K et al (2017) Transcribed ultraconserved region Uc.63+ promotes resistance to docetaxel through regulation of androgen receptor signaling in prostate cancer. Oncotarget 8:94259–94270

Senn JJ, Burel S, Henry SP (2005) Non-CpG-containing antisense 2′-methoxyethyl oligonucleotides activate a proinflammatory response independent of toll-like receptor 9 or myeloid differentiation factor 88. J Pharmacol Exp Ther 314:972–979

Shalem O, Sanjana NE, Hartenian E et al (2014) Genome-scale CRISPR-Cas9 knockout screening in human cells. Science 343:84–87

Sharma S, Mandal P, Sadhukhan T et al (2015) Bridging links between long noncoding RNA HOTAIR and HPV Oncoprotein E7 in cervical Cancer pathogenesis. Sci Rep 5:11724

Sharpless NE, Depinho RA (2006) The mighty mouse: genetically engineered mouse models in cancer drug development. Nat Rev 5:741–754

Shtivelman E, Bishop JM (1989) The PVT gene frequently amplifies with MYC in tumor cells. Mol Cell Biol 9:1148–1154

Singh M, Murriel CL, Johnson L (2012) Genetically engineered mouse models: closing the gap between preclinical data and trial outcomes. Cancer Res 72:2695–2700

Skarnes WC, Von Melchner H, Wurst W et al (2004) A public gene trap resource for mouse functional genomics. Nat Genet 36:543–544

Somarowthu S, Legiewicz M, Chillón I et al (2013) HOTAIR forms an intricate and modular secondary structure. Mol Cell 58:353–361

Staton AA, Giraldez AJ (2011) Use of target protector morpholinos to analyze the physiological roles of specific miRNA–mRNA pairs *in vivo*. Nat Protoc 6:2035–2049

Stein CA, Castanotto D (2017) FDA-approved oligonucleotide therapies in 2017. Mol Ther 25:1069–1075

Stein CA, Subasinghe C, Shinozuka K et al (1988) Physicochemical properties of phosphorothioate oligodeoxynucleotides. Nucleic Acids Res 16:3209–3221

Sun M, Liu XH, Wang KM et al (2014) Downregulation of BRAF activated non-coding RNA is associated with poor prognosis for non-small cell lung cancer and promotes metastasis by affecting epithelial-mesenchymal transition. Mol Cancer 13:68

Taft RJ, Pang KC, Mercer TR et al (2010) Non-coding RNAs: regulators of disease. J Pathol 220:126–139

Terracciano D, Terreri S, de Nigris F et al (2017) The role of a new class of long noncoding RNAs transcribed from ultraconserved regions in cancer. Biochim Biophys Acta Rev Cancer 1868:449–455

Terreri S, Durso M, Colonna V et al (2016) New cross-talk layer between Ultraconserved non-coding RNAs, MicroRNAs and Polycomb protein YY1 in bladder Cancer. Genes 7:127

Tichon A, Gil N, Lubelsky Y et al (2016) A conserved abundant cytoplasmic long noncoding RNA modulates repression by Pumilio proteins in human cells. Nat Commun 7:1–10

To C, Epp T, Reid T et al (2004) The Centre for Modeling Human Disease Gene Trap resource. Nucleic Acids Res 32:D557–D559

Trimarchi T, Bilal E, Ntziachristos P et al (2014) Genome-wide mapping and characterization of notch-regulated long noncoding RNAs in acute leukemia. Cell 158:593–606

Tsai MC, Manor O, Wan Y et al (2010) Long noncoding RNA as modular scaffold of histone modification complexes. Science 329:689–693

Tsang WP, Ng EK, Ng SS et al (2009) Oncofetal H19-derived miR-675 regulates tumor suppressor RB in human colorectal cancer. Carcinogenesis 31:350–358

Tseng YY, Moriarity BS, Gong W et al (2014) PVT1 dependence in cancer with MYC copy-number increase. Nature 512:82–86

Van Dyke T, Jacks T (2002) Cancer modeling in the modern era: progress and challenges. Cell 108:135–144

Vannini I, Wise PM, Challagundla KB et al (2017) Transcribed ultraconserved region 339 promotes carcinogenesis by modulating tumor suppressor microRNAs. Nat Commun 8:1801

Vendramin R, Verheyden Y, Ishikawa H et al (2018) SAMMSON fosters cancer cell fitness by concertedly enhancing mitochondrial and cytosolic translation. Nat Struct Mol Biol 25:1035–1046

Vollmer J, Weeratna RD, Jurk M et al (2004) Oligodeoxynucleotides lacking CpG dinucleotides mediate toll-like receptor 9 dependent T helper type 2 biased immune stimulation. Immunology 113:212–223

Walrath JC, Hawes JJ, Van Dyke T et al (2010) Genetically engineered mouse models in cancer research. Adv Cancer Res 106:113–164

Wang P, Ren Z, Sun P (2012) Overexpression of the long non-coding RNA MEG3 impairs *in vitro* glioma cell proliferation. J Cell Biochem 113:1868–7184

Weidle UH, Birzele F, Kollmorgen G et al (2017) Long non-coding RNAs and their role in metastasis. Cancer Genomics Proteomics 14:143–160

Wilkinson KA, Merino EJ, Weeks KM (2006) Selective 2′-hydroxyl acylation analyzed by primer extension (SHAPE): quantitative RNA structure analysis at single nucleotide resolution. Nat Protoc 1:1610–1616

Xue S, Jiang SQ, Li QW et al (2018) Decreased expression of BRAF-activated long non-coding RNA is associated with the proliferation of clear cell renal cell carcinoma. BMC Urol 18:79

Yan X, Hu Z, Feng Y et al (2015) (2015) comprehensive genomic characterization of long non-coding RNAs across human cancers. Cancer Cell 28:529–540

Yang Z, Zhou L, Wu LM et al (2011) Overexpression of long non-coding RNA HOTAIR predicts tumor recurrence in hepatocellular carcinoma patients following liver transplantation. Ann Surg Oncol 18:1243–1250

Yang F, Abdelmohsen K, Srikantan S et al (2014) Reciprocal regulation of HIF-1alpha and lincRNA-p21 modulates the Warburg effect. Mol Cell 53:88–100

Yoneyama M, Kikuchi M, Natsukawa T et al (2004) The RNA helicase RIG-I has an essential function in double-stranded RNA-induced innate antiviral responses. Nat Immunol 5:730–737

Yoo BH, Bochkareva E, Bochkarev A et al (2004) 2'-O-methyl-modified phosphorothioate antisense oligonucleotides have reduced non-specific effects *in vitro*. Nucleic Acids Res 32:2008–2016

Yoon J-H, Abdelmohsen K, Srikantan S et al (2012) LincRNA-p21 suppresses target mRNA translation. Mol Cell 47:648–655

Yoshimizu T, Miroglio A, Ripoche MA et al (2008) The H19 locus acts in vivo as a tumor suppressor. Proc Natl Acad Sci U S A 105:12417–12422

Zamecnik PC, Stephenson ML (1978) Inhibition of Rous sarcoma virus replication and cell transformation by a specific oligodeoxynucleotide. Proc Natl Acad Sci U S A 75:280–284

Zhai H, Fesler A, Schee K et al (2013) Clinical significance of long intergenic noncoding RNA-p21 in colorectal cancer. Clin Colorectal Cancer 12:261–266

Zhang X, Zhou Y, Mehta KR et al (2003) A pituitary-derived MEG3 isoform functions as a growth suppressor in tumor cells. J Clin Endocrinol Metab 88:5119–5126

Zhang X, Gejman R, Mahta A et al (2010) Maternally expressed gene 3, an imprinted noncoding RNA gene, is associated with meningioma pathogenesis and progression. Cancer Res 70:2350–2358

Zhang B, Arun G, Mao YS et al (2012) The lncRNA Malat1 is dispensable for mouse development but its transcription plays a cis-regulatory role in the adult. Cell Rep 2:111–123

Zhang D, Zhang G, Hu X et al (2017) Oncogenic RAS regulates long non-coding RNA Orilnc1 in human cancer. Cancer Res 77:3745–3757

Zhou Y, Zhong Y, Wang Y et al (2007) Activation of p53 by MEG3 non-coding RNA. J Biol Chem 282:24731–24742

Zhou Y, Zhang X, Klibanski A (2012) MEG3 noncoding RNA: a tumor suppressor. J Mol Endocrinol 48:R45–R53

Zhu S, Li W, Liu J et al (2016) Genome-scale deletion screening of human long non-coding RNAs using a paired-guide RNA CRISPR-Cas9 library. Nat Biotechnol 34:1279–1286

Long Noncoding RNAs in Non-Small Cell Lung Cancer: State of the Art

Panagiotis Paliogiannis, Valentina Scano, Arduino Aleksander Mangoni, Antonio Cossu, and Giuseppe Palmieri

Contents

Abstract Lung cancer is one of the most common malignancies worldwide. Despite a significant amount of basic and clinical research, mortality rates remain extremely high, especially for patients affected by advanced stage disease. Recently, new molecules playing several roles in the pathogenesis, diagnosis, and, potentially,

P. Paliogiannis (✉) · V. Scano · A. Cossu
Department of Medical, Surgical, and Experimental Sciences, University of Sassari, Sassari, Italy
e-mail: ppaliogiannis@uniss.it; cossu@uniss.it

A. A. Mangoni
Discipline of Clinical Pharmacology, College of Medicine and Public Health, Flinders University and Flinders Medical Centre, Flinders Drive, Bedford Park, SA, Australia

Medizinische Fakultät Carl Gustav Carus, Technische Universität Dresden, Dresden, Germany
e-mail: arduino.mangoni@flinders.edu.au

G. Palmieri
Unit of Cancer Genetics, Institute Biomolecular Chemistry, CNR, Sassari, Italy

S. Jurga, J. Barciszewski (eds.), *The Chemical Biology of Long Noncoding RNAs*, RNA Technologies 11, https://doi.org/10.1007/978-3-030-44743-4_12

clinical management of lung cancer are under investigation, including noncoding fragments of the human genome, also known as noncoding RNAs (ncRNAs). NcRNAs are commonly divided into two categories according to their size. The first category includes small ncRNAs, such as the recently discovered miRNAs, siRNAs, and the classical cellular RNAs (ribosomal, transfer, and other RNAs). Noncoding RNAs greater than 200 nucleotides represent a further category that includes long noncoding RNAs (lncRNAs). LncRNAs have numerous biological and pathophysiological effects. Numerous studies have recently investigated their involvement in the oncogenesis and the progression of pulmonary malignancies. In this chapter, we summarize the current knowledge regarding the role of lncRNAs in the pathogenesis, diagnosis, and clinical management of non-small cell lung cancer.

Keywords Lung · Cancer · NSCLC · LncRNA · Biomarkers

1 Introduction

Long noncoding RNAs (lncRNAs) are defined as autonomously transcribed non-coding RNAs that are longer than 200 nucleotides and have minimal coding potential. By contrast, noncoding transcripts with less than 200 nucleotides are defined as small noncoding RNAs (sncRNAs), which include micro-RNAs (miRNAs), small interfering RNAs (siRNAs), transfer RNAs (tRNAs), ribosomal RNAs (rRNAs), and other classes of small RNAs. For a long time, the DNA sequence of lncRNAs was considered "junk DNA." Since the early 1990s, thousands of lncRNAs have been discovered and investigated; however, their exact number and role in human physiology and pathology are poorly understood. Recently, Hon et al. (2017) integrated multiple transcript collections using FANTOM5 cap analysis of gene expression (CAGE) data to generate a comprehensive atlas of 27,919 human lncRNA genes with high-confidence 5′ ends and expression profiles across 1829 samples from key human primary cell types and tissues.

On the basis of their extension and relation with coding genes on the DNA strands, lncRNAs can be broadly classified as genic (overlapping a protein-coding transcript at one or more nucleotides), nested (contained entirely within protein-coding transcripts), and intergenic (not overlapping a protein-coding transcript); other subtypes describe particular conditions (containing, overlapping, multiple relationships, etc.) (Ransohoff et al. 2018). In accordance with their level of activity, lncRNAs act at the transcriptional, posttranscriptional, and epigenetic level. At the transcriptional level, they have several functions including acting as decoys to disrupt the binding of transcriptional factors with promoters of target genes, altering the localization of transcriptional factors in the genome, competing with endogenous RNA, and forming scaffolds with DNA and proteins. At the posttranscriptional level, they modulate directly or indirectly the effects of micro-RNAs (miRNAs) on target genes and regulate the alternative splicing of mRNA. At the epigenetic level, they interact with proteins involved in histone modifications, regulate DNA methylation in promoter regions, and interact with chromatin modification complexes (Wei and Zhou 2016).

Finally, on the basis of their specific molecular functions, lncRNAs act as molecular signal transducers, decoys, guides for ribonucleoprotein complex, scaffolds, and as sponge to sequester miRNAs (Peng et al. 2018). As signaling molecules, they serve as spatiotemporal indicators of gene regulation that reflect the biological effects of transcription factors (TFs) or signaling pathways; as decoys, they sequester TFs and other proteins away from chromatin or into nuclear subdomains; as guides, they recruit RNA-binding proteins to target genes; and as scaffolds, they recruit several proteins to form complexes with specific biological roles.

The reported functions have been described in several physiologic conditions and pathologies, including cancer (Palmieri et al. 2017). In particular, during the last decade, lncRNAs have been studied in the context of lung cancer, in order to better understand their roles in pulmonary carcinogenesis, and their potential application either as new therapeutic targets or as biomarkers for early diagnosis, prognosis, and therapy monitoring. This chapter provides an overview of the state of the art of current research on lncRNAs in the pathophysiology and clinical management of non-small cell lung cancer.

2 Lung Cancer: Current Status

Lung cancer is one of the most common malignancies and the leading cause of cancer-related deaths worldwide. In 2018, the International Agency for Research on Cancer (IARC) observatory estimated approximately 2,100,000 new cases and more than 1,700,000 deaths, a significant increase in comparison with previous estimations (Paliogiannis et al. 2013). The narrow gap between incidence and mortality rates highlights the challenges with early diagnosis and improving survival rates, especially in patients with advanced stage disease. Cancer Research UK reported recently that the 1-year overall survival rate is 32% for lung cancer patients, while the 5-year survival rate is around 10%. These figures suggest that there is a long way, in terms of basic and clinical research, to improve lung cancer survival rates.

Lung cancer includes a wide range of malignancies, which are broadly divided in small-cell lung cancer (SCLC) and non-small cell lung cancer (NSCLC). The former is biologically, pathologically, and clinically different from other subtypes as it is characterized by aggressive behavior, early lymphatic and distant metastasis, and a high responsiveness to chemotherapy (Fig. 1). NSCLC, the focus of this chapter, comprises several further subtypes; the most common are adenocarcinoma (~50%), squamous cell carcinoma (~25%), and large cell carcinomas (~10%). Squamous cell carcinoma was the most common histotype until the 1980s when it was superseded

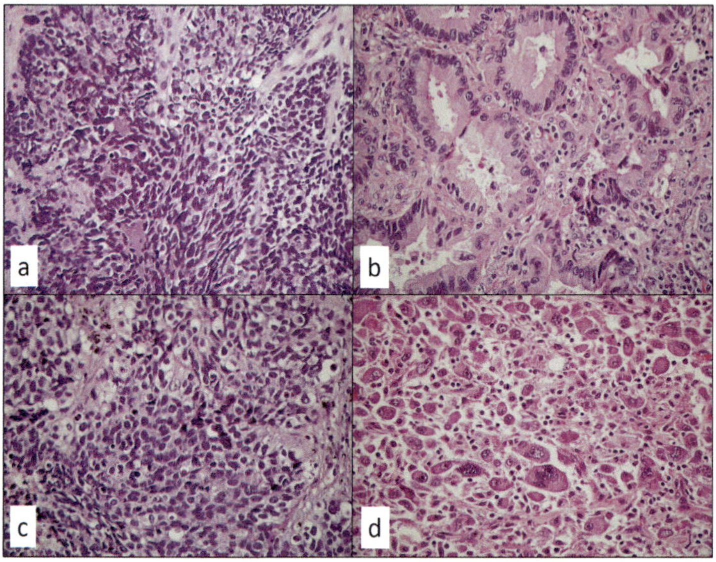

Fig. 1 The main histological subtypes of lung cancer. (**a**) Small-cell lung cancer, (**b**) adenocarcinoma, (**c**) squamous cell carcinoma, and d) large cell carcinoma. All samples are stained with hematoxylin and eosin and magnified at 40x

by adenocarcinoma. This might be explained by changes in smoking habits, particularly changes in the characteristics of cigarettes, increased puff volume, increased nitrate levels, and higher smoking incidence in women. Other than their morphological differences (Fig. 1), NSCLC subtypes show consistent differences in biological and clinical behavior, as well as different responses to current therapies, such as the recently introduced targeted agents.

Targeted therapies represent the most important innovation in the treatment of lung cancer over the last few years, also considering that surgical resection is an option in relatively a few cases. In 2004, activating mutations within the kinase domain of the epidermal growth factor receptor (EGFR) gene were discovered in lung adenocarcinomas, suggesting that these tumors were highly sensitive to tyrosine kinase inhibitors (TKIs), a class of agents that selectively inhibits the *EGFR* molecular pathway (Paliogiannis et al. 2015). TKIs such as gefitinib and erlotinib have been shown to significantly improve the clinical outcomes of approximately 40–60% Asian and 12–16% Caucasian patients with adenocarcinomas harboring *EGFR* mutations, with survival rates that were nearly twofold when compared to traditional chemotherapy (Shi et al. 2015). However, the increased frequency of resistance to TKIs reduced the initial enthusiasm around these agents, prompting at the same time further research to discover novel molecular targets and molecules with greater therapeutic efficacy and lower resistance rates. For example, additional agents, such as osimertinib, alectinib, and crizotinib, that target genetic alterations of the ALK and ROS1 genes have been introduced in clinical practice; these drugs show a better ability to overcome resistance mechanisms in comparison to older medications. Other molecular targets such as KRAS, BRAF, HER2, MET, and RET, as well as their molecular pathways, are currently being investigated for the development of novel targeted agents (Colombino et al. 2019).

Immunotherapy with immune checkpoint inhibitors (ICIs) has been introduced more recently in NSCLC, further improving survival outcomes in patients with adenocarcinomas a well as in other NSCLC subtypes. Nivolumab, a monoclonal antibody targeting programmed death 1 protein (PD-1), was the first drug approved for advanced NSCLC not responding to platinum-based chemotherapy. Nivolumab showed durable responses in 10 (37%) of 27 confirmed responders with squamous NSCLC and 19 (34%) of 56 with non-squamous NSCLC that had ongoing response after a minimum follow-up of 2 years (Horn et al. 2017). Other agents, pembrolizumab and atezolizumab, are also gradually replacing standard chemotherapy as second- and first-line treatment in pan-negative advanced NSCLC. For ICIs, the expression of PD-L1 is currently considered as a major predictive factor for immunotherapy, despite some limitations in accurately selecting patients who would respond to treatment. Active research is ongoing to overcome such limitations, including the combination of immunotherapy drugs with different mechanisms of action, and their associations with targeted therapies.

3 LncRNAs in the Pathogenesis of NSCLC

A large number of studies reported abnormal patterns of expression of lncRNAs in NSCLC (Table 1). Specifically, these alterations promote molecular pathways that are involved in proliferation and migration, either by influencing second messengers' activation or triggering the transcription of growth factors. In this setting, lncRNAs are upregulated in neoplastic tissue and, in some studies, also in blood or plasma; their concentration may vary depending on the stage of the disease (presence of metastasis or tumor size). On the other hand, downregulated lncRNAs have also been described in tumor tissue. These lncRNAs promote cellular apoptosis through activation of proapoptotic genes or sequestration of pro-oncogenic molecules. Therefore, similar to protein-coding genes, lncRNAs can be classified as oncogenic or tumor suppressors.

3.1 Main LncRNAs with Oncogenic Functions

3.1.1 MALAT-1

Metastasis-associated lung adenocarcinoma transcript 1 (*MALAT-1*) is one of the first lncRNAs that was associated with lung cancer, especially with lung adenocarcinoma. *MALAT-1* is an 8.7 kB intergenic lncRNA (lincRNA) located on

Table 1 Biological activity of the main lncRNAs involved in the pathophysiology and clinical management of NSCLC

LncRNA	Main functions and effects
MALAT-1	Oncogenic: Alternative splicing of pre-mRNAs and chromobox homolog 4 (CXB4) regulation. Induces EMT. Worst prognosis and higher circulating levels in metastatic patients. Resistance to TKIs
HOTAIR	Oncogenic: Promotes gelatinases and suppresses metalloproteinases. Lower disease-free survival
CCAT2	Oncogenic: WNT signaling pathway. Induces Pokemon and suppresses p21. Increased levels in adenocarcinomas. Resistance to chemotherapies and radiotherapies
H19	Oncogenic: Induces the expression of JNK1/2. Poorer survival. Resistance to platinum-based chemotherapies
ANRIL	Oncogenic: ANRIL binds PCR2 and suppresses INK4A-ARF-INK4B. Higher expression in NSCLC tissues and in higher disease stages. Poorer survival
LUACT1	Oncogenic: Affects cellular growth. Correlates with cigarette smoking
SOX2-OT	Oncogenic: Affects expression of EZH2. Increased in squamous cell cancer
MEG3	Onco-suppressor: Induces cell apoptosis and impedes tumorigenesis. Upregulates p53. Low levels associated with bad prognosis and bad response to chemotherapy
TUG1	Onco-suppressor: Suppresses HOXB7 through PCR2. Resistance to paclitaxel
SPRY4-IT1	Onco-suppressor: Downregulated by EZH2. Lower levels in lung adenocarcinoma
GAS5	Onco-suppressor: Induces apoptosis. Low circulating levels in NSCLC patients. Predicts radiosensitivity
BANCR	Onco-suppressor: Inhibits EMT proteins. Decreased in NSCLC tissues. Correlations with stage and prognosis
TARID	Onco-suppressor: Activates TCF21 via GADD45A. Low levels in NSCLC
SCAL1	Induced by cigarette smoking. Regulated by NRF2. Upregulated in NSCLC
DQ786227	Malignant transformation in respiratory cells exposed to benzo(a)pyrene
LOC728228	Malignant transformation in respiratory cells exposed to benzo(a)pyrene

(continued)

Table 1 (continued)

LncRNA	Main functions and effects
CAR-10	Induced by air pollution. Upregulated by dibenz [a,h]anthracene, bounds (YB-1), and upregulates EGFR
NR-026689, XIST, HIF1A-AS1, lncRNA16, UCA1, RP11-397D12.4, AC007403.1, and ERICH1-AS1	Increased levels in NSCLC
LINC00313	Stage and histotype specific (squamous cell carcinomas)
RP11-21 L23.2, GPR158-AS1, RP11-701P16.5, and RP-11379F4.4	Poor prognosis
CTD-2358C21.4, RP11-94 L15.2, KCNK15. AS1, and AC104134.2	Better prognosis
AK126698 and ROR	Resistance to cisplatin chemotherapies. ROR induces resistance to radiotherapy
CNQ1OT1	Higher expression in paclitaxel-sensitive patients
BC087858	Activates PI3K/AKT and MERK/ERK. Induces EMT resistance to TKIs
pR-lncRNA-1, LINC-PINT, and TUSC7	Induce radioresistance
PVT1	Predicts radiosensitivity

chromosome 11q13. Silencing of *MALAT-1* in vitro reduces the mobility of lung adenocarcinoma cells. In alveolar basal epithelial cells (A549 cell line) and human NSCLC cells derived from lymph nodes (H1299 cell line), *MALAT-1* sponges miR206 promoting cellular invasion and migration (Tang et al. 2018). *MALAT-1* upregulation in lung cancer tissue is associated with metastasis progression and poor prognosis, particularly in early stage NSCLC patients with metastasis (Tano et al. 2010). Indeed, in patients with NSCLC and bone or brain metastases, the circulating concentrations of *MALAT-1* were found to be higher than in healthy individuals.

MALAT-1 transcription is regulated by p53, which is able to sequester its promoter. Although the activity of *MALAT-1* in lung carcinogenesis is not fully understood, two potential mechanisms of action have been described: (a) contribution to an alternative splicing of pre-mRNAs that produces an aberrant expression of genes such as B-MYB transcription factor, and (b) interaction with the demethylated chromobox homolog 4 (CXB4) that controls the relocation of growth-related genes in interchromatin granules. In addition, it has been reported that the CXC motif chemokine ligand 5 (CXCL5), as a downstream gene of *MALAT-1*, mediated the effects of *MALAT-1* on NSCLC migration and invasion (Guo et al. 2015).

3.1.2 HOTAIR

The HOX transcript antisense RNA (*HOTAIR*), a 2.4 kB antisense lncRNA located in chromosome 12, exhibits altered expression in several human cancers. In NSCLC cancer cells knocked out for *HOTAIR* in vitro, a decrease in proliferation and metastasis progression has been described. Experiments in fibroblasts showed that *HOTAIR* acts on homeobox D (a family of transcription factors) genes through epigenetic silencing: these genes are located in chromosome 2, where *HOTAIR* is driven to by polycomb repressive complex 2 (PCR2), a protein complex able to silence chromatin by methylation of histone 3 on lysine 27 (H3K27). *HOTAIR* acts as a scaffold binding to PCR2 and to lysine-specific demethylase 1 LSD-1/CoREST/REST complex. The LSD-1/CoREST/REST complex demethylates the lysine 4 of histone 3, resulting globally in chromatin rearrangement (Rinn et al. 2007). Furthermore, it has been demonstrated that *HOTAIR* promotes the expression of gelatinases and represses the expression of cell-adhesion proteins and metalloproteinases, promoting neoplastic cell motility and metastasis (Zhao et al. 2014); *HOTAIR* also suppresses $p21^{waf1}$ (discussed later) and HOXA5, a further protein involved in NSCLC cell migration and invasion. Taken together, these findings clearly support an important role of *HOTAIR* in NSCLC progression and metastasis. Nakagawa et al. (2013) confirmed that this lncRNA is associated with a shorter disease-free survival in patients affected by NSCLC.

3.1.3 CCAT2

The role of colon cancer-associated transcript 2 (*CCAT2*), an lncRNA first described in colorectal cancer, in carcinogenesis is not fully understood (Palmieri et al. 2017). *CCAT2* is a 1.7 kb intergenic lncRNA located on chromosome 8q24 that is involved in the WNT signaling pathway through an interaction with the transcription factor 7-like 2 (TCF7L2). This results in the upregulation of the expression of *MYC* and some miRNAs, such as miR-20a, known to regulate cell proliferation. A single nucleotide polymorphism (SNP) of *CCAT2*, rs6983267, has been associated with *CCAT2* overexpression. Zhao et al. (2018) showed that knockdown of *CCAT2* in NSCLC cells limited malignant growth and invasion, while artificial overexpression of *CCAT2* led to opposite effects. In addition, CCAT2 knockdown significantly decreased the expression of POK erythroid myeloid ontogenic factor (Pokemon) and induced the expression of the p21 tumor suppressor; this suggests that Pokemon overexpression could reverse the decrease of cell viability and cell invasion triggered by *CCAT2* silencing. In addition, *CCAT2* overexpression has been significantly associated with lung adenocarcinoma but not with squamous cell cancer. Silencing CCAT2 by siRNA has led to inhibition of proliferation and invasion in NSCLC cell lines in vitro (Qiu et al. 2014).

3.1.4 H19

Overexpression of *H19*, a 2.3 kb lncRNA located on chromosome 11p15 that is expressed only in the maternally inherited chromosome (imprinting phenomenon), is associated with poor prognosis in various cancers. *H19* overexpression is associated with hypomethylation of its promoter region; in NSCLC, this characteristic is frequently associated with loss of imprinting. *H19* induces cellular proliferation by stimulating the expression of c-Jun and c-Jun N-terminal kinase 1/2 (JNK1/2), and it acts as a sponge by sequestering let-7, a miRNA able to inhibit carcinogenesis by downregulating tumor-promoting proteins (e.g., RAS and MYC) (Kallen et al. 2013). Other pathophysiological mechanisms in lung oncogenesis have been also identified: miR-17/STAT3, miR-484/ROCK2, and miR-196b/LIN28B regulation, SAHH interaction and attenuation, and BPDE-DNA adduct formation. Several studies confirmed that the higher expression of *H19* was positively correlated with advanced tumor stage and tumor size, as well as that *H19* expression is an independent prognostic factor for overall survival of NSCLC (Chen et al. 2013).

3.1.5 ANRIL

Noncoding RNA in the INK4 locus (*ANRIL*) is a recently characterized lncRNA that is functionally correlated with the phospholipase D (PLD): the overexpression of *ANRIL* is associated with inhibition of PLD, with consequent anti-tumorigenic effects, while knockdown of *ANRIL* suppresses PLD inhibition-induced apoptosis (Kang et al. 2015). *ANRIL* is a 3.8 bp antisense lncRNA located on chromosome 9p21 which is transcribed from the INK4b-ARF-INK4a gene cluster and has been proven to be upregulated in multiple cancers, such as breast cancer, cervical cancer, nasopharyngeal carcinoma, and thyroid cancer. In tumorigenesis, *ANRIL* binds to PCR2 and causes a chromatin rearrangement and consecutive silencing of the INK4A-ARF-INK4B gene cluster, which contains tumor-suppressor genes also known as p16, p14, and p15. In addition, *ANRIL* has been proven to inhibit the expression of P21 and KLF2 and attenuate the transforming growth factor β (TGF-β)/Smad signaling pathway, promoting cancer invasion and metastasis. In recent studies, the expression level of *ANRIL* was higher in NSCLC tissues and lung cancer cells than in adjacent non-tumor tissues and normal human bronchial epithelial cells (Lu et al. 2016). The higher expression levels of *ANRIL* in NSCLC were positively correlated with advanced tumor-node-metastasis stage and had negative prognostic implications. Moreover, knockdown of *ANRIL* expression could inhibit lung cancer cell proliferation, migration, and invasion in vitro. For this reason, *ANRIL* has been recently investigated as part of promising panels for the diagnosis of NSCLC, which include other lncRNAs and traditional biomarkers such as the carcinoembryonic antigen (CEA) and the cytokeratin 19 fragment (CYFRA 21-1).

3.1.6 LUACT1 (SCAL1)

Lung cancer-associated transcript 1 is known also as smoke- and cancer-related long-chain noncoding RNA 1 (SCAL-1). It is located on chromosome 5 and it is typically tobacco-induced. The transcript includes four exons and three introns. *LUACT1* expression is transcriptionally regulated by nuclear factor erythroid 2-related factor (NRF2) and is determined by knockdown through siRNA in NRF2 and kelch-like ECH-associated protein 1 (KEAP1). Induction of *LUACT1* has been shown both in vitro and in vivo. It can play the downstream role of NRF2 in regulation of gene expression and intermediate in protection against oxidation stress in epithelial cells of the respiratory system. In a recent study, the lncRNA landscape in lung cancer has been characterized using publicly available transcriptome sequencing data from a cohort of 567 adenocarcinoma and squamous cell carcinoma tumors; functional validation, using both knockdown and overexpression, shows that the most differentially expressed lncRNA was *LUACT1* that was sufficient to affect cellular growth independently of other common cancer mutations (White et al. 2014).

3.1.7 SOX2-OT

SOX2 overlapping transcript (*SOX2-OT*) is an overlapping lncRNA located on chromosome 3 that is highly expressed in embryonic stem cells. Dysregulation of *SOX2-OT* has been observed in various tumors, including gastric cancer, esophageal cancer, breast cancer, hepatocellular carcinoma, ovarian cancer, pancreatic ductal adenocarcinoma, laryngeal squamous cell carcinoma, cholangiocarcinoma, osteosarcoma, nasopharyngeal carcinoma, glioblastoma, and lung cancer, wherein it typically functions as an oncogene and possibly as a tumor-suppressor gene. Hou et al. (Hou et al. 2014) showed that the expression level of *SOX2-OT* in 53.01% of human primary lung cancer was twofold higher than that in pair-matched adjacent non-tumor samples. Compared to adenocarcinomas, *SOX2-OT* expression was significantly higher in squamous cell carcinoma of the lung. Knockdown of *SOX2-OT* inhibited cell proliferation by decreasing the number of cells in S phase and inducing G2/M arrest. The protein expressions of EZH2 and cyclin B1 and Cdc2 were reduced, and ectopic expression of EZH2 restored the G2/M transition and cyclin B1 and Cdc2 protein expression (Hou et al. 2014). Further studies demonstrated that *SOX2-OT* expression was obviously higher in NSCLC tissues and serum samples than in normal controls and that *SOX2-OT* overexpression was associated with poor survival in patients with lung cancer.

3.2 Main LncRNAs with Onco-suppressive Functions

3.2.1 MEG3

Low concentrations of the lncRNA maternally expressed 3 (*MEG3*) correlate with poor prognosis in NSCLC (Zhou et al. 2012). *MEG3* is a 6.9 kb lncRNA located on chromosome 14q32, expressed only in the maternal-inherited chromosome. It is expressed in normal human tissues, especially in brain and the pituitary, and is thought to be a tumor suppressor. Recent studies showed that *MEG3* expression is disrupted in various human cancers, such as bladder cancer, glioma, and hepatocellular carcinoma. In lung cancer, *MEG3* upregulates p53 expression, inhibiting the exon 3 ubiquitin ligase from preventing p53 transcription. It can also act as a guide for PCR2, bringing it to the regulatory regions of target genes. Interestingly, its overexpression has different effects in in vitro and in vivo experiments: in vitro overexpression induces cell apoptosis, whereas in vivo overexpression inhibits tumorigenesis. A recent report demonstrated that expression of *MEG3* in NSCLC cell lines was negatively correlated with miR-205-5p, which enhances cell proliferation and represses apoptosis through targeting low-density lipoprotein (LDL) receptor-related protein-1 (LRP1) (Wang et al. 2017a). Other reports showed that *MEG3* expression was decreased in NSCLC tumor tissues compared with normal tissues and associated with advanced pathological stage and tumor size. Moreover, patients with lower levels of *MEG3* expression had a relatively poor prognosis.

3.2.2 TUG1

The concentration of the lncRNA taurine-upregulated gene 1 (*TUG1*) in squamous carcinoma and adenocarcinoma is negatively associated with advanced disease stage and shorter overall survival. *TUG1* is a 5.6 kb intergenic lncRNA located on 22q12 chromosome that was originally identified in a genomic screen of taurine-treated mouse retinal cells. *TUG1* has been demonstrated to serve crucial regulatory roles in various cancer-associated biological processes. It binds to PRC2 in the promoter region of CELF1 and negatively regulates CELF1 expression. It guides PCR2 into the homeobox B7 region (HOXB7), an oncogene responsible for activating both the PI3K/ERK and MAPK pathways, resulting in an increase in cellular proliferation (Zhang et al. 2014). PCR2 suppresses the expression of HOXB7. *TUG1* expression is mediated by wild-type *p53*; this effect is lost in cases of *p53* mutations with R175H missense substitution. *TUG1* is found to exhibit aberrant expression in a variety of malignancies. Dysregulation of *TUG1* has been shown to contribute to proliferation, migration, cell cycle changes, inhibited apoptosis, and drug resistance of cancer cells which revealed an oncogenic role for this lncRNA, but some reports have shown downregulation of *TUG1* in lung cancer samples compared with noncancerous samples. Interestingly, in NSCLC patients, *TUG1* downregulation correlated with sex, smoking status, and tumor differentiation grade.

3.2.3 SPRY4-IT1

Sprouty homolog 1 4 intronic transcript 1 (*SPRY4-IT1*) is derived from an intron of the SPRY4 gene located in chromosome 5q31.3. Its downregulation, due to transcriptional repression mediated by EZH2, a histone methyltransferase able to induce a H3K27 modification, favors cell migration and invasion in vitro. By contrast, its upregulation induces apoptosis and, in mice, reduces metastasis. In a recent study, *SPRY4-IT1* expression was observed to be significantly lower, and the expression of EZH2 significantly higher, in lung adenocarcinoma tissues when compared to the adjacent normal tissues (Wen et al. 2018). *SPRY4-IT1*-suppressed expression in NSCLC is correlated with larger tumor size and lymph node metastasis.

3.2.4 GAS5

Growth arrest specific 5 (*GAS5*) is an lncRNA located in chromosome 1q25, involved in inducing apoptosis. In recent studies, the expression pattern of *GAS5* was investigated in NSCLC specimens and healthy tissues, and its biological functions in the development and progression of NSCLC were assessed. *GAS5* expression was downregulated in cancerous tissues compared to adjacent noncancerous tissues and was highly related to tumor size and stage. Furthermore, *GAS5* overexpression increased tumor cell growth arrest and induced apoptosis in vitro and in vivo. In addition, siRNA-mediated knockdown of *GAS5* promoted tumor cell growth. It has been demonstrated that the ectopic expression of *GAS5* significantly upregulates p53 expression and downregulates transcription factor E2F1 expression. In addition, upregulation of GAS5 in NSCLC cells was able to suppress their growth, migration, and invasion via the miR-205/PTEN axis. Therefore, GAS5 is a tumor suppressor in NSCLC which acts through p53-dependent and p53-independent pathways. A recent study showed that *GAS5* circulating concentrations are reduced in NSCLC patients but tend to normalize after surgical resection of the tumor (Liang et al. 2016). The authors found that *GAS5* expression levels could distinguish NSCLC patients from control patients with 82.2% sensitivity and 72.7% specificity and that the combination of the *GAS5* and carcinoembryonic antigen could produce an area of 0.909 (95% confidence interval 0.857–0.962) under the receiver-operating characteristic curve in distinguishing NSCLC patients from control subjects.

3.2.5 BANCR

BRAF-activated noncoding RNA (*BANCR*) is a 693-bp lncRNA on chromosome 9 that is overexpressed in melanoma cells and crucial for melanoma cell migration. In a recent study, overexpression of BANCR was found to play a key role in epithelial-mesenchymal transition (EMT) through the regulation of E-cadherin,

N-cadherin, and vimentin expression (Sun et al. 2014). In this study, BANCR expression was significantly decreased in 113 NSCLC tumor tissues compared with normal tissues. Additionally, reduced BANCR expression was associated with larger tumor size, advanced pathological stage, metastasis distance, and shorter overall survival of NSCLC patients. Finally, reduced BANCR expression was found to be an independent prognostic factor for NSCLC.

3.2.6 TARID

TCF21 antisense RNA inducing demethylation (TARID) has been demonstrated to activate TCF21 expression by inducing promoter demethylation. This occurs because TARID interacts with both the TCF21 promoter and GADD45A (growth arrest and DNA-damage-inducible, alpha), a regulator of DNA demethylation. In a pilot study, TARID was downregulated in NSCLC cells and tissues, but its potential pathophysiological and clinical roles need to be further elucidated (Arab et al. 2014).

3.3 LncRNAs and Endothelial to Mesothelial Transition (EMT)

EMT is described as a reversible phenomenon in which an epithelial cell loses its distinctive characteristics and becomes a mesenchymal cell through the activation of different pathways that culminates in loss of E-cadherin. EMT plays a crucial role in the pathogenesis of NSCLC. It is believed that EMT represents one of the mechanisms of resistance to target therapies in NSCLC patients; moreover, EMT and subsequently mesothelial to endothelial transition (MET) could represent a key process that allows lung cancer cells to metastasize. Numerous lncRNAs have been demonstrated to be involved in EMT, particularly *MALAT-1*, *HOTAIR,* and *SPRY4-IT1*. As discussed earlier, *MALAT-1* concentrations in peripheral blood are higher in patients with brain metastases, suggesting a possible role in inducing EMT. In particular, *MALAT-1* upregulation increases ZEB1/2 and decreases E-cadherin levels concentrations in these patients.

HOTAIR is able to bind to PCR2, a protein complex needed for H3K27 trimethylation that contains the histone methyltransferase EZH2, which in turn represses E-cadherin gene by H3K27 methylation (Cao et al. 2008). *SPRY4-IT1* and *BANCR* act in similar ways as previously discussed.

4 LncRNAs and Lung Cancer Risk Factors

Altered concentrations of lncRNAs have been described in relation to well-known risk factors for lung cancer, especially cigarette smoking. The most studied is smoking cancer-associated lncRNA 1 (*SCAL-1*) which is induced by cigarette smoking and is upregulated in NSCLC cell lines (Thai et al. 2013). This lncRNA, located in chromosome 5, is regulated transcriptionally by nuclear factor erythroid 2-related factor (NRF2). As previously discussed, cigarette smoking initially induces upregulation of the active H19 allele. This is likely to progress to loss of imprinting as the burden of smoking increases and as the epithelium undergoes transition from normal to neoplastic. The *lncLUACT1*, previously described, also correlates with cigarette smoking, while lncRNA *DQ786227* and lncRNA *LOC728228* are involved in malignant transformation in respiratory cells exposed to benzo(a)pyrene (Gao et al. 2013; Hu et al. 2015). Finally, lncRNA *CAR-10* is upregulated in NSCLCs due to a different risk factor for NSCLC—air pollution. This lncRNA is upregulated by a polycyclic aromatic hydrocarbon, dibenz[a,h]anthracene (a pollutant of smoke and oils), that increases the expression of FoxF2. CAR-10 bounds and stabilizes transcription factor Y-box-binding protein 1 (YB-1), leading to upregulation of the EGFR and proliferation of lung cancer cells (Wei et al. 2016).

5 LncRNAs as Diagnostic and Prognostic Biomarkers

5.1 LncRNAs as Diagnostic Biomarkers

The high interest gained by lncRNAs in cancer is also related to their putative role as diagnostic biomarkers for early diagnosis of tumors and/or for differential diagnosis of specific malignancies. This would be particularly relevant in NSCLC in order to improve the currently low survival rates. From this perspective, lncRNAs have some features that make them suitable as potential biomarkers: as previously discussed, the tissue or blood concentrations of some of them change in relation to either the presence of the tumor, its stage, or prognosis; others are tissue-specific, and their concentrations change in particular NSCLC histotypes. Furthermore, they can also be determined in other biological fluids, particularly pleural effusions, frequent in NSCLC patients. The most relevant issue against the implementation of specific lncRNAs detection in clinical practice is represented by their generally low concentrations in biological fluids, which prevents easy determination with standard analytical methods. Concentrations and types of lncRNA detected also vary depending on the biological sample (whole blood, serum, and plasma); to date, there are no NSCLC-associated lncRNA recognized in sputum.

Whole blood concentrations of several lncRNAs are altered in NSCLC patients (Table 1). For the discrimination of NSCLC patients from cancer-free controls, MALAT-1 showed a sensitivity of 56% and a specificity of 96% in cellular fractions of whole blood (Weber et al. 2013). Hu et al. (2015) reported that circulating SPRY4-IT1, ANRIL, and NEAT1 were significantly increased in plasma samples of NSCLC patients. Receiver operating characteristic curve (ROC) analysis revealed that plasma ANRIL provided the highest diagnostic performance with an area under ROC curve value (AUC) of 0.798. Combination of the three factors further increased the diagnostic performance (AUC, 0.876; sensitivity, 82.8%; specificity, 92.3%). Other lncRNAs, particularly *NR-026689*, *XIST*, *HIF1A-AS1*, *lncRNA16*, *UCA1*, *RP11-397D12.4*, *AC007403.1*, and *ERICH1-AS1*, have been shown to be increased in NSCLC patients. On the other hand, blood concentrations of onco-suppressive lncRNAs, particularly GAS5, BANCR, and TARID, are generally lower in NSCLC patients.

Currently, none of the molecules described is used in clinical practice due to the technical reasons mentioned above and the need to better establish their predictive capacity (sensibility, specificity, positive and negative predictive values) through adequately designed clinical trials. In some cases, combination of lncRNAs with traditional biomarkers may be effective. Qiu et al. (2014) showed that CCAT2 combined with CEA could predict lymph node metastasis in NSCLC patients. Currently, one clinical trial conducted in China is recruiting patients to evaluate the role of lncRNAs as potential biomarkers for lung cancer diagnosis [NCT03830619].

The association between blood concentrations of some lncRNAs and stage of the disease is also of interest. For example, *LINC00313*, an intergenic lncRNA, can be detected in serum of patients affected with T2N1-stage lung adenocarcinoma (Li et al. 2015). Another interesting feature of some lncRNAs is their histotype specificity. Even if NSCLC subtypes present specific morphologic, immunohistochemical, and molecular characteristics, the exact diagnosis may be difficult in some cases. Biomarkers detectable in serum specific for NSCLC histotypes would be useful in this context. In a study that compared lncRNAs expressed in lung adenocarcinoma and lung squamous cell carcinoma tissues, *LINC01133* was upregulated in lung squamous cell carcinomas, but not in adenocarcinomas (Zhang et al. 2015). Conversely, Qiu et al. (2014) reported that *CCAT2* overexpression is significantly associated with lung adenocarcinoma, but not with squamous cell cancer.

5.2 LncRNAs as Prognostic Markers

Most of the lncRNAs described have been found to be related to the prognosis of the disease, despite their obscure role in NSCLC pathogenesis. For example, *RP11-21 L23.2*, *GPR158-AS1*, *RP11-701P16.5*, and *RP-11379F4.4* were correlated with poor overall survival, while *CTD-2358C21.4*, *RP11-94 L15.2*, *KCNK15.AS1*, and *AC104134.2* were related to better overall survival (Zhou et al. 2015). Also, low expression of some lncRNAs is associated with prognosis: for example, low concentrations of MEG3, GAS-6AS1, and other onco-suppressing molecules correlate with poor overall survival (Han et al. 2013).

6 LncRNAs in NSCLC Therapy

Some lncRNAs have shown to be implied in lung cancer therapy, both for their implications in acquired or non-acquired therapy resistance and for their possible use as therapeutic targets.

6.1 *Role of LncRNAs in Resistance to Therapy*

6.1.1 Resistance to Chemotherapy

The development of resistance to chemotherapy, for example, cisplatin, is commonly observed in lung cancer. A different expression profile of 1380 lncRNAs was found in vitro in A549 cells and cisplatin-resistant A549/CDDP cells, suggesting that lncRNAs are involved in chemotherapy resistance mechanisms (Yang et al. 2013). Cisplatin acts by inhibiting DNA replication and damaging cell membrane, leading to apoptosis. Resistance may arise because of altered expression of lncRNAs, which can reactivate proliferation pathways and/or repair cisplatin-induced damage. *AK126698* targets Wnt, while *lncROR* targets PI3K/AKT/mTOR, increasing sensitivity to cisplatin-based therapies (Shi et al. 2017). HOTAIR hyper-expression induces cisplatin resistance by downregulating the cyclin-dependent kinase inhibitor 1, a protein associated with cell cycle arrest, and upregulating the expression of stem cell-related biomarkers such as Klf4 (Liu et al. 2016). Furthermore, H19 plays a role in platinum therapy resistance by regulating apoptosis proteins such as BAX, BAK, and FAS (Wang et al. 2017b). Furthermore, patients with downregulation of MEG3 exhibit reduced response to cisplatin therapy, probably because MEG3 is able to modulate the expression of p53, activate the Wnt/β-catenin pathway, and sponge regulatory miRNAs.

Two lncRNAs were found to be associated with altered response to paclitaxel: *CNQ1OT1*, which exhibits increased expression in lung adenocarcinoma cells that are paclitaxel-sensitive (Ren et al. 2017), and TUG1, by influencing EZH2 in lung squamous cell carcinoma (Niu et al. 2017).

6.1.2 Resistance to Targeted Therapy

Some lncRNAs can affect the efficacy of anti-EGFR TKIs by inducing alterations which allow cancer cells to "escape" their effects, resulting in an acquired resistance to therapy and disease progression. *UCA1* expression, upregulated in patients with EGFR-TKIs resistance, activates the AKT-mTOR pathway and stimulates EMT (Cheng et al. 2015). The lncRNA *BC087858* also induces EGFR-TKIs resistance by activating the PI3K/AKT and MERK/ERK pathways, as well as inducing EMT (Pan et al. 2016). *lncBC0587858* induces EMT by upregulating FOXC1, leading to

E-cadherin inhibition and induction of EMT (Xia et al. 2013). *MALAT-1* promotes EMT through EMT-associated transcription genes and activation of the Wnt pathway (Samatov et al. 2013).

6.1.3 Resistance to Radiotherapy

Mechanisms of radioresistance are still largely unknown. Some studies suggested a modulation by noncoding RNAs principally in response to DNA damage, radiation-associated cell death, hypoxia, and activation of cancer stem cells. The ncRNAs responsible for these events are predominantly miRNAs; however, some lncRNAs are also involved, particularly *lncROR*, *pR-lncRNA-1*, *LINC-PINT*, and *TUSC7*. In mice lung cancer models, upregulation of *HOTAIR* is associated with a decreased radiosensitivity by the inactivation of the β-catenin pathway (Chen et al. 2015). LncRNAs may be also suitable as markers of response to radiotherapy: the lncRNA plasmacytoma variant transcript 1 (*PVT1*) and the *lncGAS5* have been demonstrated to serve as putative biomarkers in predicting radiosensitivity (Wu et al. 2017; Xue et al. 2017).

6.2 LncRNAs as New Therapeutic Targets

The comprehension of the pathogenic roles of lncRNAs in lung carcinogenesis is essential in order to investigate and establish new therapeutic targets. Numerous studies have been published to date (the most important are summarized in this chapter), and several interactions between lncRNAs and the main molecular pathways involved in lung carcinogenesis have been explored, evidencing opportunities for novel therapies. It is in principle possible to modulate the action of lncRNAs by blocking these interactions with siRNAs, antisense oligonucleotides, ribozyme, and aptamers. As previously discussed, these methods have been employed to inhibit oncogenic lncRNAs in several studies, with encouraging results. For example, experimental silencing of *MALAT-1* with antisense oligonucleotides in mouse models reduced lung cancer metastasis (Gutschner et al. 2013). Nevertheless, there are currently no clinical trials testing lncRNA-targeting agents suggesting that additional research is warranted.

7 Conclusions and Future Perspectives

NSCLC is currently the leading cause of cancer-related death worldwide. Despite recent advances in the surgical and clinical management, mortality rates remain extremely high and close to incidence rates. Therefore, further research is warranted to improve survival, especially in the advanced stages of the disease. LncRNAs represent an emerging class of noncoding RNAs, which show encouraging results and potential applications in the diagnosis of NSCLC and in predicting the prognosis in subgroups of patients. In particular, numerous lncRNAs have been evaluated as biomarkers for early diagnosis, differential diagnosis, and stage stratification of NSCLC patients with encouraging results. Nevertheless, their clinical applicability and their predictive potential have to be tested with methodologically tailored studies. Furthermore, recent evidence strongly suggests that some lncRNAs or their combinations can predict either sensitivity or resistance to cisplatinum-based chemotherapy and TKI-based treatments, which further impacts prognosis. In this context, lncRNAs might be proposed as biomolecular markers for patient selection and implementation of personalized oncological treatments and for establishing alternative therapeutic strategies in cases of prediction of resistance to these treatments.

Currently, there are no available biomarkers to implement such a task, with the exception of some somatic mutations, such as the T790M *EGFR* mutation which determines resistance to TKIs in patients with lung adenocarcinoma. These mutations can be absent in the initial diagnosis and develop subsequently during treatment. This dictates the need to monitor the mutational status of driver and resistance-conferring genes during the course of the disease. Liquid biopsy methods are being developed to this regard, but they harbor several limitations, mainly due to technical reasons. In this setting, the use of lncRNAs might be useful for the prediction of sensitivity/resistance to therapies; however, their detection in biological fluids is challenging due to instability, which imposes the use of novel techniques.

Finally, lncRNAs can be molecular therapeutic targets themselves, considering their involvement in several pathophysiological mechanisms of lung cancer. Unfortunately, these small molecules are involved in numerous complex physiological and pathological processes, which currently limits their potential as targets. Further research is warranted to understand the interactions between lncRNAs and other classes of molecules to better elucidate their potential implication in the treatment of NSCLC.

References

Arab K, Park YJ, Lindroth AM et al (2014) Long noncoding RNA TARID directs demethylation and activation of the tumor suppressor TCF21 via GADD45A. Mol Cell 55:604–614

Cao Q, Yu J, Dhanasekaran SM et al (2008) Repression of E-cadherin by the polycomb group protein EZH2 in cancer. Oncogene 27:7274–7284

Chen B, Yu M, Chang Q et al (2013) Mdig de-represses H19 large intergenic non-coding RNA (lincRNA) by down-regulating H3K9me3 and heterochromatin. Oncotarget 4:1427–1437

Chen J, Shen Z, Zheng Y et al (2015) Radiotherapy induced Lewis lung cancer cell apoptosis via inactivating β-catenin mediated by upregulated HOTAIR. Int J Clin Exp Pathol 8:7878–7886

Cheng N, Cai W, Ren S et al (2015) Long non-coding RNA UCA1 induces non-T790M acquired resistance to EGFR-TKIs by activating the AKT/mTOR pathway in EGFR-mutant non-small cell lung cancer. Oncotarget 6:23582–23593

Colombino M, Paliogiannis P, Cossu A et al (2019) EGFR, KRAS, BRAF, ALK, and cMET genetic alterations in 1440 Sardinian patients with lung adenocarcinoma. BMC Pulm Med 19:209

Gao L, Mai A, Li X et al (2013) LncRNA-DQ786227-mediated cell malignant transformation induced by benzo(a)pyrene. Toxicol Lett 223:205–210

Guo F, Guo L, Li Y et al (2015) MALAT1 is an oncogenic long non-coding RNA associated with tumor invasion in non-small cell lung cancer regulated by DNA methylation. Int J Clin Exp Pathol 8:15903–15910

Gutschner T, Hämmerle M, Eissmann M et al (2013) The noncoding RNA MALAT1 is a critical regulator of the metastasis phenotype of lung cancer cells. Cancer Res 73:1180–1189

Han L, Kong R, Yin DD et al (2013) Low expression of long noncoding RNA GAS6-AS1 predicts a poor prognosis in patients with NSCLC. Med Oncol 30:694

Hon CC, Ramilowski JA, Harshbarger J et al (2017) An atlas of human long non-coding RNAs with accurate 5′ ends. Nature 543:199–204

Horn L, Spigel DR, Vokes EE et al (2017) Nivolumab versus docetaxel in previously treated patients with advanced non-small cell lung cancer: two-year outcomes from two randomized, open label, phase III trials (CheckMate 017 and CheckMate 057). J Clin Oncol 35:3924–3933

Hou Z, Zhao W, Zhou J et al (2014) A long noncoding RNA Sox2ot regulates lung cancer cell proliferation and is a prognostic indicator of poor survival. Int J Biochem Cell Biol 53:380–388

Hu G, Yang T, Zheng J et al (2015) Functional role and mechanism of lncRNA LOC728228 in malignant 16HBE cells transformed by anti-benzopyrene-trans-7,8-dihydrodiol-9,10-epoxide. Mol Carcinog 54:192–204

Kallen AN, Zhou XB, Xu J et al (2013) The imprinted H19 lncRNA antagonizes let-7 microRNAs. Mol Cell 52:101–112

Kang YH, Kim D, Jin EJ (2015) Down-regulation of phospholipase D stimulates death of lung cancer cells involving up-regulation of the long ncRNA ANRIL. Anticancer Res 35:2795–2803

Li M, Qiu M, Xu Y et al (2015) Differentially expressed protein-coding genes and long noncoding RNA in early-stage lung cancer. Tumour Biol 36:9969–9978

Liang W, Lv T, Shi X et al (2016) Circulating long noncoding RNA GAS5 is a novel biomarker for the diagnosis of nonsmall cell lung cancer. Medicine (Baltimore) 95:e4608

Liu MY, Li XQ, Gao TH et al (2016) Elevated HOTAIR expression associated with cisplatin resistance in non-small cell lung cancer patients. J Thorac Dis 8:3314–3322

Lu Y, Zhou X, Xu L et al (2016) Long noncoding RNA ANRIL could be transactivated by c-Myc and promote tumor progression of non-small-cell lung cancer. Onco Targets Ther 9:3077–3084

Nakagawa T, Endo H, Yokoyama M et al (2013) Large noncoding RNA HOTAIR enhances aggressive biological behavior and is associated with short disease-free survival in human non-small cell lung cancer. Biochem Biophys Res Commun 436:319–324

Niu Y, Ma F, Huang W et al (2017) Long non-coding RNA TUG1 is involved in cell growth and chemoresistance of small cell lung cancer by regulating LIMK2b via EZH2. Mol Cancer 16:5

Paliogiannis P, Attene F, Cossu A et al (2013) Lung cancer epidemiology in North Sardinia, Italy. Multidiscip Respir Med 8:45

Paliogiannis P, Attene F, Cossu A et al (2015) Impact of tissue type and content of neoplastic cells of samples on the quality of epidermal growth factor receptor mutation analysis among patients with lung adenocarcinoma. Mol Med Rep 12:187–191

Palmieri G, Paliogiannis P, Sini MC et al (2017) Long non-coding RNA CASC2 in human cancer. Crit Rev Oncol Hematol 111:31–38

Pan H, Jiang T, Cheng N et al (2016) Long non-coding RNA BC087858 induces non-T790M mutation acquired resistance to EGFR-TKIs by activating PI3K/AKT and MEK/ERK pathways and EMT in non-small-cell lung cancer. Oncotarget 7:49948–49960

Peng W, Wang J, Shan B et al (2018) Diagnostic and prognostic potential of circulating long non-coding RNAs in non small cell lung cancer. Cell Physiol Biochem 49:816–827
Qiu M, Xu Y, Yang X et al (2014) CCAT2 is a lung adenocarcinoma-specific long non-coding RNA and promotes invasion of non-small cell lung cancer. Tumour Biol 35:5375–5380
Ransohoff JD, Wei Y, Khavari PA (2018 Mar) The functions and unique features of long intergenic non-coding RNA. Nat Rev Mol Cell Biol 19(3):143–157
Ren K, Xu R, Huang J et al (2017) Knockdown of long non-coding RNA KCNQ1OT1 depressed chemoresistance to paclitaxel in lung adenocarcinoma. Cancer Chemother Pharmacol 80:243–250
Rinn JL, Kertesz M, Wang JK et al (2007) Functional demarcation of active and silent chromatin domains in human HOX loci by noncoding RNAs. Cell 129:1311–1323
Samatov TR, Tonevitsky AG, Schumacher U (2013) Epithelial-mesenchymal transition: focus on metastatic cascade, alternative splicing, non-coding RNAs and modulating compounds. Mol Cancer 12:107
Shi Y, Li J, Zhang S et al (2015) Molecular epidemiology of EGFR mutations in asian patients with advanced non-small-cell lung cancer of adenocarcinoma histology - mainland China subset analysis of the PIONEER study. PLoS One 10:e0143515
Shi H, Pu J, Zhou XL et al (2017) Silencing long non-coding RNA ROR improves sensitivity of non-small-cell lung cancer to cisplatin resistance by inhibiting PI3K/Akt/mTOR signaling pathway. Tumour Biol 39:1010428317697568
Sun M, Liu XH, Wang KM et al (2014) Downregulation of BRAF activated non-coding RNA is associated with poor prognosis for non-small cell lung cancer and promotes metastasis by affecting epithelial-mesenchymal transition. Mol Cancer 13:68
Tang Y, Xiao G, Chen Y et al (2018) LncRNA MALAT1 promotes migration and invasion of non-small-cell lung cancer by targeting miR-206 and activating Akt/mTOR signaling. Anti-Cancer Drugs 29:725–735
Tano K, Mizuno R, Okada T et al (2010) MALAT-1 enhances cell motility of lung adenocarcinoma cells by influencing the expression of motility-related genes. FEBS Lett 584:4575–4580
Thai P, Statt S, Chen CH et al (2013) Characterization of a novel long noncoding RNA, SCAL1, induced by cigarette smoke and elevated in lung cancer cell lines. Am J Respir Cell Mol Biol 49:204–211
Wang P, Chen D, Ma H et al (2017a) Long non-coding RNA MEG3 regulates proliferation and apoptosis in non-small cell lung cancer via the miR-205-5p/LRP1 pathway. RSC Adv 7:49710–49719
Wang Q, Cheng N, Li X et al (2017b) Correlation of long non-coding RNA H19 expression with cisplatin-resistance and clinical outcome in lung adenocarcinoma. Oncotarget 8:2558–2567
Weber DG, Johnen G, Casjens S et al (2013) Evaluation of long noncoding RNA MALAT1 as a candidate blood-based biomarker for the diagnosis of non-small-cell lung cancer. BMC Res Notes 6:518
Wei MM, Zhou GB (2016) Long non-coding RNAs and their roles in non-small-cell lung Cancer. Genomics Proteomics Bioinformatics 14:280–288
Wei MM, Zhou YC, Wen ZS et al (2016) Long non-coding RNA stabilizes the Y-box-binding protein 1 and regulates the epidermal growth factor receptor to promote lung carcinogenesis. Oncotarget 7:59556–59571
Wen X, Han XR, Wang YJ et al (2018) Effects of long noncoding RNA SPRY4-IT1-mediated EZH2 on the invasion and migration of lung adenocarcinoma. J Cell Biochem 119:1827–1840
White NM, Cabanski CR, Silva-Fisher JM et al (2014) Transcriptome sequencing reveals altered long intergenic non-coding RNAs in lung cancer. Genome Biol 15:429
Wu D, Li Y, Zhang H et al (2017) Knockdown of Lncrna PVT1 enhances Radiosensitivity in non-small cell lung Cancer by sponging Mir-195. Cell Physiol Biochem 42:2453–2466
Xia L, Huang W, Tian D et al (2013) Overexpression of forkhead box C1 promotes tumor metastasis and indicates poor prognosis in hepatocellular carcinoma. Hepatology 57:610–624

Xue Y, Ni T, Jiang Y et al (2017) Long noncoding RNA GAS5 inhibits tumorigenesis and enhances radiosensitivity by suppressing miR-135b expression in non-small cell lung cancer. Oncol Res 25:1305–1316

Yang Y, Li H, Hou S et al (2013) The noncoding RNA expression profile and the effect of lncRNA AK126698 on cisplatin resistance in non-small-cell lung cancer cell. PLoS One 8:e65309

Zhang EB, Yin DD, Sun M et al (2014) P53-regulated long non-coding RNA TUG1 affects cell proliferation in human non-small cell lung cancer, partly through epigenetically regulating HOXB7 expression. Cell Death Dis 5:e1243

Zhang J, Zhu N, Chen X (2015) A novel long noncoding RNA LINC01133 is upregulated in lung squamous cell cancer and predicts survival. Tumour Biol 36:7465–7471

Zhao W, An Y, Liang Y et al (2014) Role of HOTAIR long noncoding RNA in metastatic progression of lung cancer. Eur Rev Med Pharmacol Sci 18:1930–1936

Zhao C, Qiao C, Zong L et al (2018) Long non-coding RNA-CCAT2 promotes the occurrence of non-small cell lung cancer by regulating the Wnt/β-catenin signaling pathway. Oncol Lett 16:4600–4606

Zhou Y, Zhang X, Klibanski A (2012) MEG3 noncoding RNA: a tumor suppressor. J Mol Endocrinol 48:R45–R53

Zhou M, Guo M, He D et al (2015) A potential signature of eight long non-coding RNAs predicts survival in patients with non-small cell lung cancer. J Transl Med 13:231

Long Noncoding RNAs in Cardiovascular Diseases

Laura Schoppe, Tim Meinecke, Patrick Hofmann, Ulrich Laufs, and Jes-Niels Boeckel

Contents

Abstract A major part of the human transcriptome consists of long noncoding RNAs (lncRNAs). These are RNA molecules longer than 200 nucleotides that do not contain open reading frames but can have a multitude of regulatory functions within cells. LncRNAs were found to be regulators of and to be regulated in several cardiovascular disease (CVD) entities. LncRNAs were found in various cell types

---

Authors Laura Schoppe, Tim Meinecke, and Patrick Hofmann equally contributed to this chapter.

L. Schoppe
University of Heidelberg, Heidelberg, Germany

T. Meinecke · P. Hofmann · U. Laufs · J.-N. Boeckel (✉)
Klinik und Poliklinik für Kardiologie, University Hospital of Leipzig, Leipzig, Germany
e-mail: Jes-Niels.Boeckel@medizin.uni-leipzig.de

S. Jurga, J. Barciszewski (eds.), *The Chemical Biology of Long Noncoding RNAs*, RNA Technologies 11, https://doi.org/10.1007/978-3-030-44743-4_13

and tissues comprising the heart such as cardiomyocytes, endothelial cells, and vascular smooth muscle cells but also in non-cardiovascular cells of the blood. Functional characterizations suggest a pathophysiological importance of lncRNAs in development and course of CVD. Interestingly, several lncRNAs appear to be predominantly regulated in circulating immune cell populations. Recent studies showed that lncRNA expression profiles are able to discriminate different pathologies of heart failure and that their expression is further altered in response to therapy. Therefore, lncRNAs have the potential to serve as biomarkers for CVD and therapy outcome and as targets for therapy of CVD. We discuss the regulation of lncRNAs in the heart, the vasculature, and the blood in cardiovascular disease in this chapter.

Keywords LncRNA · Cardiovascular disease · Blood · Myocardium · Vessels · Myocardial infarction · Atherosclerosis · Coronary artery disease · Cardiac hypertrophy

1 Introduction

Cardiovascular diseases, including atherosclerosis causing coronary artery disease, myocardial infarction, and ischemic cardiomyopathy, remain the leading cause of morbidity and death in the western world. Endothelial cells (EC) lined within the interior surface of vessels have anti-adhesive properties, forming a barrier between the circulating blood and underlying tissues. Dysfunction or injury of ECs contributes to formation of arteriosclerosis which is a hallmark of coronary artery disease (CAD) (Leistner et al. 2016). The formation of arteriosclerotic plaques is a multifactorial process characterized by pathophysiological thickening of the vessel wall resulting from infiltration of immune cells and proliferation of vascular smooth muscle cells. Pro-inflammatory stimuli can cause impairment of ECs, enhancing atherosclerotic plaque formation. Several risk factors such as hypertension, hyperlipidemia, and diabetes mellitus further promote atherosclerotic plaque formation, thereby contributing to development and progression of atherosclerosis.

Heart failure caused by myocardial ischemia or cardiomyopathies (CMPs) accounts for more than 400,000 deaths per year globally. Therefore, CMPs are one of the leading causes of death in central Europe and North America and represent the leading cause for heart transplantation in patients below the age of 55. CMPs are myocardial disorders in which the heart muscle structurally and functionally deteriorates, without being primarily caused by CAD, hypertension, alcohol, viral infections, and valvular or congenital heart diseases. The heterogeneous pathophysiological aspects leading to CMPs and the poor outcome in prevalent heart failure treatment result in the need to further understand the underlying molecular mechanisms of this disease.

Although improvement in the treatment of systolic heart failure was made within recent decades using blockage of neuroendocrine activation, there has only been marginal improvement in treatment of distinct sub-entities, such as heart failure-induced cachexia or progressing cardiac fibrosis, underlining the need for innovative new treatment approaches.

The majority of the human transcriptome, which covers up to 75% of the genome, according to the ENCODE project, is made up of noncoding RNAs (ncRNAs) rather than protein-coding RNAs (Carninci et al. 2005). The population of ncRNAs is heterogeneous. NcRNAs are subdivided into circular and linear RNAs, the latter further into short (e.g., microRNAs (miRNAs)) and long noncoding RNAs (lncRNAs). LncRNAs are defined as RNA molecules longer than 200 nucleotides that are not translated into proteins.

LncRNAs have been classified into several subgroups by mode of action and depending on where they are encoded in relation to protein-coding genes in the genome. Interestingly, comparing different organisms, the number of ncRNAs varies, while in the human genome the class of lncRNAs was found to be the largest class of ncRNAs (Cabili et al. 2011; Derrien et al. 2012).

LncRNAs are transcribed by RNA polymerase II (Pol II), are posttranscriptionally modified, and can be polyadenylated (Ravasi et al. 2006; Wu et al. 2008; Khalil et al. 2009; Ramsköld et al. 2009). The genomic loci of lncRNAs have been used to classify them: they can be derived from pseudogenes, introns, and enhancers, have a stand-alone locus, or can be the antisense of a coding RNA (Kung et al. 2013). Several lncRNAs are regulated during emryonic development and their expression was found to be specific for certain differentation stages and specialised cell and tissue types, which suggests important biological roles and has made them a focus of preclinical research (Djebali et al. 2012). Although lncRNA sequences are poorly conserved across vertebrates, it is thought that their secondary and tertiary structures, as well as their regulatory sequence elements, exhibit a higher level of conservation, thereby maintaining their evolutionary conserved function (Diederichs 2014; Noviello et al. 2018). This, however, is an added challenge in lncRNA research using animal models. In general, expression levels of lncRNAs, similiar to other ncRNA classes, are rather low compared to usual mRNA expression levels (Ravasi et al. 2006).

The functions of lncRNAs are diverse and divided, *inter alia*, based on their presence in the cytoplasm or nucleus. LncRNAs located in the nucleus can affect epigenetic processes by forming a scaffold, stabilising the assembly of protein complexes at certain genomic loci, recruit protein factors, tether them to the DNA, or modulate the formation of nuclear compartments (Kung et al. 2013; Hanly et al. 2018). Furthermore, nuclear-located lncRNAs play an important role in the regulation of transcription and posttranscriptional processing by altering the binding of transcription factors and even affecting the binding of Pol II to the DNA (Kung et al. 2013). In contrast, cytoplasmic lncRNAs can interfere with the translation of mRNAs by competitively binding and thereby occupying the shared binding sequences of miRNAs, thereby inhibiting the binding of lncRNAs to their target mRNAs (Huang 2018). LncRNA functions are heterogeneous, thereby offering several potential targeting strategies for therapy, which warrants further study regarding their individual mechanisms of actions in different diseases. The three major cellular compartments involved in cardiovascular disease are the blood, the heart muscle, and the vasculature. Therefore, involvement of lncRNAs in CVD will be discussed based on this classification in the following article.

2 Molecular Mechanisms of Cardiovascular Diseases

Inflammation is a major trigger in the development of cardiovascular diseases (Ruparelia et al. 2017). The pro-inflammatory cytokine TNF-α is released in the acute phase of the inflammation process by activated macrophages (Chu 2013). The binding of TNF-α to its receptors activates the transcription factor NF-κB and the signaling of the MAPK pathway. Subsequently, NF-κB initiates the orchestrated transcription of several genes regulating cell proliferation and inflammation, as well as anti-apoptotic factors preliminarily fostering cell survival, implemented by activation of JNK signaling, resulting in the induction of transcription factors that drive transcription of pro-apoptotic factors and genes involved in cell differentiation and proliferation(Sedger and McDermott 2014).

Blood flow applies a constant force to the endothelium, this so called shear stress, can be laminar or disturbed. Vascular bifurcations cause disturbed shear stress having a pro-inflammatory effect on the vessel, while laminar shear stress has anti-inflammatory effects (Cunningham and Gotlieb 2005). Transcription factors of the KLF family are responsible for the atheroprotective effect of laminar shear stress, whereas NF-κB mediates the pro-inflammatory effect of disturbed shear stress (Hahn and Schwartz 2009). Inflammatory activation of the endothelium leads to expression of adhesion molecules, mainly ICAM-1, VCAM-1, and E-selectin, and their presentation on the endothelial cell surface. Circulating immune cells of the blood can bind to these adhesion molecules via integrins and transmigrate through the endothelial monolayer (Hartman and Frishman 2014).

Inflammatory activation of cells can be induced by cellular senescence or bacterial infection. Senescent cells secrete pro-inflammatory cytokines, thereby creating an inflammation-promoting environment (Rodier et al. 2009; Coppé et al. 2010). Senescent cells do not proliferate or undergo apoptosis, but remain metabolically and transcriptionally active. Senescence is initiated by two different pathways which are both induced upon DNA damage or other harmful stimuli. The p53/p21 pathway is the major pathway induced by senescence, being activated via ATM, a sensor protein for DNA damage. ATM regulates p53, which in turn regulates p21, which can control cell cycle arrest in G1 phase. The second pathway is driven by induction of p16, which also can result in a G1 phase arrest. All of the mechanisms mentioned above regulates the inflammatory activation of blood vessels and thereby are associated with the development of CVD.

Other important activators of inflammation are the pattern recognition receptors (PRRs). These include PAMPS (pathogen-associated molecular patterns), DAMPs (damage-associated molecular patterns), SR (scavenger receptors), mannose receptors, TLRs (Toll-like receptors), intracellular NOD proteins, and collectins. First in line to detect damage and pathogens via PRRs in the cardiovascular system are endothelial cells and leukocytes. The endothelium and leukocytes respond differently to activation by PRRs, caused by the different expression of receptor types and differences in activation of downstream pathways (Mitchell et al. 2007). Changes in activation of PRRs, such as chronic activation, can further enhance the development of atherosclerosis (Herwald and Egesten 2011).

NF-κB is the main intracellular relay of inflammation, being activated among others by the NOD pathway, TLR pathway, IL-1, and TNF as well as being a oxygen-sensitive transcription factor. NF-κB was first described in B lymphocytes but exists in nearly every human cell. NF-κB binds the κB-motive, which appears in numerous regulatory genomic loci, binding of NF-κB enhances the transcription of genes regulating both the innate and the adaptive immune response (Smith et al. 2006).

HIF (hypoxia-inducible factor) is a hypoxia-induced transcription factor that can also alter the activity of the inflammatory response (Taylor et al. 2016), being frequently induced in CVD (Abe et al. 2017).

Interleukins are key molecules in the modulation and activation of the immune response. For example, interleukin-6, which triggers the secretion of acute-phase proteins, was reported to play a role in the pathogenesis of atherosclerosis and cardiac dysfunction (Kanda and Takahashi 2004). Plasma levels of several interleukins were shown to correlate with severity and mortality of chronic heart failure (Rauchhaus et al. 2000).

During the development and progression of atherosclerosis, endothelial cell activation is characterized by the induction of ICAM-1, PCAM, E-selectin, and VCAM-1. Their elevated expressions enable the binding and activation of monocytes to the activated endothelium. Classical monocytes may then migrate through the endothelium and differentiate into macrophages in the tissue, at sites of inflammatory initiation. In the tissue, M2 macrophages can incorporate low-density lipoprotein via the scavenger receptor SR-A and SR-B, being activated by a variety of ligands, such as oxidized lipoproteins and chemically altered proteins (Kelley et al. 2014). The expression of SR-A and SR-B is induced on the cellular level by NF-κB.

PPARβ/δ (peroxisome proliferator-activated receptor-β/δ) expression is elevated in PBMCs of CAD patients (Barbosa et al. 2019). PPARβ/δ suppresses the activities of several transcription factors including NF-κB (Palomer et al. 2018) and further promotes the resolution of inflammation (Quintela et al. 2012).

The functions of lncRNAs in these molecular mechanisms underlying cardiovascular diseases are as heterogeneous as the lncRNAs themselves. This review focuses on the pathways affected in the three major compartments of the cardiovascular system, which are the blood, the heart muscle, and the vasculature. Each of these compartments shows different pathologies in which the pathological mechanisms described above play a part. The study of lncRNAs in the context of these pathologies offers a new dimension of understanding the development of these diseases and presents potential for the development of new treatment startegies.

3 LncRNA in Human Blood in Cardiovascular Diseases

The investigation of lncRNA expression and regulation on a tissue-specific level may improve our understanding of the molecular mechanisms underlying cardiovascular diseases. To the present day, measurement of lncRNAs in the blood of CVD patients is less invasive than other options when thinking about diagnostic and prognostic purposes. LncRNAs in the blood are easier to obtain as blood sampling has less impact on the patient compared to tissue sampling necessary when analyzing lncRNAs exclusively expressed in the heart or vasculature. However, unlike miRNAs, lncRNAs seem to be much more vulnerable to degradation by RNases in the cell-free compartment of the blood. Therefore, lncRNAs expressed in blood cells might serve as a more stable class of biomarkers compared to lncRNAs in the cell-free compartment of the blood (Devaux 2017). Nevertheless, some lncRNAs seem to be resistant to degradation and were recently measured in serum and plasma. Expressionally changed lncRNAs in the left ventricle were over 70% originated from the mitochondria, suggesting that circulating mitochondrial lncRNA in patients suffering from heart failure might be released from the heart tissue (Yang et al. 2014).

3.1 *Blood LncRNAs in Coronary Artery Disease*

Coronary artery disease (CAD) is the most common vascular disease (GBD 2013 Mortality and Causes of Death Collaborators 2015). CAD can lead to stable and instable angina, myocardial infarction, and sudden cardiac death. Risk factors include smoking, lack of exercise, obesity, hypertension, elevated glucose levels and elevated cholesterol levels (Wong 2014). In the past years, several studies aimed to identify lncRNAs circulating in blood that might be used as disease biomarkers or to predict the outcome of CAD.

3.1.1 HOTAIR

One of the most extensively studied lncRNA is **HOTAIR** (HOX antisense intergenic RNA). **HOTAIR** was shown to be upregulated in patients with CAD (Fig. 1, Table 1) (Avazpour et al. 2018). **HOTAIR** was detected in peripheral blood mononuclear cells (PBMCs) as well as plasma. HOTAIR is encoded in the HOXC gene cluster on chromosome 12 and functions by repression of Homeobox D cluster (HOXD) transcription (Cao 2014).

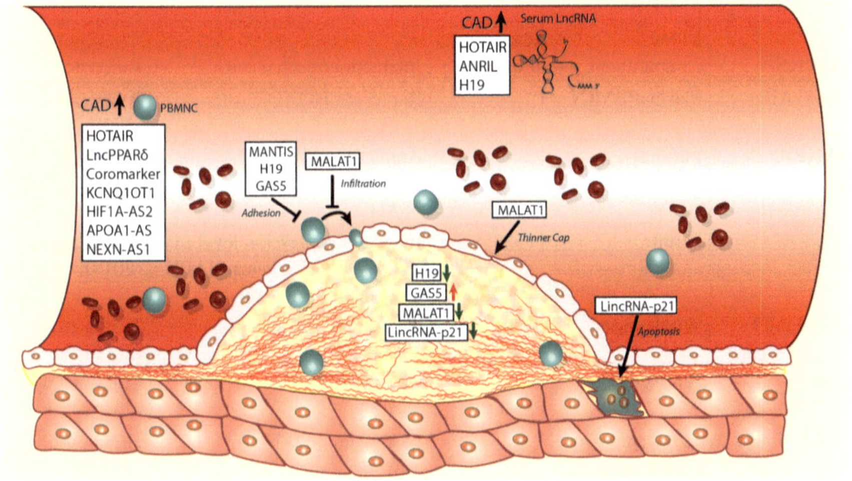

Fig. 1 Role of lncRNAs in atherosclerosis and plaque formation. LncRNAs in the blood appear to play an important role in atherosclerosis. The lncRNAs MANTIS, H19, and GAS5 in endothelial cells were shown to lower the adhesion potential of immune cells during the inflammtory response. Deletion of MALAT1 leads to increased infiltration of leukocytes and a thinner plaque cap. Levels of H19, MALAT1, and LincRNA-p21 are reduced in atherosclerotic plaque while GAS5 level is elevated. LincRNA-p21 inhibits cell proliferation and induces apoptosis in vascular smooth muscle cells. HOTAIR, ANRIL, and H19 are elevated in blood serum in coronary artery disease (CAD). HOTAIR, LncPPARδ, CoroMarker, KCNQ1OT1, HIF1A-AS2, APOA1-AS, and NEXN-AS1 are enriched in peripheral blood monocytes in CAD

3.1.2 ANRIL

ANRIL (antisense noncoding RNA in the INK4 locus) has already been associated with several human diseases such as CAD, intracranial aneurysm, gliomas, basal cell carcinomas, and type 2 diabetes. Initially ANRIL was identified as a major disease hot spot in genome-wide association studies (GWAS) searching for genetic causes of CAD (Pasmant et al. 2011). **ANRIL** was found to be elevated in patients with CAD (Fig. 1) and has further been correlated with a worse outcome of the disease (Hu et al. 2019c). Several mechanisms of action were described for ANRIL, such as NF-κB pathway activation leading to increased inflammation (Zhou et al. 2016), $p15^{INK4b}$ methylation (Zhuang et al. 2012), and induction of methylthioadenosine phosphorylase (MTAP), which is found in arteriosclerotic plaques, damaging plaque stability and potentially inducing thrombogenesis (Holdt et al. 2011). Interestingly, alterations in splicing and polymorphisms of ANRIL are suspected to increase the individual risk for CAD as well as other diseases (Pasmant et al. 2011; Yari et al. 2018; Xu et al. 2018).

3.1.3 MIAT

MIAT (myocardial infarction-associated transcript) and **MALAT1** (metastasis-associated lung adenocarcinoma transcript 1) have been proposed as potential

Table 1 Blood lncRNAs in coronary artery disease

LncRNA	Blood compartment	Regulation in CAD	Interaction/mechanism	References
HOTAIR	PlasmaPBMCs	↑	Represses HOXD cluster	Avazpour et al. (2018)
ANRIL	Plasma	↑	Activates NF-κB pathway and promotes expression of MTAP	Hu et al. (2019c)
H19	Serum	↑	Increases expression level of TGF-β1 and regulates MAPK and NF-κB pathway	Yao et al. (2019)
MALAT1	Whole blood	↑	Unknown	Toraih et al. (2019)
MIAT	Whole blood	↑	Unknown	Toraih et al. (2019)
NEXN-AS1	Monocytes (THP1) and macrophages	↑	Modulates NEXN	Hu et al. (2019b)
LncPPARδ	Monocytes	↑	Downregulates PPARδ, ADRP, and ANGPTL4	Cai et al. (2016a)
CoroMarker	PBMCs	↑	Pro-inflammatory	Cai et al. (2016b)
KCNQ1OT1	PBMCs	↑	Silencing of large genomic regions via formation of repressive chromatin	Pandey et al. (2008) and Zhang et al. (2019d)
HIF1A-AS2	PBMCs	↑	Antisense to HIF1-α mRNA	Zhang et al. (2019d)
APOA1-AS	PBMCs	↑	Antisense to APOA1 mRNA	Zhang et al. (2019d)

biomarkers for CVD, indeed both are found to be significantly elevated in CAD. MIAT was tested as a biomarker for detecting the presence of significant CAD (Toraih et al. 2019). A GWAS found two single nucleotide polymorphisms (SNPs) (rs4102217, rs619586) in MALAT1 that increase the risk of CAD (Hu et al. 2019a).

3.1.4 H19

H19 has an important role in embryogenesis and is postnatally repressed. However after birth, it can still be induced by hypoxia (Yoshimizu et al. 2008) and has been suggested as a biomarker that is elevated in the blood of patients with long-term CVD (Fig. 1), additionally being positively correlated with elevated TGF-β1 level in serum. In vitro experiments revealed that overexpression of H19 triggers TGF-β1 expression, suggesting a role in inflammation and fibrosis (Yao et al. 2019).

3.1.5 KCNQ1OT1, HIF1A-AS, and APOA1-AS

KCNQ1OT1 (KCNQ1 opposite strand/antisense transcript 1), **HIF1A-AS2** (hypoxia-inducible factor 1 alpha-antisense RNA 2), and **APOA1-AS** (apolipoprotein A-1 antisense RNA) were significantly upregulated in CAD (Fig. 1), showing individual but especially combined diagnostic value as a marker for CAD (AUC: 0.990). APOA1-AS was positively correlated with NT-proBNP, CKMB, MYO, and HsTnT, whereas HIF1A-AS2 was correlated with NT-proBNP and HsTnT (Zhang et al. 2019d). KCNQ1OT1 was reported to mediate silencing of large genomic regions via formation of repressive chromatin structures (Pandey et al. 2008).

3.1.6 NEXN-AS1

NEXN is an actin-binding protein that has protective functions in atherosclerosis. LncRNA **NEXN-AS1** (nexilin F-actin binding protein antisense RNA 1) is encoded complementarily to NEXN mRNA in the genome. Both exhibit significantly lower expression in human atherosclerotic plaques in endothelial cells, macrophages, and vascular smooth muscle cells as compared to healthy arteries. NEXN-AS1 has moreover been detected in THP1 monocytes, where elevated NEXN-AS1 level resulted in decreased ability of monocytes to adhere to endothelial cells. This effect was offset after NEXN knockdown, indicating that NEXN-AS1 functions mainly via modulating NEXN (Hu et al. 2019b).

3.1.7 LncPPARδ

LncPPARδ, also known as NONHSAT112178, can affect the expression of the neighboring protein-coding gene peroxisome proliferator-activated receptor δ (PPARδ) and its direct target genes, such as adipose differentiation-related protein (ADRP) and angiopoietin-like 4 (ANGPTL4). Due to its higher expression in PBMCs of CAD patients (Fig. 1) and its potential effect on PPARδ, it has been also proposed as a diagnostic marker for CAD (Cai et al. 2016a).

A recent study proposed the lncRNA **CoroMarker** (OTTHUMT00000387022) as a biomarker for diagnosis of CAD (Fig. 1) able to independently predict presence of CAD. CoroMarker knockdown in vitro decreased expression levels of the inflammatory markers IL-1β, IL-6, and TNF-α (Cai et al. 2016b).

3.2 *Blood LncRNAs in Cardiac Hypertrophy and Heart Failure*

Heart failure (HF) affects around 1–2% of adult population in the modern world and is one of the main causes of death in Europe (Lozano et al. 2012). It is a multifactorial and progressive disease and can occur with reduced (HFrEF) and preserved (HFpEF) ejection fraction (Tanai and Frantz 2015).

The currently most-noticed lncRNA associated with HF is **LIPCAR** (long intergenic noncoding RNA predicting cardiac remodeling), interestingly encoded in the mitochondrial DNA (Table 2) (Kumarswamy et al. 2014). LIPCAR was found to be downregulated shortly after myocardial infarction (MI). However, LIPCAR levels were also found to be increased in serum during post-MI-LV-remodeling. In chronic HF, LIPCAR levels are elevated and are further associated with mortality. This lncRNA transcript was therefore suggested as a marker for cardiac remodeling and HF (Kumarswamy et al. 2014). In patients with diabetes mellitus type 2, circulating LIPCAR was also shown to be negatively correlated to diastolic function while being positively correlated to grade I diastolic dysfunction (de Gonzalo-Calvo et al. 2016).

Heat2 (heart failure-associated transcript 2) has been identified using next-generation sequencing being regulated in PBMCs of HFrEF patients. Heat2 is enriched in the nuclei of basophils and eosinophils (Boeckel et al. 2018). Eosinophils and basophils are the least studied immune cells in CVD, but eosinophils are known to promote cardiac fibrosis by secreting TGF-β after infiltrating into the myocardium (Jacobsen et al. 2012).

ENST00000507296 is an lncRNA proposed as a biomarker for the diagnosis of dilated cardiomyopathy (DCM)-HF and cardiovascular events in DCM-HF. Its molecular function remains unknown (Zhang et al. 2019c).

Table 2 Blood lncRNAs in heart failure

LncRNA	Blood compartment	Regulation in HF	Interaction	References
LIPCAR	Serum	↑	Regulates mitochondrial pathways	Kumarswamy et al. (2014)
Heat2	Whole blood PBMCS Eosinophils Basophils	↑	Unknown	Boeckel et al. (2018)
ENST00000507296	Plasma	↑ (DCM)	Unknown	Zhang et al. (2019c)
NRON	Whole blood	↑	Unknown	Xuan et al. (2017)
MHRT	Whole blood	↑	Unknown	Xuan et al. (2017)

NRON (noncoding repressor of NFAT) and ***MHRT*** (myosin heavy-chain-associated RNA transcripts) could be verified as independent markers for HF. MHRT was positively associated with LDH and AST, while NRON was negatively associated to HDL and positively to LDL (Xuan et al. 2017).

3.3 Blood LncRNAs in Acute Myocardial Infarction

Acute myocardial infarction (AMI) is the blockage of coronary blood flow, which can result in malfunction and damage of the heart muscle. This is mostly due to thrombus formation in the coronary vessels.

The previously introduced lncRNAs **H19**, **MIAT**, and **MALAT1** have also been described as potential markers for detection of AMI in PBMCs, with a combined AUC of 0.756 (Table 3) (Wang et al. 2019c).

LncRNA **ZFAS1** (zinc finger antisense 1) was significantly lower in whole blood 6–24 h after AMI (Zhang et al. 2016b). ZFAS1 is antisense to the mRNA Znfx1 (zinc finger NFX-1-type containing).

LIPCAR, KCNQ1OT1, and **HIF1A-AS2** were found to be significantly increased in patients with STEMI (ST-elevation myocardial infarction). In this regard, LIPCAR was also predicting the occurence of STEMI, while after percutaneous coronary intervention, LIPCAR expression levels were reduced and were found to be abel to predict further cardiac events after AMI (Li et al. 2018d).

Table 3 Blood lncRNAs in acute myocardial infarction

LncRNA	Blood compartment	Regulation in AMI	Interaction	References
H19	PBMCs	↑	Unknown	Wang et al. (2019c)
MIAT	PBMCs	↑	Unknown	Wang et al. (2019c)
MALAT1	PBMCs	↑	Unknown	Wang et al. (2019c)
ZFAS1	Whole blood	↓	Antisense to Znfx1 mRNA	Zhang et al. (2016b)
LIPCAR	Plasma	↑	Regulates mitochondrial signaling	Li et al. (2018d)
KCNQ1OT1	Plasma	↑	Promotes formation of repressive chromatin structures	Li et al. (2018d)
HIF1A-AS2	Plasma	↑	Antisense to HIF-1α mRNA	Li et al. (2018d)

3.4 Blood LncRNAs in Acute Ischemic Stroke

Acute ischemic stroke (AIS) is the sudden shortage of blood flow to cerebral tissue. It can be caused by embolism, dissection, stenosis, and vasculitis. Symptoms vary according to the affected area of the brain.

Two studies examined level of circulating **ANRIL** in AIS. Interestingly, one study reported ANRIL levels in serum to be significantly higher alongside with hs-CRP and MMP-9, suggesting a diagnostic value for incidence and severity of AIS (Zhang et al. 2019b). Reduction of ANRIL in plasma was further associated with higher risk for AIS, as well as a worse outcome, and was further associated with higher levels of inflammatory marker gene expression (Feng et al. 2019).

MIAT was reported to be able to discriminate AIS patients from controls with an AUC of 0.842. Patients with higher MIAT had poor prognoses compared to those individuals with lower levels (Zhu et al. 2018). After middle cerebral artery occlusion in mouse model, the hypoxia-inducible lncRNA **H19** was shown to be elevated in brain tissue, leukocytes, and plasma and was further positively correlated with the levels of TNF indicating the induction of inflammation.

H19 knockdown in mice resulted in neurological deficits and further led to reduction of brain tissue after AIS (Wang et al. 2017). Neurogenesis might be disturbed by H19 through inhibition of p53 and decreased Notch1 transcription (Wang et al. 2019a) (Table 4).

3.5 Blood LncRNAs in Kawasaki Disease

Kawasaki disease (KD), an acute vasculitis in children, that can lead to long-term cardiovascular damage if untreated. It is the most frequent cause of cardiac disease at a young age in developed countries (Burns and Glodé 2004).

LncRNA **XLOC_006277** was identified as a marker for progression of KD. XLOC_006277 is elevated in all types of KD, higher during acute stage and even higher in patients which develops complications, such as coronary artery aneurysms. In vitro knockdown of XLOC_006277 resulted in reduction of matrix-metallo-protease-8 and matrix-metallo-protease-9 expression (Ko et al. 2019).

Table 4 LncRNA in acute ischemic stroke

LncRNA	Blood compartment	Regulation in AIS	Interaction	References
ANRIL	Plasma/serum	↓/↑	Unknown	Zhang et al. (2019b) and Feng et al. (2019)
MIAT	Whole blood	↑	Unknown	Zhu et al. (2018)
H19	Plasma	↑	Inhibition of p53/Notch1 pathway	(Wang et al. 2017, 2019a)

THRIL (TNF-α and hnRNPL-related immunoregulatory lincRNA) was found to be elevated in whole blood of patients in acute stage of KD as well as after 2–3 months in the recovery stage. Therefore, THRIL was associated with stage and severity of KD. Mechanistically, THRIL was shown to form a complex with heterogeneous nuclear ribonucleoprotein L (hnRNPL) that is known to control the promoter of TNF (Li et al. 2014).

4 LncRNAs in the Myocardium and Cardiovascular Disease

LncRNAs present in the myocardium are emerging to have important roles in the development and progression of CVDs. Several lncRNAs have meanwhile been reported to have regulatory effects on cardiac ischemia, the course of reperfusion injury, myocardial infarction, and myocardial hypertrophy. Less is currently known about their effects in cardiomyopathies such as dilated cardiomyopathy (DCM). Besides their potential use as biomarkers for CVD, increasing knowledge of the respective mechanisms of action by which lncRNAs in the myocardium can aggravate or alleviate disease is of great importance for the development of future treatment options.

4.1 Myocardial LncRNAs in Cardiac Ischemia and Reperfusion Injury

Cardiac ischemia describes a pathological condition where blood flow to the myocardium is reduced or restricted, either through narrowing of coronaries due to CAD or by AMI. Blood supply becomes insufficient for the tissue's requirements, leading to hypoxia, insufficient metabolite supply, as well as the accumulation of metabolic waste, ultimately leading to apoptosis or necrosis of cells within the tissue. Reperfusion injury can occure when blood flow is restored in a previously ischemic tissue. Reactive oxygen species are generated, and inflammatory signaling is activated in the reperfused tissue, exacerbating tissue damage (Kalogeris et al. 2012). Several lncRNAs have been shown to heighten or reduce cardiomyocyte apoptosis in this context.

The lncRNA **FTX** (Five prime to Xist) has been shown to be downregulated in cardiomyocytes upon ischemia/reperfusion (I/R) injury (Fig. 2, Table 5). It was further reported that increased expression of FTX in vitro can inhibit hydrogen peroxide-induced cardiomyocyte apoptosis. Mechanistically, the lncRNA FTX regulates the expression of anti-apoptotic protein BCL2L2 by functioning as an endogenous sponge for miR-29b-1-5p (Long et al. 2018).

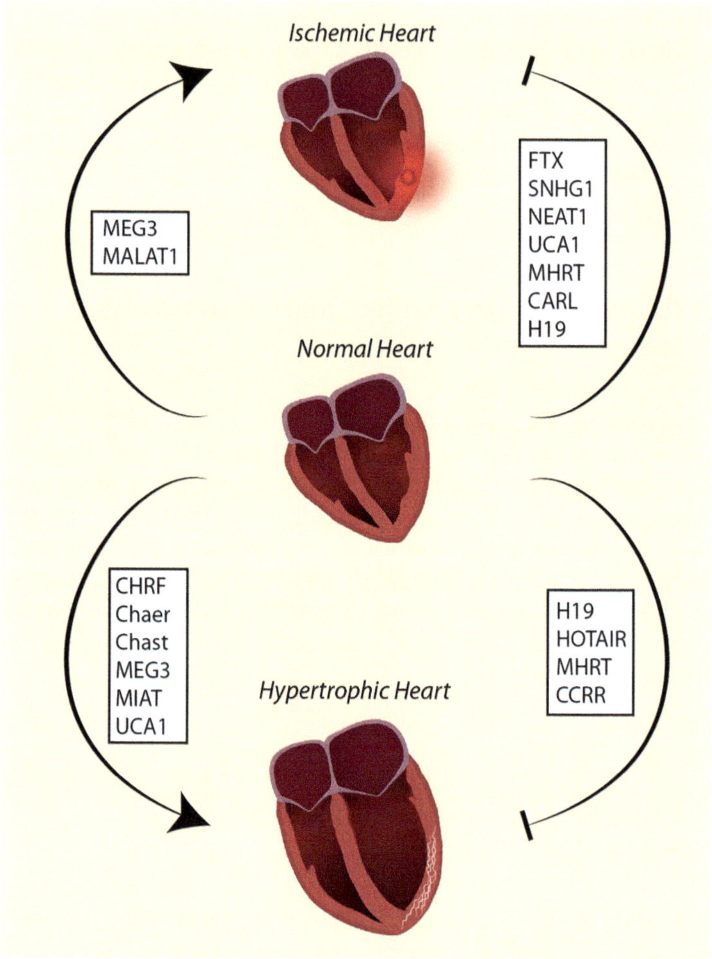

Fig. 2 LncRNA affecting myocardial diseases. Two major cardiac pathologies in which lncRNAs have been shown to play a decisive role are the development of ischemic and hypertrophic heart disease. High lncRNA expression can have disease-inhibiting or disease-promoting effects. Increased expression of MEG3 and MALAT1 in cardiomyocytes was demonstrated to have an adverse effect in ischemic heart disease, promoting apoptosis. FTX, SNHG1, NEAT1, UCA1, MHRT, CARL, and H19, on the other hand, had shown anti-apoptotic effects, protecting cardiomyocytes. In hypertrophic cardiomyopathy, CHRF, Chaer, Chast, MEG3, MIAT, and UCA1 have been shown to aggravate disease, whereas H19, HOTAIR, MHRT, and CCRR have cardioprotective effects in the context of hypertrophy

SNHG1 (small nucleolar RNA host gene 1) has similarly been shown to alleviate human cardiomyocyte apoptosis after hydrogen peroxide treatment by functioning as a competing endogenous RNA (ceRNA) for miR-195 (Fig. 2). SNHG1 thereby positively regulates BCL2L2 expression on mRNA and protein level. Although downregulated upon hydrogen peroxide treatment, SNHG1 overexpression was able to improve cellular viability (Zhang et al. 2018a). Indeed, SNHG1

Table 5 Myocardial lncRNAs in cardiac ischemia and reperfusion injury

LncRNA	Cell type	Regulation in I/R	Interaction	References
FTX	Cardiomyocyte	↓	Competing endogenous RNA (ceRNA) for miR-29b-1-5p, thereby regulating BCL2L2	Long et al. (2018)
SNHG1	Cardiomyocyte	↓	CeRNA for miR-195, thereby positively regulating BCL2L2	Zhang et al. (2018a) and Chen et al. (2019b)
NEAT1	Cardiomyocyte	↓/↑	Stimulates BCL2L12 expression and inhibits miR-125a-5p/miR-520a interaction	Yan et al. (2019) and Wu et al. (2019)
UCA1	Cardiomyocyte	↓	Downregulation of UCA1 causes increased p27 expression.	Liu et al. (2015)

overexpression also protected cardiomyocytes from the toxic effects of the chemotherapeutic drug doxorubicin (Chen et al. 2019b).

The lncRNA **NEAT1** (nuclear paraspeckle assembly transcript 1) received its name due to its involvement in the formation of nuclear paraspeckles (Clemson et al. 2009), and its dysregulation was first reported in several tumor types (Zeng et al. 2014; Qi et al. 2018). In the context of CVD, NEAT1 was found to play a role in skeletal muscle myogenesis (Wang et al. 2019b). Furthermore, NEAT1 is downregulated in murine cardiomyocytes in vivo upon I/R, whereas NEAT1 overexpression was found to suppress apoptosis via stimulation of BCL2L12 expression (Fig. 2). This effect was attributable to the direct interaction of NEAT1 with BCL2L12, thereby abolishing the inhibitory effect of miR-125a-5p (Yan et al. 2019). However, other research has found NEAT1 to be induced in cardiomyocytes of rats in an I/R model. In line, the same study further demonstrated that knockdown of NEAT1 inhibited cardiomyocyte apoptosis. The mechanism behind this regulation involved direct interaction with miR-520a, affecting regulation of pro- and anti-apoptotic factors (Wu et al. 2019).

UCA1 (urothelial carcinoma associated 1), originally identified in bladder cell transitional carcinoma (Wang et al. 2006), was, in contrast to NEAT1, recently demonstrated to be downregulated in the hearts of rats upon cardiac I/R injury exacerbating cardiomyocyte apoptosis (Fig. 2). This downregulation was shown to lead to an increase in pro-apoptotic p27 protein expression. Therefore, UCA1 seems to fulfill anti-apoptotic function in the ischemic heart (Liu et al. 2015).

4.2 *Myocardial LncRNAs in Myocardial Infarction*

Myocardial infarction (MI) is a frequent consequence of myocardial ischemia and coronary artery disease. Many of the lncRNAs mentioned previously as circulating biomarkers in the blood are also expressed in myocardial cells under MI conditions.

Unlike many of the previously described lncRNAs, **MHRT** (myosin heavy-chain-associated RNA transcript) has a protective effect against cardiomyocyte apoptosis (Fig. 2). MHRT is upregulated in patients with acute MI, and its functions were studied in hydrogen peroxide stimulated neonatal rat cardiac myocytes. Expression of MHRT was heart specific. Knockdown of MHRT in vitro resulted in increased cardiomyocyte apoptosis upon I/R injury (Zhang et al. 2016a).

CARL (cardiac apoptosis-related lncRNA) likewise has cardioprotective function by acting as an endogenous sponge for miR-539 to regulate PHB2 expression, thereby inhibiting mitochondrial fission and apoptosis (Fig. 2) (Wang et al. 2014b).

CPR (cardiomyocyte proliferation regulator) was demonstrated to negatively regulate cardiomyocyte proliferation that is upregulated in adult versus developing hearts. Knockout improved cardiac repair after MI, which is decisive for recovery. Mechanistically, CPR causes epigenetic changes by recruiting DNMT3A to the promoter of microchromosome maintenance 3, thereby silencing its expression (Ponnusamy et al. 2019).

CRRL (cardiomyocyte regeneration-related lncRNA) is likewise upregulated in the adult heart. It acts as a ceRNA for miR-199a-3p, thereby increasing expression of Hopx. Via this mechanism, CRRL causes negative regulation of cardiomyocyte proliferation. CRRL loss promotes regeneration of the myocardium after damage and prevents detrimental cardiac remodeling (Chen et al. 2018).

CAREL's (cardiac regeneration-related long noncoding ribonucleic acid) upregulation during development also causes loss of cardiac regenerative capacity. CAREL is a ceRNA for miR-296, releasing expression of its targets Trp53inp1 and Itm2a (Cai et al. 2018).

Sirt1-AS (silent information regulatory factor 2-related enzyme 1 antisense lncRNA) on the other hand was demonstrated to be a positive regulator of cardiomyocyte proliferation that becomes upregulated throughout heart development. It was shown to reduce infarct size, alleviate apoptosis, and improve heart function in mice (Li et al. 2018a).

4.3 *Myocardial LncRNAs in Cardiac Hypertrophy and Heart Failure*

Cardiac hypertrophy (CH) describes thickening of the heart muscle in response to pathophysiological events such as hemodynamic stress. Although this initially allows the heart to compensate the pathological conditions, it ultimately leads to heart failure (Carreño et al. 2006).

4.3.1 Pro-hypertrophic LncRNAs

CHRF (cardiac hypertrophy-related factor), also known as AK048451, is a regulatory factor in CH. CHRF functions as an endogenous sponge for anti-hypertrophic miR-489, which it binds directly. This leads to an upregulation of Myd88 expression and amplified hypertrophy (Fig. 2, Table 7) (Wang et al. 2014a).

Chaer (cardiac hypertrophy-associated epigenetic regulator) is an important epigenetic regulator enriched in cardiac tissue under stress conditions and was shown to be a necessary factor in the development of CH (Fig. 2). Chaer interaction with polycomb repressor complex 2 (PRC2) inhibits its binding to the genome, ultimately preventing the inactivating H3K27 methylation at promoter regions of hypertrophy genes. Furthermore, it was shown that inhibition of Chaer before pressure overload could have a beneficial effect on heart function (Wang et al. 2016b).

Chast (cardiac hypertrophy-associated transcript) was shown to be upregulated in TAC mice, and its human homolog CHAST was upregulated in hypertrophic patients' hearts (Fig. 2). Lentiviral overexpression of Chast in phenylephrine (PE) and isoproterenol (ISO) treated cardiomyocytes induced hypertrophy, while silencing of the lncRNA reduced hypertrophic response (Viereck et al. 2016).

Table 7 Myocardial lncRNAs in cardiac hypertrophy

LncRNA	Cell type	Regulation in CH	Interaction	References
CHRF	Cardiomyocyte	↓	Sponge for miR-489 and Myd88 upregulation	Wang et al. (2014a)
Chaer	Cardiomyocyte	↑	PRC2 inhibition	Wang et al. (2016b)
CHAST	Cardiomyocyte	↑	Unknown	Viereck et al. (2016)
MEG3	Cardiomyocyte	↑	Regulates miR-361-5p and HDAC9 as a ceRNA	Zhang et al. (2019a)
MIAT	Cardiomyocyte	↑	CeRNA for miR-93, thereby promoting TLR4 expression	Li et al. (2018e, f)
UCA1	Cardiomyocyte	↑	Downregulates miR-184, thereby increasing HOXA9 expression	Zhou et al. (2018)
H19	Cardiomyocyte	↑	MiR-675 CaMKIIδ binding	Liu et al. (2016)
HOTAIR	Cardiomyocyte	↓	CeRNA for miR-19 and disinhibition of PTEN	Lai et al. (2017)
MHRT	Cardiomyocyte	↓	Antagonizes Brg1	Han et al. (2014)
CCRR	Cardiomyocyte	↓	Inhibits Connexin43 endocytosis	Zhang et al. (2018b)

MEG3 was also shown to promote CH (Fig. 2). STAT3 transcription factor upregulates the expression of MEG3 in TAC mice as well as angiotensin II (AngII)-stimulated cardiomyocytes and chronic heart failure tissue. MEG3 lncRNA regulates miR-361-5p and HDAC9 as a ceRNA (Zhang et al. 2019a).

MIAT also contributes to CH by functioning as a ceRNA for miR-93, thereby promoting TLR4 expression (Fig. 2). MIAT was upregulated in AngII-treated cardiomyocytes, and its knockdown reduced cell surface area and fetal gene expression (ANF and β-MHC) related to cardiac remodeling (Li et al. 2018e). A different study additionally demonstrated that MIAT targets the miR-150/P300 axis as a positive regulator of hypertrophy (Li et al. 2018f).

UCA1 (urothelial carcinoma associated 1) is highly expressed in TAC mice as well as PE-treated cardiomyocytes (Fig. 2). It was demonstrated to downregulate miR-184, thereby increasing HOXA9 expression. Knockdown of UCA1 reduced cell surface and fetal gene expression (ANP, BNP) (Zhou et al. 2018).

4.3.2 Anti-hypertrophic LncRNAs

H19 is upregulated in CH and heart failure (Fig. 2). Its effect was shown be anti-hypertrophic via its encoded miR-675. Furthermore, this effect was demonstrated to be due to the direct binding of miR-675 to CaMKIIδ (Liu et al. 2016).

HOTAIR expression is downregulated in TAC mice and AngII-stimulated cardiomyocytes (Fig. 2). Its expression was shown to negatively correlate with that of miR-19, indicating its role as a ceRNA, causing disinhibition of PTEN. Overexpression of HOTAIR led to a reduction in cell surface area as well as reduced expression of fetal genes (ANP, BNP, β-MHC) associated with hypertrophic development in response to AngII-stimulation (Lai et al. 2017).

Mhrt (myosin heavy-chain-associated RNA transcripts) are a cluster of transcripts from Myh7 loci that are heart-specifically abundantly expressed in adult mice. Stress induces repression of Mhrt expression, and restoration of Mhrt expression offers protection from hypertrophy and heart failure (Fig. 2). The mechanism behind this is the antagonistic effect Mhrt has on chromatin remodeling factor Brg1. Mhrt prevents Brg1's binding to its genomic target regions by sequestering it. The human variant of MHRT is also repressed in cardiomyopathy, and a similar role is suggested (Han et al. 2014).

CCRR (cardiac conduction regulatory RNA) was demonstrated to be downregulated in heart failure patients as well as a mouse model of heart failure (Fig. 2). CCRR silencing in mouse induced cardiac arrhythmias in otherwise healthy mice. Overexpression on the other hand improved cardiac conduction by inhibiting Connexin43 endocytosis (Zhang et al. 2018b).

Further lncRNAs, whose mechanisms and functions are not yet fully understood, have been identified in biopsies of heart failure patient tissue and show high association with the disease. **EGOT, LOC285194, RMRP, RNY5, SOX2-OT,** and **SRA1** showed significantly increased expression in HF myocardial biopsies. **CDKN2B-AS** and **TUSC7 (LOC285194)** are found to be modulated in cardiac tissue as well as the bloodstream, making them attractive potential biomarkers of disease (Greco et al. 2016).

Additionally, **SLC26A4-AS1, RP11-344E13.3**, and **MAGI1-IT1** upregulation has been associated with CH (Song et al. 2016).

5 LncRNA in Vasculature and Cardiovascular Disease

5.1 Vascular LncRNAs in Atherosclerosis

Atherosclerosis is an inflammatory disease with high prevalence in the population (Epstein and Ross 1999). The formation of atherosclerotic plaques usually starts at an early age, and plaques grow over time at varying speeds, depending on genetic and environmental factors. Under non-inflammatory conditions, endothelial cells (EC) lined within the interior surface of vessels have anti-adhesive properties. In contrast, activated endothelial cells rapidly initiate the expression of cell surface adhesion molecules such as ICAM-1, VCAM-1, and E-selectin. Circulating monocytes can adhere to these adhesion molecules and migrate through the endothelial monolayer, where they differentiate into macrophages. In the progression of plaque formation, apolipoprotein B100 (ApoB100) was reported to bind to extracellular matrix proteoglycans and thereby lead to accumulation of LDL particles in the intima, where they often undergo oxidative modifications (Weber and Noels 2011). Oxidized LDL further triggers the expression of adhesion molecules on the endothelium. A subtype of macrophages, the foam cells, are characterized by preceded ingestion of oxidized LDL particles within sites of plaque development. Foam cells are known to eventually rupture, thereby leading to accumulation of apoptotic cells, debris, and cholesterol crystals in the plaque, which in turn can result in the formation of a necrotic core. A further pro-atherosclerotic stimulus is the release of cytokines such as INF-γ or TNF-α, which trigger the proliferation of smooth muscle cells, the release of extracellular matrix (ECM) molecules, as well as remodeling of the ECM, which can result in further growth of the plaque. Late-stage plaques can rather be covered by a thin fibrous cap, which mainly consists of macrophages and is prone to rupture. Recent studies indicate that besides plaque rupture, plaque erosion also plays an important role in arterial thrombosis.

Several lncRNAs were shown to play a role in the process of atherosclerotic lesion formation. The lncRNA **MEG3** (maternally expressed gene 3) is a nuclear-located transcript, which was shown to be upregulated in senescent human umbilical vein endothelial cells (HUVECs) compared to non-senescent control cells (Fig. 3, Table 8) (Boon et al. 2016). Interestingly, MEG3 expression in HUVECs was repressed upon stimulation with the inflammatory cytokine TNF-α, although the duration of TNF-α stimulation remains unclear in this study (Wu et al. 2017b). Senescent cells are known to secrete pro-inflammatory cytokines and create a pro-inflammatory environment (Coppé et al. 2008; Rodier et al. 2009). Cytokine secretion upon senescence is maintained mainly by the pro-inflammatory NF-κB

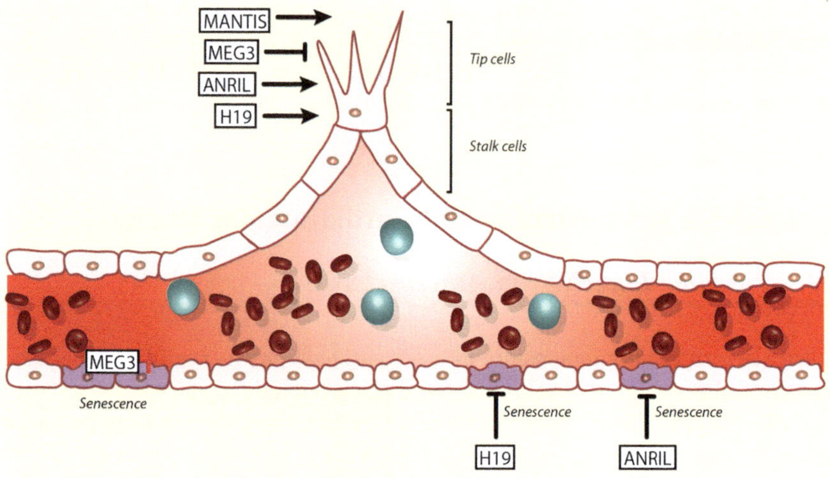

Fig. 3 Angiogenesis is the process of the formation of new blood vessels from existing vessels through proliferation of endothelial stalk cells and migration of the tip cell. Otherwise, quiescent ECs start to proliferate and migrate in response to VEGF, and senescence counteracts this process. ANRIL and H19 were shown to inhibit senescence, while the expression of MEG3 was induced in senescent ECs. MEG3 furthermore inhibited angiogenesis, while H19, ANRIL, and MANTIS had pro-angiogenic effects

transcription factor family (Chien et al. 2011), and NF-κB in turn is mainly activated via TNF cytokines. Overexpression of MEG3 in HUVECs delayed proliferation and angiogenesis (Chien et al. 2011; He et al. 2017), supporting the finding that MEG3 plays a role in senescence of ECs. MEG3 was shown to interact with chromatin factors EZH2 and JARID2 and thereby mediate epigenetic regulation of gene expression (Kaneko et al. 2014). Another study furthermore reported that MEG3 exerts its function by inhibiting expression of miR-21 and thereby upregulates expression of the miR-21 targets RhoB and PTEN (Chien et al. 2011). Functionally, MEG3 was also shown to bind miR-9 and thereby acts as an miRNA sponge (He et al. 2017). There are still open questions on how generally poorly expressed lncRNA can act as decoys for usually highly abundant miRNAs stoichiometry-wise (Ulitsky 2018). Pharmacological inhibition of MEG3 in HUVECs prevented senescence-mediated inhibition of angiogenic sprouting in vitro and improved perfusion in a hind limb ischemia model in mice (Boon et al. 2016). The primary mechanism of action of MEG3 in ECs remains to be elucidated given the partially contradicting mechanistic and regulatory findings.

Atherosclerotic lesion formation was often reported to originate from regions with low or disturbed shear stress like bifurcations of arteries (Moore and Tabas 2011). Shear stress-regulated lncRNAs could therefore represent potential therapeutic targets for pharmaceutical interventions. In this regard, the lncRNA **MANTIS** was reported to be upregulated by laminar shear stress in a KLF-2- and KLF-4-

Table 8 Vascular lncRNAs in atherosclerosis

LncRNA	Compartment	Regulation in atherosclerosis	Function	References
MEG3	Endothelial cells	Depends on stimulus	Interacts with EZH2 and JARID2; upregulation of RhoBP and PTEN through inhibition of miR-21; sponging of miR-9	Chien et al. (2011), Kaneko et al. (2014), and He et al. (2017)
MANTIS	Endothelial cells	↓/↑	Interaction with BRG1	Leisegang et al. (2017, 2019)
H19	Endothelial cells and myoblasts	↓	Inhibition of STAT3 activation; sponging of miRs	Kallen et al. (2013) and Hofmann et al. (2019)
GATA6-AS	Endothelial cells	↑	Interaction with LOXL2 and alteration of H3K4me3 methylation	Neumann et al. (2018)
MALAT1	Endothelial cells	↓	Sponge for miR-504	Cremer et al. (2019)
LincRNA-p21	Endothelial cells and vascular smooth muscle cells	↓	Sponging of miR-130b; interaction with MDM2	Wu et al. (2014) and He et al. (2015)
HOTAIR	Fibroblasts and endothelial cells	↓	Promotion of proliferation, migration, and survival	Peng et al. (2017)
GAS5	Endothelial cells	↑	Interaction with β-catenin	Wang et al. (2016a), Chen et al. (2016), and Shen and She (2018)
ANRIL	Endothelial cells	↑	Activating of the NF-κB signaling pathway; interaction with EZH2 and p300	Thomas et al. (2017), Zhang et al. (2017), and Arslan et al. (2017)

dependent manner. Deletion of MANTIS in ECs inhibited angiogenic sprouting and alignment of the cells in the direction of flow (Fig. 3). Moreover, an increase of ICAM-1-mediated monocyte adhesion was observed, which represents an important mechanism in the development of atherosclerotic plaques (Fig. 1). Furthermore, MANTIS was required to mediate the observed atheroprotective effect of sHMG-CoA reductase inhibitors (statins) by facilitating the atorvastatin-induced changes in EC gene expression (Leisegang et al. 2017, 2019). MANTIS might thereby represent an interesting target for pharmaceutical intervention, as the atheroprotective effect of MANTIS might be essential for regions of disturbed blood flow, where KLF expression is usually low.

Another shear stress-regulated lncRNA is **H19**, which was one of the first lncRNAs to be discovered and initially thought to encode for a protein (Han et al. 1996). H19 is upregulated by laminar shear stress in a KLF2-dependent manner and repressed by pro-inflammatory stimulation with TNF-α or by aging (Hofmann et al. 2019). H19 is highly expressed in embryonic tissues but downregulated shortly after birth in most of the analyzed tissues. H19 was further found to be elevated in human atherosclerotic lesions in a mosaic-like manner in cells of unknown origin (Han et al. 1996). H19 was clearly localized in the endothelial layer and intra-plaque blood vessels in samples of atherosclerotic plaques from human carotid arteries and was downregulated in this tissue compared to healthy human carotid arteries (Fig. 1) (Hofmann et al. 2019). Depletion of H19 in HUVECs resulted in delayed cell cycle progression, increased senescence, inhibition of angiogenic sprouting in vitro and ex vivo, and further reduced perfusion upon hind limb ischemia in vivo (Fig. 3). Depletion of H19 led to higher expression of ICAM-1 and VCAM-1 and subsequently to increased monocyte adhesion to the endothelium through elevated STAT3 phosphorylation, whereas inhibition of STAT3 phosphorylation abolished the effect of H19 depletion (Fig. 1) (Hofmann et al. 2019). H19 was further suggested to bind miRNAs of the let-7 family with four identified binding sites for let-7 miRNAs. Phenotypically, depletion of H19 in C2C12 cells resulted in accelerated muscle differentiation while this effect was abolished by simultaneous overexpression of let-7 miRNAs (Kallen et al. 2013). Furthermore, H19 were shown to be enhanced on an expressional level in many cancer cells (Matouk et al. 2007), highlighting the often cell-type-specific expression of lncRNAs and the necessity to find therapeutic approaches for targeting lncRNAs in a tissue-specific manner.

A hypoxia-regulated lncRNA antisense to the GATA6 gene, **GATA6-AS**, was shown to regulate endothelial gene expression by interacting with the epigenetic regulator LOXL2. GATA6-AS acts as a negative regulator of LOXL2 function, while silencing of GATA6-AS reduced H3K4me3 methylation of two angiogenesis-related genes. Silencing of GATA6-AS furthermore diminished TGF-β-induced endothelial-to-mesenchymal transition in vitro and angiogenesis in vivo (Neumann et al. 2018).

Expression of the hypoxia-induced lncRNA, **MALAT1**, was shown to be reduced in atherosclerotic plaques (Cremer et al. 2019), while genetic variations of the MALAT1 gene are associated with a decreased risk for CAD (Wang et al. 2018). Interestingly, deletion of MALAT1 in mice analyzed in a model for atherosclerosis showed an increased infiltration of CD45+ leukocytes into sites of atherosclerotic plaques and further resulted in a thinner plaque cap, whereas total plaque size was only modestly increased. MALAT1 acts as a sponge for miR-503, and MALAT1 deficiency in bone marrow mononuclear cells increased adhesion to endothelial cells in vitro (Cremer et al. 2019).

Expression of the nuclear lncRNA **lincRNA-p21** is reduced in human and mouse atherosclerotic plaques (Fig. 1) (Çekin et al. 2018). LincRNA-p21 is regulated by p53 and physically interacts with MDM2 in a feedback mechanism to enhance p53 transcriptional activity. Binding of lincRNA-p21 to MDM2 releases MDM2-mediated repression of p53 and enables the interaction of p53 with p300, mediating binding of p53 to promoters of its target genes. In vitro experiments further showed that lincRNA-p21 represses cell proliferation and induces apoptosis in vascular smooth muscle cells (SMCs) and mouse mononuclear macrophages. Furthermore, its inhibition promoted neointimal hyperplasia in a carotid artery injury model in vivo (Wu et al. 2014). In mice, lincNA-p21 acts as a ceRNA for miR-130b in the endothelium and thereby delays proliferation (He et al. 2015).

HOX antisense intergenic RNA (**HOTAIR**) is found in the nucleus, as well as in the cytoplasm, and its silencing reduced senescence of fibroblasts (Yoon et al. 2013; Li et al. 2018c). HOTAIR expression was reported to be reduced in ECs at sites of atherosclerotic plaques, and it further promoted proliferation and migration as well as survival in ECs in vitro (Peng et al. 2017).

The lncRNA growth arrest specific 5 (**GAS5)** is another lncRNA reported to accumulate in atherosclerotic plaques (Fig. 1). A demonstrable increase of this lncRNAs was reported for rat and human plaques, while polymorphisms in the GAS5 gene are associated with an increased risk for atherosclerosis (Chen et al. 2016; Shen and She 2018). An increase in GAS5 expression in atherosclerotic plaques suggests a detrimental role of GAS5, although the cell type in which GAS5 expression is increased is unknown. Interestingly, pharmacological inhibition of GAS5 in rats aggravates hypertension in these animals and furthermore resulted in vascular remodeling and vascular leakage (Wang et al. 2016a). GAS5 physically interacts with the β-catenin protein in human endothelial cells. More interestingly, GAS5 knockdown had a pro-inflammatory and activating effect on ECs as evidenced by increased expression of the cell surface adhesion molecules ICAM-1, VCAM-1, and E-selectin in human endothelial cells (Fig. 1) (Wang et al. 2016a). The findings from the two aforementioned studies are controversial in regard of the inflammatory modulation and thereby demonstrate in an exemplary fashion a further controversial effect of a studied lncRNA in different diseases. Nevertheless, it was not stated in which cells of atherosclerotic plaques GAS5 expression was increased, and as the endothelium only comprises a small fraction of the whole mass of an atherosclerotic plaque, a potential reduction in GAS5 expression in ECs could be well masked by an increase in a more abundant cell type such as VSMCs or else.

ANRIL is a 3.8 kb long lncRNA, and its overexpression increased VEGF expression and thereby promoted angiogenesis in a rat model of diabetes mellitus. When the same rats were furthermore subjected to cerebral infarction, ANRIL was found to promote the pro-angiogenic effect by activating the NF-κB signaling pathway (Fig. 3) (Zhang et al. 2017). ANRIL expression was upregulated in atherosclerotic plaques and was also shown to promote VEGF expression through binding to the chromatin proteins EZH2 and p300, which regulate endothelial network formation in vitro (Thomas et al. 2017; Arslan et al. 2017).

5.2 *Vascular LncRNAs in Aneurysms*

An aneurysm is an outward bulging in a blood vessel due to a weak vessel wall. Rupture of an aneurysm can, depending on its localization, lead to a life-threatening condition. Risk factors for the development of an aneurysm include, among others, age, smoking, hypertension, and atherosclerosis. Atherosclerosis might promote the formation of aneurysms by weakening of the vessel wall along with reducing its elasticity. Atherosclerotic plaques generate a pro-inflammatory environment resulting in deregulation of many lncRNAs. Inflammatory activation of the inner vessel wall was shown to promote the formation of aneurysms (Sun et al. 2007). Matrix metalloproteases (MMPs), which remodel the extracellular matrix of blood vessels, were additionally shown to be more active at sites of aneurysm formation while their inhibitors were less abundant. Several lncRNAs were proposed to play a role in these aneurysm-forming processes. Indeed, several studies did extensive work to identify the underlying pathways and mechanisms of lncRNAs in this disease condition.

The only lncRNA that was directly linked to abdominal aortic aneurysms is **H19** (Table 9). Pharmacological inhibition of H19 significantly inhibited aneurysm formation in two independent in vivo models (Fig. 4) (Li et al. 2018b). On a cellular level, the knockdown of H19 led to decreased VSMC apoptosis, most likely via a direct interaction of H19 with the hypoxia-inducible factor (HIF-) 1α in the cytoplasm and subsequent stabilization of p53 (Li et al. 2018b). H19 seems to play numerous roles in the cardiovascular system. Some of these roles are conserved across cell types, like the reduced proliferation upon its depletion in VSMCs and ECs.

Expression of the lncRNA **HOTAIR** was decreased in sporadic thoracic aortic aneurysm tissue and was further negatively correlated with the size of the aortic diameter (Fig. 4). Inhibition of HOTAIR in vitro reduced proliferation and promoted apoptosis in human aortic SMCs (Guo et al. 2017). Furthermore, expression of collagen type I and III was reduced upon depletion of HOTAIR (Guo et al. 2017). By regulating these mechanisms in aneurysm progression, HOTAIR is an important candidate for further studies.

Table 9 Vascular lncRNAs in aneurysms

LncRNA	Compartment	Regulation in aneurysms	Function	References
H19	Vascular smooth muscle cells	↑	Interaction with HIF-1α	Li et al. (2018b)
HOTAIR	Vascular smooth muscle cells	↓	Regulation of extracellular matrix genes	Guo et al. (2017)
ANRIL	Unknown	Unknown	Risk factor	Yasuno et al. (2010)
MALAT1	Vascular smooth muscle cells	Unknown	Complex formation with BRG1 and HDAC9	Lino Cardenas et al. (2018)

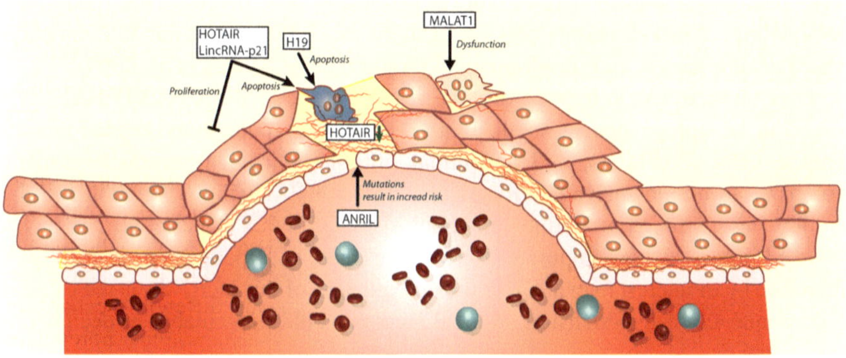

Fig. 4 Aneurysms are an outward bulging of the vessel wall caused by various reasons like smooth muscle cell apoptosis, extracellular matrix remodeling, and others. Depletion of H19 inhibited smooth muscle cell (SMC) apoptosis by stabilization of p53 trough HIF1-α and reduced aneurysm formation in vivo. HOTAIR expression was reduced in aneurysm tissue, and its depletion promoted SMC apoptosis and extracellular matrix remodeling, both factors that can influence aneurysm formation. Mutations in the ANRIL locus were identified as a risk factor for aneurysm formation, and its depletion promoted inflammatory activation of the tissue. MALAT1 was shown to have detrimental effects in aneurysm formation by altering expression of contractile genes in SMCs

Mutations in the **ANRIL** locus were identified as a risk factor for intracranial aneurysms (Fig. 4). The depletion of ANRIL promoted senescence in one study and resulted in increased expression of IL-6 and IL-8 via the NF-κB pathway (Yasuno et al. 2010; Kotake et al. 2011; Zhou et al. 2016). Even though its functional role in aneurysm formation was not analyzed so far, ANRIL shows great potential to be involved in the formation of aneurysms.

MALAT1 was shown to form a complex with the histone deacetylase HDAC9 and the chromatin remodeling enzyme BRG1. HDAC9 localization was fully dependent on the presence of MALAT1 in VSMCs (Lino Cardenas et al. 2018). The complex associated with promoters of contractile genes, and this association was dependent on MALAT1. VSMCs isolated from a mouse strain with high susceptibility towards aneurysm formation showed increased interaction of the complex compared to wild-type littermates. Genetic deletion of MALAT1 in a mouse model for aneurysm formation restored the detrimental phenotype and demonstrates that MALAT1 disrupts expression of contractile genes in VSMCs (Fig. 4) (Lino Cardenas et al. 2018).

5.3 *Vascular LncRNAs in Hypertension*

High blood pressure or hypertension is a common disease with high prevalence among the adult population. Hypertension is a risk factor for numerous CVDs like stroke, myocardial infarction, chronic kidney disease, and others (Sekar et al. 2017).

The pathophysiological causes for hypertension are diverse, but pathologic remodeling of blood vessels plays a primary role among vascular inflammation, vascular senescence, and developmental programming (Sedeek et al. 2009).

Data on lncRNAs in hypertension is sparse, but the lncRNA **PAXIP1-AS1** was found to be upregulated in small pulmonary arteries, pulmonary arterial smooth muscle cells (PASMCs), and adventitial fibroblasts from hypertensive patients compared to normotensive controls. Knockdown of PAXIP1-AS1 inhibited proliferation and migration and promoted apoptosis in PASMCs (Jandl et al. 2019). The affected processes play important roles in the onset of hypertension, and it would therefore be of interest to evaluate if the upregulation of this lncRNAs is the cause or a consequence of hypertension. Animal studies would be valuable in deciphering this question.

GAS5 is an interesting lncRNA suggested to play a role in hypertension via its regulation of EC activation and proliferation, as well as on VSMC phenotypic conversion through β-catenin signaling (Wang et al. 2016a). Nevertheless, another study did not find an association of GAS5 with hypertension based on a bioinformatic screening only (Jusic and Devaux 2019).

H19 was upregulated in serum and lung samples from rats and mice in a model of pulmonary hypertension and was associated with PASMC proliferation. Interestingly, depletion of H19 counteracted pulmonary artery remodeling and pulmonary artery hypertension, likely via sponging of miR-let7b (Su et al. 2018).

MANTIS was found repressed in lung samples from patients suffering from idiopathic pulmonary hypertension compared to healthy controls, as well as in CD31 + cells from lungs of a rat model of hypertension. Mechanistically, MANTIS interacts with the chromatin-remodeling factor BRG1 and facilitates its interaction with angiogenic genes (Leisegang et al. 2017). Interestingly, deletion of BRG1 ameliorates pulmonary hypertension (Chen et al. 2013). MANTIS might therefore serve as an interesting target for the treatment of hypertension (Table 10).

6 Future Perspectives – Translational/Therapy Options

LncRNAs are a promising new area of research with enormous potential for clinical applications in diagnosis and therapy of CVD. Circulating lncRNAs in the blood are easily assayable biomarkers. Their potential as biomarkers in the context of cardiovascular disease could improve risk stratification and allocation of treatment.

Therefore, understanding of the molecular mechanisms underlying lncRNAs in cardiovascular disease still represents an important aim of ongoing research.

Table 10 Vascular lncRNAs in hypertension

LncRNA	Compartment	Regulation in hypertension	Function	References
PAXIP1-AS1	Vascular smooth muscle cells and fibroblasts	↑	Inhibition of PAXIP1-AS1 reduces proliferation and migration and promotes apoptosis.	Jandl et al. (2019)
GAS5	Endothelial cells and vascular smooth muscle cells	Unknown	Regulation of β-catenin signaling	Wang et al. (2016a)
H19	Serum, lung, and vascular smooth muscle cells	↑	Sponging of miR-let7b	Su et al. (2018)
MANTIS	Lung and CD31+ cells	↓	Interaction with BRG1	Chen et al. (2013) and Leisegang et al. (2017)

References

Abe H, Semba H, Takeda N (2017) The roles of hypoxia signaling in the pathogenesis of cardiovascular diseases. J Atheroscler Thromb 24:884–894

Arslan S, Berkan Ö, Lalem T et al (2017) Long non-coding RNAs in the atherosclerotic plaque. Atherosclerosis 266:176–181

Avazpour N, Hajjari M, Yazdankhah S et al (2018) Circulating HOTAIR LncRNA is potentially up-regulated in coronary artery disease. Genomics Inform 16:e25

Barbosa JE, Stockler-Pinto MB, Cruz BO et al (2019) Nrf2, NF-κB and PPARβ/δ mRNA expression profile in patients with coronary artery disease. Arq Bras Cardiol. pii: S0066-782X2019005001205

Boeckel J-N, Perret MF, Glaser SF et al (2018) Identification and regulation of the long non-coding RNA Heat2 in heart failure. J Mol Cell Cardiol 126:13–22

Boon RA, Hofmann P, Michalik KM et al (2016) Long noncoding RNA Meg3 controls endothelial cell aging and function. J Am Coll Cardiol 68:2589–2591

Burns JC, Glodé MP (2004) Kawasaki syndrome. Lancet (London, England) 364:533–544

Cabili MN, Trapnell C, Goff L et al (2011) Integrative annotation of human large intergenic noncoding RNAs reveals global properties and specific subclasses. Genes Dev 25:1915–1927

Cai Y, Yang Y, Chen X et al (2016a) Circulating "LncPPARδ" from monocytes as a novel biomarker for coronary artery diseases. Medicine (Baltimore) 95:e2360

Cai Y, Yang Y, Chen X et al (2016b) Circulating 'lncRNA OTTHUMT00000387022' from monocytes as a novel biomarker for coronary artery disease. Cardiovasc Res 112:714–724

Cai B, Ma W, Ding F et al (2018) The long noncoding RNA CAREL controls cardiac regeneration. J Am Coll Cardiol 72:534–550

Cao J (2014) The functional role of long non-coding RNAs and epigenetics. Biol Proced Online 16:11

Carninci P, Kasukawa T, Katayama S et al (2005) The transcriptional landscape of the mammalian genome. Science 309:1559–1563

Carreño JE, Apablaza F, Ocaranza MP et al (2006) Cardiac hypertrophy: molecular and cellular events. Rev Española Cardiol (English Ed) 59:473–486

Çekin N, Özcan A, Göksel S et al (2018) Decreased FENDRR and LincRNA-p21 expression in atherosclerotic plaque. Anatol J Cardiol 19:131–136
Chen D, Fang F, Yang Y et al (2013) Brahma-related gene 1 (Brg1) epigenetically regulates CAM activation during hypoxic pulmonary hypertension. Cardiovasc Res 100:363–373
Chen L, Yao H, Hui JY et al (2016) Global transcriptomic study of atherosclerosis development in rats. Gene 592:43–48
Chen G, Li H, Li X et al (2018) Loss of long non-coding RNA CRRL promotes cardiomyocyte regeneration and improves cardiac repair by functioning as a competing endogenous RNA. J Mol Cell Cardiol 122:152–164
Chen L, Zhang D, Yu L et al (2019a) Targeting MIAT reduces apoptosis of cardiomyocytes after ischemia/reperfusion injury. Bioengineered 10:121–132
Chen S, Wang J, Zhou Y (2019b) Long non-coding RNA SNHG1 protects human AC16 cardiomyocytes from doxorubicin toxicity by regulating miR-195/Bcl-2 axis. Biosci Rep 39: pii:BSR20191050
Chien Y, Scuoppo C, Wang X et al (2011) Control of the senescence-associated secretory phenotype by NF-κB promotes senescence and enhances chemosensitivity. Genes Dev 25:2125–2136
Chu W-M (2013) Tumor necrosis factor. Cancer Lett 328:222–225
Clemson CM, Hutchinson JN, Sara SA et al (2009) An architectural role for a nuclear noncoding RNA: NEAT1 RNA is essential for the structure of Paraspeckles. Mol Cell 33:717–726
Coppé J-P, Patil CK, Rodier F et al (2008) Senescence-associated secretory phenotypes reveal cell-nonautonomous functions of oncogenic RAS and the p53 tumor suppressor. PLoS Biol 6:2853–2868
Coppé J-P, Desprez P-Y, Krtolica A et al (2010) The senescence-associated secretory phenotype: the dark side of tumor suppression. Annu Rev Pathol 5:99–118
Cremer S, Michalik KM, Fischer A et al (2019) Hematopoietic deficiency of the Long noncoding RNA MALAT1 promotes atherosclerosis and plaque inflammation. Circulation 139:1320–1334
Cunningham KS, Gotlieb AI (2005) The role of shear stress in the pathogenesis of atherosclerosis. Lab Investig 85:9–23
de Gonzalo-Calvo D, Kenneweg F, Bang C et al (2016) Circulating long-non coding RNAs as biomarkers of left ventricular diastolic function and remodelling in patients with well-controlled type 2 diabetes. Sci Rep 6:37354
Derrien T, Johnson R, Bussotti G et al (2012) The GENCODE v7 catalog of human long noncoding RNAs: analysis of their gene structure, evolution, and expression. Genome Res 22:1775–1789
Devaux Y (2017) Transcriptome of blood cells as a reservoir of cardiovascular biomarkers. Biochim Biophys Acta, Mol Cell Res 1864:209–216
Diederichs S (2014) The four dimensions of noncoding RNA conservation. Trends Genet 30:121–123
Djebali S, Davis CA, Merkel A et al (2012) Landscape of transcription in human cells. Nature 489:101–108
Epstein FH, Ross R (1999) Atherosclerosis — an inflammatory disease. N Engl J Med 340:115–126
Feng L, Guo J, Ai F (2019) Circulating long noncoding RNA ANRIL downregulation correlates with increased risk, higher disease severity and elevated pro-inflammatory cytokines in patients with acute ischemic stroke. J Clin Lab Anal 33:e22629
GBD 2013 Mortality and Causes of Death Collaborators G 2013 M and C of D (2015) Global, regional, and national age-sex specific all-cause and cause-specific mortality for 240 causes of death, 1990–2013: a systematic analysis for the Global Burden of Disease Study 2013. Lancet (London, England) 385:117–1171
Gong L-C, Xu H-M, Guo G-L et al (2017) Long non-coding RNA H19 protects H9c2 cells against hypoxia-induced injury by targeting MicroRNA-139. Cell Physiol Biochem 44:857–869
Gong X, Zhu Y, Chang H et al (2019) Long noncoding RNA MALAT1 promotes cardiomyocyte apoptosis after myocardial infarction via targeting miR-144-3p. Biosci Rep 39:pii: BSR20191103

Greco S, Zaccagnini G, Perfetti A et al (2016) Long noncoding RNA dysregulation in ischemic heart failure. J Transl Med 14:183
Guo X, Chang Q, Pei H et al (2017) Long non-coding RNA-mRNA correlation analysis reveals the potential role of HOTAIR in pathogenesis of sporadic thoracic aortic aneurysm. Eur J Vasc Endovasc Surg 54:303–314
Hahn C, Schwartz MA (2009) Mechanotransduction in vascular physiology and atherogenesis. Nat Rev Mol Cell Biol 10:53–62
Han DK, Khaing ZZ, Pollock RA et al (1996) H19, a marker of developmental transition, is reexpressed in human atherosclerotic plaques and is regulated by the insulin family of growth factors in cultured rabbit smooth muscle cells. J Clin Invest 97:1276–1285
Han P, Li W, Lin C-H et al (2014) A long noncoding RNA protects the heart from pathological hypertrophy. Nature 514:102–106
Hanly DJ, Esteller M, Berdasco M (2018) Interplay between long non-coding RNAs and epigenetic machinery: emerging targets in cancer? Philos Trans R Soc Lond Ser B Biol Sci 373:20170074
Hartman J, Frishman WH (2014) Inflammation and atherosclerosis. Cardiol Rev 22:147–151
He C, Ding J-W, Li S et al (2015) The role of Long intergenic noncoding RNA p21 in vascular endothelial cells. DNA Cell Biol 34:677–683
He C, Yang W, Yang J et al (2017) Long noncoding RNA MEG3 negatively regulates proliferation and angiogenesis in vascular endothelial cells. DNA Cell Biol 36:475–481
Herwald H, Egesten A (2011) The discovery of pathogen recognition receptors (PRR). J Innate Immun 3:435–436
Hofmann P, Sommer J, Theodorou K et al (2019) Long non-coding RNA H19 regulates endothelial cell aging via inhibition of STAT3 signalling. Cardiovasc Res 115:230–242
Holdt LM, Sass K, Gäbel G et al (2011) Expression of Chr9p21 genes CDKN2B (p15INK4b), CDKN2A (p16INK4a, p14ARF) and MTAP in human atherosclerotic plaque. Atherosclerosis 214:264–270
Hu W, Ding H, Ouyang A et al (2019a) LncRNA MALAT1 gene polymorphisms in coronary artery disease: a case–control study in a Chinese population. Biosci Rep 39:BSR20182213
Hu Y-W, Guo F-X, Xu Y-J et al (2019b) Long noncoding RNA NEXN-AS1 mitigates atherosclerosis by regulating the actin-binding protein NEXN. J Clin Invest 129:1115–1128
Hu Y, Hu J, Hu Y et al (2019c) Diagnostic value of circulating lncRNA ANRIL and its correlation with coronary artery disease parameters. Brazilian J Med Biol Res 52:e8309
Huang Y (2018) The novel regulatory role of lncRNA-miRNA-mRNA axis in cardiovascular diseases. J Cell Mol Med 22:5768–5775
Huang S, Zhang L, Song J et al (2019) Long noncoding RNA MALAT1 mediates cardiac fibrosis in experimental postinfarct myocardium mice model. J Cell Physiol 234:2997–3006
Ishii N, Ozaki K, Sato H et al (2006) Identification of a novel non-coding RNA, MIAT, that confers risk of myocardial infarction. J Hum Genet 51:1087–1099
Jacobsen EA, Helmers RA, Lee JJ et al (2012) The expanding role(s) of eosinophils in health and disease. Blood 120:3882–3890
Jandl K, Thekkekara Puthenparampil H, Marsh LM et al (2019) Long non-coding RNAs influence the transcriptome in pulmonary arterial hypertension: the role of PAXIP1-AS1. J Pathol 247:357–370
Jusic A, Devaux Y (2019) Noncoding RNAs in hypertension. Hypertension 74:477–492
Kallen AN, Zhou X-B, Xu J et al (2013) The imprinted H19 LncRNA antagonizes Let-7 MicroRNAs. Mol Cell 52:101–112
Kalogeris T, Baines CP, Krenz M et al (2012) Cell biology of ischemia/reperfusion injury. Int Rev Cell Mol Biol 298:229–317
Kanda T, Takahashi T (2004) Interleukin-6 and cardiovascular diseases. Jpn Heart J 45:183–193
Kaneko S, Bonasio R, Saldaña-Meyer R et al (2014) Interactions between JARID2 and noncoding RNAs regulate PRC2 recruitment to chromatin. Mol Cell 53:290–300
Kelley JL, Ozment TR, Li C et al (2014) Scavenger receptor-A (CD204): a two-edged sword in health and disease. Crit Rev Immunol 34:241–261

Khalil AM, Guttman M, Huarte M et al (2009) Many human large intergenic noncoding RNAs associate with chromatin-modifying complexes and affect gene expression. Proc Natl Acad Sci U S A 106:11667–11672

Ko T-M, Chang J-S, Chen S-P et al (2019) Genome-wide transcriptome analysis to further understand neutrophil activation and lncRNA transcript profiles in Kawasaki disease. Sci Rep 9:328

Kotake Y, Nakagawa T, Kitagawa K et al (2011) Long non-coding RNA ANRIL is required for the PRC2 recruitment to and silencing of p15 INK4B tumor suppressor gene. Oncogene 30:1956–1962

Kumarswamy R, Bauters C, Volkmann I et al (2014) Circulating Long noncoding RNA, LIPCAR, predicts survival in patients with heart failure. Circ Res 114:1569–1575

Kung JTY, Colognori D, Lee JT (2013) Long noncoding RNAs: past, present, and future. Genetics 193:651–669

Lai Y, He S, Ma L et al (2017) HOTAIR functions as a competing endogenous RNA to regulate PTEN expression by inhibiting miR-19 in cardiac hypertrophy. Mol Cell Biochem 432:179–187

Leisegang MS, Fork C, Josipovic I et al (2017) Long noncoding RNA MANTIS facilitates endothelial Angiogenic function. Circulation 136:65–79

Leisegang MS, Bibli S-I, Günther S et al (2019) Pleiotropic effects of laminar flow and statins depend on the Krüppel-like factor-induced lncRNA MANTIS. Eur Heart J 40:2523–2533

Leistner DM, Boeckel JN, Reis SM et al (2016) Transcoronary gradients of vascular miRNAs and coronary atherosclerotic plaque characteristics. Eur Heart J 37:1738–1749

Li Z, Chao T-C, Chang K-Y et al (2014) The long noncoding RNA THRIL regulates TNFα expression through its interaction with hnRNPL. Proc Natl Acad Sci U S A 111:1002–1007

Li X, Dai Y, Yan S et al (2017a) Down-regulation of lncRNA KCNQ1OT1 protects against myocardial ischemia/reperfusion injury following acute myocardial infarction. Biochem Biophys Res Commun 491:1026–1033

Li X, Zhou J, Huang K (2017b) Inhibition of the lncRNA Mirt1 attenuates acute myocardial infarction by suppressing NF-κB activation. Cell Physiol Biochem 42:1153–1164

Li B, Hu Y, Li X et al (2018a) Sirt1 antisense Long noncoding RNA promotes cardiomyocyte proliferation by enhancing the stability of Sirt1. J Am Heart Assoc 7:e009700

Li DY, Busch A, Jin H et al (2018b) H19 induces abdominal aortic aneurysm development and progression. Circulation 138:1551–1568

Li L, Zhang M, Chen W et al (2018c) LncRNA-HOTAIR inhibition aggravates oxidative stress-induced H9c2 cells injury through suppression of MMP2 by miR-125. Acta Biochim Biophys Sin Shanghai 50:996–1006

Li M, Wang Y-F, Yang X-C et al (2018d) Circulating Long noncoding RNA LIPCAR acts as a novel biomarker in patients with ST-segment elevation myocardial infarction. Med Sci Monit 24:5064–5070

Li Y, Wang J, Sun L et al (2018e) LncRNA myocardial infarction-associated transcript (MIAT) contributed to cardiac hypertrophy by regulating TLR4 via miR-93. Eur J Pharmacol 818:508–517

Li Z, Liu Y, Guo X et al (2018f) Long noncoding RNA myocardial infarction-associated transcript is associated with the microRNA-150-5p/P300 pathway in cardiac hypertrophy. Int J Mol Med 42:1265–1272

Lino Cardenas CL, Kessinger CW, Cheng Y et al (2018) An HDAC9-MALAT1-BRG1 complex mediates smooth muscle dysfunction in thoracic aortic aneurysm. Nat Commun 9:1009

Liu Y, Zhou D, Li G et al (2015) Long non coding RNA-UCA1 contributes to cardiomyocyte apoptosis by suppression of p27 expression. Cell Physiol Biochem 35:1986–1998

Liu L, An X, Li Z et al (2016) The H19 long noncoding RNA is a novel negative regulator of cardiomyocyte hypertrophy. Cardiovasc Res 111:56–65

Long B, Li N, Xu X-X et al (2018) Long noncoding RNA FTX regulates cardiomyocyte apoptosis by targeting miR-29b-1-5p and Bcl2l2. Biochem Biophys Res Commun 495:312–318

Lozano R, Naghavi M, Foreman K et al (2012) Global and regional mortality from 235 causes of death for 20 age groups in 1990 and 2010: a systematic analysis for the global burden of disease study 2010. Lancet 380:2095–2128
Matouk IJ, DeGroot N, Mezan S et al (2007) The H19 non-coding RNA is essential for human tumor growth. PLoS One 2:e845
Mitchell JA, Ryffel B, Quesniaux VFJ et al (2007) Role of pattern-recognition receptors in cardiovascular health and disease. Biochem Soc Trans 35:1449–1452
Moore KJ, Tabas I (2011) Macrophages in the pathogenesis of atherosclerosis. Cell 145:341–355
Neumann P, Jaé N, Knau A et al (2018) The lncRNA GATA6-AS epigenetically regulates endothelial gene expression via interaction with LOXL2. Nat Commun 9:237
Noviello TMR, Di Liddo A, Ventola GM et al (2018) Detection of long non–coding RNA homology, a comparative study on alignment and alignment–free metrics. BMC Bioinformatics 19:407
Palomer X, Barroso E, Pizarro-Delgado J et al (2018) PPARβ/δ: a key therapeutic target in metabolic disorders. Int J Mol Sci 19:913
Pandey RR, Mondal T, Mohammad F et al (2008) Kcnq1ot1 antisense noncoding RNA mediates lineage-specific transcriptional silencing through chromatin-level regulation. Mol Cell 32:232–246
Pasmant E, Sabbagh A, Vidaud M, Bièche I (2011) ANRIL, a long, noncoding RNA, is an unexpected major hotspot in GWAS. FASEB J 25:444–448
Peng Y, Meng K, Jiang L et al (2017) Thymic stromal lymphopoietin-induced HOTAIR activation promotes endothelial cell proliferation and migration in atherosclerosis. Biosci Rep 37: pii: BSR20170351
Ponnusamy M, Liu F, Zhang Y-H et al (2019) Long noncoding RNA CPR (Cardiomyocyte proliferation regulator) regulates cardiomyocyte proliferation and cardiac repair. Circulation 139:2668–2684
Qi L, Liu F, Zhang F et al (2018) lncRNA NEAT1 competes against let-7a to contribute to non-small cell lung cancer proliferation and metastasis. Biomed Pharmacother 103:1507–1515
Qu X, Song X, Yuan W et al (2016) Expression signature of lncRNAs and their potential roles in cardiac fibrosis of post-infarct mice. Biosci Rep 36: pii:e00337
Qu X, Du Y, Shu Y et al (2017) MIAT is a pro-fibrotic long non-coding RNA governing cardiac fibrosis in post-infarct myocardium. Sci Rep 7:42657
Quintela AM, Jiménez R, Gómez-Guzmán M et al (2012) Activation of peroxisome proliferator-activated receptor-β/−δ (PPARβ/δ) prevents endothelial dysfunction in type 1 diabetic rats. Free Radic Biol Med 53:730–741
Ramsköld D, Wang ET, Burge CB et al (2009) An abundance of ubiquitously expressed genes revealed by tissue transcriptome sequence data. PLoS Comput Biol 5:e1000598
Rauchhaus M, Doehner W, Francis DP et al (2000) Plasma cytokine parameters and mortality in patients with chronic heart failure. Circulation 102:3060–3067
Ravasi T, Suzuki H, Pang KC et al (2006) Experimental validation of the regulated expression of large numbers of non-coding RNAs from the mouse genome. Genome Res 16:11–19
Rodier F, Coppé J-P, Patil CK et al (2009) Persistent DNA damage signalling triggers senescence-associated inflammatory cytokine secretion. Nat Cell Biol 11:973–979
Ruparelia N, Chai JT, Fisher EA et al (2017) Inflammatory processes in cardiovascular disease: a route to targeted therapies. Nat Rev Cardiol 14:133–144
Sedeek M, Hébert RL, Kennedy CR et al (2009) Molecular mechanisms of hypertension: role of Nox family NADPH oxidases. Curr Opin Nephrol Hypertens 18:122–127
Sedger LM, McDermott MF (2014) TNF and TNF-receptors: from mediators of cell death and inflammation to therapeutic giants – past, present and future. Cytokine Growth Factor Rev 25:453–472
Sekar D, Shilpa BR, Das AJ (2017) Relevance of microRNA 21 in different types of hypertension. Curr Hypertens Rep 19:57

Shen Z, She Q (2018) Association between the deletion allele of Ins/Del polymorphism (Rs145204276) in the promoter region of GAS5 with the risk of atherosclerosis. Cell Physiol Biochem 49:1431–1443
Smith EM, Gregg M, Hashemi F et al (2006) Corticotropin Releasing Factor (CRF) activation of NF-κB-directed transcription in leukocytes. Cell Mol Neurobiol 26:1019–1034
Song C, Zhang J, Liu Y et al (2016) Construction and analysis of cardiac hypertrophy-associated lncRNA-mRNA network based on competitive endogenous RNA reveal functional lncRNAs in cardiac hypertrophy. Oncotarget 7:10827–10840
Su H, Xu X, Yan C et al (2018) LncRNA H19 promotes the proliferation of pulmonary artery smooth muscle cells through AT 1 R via sponging let-7b in monocrotaline-induced pulmonary arterial hypertension. Respir Res 19:254
Sun R, Zhang L (2019) Long non-coding RNA MALAT1 regulates cardiomyocytes apoptosis after hypoxia/reperfusion injury via modulating miR-200a-3p/PDCD4 axis. Biomed Pharmacother 111:1036–1045
Sun J, Sukhova GK, Wolters PJ et al (2007) Mast cells promote atherosclerosis by releasing proinflammatory cytokines. Nat Med 13:719–724
Tanai E, Frantz S (2015) Pathophysiology of heart failure. In: Comprehensive physiology. John Wiley & Sons, Inc., Hoboken, pp 187–214
Taylor CT, Doherty G, Fallon PG et al (2016) Hypoxia-dependent regulation of inflammatory pathways in immune cells. J Clin Invest 126:3716–3724
Thomas AA, Feng B, Chakrabarti S (2017) ANRIL: a regulator of VEGF in diabetic retinopathy. Investig Ophthalmol Vis Sci 58:470–480
Toraih EA, El-Wazir A, Alghamdi SA et al (2019) [PROVISIONAL] Association of long non-coding RNA MIAT and MALAT1 expression profiles in peripheral blood of coronary artery disease patients with previous cardiac events. Genet Mol Biol. pii:S1415-47572019005020101
Ulitsky I (2018) Interactions between short and long noncoding RNAs. FEBS Lett 592:2874–2883
Viereck J, Kumarswamy R, Foinquinos A et al (2016) Long noncoding RNA Chast promotes cardiac remodeling. Sci Transl Med 8:326ra22
Wang X-S, Zhang Z, Wang H-C et al (2006) Rapid identification of UCA1 as a very sensitive and specific unique marker for human bladder carcinoma. Clin Cancer Res 12:4851–4858
Wang K, Liu F, Zhou L-Y et al (2014a) The Long noncoding RNA CHRF regulates cardiac hypertrophy by targeting miR-489. Circ Res 114:1377–1388
Wang K, Long B, Zhou L-Y et al (2014b) CARL lncRNA inhibits anoxia-induced mitochondrial fission and apoptosis in cardiomyocytes by impairing miR-539-dependent PHB2 downregulation. Nat Commun 5:3596
Wang K, Liu C-Y, Zhou L-Y et al (2015) APF lncRNA regulates autophagy and myocardial infarction by targeting miR-188-3p. Nat Commun 6:6779
Wang YNZ, Shan K, Di Yao M et al (2016a) Long noncoding RNA-GAS5. Hypertension 68:736–748
Wang Z, Zhang X-J, Ji Y-X et al (2016b) The long noncoding RNA Chaer defines an epigenetic checkpoint in cardiac hypertrophy. Nat Med 22:1131–1139
Wang J, Zhao H, Fan Z et al (2017) Long noncoding RNA H19 promotes neuroinflammation in ischemic stroke by driving histone deacetylase 1–dependent M1 microglial polarization. Stroke 48:2211–2221
Wang G, Li Y, Peng Y et al (2018) Association of polymorphisms in MALAT1 with risk of coronary atherosclerotic heart disease in a Chinese population. Lipids Health Dis 17:75
Wang J, Cao B, Zhao H et al (2019a) Long noncoding RNA H19 prevents neurogenesis in ischemic stroke through p53/Notch1 pathway. Brain Res Bull 150:111–117
Wang S, Zuo H, Jin J et al (2019b) Long noncoding RNA Neat1 modulates myogenesis by recruiting Ezh2. Cell Death Dis 10:505

Wang X-M, Li X-M, Song N et al (2019c) Long non-coding RNAs H19, MALAT1 and MIAT as potential novel biomarkers for diagnosis of acute myocardial infarction. Biomed Pharmacother 118:109208

Weber C, Noels H (2011) Atherosclerosis: current pathogenesis and therapeutic options. Nat Med 17:1410–1422

Wong ND (2014) Epidemiological studies of CHD and the evolution of preventive cardiology. Nat Rev Cardiol 11:276–289

Wu Q, Kim YC, Lu J et al (2008) Poly A- transcripts expressed in HeLa cells. PLoS One 3:e2803

Wu G, Cai J, Han Y et al (2014) LincRNA-p21 regulates Neointima formation, vascular smooth muscle cell proliferation, apoptosis, and atherosclerosis by enhancing p53 activity. Circulation 130:1452–1465

Wu T, Wu D, Wu Q et al (2017a) Knockdown of Long non-coding RNA-ZFAS1 protects cardiomyocytes against acute myocardial infarction via anti-apoptosis by regulating miR-150/CRP. J Cell Biochem 118:3281–3289

Wu Z, He Y, Li D et al (2017b) Long noncoding RNA MEG3 suppressed endothelial cell proliferation and migration through regulating miR-21. Am J Transl Res 9:3326–3335

Wu H, Zhao Z-A, Liu J et al (2018) Long noncoding RNA Meg3 regulates cardiomyocyte apoptosis in myocardial infarction. Gene Ther 25:511–523

Wu H-J, Tang G-M, Shao P-Y et al (2019) Long non-coding RNA NEAT1 modulates hypoxia/reoxygenation-induced cardiomyocyte injury via targeting microRNA-520a. Exp Ther Med 18:2199–2206

Xu B, Fang Z, He S et al (2018) ANRIL polymorphism rs4977574 is associated with increased risk of coronary artery disease in Asian populations. Medicine (Baltimore) 97:e12641

Xuan L, Sun L, Zhang Y et al (2017) Circulating long non-coding RNAs *NRON* and *MHRT* as novel predictive biomarkers of heart failure. J Cell Mol Med 21:1803–1814

Yan H, Liang H, Liu L et al (2019) Long noncoding RNA NEAT1 sponges miR-125a-5p to suppress cardiomyocyte apoptosis via BCL2L12. Mol Med Rep 19:4468–4474

Yang K-C, Yamada KA, Patel AY et al (2014) Deep RNA sequencing reveals dynamic regulation of myocardial noncoding RNAs in failing human heart and remodeling with mechanical circulatory support. Circulation 129:1009–1021

Yao Y, Xiong G, Jiang X et al (2019) The overexpression of lncRNA H19 as a diagnostic marker for coronary artery disease. Rev Assoc Med Bras 65:110–117

Yari M, Bitarafan S, Broumand MA et al (2018) Association between Long noncoding RNA ANRIL expression variants and susceptibility to coronary artery disease. Int J Mol Cell Med 7:1–7

Yasuno K, Bilguvar K, Bijlenga P et al (2010) Genome-wide association study of intracranial aneurysm identifies three new risk loci. Nat Genet 42:420–425

Yoon J-H, Abdelmohsen K, Kim J et al (2013) Scaffold function of long non-coding RNA HOTAIR in protein ubiquitination. Nat Commun 4:2939

Yoshimizu T, Miroglio A, Ripoche M-A et al (2008) The H19 locus acts in vivo as a tumor suppressor. Proc Natl Acad Sci U S A 105:12417–12422

Zangrando J, Zhang L, Vausort M et al (2014) Identification of candidate long non-coding RNAs in response to myocardial infarction. BMC Genomics 15:460

Zeng C, Xu Y, Xu L et al (2014) Inhibition of long non-coding RNA NEAT1 impairs myeloid differentiation in acute promyelocytic leukemia cells. BMC Cancer 14:693

Zhang J, Gao C, Meng M et al (2016a) Long noncoding RNA MHRT protects cardiomyocytes against H2O2-induced apoptosis. Biomol Ther (Seoul) 24:19–24

Zhang Y, Sun L, Xuan L et al (2016b) Reciprocal changes of circulating Long non-coding RNAs ZFAS1 and CDR1AS predict acute myocardial infarction. Sci Rep 6:22384

Zhang B, Wang D, Ji T-F et al (2017) Overexpression of lncRNA ANRIL up-regulates VEGF expression and promotes angiogenesis of diabetes mellitus combined with cerebral infarction by activating NF-κB signaling pathway in a rat model. Oncotarget 8:17347–17359

Zhang N, Meng X, Mei L et al (2018a) The Long non-coding RNA SNHG1 attenuates cell apoptosis by regulating miR-195 and BCL2-like protein 2 in human cardiomyocytes. Cell Physiol Biochem 50:1029–1040

Zhang Y, Sun L, Xuan L et al (2018b) Long non-coding RNA CCRR controls cardiac conduction via regulating intercellular coupling. Nat Commun 9:4176

Zhang YYY, Jiao L, Sun L et al (2018c) LncRNA *ZFAS1* as a SERCA2a inhibitor to cause intracellular Ca $^{2+}$ overload and contractile dysfunction in a mouse model of myocardial infarction. Circ Res 122:1354–1368

Zhang J, Liang Y, Huang X et al (2019a) STAT3-induced upregulation of lncRNA MEG3 regulates the growth of cardiac hypertrophy through miR-361-5p/HDAC9 axis. Sci Rep 9:460

Zhang K, Qi M, Yang Y et al (2019b) Circulating lncRNA ANRIL in the serum of patients with ischemic stroke. Clin Lab 65(8). https://doi.org/10.7754/Clin.Lab.2019.190143

Zhang X, Nie X, Yuan S et al (2019c) Circulating Long non-coding RNA ENST00000507296 is a prognostic Indicator in patients with dilated cardiomyopathy. Mol Ther Nucleic Acids 16:82–90

Zhang Y, Zhang L, Wang Y et al (2019d) KCNQ1OT1, HIF1A-AS2 and APOA1-AS are promising novel biomarkers for diagnosis of coronary artery disease. Clin Exp Pharmacol Physiol 46:635–642

Zhou X, Han X, Wittfeldt A et al (2016) Long non-coding RNA ANRIL regulates inflammatory responses as a novel component of NF-κB pathway. RNA Biol 13:98–108

Zhou G, Li C, Feng J et al (2018) lncRNA UCA1 is a novel regulator in cardiomyocyte hypertrophy through targeting the miR-184/HOXA9 Axis. Cardiorenal Med 8:130–139

Zhu M, Li N, Luo P et al (2018) Peripheral blood leukocyte expression of lncRNA MIAT and its diagnostic and prognostic value in ischemic stroke. J Stroke Cerebrovasc Dis 27:326–337

Zhuang J, Peng W, Li H et al (2012) Methylation of p15INK4b and expression of ANRIL on chromosome 9p21 are associated with coronary artery disease. PLoS One 7:e47193

Long Noncoding RNAs in Cardiovascular Development and Diseases

Jiali Deng, Mengying Guo, and Junjie Xiao

Contents

Abstract Cardiovascular disease (CVD) has become one of the major diseases affecting human health worldwide. Long noncoding RNAs (lncRNAs) refer to a class of RNAs with a length of more than 200 nucleotides and a low coding potential. LncRNAs are involved in a variety of physiological and pathological processes and are reported to play an essential role in cell proliferation, differentiation, development, senescence, and apoptosis, closely related to various diseases such as tumors, endocrine diseases, and nervous system diseases. In the cardiovascular system, lncRNAs also have an important regulatory role in cardiovascular development and diseases, including pathological cardiac hypertrophy, myocardial

J. Deng
Cardiac Regeneration and Ageing Lab, Institute of Cardiovascular Sciences, School of Life Science, Shanghai University, Shanghai, China
e-mail: dengjiali@shu.edu.cn

M. Guo
School of Medicine (in preparation), Shanghai University, Shanghai, China

J. Xiao (✉)
Cardiac Regeneration and Ageing Lab, Institute of Cardiovascular Sciences, School of Life Science, Shanghai University, Shanghai, China

School of Medicine (in preparation), Shanghai University, Shanghai, China
e-mail: junjiexiao@shu.edu.cn

S. Jurga, J. Barciszewski (eds.), *The Chemical Biology of Long Noncoding RNAs*, RNA Technologies 11, https://doi.org/10.1007/978-3-030-44743-4_14

infarction, cardiac ischemia-reperfusion injury, atherosclerosis, heart failure, etc. In this chapter, we will briefly summarize the classification and characteristics of lncRNAs and focus on the latest knowledge of lncRNAs in cardiovascular development and diseases. In addition, we will discuss the molecular mechanism underlying lncRNA regulation of cardiovascular diseases and highlight the potential application of lncRNAs.

Keywords lncRNAs · Cardiovascular diseases · Development · mechanism

1 Introduction

At present, cardiovascular disease has become one of the major chronic diseases that threaten human life and health (Thomas et al. 2018). Although the increasingly developed medical technologies can relieve the symptoms and improve the life quality of patients with cardiovascular diseases after proper treatments, it is still unable to terminate the further development or reverse the pathological processes of cardiovascular diseases. Moreover, compared to past decades, unhealthy living habits of human beings have gradually become new risk factors for cardiovascular diseases (Mozaffarian et al. 2015). Thus, there is an urgent need for early diagnosis and effective treatment of cardiovascular diseases.

Under the stimulation of various etiologies, cardiac myocytes can produce compensatory hypertrophy and fight against insufficiency of ejection. Cardiac hypertrophy can finally develop to heart failure. Heart failure is a complex disease, usually caused by other diseases, characterized by a decline in the efficiency of heart to pump blood, and unable to meet the needs of the body. Myocardial infarction is one of the main death causes of cardiovascular disease, which is characterized by local damage of myocardial tissue and lack of blood supply to the heart, leading to cell death. Coronary atherosclerotic heart disease (CHD) is caused by the formation of atherosclerotic plaques. With the structural remodeling of the arterial wall, the activation of endothelial cells, and the activation of inflammatory cells, it can eventually develop into myocardial ischemia.

With the completion of the Human Genome Project, genome-wide association study (GWAS) found that the number of coding genes in the human genome decreases from expected 2 million at least to 20,500, which means that the number of protein-coding RNAs is less than 3% of the total RNAs, and most of the remaining genes do not have the function of encoding proteins (Rinn and Chang 2012). Initially, because the number, type, and function of noncoding sequences are not very clear, these noncoding transcripts (ncRNAs) are not considered to play any role in gene expression regulation. Recently, ENCODE published decoding of noncoding sequences, which account for 98% of the human genome. It is clear that at least 80% of the gene transcripts have biological functions. Among them,

1.1–1.5% of the transcripts can encode proteins, and the majority of the remaining transcripts are ncRNAs (de Souza 2012; Pazin 2015). Among ncRNAs, long-chain noncoding RNAs (lncRNAs) are those with more than 200 nts in length. So far, more than one hundred thousand eukaryotic lncRNAs have been found, among which a small part is confirmed to encode small peptide chains (Anderson et al. 2015). In recent years, lncRNAs are reported to regulate gene expression by modulating promoters and enhancers of genes by epigenetics, transcriptional regulation, and posttranscriptional regulation and then regulate proliferation, apoptosis, and autophagy, thus participating in occurrence and development of diseases (Kapranov et al. 2007; Qureshi and Mehler 2012).

Cardiovascular disease is one of the most important threats to human health worldwide. Considering the huge social burden brought by cardiovascular disease, the progress in disease management should not only be limited to the treatment and research of these diseases but also focus on the development of early detection and prevention platform for cardiovascular disease. In recent years, many biomarkers related to cardiovascular disease have been found and applied in the clinic such as cardiac calcium protein and creatine kinase (CK). However, with the development and progress of deep RNA sequencing technology, new members of the genome, including noncoding RNAs (ncRNAs), have been found gradually. They do not code proteins but can regulate the function of genes, thus regulating a series of physiological and pathological processes. According to different functions, lengths, and structures, ncRNAs can be divided into transfer RNA (tRNA), ribosomal RNA (rRNA), micronuclear RNA (snRNA), micronuclear RNA (snoRNA), guide RNA (gRNA), microRNA (miRNA), long-chain ncRNA (lncRNA), circular ncRNA, etc. The role of miRNA and lncRNA in the treatment, prognosis, and diagnosis of cardiovascular disease has been paid more attention worldwide. In addition, in recent years, the newly discovered circRNA, which is produced by reverse splicing of precursor mRNA, exists in many species, and the research results in various fields suggest that it may participate in a wide range of pathophysiological regulation. ncRNAs exist in peripheral circulation, such as blood, and some are stable in a variety of environmental conditions. It has a unique value in the early diagnosis and prognosis of a variety of diseases.

LncRNAs are potential biomarkers for disease diagnosis, treatment, and prognosis evaluation. In recent years, more and more studies have confirmed that lncRNAs play a role in the development of cardiovascular diseases, including pathological cardiac hypertrophy, myocardial infarction, cardiac ischemia-reperfusion injury, atherosclerosis, heart failure, etc., through various regulatory mechanisms (Boon et al. 2016; Kopp and Mendell 2018). This chapter reviews the latest research progress of lncRNAs in cardiovascular diseases.

2 Classification and Characteristics of LncRNAs

With the development of whole genome sequencing technology, we have a deeper understanding of the transcriptome of organisms. At present, it is believed that more than 90% of human genome noncoding RNAs (ncRNAs) play an important biological role, rather than the previously thought "transcription noise" or "transcription garbage" (Mercer et al. 2009). According to the length, noncoding RNA can be divided into two categories: small noncoding RNA (ncRNA, $<$ 200 nucleotides), including miRNA, siRNA, and piRNA, and long noncoding RNA (lncRNA, $>$ 200 NT); both have no ability to encode protein (St Laurent et al. 2015). More and more studies have shown that lncRNAs play an important role in regulating important cell biological functions, such as cell proliferation, differentiation, apoptosis, invasion, drug resistance, and immune response. In addition, the abnormal expression of lncRNAs is closely related to the occurrence and development of various malignant tumors (Hung and Chang 2010).

There are many kinds of lncRNAs based on various classification methods. Traditional classification is based on their location in genes, which mainly includes five categories: sense lncRNAs, antisense lncRNAs, bidirectional lncRNAs, intronic lncRNAs, and intergenic lncRNAs (Ponting et al. 2009). With the discovery of new lncRNAs, this traditional classification method fails to cover all kinds. LncRNAs are then classified into eight broad categories: divergent, convergent, intronic, intergenic, overlapping-sense, overlapping-antisense, enhancer RNAs, and miRNA host gene (Schmitz et al. 2016; Wu et al. 2017a, b). Compared with traditional taxonomy, the latter describes the relationship between lncRNAs and adjacent genes in more detail. However, this classification still refers to only one criterion. With the deepening of the research on lncRNAs, the classification work still needs to be improved, and more factors, such as gene splicing or the working mechanism of lncRNAs, should be considered as well.

The classification of lncRNAs is shown in Fig. 1.

The number of lncRNAs is huge, and their biological functions are also diverse (Hon et al. 2017). The functions of lncRNAs can be summarized into three categories: the first is chromosome modification and regulation of gene transcription. For example, Xist, as a cis-response element, maintains the inactivation of X chromosome by binding to X chromosome and attracting polycomb repressive complex 2 (PRC2) (Froberg et al. 2013). The inactivation of X chromosome is a genomic imprinting phenomenon, which is an important event in the development. This suggests that lncRNAs may play a key role in the development of the heart. The second is that lncRNAs can directly bind to proteins, affecting the function and the stability of proteins (Banisadr et al. 2002). LncRNAs regulate the activity of corresponding proteins by binding to specific proteins, forming nucleic acid protein complexes with proteins as structural components, and binding to specific proteins to change the cellular localization of the proteins (Tsai et al. 2010). The third category is coding function. A muscle-specific lncRNA encodes short peptides containing 34 amino acids named DWORF mainly located in the sarcoplasmic reticulum. By

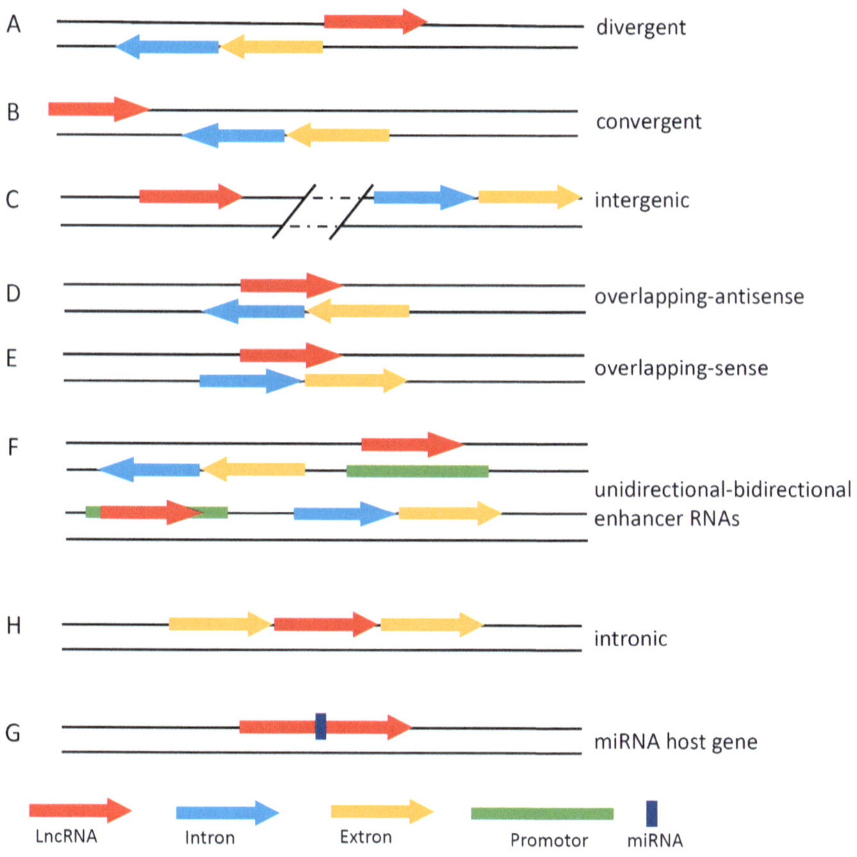

Fig. 1 Classification of lncRNAs

competitive binding to sarcoplasmic endoplasmic reticulum calcium ATPase (SERCA), DWORF blocks the binding of SERCA with phospholamban, sarcolipin, and myoregulin and enhances the activity of SERCA, thereby regulating the systolic function of mice heart (Nelson et al. 2016).

Because of lncRNA's lack of open reading frame, it does not have the function of coding protein. LncRNAs can widely regulate gene expression at the level of chromosome modification, transcription, and posttranscription and participate in tumor development by inactivating tumor suppressor or activating oncogene. The abnormal expression of lncRNAs can mediate tumor drug resistance through different ways such as targeting drug efflux pump molecules and apoptosis regulatory proteins. The intervention of abnormal expression of lncRNAs can change the drug resistance behavior of tumor, suggesting that lncRNAs play an important role in the process of mediating and regulating tumor drug resistance. Therefore, finding out the molecular mechanism of lncRNAs related to drug resistance and its regulation will help to improve the clinical efficiency and open up a new research field for exploring the theory of tumor drug resistance. The mechanism of lncRNAs involved in tumor drug resistance is complex. It can be induced by changing drug concentration in tumor cells, cell cycle change, abnormal apoptosis, and epithelial mesenchymal transition (EMT) (Chen et al. 2016; Zhou et al. 2016).

3 LncRNAs in Cardiovascular Development

Cardiac development in mammals is closely related to many kinds of stem cells and cardiac cells, as well as to the differentiation of smooth muscle cells and endothelial cells. Cardiac progenitor cells (CPC), which originated from mesoderm (lateral mesoderm), experienced the induction and specialization of cardiac progenitor cells, cardiac tube formation, cardiac cyclization and asymmetric development, chamber specialization, and growth. The whole development process was regulated by a variety of core transcription factors, such as Nkx2.5, mesp1, MEF2C, GATA4, and other genes. The imbalance of cardiac gene regulatory network will cause congenital heart disease and various acquired chronic heart diseases. The discovery of lncRNAs adds a new way of cardiac development regulation. At present, a lot of studies have shown that many lncRNAs play a key role in the heart cells of mouse mesoderm, including lncRNA Braveheart (Bvht) and lncRNA fetal-lethal noncoding developmental regulatory RNA (Fendrr) in mouse heart mesoderm. Bvht can activate the gene network of cardiovascular system and enter the stage of gastrointestinal embryogenesis and mesodermal patterning, promoting the directional differentiation of stem cells into cardiac progenitor cells with various functions. Bvht can also interact with the SUZ12 of PRC2 or form molecular scaffold structure with chromatin modifications to regulate cardiac development (Klattenhoff et al. 2013). Fendrr binds to the histone-modifying complexes PRC2 and TrxG/MLL and plays a role in the proliferation and differentiation of cardiac embryonic layer and ventral cells. Downregulation of Fendrr in mouse embryos could inhibit the expression of its target gene, resulting in lethal distortion of the heart (Grote et al. 2013). A total of 149 specific lncRNAs related to the expression of cardiac embryos has been reported. LncRNAs regulate nuclear factor kappa B and cyclic adenosine phosphate reaction protein genes to orchestrate the development of cardiac embryos (Matkovich et al. 2014). In addition, some studies have shown that lncRNAs induced by embryonic cardiac enhancer play a key role in directional differentiation of embryonic stem cells (Thum and Condorelli 2015). By studying the differentiation of embryonic tumor cells in vitro, 14 lncRNAs were found downregulated and 28 lncRNAs were upregulated in cardiomyocytes (Ounzain et al. 2014). In vitro studies also showed that lncRNA p21 could significantly inhibit the growth and proliferation of mouse monocyte macrophages and vascular smooth muscle cells and induce cell apoptosis (Wu et al. 2014).

Many lncRNAs were changed dynamically during the development of mouse heart and were significantly related to the expression of adjacent genes. LncRNA Pp1r1b can regulate the adjacent gene Tcap, and the expression ratio of the two genes can help to distinguish the type of congenital heart disease (Touma et al. 2016). Scot et al. found that with the development of the heart, the expression of Lcrna Kcnq1ot1 decreased, and the level of its target mRNA KCNQ1 increased. KCNQ1 is essential for the repolarization of cardiomyocytes, consistent with the need to enhance the contractile function of myocardium during the transition from embryo to adult rat heart (Pandey et al. 2008).

LncRNA Hoxblinc interacts with Set1/MLL complex to activate Hoxb gene and promote the differentiation of mesodermal cells (Deng et al. 2016). MAPK1 upregulates the expression of lncRNA MALAT1 and the proliferation of cardiomyocytes through PI3K/AKT signaling pathway (Zhao et al. 2015). LncRNA H19 regulates the downregulation of miR-19b, promotes cell proliferation, and inhibits apoptosis in the process of late cardiac differentiation (Han et al. 2016). Inhibition of the transcription of lncRNA UPH can block the expression of Hand2, hinder the differentiation of fibroblasts into cardiomyocytes, and lead to right ventricular hypoplasia and fetal death in mice (Anderson et al. 2016). In some studies, P19 cells were used to differentiate into cardiomyocytes to simulate the development of the heart in vitro. In this process, 28 lncRNAs were upregulated and 12 downregulated (Zhu et al. 2014). Wang et al. found that 51 lncRNAs expressed differently in the heart of embryo and adult fish (Wang et al. 2017).

Whether lncRNAs can be used as a molecular marker for early diagnosis of cardiac dysplasia and whether targeted interventions against lncRNAs can be used as a new treatment for cardiac dysplasia still need to be further investigated.

The relationship between some lncRNAs and cardiovascular development is shown in Fig. 2 and Table 1.

4 LncRNAs in Pathological Cardiac Hypertrophy

Cardiac hypertrophy is a state formed by the heart in order to maintain the original perfusion function under the changes of pressure and volume (Sundaresan et al. 2009). Pathological cardiac hypertrophy is accompanied by a series of adverse cardiovascular events, including arrhythmia, heart failure, and so on (Schiattarella and Hill 2015). Pathological myocardial hypertrophy is a common pathophysiological manifestation of various cardiovascular diseases, such as hypertension, coronary heart disease, valvular disease, and so on. Pathological myocardial hypertrophy is the early sign of heart failure and a risk factor for heart failure, stroke, coronary heart disease, and sudden death. Ventricular hypertrophy is related to the reactivation of fetal genes, including atrial natriuretic peptide (ANP), brain natriuretic peptide (BNP) and β-myosin heavy chain (β-MHC), and sarcoplasmic reticulum Ca^{2+}-ATPase (SERCA2a) and α-myosin heavy chain (α-MHC) (Depre et al. 1998). Studies have shown that chromatin remodeling and histone modification play an important role in cardiac development and heart disease, and epigenetic reprogramming is a key process of pathological gene induction in myocardial hypertrophy and remodeling, while lncRNAs may participate in this process and

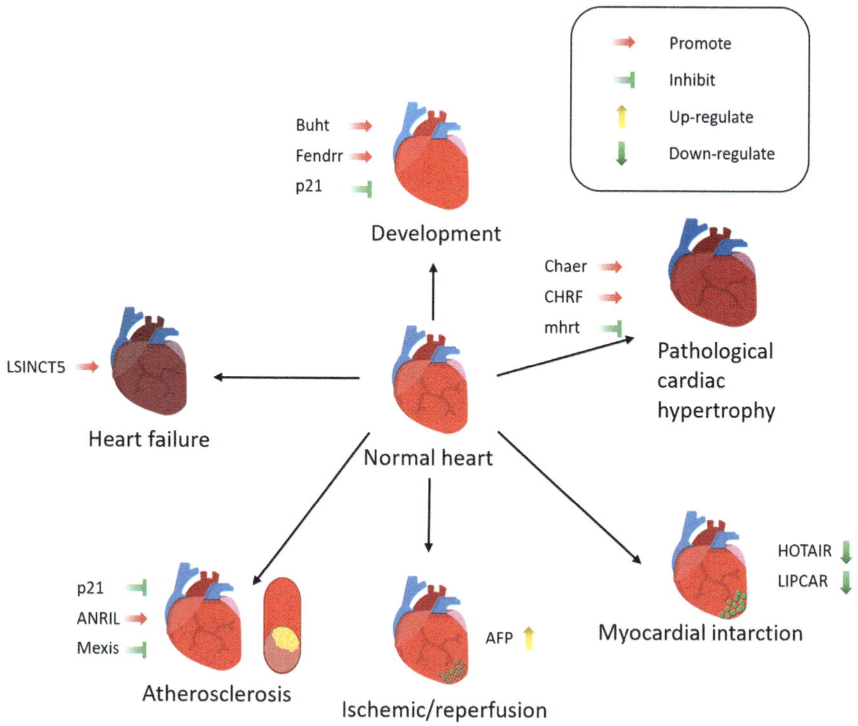

Fig. 2 The relationship between some lncRNAs and cardiovascular diseases
LncRNA Bvht and lncRNA Fendrr promote heart development, while lncRNA p21 inhibits heart development. LncRNA Chaer and CHRF promote pathological cardiac hypertrophy, while lncRNA mhrt inhibits pathological cardiac hypertrophy. LncRNA HOTAIR and LIPCAR are negatively correlated with myocardial infarction. LncRNA AFP is positively correlated with cardiac ischemia-reperfusion injury. LncRNA p21 and ANRIL induce atherosclerosis, while lncRNA MeXis inhibits atherosclerosis. LncRNA LSINCT5 induces heart failure

play an important role in cardiac pathological hypertrophy (Luther 2005). However, the potential regulatory mechanism remains to be elucidated.

LncRNA cardiac hypertrophy-associated epigenetic regulator (Chaer) enriched in the heart plays an important role in the development of cardiac hypertrophy. Chaer interacts directly with the catalytic subunit of PRC2 through a 66-mer sequence and interferes with PRC2 targeting genomic sites, thus inhibiting the methylation of histone H3 at lysine 27 in the promoter region of myocardial hypertrophy-related genes. Further studies have found that inhibiting the expression of Chaer in the heart could significantly reduce myocardial hypertrophy and dysfunction (Wang et al. 2016). LncRNA cardiac hypertrophy-related factor (CHRF) can reduce Myd88 binding to miR-489 through competitive binding miR-489, thus upregulating the expression of target gene of Myd88 and inducing myocardial hypertrophy (Wang et al. 2014a, b). Recent research found that lncRNA myosin heavy-chain-associated RNA transcripts (mhrt) can affect the acetylation of myocardin by HDAC5, thereby

Table 1 LncRNAs in cardiovascular diseases

LncRNAs	Related progress or disease	Tissues	Mechanism	References
Bvht	Development	Embryonic stem cell	Interacts with the SUZ12 of PRC2, promotes the directional differentiation of stem cells into cardiac progenitor cells	Klattenhoff et al. (2013)
Fendrr	Development	Lateral mesoderm	Binds to PRC2 and TrxG/MLL, promotes proliferation and differentiation of cardiac embryonic layer and ventral cells	Grote et al. (2013)
p21	Development, atherosclerosis	Heart	Enhances p53, inhibits the growth and proliferation, and induces cell apoptosis of mouse monocyte macrophages and vascular smooth muscle cells, inhibits atherosclerosis	Wu et al. (2014)
Chaer	Pathological cardiac hypertrophy	Heart	Interferes PRC2 targeting genomic sites, promotes myocardial hypertrophy and dysfunction	Wang et al. (2016)
CHRF	Pathological cardiac hypertrophy	Heart	Downregulates miR-489, induces myocardial hypertrophy	Wang et al. (2014a, b)
mhrt	Pathological cardiac hypertrophy	Heart	Affects the acetylation of myocardin by HDAC5, inhibits cardiac hypertrophy	Luo et al. (2018)
HOTAIR	Myocardial infarction	Blood	Negatively correlated with myocardial infarction	Gao et al. (2017)
AFP	Ischemia-reperfusion	Heart	Inhibits miR-188-3p, positively correlated with ischemia-reperfusion	Wang et al. (2015)
LIPCAR	Myocardial infarction	Blood	Downregulated after myocardial infarction and upregulated in patients with end-stage heart failure	Kumarswamy et al. (2014)
ANRIL	Atherosclerosis	Heart	Induces atherosclerosis	Holdt et al. (2010), Congrains et al. (2013), and Aguilo et al. (2016)
MeXis	Atherosclerosis	Bone marrow	Inhibits atherosclerosis	Sallam et al. (2018)
LSINCT5	Heart failure	Heart	Activates caspase-1, induces heart failure	Zhang et al. (2015)

inhibiting cardiac hypertrophy caused by cardiac myosin. In addition, cardiomyosin can directly activate mhrt transcription by binding with CarG box. Therefore, mhrt and cardiomyosin form regulatory rings in the process of cardiac hypertrophy (Luo et al. 2018).

Cardiac hypertrophic-associated transcript (Chase) is a kind of hypertrophic lncRNA, which inhibits the expression of autophagy regulator pleckstrin homology domain-containing family M member 1 (plekhml), thus hindering the autophagy of cardiac myocytes and inducing cardiac hypertrophy. The expression of Chase was partly induced by nuclear factor of activated T-cells (NFAT). In addition, Chase is related to Wnt signaling pathway, and the activation of Wnt signaling pathway can drive cardiac hypertrophy (Dawson et al. 2013; Viereck et al. 2016). The mesoderm posterior 1 (MESP1) gene plays a role in the development of heart and in the process of epithelial mesenchymal transition. As an upstream regulator of mesp1, Bvht may be involved in the regulation of cardiac hypertrophy. The interaction between Bvht and SUZ12 subunit of PRC2 results in the downregulation of SUZ12 level at the position of mesp1 promoter, hinders H3K27me3, activates the expression of mesp1 gene, and promotes the progress of cardiac hypertrophy (Lindsley et al. 2008). Myocardial infarction-associated transcript (MIAT) increased significantly in the process of myocardial hypertrophy. MIAT siRNA inhibited the upregulation of ANP, BNP, and β-MHC induced by angiotensin II in mouse models and H9c2 cells, and the mechanism may be due to the fact that MIAT promoted the progress of cardiac hypertrophy by inhibiting miR-150 (Zhu et al. 2016). H19 distributes in cytoplasm and nucleus, mainly expressed in skeletal muscle and heart of adults. In recent years, it has been found that H19 has a regulatory effect on cardiac diseases such as cardiac hypertrophy. Liu et al. treated cardiomyocytes with adrenaline and found that adenovirus overexpression of H19 inhibited the hypertrophy of cardiomyocytes; after siRNA-H19 treatment, cardiomyocytes area increased. H19 may act as a precursor of miR-675 in the process of cardiac hypertrophy (Liu et al. 2016). Calcium/calmodulin-dependent protein kinase II delta chain (CAMK2D) is a multifunctional serine/threonine protein kinase mainly existing in the heart, which can phosphorylate ion channels, transcription factors, signal molecules, and other membrane proteins that are crucial to cardiac electrical activity and structure (Bers and Grandi 2009). CAMK2D can be used as an inducer of cardiac hypertrophy. Studies have shown that CAMK2D is a direct target of miR-675 in cardiomyocytes, and miR-675 downregulates the mRNA and protein levels of CAMK2D. Therefore, H19 may regulate CAMK2D to inhibit cardiomyocyte hypertrophy through miR-675. Bernhard Herrmann's team named an lncRNA in cardiac and ventral parietal progenitors foxf1 adjacent noncoding development regulatory RNA (Fendrr). In vivo, Fendrr can combine with mixed lineage leukemic histone modification complex (MLL) of PRC2 and TrxG families to increase H3K27me3, inhibit the expression of mast gene, and inhibit cardiomyocyte hypertrophy (Grote et al. 2013). Phospholipid reptilase 4 (Plscr 4) is a member of the phospholipid reptilase gene family, which encodes a group of Ca^{2+}-dependent polypalmitoylated type II membrane proteins. Overexpression of lncRNA Plscr4 weakened angiotensin II, and TAC-induced cardiomyocyte hypertrophy, on the contrary, inhibited Plscr4-induced cardiomyocyte hypertrophy. In vitro and in vivo experiments showed that Plscr4 overexpression could downregulate miR-214, promote the expression of mitochondrion fusion protein-2 (Mfn2), and weaken cardiomyocardial hypertrophy (Lv et al. 2018).

These findings may play a vital role in revealing the complete mechanism of myocardial hypertrophy.

5 LncRNAs in Myocardial Infarction

Myocardial infarction refers to the ischemic necrosis of the myocardium, based on coronary artery disease; the blood flow of the coronary artery decreases sharply or interrupts, resulting in severe and lasting acute ischemia of the corresponding myocardium, which eventually leads to ischemic necrosis of the myocardium. Acute myocardial infarction (AMI) is the main cause of death in the world. At present, the diagnosis of AMI mainly depends on the patient's past history, chest pain symptoms, electrocardiogram (ECG), and laboratory examination. In laboratory examination items, cTnI/cTnT has been widely used in the clinic, but it will still be elevated in a variety of non-AMI patients, such as viral myocarditis, heart failure, atrial fibrillation, chronic kidney disease, sepsis, etc. Creatine kinase MB (CK-MB) is also affected by a variety of factors (Finsterer et al. 2007; Aldous 2013). In addition, not all AMI patients have ECG changes. Moreover, the clinical manifestations and past medical history of AMI are different from each other. Thus, more rapid and stable circulating biomarkers for early diagnosis of AMI are highly needed, and they can help effectively reduce mortality and save more patients' lives.

LncRNA MIAT is associated with myocardial infarction, and the expression changes of six single nucleotide polymorphisms (SNPs) in MIAT gene may give genetic susceptibility to myocardial infarction (Ishii et al. 2006). LncRNA HOX antisense intergenic RNA (HOTAIR) is associated with myocardial infarction. HOTAIR has been found to be a protective factor of myocardial cells, and its cardioprotective function was partly based on the negative regulation of miR-1. Compared with healthy controls, the expression of HOTAIR in serum of patients with acute myocardial infarction was significantly decreased. HOTAIR was downregulated in the serum of mice ligated by coronary artery and in cultured cardiomyocytes exposed to hypoxia. The plasma concentration of HOTAIR can be used as a biomarker for the diagnosis of acute myocardial infarction (Gao et al. 2017). LncRNA autophagy-promoting factor (APF) can regulate autophagic cell death by affecting the activity of miR-188-3p and miR-188-3p downstream target ATG7 (Wang et al. 2015). Autophagy is a key cellular stress response that degrades defective macromolecules and organelles and provides bioenergy intermediates during hypoxia and nutrition deficiency. It has a certain protective effect on myocardium (Bravo-San Pedro et al. 2017; Frudd et al. 2018). ATG7 is an important autophagy-promoting gene involved in myocardial injury induced by ischemia-reperfusion (I/R). Under the pathological conditions of I/R, the expression of APF is significantly increased, and the activity of miR-188-3p and the expression of ATG-7 are inhibited, which eventually leads to autophagy. The expression of lncRNAs in peripheral blood was also found to be altered after the onset of myocardial infarction. Many studies have shown that lncRNAs can participate in many biological processes through dynamic expression changes. For example, long intergenic noncoding RNA predicting cardiac remodeling (LIPCAR) is downregulated after myocardial infarction and upregulated in patients with end-stage heart failure (Kumarswamy et al. 2014).

After AMI, cardiac fibroblasts are activated, and then a large number of extracellular matrix (ECM) proteins are produced, which eventually lead to interstitial fibrosis and cardiac remodeling, leading to arrhythmia and heart failure. Therefore, inhibition of ECM oversecretion and deposition is an important treatment strategy to improve the prognosis of AMI. A recent study found that 53 kinds of lncRNAs were upregulated by two times in the peri-infarct area of mice 4 weeks after MI and 37 kinds of lncRNAs were downregulated by more than 0.5 times. The upregulation of nonmmuta022554 is the most significant, and it is positively correlated with six genes of PI3K/AKT signaling pathway and ECM receptor interaction involved in cardiac fibrosis (Qu et al. 2016). Overexpression of lncRNA H19 can lead to proliferation and fibrosis of cardiac fibroblasts (Tao et al. 2016). Using microarray analysis, the study found that MIRT1 and MIRT2 of lncRNA were upregulated 5 and 13 times after AMI, respectively, and were related to cardiac remodeling and ejection fraction (Zangrando et al. 2014).

Therefore, further research will hopefully help us to find biological indicators that can indicate the occurrence of myocardial infarction and predict the severity of myocardial infarction at an early stage. Thus, it will be of great biological significance and clinical relevance to obtain the relevant data of the indicators through the analysis of peripheral blood.

6 LncRNAs in Cardiac Ischemia-Reperfusion Injury

Myocardial infarction is an important disease endangering human health and the most common cause of death, which brings a heavy burden to the world. Reperfusion treatment after myocardial infarction, including percutaneous coronary intervention and thrombolysis, can effectively save the myocardial infarction area, but the sudden recovery of interrupted blood flow caused by revascularization will lead to a series of adverse events, such as myocardial stunning, reperfusion arrhythmia, microvascular dysfunction, etc. Although most of the blood supply recovered after myocardial ischemia makes that necrotic myocardium can be recovered, it will also cause progressive aggravation of some myocardial damage. This pathological process is called myocardial ischemia-reperfusion injury (Hausenloy and Yellon 2013). Myocardial ischemia-reperfusion injury is mainly mediated by oxidative stress, intracellular Ca^{2+} overload, increased inflammatory response, mitochondrial dysfunction, cell necrosis, apoptosis, autophagy, and other factors. A large number of studies have shown that lncRNA is closely related to myocardial ischemia-reperfusion injury (Wu et al. 2017a, b).

LncRNA H19 regulates cardiomyocyte apoptosis through miR-675 negative regulation of peroxisome proliferator-activated receptor alpha (PPARα), and inhibition of lncRNA H19 can effectively reduce apoptosis caused by ischemia-reperfusion (Luo et al. 2019). After myocardial ischemia-reperfusion, the RNA component of mitochondrion RNA processing endoribonuclease (RMRP), a kind of lncRNAs, was upregulated. LncRNA RMRP could increase cardiomyocyte apoptosis by downregulating miR-206 and then upregulating autophagy-related protein 3 (Atg3) (Kong et al. 2019). After myocardial ischemia-reperfusion, the expression of urothelial carcinoma-associated 1 (UCA1) decreased, and lncRNA UCA1 can regulate the apoptosis of cardiomyocytes by inhibiting the expression of p27 protein (Liu et al. 2015). After anoxia and reoxygenation, the expression of KCNQ1 overlapping transcription 1 (KCNQ1OT1) was upregulated, and lncrna KCNQ1OT1 regulated cardiomyocyte apoptosis by adiponectin receptor 1 (AdipoR1), p38 protein, and mitogen-activated protein kinase (MAPK)/NF-κB pathway (Li et al. 2017). LncRNA ROR expression in the plasma of patients with myocardial infarction is increased, and lncRNA ROR can activate p38/MAPK pathway and regulate cardiomyocyte apoptosis (Zhang et al. 2018a, b). The expression of lncRNA LINC00652 increased after myocardial ischemia-reperfusion in mice. LncRNA LINC00652 acts on the glucagon-like peptide 1 receptor (GLP-1R) and regulates the cyclic adenosine monophosphate/protein kinase A (cAMP/PKA) pathway to regulate cardiomyocyte apoptosis (Zhang et al. 2018a, b). After myocardial ischemia-reperfusion in diabetic rats, the expression of lncRNA myocardial infarction-related transcript 1 (MIRT1) increased, and the decreased expression of lncRNA MIRT can reduce cardiomyocyte apoptosis in diabetic rats by inhibiting the activation of NF-κ B signaling pathway (Liu et al. 2019). LncRNAs not only regulate the necrosis, apoptosis, and autophagy of myocardial cells after myocardial infarction but also regulate the arrhythmia after myocardial infarction. LncRNA MALAT1 expression was upregulated after myocardial infarction in rats. LncRNA MALAT1 could negatively regulate miR-200c, and miR-200c could negatively regulate high mobility group box 1 (HMGB1). MiR-200c/HMGB1 regulates the expression of outward potassium current and peak current density, so as to regulate the arrhythmia after myocardial infarction (Zhu et al. 2018).

Studying the expression level of lncRNAs during early ischemia-reperfusion in mice with myocardial infarction reveals the pathological mechanism of lncRNAs on myocardial ischemia-reperfusion. After myocardial ischemia-reperfusion, many lncRNAs showed abnormal expression (Wu et al. 2017a, b). It has been shown that 64 kinds of lncRNAs increased and 87 kinds of lncRNAs decreased after ischemia-reperfusion among the total 31,423 lncRNAs detected. Among them, AK035396 and AK005401 were the most obvious upregulated one, while AK080112 and AK156124 were the most obvious downregulated one. In addition, the abnormal expression of lncRNAs was more obvious in the ischemic infarct area, which might be the cause of imbalance between cell recovery and tissue necrosis (Liu et al. 2013).

LncRNA can help us to study the pathological mechanism of myocardial ischemia-reperfusion injury and find effective therapeutic target.

7 LncRNAs in Atherosclerosis

Atherosclerosis refers to the extensive atherosclerotic plaque formed by subintimal lipid deposition accompanied by the proliferation of smooth muscle cells and fibrous matrix components. An important pathogenesis of atherosclerosis is the migration and proliferation of vascular smooth muscle cells. Plaque development is the pathological basis of atherosclerosis, which is characterized by vascular stenosis and spasm. LncRNA antisense noncoding RNA in the INK4 locus (ANRIL) can induce atherosclerosis by regulating the behavior of vascular smooth muscle cells. ANRIL is located on chromosome 9p21, which does not contain any known protein-coding genes. ANRIL can promote the proliferation of vascular smooth muscle cells and the formation of atherosclerotic plaque. Besides, ANRIL can also increase proliferation and adhesion and reduce apoptosis of vascular smooth muscle cell through reverse regulation of target genes (Holdt et al. 2010). These processes are fundamental in the pathogenesis of coronary atherosclerosis. This regulation relies on a unique scattered repeat sequence, which exists in the promoter of ANRIL target gene. In addition, single nucleotide polymorphism (SNPs) in 9p21 region is closely related to the expression level of ANRIL (Congrains et al. 2013; Aguilo et al. 2016). LncRNA macrophage-expressed LXR-induced sequence (MeXis) is associated with atherosclerosis, which controls the expression of the key protein ABCA1 (Sallam et al. 2018). Abca1 is an ATP-dependent transport membrane protein. It pumps cholesterol out of cells in arterial wall, which is essential for the formation of high-density lipoprotein (HDL). The plasma HDL level is negatively correlated with atherosclerotic cardiovascular disease. The protective effect against atherosclerosis associated with HDL is by promoting the removal of cholesterol by macrophages in the arterial wall and delivering cholesterol to liver for excretion, thus achieving reverse cholesterol transport (Calabresi et al. 2015). Mice lacking MeXis are more likely to have blocked blood vessels. In addition, increasing the expression of MeXis can clear excess cholesterol more effectively. LncRNA-p21 is a key regulator of cell proliferation and apoptosis in atherosclerosis by inhibiting the ability of p53, while the inactivation of p53 can promote atherosclerosis. The expression of p21 in atherosclerotic plaque and coronary artery disease of ApoE knockout mice was significantly downregulated (Wu et al. 2014).

Because lncRNAs are expressed in many coronary artery smooth muscle cells, vascular endothelial cells, monocyte-macrophage system, and many other tissues related to coronary heart disease, we speculate that lncRNAs may play a role in the new mechanism of regulating the occurrence and development of coronary heart disease, and there are reasons to expect them to become important clues for the study of new diagnostic and preventive methods for coronary atherosclerotic heart disease.

8 LncRNAs in Heart Failure

Heart failure is the final stage of most cardiovascular diseases and one of the leading causes of death among the elderly worldwide (Granger et al. 2003). LncRNAs are involved in lipid metabolism, proliferation, and differentiation of cardiomyocytes, cardiac fibrosis, and so on. Relevant animal models and clinical studies have confirmed that lncRNA was closely related to the occurrence and development of heart failure (Kumarswamy et al. 2014; Long et al. 2017). In the context of precision medicine, lncRNA has great value in the diagnosis, risk stratification, and treatment of heart failure.

B-type brain natriuretic peptide (BNP) has been proved to be closely related to heart failure and has become one of the main biomarkers of heart failure (Ishino et al. 2008). LncRNA LSINCT5 is increased in cardiomyocytes with BNP overexpression. Interference with LSINCT5 by siRNA can reduce the apoptosis of cardiomyocytes (Prescimone et al. 2015). Previous studies have shown that caspase-1 plays an important role in cardiovascular diseases (Syed et al. 2005). The expression of LSINCT5 is associated with the expression of caspase-1 and interleukin-1 beta. This newly discovered BNP/LSINCT5/caspase-1/interleukin-1 beta signaling pathway helps us better understand the molecular mechanism behind chronic heart failure (Zhang et al. 2015). Future research may focus on whether LSINCT5 will become a diagnostic tool or even a therapeutic target for chronic heart failure.

Bace1-as is a long-chain noncoding RNA transcribed from the antisense chain of β-endocrine enzyme-1 gene (BACE1). Bace1-as increased the stability of BACE1 transcripts. Previous studies have shown that bace1-as is closely related to Alzheimer's disease. However, Greco et al. found that the expression of Bace1-as and β-amylase increased significantly in patients with heart failure and the survival ability of human embryonic stem cells and mouse embryonic stem cells treated with β-amylase decreased; in addition, by silencing BACE1, it could prevent the apoptosis of endothelial cells caused by Bace1-as overexpression. Bace1-as, BACE1, and β-amylase are involved in the pathological mechanism of heart failure (Greco et al. 2017). In doxorubicin-induced heart failure mice model, CHRF was detected by qRT-PCR, and TGF-β1 was detected by enzyme-linked immunosorbent assay. In order to verify that CHRF, TGF-β1, and caspase-3 are closely related to heart failure in mice, Chen et al. also showed the same results in vitro. The cardiac protective effect of valsartan can be achieved by regulating CHRF-TGF-β1/Smads and TGF-β1/p38. Wang et al. confirmed that CHRF acts as the endogenous skeleton of miR-489 and regulates the expression of MyD88 in vivo and in vitro (Wang et al. 2014a, b; Chen et al. 2018).

LncRNA is not only directly related to heart failure but also can affect the occurrence of heart failure by interacting with other noncoding RNAs. For example, lncRNA AFP, miR-188-3p, and Atg7 constitute a new autophagy mode. APF can target and regulate miR-188-3p, thus affecting the expression of Atg7 (key promoter and autophagy marker in autophagy), effectively reducing the area of myocardial infarction, reducing the probability of heart failure, or prolonging the survival time. Atg7 participates in the inhibition of Atg7 translation by miR-188-3p, and miR-188-3p inhibits autophagy and cell death by targeting Atg7 in vivo and in vitro. APF directly binds to miR-188-3p, thus inhibiting its activity. Regulating the level of lncRNA AFP can be used as a potential target and diagnostic tool for a new treatment strategy of myocardial infarction and heart failure (Wang et al. 2015). LncRNA CARL is involved in the pathogenesis of many diseases, including heart failure, by impairing the downregulation of miR-539-dependent PHB2 and inhibiting the mitotic abnormalities induced by hypoxia-induced mitosis of cardiomyocytes. Among them, CARL acts as a "sponge" of miRNA, and the regulation of CARL may provide a new method for interventional heart disease treatment (Wang et al. 2014a, b). Tumor necrosis factor (TNF) mediates heart failure by activating the target gene of NF-κB pathway. The results showed that TNFα stimulated the expression of various lncRNAs in mouse embryonic fibroblasts. LETHE is one of the pseudo lncRNAs. It selectively induces pro-inflammatory cytokines through NF-κB or glucocorticoid receptor agonists and plays a role in the negative feedback signal of NF-κB. In addition, LETHE can induce TNFα and IL-1β, but it is not a Toll-like receptor agonist. It is suggested that these expressed lncRNAs may be used as intervention targets in the treatment of heart failure (Rapicavoli et al. 2013). Yang et al. found 113 new types of lncRNA by studying the changes of lncRNA expression profile in left ventricular (LV) myocardium of ischemic and nonischemic heart failure heart tissue before and after the installation of left ventricular assist device (LVAD). In addition, 679 and 570 lncRNAs were differentially expressed in ischemic and nonischemic heart failure tissues, respectively, which is equivalent to about 10% of patients with improved cardiac function after LVAD installation. The change of lncRNA expression in the heart not only distinguishes non-heart failure patients in LV samples of cardiomyopathy but also distinguishes LV samples of cardiomyopathy (including ischemic and nonischemic heart failure) before and after LVAD treatment. This study not only shows that up- or downregulation of lncRNA expression profile can distinguish heart failure of different causes but also proves for the first time that lncRNA may play an important role in the pathogenesis of cardiomyopathy and help to reverse the left ventricular remodeling caused by hemodynamics after mechanical circulation support unloading (Yang et al. 2014). Danhua et al. screened the lncRNAs in the heart, whole blood, and plasma of heart failure mice by chip technology and found that 32 of them were stably expressed in the blood circulation. Further study found that these circulating lncRNA levels are not only from heart failure but also may be activated and released by other cells in the circulatory system during heart failure (Li et al. 2013). These stably expressed lncRNAs in heart failure mice may be used as biomarkers for diagnosis and treatment of heart failure.

The relationship between some lncRNAs and cardiovascular diseases is shown in Fig. 2 and Table 1.

9 Conclusions

In summary, current research has expanded the understanding of lncRNA function and regulation. Although some functions of lncRNAs are clearly revealed in the study based on human and mouse tissue samples and disease models, the current research on lncRNAs is still in the primary stage. A large number of long-chain noncoding transcriptional fragments and their functions have not been explained. Many of the data obtained by sequencing have not been effectively utilized, and not all studies can provide accurate location of lncRNAs in chromosomes. The extensive expression of lncRNAs in cardiovascular diseases shows the broad prospects of lncRNAs as a cardiovascular marker. It is believed that with further research, lncRNAs as a target for diagnosis and treatment of diseases will become a hot topic for a period of time.

Acknowledgment This work was supported by the grants from the National Natural Science Foundation of China (81722008, 91639101, and 81570362 to JJ Xiao), Innovation Program of Shanghai Municipal Education Commission (2017-01-07-00-09-E00042 to JJ Xiao), and the grant from the Science and Technology Commission of Shanghai Municipality (17010500100 to JJ Xiao).

Conflict of Interest The authors declare that they have no conflict of interest.

References

Aguilo F, Di Cecilia S, Walsh MJ (2016) Long non-coding RNA ANRIL and polycomb in human cancers and cardiovascular disease. Curr Top Microbiol Immunol 394:29–39
Aldous SJ (2013) Cardiac biomarkers in acute myocardial infarction. Int J Cardiol 164:282–294
Anderson DM, Anderson KM, Chang CL et al (2015) A micropeptide encoded by a putative long noncoding RNA regulates muscle performance. Cell 160:595–606
Anderson KM, Anderson DM, Mcanally JR et al (2016) Transcription of the non-coding RNA upperhand controls Hand2 expression and heart development. Nature 539:433–436
Banisadr G, Fontanges P, Haour F et al (2002) Neuroanatomical distribution of CXCR4 in adult rat brain and its localization in cholinergic and dopaminergic neurons. Eur J Neurosci 16:1661–1671
Bers DM, Grandi E (2009) Calcium/calmodulin-dependent kinase II regulation of cardiac ion channels. J Cardiovasc Pharmacol 54:180–187
Boon RA, Jae N, Holdt L et al (2016) Long noncoding RNAs: from clinical genetics to therapeutic targets? J Am Coll Cardiol 67:1214–1226
Bravo-San Pedro JM, Kroemer G, Galluzzi L (2017) Autophagy and Mitophagy in cardiovascular disease. Circ Res 120:1812–1824
Calabresi L, Gomaraschi M, Simonelli S et al (2015) HDL and atherosclerosis: insights from inherited HDL disorders. Biochim Biophys Acta 1851:13–18
Chen X, Liu B, Yang R et al (2016) Integrated analysis of long non-coding RNAs in human colorectal cancer. Oncotarget 7:23897–23908
Chen L, Yan KP, Liu XC et al (2018) Valsartan regulates TGF-beta/Smads and TGF-beta/p38 pathways through lncRNA CHRF to improve doxorubicin-induced heart failure. Arch Pharm Res 41:101–109

Congrains A, Kamide K, Ohishi M et al (2013) ANRIL: molecular mechanisms and implications in human health. Int J Mol Sci 14:1278–1292
Dawson K, Aflaki M, Nattel S (2013) Role of the Wnt-Frizzled system in cardiac pathophysiology: a rapidly developing, poorly understood area with enormous potential. J Physiol 591:1409–1432
De Souza N (2012) The ENCODE project. Nat Methods 9:1046
Deng C, Li Y, Zhou L et al (2016) HoxBlinc RNA recruits Set1/MLL complexes to activate Hox gene expression patterns and mesoderm lineage development. Cell Rep 14:103–114
Depre C, Shipley GL, Chen W et al (1998) Unloaded heart in vivo replicates fetal gene expression of cardiac hypertrophy. Nat Med 4:1269–1275
Finsterer J, Stollberger C, Krugluger W (2007) Cardiac and noncardiac, particularly neuromuscular, disease with troponin-T positivity. Neth J Med 65:289–295
Froberg JE, Yang L, Lee JT (2013) Guided by RNAs: X-inactivation as a model for lncRNA function. J Mol Biol 425:3698–3706
Frudd K, Burgoyne T, Burgoyne JR (2018) Oxidation of Atg3 and Atg7 mediates inhibition of autophagy. Nat Commun 9:95
Gao L, Liu Y, Guo S et al (2017) Circulating long noncoding RNA HOTAIR is an essential mediator of acute myocardial infarction. Cell Physiol Biochem 44:1497–1508
Granger CB, Mcmurray JJ, Yusuf S et al (2003) Effects of candesartan in patients with chronic heart failure and reduced left-ventricular systolic function intolerant to angiotensin-converting-enzyme inhibitors: the CHARM-alternative trial. Lancet 362:772–776
Greco S, Zaccagnini G, Fuschi P et al (2017) Increased BACE1-AS long noncoding RNA and beta-amyloid levels in heart failure. Cardiovasc Res 113:453–463
Grote P, Wittler L, Hendrix D et al (2013) The tissue-specific lncRNA Fendrr is an essential regulator of heart and body wall development in the mouse. Dev Cell 24:206–214
Han Y, Xu H, Cheng J et al (2016) Downregulation of long non-coding RNA H19 promotes P19CL6 cells proliferation and inhibits apoptosis during late-stage cardiac differentiation via miR-19b-modulated Sox6. Cell Biosci 6:58
Hausenloy DJ, Yellon DM (2013) Myocardial ischemia-reperfusion injury: a neglected therapeutic target. J Clin Invest 123:92–100
Holdt LM, Beutner F, Scholz M et al (2010) ANRIL expression is associated with atherosclerosis risk at chromosome 9p21. Arterioscler Thromb Vasc Biol 30:620–627
Hon CC, Ramilowski JA, Harshbarger J et al (2017) An atlas of human long non-coding RNAs with accurate 5′ ends. Nature 543:199–204
Hung T, Chang HY (2010) Long noncoding RNA in genome regulation: prospects and mechanisms. RNA Biol 7:582–585
Ishii N, Ozaki K, Sato H et al (2006) Identification of a novel non-coding RNA, MIAT, that confers risk of myocardial infarction. J Hum Genet 51:1087–1099
Ishino M, Takeishi Y, Niizeki T et al (2008) Risk stratification of chronic heart failure patients by multiple biomarkers: implications of BNP, H-FABP, and PTX3. Circ J 72:1800–1805
Kapranov P, Cheng J, Dike S et al (2007) RNA maps reveal new RNA classes and a possible function for pervasive transcription. Science 316:1484–1488
Klattenhoff CA, Scheuermann JC, Surface LE et al (2013) Braveheart, a long noncoding RNA required for cardiovascular lineage commitment. Cell 152:570–583
Kong F, Jin J, Lv X et al (2019) Long noncoding RNA RMRP upregulation aggravates myocardial ischemia-reperfusion injury by sponging miR-206 to target ATG3 expression. Biomed Pharmacother 109:716–725
Kopp F, Mendell JT (2018) Functional classification and experimental dissection of Long noncoding RNAs. Cell 172:393–407
Kumarswamy R, Bauters C, Volkmann I et al (2014) Circulating long noncoding RNA, LIPCAR, predicts survival in patients with heart failure. Circ Res 114:1569–1575
Li D, Chen G, Yang J et al (2013) Transcriptome analysis reveals distinct patterns of long noncoding RNAs in heart and plasma of mice with heart failure. PLoS One 8:e77938

Li X, Dai Y, Yan S et al (2017) Down-regulation of lncRNA KCNQ1OT1 protects against myocardial ischemia/reperfusion injury following acute myocardial infarction. Biochem Biophys Res Commun 491:1026–1033
Lindsley RC, Gill JG, Murphy TL et al (2008) Mesp1 coordinately regulates cardiovascular fate restriction and epithelial-mesenchymal transition in differentiating ESCs. Cell Stem Cell 3:55–68
Liu H, Song G, Zhou L et al (2013) Compared analysis of LncRNA expression profiling in pdk1 gene knockout mice at two time points. Cell Physiol Biochem 32:1497–1508
Liu Y, Zhou D, Li G et al (2015) Long non coding RNA-UCA1 contributes to cardiomyocyte apoptosis by suppression of p27 expression. Cell Physiol Biochem 35:1986–1998
Liu L, An X, Li Z et al (2016) The H19 long noncoding RNA is a novel negative regulator of cardiomyocyte hypertrophy. Cardiovasc Res 111:56–65
Liu Y, Wang T, Zhang M et al (2019) Down-regulation of myocardial infarction associated transcript 1 improves myocardial ischemia-reperfusion injury in aged diabetic rats by inhibition of activation of NF-kappaB signaling pathway. Chem Biol Interact 300:111–122
Long QQ, Wang H, Gao W et al (2017) Long noncoding RNA Kcna2 antisense RNA contributes to ventricular arrhythmias via silencing Kcna2 in rats with congestive heart failure. J Am Heart Assoc 6
Luo Y, Xu Y, Liang C et al (2018) The mechanism of myocardial hypertrophy regulated by the interaction between mhrt and myocardin. Cell Signal 43:11–20
Luo H, Wang J, Liu D et al (2019) The lncRNA H19/miR-675 axis regulates myocardial ischemic and reperfusion injury by targeting PPARalpha. Mol Immunol 105:46–54
Luther HP (2005) Role of endogenous antisense RNA in cardiac gene regulation. J Mol Med (Berl) 83:26–32
Lv L, Li T, Li X et al (2018) The lncRNA Plscr4 controls cardiac hypertrophy by regulating miR-214. Mol Ther Nucleic Acids 10:387–397
Matkovich SJ, Edwards JR, Grossenheider TC et al (2014) Epigenetic coordination of embryonic heart transcription by dynamically regulated long noncoding RNAs. Proc Natl Acad Sci U S A 111:12264–12269
Mercer TR, Dinger ME, Mattick JS (2009) Long non-coding RNAs: insights into functions. Nat Rev Genet 10:155–159
Mozaffarian D, Benjamin EJ, Go AS et al (2015) Heart disease and stroke statistics--2015 update: a report from the American Heart Association. Circulation 131:e29–e322
Nelson BR, Makarewich CA, Anderson DM et al (2016) A peptide encoded by a transcript annotated as long noncoding RNA enhances SERCA activity in muscle. Science 351:271–275
Ounzain S, Pezzuto I, Micheletti R, Burdet F et al (2014) Functional importance of cardiac enhancer-associated noncoding RNAs in heart development and disease. J Mol Cell Cardiol 76:55–70
Pandey RR, Mondal T, Mohammad F et al (2008) Kcnq1ot1 antisense noncoding RNA mediates lineage-specific transcriptional silencing through chromatin-level regulation. Mol Cell 32:232–246
Pazin MJ (2015) Using the ENCODE resource for functional annotation of genetic variants. Cold Spring Harb Protoc 2015:522–536
Ponting CP, Oliver PL, Reik W (2009) Evolution and functions of long noncoding RNAs. Cell 136:629–641
Prescimone T, D'amico A, Caselli C et al (2015) Caspase-1 transcripts in failing human heart after mechanical unloading. Cardiovasc Pathol 24:11–18
Qu X, Song X, Yuan W et al (2016) Expression signature of lncRNAs and their potential roles in cardiac fibrosis of post-infarct mice. Biosci Rep 36:e00337
Qureshi IA, Mehler MF (2012) Emerging roles of non-coding RNAs in brain evolution, development, plasticity and disease. Nat Rev Neurosci 13:528–541
Rapicavoli NA, Qu K, Zhang J et al (2013) A mammalian pseudogene lncRNA at the interface of inflammation and anti-inflammatory therapeutics. elife 2:e00762

Rinn JL, Chang HY (2012) Genome regulation by long noncoding RNAs. Annu Rev Biochem 81:145–166

Sallam T, Jones M, Thomas BJ et al (2018) Transcriptional regulation of macrophage cholesterol efflux and atherogenesis by a long noncoding RNA. Nat Med 24:304–312

Schiattarella GG, Hill JA (2015) Inhibition of hypertrophy is a good therapeutic strategy in ventricular pressure overload. Circulation 131:1435–1447

Schmitz SU, Grote P, Herrmann BG (2016) Mechanisms of long noncoding RNA function in development and disease. Cell Mol Life Sci 73:2491–2509

St Laurent G, Wahlestedt C, Kapranov P (2015) The landscape of long noncoding RNA classification. Trends Genet 31:239–251

Sundaresan NR, Gupta M, Kim G et al (2009) Sirt3 blocks the cardiac hypertrophic response by augmenting Foxo3a-dependent antioxidant defense mechanisms in mice. J Clin Invest 119:2758–2771

Syed FM, Hahn HS, Odley A et al (2005) Proapoptotic effects of caspase-1/interleukin-converting enzyme dominate in myocardial ischemia. Circ Res 96:1103–1109

Tao H, Cao W, Yang JJ et al (2016) Long noncoding RNA H19 controls DUSP5/ERK1/2 axis in cardiac fibroblast proliferation and fibrosis. Cardiovasc Pathol 25:381–389

Thomas H, Diamond J, Vieco A et al (2018) Global atlas of cardiovascular disease 2000-2016: the path to prevention and control. Glob Heart 13:143–163

Thum T, Condorelli G (2015) Long noncoding RNAs and microRNAs in cardiovascular pathophysiology. Circ Res 116:751–762

Touma M, Kang X, Zhao Y et al (2016) Decoding the Long noncoding RNA during cardiac maturation: a roadmap for functional discovery. Circ Cardiovasc Genet 9:395–407

Tsai MC, Manor O, Wan Y et al (2010) Long noncoding RNA as modular scaffold of histone modification complexes. Science 329:689–693

Viereck J, Kumarswamy R, Foinquinos A et al (2016) Long noncoding RNA Chast promotes cardiac remodeling. Sci Transl Med 8:326ra322

Wang K, Liu F, Zhou LY et al (2014a) The long noncoding RNA CHRF regulates cardiac hypertrophy by targeting miR-489. Circ Res 114:1377–1388

Wang K, Long B, Zhou LY et al (2014b) CARL lncRNA inhibits anoxia-induced mitochondrial fission and apoptosis in cardiomyocytes by impairing miR-539-dependent PHB2 downregulation. Nat Commun 5:3596

Wang K, Liu CY, Zhou LY et al (2015) APF lncRNA regulates autophagy and myocardial infarction by targeting miR-188-3p. Nat Commun 6:6779

Wang Z, Zhang XJ, Ji YX et al (2016) The long noncoding RNA Chaer defines an epigenetic checkpoint in cardiac hypertrophy. Nat Med 22:1131–1139

Wang L, Ma X, Xu X et al (2017) Systematic identification and characterization of cardiac long intergenic noncoding RNAs in zebrafish. Sci Rep 7:1250

Wu G, Cai J, Han Y et al (2014) LincRNA-p21 regulates neointima formation, vascular smooth muscle cell proliferation, apoptosis, and atherosclerosis by enhancing p53 activity. Circulation 130:1452–1465

Wu T, Wu HD, Xu ZX et al (2017a) Abnormal expression of long non-coding RNAs in myocardial infarction. Heart Vessel 32:1253–1261

Wu X, Zhu H, Zhu S et al (2017b) lncRNA expression character associated with ischemic reperfusion injury. Mol Med Rep 16:3745–3752

Yang KC, Yamada KA, Patel AY et al (2014) Deep RNA sequencing reveals dynamic regulation of myocardial noncoding RNAs in failing human heart and remodeling with mechanical circulatory support. Circulation 129:1009–1021

Zangrando J, Zhang L, Vausort M et al (2014) Identification of candidate long non-coding RNAs in response to myocardial infarction. BMC Genomics 15:460

Zhang X, Sha M, Yao Y et al (2015) Increased B-type-natriuretic peptide promotes myocardial cell apoptosis via the B-type-natriuretic peptide/long non-coding RNA LSINCT5/caspase-1/interleukin 1beta signaling pathway. Mol Med Rep 12:6761–6767

Zhang W, Li Y, Wang P (2018a) Long non-coding RNA-ROR aggravates myocardial ischemia/reperfusion injury. Braz J Med Biol Res 51:e6555

Zhang SB, Liu TJ, Pu GH et al (2018b) Suppression of Long non-coding RNA LINC00652 restores sevoflurane-induced cardioprotection against myocardial ischemia-reperfusion injury by targeting GLP-1R through the cAMP/PKA pathway in mice. Cell Physiol Biochem 49:1476–1491

Zhao J, Li L, Peng L (2015) MAPK1 up-regulates the expression of MALAT1 to promote the proliferation of cardiomyocytes through PI3K/AKT signaling pathway. Int J Clin Exp Pathol 8:15947–15953

Zhou M, Sun Y, Sun Y et al (2016) Comprehensive analysis of lncRNA expression profiles reveals a novel lncRNA signature to discriminate nonequivalent outcomes in patients with ovarian cancer. Oncotarget 7:32433–32448

Zhu S, Hu X, Han S et al (2014) Differential expression profile of long non-coding RNAs during differentiation of cardiomyocytes. Int J Med Sci 11:500–507

Zhu XH, Yuan YX, Rao SL et al (2016) LncRNA MIAT enhances cardiac hypertrophy partly through sponging miR-150. Eur Rev Med Pharmacol Sci 20:3653–3660

Zhu P, Yang M, Ren H et al (2018) Long noncoding RNA MALAT1 downregulates cardiac transient outward potassium current by regulating miR-200c/HMGB1 pathway. J Cell Biochem 119:10239–10249

Long Noncoding RNAs as Players in Breast Tumorigenesis

Cyrinne Achour and Francesca Aguilo

Contents

Abstract Comprehensive analysis of the mammalian genome uncovered the discovery of pervasive transcription of large RNA transcripts that do not code for proteins, namely, long noncoding RNAs (lncRNAs). LncRNAs play important roles in the regulation of gene expression from integration of chromatin remodeling complexes to transcriptional and posttranscriptional regulation of protein-coding genes. Application of next-generation sequencing technologies to cancer transcriptomes has revealed that aberrant expression of lncRNAs is associated with tumor progression and metastasis. Although thousands of lncRNAs have been shown to be dysregulated in different cancer types, just few of them have been fully characterized. In this book chapter, we

The original version of this chapter was revised. A correction to this chapter can be found at https://doi.org/10.1007/978-3-030-44743-4_21

C. Achour · F. Aguilo (✉)
Wallenberg Centre for Molecular Medicine (WCMM), Umeå University, Umeå, Sweden

Department of Medical Biosciences, Umeå University, Umeå, Sweden
e-mail: cyrinne.achour@umu.se; francesca.aguilo@umu.se

S. Jurga, J. Barciszewski (eds.), *The Chemical Biology of Long Noncoding RNAs*, RNA Technologies 11, https://doi.org/10.1007/978-3-030-44743-4_15

aim to highlight recent findings of the mechanistic function of lncRNAs in breast cancer and summarize key examples of lncRNAs that are misregulated during breast tumorigenesis. We have categorized breast cancer–associated lncRNA according to their contribution to tumor suppression or tumor progression based on recent studies. Because some of them are expressed in a specific molecular breast cancer subtype, we have outlined lncRNAs that can potentially serve as diagnostic and prognostic markers, in which expression is linked to chemotherapy resistance. Finally, we have discussed current limitations and perspectives on potential lncRNA targets for use in therapies against breast cancer.

Keywords Long noncoding RNA (lncRNA) · Gene expression regulation · Breast cancer · Metastasis · Chemoresistance

1 Introduction

Over the last decades, the development of high-throughput technologies has changed the perception of the central dogma stating that genetic information inscribed in DNA is transcribed into RNA and afterward translated into proteins. Indeed, transcription is not limited to protein-coding genes but is pervasive throughout the genome. Hence, tens of thousands of transcripts with biological function do not code for proteins. These transcripts are collectively known as noncoding RNA (ncRNA). Transfer RNA (tRNA) and ribosomal RNA (rRNA) were among the first ncRNAs to be identified. Both are highly abundant and mostly function in translation regulation. More recently, regulatory ncRNAs with key roles in cellular homeostasis have been described. These ncRNAs vary in localization and function, and an arbitrary size cutoff distinguishes two groups: small noncoding RNA (sncRNA) and long noncoding RNA (lncRNA). These sncRNAs include the small interfering RNA (siRNA), microRNA (miRNA), and PIWI-interacting RNA (piRNA), which have been well reviewed elsewhere (Okamura and Lai 2008; Gebert and MacRae 2019).

LncRNAs are noncoding transcripts that are more than 200 nucleotides in length. They share common characteristics with messenger RNAs (mRNAs). For instance, lncRNAs are transcribed by RNA polymerase II; many of them present a 5′ cap, undergo splicing, and are polyadenylated at their 3′ ends. LncRNAs were initially considered to be transcriptional noise. However, recent studies have shown that they are conserved in a cell-type-specific manner and can vary in response to environmental stimuli and during development, supporting their biological significance. Although lncRNAs can be found in both cytoplasm and nucleus, they are enriched in the later one, where they modulate transcription of genes *in cis*, located at or adjacent to their locus, and *in trans*, located on different genomic loci or in distal chromosomes. Furthermore, the transcription of lncRNAs can affect the transcription of neighboring genes, representing a limitation for their study as such regulation is independent of the RNA product.

Based on their position relative to the nearest protein-coding gene, lncRNAs can be classified into different groups: (a) long intergenic ncRNAs (lincRNAs), which do not overlap in close proximity to protein-coding genes; (b) antisense lncRNAs, transcribed from the antisense strand of a protein-coding gene; (c) sense-overlapping lncRNAs, which overlap with at least one intron/exon of different protein-coding genes in the sense RNA strand; and (d) bidirectional lncRNAs, transcribed from a promoter of a protein-coding gene, yet in the opposite direction. LncRNAs can also be classified based on the functionality of the genomic region by which the lncRNA is encoded. For instance, enhancer RNAs (eRNAs) are bidirectionally transcribed from active enhancers (Lai and Shiekhattar 2014).

LncRNAs can serve to recruit chromatin-modifying complexes to specific loci to achieve appropriate temporal and spatial gene regulation by activating or inactivating transcription. Hence, many lncRNAs have been shown to interact with polycomb repressive complexes 1 and 2 (PRC1 and PRC2) to silence target genes (Achour and Aguilo 2018). In addition, an increasing number of lncRNAs regulate gene expression by directly interacting with transcription factors and recruit them to their DNA target site. LncRNAs can also act as competitors for the binding of DNA response elements of transcription factors and act as inhibitors of their transcriptional activation activity. For instance, glucocorticoid receptor can bind both DNA and lncRNA *GAS5* through its DNA-binding domain (Kino et al. 2010). Others affect RNA processing events such as pre-mRNA splicing, mRNA export, localization, translation, and stability, by directly interacting to RNA-binding proteins (He et al. 2019). Such mechanism would confer an energetic and dynamic advantage to cells, as RNAs are less expensive and more rapidly produced than proteins. LncRNAs can recognize specific targets by directly binding to genomic DNA through R-loops and/or RNA-DNA triplex structures (Chu et al. 2011). In addition, recent studies have proposed that lncRNAs with common miRNA-binding sites can act as molecular "sponges," thereby activating the transcription of those miRNA targets (Salmena et al. 2011).

LncRNAs regulate cellular homeostasis, dysregulation of which may impact normal cellular function, and lead to tumor development and metastasis. Indeed, many lncRNAs have been shown to influence the different hallmarks of cancer by either sustaining proliferative signaling, evading growth suppressors, resisting cell death, enabling replicative immortality, inducing angiogenesis, or activating invasion and metastasis. In this book chapter, we have provided an overview of the current state of lncRNAs in breast cancer, with a particular focus on lncRNAs that impact cell growth, invasion, and metastasis. Based on their contribution to the activation or repression of oncogenic pathways, lncRNAs can be classified as oncogenic or tumor suppressor genes, respectively. Some key examples are summarized here. We have also highlighted the role of lncRNAs in chemotherapy resistance and the possibility of using oncogenic lncRNAs as novel prognostic markers and therapeutic targets for breast cancer.

2 LncRNAs and Breast Cancer Molecular Subtypes

Breast cancer is one of the most common malignant tumors in women and is the leading cause of morbidity and mortality for women worldwide (Torre et al. 2015). Breast cancer is a heterogeneous disease that can be classified into various subtypes based on molecular, histological, and clinical characteristics, with different clinical implications and treatment responses. Based on the presence of hormone receptors (progesterone (PR) and estrogen (ER)) and epidermal growth factor receptor type 2 (HER2) defined by immunohistochemistry staining, tumors are classified into four subtypes: luminal A, luminal B, HER2-positive (harboring an *ERBB2* amplification), and triple-negative (Table 1). Global gene expression studies of breast carcinomas distinguished five tumor subclasses, which mapped to the IHC-defined subtypes, except for the normal-like tumors, which share similar hormonal status with the luminal A but are characterized by normal breast tissue profiling and worse prognosis (Sorlie et al. 2001). Later, the PAM50 assay, in which the expression of 50 genes is assessed, was developed to further characterize the distinct subtypes (Parker et al. 2009).

Several studies have shown that lncRNAs are highly correlated with breast cancer subtypes achieving greater specificity than the protein-coding genes. Cluster analysis of unsupervised lncRNA expression retrieved from a large cohort of breast cancer patients from The Cancer Genome Atlas Portal (TCGA) revealed four subgroups that highly correlated with the defined PAM50 subtypes. Overall, 370 lncRNAs were overexpressed in the basal-like (cluster I), 220 were overexpressed in the HER2-positive (cluster II), 339 in luminal A (cluster III), and 279 in luminal B (cluster IV) breast cancer subtypes (Su et al. 2014). Another study identified a lncRNA molecular subtype-specific signature consisting of 42 lncRNAs (36 upregulated and 6 downregulated) for luminal A, 9 (8 upregulated and 1 downregulated) for luminal B, 14 (8 upregulated and 6 downregulated) for HER2$^+$, and 74 (28 upregulated and 46 downregulated) for the basal-like subtype

Table 1 Definition of breast cancer subtypes

BC subtype	Luminal A	Luminal B	HER-2	Basal
Molecular characteristics	ER/PR positive HER2/neu negative low Ki-67 (<14%)	ER/PR positive HER2/neu positive or negative high Ki-67 (≥14%)	ER/PR negative HER2/neu positive Ki-67	ER/PR negative HER2/neu negative Ki-67
Prevalence of invasive cancer	50%	20%	15%	~15%
Outcome	Favorable prognosis	Poor prognosis	Poor prognosis	Poor prognosis

Immunohistochemical phenotype of molecularly defined breast cancer subtypes based on the presence of *ER* estrogen receptor, *PR* progesterone receptor, *HER-2* human epidermal growth factor-2, and *Ki 67* proliferation index. The prevalence of invasive cancer and the outcome are also indicated

(Van Grembergen et al. 2016) (Fig. 1). In addition, overexpression of lncRNAs *CTD-2015G9.2*, *CTD-2527I21.15*, *LINC00393*, *LINC001198*, *RP11-10A14.5*, and *RP11-19E11* was highlighted in the basal-like subtype (Bradford et al. 2016). Although thousands of lncRNAs regulated by estrogen signaling have been identified in MCF7 human breast carcinoma cells, a three-lncRNA signature, including *PTPRG-AS1*, *LINC00324*, and *SNHG17*, was associated with the ER-positive and ER-negative subtypes of breast cancer (Zhao et al. 2014a).

3 Metastatic Breast Cancer

Metastasis occurs when cancer cells spread to other tissue sites and organs from the primary tumor and reach the blood circulation, the extracellular matrix, and the stroma. Invasion and metastasis of breast cancer target primarily the lungs and are generally the cause of death of breast cancer patients. LncRNAs participate in the invasion and metastasis of breast cancer by regulating cell adhesion molecules (CAM), extracellular matrix (ECM), and matrix metallo-proteinases (MMPs). For instance, in breast cancer cell lines, lncRNA *MALAT1* directly repressed the expression of tenascin Xb, an ECM protein that has been shown to have antimetastatic properties (Arun et al. 2016). Nevertheless, as discussed below, the role of lncRNA *MALAT1/Malat1* in breast tumorigenesis remains controversial. A large number of lncRNAs, including *SNHG12*, *HULC*, *ANCR*, and *BANCR* (Shi et al. 2016; Li et al. 2017b; Wang et al. 2017; Lou et al. 2018), have been shown to upregulate the expression of MMPs, thereby affecting cell-cell and cell-matrix interactions and promoting cellular migration and invasion. LncRNAs can also exert antimetastatic function through MMPs. Indeed, the overexpression of the tumor suppressor lncRNA *MEG3* decreased cell growth, invasion, and angiogenesis of breast cancer cells through downregulation of MMP-9 (Zhang et al. 2017a).

LncRNAs may influence cancer development by dysregulating multiple signaling pathways, such as p53, Bcl-2/Bax, or KLF4-KRT6/13, among others, thus affecting apoptosis, chemoresistance, radioresistance, and angiogenesis. For instance, it has been shown that lncRNA *LINP1* promoted metastasis and epithelial-to-mesenchymal transition and its overexpression in breast cancer cells correlated with the low sensitivity of the cells to 5-fluorouracil and doxorubicin through the inhibition of pro-apoptotic protein Bax (Liang et al. 2018).

During the process of epithelial-to-mesenchymal transition (EMT), epithelial cancer cells lose their defining epithelial features and gain a mesenchymal phenotype, which is critical for cell migration and invasion. The transition from epithelial to mesenchymal state is conducted by the repression of cell adhesion proteins such as E-cadherin; the increased expression of mesenchymal markers such as N-cadherin, vimentin, fibroblast-specific protein 1, and fibronectin; and the remodeling of the ECM and MMP. Transforming growth factor-β (TGF-β) signaling has been shown to be a major inducer of EMT and to facilitate breast cancer metastasis. Several studies have shown that lncRNAs can mediate TGF-β signaling.

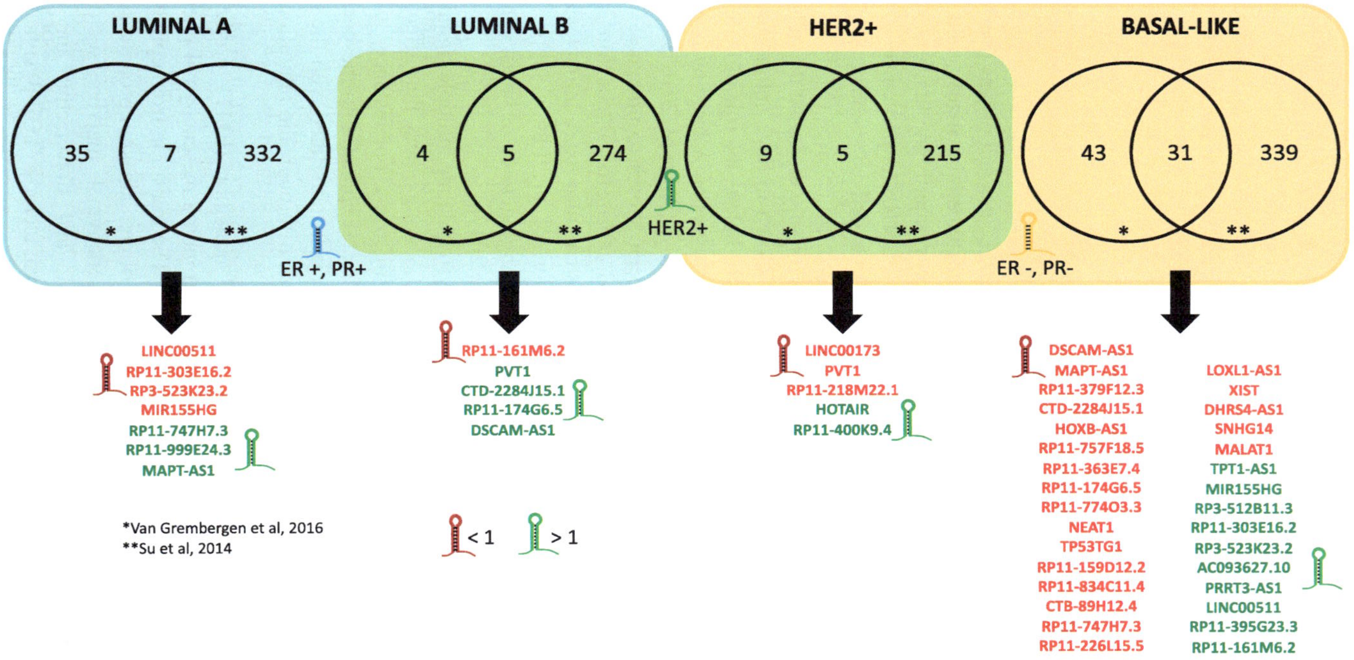

Fig. 1 Venn diagram depicting the lncRNAs that are differentially expressed in the breast cancer molecular subtypes. Data are retrieved from Van Grembergen et al. (2016) (∗) and Su et al. (2014) (∗∗). Common dysregulated lncRNAs are shown in red (downregulated) and green (upregulated). Fold change = 1.5 and FDR < 0.05. The blue, green, and orange rectangles group the subtypes based on the expression of ER, PR, and HER2. ER+/PR+ in blue (luminal A and luminal B), HER2 amplification in green (luminal B and HER2+), and ER-/PR- in orange (HER2+ and basal-like). *ER* estrogen receptor, *PR* progesterone receptor, *HER2* epidermal growth factor receptor type 2

Genome-wide lncRNA profile in mouse mammary gland epithelial cells has identified a signature of differentially expressed lncRNAs upon TGF-β induction of EMT. *LncRNA-HIT* was one of the top upregulated lncRNAs, playing a central role in TGF-β-induced EMT, cell migration, and invasion. Hence, depletion of *lncRNA-HIT* could reverse the process of EMT-associated gene expression of E-cadherin and vimentin (Richards et al. 2016). Moreover, lncRNAs can also activate TGF-β signaling pathway to promote EMT. For instance, lncRNA *ROR*, *CCAT2*, and *NORAD* have been shown to promote breast cancer metastasis by upregulating critical factors in the TGF-β signaling pathway (Cai et al. 2015; Hou et al. 2018; Zhou et al. 2019).

EMT is associated with the increased enrichment of cancer stem cells (CSCs). CSCs constitute a subpopulation of cancer cells that have the properties of self-renewal and are responsible for tumor initiation, formation, and recurrence. In breast tumors, this population of cells, exhibiting a $CD44^{+}/CD24^{-/low}$ phenotype with high ALDH activity ($ALDH^{+}$), is known as breast CSCs (BCSCs). *HOTAIR* has been shown to indirectly repress miR-7 through inhibition of the tumor suppressor *HoxD10*, maintaining the EMT process and the pool of BCSCs (Zhang et al. 2014). In addition, other lncRNAs such as *linc00617*, *ROR*, and *H19*, to name a few examples, have been described to promote the stem cell characteristics of BCSC (Hou et al. 2014).

4 Examples of LncRNAs in Breast Tumorigenesis

A large number of lncRNAs have been shown to play a role in breast cancer. However, the role and regulatory mechanisms of the vast majority of these transcripts is still elusive. We have summarized the function of some of the well-characterized oncogenic and tumor suppressor lncRNAs in breast tumorigenesis and metastasis below and in Fig. 2.

4.1 X-Inactive-Specific Transcript (XIST)

Although *X-inactive-specific transcript (XIST)* plays a central role in the initiation of X chromosome inactivation (XCI) during early development (Brown et al. 1992; Penny et al. 1996), lncRNA *XIST* is also expressed in adult females. X chromosome abnormalities contribute to the pathogenesis of breast cancer. For instance, duplication of the active X chromosome and loss of Xi has been frequently observed in the triple-negative breast cancer subtype, which leads to overexpression of a small subset of X-linked genes (Rakha et al. 2008). *XIST* expression correlated with the loss of the heterochromatic X chromosome or Barr body in somatic cells, a strong cytological marker of the inactive X chromosome (Xi) (Pageau et al. 2007). *BRCA1*, a tumor suppressor required for the cellular response to DNA damage and

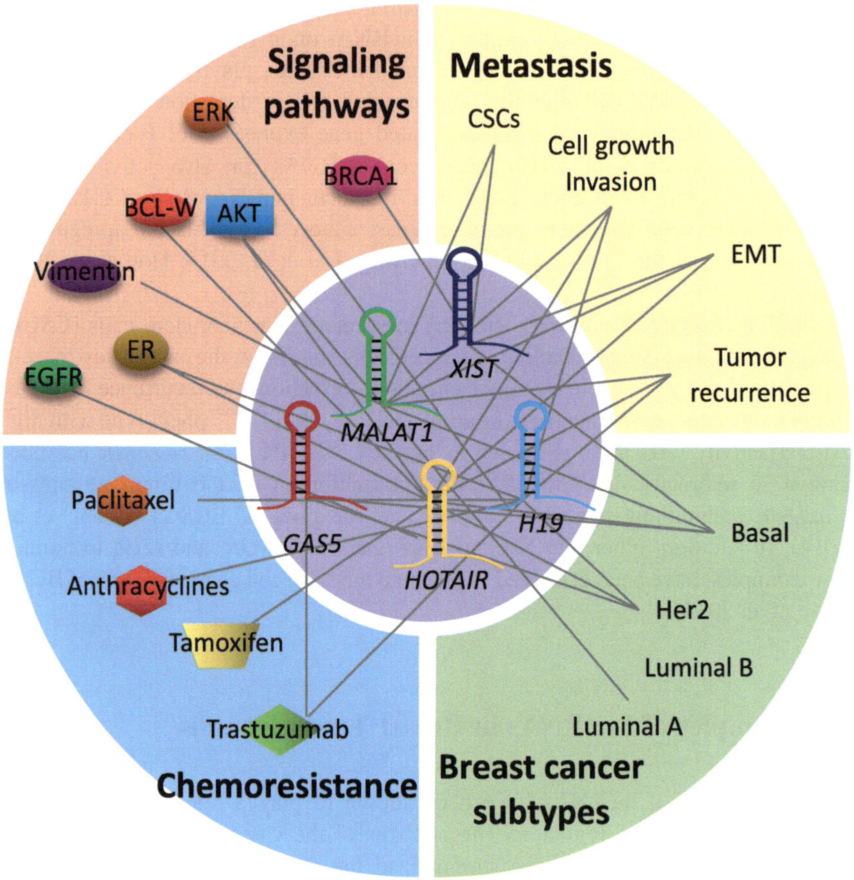

Fig. 2 Network of five common lncRNAs linked to metastasis and chemoresistance in breast cancer. LncRNAs are important factors that alter a plethora of signaling pathways. LncRNAs *XIST*, *MALAT1*, *H19*, *GAS5*, and *HOTAIR* play roles in chemoresistance and are involved in metastasis and tumor recurrence, epithelial-to-mesenchymal transition (EMT), and cancer stem cells (CSCs). The gray lines represent the association of the lncRNAs to the signaling pathways, chemoresistance, breast cancer subtypes, and metastasis

homologous recombination (Scully and Livingston 2000), physically associated with lncRNA *XIST* to maintain a proper Xi heterochromatin superstructure. Furthermore, loss of lncRNA *XIST* was associated with a *BRCA1* deficiency in sporadic basal-like cancers (Richardson et al. 2006). However, the role for BRCA in regulating the maintenance of facultative heterochromatin mediated by *XIST* has been controversial, and subsequent studies revealed that lncRNA *XIST* functioned independently of *BRCA1* in XCI (Sirchia et al. 2005; Xiao et al. 2007). Dysregulated expression levels of *XIST* have been associated with breast cancer where it has been shown to act as a tumor suppressor. Hence, depletion or overexpression of *XIST* resulted in increased or decreased levels, respectively, of AKT phosphorylation and cell viability (Huang et al. 2016). In addition, lncRNA *XIST* was significantly downregulated in brain, but not bone, metastatic tumors from patients with breast cancer. Decreased expression of *XIST* stimulated EMT and activated c-Met via MSN-mediated protein stabilization, which resulted in the promotion of stemness in the tumor cells (Xing et al. 2018).

4.2 H19

H19 encodes a capped, spliced, and polyadenylated noncoding lncRNA, which is predominantly cytoplasmic (Brannan et al. 1990). The oncofetal lncRNA *H19* is expressed in the embryo, is downregulated at birth, and then reappears in tumors, where it is actively involved in all stages of tumorigenesis, from translational deregulation and genomic instability through enhanced proliferation and metastasis. Indeed, lncRNA *H19* is overexpressed in almost every human cancer, where it is frequently associated with poor prognosis (Raveh et al. 2015). In addition, *H19* acts as molecular sponge of the tumor suppressor *let-7* microRNA, limiting its bioavailability and leading to increased cell proliferation (Kallen et al. 2013). LncRNA *H19* has been described to be overexpressed in 70% of breast cancers where it exerts tumorigenic functions through different mechanisms. For instance, *H19* overexpression conferred proliferation advantage to tumor cells via positive regulation by E2F1 (Berteaux et al. 2005). Furthermore, *H19* upregulated tyrosine kinase receptors and its downstream AKT and ERK signaling pathways to promote breast tumorigenesis (Vennin et al. 2015).

4.3 HOX Transcript Antisense Intergenic RNA (HOTAIR)

HOX transcript antisense intergenic RNA (*HOTAIR*) is transcribed from the antisense strand of the *HOXC* locus on chromosome 12. LncRNA *HOTAIR* undergoes splicing and is polyadenylated (Rinn et al. 2007). *HOX* genes encoding transcription factors are clustered in four loci *HOXA-HOXD*. During development of vertebrates, *HOXA-HOXD* expression follows a spatial and temporal order along the body and the appendicular axis (Forlani et al. 2003). LncRNA *HOTAIR* has been shown to act *in trans* and repress *HOXD* cluster of genes by recruiting PRC2. In breast cancer, *HOTAIR* is overexpressed in both primary and metastatic tumors, and its expression in primary tumors correlates with later metastases, patient prognosis, and death. LncRNA *HOTAIR* promotes the reorganization of PRC2 occupancy on target genes, including progesterone receptor and protocadherin, which repressed their expression. A specific subunit of the PRC2 complex, enhancer of zeste homolog 2 (EZH2), has been found to be associated with *HOTAIR* expression and therefore with invasion and metastasis. Overexpression of lncRNA *HOTAIR* in the highly invasive breast cancer cell line MDA-MB-231 led to lung metastasis after tail vein xenograft in mice. In contrast, knockdown of EZH2 showed an absence of metastasis to the lungs, thus confirming that EZH2 mediated the metastatic function of *HOTAIR* (Gupta et al. 2010). *HOTAIR* is an estrogen-responsive gene being its expression linked to ER in breast cancer (Bhan et al. 2013). Its promoter contains multiple estrogen response elements, which are targeted by ER and ER co-regulators in the presence of estradiol to induce *HOTAIR* transcription and contribute to breast cancer progression. In addition, a recent study has shown that overexpression of *HOTAIR* upregulated the expression of the apoptosis inhibitor Bcl-w in breast cancer. Bcl-w has been shown to promote migration, invasion, and therapy resistance of cancer cells. Mechanistically, lncRNA *HOTAIR* acted as a sponge for miR-206, which is known to inhibit Bcl-w expression, thereby facilitating the proliferation of breast cancer cells (Ding et al. 2017).

4.4 Metastasis-Associated Lung Adenocarcinoma 1 (MALAT1)

Metastasis-associated lung adenocarcinoma transcript 1 (*MALAT1*, also known as *nuclear-enriched abundant transcript 2* or *NEAT2*) is highly conserved through evolution, and is among the most abundantly expressed lncRNAs in normal tissues (Hutchinson et al. 2007). *MALAT1* is posttranscriptionally processed by RNAses at its 3′-end, generating a *MALAT1-associated small cytoplasmic RNA* (mascRNA), with unknown function, and a long nuclear transcript localized in nuclear speckles which was shown to regulate alternative splicing. However, *Malat1* knockout mice were viable and fertile and displaced no global difference in alternative splicing compared to their wild-type counterparts (Zhang et al. 2012).

MALAT1 was first discovered as a prognostic marker in non-small cell lung cancer patients, but later studies have shown that lncRNA *MALAT1* is dysregulated in a broad range of malignant tumors, including breast cancer, and its function has been directly linked to cancer progression and metastasis. *MALAT1* is highly expressed in primary breast tumors, and its expression levels have been shown to be positively correlated with lymph node size, ER expression, tumor stage, and histological grade (Gutschner et al. 2013). In addition, *MALAT1* overexpression correlated with decreased survival in lymph node-negative patients of the HER2- and triple-negative breast cancer subtypes (Jadaliha et al. 2016). Overall, these studies suggested that *MALAT1* could be used as a diagnostic and prognostic marker for breast cancer. *MALAT1* promoted cell proliferation, tumorigenesis, and metastasis of triple-negative breast cancer cells, despite having a comparatively lower expression level than ER- or HER2-positive breast cancer cells. Furthermore, downregulated *MALAT1* expression after treatment with high concentration of 17-β-estradiol inhibited cell growth, invasion, and metastasis of both ERα-positive and ERα-negative cell lines (Zhao et al. 2014b). More recently, an antisense transcript coexisting in the *MALAT1* locus termed *TALAM1* has been shown to synergize with

MALAT1 during tumorigenesis in aggressive breast cancer (Gomes et al. 2019). In support to the functional role of lncRNA *MALAT1* in regulating breast tumorigenesis, it has been reported that ablation or depletion of *Malat1* in the MMTV (mouse mammary tumor virus)-PyMT mouse mammary carcinoma model resulted in decreased tumor growth, differentiation into cystic tumors, and reduction in metastasis. Mechanistically, depletion of *Malat1* led to alterations in gene expression and splicing of signatures correlating with differentiation and pro-tumorigenic signaling pathways (Arun et al. 2016). Nonetheless, opposite observations of *MALAT1* in tumorigenesis have also been reported, and ablation of *Malall* in mice did not impact global gene expression or alternative splicing (Zhang et al. 2012). However, this work has shown that *Malat1* neighboring genes were dysregulated in *Malall* knockout mice, which raises the question to re-evaluate if the in vivo tumorigenic effect of *Malat1* lncRNA observed by other studies was a consequence to *Malat1* deletion or if the deletion of *Malat1* suppressed regulatory sequences of those genes. Indeed, later studies using targeted insertional inactivation of *Malat1* in MMTV-PyMT mice, for which lncRNA *Malat1* was directly targeted and not its neighboring genes, observed an opposite phenotype and showed that loss of *Malat1* induced metastasis. Rescue experiments and overexpression of *Malat1* in mice also validated that *Malat1* is a suppressor of breast cancer lung metastasis (Kim et al. 2018). In agreement with the tumor suppressor role of *MALAT1* lncRNA, lower expression of *MALAT1* in breast cancer patients was associated with shorter relapse-free survival. Furthermore, depletion of *MALAT1* in breast cancer cell lines induced an EMT program and enhanced cell migration and invasion by activating phosphatidylinositide-3 kinase-AKT signaling (Xu et al. 2015). Overall, several studies have reported contradictory effects of *MALAT1* in breast cancer metastasis; therefore, the function of lncRNA *MALAT1* still remains unclear and its roles must be more carefully dissected.

4.5 Growth Arrest-Specific Transcript 5 (GAS5)

Growth arrest-specific transcript 5 (*GAS5*) comprises 12 exons that due to the presence of alternative 5′-splice donor sites in exon 7 yield two mature lncRNAs, termed *GAS5a* and *GAS5b*. Within its introns, the gene encodes 10 box C/D snoRNAs which participate in the 2’-O-methylation of ribosomal RNA (rRNA) (Smith and Steitz 1998). LncRNA *GAS5* is highly expressed in growth-arrested cells from which it was isolated after screening of tumor suppressor genes that accumulate upon lack of nutrients and growth factors (Schneider et al. 1988). LncRNA *GAS5* acts as a tumor suppressor in several types of human cancer, including breast cancer, by both facilitating growth arrest and promoting apoptosis. Depletion of lncRNA *GAS5* in triple-negative and ER-positive cell lines attenuated cell responses to apoptotic stimuli, including classical chemotherapeutic agents (Pickard and Williams 2014). LncRNA *GAS5* is significantly low expressed in multiple cancers, with expression levels related to both clinicopathological characteristics and patient prognosis. For instance, lncRNA *GAS5* could serve as a novel early diagnostic biomarker for breast tumorigenesis as it is already downregulated in stages I and II of breast cancer (Mourtada-Maarabouni et al. 2009). Furthermore, *GAS5* can also be used to assess the prognosis evaluation for breast cancer patients after surgery (Han et al. 2016). In addition, downregulation of lncRNA *GAS5* leads to trastuzumab resistance in HER2-positive breast cancer, indicating that *GAS5* could be a candidate drug target for this breast cancer subtype (Li et al. 2016a).

5 LncRNAs Associated with Chemotherapy Resistance

Over the past few years, substantial advances in the treatment of breast cancer, involving the use of cytotoxic drugs, hormonal and targeted therapies, and immunotherapeutic agents, have decreased the mortality rate. Furthermore, a better understanding of the heterogeneity of breast cancer has led to the development of more effective therapies and individualized approaches to target specific molecular subtypes. Although in the majority of patients clinical responses are observed, the presence of primary and acquired resistance to chemotherapy treatment remains a significant common challenge that results in therapeutic failure. Recently, lncRNAs have emerged as key players in breast cancer drug resistance.

Endocrine therapy has proved to be one of the most effective treatment modality against ER-positive tumors. Tamoxifen, a partial nonsteroidal estrogen agonist is the prototype of the selective ER modulator family, and acts as a competitive inhibitor of estradiol in the estrogen receptor (Lumachi et al. 2011). Several lncRNAs have been shown to contribute to tamoxifen resistance by influencing different mechanisms. For instance, lncRNA *HOTAIR* is directly repressed by ER and, thus, upregulated upon the blockade of ER signaling, either by hormone deprivation or by tamoxifen treatment. Elevated lncRNA *HOTAIR* increased ER protein level and thus ER occupancy on the chromatin, enhancing ER transcriptional program (Xue et al. 2016). LncRNA *ROR* is also involved in tamoxifen resistance in ER-positive tumors. Hence, inhibition of lncRNA *ROR* reversed resistance to tamoxifen by inducing autophagy in breast cancer (Li et al. 2017a). Furthermore, downregulated lncRNA *ROR* could enhance the sensibility of breast cancer cells to tamoxifen by increasing and decreasing miR-205 and ZEB1/2, respectively (Zhang et al. 2017b). In addition, lncRNA *urothelial carcinoma-associated 1* (*UCA1*) conferred resistance to tamoxifen by activating the mTOR signaling pathway and by sponging miR-18a, a negative regulator of HIF1α (Li et al. 2016b).

Achieving a blockade of HER2 receptors with antibody-drug conjugates such as trastuzumab has been shown to improve the poor prognosis associated with HER2-positive metastatic tumors. However, trastuzumab resistance has become prevalent in recent years, and it is the leading cause of mortality in the HER2-positive breast cancer subtype. The role of TGF-β-induced EMT in trastuzumab resistance is well established. LncRNA activated by TGF-β (*lnc-ATB*), a mediator of TGF-β signaling, has been described to predispose breast cancer patients to EMT and trastuzumab resistance by competitively binding miR-200c and upregulating the oncogenes *ZEB1* and *ZNF217* (Shi et al. 2015). On the contrary, downregulation of lncRNA *GAS5* led to trastuzumab resistance in HER2-positive breast cancer. *GAS5* suppressed cancer proliferation by acting as a molecular sponge for miR-21, leading to the derepression of PTEN (Li et al. 2016a).

Cytotoxic chemotherapy is used during adjuvant systemic therapy, administrated after breast cancer surgery regardless of the breast cancer subtype, to prevent distant recurrence of the disease. Anthracyclines, such as doxorubicin and epirubicin, and/or taxanes, including paclitaxel and docetaxel, are the current cytotoxic agents used for both early and advanced stages of breast cancer (Hernandez-Aya and Gonzalez-Angulo 2013). Several studies have shown that lncRNA *H19* confers chemotherapy resistance in breast cancer. ERα upregulated the expression of *H19,* which is associated with paclitaxel resistance. Hence, in ERα-positive breast cancer cells, lncRNA *H19* attenuated cell apoptosis by inhibiting transcription of pro-apoptotic genes by recruiting polycomb factors at their promoters (Si et al. 2016). In doxorubicin-resistant breast cancer cells, *H19* was significantly upregulated, which led to the upregulated expression of *CUL4A* and *ABCB1*/MDR1, contributing to the cell resistance of multiple chemotherapeutic agents (Zhu et al. 2017).

Triple-negative breast cancer does not express conventional therapeutic targets and is the only type of malignant breast cancer for which no targeted therapy is available, with cytotoxic chemotherapy being the main treatment. Several lncRNAs have been shown to be involved in chemoresistance of these patients limiting the options for treatment. For instance, lncRNA *ferritin heavy chain 1 pseudogene 3* (*FTH1P3*) expression is enriched in paclitaxel-resistant breast cancer tissue samples, promoting cellular proliferation by targeting miR-206/ABCB1 axis (Wang et al. 2018). LncRNA *HIF1A-AS2* and *AK124454* have been shown to promote cell proliferation and invasion and were also contributing to paclitaxel resistance in triple-negative breast cancer patients (Jiang et al. 2016). Epidermal growth factor receptor (EGFR) and the non-receptor tyrosine kinase c-ABL are frequently overexpressed in triple-negative breast cancer to enhance proliferation, survival, invasion, and metastasis of cancer cells (Corkery et al. 2009). Co-treatment with clinically validated inhibitors of EGFR and c-ABL blocked β-catenin nuclear expression, which resulted in repression of the lncRNA *HOTAIR.* Consistently, *HOTAIR* expression and cellular growth could be restored by forcing the expression of β-catenin in the presence of both drugs. Furthermore, depletion of *HOTAIR* alone phenocopied the dual treatment in growth suppression (Wang et al. 2015).

6 LncRNAs in Diagnostic, Prognosis, and Therapies in Triple-Negative Breast Cancer

Due to the lack of early detection biomarkers and hormonal or targeted therapies, triple-negative breast cancer is the most aggressive subtype. As a consequence, patients with triple-negative breast cancer usually suffer from distal recurrence and, hence, have a poor prognosis. Therefore, the identification of potential biomarkers for prognoses and novel therapies is urgently needed. A number of lncRNAs have been associated with clinical diagnosis and survival outcome in patients with triple-negative breast cancer, thus serving as novel predictors for tumor diagnosis and prognosis. For instance, the expression of the lncRNAs *ANRIL*, *UCA1*, and *HIF1A-AS2* was increased in the plasma of patients with triple-negative breast cancer in comparison to patients with other breast cancer subtypes (Liu et al. 2017). LncRNA *NEAT1* was also found to be overexpressed in blood samples of patients with triple-negative breast cancer compared to the other subtypes. Knockdown of *NEAT1* led to a decrease of stemness and an increase in chemotherapy sensitivity, supporting the use of *NEAT1* as a target for triple-negative breast cancer (Shin et al. 2019). Another example is the analysis of microarray expression profiles from 473 breast cancer patients, of which 12 lncRNAs have been identified as either upregulated or downregulated in patients with tumor recurrence (Zhou et al. 2016). Therefore, those lncRNAs may be used as a prognostic signature for breast cancer patients to predict the tumor recurrence and recurrence-free survival.

Despite our current understanding of lncRNA function in breast cancer is limited, and controversies still remain in the field, several studies have highlighted the use of lncRNAs in cancer therapy. Antisense oligonucleotides (ASOs) are single-stranded oligonucleotides that are complementary to a target sequence which induces its RNA degradation by RNase H. They offer advantages over siRNA targeting, as ASOs show fewer off-target effects and higher specificity. ASOs targeting lncRNA *Malat1* in the mouse MMTV-PyMT carcinoma model resulted in a decrease of the tumor growth and lung metastasis (Arun et al. 2016). More recently, a novel tool using ASO has been developed to target *MALAT1* in lung cancer cells, but one could think of its application in breast cancer. This new technology uses functionalized gold nanoparticles conjugated with positively charged TAT peptides to stabilize the negatively charged ASO and guide it to the cell nucleus where it can specifically target *MALAT1* (Gong et al. 2019). This novel method to target lncRNA *MALAT1* showed a significant reduction of the migration capacity of cells in vitro and of the formation of tumors in a mouse model. The use of small molecules which target specific lncRNA-protein interactions to finely tune gene expression and modify chromatin state may represent a novel therapeutic approach in cancer. Hence, targeting chromatin-modifying enzymes through interacting lncRNAs would increase the specificity of such compounds and provide reversible regulation at non-catalytic domains of the protein. For instance, a small molecule compound (AC1Q3QWB) has been identified as a selective and efficient disruptor of

HOTAIR-EZH2 interaction, resulting in blocking PRC2 recruitment and increasing the expression of tumor suppressor genes in breast cancer (Li et al. 2019). Furthermore, treatment of breast cancer cell lines with a peptide nucleic acid-based approach that also blocked *HOTAIR*-EZH2 interaction decreased invasion and increased chemotherapy sensitivity (Ozes et al. 2017). Although the application of these approaches against lncRNA function is still in its infancy, it proves to be an exciting area to explore for novel therapeutic targeting of breast cancer.

7 Conclusion and Future Perspectives

Over the past few years, high-throughput sequencing analysis has highlighted the importance of lncRNAs in cellular homeostasis by playing essential roles in gene expression regulation. LncRNAs have been implicated in several processes including epigenetic control, RNA processing, stability, and translation, but knowledge of the precise mechanisms of action is still in the early stages. A limited number of well-characterized lncRNAs have given important clues about the biology of lncRNA in cancer revealing their function as tumor suppressors or pro-oncogenes, depending on how they influence tumorigenesis. In breast cancer, aberrant expression of lncRNAs has been shown to promote tumor growth and metastasis through sustaining proliferation and cell cycle progression, promoting migration, invasion, and EMT transition, and regulating cell stemness, among others. LncRNA gene signatures have been associated with different breast cancer subtypes and clinical outcomes, suggesting their potential as attractive therapeutic targets and novel molecular biomarkers for breast cancer. Therefore, a better understanding of the biology and functional mechanisms of lncRNA in cancer could pave the way for novel anticancer therapies in the future.

Acknowledgments We sincerely apologize to authors whose work could not be included due to space limitations. This work is supported by grants from the Knut and Alice Wallenberg Foundation, Umeå University, Västerbotten County Council, Kempe Foundation, Swedish Research Council, Cancerfonden, and the Cancer Research Foundation in Northern Sweden.

References

Achour C, Aguilo F (2018) Long non-coding RNA and Polycomb: an intricate partnership in cancer biology. Front Biosci (Landmark Ed) 23:2106–2132

Arun G, Diermeier S, Akerman M et al (2016) Differentiation of mammary tumors and reduction in metastasis upon Malat1 lncRNA loss. Genes Dev 30:34–51

Berteaux N, Lottin S, Monte D et al (2005) H19 mRNA-like noncoding RNA promotes breast cancer cell proliferation through positive control by E2F1. J Biol Chem 280:29625–29636

Bhan A, Hussain I, Ansari KI et al (2013) Antisense transcript long noncoding RNA (lncRNA) HOTAIR is transcriptionally induced by estradiol. J Mol Biol 425:3707–3722

Bradford JR, Cox A, Bernard P et al (2016) Consensus analysis of whole transcriptome profiles from two breast cancer patient cohorts reveals long non-coding RNAs associated with intrinsic subtype and the tumour microenvironment. PLoS One 11:e0163238
Brannan CI, Dees EC, Ingram RS et al (1990) The product of the H19 gene may function as an RNA. Mol Cell Biol 10:28–36
Brown CJ, Hendrich BD, Rupert JL et al (1992) The human XIST gene: analysis of a 17 kb inactive X-specific RNA that contains conserved repeats and is highly localized within the nucleus. Cell 71:527–542
Cai Y, He J, Zhang D (2015) Long noncoding RNA CCAT2 promotes breast tumor growth by regulating the Wnt signaling pathway. Onco Targets Ther 8:2657–2664
Chu C, Qu K, Zhong FL et al (2011) Genomic maps of long noncoding RNA occupancy reveal principles of RNA-chromatin interactions. Mol Cell 44:667–678
Corkery B, Crown J, Clynes M et al (2009) Epidermal growth factor receptor as a potential therapeutic target in triple-negative breast cancer. Ann Oncol 20:862–867
Ding W, Ren J, Ren H et al (2017) Long noncoding RNA HOTAIR modulates MiR-206-mediated Bcl-w signaling to facilitate cell proliferation in breast cancer. Sci Rep 7:17261
Forlani S, Lawson KA, Deschamps J (2003) Acquisition of Hox codes during gastrulation and axial elongation in the mouse embryo. Development 130:3807–3819
Gebert LFR, MacRae IJ (2019) Regulation of microRNA function in animals. Nat Rev Mol Cell Biol 20:21–37
Gomes CP, Nobrega-Pereira S, Domingues-Silva B et al (2019) An antisense transcript mediates MALAT1 response in human breast cancer. BMC Cancer 19:771
Gong N, Teng X, Li J et al (2019) Antisense oligonucleotide-conjugated nanostructure-targeting lncRNA MALAT1 inhibits cancer metastasis. ACS Appl Mater Interfaces 11:37–42
Gupta RA, Shah N, Wang KC et al (2010) Long non-coding RNA HOTAIR reprograms chromatin state to promote cancer metastasis. Nature 464:1071–1076
Gutschner T, Hammerle M, Diederichs S (2013) MALAT1 – a paradigm for long noncoding RNA function in cancer. J Mol Med (Berl) 91:791–801
Han L, Ma P, Liu SM et al (2016) Circulating long noncoding RNA GAS5 as a potential biomarker in breast cancer for assessing the surgical effects. Tumour Biol 37:6847–6854
He RZ, Luo DX, Mo YY (2019) Emerging roles of lncRNAs in the post-transcriptional regulation in cancer. Genes Dis 6:6–15
Hernandez-Aya LF, Gonzalez-Angulo AM (2013) Adjuvant systemic therapies in breast cancer. Surg Clin North Am 93:473–491
Hou P, Zhao Y, Li Z et al (2014) LincRNA-ROR induces epithelial-to-mesenchymal transition and contributes to breast cancer tumorigenesis and metastasis. Cell Death Dis 5:e1287
Hou L, Tu J, Cheng F et al (2018) Long noncoding RNA ROR promotes breast cancer by regulating the TGF-beta pathway. Cancer Cell Int 18:142
Huang YS, Chang CC, Lee SS et al (2016) Xist reduction in breast cancer upregulates AKT phosphorylation via HDAC3-mediated repression of PHLPP1 expression. Oncotarget 7:43256–43266
Hutchinson JN, Ensminger AW, Clemson CM et al (2007) A screen for nuclear transcripts identifies two linked noncoding RNAs associated with SC35 splicing domains. BMC Genomics 8:39
Jadaliha M, Zong X, Malakar P et al (2016) Functional and prognostic significance of long non-coding RNA MALAT1 as a metastasis driver in ER negative lymph node negative breast cancer. Oncotarget 7:40418–40436
Jiang YZ, Liu YR, Xu XE et al (2016) Transcriptome analysis of triple-negative breast cancer reveals an integrated mRNA-lncRNA signature with predictive and prognostic value. Cancer Res 76:2105–2114
Kallen AN, Zhou XB, Xu J et al (2013) The imprinted H19 lncRNA antagonizes let-7 microRNAs. Mol Cell 52:101–112
Kim J, Piao HL, Kim BJ et al (2018) Long noncoding RNA MALAT1 suppresses breast cancer metastasis. Nat Genet 50:1705–1715

Kino T, Hurt DE, Ichijo T et al (2010) Noncoding RNA gas5 is a growth arrest- and starvation-associated repressor of the glucocorticoid receptor. Sci Signal 3:ra8
Lai F, Shiekhattar R (2014) Enhancer RNAs: the new molecules of transcription. Curr Opin Genet Dev 25:38–42
Li W, Zhai L, Wang H et al (2016a) Downregulation of LncRNA GAS5 causes trastuzumab resistance in breast cancer. Oncotarget 7:27778–27786
Li X, Wu Y, Liu A et al (2016b) Long non-coding RNA UCA1 enhances tamoxifen resistance in breast cancer cells through a miR-18a-HIF1alpha feedback regulatory loop. Tumour Biol 37:14733–14743
Li Y, Jiang B, Zhu H et al (2017a) Inhibition of long non-coding RNA ROR reverses resistance to Tamoxifen by inducing autophagy in breast cancer. Tumour Biol 39:1010428317705790
Li Z, Hou P, Fan D et al (2017b) The degradation of EZH2 mediated by lncRNA ANCR attenuated the invasion and metastasis of breast cancer. Cell Death Differ 24:59–71
Li Y, Ren Y, Wang Y et al (2019) A compound AC1Q3QWB selectively disrupts HOTAIR-mediated recruitment of PRC2 and enhances Cancer therapy of DZNep. Theranostics 9:4608–4623
Liang Y, Li Y, Song X et al (2018) Long noncoding RNA LINP1 acts as an oncogene and promotes chemoresistance in breast cancer. Cancer Biol Ther 19:120–131
Liu M, Xing LQ, Liu YJ (2017) A three-long noncoding RNA signature as a diagnostic biomarker for differentiating between triple-negative and non-triple-negative breast cancers. Medicine (Baltimore) 96:e6222
Lou KX, Li ZH, Wang P (2018) Long non-coding RNA BANCR indicates poor prognosis for breast cancer and promotes cell proliferation and invasion. Eur Rev Med Pharmacol Sci 22:1358–1365
Lumachi F, Luisetto G, Basso SM et al (2011) Endocrine therapy of breast cancer. Curr Med Chem 18:513–522
Mourtada-Maarabouni M, Pickard MR, Hedge VL et al (2009) GAS5, a non-protein-coding RNA, controls apoptosis and is downregulated in breast cancer. Oncogene 28:195–208
Okamura K, Lai EC (2008) Endogenous small interfering RNAs in animals. Nat Rev Mol Cell Biol 9:673–678
Ozes AR, Wang Y, Zong X et al (2017) Therapeutic targeting using tumor specific peptides inhibits long non-coding RNA HOTAIR activity in ovarian and breast cancer. Sci Rep 7:894
Pageau GJ, Hall LL, Ganesan S et al (2007) The disappearing Barr body in breast and ovarian cancers. Nat Rev Cancer 7:628–633
Parker JS, Mullins M, Cheang MC et al (2009) Supervised risk predictor of breast cancer based on intrinsic subtypes. J Clin Oncol 27:1160–1167
Penny GD, Kay GF, Sheardown SA et al (1996) Requirement for Xist in X chromosome inactivation. Nature 379:131–137
Pickard MR, Williams GT (2014) Regulation of apoptosis by long non-coding RNA GAS5 in breast cancer cells: implications for chemotherapy. Breast Cancer Res Treat 145:359–370
Rakha EA, Reis-Filho JS, Ellis IO (2008) Basal-like breast cancer: a critical review. J Clin Oncol 26:2568–2581
Raveh E, Matouk IJ, Gilon M et al (2015) The H19 long non-coding RNA in cancer initiation, progression and metastasis – a proposed unifying theory. Mol Cancer 14:184
Richards EJ, Zhang G, Li ZP et al (2016) Long non-coding RNAs (LncRNA) regulated by transforming growth factor (TGF) beta. LncRNA-Hit-mediated TGF-induced epithelial to mesenchymal transition in mammary epithelia. J Biol Chem 291:22860
Richardson AL, Wang ZC, De Nicolo A et al (2006) X chromosomal abnormalities in basal-like human breast cancer. Cancer Cell 9:121–132
Rinn JL, Kertesz M, Wang JK et al (2007) Functional demarcation of active and silent chromatin domains in human HOX loci by noncoding RNAs. Cell 129:1311–1323
Salmena L, Poliseno L, Tay Y et al (2011) A ceRNA hypothesis: the Rosetta stone of a hidden RNA language? Cell 146:353–358

Schneider C, King RM, Philipson L (1988) Genes specifically expressed at growth arrest of mammalian cells. Cell 54:787–793
Scully R, Livingston DM (2000) In search of the tumour-suppressor functions of BRCA1 and BRCA2. Nature 408:429–432
Shi SJ, Wang LJ, Yu B et al (2015) LncRNA-ATB promotes trastuzumab resistance and invasion-metastasis cascade in breast cancer. Oncotarget 6:11652–11663
Shi F, Xiao F, Ding P et al (2016) Long noncoding RNA highly up-regulated in liver cancer predicts unfavorable outcome and regulates metastasis by MMPs in triple-negative breast cancer. Arch Med Res 47:446–453
Shin VY, Chen J, Cheuk IW et al (2019) Long non-coding RNA NEAT1 confers oncogenic role in triple-negative breast cancer through modulating chemoresistance and cancer stemness. Cell Death Dis 10:270
Si X, Zang R, Zhang E et al (2016) LncRNA H19 confers chemoresistance in ERalpha-positive breast cancer through epigenetic silencing of the pro-apoptotic gene BIK. Oncotarget 7:81452–81462
Sirchia SM, Ramoscelli L, Grati FR et al (2005) Loss of the inactive X chromosome and replication of the active X in BRCA1-defective and wild-type breast cancer cells. Cancer Res 65:2139–2146
Smith CM, Steitz JA (1998) Classification of gas5 as a multi-small-nucleolar-RNA (snoRNA) host gene and a member of the 5′-terminal oligopyrimidine gene family reveals common features of snoRNA host genes. Mol Cell Biol 18:6897–6909
Sorlie T, Perou CM, Tibshirani R et al (2001) Gene expression patterns of breast carcinomas distinguish tumor subclasses with clinical implications. Proc Natl Acad Sci U S A 98:10869–10874
Su X, Malouf GG, Chen Y et al (2014) Comprehensive analysis of long non-coding RNAs in human breast cancer clinical subtypes. Oncotarget 5:9864–9876
Torre LA, Bray F, Siegel RL et al (2015) Global cancer statistics, 2012. CA Cancer J Clin 65:87–108
Van Grembergen O, Bizet M, de Bony EJ et al (2016) Portraying breast cancers with long noncoding RNAs. Sci Adv 2:e1600220
Vennin C, Spruyt N, Dahmani F et al (2015) H19 non coding RNA-derived miR-675 enhances tumorigenesis and metastasis of breast cancer cells by downregulating c-Cbl and Cbl-b. Oncotarget 6:29209–29223
Wang YL, Overstreet AM, Chen MS et al (2015) Combined inhibition of EGFR and c-ABL suppresses the growth of triple-negative breast cancer growth through inhibition of HOTAIR. Oncotarget 6:11150–11161
Wang O, Yang F, Liu Y et al (2017) C-MYC-induced upregulation of lncRNA SNHG12 regulates cell proliferation, apoptosis and migration in triple-negative breast cancer. Am J Transl Res 9:533–545
Wang R, Zhang T, Yang Z et al (2018) Long non-coding RNA FTH1P3 activates paclitaxel resistance in breast cancer through miR-206/ABCB1. J Cell Mol Med 22:4068–4075
Xiao C, Sharp JA, Kawahara M et al (2007) The XIST noncoding RNA functions independently of BRCA1 in X inactivation. Cell 128:977–989
Xing F, Liu Y, Wu SY et al (2018) Loss of XIST in breast cancer activates MSN-c-Met and reprograms microglia via exosomal miRNA to promote brain metastasis. Cancer Res 78:4316–4330
Xu S, Sui S, Zhang J et al (2015) Downregulation of long noncoding RNA MALAT1 induces epithelial-to-mesenchymal transition via the PI3K-AKT pathway in breast cancer. Int J Clin Exp Pathol 8:4881–4891
Xue X, Yang YA, Zhang A et al (2016) LncRNA HOTAIR enhances ER signaling and confers tamoxifen resistance in breast cancer. Oncogene 35:2746–2755
Zhang B, Arun G, Mao YS et al (2012) The lncRNA Malat1 is dispensable for mouse development but its transcription plays a cis-regulatory role in the adult. Cell Rep 2:111–123

Zhang H, Cai K, Wang J et al (2014) MiR-7, inhibited indirectly by lincRNA HOTAIR, directly inhibits SETDB1 and reverses the EMT of breast cancer stem cells by downregulating the STAT3 pathway. Stem Cells 32:2858–2868
Zhang CY, Yu MS, Li X et al (2017a) Overexpression of long non-coding RNA MEG3 suppresses breast cancer cell proliferation, invasion, and angiogenesis through AKT pathway. Tumour Biol 39:1010428317701311
Zhang HY, Liang F, Zhang JW et al (2017b) Effects of long noncoding RNA-ROR on tamoxifen resistance of breast cancer cells by regulating microRNA-205. Cancer Chemother Pharmacol 79:327–337
Zhao W, Luo J, Jiao S (2014a) Comprehensive characterization of cancer subtype associated long non-coding RNAs and their clinical implications. Sci Rep 4:6591
Zhao Z, Chen C, Liu Y et al (2014b) 17beta-Estradiol treatment inhibits breast cell proliferation, migration and invasion by decreasing MALAT-1 RNA level. Biochem Biophys Res Commun 445:388–393
Zhou M, Zhong L, Xu W et al (2016) Discovery of potential prognostic long non-coding RNA biomarkers for predicting the risk of tumor recurrence of breast cancer patients. Sci Rep 6:31038
Zhou K, Ou Q, Wang G et al (2019) High long non-coding RNA NORAD expression predicts poor prognosis and promotes breast cancer progression by regulating TGF-beta pathway. Cancer Cell Int 19:63
Zhu QN, Wang G, Guo Y et al (2017) LncRNA H19 is a major mediator of doxorubicin chemoresistance in breast cancer cells through a cullin4A-MDR1 pathway. Oncotarget 8:91990–92003

Drosophila Models to Study Long Noncoding RNAs Related to Neurological Disorders

Yuuka Muraoka and Masamitsu Yamaguchi

Contents

Abstract Long noncoding RNAs (lncRNAs) are now emerging as important regulators of various neurological functions. The abnormal expression or function of lncRNAs is now known to be critical for the onset and/or progression of neurological disorders. *Drosophila* provides a useful model to gain insights into the molecular mechanisms of various human neurological disorders. In this chapter, after overviewing the lncRNAs related to human neurological disorders, the lncRNAs

Y. Muraoka · M. Yamaguchi (✉)
Department of Applied Biology and Advanced Insect Research Promotion Center, Kyoto Institute of Technology, Kyoto, Japan
e-mail: myamaguc@kit.ac.jp

S. Jurga, J. Barciszewski (eds.), *The Chemical Biology of Long Noncoding RNAs*, RNA Technologies 11, https://doi.org/10.1007/978-3-030-44743-4_16

identified and analyzed using *Drosophila* neurological disorder models are described.

Keywords *Drosophila* · lncRNA · RNA-binding protein · Neurological disorders · Rough eye phenotype · Neuromuscular junction · Genetic screening

1 Introduction

Long noncoding RNAs (lncRNAs) that are longer than 200 nucleotides do not have the capacity to code proteins but have diverse functions in a number of biological processes, such as the regulation of gene expression, modification of the chromatin structure, and genome imprinting (Lee and Bartolomei 2013; Chen 2016; Cao et al. 2018; Li et al. 2019a, b). Based on these functions, lncRNAs have been implicated in various diseases, such as neurological disorders, cancer, immunological diseases, diabetes, cardiovascular diseases, and respiratory diseases (Bian and Sun 2011; Huarte 2015; Wan et al. 2017; Li et al. 2019a, b).

LncRNAs have been identified in various organisms, including humans, mice, *Xenopus*, *Drosophila*, nematodes, *Arabidopsis*, and yeast (Au et al. 2011). In this chapter, we focus on *Drosophila* models that are used in the study of lncRNAs. *Drosophila* is a good model organism for investigating human diseases (Yamaguchi and Yoshida 2018). The functional orthologues of 75% of human disease-associated genes have been identified in *Drosophila* (Pandy and Nichols 2011; Yamamoto et al. 2014). The identity of the amino acid sequences of proteins between *Drosophila* and humans is approximately 40%. Moreover, in terms of the conserved functional domains of proteins, amino acid identity is more than 80%. Based on the Ensembl database, the proportion of lncRNAs to the number of whole annotated genes is very similar between humans and *Drosophila*, accounting for 13% on average in both species (Lo Piccolo 2018). *Drosophila* has a relatively short life span and produces a large number of progenies, which allows for reliable statistical analyses of data. Many mutants and RNAi lines are available from stock centers. There are also fewer ethical concerns than mammals because animal laws are not applicable to insects in many countries. Therefore, *Drosophila* is emerging as a useful model for investigating human diseases with the aim of discovering novel diagnosis markers or therapeutic targets for human diseases. It is also useful for the screening of candidate substances for the treatment of human diseases (Pandy and Nichols 2011).

In this chapter, after overviewing the roles of lncRNAs in human neurological disorders, *Drosophila* models targeting these disorders are described. We also discuss lncRNAs that have been identified by genetic screening with these *Drosophila* models.

2 LncRNAs in Human Neurological Disorders

LncRNAs play roles in various biological processes. A number of lncRNAs regulate gene transcription that is mediated by the modulation of chromatin states or recruitment of general transcription factors and regulatory factors. Some lncRNAs are also involved in the RNA processing, silencing, and degradation of mRNA. The dysregulation of lncRNAs and RNA toxicity with repeat expansion are related to various neurological disorders including Alzheimer's disease (AD), Parkinson's disease (PD), Huntington's disease (HD), frontotemporal dementia (FTD), autism spectrum disorder (ASD), schizophrenia (SCZ), spinocerebellar ataxia (SCA), fragile X tremor/ataxia syndrome (FXTAS), and amyotrophic lateral sclerosis (ALS) (Wan et al. 2017; Lekka and Hall 2018; Lo Piccolo 2018; Li et al. 2019a, b).

2.1 LncRNAs in AD

AD is one of the most common neurodegenerative diseases and is characterized by progressive defects in mental and behavioral functions. AD patients show abnormal structures called plaques and tangles in the brain (Lo Piccolo 2018). Plaques are deposits of a protein fragment called amyloid beta (Aβ) protein in the spaces between nerve cells. Tangles are twisted fibers of hyperphosphorylated tau protein built inside the nerve cells. Amyloid precursor protein (APP) is first cleaved by β-secretase 1 (BACE1) and then by γ-secretase to produce Aβ. Aβ peptides form soluble oligomers that induce dysfunctions in mitochondria and oxidative damage, which result in synaptic dysfunction and degeneration (Lo Piccolo 2018). Therefore, BACE1 plays a crucial role in the pathogenesis of AD. *BACE1* antisense lncRNA (*BACE1-AS* lncRNA) is expressed at relatively high levels in AD brains and elevates BACE1 protein levels by stabilizing BACE1 mRNA to finally increase Aβ42 levels (Faghihi et al. 2008). The knockdown of *BACE1-AS* lncRNA by the continuous infusion of siRNAs directly into the brain induced reductions in *BACE1-AS* lncRNA and *BACE1* levels, which resulted in the downregulation of Aβ synthesis and their aggregation in the brain (Wan et al. 2017; Li et al. 2019a, b).

Another example of an antisense lncRNA involved in AD is *LRP1-AS* lncRNA. LRP1 encodes a low-density lipoprotein receptor and appears to play a role in the clearance of Aβ. *LRP1-AS* lncRNA, which is upregulated in AD brains, may negatively regulate *LRP1* gene expression by modulating the chromatin state, suggesting that *LRP1-AS* lncRNA plays a critical role in AD pathogenesis (Yamanaka et al. 2015). Several other antisense lncRNAs have been implicated in the pathogenesis of AD (Lo Piccolo 2018; Li et al. 2019a, b). The antisense lncRNA *17A* is produced from the third intron of the human *G protein-coupled receptor 51* (*GPR51/GABBR2*) gene in antisense orientation and regulates the pre-mRNA processing of *GPR51/GABBR2* (Massone et al. 2011). Upregulation of the lncRNA *17A* causes the formation of a defective GABA signaling and may induce the secretion of the Aβ. The antisense lncRNA *51A* is generated from the first intron of the *SORL1* gene that is a risk factor for AD, playing a role in the Aβ formation (Ciarlo et al. 2013). In addition, aberrant expression of the other antisense lncRNA *NMD29* can also induce overproduction of Aβ42 (Massone et al. 2011).

2.2 LncRNAs in PD

PD is one of the most common neurodegenerative disorders. It is characterized by progressive defects in locomotive abilities caused by the loss of dopamine-producing neuronal cells (dopaminergic neurons) in the brain. Several antisense lncRNAs related to PD have been reported (Wan et al. 2017; Lo Piccolo 2018; Li et al. 2019a, b). Ubiquitin carboxyl-terminal hydrolase L1 (UCHL1) protein, which is involved in the pathogenesis of PD, has been detected in the Lewy bodies of nerve cells in the brains of PD patients. The antisense lncRNA for the *Uchl1* gene (*ASUchl1* lncRNA) has been shown to be involved in PD (Carrieri et al. 2015). In the cytoplasm, *ASUchl1* lncRNA promotes the interaction between *Uchl1* mRNA and translationally active polysomes to increase UCHL1 protein synthesis. The transcription factor, Nurr1, which plays a role in the differentiation and maintenance of dopaminergic neurons, has been shown to regulate the transcription of *ASUchl1* lncRNA, suggesting its potential as a candidate therapeutic target of *Uchl1*-related PD (Carrieri et al. 2015). *ASUchl1* lncRNA is downregulated in neurochemical models of PD, which results in the reduced translation of the UCHL1 protein, followed by the disruption of the ubiquitin-proteasome system (Carrieri et al. 2015).

PTEN-induced kinase 1 (PINK1) induces the early onset of PD when mutated. The antisense lncRNA for the *Pink1* gene (*NaPINK1*) has been shown to stabilize the expression of PINK1 in humans (Scheele et al. 2007). A similar antisense *Pink1* lncRNA was detected in the mouse brain and colocalized with *Pink1* mRNA (Chiba et al. 2009). Another lncRNA called metastasis-associated lung adenocarcinoma transcript 1 (*MALT1*) has also been implicated in the pathogenesis of PD (Liu et al. 2017). *MALT1* binds to the microRNA (*miR-124*) and downregulates *miR-124* expression to induce apoptosis in dopaminergic neurons (Liu et al. 2017).

2.3 LncRNAs in HD

HD is a hereditary neurodegenerative disease characterized by dementia, chorea, and psychiatric disturbances. HD is caused by the expansion of a CAG triplet repeat within the *huntingtin* gene, producing a mutant form of the huntingtin protein. The mutant form of huntingtin is neurotoxic and induces the aberrant nuclear accumulation of the RE1-silencing transcription factor (REST), which is involved in the repression of many target genes in neurons (Hwang and Zukin 2018). A high-throughput analysis of RNA expression in HD brains revealed that *NEAT1* (*nuclear paraspeckle assembly transcript 1*) and *TUG1*, which is required for retinal development, were both upregulated. In contrast, *DGCR5* (*DiGeorge syndrome critical region gene 5*) and *MEG3* (*maternally expressed 3*) were downregulated in HD brains (Johnson 2012). *NEAT1* is necessary for the formation of paraspeckles. Therefore, the upregulation of *NEAT1* in HD brains may play a role in altering the transcriptional status by sequestrating various paraspeckle components (Sunwoo et al. 2017). Since *TUG1* and *MEG3* both play roles in the formation of epigenetic ribonucleoprotein complexes, such as polycomb complex-2 (PRC2), altered epigenetic gene regulation in neurons may contribute to neurodegenerative diseases, including HD (Johnson 2012). *DGRC5* and *MEG3* are both direct targets of REST, and thus their downregulation may be caused by the aberrant accumulation of REST in the nuclei of HD striatal neurons (Johnson 2012; Hwang and Zukin 2018).

The antisense lncRNA for the *BDNF* gene (*BDNF-AS* lncRNA) encoding a secreted growth factor that promotes neural maturation and survival is also involved in HD (Wan et al. 2017). *BDNF-AS* lncRNA inhibits BDNF expression posttranscriptionally. BDNF levels are reduced in the HD brain, and the overexpression of BDNF in the forebrain may rescue HD-like phenotypes in model mice (Xie et al. 2010). Based on these findings, increases in BDNF via the downregulation of *BDNF-AS* lncRNA have potential as a reasonable approach for HD therapy. Several other antisense lncRNAs have been implicated in the pathogenesis of HD (Wan et al. 2017; Lo Piccolo 2018; Li et al. 2019a, b).

2.4 LncRNAs in ASD

ASD is characterized by defects in social interactions and communication as well as in restricted, repetitive, and stereotyped patterns of behaviors, interests, and activities. Comorbidities occur in some ASD cases, including language deficits, epilepsy, intellectual disability, motor abnormalities, anxiety, and gastrointestinal issues (Ueoka et al. 2019). The microarray profiles of more than 33,000 annotated lncRNAs and 30,000 mRNAs were obtained from the postmortem brain tissue of the autistic and control prefrontal cortex and cerebellum (Ziats and Rennert 2013; Li et al. 2019a, b). In the autistic brain, 222 differentially expressed lncRNAs were identified, and most were oriented in and around genes that are known to be associated with neurodevelopmental and psychiatric disorders, including the *Ubiquitin-protein ligase E3A* (*UBE3A*) gene (Ziats and Rennert 2013). The paternal normal allele of *UBE3A* is silenced by lncRNA. Loss-of-function mutations in the maternal allele of the imprinted *UBE3A* gene are responsible for many cases of Angelman syndrome, a monogenic disorder with an estimated ASD prevalence of 34% (Ueoka et al. 2019).

The antisense lncRNA for the *MSN* gene (*MSNP1AS* lncRNA) encoding the moesin protein, which is required for the neuronal architecture and immune responses, is also involved in HD (Wilkinson and Campbell 2013; Li et al.

2019a, b). *MSNP1AS* lncRNA has been found to be upregulated in a postmortem ASD temporal cortex, and its overexpression decreased *MSN*, which, as a consequence, reduced neurite numbers and the lengths of neurons, suggesting that *MSNP1AS* lncRNA increases the risk of ASD by negatively regulating *MSN* expression (Wilkinson and Campbell 2013; Li et al. 2019a, b).

2.5 LncRNAs in SCZ

SCZ is a severe mental disorder with a wide spectrum of neurocognitive defects. Combinations of genetic and environmental factors appear to contribute to the pathophysiology of SCZ (Li et al. 2019a, b). The lncRNA, *MIAT* (*myocardial infarction-associated transcript*) has been shown to be downregulated by neuronal activation in depolarized mouse primary cortical neurons. Consistent with this finding, cortical *MIAT* was also downregulated in the postmortem superior temporal gyrus of SCZ (Sun et al. 2018). *MIAT* directly binds to splicing factor 1 in order to modify the expression of several genes in neurons. *MIAT* may also function as a scaffold to modulate the alternative splicing of various SCZ-related genes (Liu et al. 2018; Sun et al. 2018). Several other lncRNAs have also been suggested as diagnosis markers and therapeutic targets of SCZ (Chen et al. 2016a, b).

2.6 LncRNAs in SCAs

SCAs are a heterogeneous group of disorders affecting the cerebellum and related pathways. The CAG repeat expansion is known to be associated with SCA1, SCA2, SCA3, SCA6, SCA7, and SCA17 (Manto 2005). *ATXNl3B* lncRNA, which interacts with *ataxin type 7* (*ATXN7*) mRNA, plays a role in SCA7 (Tan et al. 2014). The triplet expansion in the *ATXN7* gene in SCA7 disrupts these regulatory interactions, resulting in an increase in *ATXN7* expression in neurons. SCA8 is caused by a CTG repeat extension in the *ATXN8OS* gene encoding an antisense lncRNA (*ATXN8OS* lncRNA) that partially overlaps with the neighboring protein-coding gene, *KLHL1* (Chen et al. 2008). The CUG expansion in *ATXN8OS* lncRNA may exert effects on *KLH1* expression (Chen et al. 2008). Furthermore, the triplet expansion in *ATXN8OS* lncRNA may cause toxic RNA gain-of-function pathologies, because it induces accumulation in RNA foci to sequestering factors involved in alternative splicing (Daughters et al. 2009). In addition, the expanded ATTCT repeat is related to SCA10, and the expanded UGGAA repeat is responsible for SCA31 (Manto 2005; Lo Piccolo 2018).

2.7 LncRNAs in ALS and FTD

ALS is a severe neurodegenerative disease characterized by progressive neurodegeneration of both the upper and lower motor neurons. Patients rapidly become paralyzed and die 2–5 years after disease onset (Kanouchi et al. 2012). ALS is frequently associated with another disorder called FTD. Therefore, ALS and FTD are considered to form a broad continuum of neurodegenerative disorders (Murphy et al. 2007); however, FTD is a clinically diverse dementia syndrome that is accompanied by behavioral changes and language dysfunction.

The expansion of the hexanucleotide GGGGCC repeat in the protein-coding gene *C9ORF72* (*chromosome 9 ORF 72*) has been identified as a causative mutation for ALS/FTD (Renton et al. 2011; DeJesus-Hernandez et al. 2011; Paul and Gitler 2014). Loss-of-function of the C9ORF72 protein, toxic gain-of-function of expanded GGGGCC repeat RNAs, and toxic gain-of-function of dipeptide repeat (DPR) proteins generated from expanded repeat RNAs by repeat-associated non-ATG (RAN) translation may cause *C9ORF72*-ALS/FTD (Ling et al. 2013). The clinical features of FTD patients homozygous for the *C9ORF72* GGGGCC repeat expansion mutation were shown to be in the same range as those of heterozygous patients (Fratta et al. 2013). Mutations in *C9ORF72*-coding regions were consistently not found in ALS patients, excluding the possibility of a loss-of-function mechanism in *C9ORF72*-ALS/FTD (Harms et al. 2013). Moreover, knockout mice targeting the *C9ORF72* gene showed immunological defects, but no or mild neurological disorders (Koppers et al. 2015; Atanasio et al. 2016; Jiang et al. 2016; Sudria-Lopez et al. 2016), which further supports the loss-of-function not playing a critical role in the pathogenesis of *C9ORF72*-ALS/FTD. The repeat expanded *C9ORF72* is transcribed in both the sense and antisense orientations to form nuclear and cytoplasmic sense and antisense RNA foci in which various RNA-binding proteins, (RBPs), including the TAR DNA-binding protein (TDP-43) and fused in sarcoma (FUS), are recruited. Since *TDP43* and *FUS* are both ALS/FTD-causing genes, the sequestration of TDP43 and FUS by this mechanism may also play a role in the pathogenesis of *C9ORF72*-ALS/FTD (Moens et al. 2017; Lo Piccolo 2018).

TDP43 and FUS also appear to be involved in controlling the expression of various RNAs, including architectural lncRNAs (arcRNAs), such as *NEAT1* and *MALAT1*. The depletion of TDP43 or FUS increased the expression of both *NEAT1* and *MALAT1* in mouse ALS/FTD models (Lagier-Tourenne et al. 2012; Polymenidou et al. 2011). Furthermore, in the early stage of ALS pathogenesis, increases in *NEAT1* RNA foci were observed in postmortem humans as a consequence of altered TDP43 subcellular localization (Nishimoto et al. 2013).

3 *Drosophila* Models for Studying lncRNAs Related to Neurological Disorders

The functional orthologues of nearly 75% of human disease-associated genes have been estimated to exist in *Drosophila* (Pandy and Nichols 2011; Yamamoto et al. 2014). The proportion of lncRNAs with respect to the whole annotated gene number is very similar between humans and *Drosophila* (Lo Piccolo 2018). A large number of mutants and transgenic fly lines are available from stock centers such as the Bloomington Drosophila Stock Center (https://bdsc.indiana.edu/) and Kyoto Stock Center (http://www.dgrc.kit.ac.jp/). The Undiagnosed Disease Network (UDN) has been established in the USA, with the aims of improving the diagnosis and treatment of patients carrying undiagnosed diseases, facilitating research into the etiologies of undiagnosed diseases, and creating an integrated and collaborative research community (Takano-Shimizu-Kouno and Ohsako 2018). The screening centers of the UDN at the Baylor College of Medicine and the University of Oregon focus on the three model organisms, *Drosophila*, zebra fish, and *Caenorhabditis elegans*, to predict the pathogenicity of missense variants identified by the whole-exome sequencing of human patients (Takano-Shimizu-Kouno and Ohsako 2018). Thus, the screening center is now developing new *Drosophila* resources that will be useful for studying rare intractable human diseases. It is important to note that there are fewer ethical concerns for *Drosophila* because animal laws are generally not applicable to insects. Therefore, *Drosophila* is now emerging as a commonly used model organism to study various human diseases (Pandy and Nichols 2011; Yamaguchi and Yoshida 2018).

There are approximately 100,000 neurons in the adult fly brain. These neurons form discrete circuits and neuropils that mediate complex behaviors, such as learning and memory, circadian wake and sleep rhythms, feeding, grooming, and courtship and aggression. In addition, the responses of *Drosophila* to various drugs that act on the functions of the central nervous system (CNS) appear to be similar to those observed in mammals (Rothenfluh and Heberlein 2002; Satta et al. 2003; Wolf and Heberlein 2003; Nichols 2006; Andretic et al. 2008). Therefore, *Drosophila* provides a useful model for the screening of candidate therapeutic substances for various human neurological disorders. Although 1077 lncRNAs have been identified and their expression profiles reported in *Drosophila* (Chen et al. 2016a, b), information on their relationship with the homologues of human disease-causing genes remains limited.

In *Drosophila*, the GAL4-UAS targeted expression system (Brand and Perrimon 1993) is commonly used to express proteins and lncRNAs at a specific developmental stage and in specific cells, tissues, and organs (Fig. 1). When RNA interference (RNAi) is combined, it is also possible to knockdown specific protein-coding genes and lncRNAs by expressing double-stranded RNAs (dsRNAs) targeted to specific regions of mRNAs and lncRNAs using the GAL4-UAS system (Fig. 1). More sophisticated GAL4/UAS/Gal80^{ts} system has also been developed (Suster et al. 2004) (Fig. 1). The Vienna Drosophila RNAi Center (VDRC)

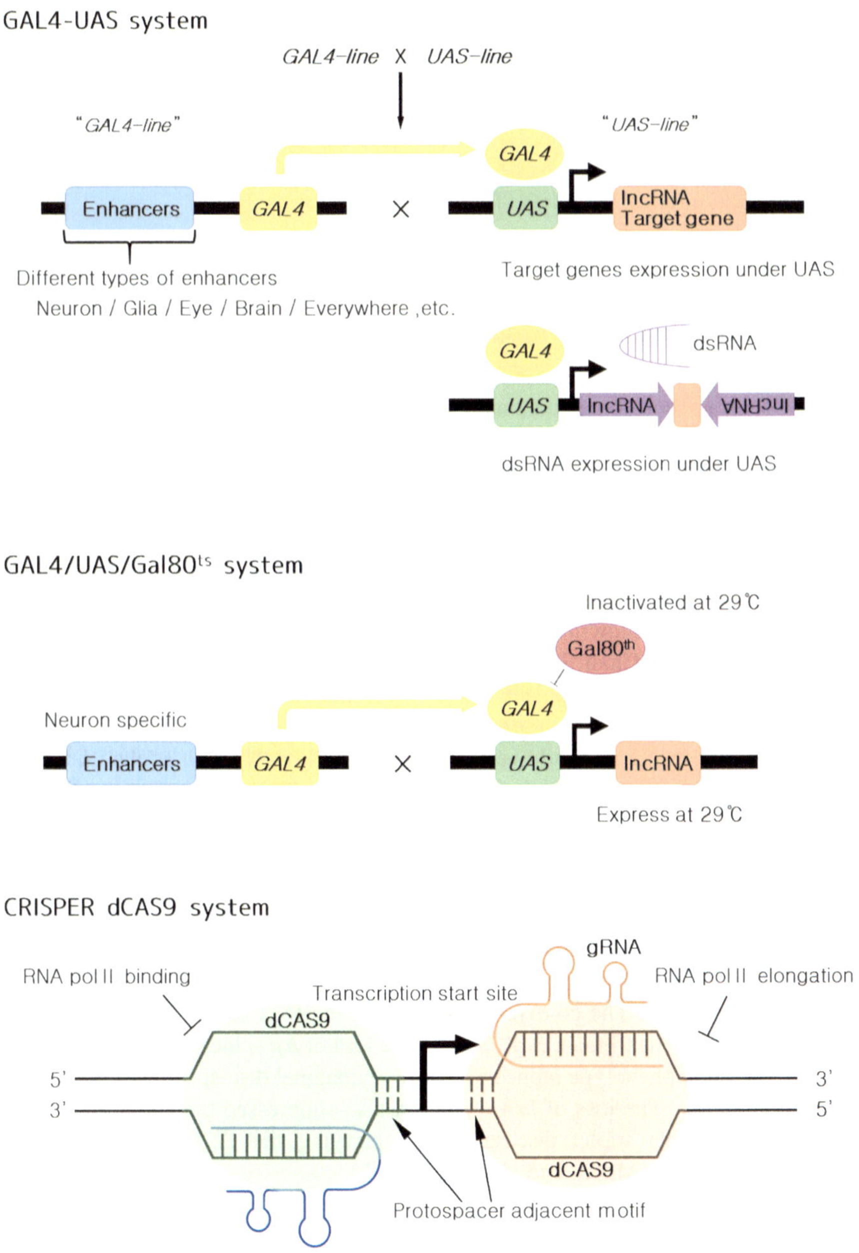

Fig. 1 The Gal4-UAS targeted expression system: this system is used to overexpress or knockdown target genes encoding proteins or lncRNAs. The GAL4 driver strain was crossed with transgenic lines carrying target genes. In the progeny, mRNA or double-stranded RNA is expressed tissue-specifically to overexpress or knockdown target genes, and the phenotype was examined. GAL4/UAS/Gal80^{ts} system: Gal80^{ts} is a Gal4 inhibitor with a temperature-sensitive mutation. The GAL4 inhibitor activity of Gal80^{ts} is inactivated by a shift in the temperature to higher than 29 °C. Thus, constitutively expressed Gal80^{ts} blocks the GAL4/UAS system until the shift in the temperature to higher than 29 °C. CRISPER dCAS9 system: in this system, the catalytically inactive Cas9

(http://stockcenter.vdrc.at/control/main) provides useful resources for this purpose; however, the RNAi fly lines targeting lncRNAs remain limited in this center. In addition, the CRISPR-dCas9 method (Port and Bullock 2016) has been developed to specifically knockdown any lncRNAs by designing specific guide RNAs that are specifically expressed by the GAL4-UAS system (Fig. 1). In this section, we describe several *Drosophila* models utilized to identify and characterize lncRNAs related to neurological disorders (Table 1).

3.1 Drosophila *SCA1 Model*

SCA1 is a rare autosomal dominant disorder characterized by neurological and cerebellar dysfunctions. SCA1 is caused by the expansion of CAG trinucleotide repeats translated into polyglutamine tracts (polyQ) in the ataxin-1 protein. *Drosophila* models for SCA1 have been created by the eye imaginal disc-specific expression of full-length human ataxin-1 carrying 30Q (wild type) and its mutant form carrying 82Q (Fernandez-Funez et al. 2000). *Drosophila* compound eyes carrying photoreceptor neurons have commonly been used as a model to study neurodegeneration, allowing for the identification of novel genetic modifiers in screens (Koon and Chan 2017). The GAL4/UAS system (Fig. 1) allows toxic RNAs or toxic proteins that are commonly found in human neurodegenerative diseases to be expressed in eye imaginal discs, and this often results in an aberrant morphology of the compound eye (the rough eye phenotype) in adults. Since the compound eye is not required for fly viability, this system allows for the effective screening of gene mutations that may suppress or enhance the rough eye (Fig. 2).

Although the overexpression of not only the mutant form but also the wild-type ataxin-1 induced the rough eye phenotype with severe degeneration in adults, the expression of the protein with 82Q induced a more severe eye degeneration phenotype (Fernandez-Funez et al. 2000). Genetic screening to detect suppressors and enhancers of the rough eye phenotype identified the gene encoding *hsrω* lncRNA as a genetic interactant. The co-expression of *hsrω* lncRNA enhanced the rough eye phenotype (Fernandez-Funez et al. 2000). The loss of *hsrω* lncRNA suppressed the eye degeneration phenotype induced by the eye imaginal disc-specific expression of the ataxin-1 82Q. The loss of *hsrω* lncRNA also suppressed the eye degeneration phenotype of other triplet disease models carrying the 127Q, MJDtr-Q78, or httex1pQ transgene (Mallik and Lakhotia 2009b).

Fig. 1 (continued) protein (dCas9) is targeted either to the template or non-template DNA strand based on the targeting sequence of the sgRNA and an adjacent PAM sequence. Binding of dCas9: the sgRNA complex upstream of the transcription start site interferes with transcription initiation by blocking the recruitment of RNA polymerase II, while its assembly at a downstream site prevents transcription elongation

Table 1 lncRNAs related to *Drosophila* models for human neurological disorders

RNA	Function	Associated genes and proteins	Associated model for human disease	References
hsrω lncRNA	Heat-shock response, formation of ω-speckles, regulation of apoptosis	hnRNPs, HP1, RNA polymerase II, ISWI, DIAP1, LAMP1, *dFIG4*	SCA1, *FUS*-ALS/FTD, *FIG4*-CMT	Fernandez-Funez et al. (2000), Mallik and Lakhotia (2009a, b), Lakhotia et al. (2012), Koon and Chan (2017), Lo Piccolo and Yamaguchi (2017), Lo Piccolo et al. (2017), and Muraoka et al. (2018)
Human *ataxin-3* RNA containing MJDtr-Q78 repeat	Production of ataxin-3 proteins carrying an expanded polyQ repeat, RNA toxicity	*Muscle-blind*	SCA3	Warrick et al. (1998), Li et al. (2008), and Saitoh et al. (2015)
Human *SCA8* lncRNA carrying expanded CUG repeats	RNA toxicity, RAN translation	*Staufen, muscle-blind, split ends, CG3249*	SCA8	Mutsuddi et al. (2004) and Zu et al. (2011)
RNA carrying the expanded UGGAA repeats	RNA foci formation, RAN translation	TDP-43, FUS, hnRNPA2/B1	SCA31, *TDP-43*-ALS/FTD, *FUS*-ALS/FTD	Ishiguro et al. (2017) and Ueyama and Nagai (2018)
RNA carrying premutation CGG repeats	MicroRNA processing, RAN translation	Pur alpha, hnRNP A2/B1, CUGBP1, DGCR8, DROSHA	FXTAS	Jin et al. (2003, 2007), Sofola et al. (2007a, b), Sellier et al. (2013), and Todd et al. (2013)
RNA with iGGGGCC repeats	RAN translation, RNA foci formation	*RanGAP, importins, exportins, Ran-GTP regulators, arginine methylase*	*C9ORF72*-ALS/FTD	Xu et al. (2013), Mizielinska et al. (2014), and Tran et al. (2015)
CR18854 long-hairpin RNA	Embryogenesis	hnRNP, Hrp59, staufen, *dFIG4*	*FIG4*-CMT, *FUS*-ALS/FTD	Laver et al. (2013) and Muraoka et al. (2018)
CR43467 lnc RNA	Unknown		*FIG4*-CMT	Shimada et al. (2020)

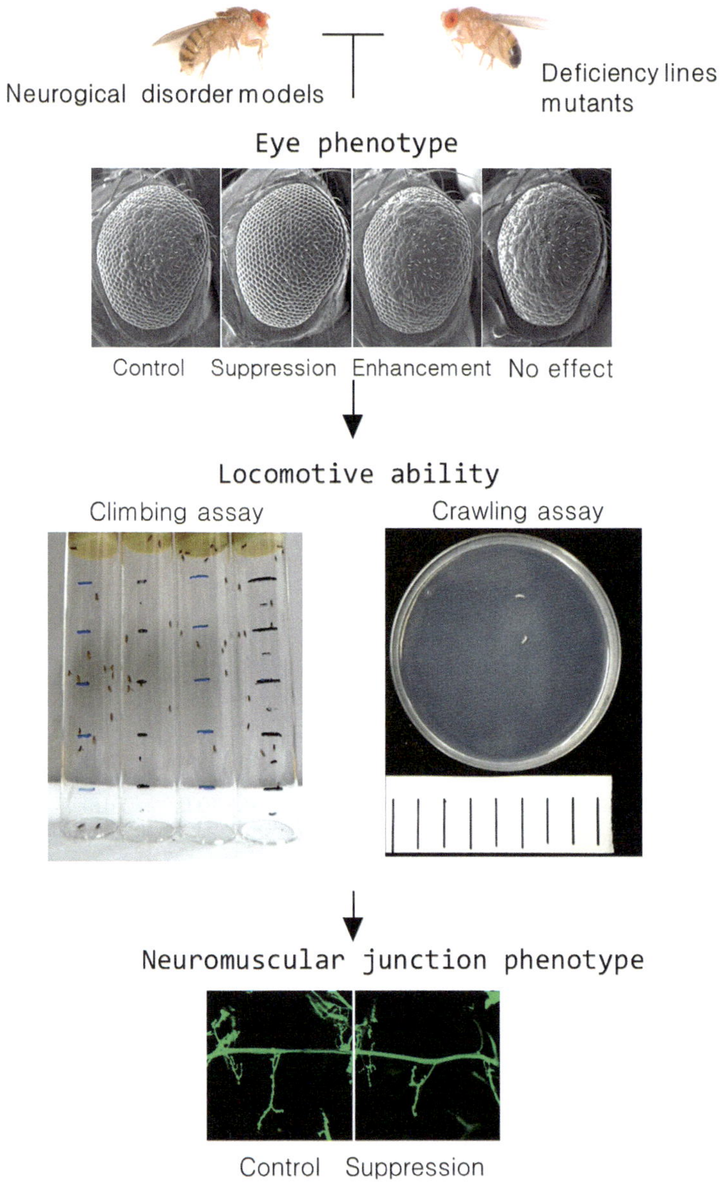

Fig. 2 Rough eye modifier screening followed by a locomotive assay and an evaluation of synapse morphology at the neuromuscular junction (NMJ). Neurological disorder model flies may be crossed with a set of deficiency lines or mutants of individual genes. In the progenies, scanning electron microscopic (SEM) images of the adult compound eyes are taken. Locomotive ability may be evaluated by a climbing assay for adult flies and a crawling assay for larvae. The synapses at the NMJ of the third instar larvae may be stained with anti-HRP IgG and inspected under a confocal laser scanning microscope

hsrω lncRNA is transcribed from the heat-shock locus 93D in the *Drosophila* chromosome. Previous studies demonstrated that *hsrω* lncRNA was responsible for restoring the proper nuclear distribution of some key regulators for various biological processes, such as several hnRNPs, HP1, and RNA polymerase II, during recovery from heat shock (Lakhotia et al. 2012). *hsrω* lncRNA plays a role in the formation of hnRNP-containing ω-speckles, specialized nuclear compartments that are built by the remodeling activity of ISWI, the catalytic subunit of several ATP-dependent chromatin remodeling complexes (Onorati et al. 2011; Prasanth et al. 2000). These ω-speckles play critical roles in the storage and sequestration of various members of the hnRNP family and other proteins required for the processing and maturation of RNA in cells under normal conditions as well as environmentally or genotoxically stressed conditions (Jolly and Lakhotia 2006; Prasanth et al. 2000). A previous study also demonstrated that *hsrω* lncRNA regulated the cellular level of *Drosophila* inhibitor of apoptosis protein 1 (DIAP1) via hnRNP Hrb57A to control apoptosis (Mallik and Lakhotia 2009a). Either or both of these distinct roles of *hsrω* lncRNA may contribute to the pathogenesis of SCA1.

3.2 **Drosophila** ***SCA3 Model***

SCA3, also known as Machado-Joseph disease (MJD), is characterized by progressive cerebellar ataxia and variable neurological disorders. It is caused by the expansion of CAG triplet repeats in the coding region of the gene encoding ataxin-3. *Drosophila* models for SCA3 have been created by expressing the C-terminal region of the ataxin-3 protein containing the normal MJDtr-Q27 repeat or disease-causing MJDtr-Q78 repeat by the GAL4-UAS system (Fig. 1). The expression of MJDtr-Q78 in eye imaginal discs induced the rough eye phenotype with the degeneration and formation of nuclear inclusion, while the expression of MJDtr-Q27 had no apparent effect (Warrick et al. 1998). Detailed analyses demonstrated that mutant ataxin-3 proteins carrying an expanded polyQ repeat caused neurodegeneration, with chaperone activity and the protein-folding pathway being implicated in this process (Saitoh et al. 2015).

In contrast, genetic modifier screening of the rough eye phenotype (Fig. 2) with SCA3 model flies identified the *muscle-blind* gene encoding a neuronally expressed RNA-binding protein (Li et al. 2008). A mutation in the *muscle-blind* gene strongly enhanced the eye degeneration phenotype in SCA3 model flies. The expression of RNA carrying the untranslated CAG repeat with a pathogenic length induced progressive neuronal defects, indicating a crucial role for RNA toxicity in the pathogenesis of SCA3 (Li et al. 2008).

3.3 Drosophila *SCA8 Model*

SCA8 is an inherited neurodegenerative disorder characterized by slowly progressive ataxia. SCA8 is caused by CUG triplet repeat expansion occurring within a transcript that is not coding a protein (Koob et al. 1999). As described above, CUG expansion in *SCA8* lncRNA may exert effects on neighboring *KLH1* gene expression (Chen et al. 2008) or cause toxic RNA gain-of-function effects (Daughters et al. 2009). *Drosophila* models for SCA8 have been created by expressing human *SCA8* lncRNA using the GAL4-UAS system (Fig. 1) (Mutsuddi et al. 2004). The eye imaginal disc-specific expression of *SCA8* lncRNA carrying CUG 9 repeats (normal) or 112 repeats (expanded) both induced late-onset, progressive neurodegeneration in the *Drosophila* retina that resulted in the rough eye phenotype. Genetic modifier screening of the rough eye phenotype (Fig. 2) identified four genes, *staufen*, *muscleblind*, *split ends*, and *CG3249*, mutations which enhanced the rough eye phenotype induced by *SCA8* lncRNA expression. All four genes encode RBPs conserved in *Drosophila* and humans and are expressed in neurons.

Bidirectional transcripts carrying the CUG repeat or antisense CAG repeat were detected in both the SCA8 mouse model and patients (Moseley et al. 2006). The CAG repeat sequence located in the noncoding region was found to be translated into repeat polypeptides without a canonical initiation codon (Zu et al. 2011). This unusual translation is designated as RAN translation. These findings suggest that toxic gain-of-function mechanisms at both the protein and RNA levels are responsible for the pathogenesis of SCA8.

3.4 Drosophila *SCA31 Model*

SCA31 is a late-onset autosomal dominant cerebellar ataxia. SCA31 is caused by the insertion of a pentanucleotide TGGAA repeat in the overlapping intron of the *brain expressed, associated with Nedd4* (*BEAN*) gene and *thymidine kinase 2* (*TK2*) gene in the antisense strand (Sato et al. 2009). RNA foci containing UGGAA repeats have been detected in the brains of SCA31 patients (Niimi et al. 2013), and, thus, a toxic gain-of-function mechanism caused by UGGAA repeat RNA has been suggested for the pathogenesis of SCA31 (Ueyama and Nagai 2018).

Drosophila models for SCA31 have been generated by expressing RNA carrying expanded UGGAA repeats using the GAL4-UAS system (Fig. 1). The eye imaginal disc-specific expression of expanded UGGAA repeats induced neurodegeneration, which was accompanied by the accumulation of RNA foci containing the expanded UGGAA repeats and pentapeptide repeat proteins produced by RAN translation, as observed in the brains of SCA31 patients (Ishiguro et al. 2017). TDP-43, FUS, and hnRNPA2/B1 have been shown to bind to RNA carrying expanded UGGAA repeats; alter the structure of RNA, as evaluated using a CD spectrum; and suppress its toxicity (Ishiguro et al. 2017; Ueyama and Nagai 2018). As described above,

TDP-43 and FUS are both well-known RBPs involved in the pathogenesis of ALS/FTD. These RBPs appear to function as RNA chaperones to regulate repeat-associated translation, which plays a crucial role in the pathogenesis of both SCA31 and ALS/FTD.

3.5 Drosophila *FXTAS Model*

FXTAS is a late-onset neurodegenerative disease characterized by kinetic tremors, gait ataxia, parkinsonism, and dementia. It is caused by a premutation expansion of CGG triplet repeats (55–200 repeat) in the 5′-UTR of the *FMR1* gene (Ueyama and Nagai 2018).

Drosophila models for FXTAS have been established by expressing RNA carrying 60 or 90 CGG repeats fused with the EGFP sequence (Jin et al. 2003). The eye imaginal disc-specific expression of RNA carrying premutation CGG repeats was shown to induce the rough eye phenotype with neurodegeneration in dose- and repeat length-dependent manners. By using this FXTAS fly model, RBPs that have affinity to CGG repeat RNA were identified from fly brain lysates. RBPs, such as Pur alpha, hnRNP A2/B1, and CUG-binding protein 1 (CUGBP1), were identified by this screen. The expression of Pur alpha suppressed neurodegeneration induced by the expression of CGG repeat RNA in FXTAS fly models, suggesting that Pur alpha plays a critical role in the pathogenesis of FXTAS (Jin et al. 2007). hnRNP A2/B1 and CUGBP1 also suppressed toxicity in FXTAS fly models (Sofola et al. 2007a, b), indicating that the sequestration of RBPs by RNAs carrying CGG repeats is one of the pathogenic mechanisms of FXTAS. The expansion of CCG repeats in the *FMR2* gene causes another type of X-linked mental retardation, FRAXE. The co-expression of CGG repeat RNA together with CCG repeat RNA decreases their toxicities independent of each other by reducing their transcript levels, possibly through a mutual RNAi pathway (Sofola et al. 2007a). A more recent study reported that the double-stranded RNA-binding protein DGCR8 bound to CGG repeats in RNA and was sequestered in CGG RNA aggregates together with its partner protein, DROSHA, resulting in a reduction in microRNA processing (Sellier et al. 2013). These findings suggest that alterations in the microRNA-processing machinery are involved in the pathogenic mechanisms of FXTAS (Ueyama and Nagai 2018).

A role has been reported for CGG repeats as a template for RAN translation to synthesize proteins carrying polyglycine. The accumulation of these polyglycine proteins in ubiquitin-positive inclusions has been observed in *Drosophila* FXTAS models as well as the brains of FXTAS patients (Todd et al. 2013). CGG repeat toxicity is suppressed by eliminating RAN translation and is enhanced by increasing polyglycine production via canonical ATG-initiated translation. These findings indicate that RAN translation producing aberrant polypeptides also contributes to neurodegeneration in FXTAS (Ueyama and Nagai 2018).

3.6 Drosophila C9ORF72-*ALS/FTD Model*

There is no orthologue of the human *C9ORF72* gene in *Drosophila*. However, *Drosophila* models for *C9ORF72*-ALS/FTD have been established by expressing the expanded 30 GGGGCC repeats with a CTCGAG interruption (iGGGGCC). Flies expressing iGGGGCC repeats in eye imaginal discs caused the rough eye phenotype with degeneration, and those in motor neurons induced locomotor defects (Xu et al. 2013). In order to distinguish between the toxic gain-of-function mechanisms of expanded repeat RNAs and DPR proteins produced by RAN translation, the following three distinct *Drosophila* models were established: flies expressing expanded pure GGGGCC repeats that produce both expanded RNAs and DPR proteins, flies expressing expanded GGGGCC repeat RNA interrupted with a stop codon that only produces expanded RNAs, and flies expressing non-GGGGCC RNAs with alternative codons that only produce DPR proteins (Mizielinska et al. 2014). Detailed analyses with these three model fly lines revealed that expanded GGGGCC repeats induced neurotoxicity mediated by DPR proteins, and RNA foci may not be a direct cause of neurodegeneration in these fly models. In addition, only poly-(glycine-arginine) and poly-(proline-arginine) proteins induced the rough eye phenotype, while poly-(glycine-alanine) and poly-(proline-alanine) proteins did not, indicating that arginine-containing DPR proteins are the main cause of neurodegeneration in *C9ORF72*-ALS/FTD fly models (Mizielinska et al. 2014). The findings of another study with a new model fly expressing 160 GGGGCC repeats flanked by human intron and exon sequences also support this conclusion (Tran et al. 2015).

Genetic modifier screening with *Drosophila* models for *C9ORF72*-ALS/FTD identified Ran GTPase-activating protein (RanGAP), a key regulator of nucleocytoplasmic transport (Zhang et al. 2015), 18 genes involved in the nuclear pore complex and nucleocytoplasmic transport (Freibaum et al. 2015), and genes encoding components of the nuclear pore complex, such as importins, exportins, Ran-GTP regulators, and arginine methylases (Boeynaems et al. 2016). These findings suggest that nucleocytoplasmic transport plays a crucial role in the pathogenesis of *C9ORF72*-ALS/FTD.

3.7 Drosophila FUS-*ALS/FTD Model*

Mutations in *FUS* are found in familial ALS/FTD and mostly inherited in an autosomal dominant manner (Picher-Martel et al. 2016). FUS carries a single RNA recognition motif and plays a role in RNA/DNA binding, RNA processing, splicing, and transcription (Tan and Manley 2010; Azuma et al. 2018). FUS normally localizes in the nucleus; however, it abnormally localizes in the cytoplasm as inclusions in ALS/FTD patients. Neurotoxicity by a gain-of-function and by a loss-of-function has been proposed to explain the pathogenesis of *FUS*-ALS/FTD, and both models have been developed (Azuma et al. 2018).

3.7.1 Loss-of-Function Model

In *Drosophila*, there is a single orthologue for human FUS, designated as *Cabeza* (*caz*). The pan-neuron-specific knockdown of *caz* caused a short life span and reduced locomotive ability in adults (Fig. 2) (Xia et al. 2012; Jäckel et al. 2015; Baldwin et al. 2016). Consistent with these findings, *caz* knockdown flies show shorter synaptic branch lengths in motor neurons than control flies, suggesting that *caz* is involved in the formation and/or maintenance of motor neurons at presynaptic terminals in the neuromuscular junction (NMJ) (Fig. 2) (Wang et al. 2011; Sasayama et al. 2012; Xia et al. 2012; Shimamura et al. 2014; Frickenhaus et al. 2015; Baldwin et al. 2016). The eye imaginal disc-specific knockdown of *caz* induced the rough eye phenotype (Fig. 2). The rough eye was accompanied by the loss-of-cone cell phenotype in pupal retinae (Shimamura et al. 2014), suggesting a role for caz in the differentiation of cone cells.

The pan-neuron- and motor neuron-specific knockdown of *hsrω* lncRNA impaired locomotive ability and the NMJ structure (Lo Piccolo and Yamaguchi 2017). Furthermore, the knockdown of *hsrω* lncRNA reduced *caz* mRNA levels and induced the abnormal cytoplasmic localization of the caz protein. Furthermore, the knockdown of *CR18854* lncRNA suppressed the locomotive defect and short synapse branch length phenotype at NMJ induced by the *caz* knockdown (Muraoka et al. 2018). Therefore, some human lncRNAs may play a critical role in the pathogenesis of *FUS*-ALS/FTD.

3.7.2 Gain-of-Function Model

Gain-of-function models for *FUS*-ALS/FTD have been established by overexpressing human FUS using the GAL4-UAS targeted expression system (Fig. 1). The expression of wild-type human FUS (*hFUSwt*) induced the rough eye phenotype, reduced life span, climbing and crawling defects, reduced synaptic bouton numbers, a decreased active zone at NMJ, axonal degeneration, and enlarged motoneurons (Fig. 2) (Chen et al. 2011; Lanson et al. 2011; Miguel et al. 2012; Xia et al. 2012; Jäckel et al. 2015; Baldwin et al. 2016). The expression of the mutant form of *hFUS* (*hFUSmut*) also induced the progressive degeneration of and functional defects in *Drosophila*, such as the rough eye phenotype, eclosion defects, climbing and crawling defects, reduced synaptic bouton numbers, and a decreased active zone at NMJ (Fig. 2) (Chen et al. 2011; Lanson et al. 2011; Machamer et al. 2014; Xia et al. 2012; Jäckel et al. 2015; Baldwin et al. 2016).

The hFUSwt expressed in *Drosophila* is mainly soluble and neurotoxic, inducing an aberrant eye morphology with the loss of pigmentation. The knockdown of *hsrω* lncRNA reduced *hFUSwt* mRNA levels and induced the formation of cytoplasmic nontoxic hFUSwt-LAMP1 insoluble inclusions (Lo Piccolo et al. 2017). The combination of these events appears to titrate soluble hFUSwt to suppress its neurotoxicity. Therefore, a lncRNA-dependent pathway may play an important role in controlling aggregation-prone hnRNPs, also suggesting that properly formed hFUSwt inclusions exhibit no toxicity in cells.

3.8 Drosophila FIG4-*CMT Model*

Charcot-Marie-Tooth disease (CMT) is an inherited group of peripheral neuropathies, showing defects in both motor and sensory neurons (Charcot and Marie 1886; Schroder 2006). In many cases, CMT shows an onset in childhood, and patients exhibit a combination of slowly progressive symptoms of sensory defects and distal muscle debility; however, the severity and progression of symptoms markedly vary among individuals (Saifi et al. 2003). More than 80 CMT-causing genes with a wide variety of functions have been identified, and 35 homologues of CMT-causing genes have been detected to date in *Drosophila*. Many are responsible for the formation of the abnormal axonal phenotypes observed in CMT. Since *Drosophila* does not have myelin sheaths or Schwann cells that surround the axon of the neuron, it is suitable only for the axonal-type CMT model (Yamaguchi and Takashima 2018).

Factor-induced-gene 4 (FIG4) carries a sac phosphatase domain composed of seven conserved motifs responsible for phosphoinositide phosphatase activity. FIG4 is responsible for the metabolism of PI(3,5)P2, which is associated with the intracellular vesicles of early and late endosomes (Sbrissa et al. 2007; Huotari and Helenius 2011). Mutations in the *FIG4* gene are associated with CMT4J (Chow et al. 2007), Yunis-Varon syndrome (YVS) (Campeau et al. 2013), and epilepsy with polymicrogyria (Baulac et al. 2014). Homozygous *FIG4*-null mice show neurodegeneration and enlarged vacuoles in neurons (Chow et al. 2007). In addition, the *FIG4* gene has been identified as a genetic susceptibility factor for ALS (Chow et al. 2009).

In *Drosophila*, there is a single orthologue for human FIG4, designated as dFIG4. *Drosophila* models for *FIG4*-CMT were established by the knockdown of *dFIG4* using the GAL4-UAS system (Fig. 1). Pan-neuron-specific *dFIG4* knockdown resulted in a short life span and reduced mobility in adults (Kyotani et al. 2016). Consistent with these findings, *dFIG4* knockdown flies showed a shorter synaptic branch length in motor neurons. The eye imaginal disc-specific knockdown of dFIG4 induced the rough eye phenotype accompanied by loss-of-cone cells (Kyotani et al. 2016). Other studies indicated that *dFIG4* null mutants were viable but showed enlarged lysosomes in muscles and neurons, accompanied by an age-dependent decline in flight ability (Bharadwaj et al. 2016).

In modifier screening of the rough eye phenotype using a set of chromosomal deficiency lines followed by the screening of mutants of individual genes (Fig. 2), the *CR18854* gene encoding a long-hairpin RNA (hpRNA) consisting of 2566 bases was identified as a genetic interactant with *dFIG4* (Fig. 3) (Muraoka et al. 2018).

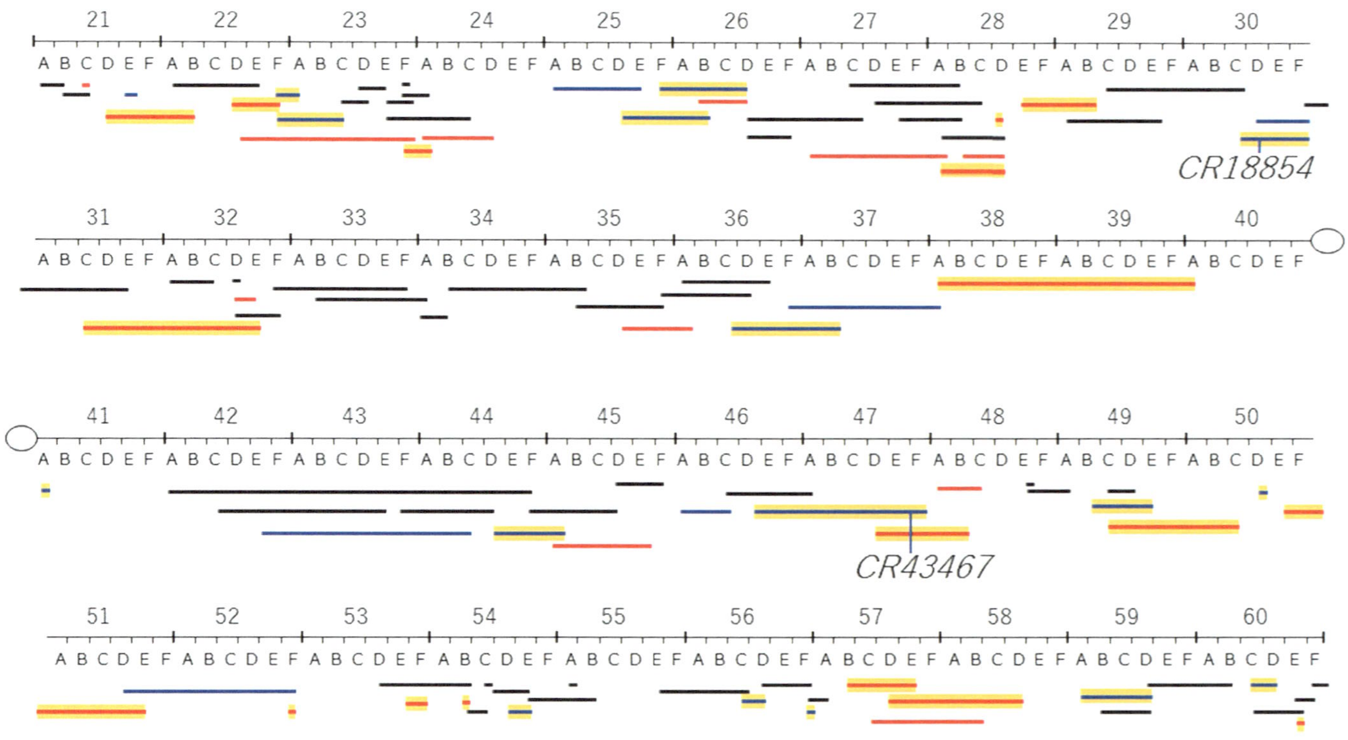

Fig. 3 Summary of rough eye modifier screening with the second chromosome deficiencies. A collection of *Drosophila* deficiency stocks on the second chromosome was crossed with the *dFIG4* knockdown fly line. The eye morphologies of their F1 progenies were inspected by SEM. The left arm of the second chromosome is divided into 20 divisions (21–40), and the right arm is also divided into 20 divisions (41–60). Each division is further divided into subdivision A to F. Lines indicate the cytological position of deficiency lines on the second chromosome. The circle indicates the position of the centromere. Black line: a genomic region whose deletion exerted no effect on the rough eye phenotype induced by the *dFIG4* knockdown. Blue line: a genomic region whose deletion suppressed the rough eye phenotype. Red line: a genomic region whose deletion enhanced the rough eye phenotype. Yellow line: a genomic region whose deletion also modified the rough eye phenotype induced by the *caz* knockdown. The positions of the genes encoding *CR18854* lncRNA and *CR43467* lncRNA are also shown

This lncRNA has been reported to bind with hnRNP, Hrp59, and RBP, staufen (Laver et al. 2013). It is important to note that *SCA8* lncRNA also genetically interacts with staufen, as described in Sect. 3.3 (Mutsuddi et al. 2004). Mutations in and the knockdown of *CR18854* suppressed both the rough eye phenotype and the loss-of-cone cell phenotype induced by the *dFIG4* knockdown (Muraoka et al. 2018). The rough eye modifier screen also identified another lncRNA, *CR43467* (Shimada et al. 2020) (Fig. 3). It has two splicing variants called *CR43467-RA* (length: 1393 nucleotides) and *CR43467-RB* (length: 2813 nucleotides). The eye disc-specific knockdown of both variants by CRISPR-dCas9 (Ghosh et al. 2016) (Fig. 1) suppressed the rough eye phenotype induced by the *dFIG4* knockdown (Shimada et al. 2020). The knockdown of *CR43467* rescued reductions in the length of synaptic branches and number of boutons at NMJ induced by the knockdown of *dFIG4* (Fig. 2). Mutations in and the knockdown of *CR18854* suppressed the enlarged vacuole phenotype induced by the fat body-specific knockdown of *dFIG4* and short synapse branch length phenotype induced by the neuron-specific *dFIG4* knockdown (Muraoka et al. 2018). The nucleotide identity of *CR43467* and *CR18854* was relatively high at 35%. Interestingly, *CR18854* genetically interacted with *caz*, while *hsrω* genetically interacted with *dFIG4*, suggesting that lncRNAs, such as *CR18854*, *CR43467*, and *hsrω*, are involved in a common pathway in the pathogenesis of CMT and ALS (Muraoka et al. 2018).

4 Perspectives

LncRNAs are now emerging as important regulators of various neurological functions, and the aberrant expression or function of lncRNAs appears to be crucial for the onset and/or progression of various neurological disorders. However, in many cases of gene diagnoses of various neurological disorders, only protein-coding exome analyses have been performed. The current knowledge on lncRNAs associated with neurological disorders strongly suggests the importance of the noncoding region of the genome, including those of lncRNAs for an accurate gene diagnosis and proper understanding of neurological disorders. In mammals, although the roles of a fraction of lncRNAs in the regulation of gene expression at the epigenetic, transcriptional, and posttranscriptional levels have been demonstrated, roles for the vast majority of lncRNAs remain unknown. The situation is similar in *Drosophila*; a limited information on roles of lncRNAs is available such as their roles in embryogenesis, neurological and gonadal functions, and stress tolerance (Li et al. 2019a, b). Moreover, bioresources for lncRNAs in *Drosophila*, such as mutants or knockdown strains, are still limited. The development of bioresources in *Drosophila*, such as effective knockdown lines for lncRNAs by CRISPR-dCas9 (Fig. 1), is needed to further clarify the roles of lncRNAs in the pathogenesis of neurological disorders.

Although the rough eye modifier screen has commonly been used to identify and characterize genes encoding lncRNAs and related RBPs, caution is needed for the following reasons. The commonly used *Drosophila* photoreceptor degeneration system provides a convenient readout but mainly reflects generic neurotoxicity instead of disease-specific selective neurotoxicity (Koon and Chan. 2017). Therefore, it is important to verify relevant phenotypes using other analytical systems that are more specific to particular types of diseases, such as inspections of CNS and NMJ morphologies (Fig. 2). In any event, *Drosophila* is now emerging as a useful and highly tractable model organism for studying human diseases, including neurological disorders. The future characterization of *Drosophila* lncRNAs associated with homologues of human neurological disorder-causing genes will expand our knowledge on human lncRNAs and contribute to the development of novel gene diagnosis markers and therapies for neurological disorders.

References

Andretic R, Kim YC, Jones FS et al (2008) *Drosophila* D1 dopamine receptor mediates caffeine-induced arousal. Proc Natl Acad Sci U S A 105:20392–20397

Atanasio A, Decman V, White D et al (2016) C9orf72 ablation causes immune dysregulation characterized by leukocyte expansion, autoantibody production, and glomerulonephropathy in mice. Sci Rep 6:23204

Au PC, Zhu QH, Dennis ES et al (2011) Long non-coding RNA-mediated mechanisms independent of the RNAi pathway in animals and plants. RNA Biol 8:404–414

Azuma Y, Mizuta I, Tokuda T et al (2018) Amyotrophic lateral sclerosis model. Adv Exp Med Biol 1076:79–96

Baldwin KR, Godena VK, Hewitt VL et al (2016) Axonal transport defects are a common phenotype in *Drosophila* models of ALS. Hum Mol Genet 25:2378–2392

Baulac A, Lenk GM, Dufresnois B et al (2014) Role of the phosphoinositide phosphatase *FIG4* gene in familial epilepsy with polymicrogyria. Neurology 82:1068–1075

Bharadwaj R, Cunningham KM, Zhang K et al (2016) FIG4 regulates lysosome membrane homeostasis independent of phosphatase function. Hum Mol Genet 25:681–692

Bian S, Sun T (2011) Functions of noncoding RNAs in neural development and neurological diseases. Mol Neurobiol 44:359–373

Boeynaems S, Bogaert E, Michiels E et al (2016) *Drosophila* screen connects nuclear transport genes to DPR pathology in c9ALS/FTD. Sci Rep 6:20877

Brand AH, Perrimon N (1993) Targeted gene expression as a means of altering cell fates and generating dominant phenotypes. Development 118:401–415

Campeau PM, Lenk GM, Lu JT et al (2013) Yunis–Varon syndrome is caused by mutations in *FIG4*, encoding a phosphoinositide phosphatase. Am J Hum Genet 92:781–791

Cao H, Wahlestedt C, Kapranov P (2018) Strategies to annotate and characterize long noncoding RNAs: advantages and pitfalls. Trends Genet 34:704–721

Carrieri C, Forrest AR, Santoro C et al (2015) Expression analysis of the long non-coding RNA antisense to Uchl1 (AS Uchl1) during dopaminergic cells' differentiation in vitro and in neurochemical models of Parkinson's disease. Front Cell Neurosci 9:114

Charcot JM, Marie P (1886) Sur une forme particuliere d'atrophie musculaire progressive, souvent familiale, debutante par les pieds et les jambes et atteignant plus tard les mains. Rev Med (Parie) 6:97–138

Chen LL (2016) Linking long noncoding RNA localization and function. Trends Biochem Sci 41:761–772

Chen WL, Lin JW, Huang HI et al (2008) SCA8 mRNA expression suggests an antisense regulation of KLHL1 and correlates to SCA8 pathology. Brain Res 1233:176–184

Chen Y, Yang M, Deng J et al (2011) Expression of human FUS protein in *Drosophila* leads to progressive neurodegeneration. Protein Cell 2:477–486

Chen B, Zhang Y, Zhang X et al (2016a) Genome-wide identification and developmental expression profiling of long noncoding RNAs during *Drosophila* metamorphosis. Sci Rep 6:23330

Chen SD, Sun XY, Niu W et al (2016b) Aberrant expression of long non-coding RNAs in schizophrenia patients. Med Sci Monitor 22:3340–3351

Chiba M, Kiyosawa H, Hiraiwa N et al (2009) Existence of Pink1 antisense RNAs in mouse and their localization. Cytogenet Genome Res 126:259–270

Chow CY, Zhang Y, Dowling JJ et al (2007) Mutation of *FIG4* causes neurodegeneration in the pale tremor mouse and patients with CMT4J. Nature 448:68–72

Chow CY, Landers JE, Bergren SK et al (2009) Deleterious variants of FIG4 a phosphoinositide phosphatase in patients with ALS. Am J Hum Genet 84:85–88

Ciarlo E, Massone S, Penna I et al (2013) An intronic ncRNA-dependent regulation of SORL1 expression affecting Abeta formation is upregulated in post-mortem Alzheimer's disease brain samples. Dis Model Mech 6:424–433

Daughters RS, Tuttle DL, Gao W et al (2009) RNA gain-of-function in spinocerebellar ataxia type 8. PLoS Genet 5:e1000600

DeJesus-Hernandez M, Mackenzie IR, Boeve BF et al (2011) Expanded GGGGCC hexanucleotide repeat in noncoding region of *C9ORF72* causes chromosome 9p-linked FTD and ALS. Neuron 72:245–256

Faghihi MA, Modarresi F, Khalil AM et al (2008) Expression of a noncoding RNA is elevated in Alzheimer's disease and drives rapid feed-forward regulation of beta-secretase. Nat Med 14:723–730

Fernandez-Funez P, Nino-Rosales ML, de Gouyon B et al (2000) Identification of genes that modify ataxin-1-induced neurodegeneration. Nature 408:101–106

Fratta P, Poulter M, Lashley T et al (2013) Homozygosity for the *C9orf72* GGGGCC repeat expansion in frontotemporal dementia. Acta Neuropathol 126:401–409

Freibaum BD, Lu Y, Lopez-Gonzalez R et al (2015) GGGGCC repeat expansion in *C9orf72* compromises nucleocytoplasmic transport. Nature 525:129–133

Frickenhaus M, Wagner M, Mallik M et al (2015) Highly efficient cell-type-specific gene inactivation reveals a key function for the *Drosophila* FUS homolog cabeza in neurons. Sci Rep 5:9107

Ghosh S, Tibbit C, Liu J-L (2016) Effective knockdown of *Drosophila* long non-coding RNAs by CRISPR interference. Nucleic Acids Res 44:e84

Harms MB, Cady J, Zaidman C et al (2013) Lack of *C9ORF72* coding mutations supports a gain of function for repeat expansions in amyotrophic lateral sclerosis. Neurobiol Aging 34(2234): e2213–e2239

Huarte M (2015) The emerging role of lncRNAs in cancer. Nat Med 21:1253–1261

Huotari J, Helenius A (2011) Endosome maturation. EMBO J 30:3481–3500

Hwang J-Y, Zukin RS (2018) REST, a master transcription regulator in neurodegenerative disease. Curr Opin Neurobiol 48:193–200

Ishiguro T, Sato N, Ueyama M et al (2017) Regulatory role of RNA chaperone TDP-43 for RNA misfolding and repeat-associated translation in SCA31. Neuron 94:108–124

Jäckel S, Summerer AK, Thömmes CM et al (2015) Nuclear import factor transportin and arginine methyltransferase 1 modify FUS neurotoxicity in *Drosophila*. Neurobiol Dis 74:76–88

Jiang J, Zhu Q, Gendron TF et al (2016) Gain of toxicity from ALS/FTD-linked repeat expansions in *C9ORF72* is alleviated by antisense oligonucleotides targeting GGGGCC-containing RNAs. Neuron 90:535–550

Jin P, Zarnescu DC, Zhang F et al (2003) RNA-mediated neurodegeneration caused by the fragile X premutation rCGG repeats in *Drosophila*. Neuron 39:739–747

Jin P, Duan R, Qurashi A et al (2007) Pur binds to rCGG repeats and modulates repeat-mediated neurodegeneration in a *Drosophila* model of fragile X tremor/ataxia syndrome. Neuron 55:556–564

Johnson R (2012) Long non-coding RNAs in Huntington's disease neurodegeneration. Neurobiol Dis 46:245–254

Jolly C, Lakhotia SC (2006) Human sat III and *Drosophila* hsr omega transcripts: a common paradigm for regulation of nuclear RNA processing in stressed cells. Nucleic Acids Res 34:5508–5514

Kanouchi T, Ohkubo T, Yokota T (2012) Can regional spreading of amyotrophic lateral sclerosis motor symptoms be explained by prion-like propagation? J Neurol Neurosurg Psychiatry 83:739–745

Koob MD, Moseley ML, Schut LJ et al (1999) An untranslated CTG expansion causes a novel form of spinocerebellar ataxia (SCA8). Nat Genet 21:379–384

Koon AC, Chan HYE (2017) *Drosophila melanogaster* as a model organism to study RNA toxicity of repeat expansion-associated neurodegenerative and neuromuscular diseases. Front Cell Neurosci 11:70

Koppers M, Blokhuis AM, Westeneng HJ et al (2015) C9orf72 ablation in mice does not cause motor neuron degeneration or motor deficits. Ann Neurol 78:426–438

Kyotani A, Azuma Y, Yamamoto I et al (2016) Knockdown of the *Drosophila FIG4* induces deficient locomotive behavior, shortening of motor neuron, axonal targeting aberration, reduction of life span and defects in eye development. Exp Neurol 277:86–95

Lagier-Tourenne C, Polymenidou M, Hutt KR et al (2012) Divergent roles of ALS-linked proteins FUS/TLS and TDP-43 intersect in processing long pre-mRNAs. Nat Neurosci 15:1488–1497

Lakhotia SC, Mallik M, Singh AK et al (2012) The large noncoding hsromega-n transcripts are essential for thermotolerance and remobilization of hnRNPs, HP1 and RNA polymerase II during recovery from heat shock in *Drosophila*. Chromosoma 121:49–70

Lanson NA Jr, Maltare A, King H et al (2011) A *Drosophila* model of FUS-related neurodegeneration reveals genetic interaction between FUS and TDP-43. Hum Mol Genet 20:2510–2523

Laver JD, Li X, Ancevicius K et al (2013) Genome-wide analysis of Staufen-associated mRNAs identifies secondary structures that confer target specificity. Nucleic Acids Res 41:9438–9460

Lee JT, Bartolomei MS (2013) X-inactivation, imprinting, and long noncoding RNAs in health and disease. Cell 152:1308–1323

Lekka E, Hall J (2018) Noncoding RNAs in disease. FEBS Lett 592:2884–2900

Li LB, Yu Z, Teng X et al (2008) RNA toxicity is a component of ataxin-3 degeneration in *Drosophila*. Nature 453:1107–1111

Li K, Tian Y, Yuan Y et al (2019a) Insights into the functions of lncRNAs in *Drosophila*. Int J Mol Sci 20:4646

Li L, Zhuang Y, Zhao X et al (2019b) Long non-coding RNA in neuronal development and neurological disorders. Front Genet 9:744

Ling SC, Polymenidou M, Cleveland DW (2013) Converging mechanisms in ALS and FTD: disrupted RNA and protein homeostasis. Neuron 79:416–438

Liu W, Zhang Q, Zhang J et al (2017) Long non-coding RNA MALAT1 contributes to cell apoptosis by sponging miR-124 in Parkinson disease. Cell Biosci 7:19

Liu YS, Rao SQ, Xu Y et al (2018) Changes in the level of long non-coding RNA gomafu gene expression in schizophrenia patients before and after antipsychotic medication. Schizophr Res 195:318–319

Lo Piccolo L (2018) *Drosophila* as a model to gain insight into the role of lncRNAs in neurological disorders. Adv Exp Med Biol 1076:119–146

Lo Piccolo L, Yamaguchi M (2017) RNAi of arcRNA hsrω affects sub-cellular localization of *Drosophila* FUS to drive neurodiseases. Exp Neurol 292:125–134

Lo Piccolo L, Jantrapirom S, Nagai Y et al (2017) FUS toxicity is rescued by the modulation of lncRNA hsrω expression in *Drosophila melanogaster*. Sci Rep 7:15660

Machamer JB, Collins SE, Lloyd TE (2014) The ALS gene FUS regulates synaptic transmission at the *Drosophila* neuromuscular junction. Hum Mol Genet 23:3810–3822

Mallik M, Lakhotia SC (2009a) The developmentally active and stress-inducible noncoding hsromega gene is a novel regulator of apoptosis in *Drosophila*. Genetics 183(3):831–852
Mallik M, Lakhotia SC (2009b) RNAi for the large non-coding hsromega transcripts suppresses polyglutamine pathogenesis in *Drosophila* models. RNA Biol 6:464–478
Manto MU (2005) The wide spectrum of spinocerebellar ataxias (SCAs). Cerebellum 4:2–6
Massone S, Vassallo I, Fiorino G et al (2011) 17A, a novel non-coding RNA, regulates GABA B alternative splicing and signaling in response to inflammatory stimuli and in Alzheimer disease. Neurobiol Dis 41:308–317
Miguel L, Avequin T, Delarue M et al (2012) Accumulation of insoluble forms of FUS protein correlates with toxicity in *Drosophila*. Neurobiol Aging 33:1008.e1-15
Mizielinska S, Gronke S, Niccoli T et al (2014) *C9orf72* repeat expansions cause neurodegeneration in *Drosophila* through arginine-rich proteins. Science 345:1192–1194
Moens TG, Partridge L, Isaacs AM (2017) Genetic models of C9orf72: what is toxic? Curr Opin Genet Dev 44:92–101
Moseley ML, Zu T, Ikeda Y et al (2006) Bidirectional expression of CUG and CAG expansion transcripts and intranuclear polyglutamine inclusions in spinocerebellar ataxia type 8. Nat Genet 38:758–769
Muraoka Y, Nakamura A, Tanaka R et al (2018) Genetic screening of genes interacting with *Drosophila* FIG4 identified a novel link between CMT-causing gene and long noncoding RNAs. Exp Neurol 310:1–13
Murphy JM, Henry RG, Langmore S et al (2007) Continuum of frontal lobe impairment in amyotrophic lateral sclerosis. Arch Neurol 64:530–534
Mutsuddi M, Marshall CM, Benzow KA et al (2004) The spinocerebellar ataxia 8 noncoding RNA causes neurodegeneration and associates with staufen in *Drosophila*. Curr Biol CB 2004 (14):302–308
Nichols CD (2006) *Drosophila melanogaster* neurobiology, neuropharmacology, and how the fly can inform central nervous system drug discovery. Pharmacol Ther 112:677–700
Niimi Y, Takahashi M, Sugawara E et al (2013) Abnormal RNA structures (RNA foci) containing a penta-nucleotide repeat (UGGAA)n in the Purkinje cell nucleus is associated with spinocerebellar ataxia type 31 pathogenesis. Neuropathology 33:600–611
Nishimoto Y, Nakagawa S, Hirose T et al (2013) The long non-coding RNA nuclear-enriched abundant transcript 1_2 induces paraspeckle formation in the motor neuron during the early phase of amyotrophic lateral sclerosis. Mol Brain 6:31
Onorati MC, Lazzaro S, Mallik M et al (2011) The ISWI chromatin remodeler organizes the hsromega ncRNA-containing omega speckle nuclear compartments. PLoS Genet 7:e1002096
Pandy UB, Nichols CD (2011) Human disease models in *Drosophila melanogaster* and its role of the fly in therapeutic drug discovery. Pharmacol Rev 63:411–436
Paul JW, Gitler AD (2014) Cell biology. Clogging information flow in ALS. Science 345:1118–1119
Picher-Martel V, Valdmanis PN, Gould PV et al (2016) From animal models to human disease: a genetic approach for personalized medicine in ALS. Acta Neuropathol Commun 4:70
Polymenidou M, Lagier-Tourenne C, Hutt KR et al (2011) Long pre-mRNA depletion and RNA missplicing contribute to neuronal vulnerability from loss of TDP-43. Nat Neurosci 14:459–468
Port F, Bullock SL (2016) Augmenting CRISPR applications in *Drosophila* with tRNA-flanked sgRNAs. Nat Methods 13:852–854
Prasanth KV, Rajendra TK, Lal AK et al (2000) Omega speckles – a novel class of nuclear speckles containing hnRNPs associated with noncoding hsr-omega RNA in *Drosophila*. J Cell Sci 113:3485–3497
Renton AE, Majounie E, Waite A et al (2011) A hexanucleotide repeat expansion in *C9ORF72* is the cause of chromosome 9p21-linked ALS-FTD. Neuron 72:257–268
Rothenfluh A, Heberlein U (2002) Drugs, flies, and videotape: the effects of ethanol and cocaine on *Drosophila* locomotion. Curr Opin Neurobiol 12:639–645

Saifi GM, Szigeti K, Snipes GJ et al (2003) Molecular mechanisms, diagnosis, and rational approaches to management of and therapy for Charcot-Marie-Tooth disease and related peripheral neuropathies. J Investig Med 51:261–283
Saitoh Y, Fujikake N, Okamoto Y et al (2015) p62 plays a protective role in the autophagic degradation of polyglutamine protein oligomers in polyglutamine disease model flies. J Biol Chem 290:1442–1453
Sasayama H, Shimamura M, Tokuda T et al (2012) Knockdown of the *Drosophila* fused in sarcoma (FUS) homologue causes deficient locomotive behavior and shortening of motoneuron terminal branches. PLoS One 7:e39483
Sato N, Amino T, Kobayashi K et al (2009) Spinocerebellar ataxia type 31 is associated with "inserted" penta-nucleotide repeats containing (TGGAA)n. Am J Hum Genet 85:544–557
Satta R, Dimitrijevic N, Manev H (2003) *Drosophila* metabolize 1,4-butanediol into γ-hydroxybutyric acid in vivo. Eur J Pharmacol 473:149–152
Sbrissa D, Ikonomov OC, Fu Z et al (2007) Core protein machinery for mammalian phosphatidylinositol 3,5-bisphosphate synthesis and turnover that regulates the progression of endosomal transport. J Biol Chem 282:23878–23891
Scheele C, Petrovic N, Faghihi MA et al (2007) The human PINK1 locus is regulated in vivo by a non-coding natural antisense RNA during modulation of mitochondrial function. BMC Genomics 8:74
Schroder JM (2006) Neuropathology of Charcot-Marie-Tooth and related disorders. Neuromol Med 8:23–42
Sellier C, Freyermuth F, Tabet R et al (2013) Sequestration of DROSHA and DGCR8 by expanded CGG RNA repeats alters microRNA processing in fragile X-associated tremor/ataxia syndrome. Cell Rep 3:869–880
Shimada S, Muraoka Y, Ibaraki K et al (2020) Identification of *CR43467* encoding a long non-coding RNA as a novel genetic interactant with *dFIG4*, a CMT-causing gene. Exp Cell Res 386:111711
Shimamura M, Kyotani A, Azuma Y et al (2014) Genetic link between Cabeza, a *Drosophila* homologue of Fused in Sarcoma (FUS), and the EGFR signaling pathway. Exp Cell Res 326:36–45
Sofola OA, Jin P, Botas J et al (2007a) Argonaute-2-dependent rescue of a *Drosophila* model of FXTAS by FRAXE premutation repeat. Hum Mol Genet 16:2326–2332
Sofola OA, Jin P, Qin Y et al (2007b) RNA-binding proteins hnRNP A2/B1 and CUGBP1 suppress fragile X CGG premutation repeat-induced neurodegeneration in a *Drosophila* model of FXTAS. Neuron 55:565–571
Sudria-Lopez E, Koppers M, de Wit M et al (2016) Full ablation of C9orf72 in mice causes immune system-related pathology and neoplastic events but no motor neuron defects. Acta Neuropathol 132:145–147
Sun C, Huang L, Li Z et al (2018) Long non-coding RNA MIAT in development and disease: a new player in an old game. J Biomed Sci 25:23
Sunwoo JS, Lee ST, Im W et al (2017) Altered expression of the long noncoding RNA NEAT1 in Huntington's disease. Mol Neurobiol 54:1577–1586
Suster ML, Seugnet L, Bate M et al (2004) Refining GAL4-driven transgene expression in *Drosophila* with a GAL80 enhancer-trap. Genesis 39:240–245
Takano-Shimizu-Kouno T, Ohsako T (2018) Humanized flies and resources for cross-species study. Adv Exp Med Biol 1076:277–288
Tan AY, Manley JL (2010) TLS inhibits RNA polymerase III transcription. Mol Cell Biol 30:186–196
Tan JY, Vance KW, Varela MA et al (2014) Cross-talking noncoding RNAs contribute to cell-specific neurodegeneration in SCA7. Nat Struct Mol Biol 21:955–961
Todd PK, Oh SY, Krans A et al (2013) CGG repeat-associated translation mediates neurodegeneration in fragile X tremor ataxia syndrome. Neuron 78:440–455

Tran H, Almeida S, Moore J et al (2015) Differential toxicity of nuclear RNA foci versus dipeptide repeat proteins in a *Drosophila* model of *C9ORF72* FTD/ALS. Neuron 87:1207–1214

Ueoka I, Pham HTN, Matsumoto K et al (2019) Autism spectrum disorder-related syndromes: modeling with *Drosophila* and rodents. Int J Mol Sci 20:4071

Ueyama M, Nagai Y (2018) Repeat expansion disease models. Adv Exp Med Biol 1076:63–78

Wan P, Su W, Zhuo Y (2017) The role of long noncoding RNAs in neurodegenerative diseases. Mol Neurobiol 54:2012–2021

Wang JW, Brent JR, Tomlinson A et al (2011) The ALS-associated proteins FUS and TDP-43 function together to affect *Drosophila* locomotion and life span. J Clin Invest 121:4118–4126

Warrick JM, Paulson HL, Gray-Board GL et al (1998) Expanded polyglutamine protein forms nuclear inclusions and causes neural degeneration in *Drosophila*. Cell 93:939–949

Wilkinson B, Campbell DB (2013) Contribution of long noncoding RNAs to autism spectrum disorder risk. Int Rev Neurobiol 113:35–59

Wolf FW, Heberlein U (2003) Invertebrate models of drug abuse. J Neurobiol 54:161–178

Xia R, Liu Y, Yang L et al (2012) Motor neuron apoptosis and neuromuscular junction perturbation are prominent features in a *Drosophila* model of Fus-mediated ALS. Mol Neurodegener 7:10

Xie Y, Hayden MR, Xu B (2010) BDNF overexpression in the forebrain rescues Huntington's disease phenotypes in YAC128 mice. J Neurosci 30:14708–14718

Xu Z, Poidevin M, Li X et al (2013) Expanded GGGGCC repeat RNA associated with amyotrophic lateral sclerosis and frontotemporal dementia causes neurodegeneration. Proc Natl Acad Sci U S A 110:7778–7783

Yamaguchi M, Takashima H (2018) *Drosophila* Charcot-Marie-Tooth disease models. Adv Exp Med Biol 1076:97–118

Yamaguchi M, Yoshida H (2018) *Drosophila* as a model organism. Adv Exp Med Biol 1076:1–10

Yamamoto S, Jaiswal M, Charng W-L et al (2014) A *Drosophila* genetic resource of mutants to study mechanisms underlying human genetic diseases. Cell 159:200–214

Yamanaka Y, Faghihi MA, Magistri M et al (2015) Antisense RNA controls LRP1 sense transcript expression through interaction with a chromatin-associated protein, HMGB2. Cell Rep 11:967–976

Zhang K, Donnelly CJ, Haeusler AR et al (2015) The *C9orf72* repeat expansion disrupts nucleocytoplasmic transport. Nature 525:56–61

Ziats MN, Rennert OM (2013) Aberrant expression of long noncoding RNAs in autistic brain. J Mol Neurosci 49:589–593

Zu T, Gibbens B, Doty NS et al (2011) Non-ATG-initiated translation directed by microsatellite expansions. Proc Natl Acad Sci U S A 108:260–265

Regulatory Roles of Long Non-coding RNAs in Skeletal Muscle Differentiation, Regeneration, and Disorders

Keisuke Hitachi and Kunihiro Tsuchida

Contents

Abstract In the past decade, long noncoding RNAs (lncRNAs), including antisense lncRNAs, promoter-associated lncRNAs, enhancer lncRNAs, and competing endogenous lncRNAs were reported to regulate several key biological functions in skeletal muscle cells. For example, lncRNAs can regulate mRNA transcription and translation, and microRNA functions in multistep programs of skeletal muscle formation. Therefore, studies of lncRNAs in skeletal muscle cells have expanded our knowledge regarding their biological significance. Although their exact molecular functions have not yet been characterized, hundreds of novel lncRNAs from skeletal muscle cells have been elucidated. Furthermore, rapid advances in genome-editing technologies have enabled examination of the in vivo regulatory roles of

K. Hitachi · K. Tsuchida (✉)
Division for Therapies against Intractable Diseases, Institute for Comprehensive Medical Science (ICMS), Fujita Health University, Toyoake, Aichi, Japan
e-mail: hkeisuke@fujita-hu.ac.jp; tsuchida@fujita-hu.ac.jp

S. Jurga, J. Barciszewski (eds.), *The Chemical Biology of Long Noncoding RNAs*, RNA Technologies 11, https://doi.org/10.1007/978-3-030-44743-4_17

lncRNAs in skeletal muscle tissues. In this chapter, we provide an overview of recent findings on the regulatory roles of lncRNAs in skeletal muscle differentiation, growth, and regeneration. In addition, we further discuss the possibility of utilizing lncRNAs as therapeutic targets for skeletal muscle disorders in humans.

Keywords Skeletal muscle differentiation · Myogenesis · Skeletal muscle hypertrophy · Skeletal muscle atrophy · Long noncoding RNA

Abbreviations

EdU	5-ethynyl-2′-deoxyuridine
ABIN-1	A20-binding inhibitor of NF-κB activation
ATRX	alpha thalassemia/mental retardation syndrome X-linked
ALS	amyotrophic lateral sclerosis
ASOs	anti-sense oligonucleotides
HuR	Human antigen R
Ash1L	ASH1-like histone lysine methyltransferase
Atrolnc-1	atrophy-related long noncoding RNA-1
bHLH	basic helix–loop–helix
BMP	bone morphogenetic protein
Bmp7	bone morphogenetic protein 7
Brca1	breast cancer 1, early onset
Cdc6	cell division cycle 6
ChIP-Seq	ChIP-sequencing
ChIRP-Seq	ChIRP-sequencing
ChIP	chromatin immunoprecipitation
ChIRP	chromatin isolation by RNA purification
ChRO1	chromatin reorganization 1
Charme	chromatin-associated muscle-specific lncRNA
CKD	chronic kidney disease
circRNAs	circular RNAs
CRISPR	clustered regularly interspaced short palindromic repeat
ceRNAs	competing endogenous lncRNAs
CE	core enhancer
Ccnd1	cyclin D1
Ccnd2	cyclin D2
Ccne1	cyclin E1
Ccne2	cyclin E2
DBE-T	DBE-Transcript
Ddx5	DEAD-box protein 5
Ddx17	DEAD-box protein 17
DAXX	death domain-associated protein
Dio3	deiodinase, iodothyronine type III
Dlk1	delta-like non-canonical Notch ligand 1

Dppa2	developmental pluripotency-associated 2
DRR	distal regulatory region
Dnmt1	DNA methyltransferase 1
Dnmt3a	DNA methyltransferase 3A
Dnmt3b	DNA methyltransferase 3B
DUM	Dppa2 upstream binding muscle lncRNA
DWORF	dwarf open reading frame
Eed	embryonic ectoderm development
eRNAs	enhancer lncRNAs
Ezh2	enhancer of zeste homologue 2
FSHD	facioscapulohumeral muscular dystrophy
Tnni2	fast skeletal muscle troponin I
Tnnt3	fast skeletal muscle troponin T
FGF2	fibroblast growth factor 2
Filip1	filamin A interacting protein 1
FISH	fluorescence in situ hybridization
FoxO1	forkhead box protein O1
Gdf5	growth differentiation factor 5
H3K4me3	histone H3 lysine 4 trimethylation
H3K27ac	histone H3 lysine 27 acetylation
PCAF	histone lysine acetyltransferase 2B
Hox-A11	homeobox A11
IMP1	IGF2-mRNA-binding protein 1
IMP2	IGF2-mRNA-binding protein 2
Irm	imprinted RNA near MEG3/Gtl2
IBM	inclusion body myositis
IRS1	insulin receptor substrate 1
IGF1	insulin-like growth factor 1
IGF1R	insulin-like growth factor 1 receptor
IGF2	insulin-like growth factor 2
IGF2R	insulin-like growth factor 2 receptor
kb	kilobase
LACTB	lactamase β
Linc-RAM	linc-RNA activator of myogenesis
lncMD	lncRNA related to muscle differentiation
linc-MD1	long noncoding MD1
lncRNAs	long noncoding RNAs
LSD1	lysine-specific demethylase 1
MAML1	mastermind-like transcriptional coactivator 1
MEG3	maternally expressed 3
lncMUMA	mechanical unloading-induced muscle atrophy-related lncRNA
Mb	megabase
Malat1	metastasis-associated lung adenocarcinoma transcript 1
miRNAs	microRNAs

Mcm4	minichromosome maintenance complex component 4
Mcm6	minichromosome maintenance complex component 6
Mcm7	minichromosome maintenance complex component 7
MAR1	muscle anabolic regulator 1
MCK	muscle creatine kinase
MDNCR	muscle differentiation-associated lncRNA
MuRF-1	muscle ring-finger 1
c-Myc	myelocytomatosis oncogene
MEF2C	myocyte enhancer factor 2C
MEF2D	myocyte enhancer factor 2D
MUNC	MyoD upstream noncoding RNA
lnc-mg	myogenesis-associated lncRNA
MyoD	myogenic differentiation 1
Myf5	myogenic factor 5
MRF4	myogenic regulatory factor 4
MRFs	myogenic regulatory factors
Myog	myogenin
Myoparr	myogenin promoter-associated myogenic regulatory anti-sense lncRNA
Mymk	myomaker
Mymx	myomixer
Mrln	myoregulin
MyHC	myosin heavy chain
Mbph1	myosin-binding protein 1
N-Ras	neuroblastoma ras oncogene
NIPBL	NIPBL cohesin loading factor
NF-κB	nuclear factor-kappa B
Neat1	nuclear-enriched abundant transcript 1
nt	nucleotides
Oct4	octamer-binding transcription factor 4
PDS5B	PDS5 cohesin-associated factor B
Pvt1	plasmacytoma variant translocation 1
Rb1	RB transcriptional corepressor 1
Rock1	Rho-associated coiled-coil containing protein kinase 1
RIP	RNA immunoprecipitation
Rian	RNA imprinted and accumulated in nucleus
RNAi	RNA interference
Pol II	RNA polymerase II
KSRP	K homology-type splicing regulatory protein
SERCA	sarcoplasmic reticulum Ca^{2+}-ATPase
SRF	serum response factor
lnc-SEMT	sheep skeletal enhanced muscularity transcript
Six1	sine oculis-related homeobox 1
Sirt1 AS lncRNA	Sirt1 antisense long noncoding RNA

Sirt1	sirtuin 1
Smad1	SMAD family member 1
Smad5	SMAD family member 5
Smad7	SMAD family member 7
siRNA	small interfering RNA
SPAR	small regulatory polypeptide of amino acid response
SRA	steroid receptor RNA activator
SMC3	structural maintenance of chromosomes 3
Suv39h1	suppressor of variegation 3–9 homolog 1
Baf60c	SWI/SNF-related, matrix-associated, actin-dependent regulator of chromatin, subfamily d, member 3
Brg1	SWI/SNF-related, matrix-associated, actin-dependent regulator of chromatin, subfamily a, member 4
SYISL	SYNPO2 intron sense-overlapping lncRNA
TDP-43	TAR DNA-binding protein 43
TA	tibialis anterior
p53	transformation-related protein 53
TGF-β	transforming growth factor-β
Tmem30a	transmembrane protein 30a
Tnrc6a	trinucleotide repeat containing 6a
Traf6	tumor necrosis factor receptor-associated factor 6
Wnt5a	wingless-type MMTV integration site family, member 5A
Wnt7b	wingless-type MMTV integration site family, member 7B
YB-1	Y box protein 1
YY1	Yin Yang 1
Yam-1	YY1-associated muscle lincRNA-1

1 Introduction

In the past few decades, two types of RNAs that do not code for proteins were discovered to play important roles in diverse biological processes. One such type of RNAs is microRNAs (miRNAs), which are small noncoding RNAs ranging in the length of 20–25 nucleotides (nt), and are highly conserved in eukaryotes. miRNAs post-transcriptionally regulate the expression of target genes via mRNA degradation and translational inhibition and play important roles in development, cell proliferation, differentiation, and various diseases (Bartel 2009). The other type is long noncoding RNAs (lncRNAs), which are longer than 200 nt. Although the roles of lncRNAs in genomic imprinting and dosage compensation have been previously shown (Bartolomei et al. 1991; Marahrens et al. 1997), the current state-of-the-art comprehensive sequencing technologies for nucleic acids have enabled detection of numerous, previously unidentified lncRNAs in cells and tissues. In humans, lncRNAs are pervasively transcribed from noncoding genomic DNA, which occupies approximately 98% of the human genome including cis-regulatory regions, untranslated regions, introns, and repetitive sequences (Chakraborty et al. 2014), and also from the coding genomic DNA in the antisense direction (Hon et al. 2017). Unlike miRNAs, lncRNAs tend not to be conserved among different species (Johnsson et al. 2014) and they act both transcriptionally and post-transcriptionally to regulate the expression of target genes by interacting with miRNAs or proteins (Ransohoff et al. 2018).

Skeletal muscle is the largest organ in human adults and typically represents 30–40% of the body weight. Skeletal muscles are necessary for both walking and maintaining posture; they also play important roles in energy consumption, glucose metabolism, and as an endocrine organ for maintaining whole-body homeostasis. Skeletal muscles in the trunk and limbs are derived from the paraxial somitic mesoderm, whereas most skeletal muscles in the head are derived from the cranial mesoderm. During embryonic development, myogenic precursor cells (myoblasts) proliferate, differentiate, and fuse together to form multinucleated myotubes, and then mature into muscle fibers to form skeletal muscle tissue. The process of skeletal muscle formation is called myogenesis. Myogenesis is highly regulated by several basic helix–loop–helix (bHLH) transcription factors, including myogenic differentiation 1 (MyoD), myogenic factor 5 (Myf5), myogenin (Myog), and myogenic regulatory factor 4 (MRF4), which are also referred to as myogenic regulatory factors (MRFs) (Tapscott 2005). The functions of these transcription factors are assisted by the myocyte enhancer factor MEF2 family of proteins (Berkes and Tapscott 2005). In vivo, myogenesis is initiated by Myf5 activation in the dorsomedial region of newly formed somites (Ott et al. 1991). Myf5 and MyoD are essential for the specification of the myogenic lineage in the somites and the determination of myoblasts (Rudnicki et al. 1993). Once the myoblasts receive a stimulus that induces myogenic differentiation, the cell cycle of proliferating myoblasts is arrested by MyoD through induction of the cyclin-dependent kinase inhibitors, p21 and p57 (Ruijtenberg and van den Heuvel 2016). Other cyclin-dependent kinase inhibitors (p18 and p27) and transformation-related protein 53 (p53) family members are also required for cell cycle arrest in myoblasts (Myers et al. 2004; Cam et al. 2006). In addition, MyoD activates the expression of early skeletal muscle-specific genes including the *Myog* gene (Tapscott 2005). Although MyoD is not sufficient for the expression of late muscle-specific genes, Myog is required to promote the entry of myoblasts into the myogenic-differentiation pathway through the activation of late muscle-specific genes (Cao et al. 2006). MyoD and Myog also activate the expression of the muscle-specific transmembrane protein myomaker (Mymk, also known as Tmem8c), which controls myoblast fusion (Millay et al. 2013). MRF4 plays a role in both the specification and differentiation of skeletal muscle cells (Kassar-Duchossoy et al. 2004). In addition to transcription factors, several miRNAs participate in the regulatory networks at different steps of skeletal muscle differentiation (Horak et al. 2016).

In adulthood, exercise induces muscle hypertrophy mainly through increases in the area and mass of muscle fibers. In contrast, immobilization, aging, and several diseases, such as muscle-wasting disorders and cancer cachexia, can decrease skeletal muscle mass; these conditions are referred to as muscle atrophy or muscle wasting. Large epidemiological surveys have revealed that muscle weakness is positively correlated with an increased overall death rate (Cooper et al. 2010). In addition to human healthcare, increasing the quantity and quality of skeletal muscle is also an important issue from the perspective of meat production.

Recent studies revealed the molecular roles of various types of lncRNAs including anti-sense lncRNAs, promoter-associated lncRNAs, enhancer lncRNAs (eRNAs), and competing endogenous lncRNAs (ceRNAs) in regulating both skeletal muscle differentiation and disorders. In addition, genome-wide studies have identified hundreds of novel lncRNAs in skeletal muscle cells and tissues. In this chapter, we first provide an overview of recent findings on the regulatory roles of lncRNAs in skeletal muscle differentiation, showing their linkages to miRNA functions. Secondly, we discuss the regulatory roles of lncRNAs as transcriptional regulators in skeletal muscle differentiation. The roles of lncRNAs in the signaling pathway that regulates skeletal muscle differentiation is summarized in Fig. 1. The molecular functions and binding partners of lncRNAs during myogenic differentiation are shown in Table 1. Thirdly, we summarize the findings of lncRNA studies in muscular disorders and discuss their potential applications for treating human muscular diseases. To this end, we discuss the current technical issues that need to be resolved before studying lncRNAs in skeletal muscle in the future.

2 Regulatory Roles of ceRNAs in Skeletal Muscle Differentiation

2.1 *CeRNAs for* miR-133 *and* miR-135

Muscle-specific miRNAs, *miR-206* and *miR-133b*, are required for the cell cycle arrest of the murine myoblast cell line C2C12 during myogenesis (Kim et al. 2006; Feng et al. 2013). Both miRNAs are located at the same genomic locus and the transcript expressed from this locus was originally identified as synapse-associated noncoding RNA 7H4 (Velleca et al. 1994). A noncoding transcript composed of 3 exons and 2 introns, which was termed long noncoding MD1 (*linc-MD1*), has been identified from the genomic locus (Cesana et al. 2011). Mouse *linc-MD1* contains *pre-miR-206* and *pre-miR-133b* in the second intron and third exon, respectively, and its expression is regulated by the MyoD protein. Later, a homologous transcript of mouse *linc-MD1* has been identified in humans (Twayana et al. 2013) (Table 1). *Linc-MD1* expression increases during the early stages of myogenic differentiation and mainly accumulates in the cytoplasm. Although *linc-MD1* is cleaved to become a source of *miR-133b*, the Human antigen R (HuR) protein favors *linc-MD1* accumulation in the cytoplasm by repressing Drosha cleavage (Legnini et al. 2014). In addition, the disease progression of Duchenne muscular dystrophy, in which

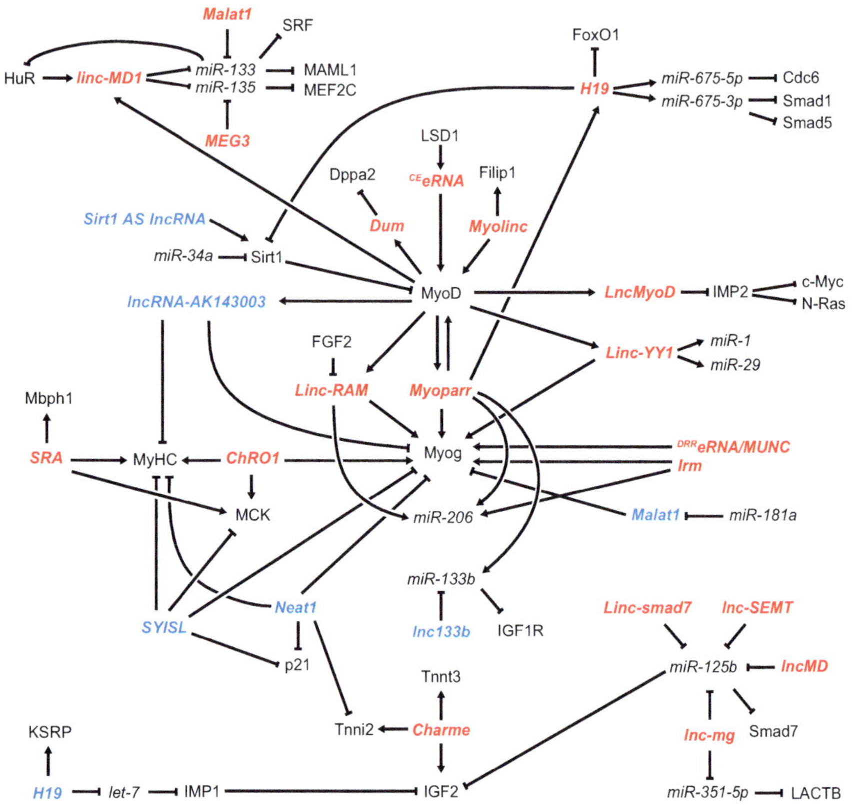

Fig. 1 Schematic representation of the lncRNA network involved in skeletal muscle differentiation

The diagram shows the network of lncRNAs and their target genes. LncRNAs promoting myogenic differentiation are represented in red. LncRNAs inhibiting myogenic differentiation are represented in blue. Owing to conflicting reports on skeletal muscle differentiation, *H19* and *Malat1* are described separately for each function

terminal differentiation of myoblasts is reduced, correlates with the expression level of *linc-MD1* (Cesana et al. 2011). During C2C12 differentiation, ectopic expression of *linc-MD1* increased the expression levels of the myogenic marker proteins, Myog and myosin heavy chain (MyHC). In contrast, RNA interference (RNAi) of *linc-MD1* decreased Myog and MyHC expression, indicating that *linc-MD1* is required for the proper myogenic differentiation (Cesana et al. 2011). Using bioinformatics analysis, it has been revealed that *linc-MD1* acts as a molecular sponge that inhibits the activities of *miR-133* and *miR-135* (Cesana et al. 2011). *miR-133* and *miR-135* target mastermind-like transcriptional coactivator 1 (MAML1) and myocyte enhancer factor 2C (MEF2C), which are transcription factors required for muscle-specific gene expression. *miR-133* also enhances myoblast proliferation by repressing the expression of serum response factor (SRF), which is a transcription

Table 1 The molecular functions and binding partners of lncRNAs important for skeletal muscle differentiation

Name	Species	Localization	Binding proteins	Binding RNAs	Function	Differentiation	Reference	PMID
m1/2-sbsRNA	Mouse	Cytoplasmic	–	SINE-containing mRNAs	mRNA decay	Inhibit	Wang et al. (2013)	23558772
^{CE}*eRNA*	Mouse	–	–	–	Transcriptional regulation	Promote	Mousavi et al. (2013)	23993744
Charme	Mouse, human	Nuclear and cytoplasmic	–	–	Transcriptional regulation	Promote	Ballarino et al. (2018)	30177572
ChRO1	Mouse	Nuclear	Histone3.3, DAXX, ATRX	–	Transcriptional regulation	Promote	Park et al. (2018)	30335163
^{DRR}*eRNA/ MUNC*	Mouse, human	Nuclear	NIPBL, SMC3, PDS5B	–	Transcriptional regulation	Promote	Mousavi et al. (2013)	23993744
							Mueller et al. (2015)	25403490
							Tsai et al. (2018)	29979962
							Cichewicz et al. (2018)	30037979
DUM	Mouse	Nuclear and cytoplasmic	Dnmt1, Dnmt3a, Dnmt3b	–	DNA methylation	Promote	Wang et al. (2015)	25686699
H19	Mouse	Cytoplasmic	KSRP	*let-7*	Sponge for miRNA, mRNA decay	Inhibit	Kallen et al. (2013)	24055342
							Giovarelli et al. (2014)	25385579
H19	Mouse, human	–	–	–	Source of *miR-675-5p* and *miR-675-3p*	Promote	Dey et al. (2014)	24532688
Irm	Mouse	Nuclear	MEF2D	–	Transcriptional regulation	Promote	Sui et al. (2019)	30792383

(continued)

Table 1 (continued)

Name	Species	Localization	Binding proteins	Binding RNAs	Function	Differentiation	Reference	PMID
linc-MD1	Mouse, human	Cytoplasmic	–	*miR-133*, *miR-135*	Sponge for miRNAs	Promote	Cesana et al. (2011)	22000014
							Twayana et al. (2013)	23863142
Linc-RAM	Mouse	Nuclear and cytoplasmic	MyoD	–	Transcriptional regulation	Promote	Yu et al. (2017)	28091529
Linc-smad7	Mouse	Nuclear and cytoplasmic	–	*miR-125b*	Sponge for miRNA	Promote	Song et al. (2018)	29912619
Linc-YY1	Mouse, human	Nuclear	YY1, Ezh2, Eed	–	Transcriptional regulation	Promote	Zhou et al. (2015)	26658965
lnc-31	Mouse	Cytoplasmic	YB-1	*Rock1* mRNA	mRNA translation	Inhibit	Dimartino et al. (2018)	29669280
lnc-SEMT	Sheep	–	–	*miR-125b*	Sponge for miRNA	Promote	Wei et al. (2018)	30153668
lnc133b	Bovine	Nuclear and cytoplasmic	–	*miR-133b*	Sponge for miRNA	Inhibit	Jin et al. (2017)	28757453
lncIRS1	Chicken	Nuclear and cytoplasmic	–	*miR-15a*, *miR-15b-5p*, *miR-15c-5p*	Sponge for miRNAs	Promote	Li et al. (2019)	30701698
lnc-mg	Mouse	Nuclear and cytoplasmic	–	*miR-125b*, *miR-351-5p*	Sponge for miRNAs	Promote	Zhu et al. (2017)	28281528
							Du et al. (2019)	30216112
lncMD	Bovine	Nuclear	–	*miR-125b*	Sponge for miRNA	Promote	Sun et al. (2016)	27589905
lncMUMA	Mouse	–	–	*miR-762*	Sponge for miRNA	Promote	Zhang et al. (2018b)	30415659
LncMyoD	Mouse	Nuclear	IMP2	-	mRNA translation	Promote	Gong et al. (2015)	26143994

lncRNA AK017368	Mouse	Nuclear and cytoplasmic	–	*miR-30c*	Sponge for miRNA	Inhibit	Liang et al. (2018)	28904016
lncRNA-AK143003	Mouse	Nuclear and cytoplasmic	–	–	Transcriptional regulation	Inhibit	Guo et al. (2017)	28588232
lncRNA-Six1	Chicken	–	–	*miR-1611*	Sponge for miRNA	Promote	Ma et al. (2018)	30518151
Malat1	Mouse	Nuclear and cytoplasmic	–	*miR-133*	Sponge for miRNA	Promote	Han et al. (2015)	25868726
Malat1	Mouse	Nuclear	Suv39h1	–	Transcriptional regulation	Inhibit	Chen et al. (2017)	28326190
MAR1	Mouse	–	–	*miR-487b*	Sponge for miRNA	Promote	Zhang et al. (2018a)	29512357
MDNCR	Bovine	–	–	*miR-133a*	Sponge for miRNA	Promote	Li et al. (2018)	30195797
MEG3	Bovine	Cytoplasmic	–	*miR-135*	Sponge for miRNA	Promote	Liu et al. (2019)	30887511
Myolinc	Mouse	Nuclear	TDP-43	–	Transcriptional regulation	Promote	Militello et al. (2018)	29618024
Myoparr	Mouse	Nuclear	Ddx17(p72)	–	Transcriptional regulation	Promote	Hitachi et al. (2019a)	30622218
Neat1	Mouse	Nuclear	Ezh2	–	Transcriptional regulation	Inhibit	Wang et al. (2019)	31243262
Sirt1 AS lncRNA	Mouse	Nuclear and cytoplasmic	–	*Sirt*1 mRNA	mRNA translation	Inhibit	Wang et al. (2016)	26902620
SRA	Mouse	–	Ddx5(p68), MyoD	–	Transcriptional regulation	Promote	Caretti et al. (2006)	17011493
SYISL	Mouse	Nuclear and cytoplasmic	Ezh2	–	Transcriptional regulation	Inhibit	Jin et al. (2018)	30279181
Yam-1	Mouse	Nuclear and cytoplasmic	–	–	Transcriptional regulation	Inhibit	Lu et al. (2013)	23942234

factor that controls skeletal muscle proliferation and differentiation. Intriguingly, *linc-MD1* increases the expression of HuR, which is repressed by *miR-133*, indicating the presence of a feedforward regulatory loop between *linc-MD1* and HuR (Legnini et al. 2014). Thus, *linc-MD1* promotes myogenic differentiation by increasing the expression levels of *miR-133*- and *miR-135*-target genes (Fig. 1). This finding was the first in a series of reports demonstrating how lncRNAs act as ceRNAs for miRNAs to control muscle differentiation.

In addition to *linc-MD1*, the lncRNA maternally expressed 3 (*MEG3*, also known as *Gtl2*), which is located in the imprinted genomic region between the delta-like non-canonical Notch ligand 1 (*Dlk1*) and deiodinase, iodothyronine type III (*Dio3*) genes, acts as a ceRNA for *miR-135*. Although maternal deletion of *MEG3* resulted in perinatal death with delayed skeletal muscle-fiber development in mice (Zhou et al. 2010), the role of *MEG3* in skeletal muscle development was not fully elucidated. During myogenic differentiation of primary bovine myoblasts, *MEG3* expression gradually increased and accumulated in the cytoplasm. The expression levels of the myogenic marker genes, *Myog* and *MyHC*, increased significantly after *MEG3* overexpression and decreased markedly after *MEG3* knockdown in primary bovine myoblasts. In contrast, the gain or loss of *MEG3* function did not affect myoblast proliferation and apoptosis (Liu et al. 2019). Using dry and wet lab analyses, it has also been found that *MEG3* functioned as a ceRNA for *miR-135*, which targets MEF2C (Table 1). Thus, *MEG3* is required for bovine skeletal muscle differentiation because it acts as a ceRNA for *miR-135* (Fig. 1).

Multiple ceRNAs for *miR-133* have also been reported. Metastasis-associated lung adenocarcinoma transcript 1 (*Malat1*), which was identified as an upregulated lncRNA in metastatic carcinoma cells, influences many aspects of cancer pathogenesis. *Malat1* is expressed in various tissues, including skeletal muscle. It has been found that knocking down *Malat1* decreased *Myog* expression in C2C12 cells (Watts et al. 2013). Later, Han et al. (2015) rediscovered that silencing *Malat1* retarded myogenic differentiation, as indicated by reduced expression of MyHC and Myog protein. In addition, *Malat1* knockdown also drastically reduced SRF expression at both the mRNA and protein levels. Intriguingly, *Malat1* and SRF mutually coregulate the expression of each other in cytoplasm during myogenesis. Because the effect of *Malat1* knockdown on SRF was only observed in the presence of *miR-133*, which targets SRF, Han et al. (2015) concluded that *Malat1* controls SRF expression by acting as a ceRNA for *miR-133* (Table 1). Thus, the competing activity of *Malat1* for *miR-133* promoted myogenic differentiation (Fig. 1). However, as described below, *Malat1* can also negatively regulate skeletal muscle differentiation, independently of *miR-133*.

Moreover, lncRNAs functioning as ceRNAs for *miR-133a* and *miR-133b* have also been found in bovine myoblasts. From the longissimus muscles of fetal and adult cattle, 13580 lncRNAs have been found by high-throughput sequencing (Li et al. 2018). Among them, expression of the muscle differentiation-associated lncRNA (*MDNCR*) was higher in fetuses than in adults. Vector-based overexpression of *MDNCR* in bovine primary myoblasts indicated that *MDNCR* inhibits myoblast proliferation and promotes myotube formation, as shown by increased MyHC expression. *MDNCR* has 32 putative *miR-133a*-binding sites, and direct binding between *MDNCR* and *miR-133a* was shown by RNA immunoprecipitation (RIP) and pull-down assays (Table 1). It has also been found that binding of *MDNCR* to *miR-133a* relieved the miRNA-inhibiting effect on GosB, a member of the activator protein 1 family of transcription factors. Thus, *MDNCR* promotes bovine myoblast differentiation by competing with *miR-133a*.

In bovine skeletal muscle tissues, a noncoding RNA, *lnc133b*, with sequences completely complementary to *miR-133b*, was identified through high-throughput sequencing data (Jin et al. 2017). Although *lnc133b* expression is widely detected in skeletal and cardiac muscle, fat, liver, spleen, lung, kidney, rumen, reticulum, omasum, and abomasum tissues in cattle, its expression gradually increases during bovine myogenesis. Curiously, both *lnc133b* overexpression and knockdown promoted myoblast proliferation and inhibited the terminal differentiation of myoblasts. Further analyses revealed that *lnc133b* positively regulates the expression of insulin-like growth factor 1 receptor (IGF1R), a target of *miR-133*, by directly binding to *miR-133b* (Table 1). Thus, the *lnc133b*–*miR-133b*–IGF1R axis regulates bovine myoblast differentiation (Fig. 1).

2.2 *CeRNAs for* miR-125b

Another miRNA that serves as a hub in the lncRNA network is *miR-125b*. During myogenesis, *miR-125b* expression gradually declines and *miR-125b* negatively regulates myogenic differentiation both in vitro and in vivo by targeting insulin-like growth factor 2 (IGF2) (Ge et al. 2011), which initiates myoblast differentiation. To date, four lncRNAs have been reported to act as ceRNAs for *miR-125b* (Sun et al. 2016; Zhu et al. 2017; Song et al. 2018; Wei et al. 2018).

Four hundred and one lncRNAs that were differentially expressed among embryonic, neonatal, and adult bovine skeletal muscles have been identified by high-throughput sequencing (Sun et al. 2016). One of them, lncRNA related to muscle differentiation (*lncMD*), whose expression showed gradual upregulation during bovine myoblast differentiation, promotes bovine myoblast differentiation (Table 1). Mechanistically, *lncMD* maintains the *IGF2* mRNA level by inhibiting *miR-125b* activity as a ceRNA (Fig. 1).

Myogenesis-associated lncRNA (*lnc-mg*) was identified in mouse skeletal muscle and was shown to promote myoblast differentiation by functioning as a ceRNA for *miR-125b* (Zhu et al. 2017) (Table 1). During C2C12 differentiation, *lnc-mg* also inhibits *miR-351-5p* activity and maintains the expression of lactamase β (LACTB), which enhances C2C12 myoblast differentiation (Du et al. 2019) (Fig. 1). Although the *lnc-mg*-knockout mice showed a decreased myofiber size, the role of *lnc-mg* in in vivo skeletal muscle formation during embryogenesis remains unknown.

In addition to cattle and mice, a lncRNA acting as an *miR-125b* ceRNA has been found in sheep skeletal muscle. Sheep skeletal enhanced muscularity transcript (*lnc-SEMT*) that highly and specifically expressed in skeletal muscle tissues in sheep has been identified by high-throughput sequencing (Wei et al. 2018). The expression of *lnc-SEMT* was upregulated during myoblast differentiation. Knockdown of *lnc-SEMT* with a small interfering RNA (siRNA) in sheep myoblasts decreased the expression levels of *MyoD1* and *Myog*. In contrast, *lnc-SEMT* overexpression increased *MyoD1* and *Myog* expression in sheep myoblasts. Thus, *lnc-SEMT* accelerates sheep myoblast differentiation (Table 1). Subsequent bioinformatics analysis revealed that *lnc-SEMT* regulates IGF2 expression by binding to *miR-125b* to function as a ceRNA (Fig. 1). Interestingly, sheep genetically modified to express *lnc-SEMT* as a transgene showed increased body weight and muscle mass. Future applications of *lnc-SEMT* for increasing skeletal muscle mass and meat production in livestock animals are expected.

The fourth lncRNA that acts as a ceRNA for *miR-125b* is *Linc-smad7*, which is located near the SMAD family member 7 (*Smad7*) gene and is enriched in myotubes (Song et al. 2018). The sequence of *Linc-smad7* differs from that of *lncRNA-Smad7*, which showed an anti-apoptotic effect upon stimulation by transforming growth factor (TGF)-β in the murine breast cancer cell line JygMC(A) (Arase et al. 2014). *Linc-smad7* expression was higher in skeletal muscle than in other tissues and gradually increased during the development of mouse hind limb muscles. In vitro studies showed that *Linc-smad7* overexpression inhibited DNA replication and induced myoblast differentiation. Of note, *Linc-smad7* overexpression promoted skeletal muscle regeneration in mice after snake venom, cardiotoxin, injection. Although *Linc-smad7* is present in both the nucleus and cytoplasm, *Linc-smad7* binds to *miR-125b* and inhibits its activity against IGF2 and Smad7 (Table 1), both of which are fundamental regulators of differentiation in skeletal muscle cells. Thus, *Linc-smad7* promotes myoblast differentiation (Fig. 1). Taken together, these studies indicate that the molecular network of lncRNAs centered on *miR-125b* is an important regulator of skeletal muscle differentiation.

2.3 CeRNAs for Other miRNAs

In addition to the lncRNAs described above, several other lncRNAs have been reported to act as ceRNAs for miRNAs during skeletal muscle differentiation. Sirtuin 1 (Sirt1) is an NAD-dependent deacetylase that inhibits myogenic differentiation by deacetylating MyoD. Sirt1 antisense long noncoding RNA (*Sirt1 AS lncRNA*) was identified in C2C12 cells by sequencing and bioinformatic analysis (Wang et al. 2014). Interestingly, treatment of actinomycin D, a widely used transcription inhibitor, showed that the half-life of *Sirt1 AS lncRNA* (10 h) was longer than that of *Sirt1* mRNA (6 h). Although Sirt1 protein expression was decreased by *miR-34a*, *Sirt1 AS lncRNA* preferably interacted with *Sirt1* mRNA to promote Sirt1 translation (Wang et al. 2016) (Table 1). *Sirt1 AS lncRNA* and *Sirt1* mRNA expression both decreased during C2C12 differentiation; therefore, *Sirt1 AS lncRNA* could inhibit myogenic differentiation by competing with *miR-34a* (Fig. 1).

In addition, *lncRNA-Six1* promotes chicken myogenesis by acting as a ceRNA for *miR-1611*, which targets sine oculis-related homeobox 1 (Six1) (Ma et al. 2018) (Table 1). The lncRNA muscle anabolic regulator 1 (*MAR1*) also promotes C2C12 differentiation by acting as a ceRNA for *miR-487b*, which targets the wingless-type MMTV integration site family, member 5A (Wnt5a) (Zhang et al. 2018a) (Table 1). The lncRNA *AK017368* promotes myoblast proliferation by attenuating the function of *miR-30c*, which targets trinucleotide repeat containing 6a (Tnrc6a), and thus suppresses C2C12 differentiation (Liang et al. 2018) (Table 1). In chickens, *lncIRS1* promotes myogenic differentiation and induces skeletal muscle hypertrophy by acting as a ceRNA for *miR-15a*, *miR-15b-5p*, and *miR-15c-5p*, which target insulin receptor substrate 1 (IRS1) (Li et al. 2019) (Table 1). Moreover, several circular RNAs (circRNAs), which are generated through back-splicing of pre-mRNAs, are involved in skeletal muscle differentiation.

2.4 H19 *lncRNA*

Among the lncRNAs whose functions are associated with miRNAs, the role of *H19* in skeletal muscle differentiation is controversial. *H19*, a lncRNA that is ubiquitously expressed in almost every tissue, is expressed from the maternally inherited allele in mammals. It has been reported that *H19* harbors binding sites for the *let-7* family of miRNAs (Kallen et al. 2013), which play important roles in development, aging, and metabolism. Gain- and loss-of-function experiments and subsequent genome-wide transcriptome analysis revealed that *H19* modulates the activity of *let-7* as a ceRNA. Although *H19* expression was strongly induced in C2C12 cells after myogenic induction, *H19* knockdown promoted myogenic differentiation as indicated by increased Myog and MyHC protein expression. Intriguingly, overexpression of *let-7* phenocopied *H19* down-modulation by downregulating an IGF2-mRNA-binding protein 1 (IMP1), which binds to *IGF2* mRNA and decreases its expression. Thus, *H19* inhibited myogenic differentiation by acting as a ceRNA for *let-7* (Fig. 1). Although 50% of satellite cells, which are adult skeletal muscle stem cells, were lost in *H19*-knockout mice, the mice showed better regeneration potential with enhanced insulin-like growth factor 1 (*IGF1*), *IGF2*, *IGF1R*, and insulin-like growth factor 2 receptor (*IGF2R*) expression (Martinet et al. 2016). It has also been found that *H19* maintained C2C12 myoblasts in an undifferentiated state by supporting RNA-binding protein K homology-type splicing regulatory protein (KSRP)-mediated decay of *Myog* mRNA (Giovarelli et al. 2014) (Table 1). These studies indicated the inhibitory role of *H19* in myogenic differentiation (Fig. 1).

However, conflicting results have been reported regarding the role of *H19* (Dey et al. 2014). Using human myoblasts with knocked down *H19* expression and *H19*-knockout murine myoblasts, it has been found that *H19* promoted myogenic differentiation; upregulated the expression of myogenic differentiation-specific genes including *MyoD1*, *Myog*, *MyHC*, *MEF2C*, and *IGF2*; and downregulated the expression of cell-cycle related genes such as Cyclin D1, D2, E1, E2 (*Ccnd1*, *Ccnd2*, *Ccne1*, and *Ccne2*), breast cancer 1, early onset (*Brca1*), minichromosome maintenance complex component 4, 6, 7 (*Mcm4*, *Mcm6*, and *Mcm7*), and cell division cycle 6 (*Cdc6*). Indeed, unlike the results of Martinet et al. (2016), *H19*-mutant mice exhibited defects in skeletal muscle regeneration after injection of cardiotoxin. Dey et al. (2014) also showed that this role of *H19* was mediated by *miR-675-3p* and *miR-675-5p* (Fig. 1), which are encoded in the first exon of *H19* and target SMAD family member 1 and 5 (Smad1 and Smad5) and Cdc6. These results indicate that *H19* promoted skeletal muscle differentiation as a source of functional miRNAs (Table 1). In accordance with this finding, it has been reported that *H19* promoted bovine myoblast differentiation by suppressing Sirt1 and Fork head box protein O1 (FoxO1) (Xu et al. 2017), both of which inhibit myogenic differentiation. Therefore, although it is possible that the role of *H19* in myogenesis could be cell context-dependent, more detailed investigations are required to reveal whether *H19* promotes or inhibits the myogenic differentiation of myoblasts.

3 Regulatory Roles of lncRNAs in Transcriptional Regulation During Myogenic Differentiation

3.1 LncRNAs Binding to the RNA helicases Ddx5 and Ddx17

Steroid receptor RNA activator (*SRA*) was first identified as a coactivator of steroid hormone receptors. Later, MyoD-binding proteins were screened from C2C12 cells by tandem affinity purification and two RNA helicases, DEAD-box protein 5 and 17 (Ddx5 and Ddx17, also known as p68 and p72, respectively), were identified as authentic MyoD-associated proteins (Caretti et al. 2006). Ddx5 and Ddx17 were shown to be required for C2C12 differentiation because they enhance the transcriptional transactivation activity of the MyoD protein. Intriguingly, RIP assays showed that *SRA* associated with endogenous MyoD protein in skeletal muscle cells, probably via the Ddx5 and Ddx17 proteins, which interact with *SRA* (Table 1). Thus, the functional relevance of *SRA* to skeletal muscle gene transcription was examined and *SRA* was shown to be required for the transcriptional activity of the MyoD protein at the muscle creatine kinase (*MCK*) promoter. Indeed, *SRA* knockdown inhibited C2C12 differentiation as shown by the reduced expression of *MyHC*, *MCK*, and myosin-binding protein 1 (*Mbph1*) mRNA (Fig. 1). Therefore, *SRA* was identified as the first lncRNA capable of regulating skeletal muscle differentiation at the transcriptional level.

The *Myog* gene, which encodes a bHLH-type transcription factor, is indispensable for skeletal muscle development; *Myog*-deficient mice showed severe skeletal muscle defects that resulted in perinatal lethality (Faralli and Dilworth 2012). *Myog* expression is restricted to skeletal muscle tissues and increases during myogenesis. MyoD and several other transcription factors can regulate the spatiotemporal expression of *Myog* in skeletal muscle through the approximately 200-base pair promoter region of the *Myog* gene (Faralli and Dilworth 2012). Although the functions of any associated DNA elements have not been elucidated, transgenic studies indicated that a region of up to 1.5 kilobase (kb) upstream of the *Myog* gene was necessary to recapitulate high-level *Myog* expression in skeletal muscles (Faralli and Dilworth 2012). Our group recently identified a *myogenin* promoter-associated myogenic regulatory anti-sense lncRNA (*Myoparr*), which was expressed from the upstream region of the *Myog* gene in the opposite direction and was revealed to serve a pivotal role in skeletal muscle differentiation (Hitachi et al. 2019a). *Myoparr*, which is a 1172-nt, single-exon lncRNA, mainly localizes in the nuclei of C2C12 cells. We also identified human *Myoparr* counterparts, one of which is spliced and the other is unspliced. During skeletal muscle differentiation, *Myoparr* and *Myog* expression gradually increase and are mutually correlated. Searching for the *Myoparr* promoter revealed that *Myoparr* shared the same promoter region with *Myog*. Through this region, *Myoparr* expression was regulated by both the MyoD protein and TGF-β signaling. Surprisingly, knocking down *Myoparr* expression with siRNAs and anti-sense oligonucleotides (ASOs) decreased *Myog* transcription and drastically inhibited C2C12 differentiation. Thus, we examined the molecular mechanisms involved in *Myoparr*-mediated transcriptional activation of *Myog*.

Chromatin isolation by RNA purification (ChIRP) and chromatin immunoprecipitation (ChIP) analyses revealed that *Myoparr* directly bound the *Myog* promoter and that *Myoparr* knockdown significantly decreased RNA polymerase II (Pol II), histone H3 lysine 4 trimethylation (H3K4me3), and histone H3 lysine 27 acetylation (H3K27ac) levels at the *Myog* locus. Intriguingly, promoter-associated lncRNAs altered the DNA-methylation status of neighboring genes (Hamazaki et al. 2015), whereas *Myoparr* knockdown caused little effect on the methylation of the *Myog* gene. Of note, *Myoparr* knockdown decreased *MyoD1* expression at the transcription level. LncRNAs can function by binding to multiple proteins and altering their functions (Ransohoff et al. 2018). Therefore, it was necessary to identify *Myoparr*-binding proteins to elucidate the molecular mechanism by which *Myoparr* activates *Myog* expression during skeletal muscle differentiation. We identified an RNA helicase, Ddx17, as a *Myoparr*-specific binding protein from differentiating C2C12 cells by the RiboTrap method and differential proteomics analysis (Table 1). Subsequent promoter assays showed that deletion of the genomic region (corresponding to the 341-nt binding region of *Myoparr* for Ddx17) and *Myoparr* knockdown largely decreased the promoter activity of *Myog*. Mechanistically, *Myoparr* promotes the protein–protein interaction between Ddx17 and histone lysine acetyltransferase 2B (PCAF), which is required for the high transcriptional activity of Ddx17 (Shin and Janknecht 2007). Intriguingly, *Myoparr* was dispensable for Ddx17 and PCAF binding to the *Myog* promoter. Thus, *Myoparr* activates *Myog* transcription, presumably by promoting the interaction between Ddx17 and PCAF on the *Myog* promoter. Furthermore, using high-throughput sequencing analysis, we revealed that *Myoparr* and Ddx17 promote myoblast cell cycle withdrawal by activating *pri-miR-206*, *pri-miR-133b*, and *H19* expression because these noncoding RNAs facilitate myoblast cell cycle withdrawal (Kim et al. 2006; Feng et al. 2013; Dey et al. 2014). Although exogenous *Myog* expression appears to suffice for inducing cell cycle exit in myoblasts (Liu et al. 2012), our experiments revealed that *Myog* gene expression did not affect 5-ethynyl-2′-deoxyuridine (EdU) uptake by differentiating myoblasts, the expression of cell cycle-related genes except for RB transcriptional corepressor 1 (*Rb1*), or the expression of *pri-miR-206*, *pri-miR-133b*, or *H19* (Hitachi et al. 2019a). Thus, *Myoparr* and Ddx17 regulate myoblast cell cycle withdrawal in a *Myog*-independent manner. Taken together, these findings indicate that *Myoparr* specifies the differentiation lineages of myoblasts by stimulating *Myog* expression and promotes myoblast cell cycle withdrawal by activating the expression of myogenic miRNAs during skeletal muscle differentiation (Fig. 1). However, the molecular mechanism whereby relatively few copies of *Myoparr* (approximately 5 copies expressed per cell) can act *in trans* to control the expression of *pri-miR-206*, *pri-miR-133b*, and *H19* located at distal loci remains unclear. Furthermore, we have identified *myogenin*-associated sense lncRNAs that are expressed in the sense orientation at the *Myog* locus. However, the molecular functions of these lncRNAs in skeletal muscle differentiation remain unknown.

3.2 LncRNAs Expressed from the Cis-Regulatory Region of MyoD1

A region upstream of *MyoD1* (up to approximately 24 kb in size) is required for spatiotemporal *MyoD1* expression in the skeletal muscles during embryogenesis (Goldhamer et al. 1992). This region contains a core enhancer (CE), which is located approximately 20 kb upstream of the *MyoD1* gene, and the distal regulatory region (DRR), which is found approximately 5 kb upstream of *MyoD1* gene. Both regions are evolutionarily conserved in mammals and are required for the specification of *MyoD1* expression in the muscle lineage. Two potentially polyadenylated lncRNAs, which are expressed from the enhancer regions of *MyoD1*, were identified using ChIP-sequencing (ChIP-Seq) data related to the genome-wide occupancies of MyoD and Myog in C2C12 cells (Mousavi et al. 2013). These lncRNAs were designated as ^{CE}*eRNA* and ^{DRR}*eRNA* based on the name of the associated enhancer region (Table 1). The expression of ^{CE}*eRNA* is under the direct control of lysine-specific demethylase 1 (LSD1) (Scionti et al. 2017). Knocking down ^{CE}*eRNA* expression severely decreased *MyoD1* transcription by reducing the Pol II occupancy and DNase I accessibility at the proximal upstream regions of *MyoD1*, indicating that ^{CE}*eRNA* is required for *MyoD1* expression during myogenic differentiation (Fig. 1). However, the detailed molecular mechanism remains unexplored.

The molecular function of ^{DRR}*eRNA* has been analyzed in more detail than that of ^{CE}*eRNA*. ^{DRR}*eRNA* is potentially polyadenylated, spliced, and mainly localized in the nucleus (Tsai et al. 2018). Conforming to the classification of a unidirectional eRNA, ^{DRR}*eRNA* was unidirectionally transcribed. ^{DRR}*eRNA* expression increases during myogenic differentiation (both in C2C12 cells and murine primary myoblasts) and is regulated by MyoD, as ^{DRR}*eRNA* expression decreased drastically after *MyoD1* knockdown. Intriguingly, functional analysis of ^{DRR}*eRNA* revealed that ^{DRR}*eRNA* knockdown did not significantly affect MyoD expression, at either the transcript or protein level. These results are consistent with a previous report demonstrating that the DRR region is a classical enhancer that is dispensable for endogenous *MyoD1* expression during myogenic differentiation in embryos, but is required in mature adult skeletal muscles (Chen et al. 2002). Instead, the siRNA-mediated knockdown of ^{DRR}*eRNA* drastically decreased *Myog* mRNA expression and inhibited myogenic differentiation. Transcriptional repression of ^{DRR}*eRNA* using the clustered regularly interspaced short palindromic repeat (CRISPR) system, dCas9-KRAB, and DRR-sgRNAs also decreased *Myog* expression without affecting *MyoD1* expression (Tsai et al. 2018). Retrovirus-mediated overexpression of ^{DRR}*eRNA* in C2C12 myoblasts showed that 1.2 and 2.0 kb fragments of ^{DRR}*eRNA*, which localized to the nucleus, activated *Myog* expression without affecting *MyoD1* expression (Mousavi et al. 2013; Tsai et al. 2018), indicating the 5′ half of ^{DRR}*eRNA* was necessary for *Myog* expression. It has also been found that ^{DRR}*eRNA* knockdown reduced Pol II and MyoD occupancy and DNase I accessibility at the promoter region of the *Myog* gene. Moreover, two independent approaches, namely ChIRP-sequencing (ChIRP-Seq) and single-molecule fluorescence in situ hybridization (FISH), revealed ^{DRR}*eRNA* localization at the *Myog* locus. Mechanistically, ^{DRR}*eRNA* binds to the cohesin complex, including NIPBL cohesin loading factor (NIPBL), structural maintenance of chromosomes 3 (SMC3), and PDS5 cohesin-associated factor B (PDS5B), which remodels chromatin, and recruits it to the *Myog* locus (Table 1). Thus, ^{DRR}*eRNA* activates *Myog* gene expression *in trans* by modulating the chromatin conformation at the *Myog* locus (Fig. 1).

Later, MyoD upstream noncoding RNA (*MUNC*), whose expression is upregulated during C2C12 differentiation, has been independently identified from the region corresponding to ^{DRR}*eRNA* by high-throughput sequencing (Mueller et al. 2015). *MUNC* has a 5′ cap structure but is not polyadenylated, both of which are features of unstable eRNAs. *MUNC* exists in both a spliced and an unspliced form. Very low levels of *MUNC* transcripts were present in the polyribosome-containing fraction and *MUNC* was predicted to have low coding potential; therefore, *MUNC* is a noncoding RNA. Knocking down *MUNC* decreased *Myog* and *MyHC* expression and inhibited myogenic differentiation of C2C12 cells, as shown by a drastically reduced fusion index, which is the degree of myotube formation by MyHC-positive multinucleated cells (Fig. 1). Surprisingly, *MUNC* knockdown affected *MyoD1* expression at the transcriptional level. In addition, primary myoblasts isolated from the mice whose *MUNC* locus was deleted (Chen et al. 2002) showed approximately 50% less *MyoD1* and *Myog* expression. CRISPR/Cas9-mediated deletion of *MUNC* in C2C12 cells showed a complete loss of *MyoD1* expression and resulted in 80% decreases in *Myog* and *MyHC* expression. Moreover, MyoD binding decreased both at the DRR and the *Myog* promoter following *MUNC* knockdown. Of note, although overexpression of both spliced and unspliced *MUNC* resulted in 20-fold greater *MyoD1* mRNA expression and 15-fold greater *Myog* mRNA expression, *MUNC* overexpression did not increase the MyoD and Myog protein expression levels. *MUNC* acts in a MyoD-independent manner considering that *MUNC* overexpression induced *Myog* and *MyHC* expression in *MyoD1*-depleted C2C12 cells (Cichewicz et al. 2018). To complement these results, knocking down a human counterpart of *MUNC*, whose expression was upregulated during myogenic differentiation, decreased the expression levels of *Myod1*, *Myog*, and *MyHC* in an immortalized human myoblast cell line (LHCN myoblasts) (Table 1). Furthermore, the role of *MUNC* in adult skeletal muscle regeneration was also examined using *MUNC* siRNA in the tibialis anterior (TA) muscles of mice. Fifteen days after cardiotoxin injection, *MUNC* knockdown impaired skeletal muscle regeneration, as shown by both a decreased myofiber diameter and decreased myogenic marker-expression levels. Thus, *MUNC* activates *MyoD1* expression *in cis* and regulates *Myog* expression *in trans* in skeletal muscle cells.

When comparing the functions of $^{DRR}eRNA$ and *MUNC*, it is clear that the lncRNAs expressed from the DRR of *MyoD1* activate *Myog* expression *in trans*. However, the biggest difference between $^{DRR}eRNA$ and *MUNC* is the effect on *MyoD1* expression. Inhibition of $^{DRR}eRNA$ did not affect *MyoD1* expression in C2C12 cells (Mousavi et al. 2013; Tsai et al. 2018). In contrast, inhibition of *MUNC* largely decreased *MyoD1* expression in C2C12 cells, murine primary myoblasts, and human LHCN myoblasts (Mueller et al. 2015; Cichewicz et al. 2018). In addition, in vivo knockdown of *MUNC* decreased *MyoD1* expression during skeletal muscle regeneration in mice (Mueller et al. 2015). Because both groups used a variety of inhibition methods, such as knockdown with siRNAs and ASOs, transcriptional repression by CRISPR/dCas9, and genetic knockout, the different results might have been caused by the use of different experimental methods. Thus, further studies, including results from other research groups, are needed to confirm the effects of $^{DRR}eRNA$ and *MUNC* on regulating *MyoD1* expression.

3.3 LncRNAs Acting as Scaffold RNAs for Transcriptional Activators

During myogenic differentiation, several lncRNAs can activate *MyoD1* and *Myog* expression by acting as scaffold RNAs for transcriptional activators. LncRNA *Myolinc* (*AK142388*), which is located approximately 8 kb downstream of the transmembrane protein 30a (*Tmem30a*) gene and approximately 11 kb upstream of the filamin A interacting protein 1 (*Filip1*) gene, is enriched in heart and skeletal muscles. *Myolinc* is predominately localized in the nucleus and its expression levels gradually increased during C2C12 differentiation. *Myolinc* knockdown impaired the formation of multi-nucleated myotubes with decreased *MyoD1* expression in C2C12 cells. *Myolinc* was also required for expression of the neighboring *Filip1* gene, but not for *Tmem30a* expression. Intriguingly, *Filip1* was required for myogenic differentiation, as well as *MyoD1* and *Myog* expression. Moreover, siRNA-mediated in vivo knockdown of *Myolinc* increased the number of small myofibers in TA muscles after cardiotoxin injection. Mechanistically, *Myolinc* interacts with TAR DNA-binding protein 43 (TDP-43), a transcriptional regulator, and recruits it to the promoter regions of *MyoD1* and *Filip1* to activate their expression (Table 1). Thus, *Myolinc* promotes myogenic differentiation by activating *MyoD1* and *Filip1* expression as a scaffold RNA for the TDP-43 protein (Militello et al. 2018) (Fig. 1).

Imprinted RNA near *MEG3/Gtl2* (*Irm*) is an imprinted lncRNA (approximately 2500 nt long) that resides near *MEG3* gene and represents an alternative splicing variant of the RNA imprinted and accumulated in nucleus (*Rian*) gene, which is located in the *Dlk1-Dio3* locus. Although *Rian* expression was not affected, the expression of *Irm*, which is localized in the nucleus, gradually increased during skeletal muscle differentiation of C2C12 cells and murine primary myoblasts. *Irm* expression also increased in TA muscles during regeneration after cardiotoxin injection (Sui et al. 2019). *Irm* overexpression resulted in accelerated C2C12 differentiation, as shown by increased expression of both *Myog* and *miR-206*. Conversely, *Irm* knockdown by two independent siRNAs inhibited C2C12 cell differentiation with a decreased fusion index and decreased the promoter activities of *Myog* and *miR-206*. Thus, *Irm* promotes myogenic differentiation by regulating *Myog* and *miR-206* expression (Fig. 1). In addition, knocking down *Irm* impaired muscle regeneration after cardiotoxin injection in mice. RNA-pulldown and RIP assays from C2C12 cell lysates showed that *Irm* directly binds to the myocyte enhancer factor 2D (MEF2D) protein through its last 3′ 590-nt region. Intriguingly, MEF2D knockdown significantly abolished upregulated *Myog* and *miR-206* expression induced by ectopic *Irm* expression. ChIRP and ChIP assays revealed that *Irm* bound the promoter region of *Myog* and *miR-206* and recruited the MEF2D and MyoD proteins to these loci. Thus, *Irm* might promote skeletal muscle differentiation by acting as a scaffold RNA for the MEF2D and MyoD proteins (Table 1).

In addition, a muscle-specific lncRNA, chromatin reorganization 1 (*ChRO1*), which specifically localizes to the chromocenters in C2C12 myotubes, was shown to be required for *Myog*, *MCK*, and *MyHC* expression by recruiting and stabilizing the alpha thalassemia/mental retardation syndrome X-linked (ATRX)–death domain-associated protein (DAXX)–histone 3.3 complex at chromocenters (Park et al. 2018) (Fig. 1).

3.4 LncRNAs Acting as Scaffold RNAs for Transcriptional Repressors

Above, we mentioned that *Malat1* promotes myoblast differentiation by acting as a ceRNA for *miR-133*. However, Chen et al. (2017) demonstrated that *Malat1* suppresses muscle differentiation by regulating *Myog* gene transcription. They used *Malat1*-knockout mice and knocked down *Malat1* in vitro to study the function of *Malat1* in skeletal muscle differentiation. They first examined *Malat1*-expression changes during skeletal muscle differentiation. Interestingly, *Malat1* expression decreased at early time points in differentiating C2C12 cells and murine primary myoblasts. In addition, *Malat1* expression rapidly increased at the early regeneration stage after cardiotoxin injection in TA muscles, whereas it gradually decreased as the satellite cells differentiated. This rapid reduction of *Malat1* expression might be regulated by *miR-181a*, which promotes myogenic differentiation by targeting homeobox A11 (Hox-A11). Gain- and loss-of-function analyses indicated that *Malat1* accelerated C2C12 myoblast proliferation, while it inhibited C2C12 differentiation by regulating the expression of *Myog* and other terminal differentiation markers, suggesting that *Malat1* acts as an anti-myogenic factor. In accordance with previous findings in HeLa cells (Tripathi et al. 2010), *Malat1* was located in the nuclear speckles of C2C12 myoblasts and myotubes (Chen et al. 2017). Functional analyses revealed that *Malat1* inhibited the transcriptional activity of MyoD protein by recruiting suppressor of variegation 3–9 homolog 1 (Suv39h1), a methyltransferase for histone 3 lysine 9, to the MyoD-binding loci. The 3′ region of *Malat1* (nts 5185–6982) was sufficient for binding with Suv39h1. Moreover, the molecular function of *Malat1* in skeletal muscle was examined using *Malat1*-knockout mice. Although *Malat1*-knockout mice showed no obvious developmental phenotype or body-size change, Chen et al. (2017) found that the knockout mice showed accelerated muscle regeneration after cardiotoxin injection. The accelerated muscle regeneration was mediated by the accelerated myogenic potential of satellite cells. Further, *Malat1*-knockout mice in an mdx background, which is a mouse model of Duchenne muscular dystrophy, showed enhanced skeletal muscle regeneration. Thus, *Malat1* inhibited myogenic differentiation and muscle regeneration and modulated the transcriptional transactivation activity of the MyoD protein by acting as a scaffold RNA for Suv39h1. Although further studies are needed to conclude whether *Malat1* positively or negatively regulates skeletal muscle differentiation, considering that *Malat1* is localized to the nucleus and that *Malat1*-deficient mice exhibit enhanced muscle regeneration, *Malat1* likely inhibits skeletal muscle differentiation.

Nuclear-enriched abundant transcript 1 (*Neat1*) is one of several architectural lncRNAs, which can serve as scaffolding molecules for a subset of nuclear bodies. *Neat1* expression is gradually increased during the C2C12 cell differentiation and skeletal muscle regeneration (Wang et al. 2019). Knocking down *Neat1* expression significantly increased p21 protein expression, whereas *Neat1* overexpression decreased p21 protein expression in C2C12 cells. In addition, *Neat1* knockdown significantly reduced myoblast growth, whereas *Neat1* overexpression enhanced myoblast proliferation. Moreover, *Neat1* knockdown increased the expression levels of *Myog*, *MyHC*, fast skeletal muscle troponin I (*Tnni2*), and *Mymk*. Thus, *Neat1* promotes myoblast proliferation by regulating p21 expression, but inhibits myogenic differentiation by regulating *Myog*, *MyHC*, *Tnni2*, and *Mymk* expression (Fig. 1). Intriguingly, in vivo knockdown of *Neat1* led to increased skeletal muscle mass in mice, indicating that *Neat1* prevents muscle growth. Moreover, *Neat1* knockdown impaired skeletal muscle regeneration after cardiotoxin injection, which was possibly a consequence of reduced satellite cell numbers. Wang et al. (2019) showed that *Neat1* physically interacts with enhancer of zeste homologue 2 (Ezh2), a core component of polycomb repressive complex 2, through the core binding region (nts 1001–1540) and recruits it to the promoter region of target genes, including *Myog* and *p21* (Table 1). Thus, the architectural lncRNA *Neat1* regulates skeletal muscle differentiation both in vitro and in vivo as a scaffold RNA for the transcriptional repressor Ezh2.

As with *Neat1*, lncRNA *SYISL/AK004418* (*SYNPO2* intron sense-overlapping lncRNA) also interacts with the Ezh2 protein and recruits it to the *Myog* and *p21* promoters (Jin et al. 2018). *SYISL* is expressed from the fourth intron of the *SYNPO2* gene and is localized both in the nucleus and cytoplasm of C2C12 myoblasts. *SYISL* was shown to be a downstream lncRNA of MyoD; therefore, its expression increased during C2C12 differentiation, as did other target genes of MyoD. Knocking down *SYISL* in C2C12 cells significantly decreased the percentage of S-phase cells, and the opposite effect was observed after *SYISL* overexpression. In addition, p21 protein expression significantly decreased following *SYISL* overexpression and p21 protein expression increased after *SYISL* knockdown. *SYISL* knockdown also promoted the myogenic differentiation of C2C12 cells, as demonstrated by increased Myog and MyHC expression at both the RNA and protein levels. Consequently, *SYISL* overexpression significantly decreased Myog and MyHC expression at both the RNA and protein levels, but did not alter MyoD protein expression. Intriguingly, *SYISL* knockdown and overexpression decreased and increased *Mymk* expression, respectively. Thus, *SYISL* promotes myoblast proliferation and fusion, while it inhibits myogenic differentiation in C2C12 cells by regulating *Myog*, *p21*, *MyHC*, and *Mymk* expression, but not MyoD protein expression (Fig. 1). Using the CRISPR/Cas9 system, Jin et al. (2018) generated *SYISL*-knockout mice, in which the genomic region containing most of *SYISL* was deleted. Satellite cells isolated from the *SYISL*-knockout mice showed impaired proliferation and enhanced myogenic differentiation, as observed in *SYISL*-knockdown experiments performed with C2C12 cells. In contrast, although *SYISL*-knockout mice showed no significant changes in body weights and growth rates, *SYISL* deletion resulted in increased muscle fiber numbers and muscle mass, accelerated muscle regeneration, and decreased fiber sizes. Regarding the opposite effects of *SYISL* in vitro and in vivo, it was proposed that the smaller myofibers observed in *SYISL*-knockout mice may be due to faster regeneration rate than occurs in wild-type mice. Mechanistically, *SYISL* directly interacted with Ezh2 to repress the downstream target genes, such as *Myog*, *MyHC*, *MCK*, and *p21*. Thus, *SYISL* regulates myogenic differentiation by recruiting Ezh2 to the target promoter regions (Table 1).

Linc-YY1, a promoter-associated lncRNA of the Yin Yang 1 (*YY1*) gene (which encodes a transcriptional repressor of multiple genes in myoblasts that recruits Ezh2 to target loci), is a downstream lncRNA of the MyoD protein. In contrast to *Neat1* and *SYISL*, *Linc-YY1* activates *Myog*, *miR-1*, and *miR-29* expression *in trans* by removing YY1, Ezh2, and embryonic ectoderm development (Eed) at their promoter regions by directly binding to the YY1 protein (Zhou et al. 2015) (Fig. 1).

Developmental pluripotency-associated 2 (Dppa2) is a regulator of octamer-binding transcription factor 4 (*Oct4*) expression that inhibits skeletal muscle cell differentiation. It was previously reported that *Dppa2* upstream binding muscle lncRNA (*DUM*, also known as *5330426P16Rik*) was expressed from the region located 2-megabase (Mb) downstream of the *Dppa2* gene (Wang et al. 2015). RNA–FISH and cellular-fractionation assays revealed that *Dum* is localized both in the nucleus and cytoplasm of C2C12 myoblasts. *Dum* expression was robustly upregulated by MyoD during the early stage of C2C12 differentiation. In vivo, *Dum* expression was highly induced at an early stage of muscle regeneration in mice. In addition, *Dum* was highly expressed in the skeletal muscles of mdx mice, when compared to those of wild-type mice. These results suggest that *Dum* is involved in the myogenic differentiation program. Indeed, both in vitro- and in vivo-knockdown analyses showed that *Dum* promotes myoblast differentiation of C2C12 cells and is required for injury-induced muscle regeneration through the regulation of *Dppa2* and *Oct4* expression. Intriguingly, *Dum* knockdown reduced DNA methylation at the promoter region of *Dppa2*. Mechanistically, *Dum* represses *Dppa2* expression *in cis* by recruiting DNA methyltransferase 1 (Dnmt1), DNA methyltransferase 3A (Dnmt3a), and DNA methyltransferase 3B (Dnmt3b) to the *Dppa2* locus (Table 1). However, for *Dum* to bind the *Dppa2* promoter, intra-chromosomal looping between the *Dum* locus and *Dppa2* promoter was necessary. Thus, *Dum* promotes skeletal muscle differentiation and accelerates muscle regeneration by repressing *Dppa2* as a scaffold for the DNA methyltransferase family (Fig. 1).

4 Other LncRNAs Regulating Skeletal Muscle Differentiation

A murine chromatin-associated muscle-specific lncRNA (*Charme*) is a novel isoform of the *5430431A17Rik* transcript that was first identified via transcriptome analysis in C2C12 cells (Ballarino et al. 2015). Ballarino et al. (2018) analyzed the function of this lncRNA in myogenic differentiation and named it as *Charme*. *Charme* was first expressed in C2C12 cells 1 day after inducing myogenic differentiation. Although the fully spliced *Charme* (*mCharme*) was mainly present in the cytoplasm, *pCharme*, an isoform still having intron 1, was predominantly associated with the chromatin. Knocking down *Charme* expression with a locked nucleic acid GAPmer impaired the later stages of differentiation in C2C12 cells, as shown by the 50%-reduced fusion index and the increased number of mononucleated MyHC-positive cells. In addition, *Charme* depletion affected the expression of over 800 genes involved in muscle function and contraction. Following ChIRP-Seq and FISH analyses revealed that *pCharme* mediated interactions between the *Charme* locus and chromatin containing the *IGF2*, *Tnni2*, and fast skeletal muscle troponin T (*Tnnt3*) genes, which was required to activate the expression of relevant myogenic genes (Fig. 1). *Charme*-knockout mice, which were generated by inserting the polyadenylation signal from the β-globin gene into exon 2 of *Charme*, were viable and fertile, but showed a significantly reduced myofiber size. In addition,

Charme-knockout mice exhibited an altered cardiac function. However, this result is not surprising because *Charme* expression was detected both in skeletal and cardiac muscle (Ballarino et al. 2015). The most noteworthy point is that *Charme*-knockout mice showed a shorter life span; the life span of *Charme*-knockout mice was approximately one-third of that of wild-type mice. A human *Charme* counterpart has also been identified (Table 1); therefore, further study on the relationship between *Charme* and human muscular diseases is expected.

As mentioned previously, YY1 represses multiple genes in myoblasts. However, the ChIP-Seq analysis of YY1 in myoblasts revealed that YY1 also activates many loci (Lu et al. 2013). The YY1-associated muscle lincRNA-1 (*Yam-1*) acts downstream of YY1, and its expression is positively regulated by YY1. During C2C12 differentiation, *Yam-1* expression is downregulated. Loss of *Yam-1* function revealed that *Yam-1* inhibits myogenic differentiation through the transcriptional regulation of *miR-715*, which targets wingless-type MMTV integration site family, member 7B (*Wnt7b*) mRNA. In addition, in vivo knockdown of *Yam-1* by siRNA improved skeletal muscle regeneration after cardiotoxin injection. Thus, a YY1–*Yam-1*–*miR-715* cascade regulates skeletal muscle differentiation (Table 1).

It was also reported that the lncRNA *LncMyoD*, a direct target of MyoD, promotes myogenic differentiation by regulating IGF2-mRNA-binding protein 2 (IMP2)-mediated translation of the proliferation-associated genes, neuroblastoma ras oncogene (*N-Ras*) and myelocytomatosis oncogene (*c-Myc*) (Gong et al. 2015) (Fig. 1). LncRNA *m1/2-sbsRNA* regulates myogenic differentiation by promoting degradation of tumor necrosis factor receptor-associated factor 6 (*Traf6*) mRNA, which is required for myogenesis, via Staufen1-mediated mRNA decay (Wang et al. 2013) (Table 1). LncRNA *lnc-31* promotes the translation of Rho-associated coiled-coil-containing protein kinase 1 (*Rock1*) mRNA, an inhibitor of differentiation, by stabilizing the translational activator Y box protein 1 (YB-1) and, thus, *lnc-31* is required for myoblast proliferation (Ballarino et al. 2015; Dimartino et al. 2018) (Table 1). *LncRNA-AK143003*, which acts as a downstream lncRNA of the MyoD protein, might inhibit skeletal muscle differentiation by regulating *Myog* and *MyHC* expression (Guo et al. 2017) (Fig. 1). However, associated miRNAs and proteins remain unidentified.

5 Regulatory Roles of LncRNAs in Skeletal Muscle Disorders

Several lncRNAs have been implicated in skeletal muscle disorders. It has been found that various novel nuclear lncRNAs were produced from a 2.2-Mb-long dystrophin gene (a causative gene for Duchenne muscular dystrophy) (Bovolenta et al. 2012). Overexpression of these lncRNAs downregulated dystrophin mRNA expression. Although the detailed molecular mechanism remains unknown, the data generated in that study indicate that lncRNAs regulated expression of the dystrophin gene.

Facioscapulohumeral muscular dystrophy (FSHD) is an autosomal-dominant disease that shows progressive muscle weakness and atrophy in the face (eyes and mouth), upper arms, and shoulder girdle, and is associated with a reduced copy number of the D4Z4 repeat, which maps to the q35 region of chromosome 4. Although the D4Z4 deletion is associated with de-repression of genes in 4q35, the underlying molecular mechanism was unknown. Cabianca et al. (2012) identified a chromatin-associated lncRNA, DBE-Transcript (*DBE-T*), which was selectively produced in FSHD patients, and showed that *DBE-T* coordinates de-repression of 4q35 genes by recruiting the Trithorax group protein ASH1-like histone lysine methyltransferase (Ash1L) to the FSHD locus.

Using high-throughput sequencing analysis, Hamann et al. (2017) identified 85 lncRNAs including *H19*, *LncMyoD*, *Neat1*, *MEG3*, and plasmacytoma variant translocation 1 (*Pvt1*), which showed differential expression in patients with inclusion body myositis (IBM) and anti-Jo-1-associated myositis, when compared with age-matched control subjects.

In addition, it is becoming clear that several muscular atrophies or wasting conditions alter the expression of selected lncRNAs. Sarcopenia is a condition characterized by reduced muscle mass and impaired skeletal muscle function with aging. In general, mice older than 2 years are used as a model for sarcopenia. Upregulated and downregulated expression of *Chronos* and *MAR1* were observed in aged mice, respectively (Neppl et al. 2017; Zhang et al. 2018a). *Chronos* regulates bone morphogenetic protein (BMP) signaling, which positively regulates skeletal muscle mass (Sartori et al. 2013), through the repression of bone morphogenetic protein 7 (*Bmp7*) expression (Neppl et al. 2017). Thus, *Chronos* inhibition resulted in muscle hypertrophy in mice. *MAR1* increases skeletal muscle mass and strength by increasing Wnt5a protein expression in mice as a ceRNA for *miR-487b*; therefore, *MAR1* overexpression attenuated muscle atrophy induced by aging and hindlimb suspension (Zhang et al. 2018a).

Increased *Myoparr* expression was observed during both skeletal muscle differentiation and muscle atrophy caused by denervation (Hitachi et al. 2019a, 2019b). *Myoparr* promotes muscle atrophy caused by denervation by activating *Myog* expression, which induces muscle atrophy (Moresi et al. 2010). *Myoparr* also represses BMP signaling by decreasing the expression of growth differentiation factor 5 (*Gdf5*, also known as *Bmp14*), which inhibits muscle atrophy (Sartori et al. 2013). Thus, *Myoparr* inhibition significantly attenuated skeletal muscle loss induced by denervation treatment (Hitachi et al. 2019a, 2019b).

In addition, increased *Pvt1* expression was observed in amyotrophic lateral sclerosis (ALS)-model mice. Silencing *Pvt1* expression attenuated mitochondrial fragmentation, apoptosis, autophagy, and atrophy in denervated muscles (Alessio et al. 2019). The expression of mechanical unloading-induced muscle atrophy-related lncRNA (*lncMUMA*) is induced by tail suspension. *LncMUMA* promotes myogenic differentiation by acting as a ceRNA for *miR-762* to increase MyoD protein expression (Zhang et al. 2018b). Upregulated expression of atrophy-related long noncoding RNA-1 (*Atrolnc-1*) was observed in mouse models of skeletal muscle atrophy induced by chronic kidney disease (CKD), starvation, and cancer (Sun et al. 2018). *Atrolnc-1* interacted with the A20-binding inhibitor of nuclear factor-kappa B (NF-κB) activation (ABIN-1) protein to enhance NF-κB activity and activate the expression of muscle ring-finger 1 (*MuRF-1*), which is an E3 ubiquitin ligase expressed in skeletal and cardiac muscle (Bodine et al. 2001). Therefore, inhibition of *Atrolnc-1* prevented muscle atrophy in mice with CKD.

Taken together, the lncRNAs described above might serve as novel therapeutic targets for treating human muscular diseases or muscular atrophy. However, for future clinical application, it should be very important to carefully confirm whether the sequences and functions of these lncRNAs are evolutionarily conserved in humans skeletal muscle tissues.

6 LncRNAs Encoding Micropeptides

Recently, the surprising discovery was made that some lncRNAs are not actually noncoding RNAs and encode micropeptides. During skeletal muscle differentiation, the expression of Linc-RNA activator of myogenesis (*Linc-RAM*) was positively and negatively regulated by the MyoD protein and fibroblast growth factor 2 (FGF2) signaling, respectively (Yu et al. 2017; Zhao et al. 2018). Knocking down *Linc-RAM* in C2C12 cells markedly decreased myogenic differentiation as evidenced by fewer MyHC-positive cells and decreased expression of *Myog* and *miR-206*. Conversely, *Linc-RAM* overexpression increased the fusion index, with increased expression of *Myog* and *miR-206*. Thus, *Linc-RAM* enhances C2C12 differentiation by regulating *Myog* and *miR-206* expression (Fig. 1). In addition, *Linc-RAM*-knockout mice exhibited delayed muscle regeneration, indicating that *Linc-RAM* promotes myogenic differentiation. Mechanistically, *Linc-RAM* acts as a scaffold to facilitate formation of the MyoD–SWI/SNF-related, matrix-associated, actin-dependent regulator of chromatin, subfamily d, member 3 (Baf60c)–SWI/SNF-related, matrix-associated, actin-dependent regulator of chromatin, subfamily a, member 4 (Brg1) complex (Table 1). Unexpectedly, this lncRNA encodes a 46-amino acid micropeptide, known as Myoregulin (Mrln) (Anderson et al. 2015). Mrln is a transmembrane alpha-helix protein that shows skeletal muscle-specific expression. Mrln interacted with the sarcoplasmic reticulum Ca^{2+}-ATPase (SERCA), which is necessary for muscle relaxation by regulating Ca^{2+} uptake. *Mrln*-knockout mice showed improved exercise performance and enhanced Ca^{2+} mobilization in skeletal muscles. Thus, Mrln regulates skeletal muscle performance by inhibiting SERCA activity. Inspired by that work, recent studies showed additional examples of lncRNAs encoding micropeptides such as dwarf open reading frame (DWORF), Myomixer (Mymx), and small regulatory polypeptide of amino acid response (SPAR), which play important roles in muscle contractility, muscle formation, and muscle regeneration, respectively (Nelson et al. 2016; Bi et al. 2017; Matsumoto et al. 2017). Thus, the possibility that lncRNAs encode micropeptides should be explored in future studies.

7 Conclusion and Perspectives

As mentioned above, we introduced the regulatory roles of lncRNAs in skeletal muscle differentiation, growth, regeneration, and disorders. Several of these lncRNAs have potential applications for treating human muscle disorders. Moreover, there is no doubt that novel lncRNAs regulating myogenic differentiation, muscle size, and muscular diseases will be identified and characterized in the future.

Although studying lncRNAs is a new frontier in skeletal muscle biology, analyzing the functions of lncRNAs will likely be more difficult than studying protein-coding RNAs because lncRNAs may not be evolutionarily conserved, may be expressed at low abundances, and may have unexpected functions. Therefore, the following points should be taken into consideration for further lncRNA studies. First, it is important to investigate lncRNA co-localization with target chromosomes, miRNAs, and proteins in conjunction with in situ hybridization and cell fractionation. For example, lncRNAs acting as ceRNAs should localize with target miRNAs in the cytoplasm because most miRNAs are preferentially located in the cytoplasm versus the nucleus (Zhang et al. 2017). Second, it is important to determine the number of lncRNAs to be analyzed in a cell. Although lncRNAs working *in cis* generally tend to be expressed at much lower levels than protein-coding RNAs (Derrien et al. 2012), the number of lncRNAs functioning as ceRNAs should be equal to or greater than the target miRNAs. Third, it is important to confirm the intracellular binding of endogenous lncRNA with target factors such as miRNAs, proteins, and genomic loci by RIP, ChIRP, or other appropriate methods. No optimized methods have been developed for suppressing the expression and function of diverse lncRNAs. It has also been reported that overexpressing some functional lncRNAs has no biological effect (Wang et al. 2011). Thus, it is important to investigate lncRNA functions using multiple approaches such as RNAi, ASO, CRISPR-mediated activation or interference, and gene knockdown/knockout in cells. In addition, using knockout animals to verify the in vivo functions of lncRNAs increases the reliability. Finally, it is also important to consider the possibility that the micropeptides encoded by some lncRNAs could be functional. We hope that the fundamental functions of lncRNAs in skeletal muscle will become clearer in the future by paying attention to the above-mentioned pitfalls and points of interest.

Acknowledgments This work was supported in part by JSPS KAKENHI (grant numbers 17 K08646 and 19H03427) and an Intramural Research Grant (grant numbers 29-4 and 2-5) for Neurological and Psychiatric Disorders of ncnp.

References

Alessio E, Buson L, Chemello F et al (2019) Single cell analysis reveals the involvement of the long non-coding RNA Pvt1 in the modulation of muscle atrophy and mitochondrial network. Nucleic Acids Res 47:1653–1670

Anderson DM, Anderson KM, Chang CL et al (2015) A micropeptide encoded by a putative long noncoding RNA regulates muscle performance. Cell 160:595–606

Arase M, Horiguchi K, Ehata S et al (2014) Transforming growth factor-β-induced lncRNA-Smad7 inhibits apoptosis of mouse breast cancer JygMC(A) cells. Cancer Sci 105:974–982

Ballarino M, Cazzella V, D'Andrea D et al (2015) Novel long noncoding RNAs (lncRNAs) in myogenesis: a miR-31 overlapping lncRNA transcript controls myoblast differentiation. Mol Cell Biol 35:728–736

Ballarino M, Cipriano A, Tita R et al (2018) Deficiency in the nuclear long noncoding RNA *Charme* causes myogenic defects and heart remodeling in mice. EMBO J 37:e99697

Bartel DP (2009) MicroRNAs: target recognition and regulatory functions. Cell 136:215–233

Bartolomei MS, Zemel S, Tilghman SM (1991) Parental imprinting of the mouse H19 gene. Nature 351:153–155

Berkes CA, Tapscott SJ (2005) MyoD and the transcriptional control of myogenesis. Semin Cell Dev Biol 16:585–595

Bi P, Ramirez-Martinez A, Li H et al (2017) Control of muscle formation by the fusogenic micropeptide myomixer. Science 356:323–327

Bodine SC, Latres E, Baumhueter S et al (2001) Identification of ubiquitin ligases required for skeletal muscle atrophy. Science 294:1704–1708

Bovolenta M, Erriquez D, Valli E et al (2012) The DMD locus harbours multiple long non-coding RNAs which orchestrate and control transcription of muscle dystrophin mRNA isoforms. PLoS One 7:e45328

Cabianca DS, Casa V, Bodega B et al (2012) A long ncRNA links copy number variation to a polycomb/trithorax epigenetic switch in FSHD muscular dystrophy. Cell 149:819–831

Cam H, Griesmann H, Beitzinger M et al (2006) p53 family members in myogenic differentiation and rhabdomyosarcoma development. Cancer Cell 10:281–293

Cao Y, Kumar RM, Penn BH et al (2006) Global and gene-specific analyses show distinct roles for Myod and Myog at a common set of promoters. EMBO J 25:502–511

Caretti G, Schiltz RL, Dilworth FJ et al (2006) The RNA helicases p68/p72 and the noncoding RNA SRA are coregulators of MyoD and skeletal muscle differentiation. Dev Cell 11:547–560

Cesana M, Cacchiarelli D, Legnini I et al (2011) A long noncoding RNA controls muscle differentiation by functioning as a competing endogenous RNA. Cell 147:358–369

Chakraborty S, Deb A, Maji RK et al (2014) LncRBase: an enriched resource for lncRNA information. PLoS One 9:e108010

Chen JCJ, Ramachandran R, Goldhamer DJ (2002) Essential and redundant functions of the MyoD distal regulatory region revealed by targeted mutagenesis. Dev Biol 245:213–223

Chen X, He L, Zhao Y et al (2017) *Malat1* regulates myogenic differentiation and muscle regeneration through modulating MyoD transcriptional activity. Cell Discov 3:17002

Cichewicz MA, Kiran M, Przanowska RK et al (2018) MUNC, an enhancer RNA upstream from the *MYOD* gene, induces a subgroup of myogenic transcripts in *trans* independently of MyoD. Mol Cell Biol 38:e00655–e00617

Cooper R, Kuh D, Hardy R et al (2010) Objectively measured physical capability levels and mortality: systematic review and meta-analysis. BMJ 341:c4467

Derrien T, Johnson R, Bussotti G et al (2012) The GENCODE v7 catalog of human long noncoding RNAs: analysis of their gene structure, evolution, and expression. Genome Res 22:1775–1789

Dey BK, Pfeifer K, Dutta A (2014) The H19 long noncoding RNA gives rise to microRNAs miR-675-3p and miR-675-5p to promote skeletal muscle differentiation and regeneration. Genes Dev 28:491–501

Dimartino D, Colantoni A, Ballarino M et al (2018) The long non-coding RNA *lnc-31* interacts with *Rock1* mRNA and mediates its YB-1-dependent translation. Cell Rep 23:733–740

Du J, Zhang P, Zhao X et al (2019) MicroRNA-351-5p mediates skeletal myogenesis by directly targeting lactamase-β and is regulated by lnc-mg. FASEB J 33:1911–1926

Faralli H, Dilworth FJ (2012) Turning on myogenin in muscle: a paradigm for understanding mechanisms of tissue-specific gene expression. Comp Funct Genomics 2012:836374

Feng Y, Niu LL, Wei W et al (2013) A feedback circuit between miR-133 and the ERK1/2 pathway involving an exquisite mechanism for regulating myoblast proliferation and differentiation. Cell Death Dis 4:e934

Ge Y, Sun Y, Chen J (2011) IGF-II is regulated by microRNA-125b in skeletal myogenesis. J Cell Biol 192:69–81

Giovarelli M, Bucci G, Ramos A et al (2014) H19 long noncoding RNA controls the mRNA decay promoting function of KSRP. Proc Natl Acad Sci U S A 111:E5023–E5028

Goldhamer DJ, Faerman A, Shani M et al (1992) Regulatory elements that control the lineage-specific expression of myoD. Science 256:538–542

Gong C, Li Z, Ramanujan K et al (2015) A long non-coding RNA, *LncMyoD*, regulates skeletal muscle differentiation by blocking IMP2-mediated mRNA translation. Dev Cell 34:181–191

Guo Y, Wang J, Zhu M et al (2017) Identification of MyoD-responsive transcripts reveals a novel long non-coding RNA (lncRNA-AK143003) that negatively regulates myoblast differentiation. Sci Rep 7:2828

Hamann PD, Roux BT, Heward JA et al (2017) Transcriptional profiling identifies differential expression of long non-coding RNAs in Jo-1 associated and inclusion body myositis. Sci Rep 7:8024

Hamazaki N, Uesaka M, Nakashima K et al (2015) Gene activation-associated long noncoding RNAs function in mouse preimplantation development. Development 142:910–920

Han X, Yang F, Cao H, Liang Z (2015) *Malat1* regulates serum response factor through miR-133 as a competing endogenous RNA in myogenesis. FASEB J 29:3054–3064

Hitachi K, Nakatani M, Takasaki A et al (2019a) *Myogenin* promoter-associated lncRNA *Myoparr* is essential for myogenic differentiation. EMBO Rep 20:e47468

Hitachi K, Nakatani M, Tsuchida K (2019b) Long non-coding RNA *Myoparr* regulates GDF5 expression in denervated mouse skeletal muscle. Noncoding RNA 5:E33

Hon CC, Ramilowski JA, Harshbarger J et al (2017) An atlas of human long non-coding RNAs with accurate 5′ ends. Nature 543:199–204

Horak M, Novak J, Bienertova-Vasku J (2016) Muscle-specific microRNAs in skeletal muscle development. Dev Biol 410:1–13

Jin CF, Li Y, Ding XB et al (2017) lnc133b, a novel, long non-coding RNA, regulates bovine skeletal muscle satellite cell proliferation and differentiation by mediating miR-133b. Gene 630:35–43

Jin JJ, Lv W, Xia P et al (2018) Long noncoding RNA *SYISL* regulates myogenesis by interacting with polycomb repressive complex 2. Proc Natl Acad Sci U S A 115:E9802–E9811

Johnsson P, Lipovich L, Grandér D et al (2014) Evolutionary conservation of long non-coding RNAs; sequence, structure, function. Biochim Biophys Acta 1840:1063–1071

Kallen AN, Zhou XB, Xu J et al (2013) The imprinted H19 lncRNA antagonizes Let-7 microRNAs. Mol Cell 52:101–112

Kassar-Duchossoy L, Gayraud-Morel B, Gomès D et al (2004) Mrf4 determines skeletal muscle identity in *Myf5:Myod* double-mutant mice. Nature 431:466–471

Kim HK, Lee YS, Sivaprasad U et al (2006) Muscle-specific microRNA miR-206 promotes muscle differentiation. J Cell Biol 174:677–687

Legnini I, Morlando M, Mangiavacchi A et al (2014) A feedforward regulatory loop between HuR and the long noncoding RNA linc-MD1 controls early phases of myogenesis. Mol Cell 53:506–514

Li H, Yang J, Jiang R et al (2018) Long non-coding RNA profiling reveals an abundant MDNCR that promotes differentiation of myoblasts by sponging miR-133a. Mol Ther Nucleic Acids 12:610–625
Li Z, Cai B, Abdalla BA et al (2019) *LncIRS1* controls muscle atrophy via sponging miR-15 family to activate IGF1-PI3K/AKT pathway. J Cachexia Sarcopenia Muscle 10:391–410
Liang T, Zhou B, Shi L et al (2018) lncRNA *AK017368* promotes proliferation and suppresses differentiation of myoblasts in skeletal muscle development by attenuating the function of miR-30c. FASEB J 32:377–389
Liu QC, Zha XH, Faralli H et al (2012) Comparative expression profiling identifies differential roles for Myogenin and p38α MAPK signaling in myogenesis. J Mol Cell Biol 4:386–397
Liu M, Li B, Peng W et al (2019) LncRNA-MEG3 promotes bovine myoblast differentiation by sponging miR-135. J Cell Physiol 234:18361–18370
Lu L, Sun K, Chen X et al (2013) Genome-wide survey by ChIP-seq reveals YY1 regulation of lincRNAs in skeletal myogenesis. EMBO J 32:2575–2588
Ma M, Cai B, Jiang L et al (2018) lncRNA-Six1 is a target of miR-1611 that functions as a ceRNA to regulate Six1 protein expression and fiber type switching in Chicken myogenesis. Cells 7: E243
Marahrens Y, Panning B, Dausman J et al (1997) *Xist*-deficient mice are defective in dosage compensation but not spermatogenesis. Genes Dev 11:156–166
Martinet C, Monnier P, Louault Y et al (2016) *H19* controls reactivation of the imprinted gene network during muscle regeneration. Development 143:962–971
Matsumoto A, Pasut A, Matsumoto M et al (2017) mTORC1 and muscle regeneration are regulated by the LINC00961-encoded SPAR polypeptide. Nature 541:228–232
Militello G, Hosen MR, Ponomareva Y et al (2018) A novel long non-coding RNA *Myolinc* regulates myogenesis through TDP-43 and Filip1. J Mol Cell Biol 10:102–117
Millay DP, O'Rourke JR, Sutherland LB et al (2013) Myomaker is a membrane activator of myoblast fusion and muscle formation. Nature 499:301–305
Moresi V, Williams AH, Meadows E et al (2010) Myogenin and class II HDACs control neurogenic muscle atrophy by inducing E3 ubiquitin ligases. Cell 143:35–45
Mousavi K, Zare H, Dell'orso S et al (2013) eRNAs promote transcription by establishing chromatin accessibility at defined genomic loci. Mol Cell 51:606–617
Mueller AC, Cichewicz MA, Dey BK et al (2015) MUNC, a long noncoding RNA that facilitates the function of MyoD in skeletal myogenesis. Mol Cell Biol 35:498–513
Myers TK, Andreuzza SE, Franklin DS (2004) p18INK4c and p27KIP1 are required for cell cycle arrest of differentiated myotubes. Exp Cell Res 300:365–378
Nelson BR, Makarewich CA, Anderson DM et al (2016) A peptide encoded by a transcript annotated as long noncoding RNA enhances SERCA activity in muscle. Science 351:271–275
Neppl RL, Wu CL, Walsh K (2017) lncRNA Chronos is an aging-induced inhibitor of muscle hypertrophy. J Cell Biol 216:3497–3507
Ott MO, Bober E, Lyons G et al (1991) Early expression of the myogenic regulatory gene, myf-5, in precursor cells of skeletal muscle in the mouse embryo. Development 111:1097–1107
Park J, Lee H, Han N et al (2018) Long non-coding RNA ChRO1 facilitates ATRX/DAXX-dependent H3.3 deposition for transcription-associated heterochromatin reorganization. Nucleic Acids Res 46:11759–11775
Ransohoff JD, Wei Y, Khavari PA (2018) The functions and unique features of long intergenic non-coding RNA. Nat Rev Mol Cell Biol 19:143–157
Rudnicki MA, Schnegelsberg PN, Stead RH et al (1993) MyoD or Myf-5 is required for the formation of skeletal muscle. Cell 75:1351–1359
Ruijtenberg S, van den Heuvel S (2016) Coordinating cell proliferation and differentiation: antagonism between cell cycle regulators and cell type-specific gene expression. Cell Cycle 15:196–212
Sartori R, Schirwis E, Blaauw B et al (2013) BMP signaling controls muscle mass. Nat Genet 45:1309–1318

Scionti I, Hayashi S, Mouradian S et al (2017) LSD1 controls timely *MyoD* expression via MyoD core enhancer transcription. Cell Rep 18:1996–2006
Shin S, Janknecht R (2007) Concerted activation of the Mdm2 promoter by p72 RNA helicase and the coactivators p300 and P/CAF. J Cell Biochem 101:1252–1265
Song C, Wang J, Ma Y et al (2018) Linc-smad7 promotes myoblast differentiation and muscle regeneration via sponging miR-125b. Epigenetics 13:591–604
Sui Y, Han Y, Zhao X et al (2019) Long non-coding RNA *Irm* enhances myogenic differentiation by interacting with MEF2D. Cell Death Dis 10:181
Sun X, Li M, Sun Y et al (2016) The developmental transcriptome sequencing of bovine skeletal muscle reveals a long noncoding RNA, *lncMD*, promotes muscle differentiation by sponging miR-125b. Biochim Biophys Acta 1863:2835–2845
Sun L, Si M, Liu X et al (2018) Long-noncoding RNA Atrolnc-1 promotes muscle wasting in mice with chronic kidney disease. J Cachexia Sarcopenia Muscle 9:962–974
Tapscott SJ (2005) The circuitry of a master switch: Myod and the regulation of skeletal muscle gene transcription. Development 132:2685–2695
Tripathi V, Ellis JD, Shen Z et al (2010) The nuclear-retained noncoding RNA MALAT1 regulates alternative splicing by modulating SR splicing factor phosphorylation. Mol Cell 39:925–938
Tsai PF, Dell'orso S, Rodriguez J et al (2018) A muscle-specific enhancer RNA mediates cohesin recruitment and regulates transcription in *trans*. Mol Cell 71:129–141
Twayana S, Legnini I, Cesana M et al (2013) Biogenesis and function of non-coding RNAs in muscle differentiation and in Duchenne muscular dystrophy. Biochem Soc Trans 41:844–849
Velleca MA, Wallace MC, Merlie JP (1994) A novel synapse-associated noncoding RNA. Mol Cell Biol 14:7095–7104
Wang KC, Yang YW, Liu B et al (2011) A long noncoding RNA maintains active chromatin to coordinate homeotic gene expression. Nature 472:120–124
Wang J, Gong C, Maquat LE (2013) Control of myogenesis by rodent SINE-containing lncRNAs. Genes Dev 27:793–804
Wang Y, Pang WJ, Wei N et al (2014) Identification, stability and expression of *Sirt1* antisense long non-coding RNA. Gene 539:117–124
Wang L, Zhao Y, Bao X et al (2015) LncRNA *Dum* interacts with Dnmts to regulate *Dppa2* expression during myogenic differentiation and muscle regeneration. Cell Res 25:335–350
Wang GQ, Wang Y, Xiong Y et al (2016) *Sirt1* AS lncRNA interacts with its mRNA to inhibit muscle formation by attenuating function of miR-34a. Sci Rep 6:21865
Wang S, Zuo H, Jin J et al (2019) Long noncoding RNA *Neat1* modulates myogenesis by recruiting Ezh2. Cell Death Dis 10:505
Watts R, Johnsen VL, Shearer J et al (2013) Myostatin-induced inhibition of the long noncoding RNA Malat1 is associated with decreased myogenesis. Am J Physiol Cell Physiol 304:C995–C1001
Wei C, Wu M, Wang C et al (2018) Long noncoding RNA Lnc-SEMT modulates IGF2 expression by sponging miR-125b to promote sheep muscle development and growth. Cell Physiol Biochem 49:447–462
Xu X, Ji S, Li W et al (2017) LncRNA H19 promotes the differentiation of bovine skeletal muscle satellite cells by suppressing Sirt1/FoxO1. Cell Mol Biol Lett 22:10
Yu X, Zhang Y, Li T et al (2017) Long non-coding RNA Linc-RAM enhances myogenic differentiation by interacting with MyoD. Nat Commun 8:14016
Zhang T, Tan P, Wang L et al (2017) RNALocate: A resource for RNA subcellular localizations. Nucleic Acids Res 45:D135–D138
Zhang ZK, Li J, Guan D et al (2018a) A newly identified lncRNA MAR1 acts as a miR-487b sponge to promote skeletal muscle differentiation and regeneration. J Cachexia Sarcopenia Muscle 9:613–626
Zhang ZK, Li J, Guan D et al (2018b) Long noncoding RNA lncMUMA reverses established skeletal muscle atrophy following mechanical unloading. Mol Ther 26:2669–2680

Zhao Y, Cao F, Yu X et al (2018) Linc-RAM is required for FGF2 function in regulating myogenic cell differentiation. RNA Biol 15:404–412

Zhou Y, Cheunsuchon P, Nakayama Y et al (2010) Activation of paternally expressed genes and perinatal death caused by deletion of the *Gtl2* gene. Development 137:2643–2652

Zhou L, Sun K, Zhao Y et al (2015) Linc-YY1 promotes myogenic differentiation and muscle regeneration through an interaction with the transcription factor YY1. Nat Commun 6:10026

Zhu M, Liu J, Xiao J et al (2017) Lnc-mg is a long non-coding RNA that promotes myogenesis. Nat Commun 8:14718

Long Noncoding RNAs in Substance Use Disorders

Changhai Tian and Guoku Hu

Contents

Abstract Substance abuse (including illicit and prescription drugs) has become a significant public health and economic burden. Acute and chronic substance use often leads to a pathologic pattern of behaviors as well as physiologic manifestations, such as tolerance and withdrawal, collectively called substance use disorders (SUDs). Clinical and animal studies have demonstrated that SUDs are heritable in both humans and rodents. The exact molecular mechanism(s), however, is mostly unclear. Long noncoding RNAs (lncRNAs) have emerged as pivotal epigenetic regulators of gene expression in various human diseases, including SUDs. This chapter summarizes current findings and thoughts on the dysregulation as well as

C. Tian
Department of Cellular and Integrative Physiology, University of Nebraska Medical Center, Omaha, NE, USA

Department of Pharmacology and Experimental Neuroscience, University of Nebraska Medical Center, Omaha, NE, USA
e-mail: ctian@unmc.edu

G. Hu (✉)
Department of Pharmacology and Experimental Neuroscience, University of Nebraska Medical Center, Omaha, NE, USA
e-mail: guoku.hu@unmc.edu

S. Jurga, J. Barciszewski (eds.), *The Chemical Biology of Long Noncoding RNAs*, RNA Technologies 11, https://doi.org/10.1007/978-3-030-44743-4_18

the role of lncRNAs in the context of the use of substances, such as morphine, cocaine, methamphetamine, alcohol, etc. The potential applications of lncRNAs in SUDs will also be discussed in this chapter.

Keywords Substance use disorders · Long noncoding RNAs · Opioids · Morphine · Fentanyl · Cocaine · Methamphetamine · Marijuana · Alcohol · Tobacco/nicotine

1 Introduction

Abused drugs are threatening human health, and many drugs can alter a person's thinking and judgment, leading to health risks including addiction, drugged driving, and infectious disease. The use of substance(s) causes significant clinical and functional impairments which is a medical condition called substance use disorder (SUD), also known as drug use disorder or substance dependence (Goplerud et al. 2017). In terms of information released from the National Institute on Drug Abuse, the commonly abused drugs include alcohol, ayahuasca, central nervous system depressants, cocaine, heroin, inhalants, ketamine, marijuana (cannabis), methamphetamine, prescription stimulants, prescription opioids, psilocybin, Rohypnol® (flunitrazepam), *Salvia*, steroids (anabolic), synthetic cannabinoids, synthetic cathinones ("bath salts"), and tobacco. Currently, SUDs are among the most common and costly health conditions affecting the population around the world, and rates of substance use disorders vary by nation and by substance, but the overall prevalence is high. On a global level, men and younger individuals are affected at a much higher rate than women and older adults, respectively (Kuerbis et al. 2014; Morisano et al. 2014). Strikingly, drug overdose is one of the leading causes of death of young people in the United States. According to the National Center for Health Statistics at the Centers for Disease Control and Prevention, national deaths from drug overdose have increased significantly from 16,849 in 1999 to over 70,237 in 2017.

The symptoms of SUDs include, but not limited to, behavioral changes, physical changes, and social changes characterized by a series of mental, physical, and behavioral signs and symptoms (Schulte and Hser 2014). SUDs impose an enormous medical, financial, and emotional burden on society in the form of overdose and health complications, family disintegration, loss of employment, and crime. The National Institute on Drug Abuse (NIDA) estimates that the total cost of drug abuse in the United States exceeds $600 billion annually. Additionally, a sharp increase in the abuse of prescription drugs and drug abuse by teenagers has also been observed. These data suggest the urgent need for research into the effects of drugs of abuse and the mechanisms of addiction, in the expectation of uncovering novel targets for treating and preventing SUDs.

Increasing evidence from clinical and animal studies has demonstrated that SUDs are heritable in both humans and rodents, which are a result of the interplay of multiple genetic and environmental factors (Kendler et al. 2007; Mayfield et al. 2008; Wong and Schumann 2008; Zahr and Sullivan 2008; Agrawal et al. 2012; He and Wang 2012; Volkow and Muenke 2012; Verhulst et al. 2015). Indeed, genetic factors have demonstrated a major role in a person's risk of developing SUD (Prom-Wormley et al. 2017; Demontis et al. 2019; Palmer et al. 2019). Although the genetic contribution to the risk for substance abuse is 40–60% (He and Wang 2012), the fact that chronic exposure of drugs induces structural and functional changes of the brain suggests that dysregulation of gene expression could contribute to the addictive phenotype (Robison and Nestler 2011). Currently, many SUD-associated genes have been identified and shown to be dysregulated in the context of substance abuse. The expression of SUD-associated genes is regulated at various levels, including the epigenetic, transcriptional, and posttranscriptional levels (Robison and Nestler 2011; Nestler 2014). The epigenetic regulation of addiction is mainly through three different mechanisms such as posttranslational modifications of histone proteins (such as acetylation, ubiquitination, SUMOylation, methylation, phosphorylation, and ADP-ribosylation) and DNA methylation (Robison and Nestler 2011; Nestler 2014). Interestingly, the transcriptional mechanism is classically through the actions of transcription factors controlling gene expression in addiction. The transcription factors, including DeltaFosB (Hiroi et al. 1997; Nestler 2008; Perrotti et al. 2008), CREB (Carlezon et al. 1998; Barrot et al. 2002; Pluzarev and Pandey 2004; Edwards et al. 2007), NF-κB (Sullivan et al. 2004; Russo et al. 2009), and MEF2 (Pulipparacharuvil et al. 2008), have been directly linked to addiction-related behaviors, neuronal excitability, drug tolerance, and dependence. Recently, emerging evidence suggests that noncoding RNAs, especially long noncoding RNAs (lncRNAs), function as additional regulatory machinery of the genome controlling both transcriptional and posttranscriptional events in progression of SUDs.

2 Long Noncoding RNAs

It is well known that DNA sequences in the human genome corresponding to protein-coding exons and untranslated regions all together comprise only 1.5–2% (Alexander et al. 2010). Increasing evidence suggests that the non-protein-coding portion of the genome is not junk but has functional importance. With the development of advanced deep sequencing and other analysis technologies, the detection of ncRNAs has facilitated the in-depth research of different classes of ncRNAs. In terms of their length, localization, and function, these ncRNAs are divided into two groups: small ncRNAs (sncRNAs), such as microRNAs which have demonstrated predominating roles in different biological processes of higher eukaryotes, and numerous classes of long ncRNAs which have been broadly categorized as long intergenic ncRNAs, long intronic ncRNAs, telemetric ncRNAs transcribed from subtelomeric promoters, pseudogene transcripts, circular RNAs, enhancer RNAs, and transcribed ultraconserved regions (Kumarswamy and Thum 2013; Kung et al. 2013). lncRNAs are generally greater than 200 nucleotides in length, and numerous studies have revealed a huge number of long noncoding RNAs (lncRNAs) coded by the genome (Quinn and Chang 2016; Chang et al. 2017), which have been recently discovered to mediate diverse biological processes such as in the central nervous system (CNS) during normal development and under various pathological conditions (DiStefano 2018; Cao et al. 2019a; Fernandes et al. 2019).

The biogenesis of lncRNAs shares similar features with that of mRNA, such as transcribed by RNA polymerase II, processed by RNA splicing, and polyadenylated (Brannan et al. 1990). Differently, the lncRNAs can be transcribed with or without polyadenylation, and alternative cleavage, polyadenylation, and splicing also lead to different isoforms from the same locus (Fernandes et al. 2019). In addition, similar to the regulation of coding genes, some of lncRNAs, such as large intervening noncoding RNAs (lincRNAs), can also be transcriptionally regulated by key transcription factors, such as p53, NF-κB, Sox2, Oct4, and Nanog, and these lncRNAs are functional, highly conserved, and implicated in diverse biological processes (Guttman et al. 2009). The first described and identified mammalian lncRNA was H19 which was found in 1990 during liver development and shares the common features with mRNAs. Up to now, bioinformatic studies have revealed that tens of thousands of lncRNAs do not code for proteins and can also be mapped to the genome (Rinn and Chang 2012; Engreitz et al. 2016; Lee et al. 2016; Quinn and Chang 2016; Ransohoff et al. 2018). Although lncRNAs have emerged as regulators of diverse biological processes, the functional relevance of most lncRNAs remains unknown. Currently, lncRNAs have been found to involve in diverse biological processes, including X-chromosome inactivation, such as Xist and Tsix (Brown et al. 1991; Lee et al. 1999); pluripotency and reprogramming (Guttman et al. 2011; Yan et al. 2017); maintenance of genomic stability, such as LINC00657 (NORAD) and ionizing radiation-inducible lncRNA (lnc-RI) (Arainga et al. 2016; Agrahari et al. 2019); imprinting, such as H19 and air (Sleutels et al. 2002); lineage commitment and apoptosis, such as HoxA-AS3 and lncRNA MEG3 (maternally expressed gene 3) (Zhu et al. 2016; Li et al. 2019b); mammalian cell differentiation, such as LncRNA HOTAIRM1 and LncKdm2b (Fu et al. 2019; Li et al. 2019a); and adult neurogenesis, such as RMST and Sox2OT (Ng et al. 2012; Tosetti et al. 2017). A recent breakthrough in identifying the roles of lncRNAs during the development of SUDs underscores the importance of lncRNAs in SUDs.

3 LncRNAs in SUDs

SUDs are a typical behavioral condition resulted from structural and functional alterations in various brain regions of drug abusers (Robinson and Kolb 2004; Zhu et al. 2012). Drugs involved in SUDs include alcohol, opioids, methamphetamine, cocaine, tobacco/nicotine, etc. Genome-wide computational analyses have revealed that lncRNA genes from the human genome are one of the most abundant classes of ncRNAs (Jia et al. 2010) and are involved in diverse biological processes as described above. Additionally, the analyses using a transcriptomic atlas of mouse neocortical layers and in situ hybridization data from the Allen Brain Atlas have revealed that lncRNAs are highly abundant in the brain (Mercer et al. 2008; Belgard et al. 2011). Although the roles of lncRNAs, in addition, remain unclear, increasing evidence has suggested that lncRNAs play important roles in the development of SUDs. We here summarize the regulatory mechanisms of lncRNAs in SUDs.

3.1 LncRNAs in Alcohol Use Disorders

Alcohol addiction has become the second most impactful psychiatric disorder, which is characterized as compulsive alcohol seeking and taking despite adverse consequences (Hyman and Malenka 2001; Koob and Le Moal 2008; Collins et al. 2011). In addition, alcohol addiction can also cause various somatic diseases and is responsible for a mortality rate of about 6% worldwide (Friedmann 2013). Using human and animal models, the widespread alterations in brain gene expression have been observed during chronic alcohol exposure, and epigenetic regulations including histone methylation, histone acetylation, and ncRNA expression have been found to be involved in the expression of multiple genes in various types of brain cells contributing to the pathogenesis of brain related with alcohol abuse and dependence (Ponomarev 2013). Increasing evidence has suggested that lncRNAs play important roles in the synaptic plasticity alterations during drug abuse and dependence. Thus, the role of lncRNAs in the regulation of disease progression of alcohol use disorder (AUD) is gaining attention (Table 1). Although more than 130 differentially expressed lincRNAs have been observed with significant alterations in the postmortem alcoholic brain tissue across different brain regions (Mayfield 2017), the precise role and mechanism of these lincRNAs in alcohol abuse and dependence remain unclear. However, several lncRNAs have been identified with the dysregulation during chronic alcohol exposure. For example, nuclear-enriched abundant transcript 2 (NEAT2), also known as MALAT1, was significantly increased in specific regions including the cerebellum, hippocampus, and brain stem of the human alcoholic brain rather than in other regions, such as frontal or motor cortices. Differently, there was no significant increase of NEAT2 in cerebellum and cortex of alcohol-treated rats; however, NEAT2 was significantly upregulated in rat cortex following alcohol withdrawal (Kryger et al. 2012), suggesting that the dysregulation of lncRNAs induced by chronic alcohol exposure contributes to alcohol actions in the CNS. In addition, H19 as the first identified lncRNA is located at specific loci with normal hypermethylation and contributes to cell proliferation and imprinting. Increased demethylation in the offspring after prenatal exposure to alcohol has been demonstrated (Ouko et al. 2009; Stouder et al. 2011; Liang et al. 2014), suggesting that chronic alcohol use regulates some epigenetic modifications of genomic regions that are critical for embryonic development. Genome-wide association studies of alcohol dependence also revealed that another large antisense-overlapping lncRNA, called LOC100507053, was found to be located within an alcohol dehydrogenase (ADH) cluster and cover multiple loci for ADH genes and form a risk-genomic region for

Table 1 lncRNAs in alcohol use disorders (AUDs)

SUD	LncRNA	Research model	Regulation	Functions	Mechanism of action	Target	References
Alcohol	MALAT1 (NEAT2)	Human subjects	Upregulation	Epigenetic regulation; pro-angiogenesis	Alternative splicing; gene expression	VEGFR2, etc.	Mayfield (2017), Kryger et al. (2012), Lamichhane et al. (2017), Zhang et al. (2018) and Patel et al. (2019)
Alcohol	H19	Mice	Hypomethylation	Cell proliferation and imprinting	Interaction with the methyl-CpG-binding domain protein 1 MBD1	IGF2	Stouder et al. (2011), Liang et al. (2014) and Monnier et al. (2013)
Alcohol	MEG3	Mice	Upregulation	miRNA sponge contributing to EtOH-induced hepatic steatosis and apoptosis	A sponge for miR-let-7c-5p	NLRC5	Wang et al. (2018b)
Alcohol	HOTAIR	Endothelial cells	Upregulation	Pro-angiogenesis	Alternative splicing; gene expression	VEGFA and Ang2	Lamichhane et al. (2017)
Alcohol	BDNF-AS	Human subjects	Upregulation	Epigenetic regulator in the amygdala in early-onset AUD	Methylations at regulatory regions of BDNF	BDNF	Bohnsack et al. (2019) and Patel et al. (2019)
Alcohol	PSD4-1 and NETO-1	Human subjects	Upregulation	Alcohol-associated cancers	Epigenetic regulations	Unknown	Yu et al. (2016)

NLRC5 nucleotide-binding domain and leucine-rich repeat-containing receptor 5, *MEG3* maternally expressed gene 3, *MALAT1* metastasis-associated lung adenocarcinoma transcript 1, *NEAT2* nuclear-enriched abundant transcript 2, *HOTAIR* HOX transcript antisense RNA

alcohol dependence and has a significant impact on cellular responses to alcohol (Zuo et al. 2014; Xu et al. 2015).

The liver is the principal organ for the metabolism of alcohol. Alcohol-induced hepatic steatosis and apoptosis play a primary role in direct alcohol-related morbidity. Using a mouse model of chronic plus single binge ethanol gavage to mimic the chronic alcohol exposure, one study has demonstrated that lncRNA MEG3 was significantly increased in either mouse with alcohol exposure or alcohol-treated in vitro hepatocytes, and further mechanistic studies revealed that lncRNA MEG3 serves as a sponge for let-7c-5p which in turn enhances the expression of NLRC5, a target of let-7c-5p, and ultimately induces hepatic steatosis and apoptosis (Wang et al. 2018b).

In addition to liver, alcohol consumption also contributes to many other diseases related to extrahepatic tissues. Recent evidence suggests that alcohol-induced upregulation of lncRNAs-HOTAIR and MALAT1 in endothelial extracellular vesicles (EVs) plays a significant role in mediating pro-angiogenic effects of these vesicles (Lamichhane et al. 2017; Patel et al. 2019). In these studies, the authors demonstrated that ethanol conditioning upregulates the vascularization bioactivity of endothelial cell-derived EVs. Interestingly, both HOTAIR and MALAT1 were found to be significantly upregulated in EVs isolated from ethanol-exposed HUVECs (human umbilical vein endothelial cells) compared with control HUVECs (Lamichhane et al. 2017). To determine the specific roles of these lncRNAs, the authors transfected HUVECs with siRNAs targeting HOTAIR and MALAT1, or both HOTAIR and MALAT1, and then assessed the vascularization bioactivity of the EVs isolated from the transfected HUVECs using endothelial gap closure assays. The findings from this study demonstrated that downregulation of MALAT1 alone and MALAT1 and HOTAIR together significantly abrogated the ethanol-induced vascularization bioactivity in HUVEC EVs (Lamichhane et al. 2017), indicating that lncRNAs in endothelial EVs may contribute to the pathogenesis of remote organs during the chronic alcohol exposure.

Chronic alcohol exposure or alcohol abuse also contributed to the pathogenesis of cancers. Recent RNA-seq analysis of clinical data has identified another two alcohol-associated lncRNAs – lnc-PSD4-1 and lnc-NETO-1 (Yu et al. 2016). The dysregulation of these two lncRNAs has been implicated in the pathogenesis of head and neck squamous cell carcinoma. Specifically, the authors demonstrated that low expression of lnc-PSD4-1:14, an isoform of lnc-PSD4-1, showed a strong correlation with high survival rates in a Cox proportional hazards regression model (Yu et al. 2016). Increasing evidence has suggested that lncRNAs play critical roles in brain development and synaptic plasticity (Wu et al. 2013; Karpova et al. 2017; Wang et al. 2017). It was well known that neurotrophic factors (NTFs) play important roles not only in the neuronal survival of the peripheral nervous system but also in the synaptic plasticity of the brain. Brain-derived neurotrophic factor (BDNF), one of the most studied NTFs, is a central neurotrophin involved in many neuronal processes and has been linked to several psychiatric diseases as well as addictive disorders (Koskela et al. 2017). Increasing evidence from human subjects and experimental animal models suggested that the BDNF level is negatively correlated with adverse phenotypes associated with harmful alcohol consumption. For example, the alcohol addiction or the mental illness and compulsive behavior resulting from alcohol dependency are associated with the low plasma levels of BDNF (Joe et al. 2007; Zanardini et al. 2011). Consistent with the clinical data, the low levels of BDNF in the prefrontal cortex of mice and in the nucleus of the stria terminalis and in the central extended and medial amygdale of rats also contribute to the development of alcohol dependence compared with animals without alcohol addiction (Prakash et al. 2008; Logrip et al. 2009).

Furthermore, the reduction of BDNF expression or inhibition of BDNF signaling pathway in mice will increase the consumption of alcohol (Hensler et al. 2003; Jeanblanc et al. 2006; Logrip et al. 2015), whereas the activation of BDNF signaling pathway will decrease the consumption of alcohol (Pandey et al. 2006; Jeanblanc et al. 2009, 2013; Warnault et al. 2016). These studies revealed that BDNF is a critical determinant of compulsive alcohol drinking. Interestingly, one lncRNA called BDNF-AS (BDNF antisense lncRNA) has recently been shown to inhibit BDNF expression via epigenetic modifications at regulatory regions in the BDNF gene in the early-onset alcohol use disorder (AUD) group rather than in late-onset AUD group. Further molecular mechanism studies revealed that alcohol drinking during adolescence increases the BDNF-AS expression in the amygdala of AUD group, which results in increased recruitment of EZH2 and subsequently the increase of repressive H3K27 trimethylation (H3K27me3) at regulatory regions in the BDNF gene contributing to a significant decrease in BDNF expression and signaling. On the other hand, adolescent alcohol drinking also decreases the expression of the activity-regulated cytoskeleton-associated protein (ARC) and signaling by increased EZH2 deposition of repressive H3K27me3 at the ARC synaptic activity response element. These BDNF-AS-mediated epigenetic mechanisms contribute to decreased synaptic plasticity, higher risk of developing alcohol use disorder in adulthood, and the increased comorbidity with other psychiatric disorders (Bohnsack et al. 2019).

3.2 LncRNAs in Opioid Use Disorders

Opioids are a class of drugs including illegal drug heroin, synthetic opioids such as fentanyl, and prescription pain relievers, such as oxycodone, hydrocodone, codeine, morphine, and many others. A computational and annotation pipeline has been developed to identify the alteration of lncRNA transcripts in the human nucleus accumbens (hNAcc) of heroin abusers (Michelhaugh et al. 2011). In this study, Michelhaugh et al. developed a computational and annotation pipeline to identify lncRNA transcripts represented on Affymetrix U133 arrays followed by assessing a previously published dataset derived from hNAcc of heroin abusers and controls. The authors demonstrated that the expression of five lncRNAs – MIAT, MEG3, NEAT1, NEAT2, and EMX2OS – was upregulated in heroin abusers compared to matched, drug-free controls (Table 2). Myocardial infarction-associated transcript (*MIAT*) is a lncRNA predominantly expressed in heart and fetal brain tissues.

Table 2 lncRNAs in opioid use disorders

SUD	LncRNA	Research model	Regulation	Functions	Mechanism of action	Target	References
Heroin	MIAT, MEG3, NEAT1, NEAT2	Human subjects	Upregulated in NAc of heroin abusers	Unknown	Unknown	Unknown	Michelhaugh et al. (2011)
Fentanyl	MIAT	Mice	Downregulation	Pro-apoptosis of cardiomyocytes	A miRNA sponge	miR-145	Zhang et al. (2016b) and Zhao et al. (2017)
Morphine	lncRNA-Cox2	Cell and mice	Upregulation	Microglial phagocytosis	Negative regulation of phagocytosis	Phagocytosis-related genes	Hu et al. (2018)
Opiate	BDNF-AS	Human subjects	Upregulation	Inhibition of BDNF expression	Epigenetic modifications at regulatory regions in the BDNF gene	BDNF	Palma-Alvarez et al. (2017)

MIAT myocardial infarction-associated transcript

Increasing evidence suggests that MIAT functions as a competing endogenous RNA (ceRNA) and also acts as a sponge for miRNAs in mediating microvascular dysfunction, diabetic cardiomyopathy, and advanced atherosclerosis (Yan et al. 2015; Zhou et al. 2017; Ye et al. 2019). Recently, MIAT has been found to play cardioprotective effects of fentanyl in myocardial ischemia-reperfusion injury by negatively regulating miRNA-145/Bnip3 pathway axis (Zhang et al. 2016b; Zhao et al. 2017). In addition, neuronal enriched lncRNA MEG3 is a maternally imprinted lncRNA and plays an essential role in GABA neuron neurogenesis. Studies have demonstrated that MEG3 regulates the formation of brain microvessels through controlling the expression of angiogenesis genes (McLaughlin et al. 2006; Gordon et al. 2010; Mercer et al. 2010), which in turn modulates the vulnerability to heroin addiction (Nielsen et al. 2008). Interestingly, Gao et al. demonstrated that MEG3 is upregulated in morphine-exposed mouse hippocampal neuronal HT22 cells compared with control cells (Gao et al. 2019). Furthermore, the authors demonstrated that knockdown of MEG3 reduced morphine-mediated upregulation of autophagy through inactivating ERK pathway in HT22 cells (Gao et al. 2019). Brain-enriched lncRNAs – NEAT1/2 and EMX2OS – play a pivotal role in controlling the expression of synaptic genes at transcriptional level as well as posttranslational level such as mRNA splicing, which subsequently regulates synaptic density (Hutchinson et al. 2007; Bond and Fox 2009; Bernard et al. 2010; Spigoni et al. 2010). Although these lncRNAs have been associated with heroin abusers, the mechanism of the action and targets in the brain remain, however, to be elucidated.

Opioids, such as morphine, are used extensively in the clinical setting for pain management owing to their beneficial effects. However, opioid use often leads to tolerance and dependence on the drug. Recently, two independent groups have identified differentially altered lncRNAs in the spinal cord using a rat morphine tolerance model. Shao et al. performed lncRNA microarray analysis and demonstrated that the expression of 136 lncRNAs was significantly altered in the spinal cord of morphine tolerance rats compared with saline controls (Shao et al. 2018). In another study, Qiu et al. demonstrated that the expression of five LncRNAs, XR_006440, XR_009493, AF196267, MRAK150340, and MRAK037188, was downregulated and the expression of five lncRNAs, MRAK046606, XR_005988, DQ266361, uc.167−, and uc.468+, was upregulated in the spinal cord of morphine tolerance rats compared with controls (Qiu et al. 2019). These results suggest that lncRNAs could play a pivotal role in the development of morphine tolerance, although the function and the underlying mechanisms of the lncRNAs remain largely unknown.

Opioid use also manifests deleterious side effects in the brain including, but not limited to, neuroinflammation. Recent studies suggest that the long intergenic noncoding RNA (lincRNA)-Cox2 (proximal to the Cox2 gene with no overlap) plays an important role in the pathogenesis of neuroinflammation-dependent diseases including auto-inflammatory and neurodegenerative diseases. Upon immune stimulation, lincRNA-Cox2 is upregulated via the NF-kB signaling pathway and is required for the transcription of inflammatory genes (Carpenter et al. 2013; Hu et al. 2016; Liao et al. 2019; Xue et al. 2019). In addition, microglial dysfunctions, such as impaired phagocytosis and immune responses, also contribute to the pathogenesis of neurodegenerative diseases. Recently, lincRNA-Cox2 has been found to be involved in morphine-mediated neurodegeneration through regulating microglial phagocytosis activity (Agrahari et al. 2019). In response to morphine stimulation, astrocytes will release EVs that are taken up by microglial endosomes resulting in the upregulation of lincRNA-Cox2 in microglia by activating Toll-like receptor 7 (TLR7) and the impairment of microglial phagocytic activity, suggesting that lncRNA-Cox2 may indirectly contribute to the pathogenesis of SUDs.

As discussed above, BDNF is a critical determinant of compulsive alcohol drinking. Interestingly, clinical study also suggests that the decreased plasma level of BDNF is correlated with the increased risk of compulsive behavior and BDNF Val66Met polymorphism in heroin users which has been proved to contribute to the increased compulsive alcohol drinking in animals (Warnault et al. 2016; Rovis et al. 2018). Moreover, it has been reported that opiate-induced neurotoxicity in the CNS may be correlated with modifications in BDNF expression, and peripheral BDNF level may also be correlated with opiate use disorders in humans (Palma-Alvarez et al. 2017), suggesting that lncRNA BDNF-AS mediated the epigenetic regulation of BDNF that may also contribute to not only alcohol use disorder but also opiate addiction.

3.3 LncRNAs in Cocaine Use Disorders

Cocaine is a CNS stimulant that affects the brain by increasing the levels of dopamine, a brain chemical associated with pleasure and reward. Studies have demonstrated that cocaine negatively affects every part of the body along with many severe long-term effects. Cocaine use leads to alterations of gene expression in CNS cells, in addition to other permanent effects including talkativeness, excitement, alertness, anxiety, and overconfidence (Novikova et al. 2005; Repantis et al. 2010). Using cocaine-conditioned mice, a transcriptome analysis of lncRNAs of the nucleus accumbens (NAc) has been performed and revealed that cocaine administration results in genome-wide alterations of lncRNAs in the NAc, suggesting these RNA transcripts may play roles in cocaine-induced neural plasticity and addiction (Bu et al. 2012). Data analysis using Affymetrix arrays revealed that four brain-enriched lncRNAs – MIAT, MEG3, NEAT2, and EMX2OS – were significantly upregulated in the NAc of human cocaine abusers compared to matched drug-free control subjects (Sartor et al. 2012). Although upregulation of these lncRNAs has been observed in the NAc of human cocaine abusers, the potential targets and precise mechanisms remain unknown.

Increasing evidence suggests that these lncRNAs play important roles in the cardiovascular dysfunctions (Yan et al. 2015; Zhou et al. 2017; Ye et al. 2019), neurogenesis (McLaughlin et al. 2006), angiogenesis (Gordon et al. 2010), and synaptic density and gene transcription (Hutchinson et al. 2007; Bond and Fox 2009; Bernard et al. 2010; Spigoni et al. 2010); however, the mechanism of action and targets in the brain of cocaine abusers remain largely unknown. Interestingly, similar aberrations in lncRNA expression in response to cocaine and heroin abuse have been observed in human abusers (Michelhaugh et al. 2011; Bannon et al. 2015). For example, Bannon et al. performed a custom lncRNA microarray analysis of postmortem human midbrain specimens from chronic cocaine abusers and control subjects (Bannon et al. 2015). This study demonstrated a group of lncRNAs dysregulated in chronic cocaine abusers. Importantly, using in situ hybridization histochemistry (ISHH), the researchers have demonstrated that the cocaine-associated lncRNAs exhibit dopamine cell-specific expression, different subcellular distributions, and covariance of expression with known cocaine-regulated protein-coding genes. Specifically, the expression of LINC00162 and TRAF3IP2-AS1 transcripts was visualized nearly exclusively in dopamine neurons. Furthermore, the results of qPCR and Pearson's correlation analysis suggest that TRAF3IP2-AS1 transcript abundance positively correlated with the levels of TRAF3IP2 protein-coding transcript, indicating a potential role of TRAF3IP2-AS1 in regulating the expression of its cognate protein-coding gene (Bannon et al. 2015). In addition to TRAF3IP2-AS1, the authors also demonstrated that the expression of 32 well-annotated lncRNAs was differentially expressed in the midbrains of human cocaine abusers, such as upregulated lncRNAs HOTAIRM1, RPPH1, and WDR11-AS1 as well as downregulated lncRNAs RNF219-AS1, PRKCQ-AS1, STX18-AS1, LINC00540, and LINC00403 (Bannon et al. 2015) (Table 3). These lncRNAs may mediate broader downstream changes in gene expression arising within the dopamine neurons of chronic cocaine abusers through epigenetic alterations of chromatin state at gene loci. For example, as discussed by Bannon et al. (2015), given that both TRAF3IP2 and PRKCQ proteins are master mediators in regulating the activity of the transcription factor nuclear factor kappa-light-chain-enhancer of activated B cells (NF-kB), TRAF3IP2-AS1 and PRKCQ-AS1 lncRNAs could play a key role in NF-kB signaling in cocaine abuse.

BDNF is one of the neurotrophic factors and also plays a critical role in the growth, survival, and differentiation of developing neurons. It has been reported that cocaine exposure induced the synthesis and release of BDNF from NAc neurons that contributed to addictive phenotypes (Graham et al. 2007). Moreover, chronic cocaine use also induces upregulation of NAc TrkB, a high-affinity receptor of BDNF, which in turn leads to the transition to more addicted biological states (Graham et al. 2007, 2009). Both in vitro and in vivo studies have demonstrated that lncRNA BDNF-AS controls the expression of BDNF at both transcriptional and posttranscriptional levels (Modarresi et al. 2012), suggesting that lncRNA BDNF-AS could be a master regulator and controls drug-seeking behaviors in cocaine abusers.

Table 3 lncRNAs in cocaine use disorders

SUD	LncRNA	Research model	Regulation	Functions	Mechanism of action	Target	References
Cocaine	NEAT2, MIAT, MEG3, and EMX2OS	Human subjects	Upregulated in NAc of heroin abusers	Unknown	Unknown	Unknown	Michelhaugh et al. (2011)
Cocaine	BDNF-AS	Human subjects	Upregulation	Inhibition of BDNF expression	Epigenetic modifications at regulatory regions in the BDNF gene	BDNF	Palma-Alvarez et al. (2017)
Cocaine	TRAF3IP2-AS1	Human subjects	Upregulation	Inhibition of TRAF3IP2 expression	Epigenetic effects on gene expression through alterations of chromatic state	TRAF3IP2	Bannon et al. (2015)
Cocaine	PRKCQ-AS1	Human subjects	Downregulation	Inhibition of PRKCQ expression	Unknown	PRKCQ	Bannon et al. (2015)

MIAT myocardial infarction-associated transcript, *MEG3* maternally expressed 3, *NEAT2* nuclear paraspeckle assembly transcript 2, *EMX2OSE* MX2 opposite strand/antisense RNA, *PRKCQ-AS1* protein kinase C theta-antisense 1, *TRAF3IP2-AS1* tumor necrosis factor receptor-associated factor 3-interacting protein 2-antisense 1, *BDNF* brain-derived neurotrophic factor

3.4 LncRNAs in Other Drug Use Disorders

Repeated exposure to addictive drugs, including methamphetamine (METH), marijuana, or tobacco/nicotine, elicits long-lasting cellular and molecular changes. METH is a highly addictive psychostimulant (Sim et al. 2017). Zhu et al. have examined the lncRNA expression profile in the NAc of METH-sensitized mice using high-throughput strand-specific complementary RNA sequencing technology (ssRNA-seq) (Zhu et al. 2015), and demonstrated that the expression levels of 14,623 lncRNAs and 65 lncRNAs were significantly decreased and increased, respectively, in the NAc of METH-sensitized mice compared with saline controls. To further evaluate the potential functionality of these lncRNAs, the authors identified the cis- and trans-regulatory target genes of the lncRNAs followed by Gene Ontology (GO) and KEGG pathway enrichment analyses. The findings suggest that METH elicits global changes in lncRNA expression in the NAc of sensitized mice and the METH-induced alteration of the lncRNAs could play important roles in neuronal development, neuronal plasticity, learning and memory, and reward as well as addiction (Zhu et al. 2015). Indeed, five of METH-regulated lncRNAs, including Kcnq1ot1, Zfhx2as, Neat1, Neat2, and Miat, have been reported to interact with targeted loci in either *cis* or *trans* manner (Komine et al. 2006; Lewis et al. 2006; Hutchinson et al. 2007; Bernard et al. 2010) (Table 4). In addition, studies have also demonstrated that these lncRNAs modulate the LTP of hippocampus and behavioral abnormalities through various mechanisms, such as regulating the expressions of their sense partners (Komine et al. 2006; Lewis et al. 2006), function as cofactors for pre-mRNA splicing by interacting with the splicing factors (Hutchinson et al. 2007; Tollervey et al. 2011; Tsuiji et al. 2011), or regulate neuronal plasticity by modulating the expressions of multiple synaptic genes (Bernard et al. 2010). Although the precise regulatory mechanism remains largely unknown, these studies suggest that lncRNA-related nuclear modifications could play a role in METH addiction through rapid posttranscriptional changes of gene expression.

METH can also damage neurons in the brain. To understand the role of METH in the expression of lncRNAs in neurons, Xiong et al. performed microarray analysis in METH-exposed primary rat prefrontal cortical neurons (Xiong et al. 2017). This study revealed that 280 lncRNAs were upregulated, and 393 lncRNAs were downregulated in the METH-exposed neurons compared with control neurons. Results from the gene ontology (GO) and pathway analysis suggested that METH-mediated lncRNA alteration could play a role in ER stress and p53-mediated apoptosis in neurons. Furthermore, the upregulation of three lncRNAs, NR_110713, NR_027943, and growth arrest-specific transcript 5 (GAS5) in METH-stimulated neurons, was validated using qPCR analysis (Xiong et al. 2017). Moreover, GSA5 has been reported to regulate cell apoptosis through p53 signaling pathway by posttranscriptional regulation (Shi et al. 2015), which thus suggests a regulatory role of GAS5 on the p53 signaling pathway in METH-induced neurotoxicity.

Table 4 lncRNAs in other drug use disorders

SUD	LncRNA	Research model	Regulation	Functions	Mechanism of action	Target	References
METH	Kcnq1ot1, Zfhx2as, Neat1, Neat2, and Miat	Mice	Down (Neat1/2, Miat); up (Kcnq1ots and Zfhx2as)	Gene expression involved in the neuronal plasticity, learning, and memory	Interaction with targeted foci in cis or trans manner	Kcnq1ots and Zfhx2as partners; splicing factors; synaptic genes	Zhu et al. (2015)
METH	GAS5	Rat neurons	Upregulation	Neurotoxicity	Posttranscriptional regulation	P53 signaling pathway	Xiong et al. (2017)
METH	lncRNA NR_110713 and NR_027943	Rat neurons	Upregulation	Unknown	Unknown	Unknown	Xiong et al. (2017)
Nicotine	lncRNA H19	Human subjects and cells	Upregulation	Unknown	Unknown	Unknown	Kaplan et al. (2003)
Nicotine	LINC00599, LINC01362, LINC00824, <u>LINC01624</u>, RP11-563D10.1, RP11-98G13.1, AC004791.2	Human subjects	Six upregulation and one downregulation (underlined)	Unknown	Unknown	Unknown	Parker et al. (2017)

METH methamphetamine, *Kcnq1ot1* potassium voltage-gated channel subfamily Q member 1 overlapping transcript 1, *Zfhx2as* zinc finger homeobox 2 antisense, *GAS5* growth arrest-specific 5

Nicotine dependence (ND) is a type of psychiatric disorder. In 2010, Silva et al. analyzed the transcription across the entire genome in normal human bronchial epithelial cells (NHBE) with exposure to the tobacco carcinogen NNK (nicotine-derived nitrosamine ketone), utilizing whole-genome tiling arrays (Silva et al. 2010). This study identified 12 lncRNAs that were upregulated in NNK-stimulated NHBE cells (Silva et al. 2010). Northern blot and qPCR analyses further demonstrated that these lncRNAs are upregulated in NNK-exposed NHBE as well as in a number of lung cancer cell lines and breast cancer cell lines (Silva et al. 2010). The roles of these lncRNAs in cellular stress and in cancer development, however, are still under investigation.

Kaplan et al. investigated the expression of the paternally imprinted gene, lncRNA H19, in the airway epithelium of individuals with a history of mild to moderate cigarette smokers, and demonstrated that lncRNA H19 was significantly upregulated in the airway epithelium compared with that of nonsmokers (Kaplan et al. 2003). The authors further demonstrated that the smoking-induced upregulation of H19 does not result from an alteration of the normal imprinting pattern (Kaplan et al. 2003). Previous studies have demonstrated that frequent loss of imprinting of H19 is associated with lung cancer (Kondo et al. 1995), indicating an important role of H19, and the assessment of its expression, and/or loss of imprinting could serve as early a biomarker in the progression of airway epithelium toward lung cancer (Kaplan et al. 2003).

Recently, Parker et al. performed whole-blood RNA sequencing in 229 current and 286 former smokers, and identified seven smoking-associated lncRNAs, LINC00599, LINC01362, LINC00824, LINC01624, RP11-563D10.1, RP11-98G13.1, AC004791.2, which were significantly altered (Parker et al. 2017). Among them, LINC01624 was downregulated, and the left lncRNAs were significantly upregulated in current smokers compared with former smokers (Table 4). Although the mechanism of action and targets of these lncRNAs remain to be further investigated, these findings imply a role for smoking-associated lncRNAs in nicotine-mediated pathological effects.

4 The Action Mechanisms of LncRNAs in Various SUDs

Despite noncoding molecules, lncRNAs are multifunctional. Increasing evidence has demonstrated numerous functions for lncRNAs in many cellular processes, including gene imprinting (Kanduri 2016), stress response (Valadkhan and Valencia-Hipolito 2016), differentiation and development (Fatica and Bozzoni 2014), speciation and species-specific development, and vernalization in plants (Heo and Sung 2011). The addictive phenotype involving in brain structure and function was attributed to the alterations of gene regulation caused by chronic exposure to drugs of abuse. lncRNAs may function by regulating gene expression through multidimensional mechanisms, including transcriptional and epigenetic mechanisms (Kanduri 2016; Cao et al. 2019b). Recently, lncRNAs have been shown to play a crucial role in the regulation of alternative splicing in response to several stimuli or during disease (summarized in Romero-Barrios et al. 2018). Moreover, increasing evidence now indicates that chromatin-modifying complexes are directed to their sites of action by lncRNAs. Therefore, it is possible that lncRNAs play an important role in SUDs by regulating epigenetic processes. The epigenome consists of DNA methylation and histone modifications including acetylation, methylation, and phosphorylation. It has been reported that specific enzymes responsible for histone acetylation, methylation, and DNA methylation in reward-related brain areas are critically involved in cocaine addiction (LaPlant et al. 2010; Maze et al. 2010). For example, two lncRNAs – Air and Kcnq1ot1 antisense – have been found to play a role in histone methylation by interacting with histone methyltransferase G9a which is downregulated in the NAc following cocaine exposure and is important in cocaine-related behaviors (Nagano et al. 2008; Pandey et al. 2008; Maze et al. 2010). In addition, lncRNA-mediated regulation of histone acetylation, methylation, and DNA methylation, as well as other important cocaine-mediated modifications, has been reviewed (Maze and Nestler 2011). Collectively, lncRNAs exert their function at multidimensional levels in various SUDs.

5 The Potential Applications of LncRNAs in SUDs

Currently, lncRNAs have garnered attention as another species of regulatory RNA with key roles in development, epigenetics, regulation of transcription, and other essential biological processes. Aberrant expressions of lncRNAs play critical roles in the progression and development of various SUDs, some of which may be further evaluated as highly sensitive biomarkers as well as potential therapeutic targets. It has been reported that BDNF-AS lncRNA is associated with malignant status and shows the potential as a prognostic biomarker and therapeutic target for some cancers (Huang et al. 2018; Shang et al. 2018). Interestingly, BDNF-AS has also been found to be dysregulated in response to chronic cocaine, alcohol, and opiate addiction (Modarresi et al. 2012; Palma-Alvarez et al. 2017; Bohnsack et al. 2019). Moreover, BDNF-AS as a conserved noncoding antisense RNA transcript regulates the expression of BDNF at both transcriptional and posttranscriptional levels, which in turn is involved in many neuronal processes including the synaptic plasticity in the amygdala of AUD (Bohnsack et al. 2019), drug-seeking behaviors in chronic cocaine use (Graham et al. 2007), and neurotoxicity in the CNS during opiate abuse (Angelucci et al. 2007), suggesting BDNF-AS has the potential as a therapeutic target of these SUDs.

Chronic exposure to drugs of abuse in vulnerable populations causes long-term behavioral abnormalities, which involve a complex interplay between gene expression and environmental factors (Walker and Nestler 2018). Increasing evidence suggests that the alterations of gene expression in the reward circuitry of the brain, particularly in the mesolimbic dopaminergic systems, contribute to the persistent behavioral abnormalities (Alcaro et al. 2007; Walker and Nestler 2018). It has been reported that TRAF3IP2-AS1 lncRNA was upregulated, whereas PRKCQ-AS1 was downregulated in the midbrain of cocaine abusers (Table 3), which were involved in broader downstream gene alterations in dopamine neurons through epigenetic alterations of chromatin state at gene loci (Bannon et al. 2015). These studies suggest that lncRNAs are master regulators in SUDs and thus could be potential targets for therapy. Intriguingly, recent study has demonstrated that lincRNA-Cox2 contributes to morphine-mediated neurodegeneration through impairing microglial phagocytosis, and intranasal delivery of lincRNA-Cox2 small interfering RNA (siRNA) can restore microglial phagocytic activity in morphine-administered mice (Hu et al. 2018), suggesting that EV-loaded lincRNA-Cox2 siRNA can be used as therapeutic strategy for a multitude of neurodegenerative disorders associated with opiate abuse. Recently, the following study has been carried out to investigate the effects of EV-loaded lincRNA-Cox2 siRNA on the LPS-induced microglial proliferation which has been implicated in the pathogenesis of various neurodegenerative disorders, such as Alzheimer's disease (AD) (Spangenberg et al. 2016; Liao et al. 2019). The results demonstrated that intranasal delivery of EV-loaded lincRNA-Cox2 siRNA can efficiently pass through the blood-brain barrier (BBB) to target activated microglia in the brain where lincRNA-Cox2 siRNA significantly decreases the expression of lincRNA-Cox2 in the microglia contributing to the inhibition of LPS-induced microglial proliferation by controlling the expression of a set of cell cycle genes (Liao et al. 2019). These findings indicate that the EV-based delivery strategy targeting specific lncRNAs could be used to develop the therapeutics for the treatment of CNS disorders, such as SUDs.

Moreover, lncRNAs (or lncRNA fragments) can also be packaged into EVs and released into body fluids such as blood (Wang et al. 2018a; Kitagawa et al. 2019). These findings thus raise the possibility that abnormal levels of lncRNAs in EVs could serve as biomarkers for the prognosis of human diseases (Vausort et al. 2014; Zhang et al. 2016a, 2019; Dai et al. 2017; Yuan et al. 2017; Jiang et al. 2018). Recently, lncRNAs have also been found to be dysregulated and incorporated into EVs in the context of drug abuse, for example, alcohol-induced upregulation of lncRNAs HOTAIR and MALAT1 in endothelial EVs (Lamichhane et al. 2017), which may contribute to the transfer of pathophysiological phenotypes in the remote organs through the lncRNA-enriched circulating EVs. While more reliable EV lncRNAs for SUD biomarkers are to be determined, efforts aimed at deciphering the correlation between the expression levels of the EV lncRNAs and disease severity warrant investigation.

6 Conclusions and Perspectives

Although a growing number of lncRNAs have been identified in addiction-related neuroadaptations, their function and the underlying mechanism(s) in SUDs, however, remain to be investigated. Recently, many new methodologies and technologies have been proposed for the discovery and functional analysis of lncRNAs (reviewed in Kashi et al. 2016). The development and application of new tools will enhance our understanding of the roles of lncRNAs in SUDs. Additionally, a novel method has recently been developed to identify potential lncRNA-disease associations by Integrating Diverse Heterogeneous Information sources with positive pointwise Mutual Information and Random Walk with restart algorithm (named IDHI-MIRW) (Fan et al. 2019), which will be very useful for studying the relationship between lncRNAs and drug addiction. Thus, the identification of lncRNA-SUD associations will also gain insights into SUD-related lncRNAs and benefit SUD diagnoses and treatment. Furthermore, given the fact that lncRNAs often serve as master regulators in various biological processes, they could be ideal therapeutic targets for preventing and treating SUDs. Moreover, many gender-associated lncRNAs have also been identified and could contribute to the gender differences in SUDs, which indicates that it could be crucial to develop lncRNA-based gender-specific therapeutic strategies.

References

Agrahari V, Agrahari V, Burnouf PA et al (2019) Extracellular microvesicles as new industrial therapeutic frontiers. Trends Biotechnol 37:707–729

Agrawal A, Verweij KJ, Gillespie NA et al (2012) The genetics of addiction-a translational perspective. Transl Psychiatry 2:e140

Alcaro A, Huber R, Panksepp J (2007) Behavioral functions of the mesolimbic dopaminergic system: an affective neuroethological perspective. Brain Res Rev 56:283–321

Alexander RP, Fang G, Rozowsky J et al (2010) Annotating non-coding regions of the genome. Nat Rev Genet 11:559–571

Angelucci F, Gruber SH, El Khoury A et al (2007) Chronic amphetamine treatment reduces NGF and BDNF in the rat brain. Eur Neuropsychopharmacol 17:756–762

Arainga M, Su H, Poluektova LY et al (2016) HIV-1 cellular and tissue replication patterns in infected humanized mice. Sci Rep 6:23513

Bannon MJ, Savonen CL, Jia H et al (2015) Identification of long noncoding RNAs dysregulated in the midbrain of human cocaine abusers. J Neurochem 135:50–59

Barrot M, Olivier JD, Perrotti LI et al (2002) CREB activity in the nucleus accumbens shell controls gating of behavioral responses to emotional stimuli. Proc Natl Acad Sci U S A 99:11435–11440

Belgard TG, Marques AC, Oliver PL et al (2011) A transcriptomic atlas of mouse neocortical layers. Neuron 71:605–616

Bernard D, Prasanth KV, Tripathi V et al (2010) A long nuclear-retained non-coding RNA regulates synaptogenesis by modulating gene expression. EMBO J 29:3082–3093

Bohnsack JP, Teppen T, Kyzar EJ et al (2019) The lncRNA BDNF-AS is an epigenetic regulator in the human amygdala in early onset alcohol use disorders. Transl Psychiatry 9:34

Bond CS, Fox AH (2009) Paraspeckles: nuclear bodies built on long noncoding RNA. J Cell Biol 186:637–644

Brannan CI, Dees EC, Ingram RS et al (1990) The product of the H19 gene may function as an RNA. Mol Cell Biol 10:28–36

Brown CJ, Ballabio A, Rupert JL et al (1991) A gene from the region of the human X inactivation centre is expressed exclusively from the inactive X chromosome. Nature 349:38–44

Bu Q, Hu Z, Chen F et al (2012) Transcriptome analysis of long non-coding RNAs of the nucleus accumbens in cocaine-conditioned mice. J Neurochem 123:790–799

Cao M, Li H, Zhao J et al (2019a) Identification of age- and gender-associated long noncoding RNAs in the human brain with Alzheimer's disease. Neurobiol Aging 81:116–126
Cao M, Zhao J, Hu G (2019b) Genome-wide methods for investigating long noncoding RNAs. Biomed Pharmacother 111:395–401
Carlezon WA Jr, Thome J, Olson VG et al (1998) Regulation of cocaine reward by CREB. Science 282:2272–2275
Carpenter S, Aiello D, Atianand MK et al (2013) A long noncoding RNA mediates both activation and repression of immune response genes. Science 341:789–792
Chang J, Tan W, Ling Z et al (2017) Genomic analysis of oesophageal squamous-cell carcinoma identifies alcohol drinking-related mutation signature and genomic alterations. Nat Commun 8:15290
Collins PY, Patel V, Joestl SS et al (2011) Grand challenges in global mental health. Nature 475:27–30
Dai Y, Liang Z, Li Y et al (2017) Circulating long noncoding RNAs as potential biomarkers of sepsis: a preliminary study. Genet Test Mol Biomarkers 21:649–657
Demontis D, Rajagopal VM, Thorgeirsson TE et al (2019) Genome-wide association study implicates CHRNA2 in cannabis use disorder. Nat Neurosci 22:1066–1074
DiStefano JK (2018) The emerging role of Long noncoding RNAs in human disease. Methods Mol Biol 1706:91–110
Edwards S, Graham DL, Bachtell RK et al (2007) Region-specific tolerance to cocaine-regulated cAMP-dependent protein phosphorylation following chronic self-administration. Eur J Neurosci 25:2201–2213
Engreitz JM, Haines JE, Perez EM et al (2016) Local regulation of gene expression by lncRNA promoters, transcription and splicing. Nature 539:452–455
Fan XN, Zhang SW, Zhang SY et al (2019) Prediction of lncRNA-disease associations by integrating diverse heterogeneous information sources with RWR algorithm and positive pointwise mutual information. BMC Bioinf 20:87
Fatica A, Bozzoni I (2014) Long non-coding RNAs: new players in cell differentiation and development. Nat Rev Genet 15:7–21
Fernandes JCR, Acuna SM, Aoki JI et al (2019) Long non-coding RNAs in the regulation of gene expression: physiology and disease. Non-coding RNA 5:pii: E17
Friedmann PD (2013) Clinical practice. Alcohol use in adults. N Engl J Med 368:365–373
Fu L, Peng S, Wu W et al (2019) LncRNA HOTAIRM1 promotes osteogenesis by controlling JNK/AP-1 signalling-mediated RUNX2 expression. J Cell Mol Med 23:7517–7524
Gao S, Li E, Gao H (2019) Long non-coding RNA MEG3 attends to morphine-mediated autophagy of HT22 cells through modulating ERK pathway. Pharm Biol 57:536–542
Goplerud E, Hodge S, Benham T (2017) A substance use cost calculator for US employers with an emphasis on prescription pain medication misuse. J Occup Environ Med 59:1063–1071
Gordon FE, Nutt CL, Cheunsuchon P et al (2010) Increased expression of angiogenic genes in the brains of mouse meg3-null embryos. Endocrinology 151:2443–2452
Graham DL, Edwards S, Bachtell RK et al (2007) Dynamic BDNF activity in nucleus accumbens with cocaine use increases self-administration and relapse. Nat Neurosci 10:1029–1037
Graham DL, Krishnan V, Larson EB et al (2009) Tropomyosin-related kinase B in the mesolimbic dopamine system: region-specific effects on cocaine reward. Biol Psychiatry 65:696–701
Guttman M, Amit I, Garber M et al (2009) Chromatin signature reveals over a thousand highly conserved large non-coding RNAs in mammals. Nature 458:223–227
Guttman M, Donaghey J, Carey BW et al (2011) lincRNAs act in the circuitry controlling pluripotency and differentiation. Nature 477:295–300
He Y, Wang ZJ (2012) Let-7 microRNAs and opioid tolerance. Front Genet 3:110
Hensler JG, Ladenheim EE, Lyons WE (2003) Ethanol consumption and serotonin-1A (5-HT1A) receptor function in heterozygous BDNF (+/−) mice. J Neurochem 85:1139–1147
Heo JB, Sung S (2011) Vernalization-mediated epigenetic silencing by a long intronic noncoding RNA. Science 331:76–79

Hiroi N, Brown JR, Haile CN et al (1997) FosB mutant mice: loss of chronic cocaine induction of Fos-related proteins and heightened sensitivity to cocaine's psychomotor and rewarding effects. Proc Natl Acad Sci U S A 94:10397–10402

Hu G, Gong AY, Wang Y et al (2016) LincRNA-Cox2 promotes late inflammatory gene transcription in macrophages through modulating SWI/SNF-mediated chromatin remodeling. J Immunol 196:2799–2808

Hu G, Liao K, Niu F et al (2018) Astrocyte EV-induced lincRNA-Cox2 regulates microglial phagocytosis: implications for morphine-mediated neurodegeneration. Mol Ther Nucleic Acids 13:450–463

Huang Q, Yang J, He X et al (2018) LncRNA BDNF-AS is associated with the malignant status and regulates cell proliferation and apoptosis in osteosarcoma. Biosci Rep 38:BSR20181498

Hutchinson JN, Ensminger AW, Clemson CM et al (2007) A screen for nuclear transcripts identifies two linked noncoding RNAs associated with SC35 splicing domains. BMC Genomics 8:39

Hyman SE, Malenka RC (2001) Addiction and the brain: the neurobiology of compulsion and its persistence. Nat Rev Neurosci 2:695–703

Jeanblanc J, He DY, McGough NN et al (2006) The dopamine D3 receptor is part of a homeostatic pathway regulating ethanol consumption. J Neurosci 26:1457–1464

Jeanblanc J, He DY, Carnicella S et al (2009) Endogenous BDNF in the dorsolateral striatum gates alcohol drinking. J Neurosci 29:13494–13502

Jeanblanc J, Logrip ML, Janak PH et al (2013) BDNF-mediated regulation of ethanol consumption requires the activation of the MAP kinase pathway and protein synthesis. Eur J Neurosci 37:607–612

Jia H, Osak M, Bogu GK et al (2010) Genome-wide computational identification and manual annotation of human long noncoding RNA genes. RNA 16:1478–1487

Jiang N, Meng X, Mi H et al (2018) Circulating lncRNA XLOC_009167 serves as a diagnostic biomarker to predict lung cancer. Clin Chim Acta 486:26–33

Joe KH, Kim YK, Kim TS et al (2007) Decreased plasma brain-derived neurotrophic factor levels in patients with alcohol dependence. Alcohol Clin Exp Res 31:1833–1838

Kanduri C (2016) Long noncoding RNAs: lessons from genomic imprinting. Biochim Biophys Acta 1859:102–111

Kaplan R, Luettich K, Heguy A et al (2003) Monoallelic up-regulation of the imprinted H19 gene in airway epithelium of phenotypically normal cigarette smokers. Cancer Res 63:1475–1482

Karpova NN, Sales AJ, Joca SR (2017) Epigenetic basis of neuronal and synaptic plasticity. Curr Top Med Chem 17:771–793

Kashi K, Henderson L, Bonetti A et al (2016) Discovery and functional analysis of lncRNAs: methodologies to investigate an uncharacterized transcriptome. Biochim Biophys Acta 1859:3–15

Kendler KS, Myers J, Prescott CA (2007) Specificity of genetic and environmental risk factors for symptoms of cannabis, cocaine, alcohol, caffeine, and nicotine dependence. Arch Gen Psychiatry 64:1313–1320

Kitagawa T, Taniuchi K, Tsuboi M et al (2019) Circulating pancreatic cancer exosomal RNAs for detection of pancreatic cancer. Mol Oncol 13:212–227

Komine Y, Nakamura K, Katsuki M et al (2006) Novel transcription factor zfh-5 is negatively regulated by its own antisense RNA in mouse brain. Mol Cell Neurosci 31:273–283

Kondo M, Suzuki H, Ueda R et al (1995) Frequent loss of imprinting of the H19 gene is often associated with its overexpression in human lung cancers. Oncogene 10:1193–1198

Koob GF, Le Moal M (2008) Addiction and the brain antireward system. Annu Rev Psychol 59:29–53

Koskela M, Back S, Voikar V et al (2017) Update of neurotrophic factors in neurobiology of addiction and future directions. Neurobiol Dis 97:189–200

Kryger R, Fan L, Wilce PA et al (2012) MALAT-1, a non protein-coding RNA is upregulated in the cerebellum, hippocampus and brain stem of human alcoholics. Alcohol (Fayetteville, NY) 46:629–634

Kuerbis A, Sacco P, Blazer DG et al (2014) Substance abuse among older adults. Clin Geriatr Med 30:629–654
Kumarswamy R, Thum T (2013) Non-coding RNAs in cardiac remodeling and heart failure. Circ Res 113:676–689
Kung JT, Colognori D, Lee JT (2013) Long noncoding RNAs: past, present, and future. Genetics 193:651–669
Lamichhane TN, Leung CA, Douti LY et al (2017) Ethanol induces enhanced vascularization bioactivity of endothelial cell-derived extracellular vesicles via regulation of microRNAs and long non-coding RNAs. Sci Rep 7:13794
LaPlant Q, Vialou V, Covington HE 3rd et al (2010) Dnmt3a regulates emotional behavior and spine plasticity in the nucleus accumbens. Nat Neurosci 13:1137–1143
Lee JT, Davidow LS, Warshawsky D (1999) Tsix, a gene antisense to Xist at the X-inactivation centre. Nat Genet 21:400–404
Lee S, Kopp F, Chang TC et al (2016) Noncoding RNA NORAD regulates genomic stability by sequestering PUMILIO proteins. Cell 164:69–80
Lewis A, Green K, Dawson C et al (2006) Epigenetic dynamics of the Kcnq1 imprinted domain in the early embryo. Development 133:4203–4210
Li W, Shen W, Zhang B et al (2019a) Long non-coding RNA LncKdm2b regulates cortical neuronal differentiation by cis-activating Kdm2b. Protein Cell 11:161. https://doi.org/10.1007/s13238-019-0650-z
Li X, Zhao J, Geng J et al (2019b) Long non-coding RNA MEG3 knockdown attenuates endoplasmic reticulum stress-mediated apoptosis by targeting p53 following myocardial infarction. J Cell Mol Med 23:8369. https://doi.org/10.1111/jcmm.14714
Liang F, Diao L, Liu J et al (2014) Paternal ethanol exposure and behavioral abnormities in offspring: associated alterations in imprinted gene methylation. Neuropharmacology 81:126–133
Liao K, Niu F, Dagur RS et al (2019) Intranasal delivery of lincRNA-Cox2 siRNA loaded extracellular vesicles decreases lipopolysaccharide-induced microglial proliferation in mice. J NeuroImmune Pharmacol. https://doi.org/10.1007/s11481-019-09864-z
Logrip ML, Janak PH, Ron D (2009) Escalating ethanol intake is associated with altered corticostriatal BDNF expression. J Neurochem 109:1459–1468
Logrip ML, Barak S, Warnault V et al (2015) Corticostriatal BDNF and alcohol addiction. Brain Res 1628:60–67
Mayfield RD (2017) Emerging roles for ncRNAs in alcohol use disorders. Alcohol (Fayetteville, NY) 60:31–39
Mayfield RD, Harris RA, Schuckit MA (2008) Genetic factors influencing alcohol dependence. Br J Pharmacol 154:275–287
Maze I, Nestler EJ (2011) The epigenetic landscape of addiction. Ann N Y Acad Sci 1216:99–113
Maze I, Covington HE 3rd, Dietz DM et al (2010) Essential role of the histone methyltransferase G9a in cocaine-induced plasticity. Science 327:213–216
McLaughlin D, Vidaki M, Renieri E et al (2006) Expression pattern of the maternally imprinted gene Gtl2 in the forebrain during embryonic development and adulthood. Gene Expr Patterns 6:394–399
Mercer TR, Dinger ME, Sunkin SM et al (2008) Specific expression of long noncoding RNAs in the mouse brain. Proc Natl Acad Sci U S A 105:716–721
Mercer TR, Qureshi IA, Gokhan S et al (2010) Long noncoding RNAs in neuronal-glial fate specification and oligodendrocyte lineage maturation. BMC Neurosci 11:14
Michelhaugh SK, Lipovich L, Blythe J et al (2011) Mining Affymetrix microarray data for long non-coding RNAs: altered expression in the nucleus accumbens of heroin abusers. J Neurochem 116:459–466
Modarresi F, Faghihi MA, Lopez-Toledano MA et al (2012) Inhibition of natural antisense transcripts in vivo results in gene-specific transcriptional upregulation. Nat Biotechnol 30:453–459

Monnier P, Martinet C, Pontis J et al (2013) H19 lncRNA controls gene expression of the imprinted gene network by recruiting MBD1. Proc Natl Acad Sci U S A 110:20693–20698
Morisano D, Babor TF, Robaina KA (2014) Co-occurrence of substance use disorders with other psychiatric disorders: implications for treatment services. Nordic Stud Alcohol Drugs 31:5–25
Nagano T, Mitchell JA, Sanz LA et al (2008) The Air noncoding RNA epigenetically silences transcription by targeting G9a to chromatin. Science 322:1717–1720
Nestler EJ (2008) Review. Transcriptional mechanisms of addiction: role of DeltaFosB. Philos Trans R Soc Lond Ser B Biol Sci 363:3245–3255
Nestler EJ (2014) Epigenetic mechanisms of drug addiction. Neuropharmacology 76 (Pt B):259–268
Ng SY, Johnson R, Stanton LW (2012) Human long non-coding RNAs promote pluripotency and neuronal differentiation by association with chromatin modifiers and transcription factors. EMBO J 31:522–533
Nielsen DA, Ji F, Yuferov V et al (2008) Genotype patterns that contribute to increased risk for or protection from developing heroin addiction. Mol Psychiatry 13:417–428
Novikova SI, He F, Bai J et al (2005) Cocaine-induced changes in the expression of apoptosis-related genes in the fetal mouse cerebral wall. Neurotoxicol Teratol 27:3–14
Ouko LA, Shantikumar K, Knezovich J et al (2009) Effect of alcohol consumption on CpG methylation in the differentially methylated regions of H19 and IG-DMR in male gametes: implications for fetal alcohol spectrum disorders. Alcohol Clin Exp Res 33:1615–1627
Palma-Alvarez RF, Ros-Cucurull E, Amaro-Hosey K et al (2017) Peripheral levels of BDNF and opiate-use disorder: literature review and update. Rev Neurosci 28:499–508
Palmer RHC, McGeary JE, Knopik VS et al (2019) CNR1 and FAAH variation and affective states induced by marijuana smoking. Am J Drug Alcohol Abuse 45:514–526
Pandey SC, Zhang H, Roy A et al (2006) Central and medial amygdaloid brain-derived neurotrophic factor signaling plays a critical role in alcohol-drinking and anxiety-like behaviors. J Neurosci 26:8320–8331
Pandey RR, Mondal T, Mohammad F et al (2008) Kcnq1ot1 antisense noncoding RNA mediates lineage-specific transcriptional silencing through chromatin-level regulation. Mol Cell 32:232–246
Parker MM, Chase RP, Lamb A et al (2017) RNA sequencing identifies novel non-coding RNA and exon-specific effects associated with cigarette smoking. BMC Med Genet 10:58
Patel DB, Luthers CR, Lerman MJ et al (2019) Enhanced extracellular vesicle production and ethanol-mediated vascularization bioactivity via a 3D-printed scaffold-perfusion bioreactor system. Acta Biomater 95:236–244
Perrotti LI, Weaver RR, Robison B et al (2008) Distinct patterns of DeltaFosB induction in brain by drugs of abuse. Synapse 62:358–369
Pluzarev O, Pandey SC (2004) Modulation of CREB expression and phosphorylation in the rat nucleus accumbens during nicotine exposure and withdrawal. J Neurosci Res 77:884–891
Ponomarev I (2013) Epigenetic control of gene expression in the alcoholic brain. Alcohol Res 35:69–76
Prakash A, Zhang H, Pandey SC (2008) Innate differences in the expression of brain-derived neurotrophic factor in the regions within the extended amygdala between alcohol preferring and nonpreferring rats. Alcohol Clin Exp Res 32:909–920
Prom-Wormley EC, Ebejer J, Dick DM et al (2017) The genetic epidemiology of substance use disorder: a review. Drug Alcohol Depend 180:241–259
Pulipparacharuvil S, Renthal W, Hale CF et al (2008) Cocaine regulates MEF2 to control synaptic and behavioral plasticity. Neuron 59:621–633
Qiu Y, Meng LB, Di CY et al (2019) Exploration of the differentially expressed long noncoding RNAs and genes of morphine tolerance via bioinformatic analysis. J Comput Biol 26:1379. https://doi.org/10.1089/cmb.2019.0188
Quinn JJ, Chang HY (2016) Unique features of long non-coding RNA biogenesis and function. Nat Rev Genet 17:47–62

Ransohoff JD, Wei Y, Khavari PA (2018) The functions and unique features of long intergenic non-coding RNA. Nat Rev Mol Cell Biol 19:143–157
Repantis D, Schlattmann P, Laisney O et al (2010) Modafinil and methylphenidate for neuroenhancement in healthy individuals: a systematic review. Pharmacol Res 62:187–206
Rinn JL, Chang HY (2012) Genome regulation by long noncoding RNAs. Annu Rev Biochem 81:145–166
Robinson TE, Kolb B (2004) Structural plasticity associated with exposure to drugs of abuse. Neuropharmacology 47(Suppl 1):33–46
Robison AJ, Nestler EJ (2011) Transcriptional and epigenetic mechanisms of addiction. Nat Rev Neurosci 12:623–637
Romero-Barrios N, Legascue MF, Benhamed M et al (2018) Splicing regulation by long noncoding RNAs. Nucleic Acids Res 46:2169–2184
Rovis D, Cernelic Bizjak M, Vasiljev Marchesi V et al (2018) Increased risk-taking behaviour and brain-derived neurotrophic factor Val66Met polymorphism correlates to decreased serum brain-derived neurotrophic factor level in heroin users. Eur Addict Res 24:189–200
Russo SJ, Wilkinson MB, Mazei-Robison MS et al (2009) Nuclear factor kappa B signaling regulates neuronal morphology and cocaine reward. J Neurosci 29:3529–3537
Sartor GC, St Laurent G 3rd, Wahlestedt C (2012) The emerging role of non-coding RNAs in drug addiction. Front Genet 3:106
Schulte MT, Hser YI (2014) Substance use and associated health conditions throughout the lifespan. Public Health Rev 35. https://web-beta.archive.org
Shang W, Yang Y, Zhang J et al (2018) Long noncoding RNA BDNF-AS is a potential biomarker and regulates cancer development in human retinoblastoma. Biochem Biophys Res Commun 497:1142–1148
Shao J, Wang J, Huang J et al (2018) Identification of lncRNA expression profiles and ceRNA analysis in the spinal cord of morphine-tolerant rats. Mol Brain 11:21
Shi X, Sun M, Liu H et al (2015) A critical role for the long non-coding RNA GAS5 in proliferation and apoptosis in non-small-cell lung cancer. Mol Carcinog 54(Suppl 1):E1–E12
Silva JM, Perez DS, Pritchett JR et al (2010) Identification of long stress-induced non-coding transcripts that have altered expression in cancer. Genomics 95:355–362
Sim MS, Soga T, Pandy V et al (2017) MicroRNA expression signature of methamphetamine use and addiction in the rat nucleus accumbens. Metab Brain Dis 32:1767–1783
Sleutels F, Zwart R, Barlow DP (2002) The non-coding Air RNA is required for silencing autosomal imprinted genes. Nature 415:810–813
Spangenberg EE, Lee RJ, Najafi AR et al (2016) Eliminating microglia in Alzheimer's mice prevents neuronal loss without modulating amyloid-beta pathology. Brain 139:1265–1281
Spigoni G, Gedressi C, Mallamaci A (2010) Regulation of Emx2 expression by antisense transcripts in murine cortico-cerebral precursors. PLoS One 5:e8658
Stouder C, Somm E, Paoloni-Giacobino A (2011) Prenatal exposure to ethanol: a specific effect on the H19 gene in sperm. Reprod Toxicol (Elmsford, NY) 31:507–512
Sullivan PF, Neale BM, van den Oord E et al (2004) Candidate genes for nicotine dependence via linkage, epistasis, and bioinformatics. Am J Med Genet B Neuropsychiatr Genet 126B:23–36
Tollervey JR, Curk T, Rogelj B et al (2011) Characterizing the RNA targets and position-dependent splicing regulation by TDP-43. Nat Neurosci 14:452–458
Tosetti V, Sassone J, Ferri ALM et al (2017) Transcriptional role of androgen receptor in the expression of long non-coding RNA Sox2OT in neurogenesis. PLoS One 12:e0180579
Tsuiji H, Yoshimoto R, Hasegawa Y et al (2011) Competition between a noncoding exon and introns: Gomafu contains tandem UACUAAC repeats and associates with splicing factor-1. Genes Cells 16:479–490
Valadkhan S, Valencia-Hipolito A (2016) lncRNAs in stress response. Curr Top Microbiol Immunol 394:203–236
Vausort M, Wagner DR, Devaux Y (2014) Long noncoding RNAs in patients with acute myocardial infarction. Circ Res 115:668–677

Verhulst B, Neale MC, Kendler KS (2015) The heritability of alcohol use disorders: a meta-analysis of twin and adoption studies. Psychol Med 45:1061–1072

Volkow ND, Muenke M (2012) The genetics of addiction. Hum Genet 131:773–777

Walker DM, Nestler EJ (2018) Neuroepigenetics and addiction. Handb Clin Neurol 148:747–765

Wang A, Wang J, Liu Y et al (2017) Mechanisms of long non-coding RNAs in the assembly and plasticity of neural circuitry. Front Neural Circuits 11:76

Wang J, Yang K, Yuan W et al (2018a) Determination of serum exosomal H19 as a noninvasive biomarker for bladder cancer diagnosis and prognosis. Med Sci Monit 24:9307–9316

Wang Q, Li M, Shen Z et al (2018b) The long non-coding RNA MEG3/miR-let-7c-5p axis regulates ethanol-induced hepatic steatosis and apoptosis by targeting NLRC5. Front Pharmacol 9:302

Warnault V, Darcq E, Morisot N et al (2016) The BDNF valine 68 to methionine polymorphism increases compulsive alcohol drinking in mice that is reversed by tropomyosin receptor kinase B activation. Biol Psychiatry 79:463–473

Wong CC, Schumann G (2008) Review. Genetics of addictions: strategies for addressing heterogeneity and polygenicity of substance use disorders. Philos Trans R Soc Lond Ser B Biol Sci 363:3213–3222

Wu P, Zuo X, Deng H et al (2013) Roles of long noncoding RNAs in brain development, functional diversification and neurodegenerative diseases. Brain Res Bull 97:69–80

Xiong K, Long L, Zhang X et al (2017) Overview of long non-coding RNA and mRNA expression in response to methamphetamine treatment in vitro. Toxicol in Vitro 44:1–10

Xu K, Kranzler HR, Sherva R et al (2015) Genomewide association study for maximum number of alcoholic drinks in European Americans and African Americans. Alcohol Clin Exp Res 39:1137–1147

Xue Z, Zhang Z, Liu H et al (2019) lincRNA-Cox2 regulates NLRP3 inflammasome and autophagy mediated neuroinflammation. Cell Death Differ 26:130–145

Yan B, Yao J, Liu JY et al (2015) lncRNA-MIAT regulates microvascular dysfunction by functioning as a competing endogenous RNA. Circ Res 116:1143–1156

Yan P, Luo S, Lu JY et al (2017) Cis- and trans-acting lncRNAs in pluripotency and reprogramming. Curr Opin Genet Dev 46:170–178

Ye ZM, Yang S, Xia YP et al (2019) LncRNA MIAT sponges miR-149-5p to inhibit efferocytosis in advanced atherosclerosis through CD47 upregulation. Cell Death Dis 10:138

Yu V, Singh P, Rahimy E et al (2016) RNA-seq analysis identifies key long non-coding RNAs connected to the pathogenesis of alcohol-associated head and neck squamous cell carcinoma. Oncol Lett 12:2846–2853

Yuan W, Sun Y, Liu L et al (2017) Circulating LncRNAs serve as diagnostic markers for hepatocellular carcinoma. Cell Physiol Biochem 44:125–132

Zahr NM, Sullivan EV (2008) Translational studies of alcoholism: bridging the gap. Alcohol Res Health 31:215–230

Zanardini R, Fontana A, Pagano R et al (2011) Alterations of brain-derived neurotrophic factor serum levels in patients with alcohol dependence. Alcohol Clin Exp Res 35:1529–1533

Zhang K, Luo Z, Zhang Y et al (2016a) Circulating lncRNA H19 in plasma as a novel biomarker for breast cancer. Cancer Biomark 17:187–194

Zhang M, Gu H, Xu W et al (2016b) Down-regulation of lncRNA MALAT1 reduces cardiomyocyte apoptosis and improves left ventricular function in diabetic rats. Int J Cardiol 203:214–216

Zhang X, Tang X, Hamblin MH et al (2018) Long non-coding RNA Malat1 regulates angiogenesis in hindlimb ischemia. Int J Mol Sci 19:pii: E1723. https://doi.org/10.3390/ijms19061723

Zhang X, Nie X, Yuan S et al (2019) Circulating long non-coding RNA ENST00000507296 is a prognostic indicator in patients with dilated cardiomyopathy. Mol Ther Nucleic Acids 16:82–90

Zhao ZH, Hao W, Meng QT et al (2017) Long non-coding RNA MALAT1 functions as a mediator in cardioprotective effects of fentanyl in myocardial ischemia-reperfusion injury. Cell Biol Int 41:62–70

Zhou X, Zhang W, Jin M et al (2017) lncRNA MIAT functions as a competing endogenous RNA to upregulate DAPK2 by sponging miR-22-3p in diabetic cardiomyopathy. Cell Death Dis 8: e2929

Zhu J, Chen Y, Zhao N et al (2012) Distinct roles of dopamine D3 receptors in modulating methamphetamine-induced behavioral sensitization and ultrastructural plasticity in the shell of the nucleus accumbens. J Neurosci Res 90:895–904

Zhu L, Zhu J, Liu Y et al (2015) Methamphetamine induces alterations in the long non-coding RNAs expression profile in the nucleus accumbens of the mouse. BMC Neurosci 16:18

Zhu XX, Yan YW, Chen D et al (2016) Long non-coding RNA HoxA-AS3 interacts with EZH2 to regulate lineage commitment of mesenchymal stem cells. Oncotarget 7:63561–63570

Zuo L, Lu L, Tan Y et al (2014) Genome-wide association discoveries of alcohol dependence. Am J Addict 23:526–539

The Multifaceted Roles of LncRNAs in Diabetic Complications: A Promising Yet Perplexing Paradigm

Saumik Biswas and Subrata Chakrabarti

Contents

Abstract As the global epidemic of diabetes continues to grow at an alarming rate, the risk for developing diabetes-associated comorbidities will remain at an all-time high. Diabetic complications, which are separated into macro- and microvascular complications, are not only debilitating for the patient, but these chronic complications also pose significant treatment challenges for healthcare providers. In order to reduce the burden of diabetes and expand the clinical armamentarium for diagnosis and treatment, better understanding of the molecular mechanisms underlying the

S. Biswas · S. Chakrabarti (✉)
Department of Pathology and Laboratory Medicine, Schulich School of Medicine & Dentistry, Western University, London, ON, Canada
e-mail: sbiswas7@uwo.ca; Subrata.Chakrabarti@lhsc.on.ca

S. Jurga, J. Barciszewski (eds.), *The Chemical Biology of Long Noncoding RNAs*, RNA Technologies 11, https://doi.org/10.1007/978-3-030-44743-4_19

pathogenesis of diabetic complications becomes fundamental. Previously, a growing body of evidence has documented key cellular events that contribute to the progression of diabetic complications: angiogenesis, fibrosis, inflammation, and oxidative stress. However, in the last decade, recent advances in genomic technology have identified several novel epigenetic alterations that are also implicated in the progression of diabetes and its complications—alluding to the complex nature of this disease. Among these epigenetic processes, long noncoding RNAs (lncRNAs) are key regulators that are involved in altering gene expression without modifying the underlying nucleotide composition of the genome. Despite possessing no protein-coding potential and being greater than 200 nucleotides in length, lncRNAs can alter chromatin configuration by interacting with several enzymes that facilitate chromatin remodeling and gene regulation. In this chapter, we will provide a comprehensive overview of the well-documented lncRNAs that are known to impact the key cellular events implicated in the progression of diabetic complications.

Keywords Diabetic complications · Epigenetics · lncRNAs · Inflammation · Angiogenesis · Fibrosis · Oxidative stress

1 Introduction

In the next 20 years, diabetes is projected to impact over ~600 million people globally (Cho et al. 2018). The rising prevalence of diabetes becomes highly problematic, since the current burden of diabetes and its complications is already imposing considerable economic challenges on society, including intangible costs attributed to pain and suffering. A recent report estimates that the absolute global economic burden will increase from $1.3 trillion (USD) in 2015 to over $2 trillion by 2030 (Bommer et al. 2018). Undoubtedly, in order to mitigate the effects of diabetes, a sense of urgency is warranted for diabetes research, as exploratory studies will provide novel insights into the pathobiology of diabetic complications and open new avenues for better diagnosis and targeted therapies.

Hyperglycemia is the main initiator of diabetic tissue damage. Findings from early prospective clinical trials (Diabetes Control and Complications Trial [DCCT] and its follow-up, Epidemiology of Diabetes Interventions and Complications [EDIC] trial) have firmly established that the inability to maintain physiological blood glucose levels in diabetics is strongly associated with heightened risks for numerous diabetic microvascular complications, including diabetic retinopathy (DR), nephropathy (DN), and neuropathy (DNP), and macrovascular-related complications (i.e., carotid intima-media thickness and cardiovascular disease), while intensive glucose control was shown to profoundly reduce the progression of these complications in type 1 diabetic patients (Nathan 2014). In fact, 6.5 years into the DCCT study, the beneficial effects of intensive glucose control were so compelling in the intensive therapy group that all patients, who were originally placed on the conventional treatment regimen, were immediately placed onto intensive therapy—subsequently leading to the termination of the study. Interestingly, the findings from the follow-up EDIC trial, which used the same patient population from the DCCT, ultimately uncovered that a higher incidence of diabetic micro- and macrovascular complications greatly prevailed in the patients who were switched from conventional therapy to intensive therapy compared to their counterparts who were originally on intensive glycemic therapy during the DCCT—alluding to the concept of "metabolic memory," where target organs can remember early glycemic environments. As well, the mean glycated hemoglobin (HbA1c) levels were nearly equivalent between the groups (Nathan 2014), which further suggests that other inflammatory and vasoactive mechanisms can still be active despite the normalization of glucose levels. Of note, this phenomenon is not only exclusive to type 1 diabetes, but such a memory is also evident in type 2 diabetes, which has been demonstrated by the follow-up findings from another large clinical trial known as the UK Prospective Diabetes Study (UKPDS) (Holman et al. 2008).

Following the clinical demonstration of metabolic memory, decades of research have subsequently identified several biochemical aberrations that occur in a hyperglycemic environment, some of which include the formation of advanced glycation end products and its receptors (AGE/RAGE) and the activation of polyol, hexosamine, and protein kinase C (PKC) pathways (these processes have been reviewed in references (Chakrabarti et al. 2000; Lorenzi 2007; Buse 2006)). For the purposes of this chapter, we will not cover the biochemical pathways in great detail; however, we will briefly address the major pathological processes (i.e., oxidative stress, inflammation, angiogenesis, and fibrosis) involved in the progression of diabetic complications. Afterward, we will introduce epigenetics and then discuss the critical implications of long noncoding RNAs (lncRNAs) in each of these processes.

1.1 Oxidative Stress

Under homeostatic conditions in a cell, the electron transport chain (ETC) is strategically organized in such a way, where several protein complexes (termed as complexes I, II, III, and IV) are directing the flow of electrons in order to precisely regulate the levels of adenosine triphosphate (ATP). Following the metabolism of glucose in the tricarboxylic acid (TCA) cycle, NADH and $FADH_2$ (electron donors) are generated, which donate their electrons to complexes I and II, respectively. Electrons from these complexes are subsequently transferred to coenzyme Q, which are then passed to complex III, cytochrome C, complex IV, and finally molecular oxygen, where the electrons are reduced to water. During the transferring process, some of the energy generated from the electrons is used to pump protons across the mitochondrial membrane (particularly at complexes I, III, and IV), which assist in the formation of an electrochemical gradient (Giacco and Brownlee 2010).

The energy from the electrochemical gradient then drives ATP synthesis with the help of ATP synthase. In chronic hyperglycemic environments, however, the elevated presence of glucose allows for more glucose to be metabolized inside the cell, consequently heightening the level of electron donors that are directed to the ETC (Giacco and Brownlee 2010). The increase in electron shuttling and the ensuing efflux of protons then causes the mitochondrial electrochemical gradient to surpass a critical threshold, which ultimately causes disruptions in the ETC. In particular, once the threshold is reached, the electron-transferring capabilities of complex III are blocked, and the subsequent buildup of electrons at coenzyme Q is then individually donated to molecular oxygen in order to generate superoxide (an oxygen free radical). Once these reactive oxygen species are formed, these highly unstable molecules are capable of provoking cellular damage through the induction of DNA strand breaks (Pacher and Szabó 2005), which can then activate DNA repair enzymes (i.e., poly(ADP-ribose) polymerase; PARP) that further inhibit the activity of certain glycolytic enzymes (GAPDH, glyceraldehyde 3-phosphate dehydrogenase) and other nuclear proteins through direct chemical modifications (i.e., poly (ADP-ribosyl)ation). As in the case of GAPDH (Giacco and Brownlee 2010; Pacher and Szabó 2005), following its modification by PARP and export to the cytosolic compartment, the reduced activity of GAPDH leads to a buildup of metabolites that are upstream of GAPDH during glycolysis, where these glycolytic intermediates further trigger the damaging biochemical pathways (AGE/RAGE, PKC, hexosamine, and polyol) mentioned earlier.

Based on the array of evidences that have confirmed the central role of ROS in the pathogenesis of diabetic complications, an important question remains: Despite possessing short half-lives within the cellular milieu (Giacco and Brownlee 2010; Pacher and Szabó 2005), can highly reactive oxidants still have implications in the long-lasting metabolic memory even after the normalization of hyperglycemia? One possible answer could be that once ROS modifies certain biological molecules (i.e., nucleic acids, lipoproteins, and proteins), these modified molecules can then carry out their altered functions over a prolonged period of time, even after the normalization of glucose. Indeed, it has been shown that the re-establishment of good glycemic control following long-term poor glycemic control can still maintain elevated levels of retinal oxidative and nitrative stress in the diabetic retinas of animals (Kowluru 2003). Specifically, using a streptozotocin (STZ)-induced diabetes model involving Wistar rats, diabetic rats were subjected to 2 or 6 months of poor glycemic control (glycated hemoglobin levels greater than 11%) and then followed by 7 months of good glycemic control (glycated hemoglobin levels less than 5.5%). Interestingly, the reconstitution of good glycemic control after 2 months of poorly controlled diabetes reduced diabetes-induced elevations of retinal lipid peroxides and nitric oxide levels by ~50% but was unable to elicit beneficial effects on nitrotyrosine formation (a product of nitrative stress which is created between highly reactive intermediates, such as peroxynitrite, and tyrosine residues in a protein) in the diabetic retina. Conversely, in the diabetic rats that were initially subjected to 6 months of poor glucose control, 7 months of good glycemic control did not have significant effects on oxidative stress and nitric oxide levels, and the expressions of inducible nitric oxide synthase and nitrotyrosine still remained at heightened levels (greater than 80%) in the retinas of these rats compared to control rats. Similarly, in a separate study, we additionally demonstrated that NF-κB (an important nuclear transcription factor involved in inflammation and oxidative stress) and caspase-3 (a key protease responsible for the execution of apoptosis) levels were not easily reversed by the reinstitution of good glycemic control (for 7 months) following prolonged duration of diabetes (6 months) in the retinas of diabetic rats—alluding to the long-term impact of ROS on several types of molecules during the progression of DR (Kowluru et al. 2004). Nevertheless, studies by other investigators have also confirmed the ability of ROS to mediate the persistence of diabetic vascular complications after glucose normalization (Ihnat et al. 2007), further solidifying the critical impact of the mitochondria during the progression of diabetes and its complications.

1.2 Inflammation

Along with ROS, inflammation also plays a pivotal role in the development and progression of diabetes and its complications. Analysis of several inflammatory blood biomarkers showed that a chronic hyperglycemic state confers an increased risk for heightened inflammation in both type 1 (T1D) and type 2 (T2D) diabetic patients (Devaraj et al. 2006, 2007; Devaraj and Jialal 2000). Specifically, serum concentrations of C-reactive protein (CRP), soluble cell adhesion molecules (sICAM, sE-selectin, and sP-selectin), soluble CD40 ligand (sCD40L, a marker of elevated plaque burden), and nitrotyrosine levels were significantly elevated in T1D patients (without macrovascular disease) compared to control patients (Devaraj et al. 2006). Additionally, monocytes of diabetic patients were also shown to have significantly higher secretions of IL-6 and superoxide anion radicals compared to monocytes from controls—suggesting that diabetes can stimulate pro-inflammatory monocyte function. Another study showed that T1D patients with microvascular complications (DR, DN, and DNP) exhibited significant increases in CRP, nitrotyrosine, vascular cell adhesion molecules, and monocytic release of superoxide anions and IL-1β compared to diabetic patients without microvascular complications (Devaraj et al. 2007). To decipher potential pathways involved in the elevated expressions of inflammatory markers, it has been demonstrated that the NF-κB and MAPK pathways were significantly enhanced in the monocytes of diabetic patients. Interestingly, diabetic patients who presented with microvascular complications had greater activities of pp65 and pp38MAPK compared to diabetic patients without complications (Devaraj et al. 2007). Moreover, in T2D patients, similar diabetes-associated inflammatory profiles also exist where increased levels of CRP and cellular adhesion molecules and increased secretions of superoxide anions, IL-6, IL-1β, IL-8, and TNF-α by monocytes have been reported (Devaraj and Jialal 2000; Giulietti et al. 2007). Not only are inflammatory processes heightened in diabetic patients, but observations from a prospective nested case-control study, involving middle-aged women in the United States, suggest that certain inflammatory markers can also predict the future risk of acquiring type 2 diabetes. Notably, over a 4-year period, increasing plasma levels of IL-6 and CRP were associated with a higher risk of developing future diabetes, where a 28% (95% confidence interval [CI]: -1% to 65%, $P = 0.07$) and a 64% (95% CI: 22–218%, $P = 0.001$) increase in relative risk (for future diabetes) were observed per quartile increase in baseline IL-6 and CRP levels, respectively (Pradhan et al. 2001). Nevertheless, since inflammation is also implicated in insulin resistance (Shoelson et al. 2006) and many other pathological mechanisms, it becomes apparent that the process of inflammation is an important contributor to the pathogenesis of diabetes and its complications.

1.3 Angiogenesis

Physiological angiogenesis is a dynamic cellular process that is central to development, reproduction, and repair. Through the formation of new blood vessels from preexisting vasculature, angiogenesis is responsible for providing sufficient blood flow, nutrients, and oxygen to growing tissues. Based on these essential functions, it is no surprise that angiogenesis is implicated in an array of processes, ranging from fetal development to wound healing (Risau 1997). Angiogenesis is regulated by a delicate balance between angiogenic and antiangiogenic/angiostatic factors, and any shifts in this balance will ultimately govern whether the vascular network is decreased, increased, or stabilized. For instance, when certain stimuli, such as hypoxia, activate angiogenesis, various signal transduction cascades are altered that promote the expressions of many pro-angiogenic genes, leading to greater concentrations of angiogenic factors (including VEGFs, angiopoietins, TGF-β, PDGF, FGFs, nitric oxides, and various integrins) than angiostatic factors (i.e., angiostatin, thrombospondin, and others). Following upregulation, these angiogenic factors initiate the migration and proliferation of endothelial cells (ECs, cells which line the vasculature) in preexisting blood vessels with the help of various molecular mediators and cell types, which ultimately contribute to the formation of hollow capillary tubes and eventual stabilization and maturation of the newly formed blood vessels (Potente et al. 2011; Arroyo and Iruela-Arispe 2010). Under normal physiological conditions, the multistep process of angiogenesis ceases following the resolution of the initial angiogenic stimulus and the subsequent reinstitution of homeostatic mechanisms that sustain the delicate balance between positive and negative angiogenic molecules (Arroyo and Iruela-Arispe 2010). However, as a result of chronic diabetes, pathological angiogenesis occurs and continuously stimulates the neovascularization pathways in order to promote angiogenic dysfunction, which significantly alters both microvascular and macrovascular beds and contributes to the vascular abnormalities observed in diabetic patients. In particular, impairments in angiogenesis have been well documented in DR (Chakrabarti et al. 2000; Cukiernik et al. 2004), DN (De Vriese et al. 2001), DNP (Schratzberger et al. 2001), diabetic cardiomyopathy (DCM) (Feng et al. 2016; Yoon et al. 2005), and abnormal wound healing (Galiano et al. 2004). Furthermore, among the regulatory angiogenic molecules, VEGF is a potent angiogenic factor expressed by numerous ECs and non-ECs. VEGF has been studied extensively over several decades, where VEGF expressions are known to be stimulated by several pathological processes, such as inflammation (Angelo and Kurzrock 2007), AGEs (Okamoto et al. 2002), hypoxia (Levy et al. 1995), and oxidative stress (Wang et al. 2011), through a transcriptional regulation involving a complex milieu of transcription factors (i.e., HIF), mediator complexes (i.e., p300, a histone acetyltransferase), and noncoding RNAs (which will be discussed later). Nevertheless, although differential VEGF expressions have been documented between diabetic complications (Chakrabarti et al. 2000; Cukiernik et al. 2004; De Vriese et al. 2001; Schratzberger et al. 2001; Feng et al. 2016; Yoon et al. 2005), the collective findings from these studies allude to the dynamic nature of VEGF and its cellular networks in different organ systems during chronic diabetes—necessitating further exploration of novel molecular pathways.

1.4 Fibrosis

There is no doubt that constant hyperglycemic insults, under a diabetic milieu, evoke functional and structural alterations in the vasculature (Khan and Chakrabarti 2012; Biswas and Chakrabarti 2017). Some of the earliest changes in all diabetes-affected organs involve alterations in extracellular matrix (ECM) proteins and thickening of the basement membrane (BM) (Biswas et al. 2019). The continual progression of biochemical and morphological disturbances of the ECM eventually leads to organ fibrosis (a pathological response to tissue injury) and subsequent loss of organ function (Russo and Frangogiannis 2016). Generally, our ECM is comprised of an insoluble network of elastins, collagens, fibronectin (FN), proteoglycan, hyaluronans, integrins, and structural glycoproteins that provide structural support to cells and help mediate several critical events including cell growth, migration, proliferation, survival, adhesion, differentiation, and interactions between cells and between cells and the ECM of vascular tissues (Hayden et al. 2005). Although tissue-to-tissue variations exist for extracellular matrices, the homeostatic mechanisms that ensure appropriate ECM maintenance are quite universal across tissues. For example, in order for tissues and organs to possess normal structure and function, ECM remodeling is a fundamental mechanism, employed by cells, that involves the constant balance between the synthesis and degradation of ECM components (Lu et al. 2011). Several growth factors, such as TGF-β (Li et al. 2003), FGF (fibroblast growth factor) (Ingber and Folkman 1989), CTGF (connective tissue growth factor) (Heng et al. 2006), PDGF (platelet-derived growth factor) (Luttenberger et al. 2000), and VEGF (Kuiper et al. 2007), are known to stimulate the synthesis and deposition of ECM proteins, while, conversely, MMPs (matrix metalloproteinases) and serine proteases and their respective inhibitors, TIMPs (tissue inhibitors of metalloproteinases) and PAI-1 (plasminogen activator inhibitor-1), are the enzymes that are responsible for the degradation and remodeling of the ECM (Stamenkovic 2003; Ghosh and Vaughan 2012). Interestingly, over several decades of research, chronic hyperglycemia has been documented to alter numerous biochemical pathways and transcriptional mechanisms that disrupt the ECM turnover process, where the production of ECM proteins is further exacerbated and the degradation process of these ECM proteins is attenuated—leading to significant accumulation (McLennan et al. 1998; Khan and Chakrabarti 2012). In fact, considerable alterations in BM composition have been observed in the eyes of DR patients including increased depositions of FN (both total and cellular) and collagen (types I, III, IV [α1, α2], and V) in the inner limiting membrane (retinal BM); elevated levels of oncofetal FN (fibronectin isoform which contains the ED-B domain), tenascin, and type XII collagen in the BM of newly formed blood vessels located in the neovascularization zone (Ljubimov et al. 1996); and increased mRNA levels of *TGF-β1*, *VEGF*, *PlGF* (placental growth factor), *tenascin-C*, and *FN* in the retinal tissues affected by DR (Spirin et al. 1999). Of note, our laboratory has also reported elevations of several ECM- and angiogenic-related molecules present in the epiretinal membrane and vitreous humor of DR patients (Khan et al. 2004; Khan and Chakrabarti 2012; George et al. 2009).

Moreover, in DN, diabetes-related ECM alterations in the renal glomerulus are manifested as mesangial matrix expansion, tubulointerstitial fibrosis, glomerulosclerosis, and capillary basement membrane thickening (Fioretto and Mauer 2007). Interestingly, excessive ECM protein deposition can also occur in the early stages of DN, where immunohistochemical findings have shown increased expressions of type IV collagen and CTGF and decreased levels of MMP-2 in the kidneys of 5-month-old T2D Chinese Bama minipigs (Liu et al. 2007). In a similar manner, enlargements of the mesangial matrix in diabetic patients with pronounced diffuse glomerulosclerosis coincide with increased expressions of collagen (types IV and V), *FN*, and laminin, while the expression of BM-associated heparan sulfate proteoglycan is markedly reduced (Nerlich and Schleicher 1991); of note, distinct (and differential) alterations in the glomerular matrix can also result depending on the stages of legion progression in DN (Nerlich and Schleicher 1991; Mason and Wahab 2003).

Furthermore, in the myocardium, biosynthetic changes in the ECM present as interstitial and perivascular fibrosis during DCM (Asbun and Villarreal 2006). When examining these changes at the molecular level, previous studies from our laboratory have confirmed that collagen (Col1α1 and Col4α1), FGF, TGF-β1, *FN*, and inflammatory (IL-6, TNF-a, IL-1B, MCP-1, NF-KB) and angiogenic molecules (angiotensinogen and endothelin-1) are significantly augmented in the cardiac tissues of STZ-induced diabetic mice, and such upregulations accompanied focal cardiac fibrosis at 2 months (Chen et al. 2014; Feng et al. 2017). Similarly, in patients with non-insulin-dependent diabetes, the percentage of total interstitial fibrosis in the myocardium and differences in collagen phenotypes (specifically, collagen type III) are significantly higher compared to endomyocardial biopsies obtained from nondiabetic patients (Shimizu et al. 1993). Nevertheless, although experimental data supports that hyperglycemia activates several downstream biochemical pathways (i.e., AGEs and ROS (Giacco and Brownlee 2010; Chakrabarti et al. 2000)) and other pathological processes (i.e., inflammation (Kanasaki et al. 2013; Roy et al. 2016)) that contribute to ECM alterations and subsequent fibrosis of target organs affected by diabetes, accumulating research within the last three decades is beginning to demonstrate a strong role of epigenetics in such processes, which will be elaborated below (Chen et al. 2014; Feng et al. 2017; Biswas and Chakrabarti 2017, 2019; Biswas et al. 2018c).

2 Epigenetics

The effective management of diabetes and its complications has been greatly hindered with the presence of metabolic memory, which necessitates the need to discover and thoroughly investigate novel molecular paradigms that will open novel diagnostic and therapeutic avenues. In recent years, emerging mechanistic studies are beginning to demonstrate pivotal roles for epigenetics in day-to-day cellular functions and in the pathogenesis of many diseases (Zoghbi and Beaudet 2016; Biswas et al. 2018c, 2019). The term "epigenetics" includes any process that alters gene activity without changing the underlying DNA sequence, and the ensuing modifications can be heritable (Felsenfeld 2014). Epigenetic mechanisms can be governed by external determinants, such as environmental and lifestyle factors, in children and adults that contribute to aberrant gene expressions involved in metabolic disorders (Ling and Groop 2009). Furthermore, the perpetual and self-propagating nature of epigenetic modifications makes it one of the ideal mechanisms underlying metabolic memory, as several studies have demonstrated that epigenetic mechanisms not only persist long term but can also influence genes involved in inflammation, oxidative stress, and fibrosis despite the normalization of glucose (Roy et al. 1990; Zhong and Kowluru 2011; El-Osta et al. 2008). Although it is likely that the identification of novel epigenetic processes will continue to surface as the research progresses, given the existing technologies, three distinct epigenetic mechanisms are extensively studied: DNA methylation, histone post-translational modifications, and the activity of noncoding RNAs. Due to the breadth of information currently available for these epigenetic mechanisms, in the next section, we will introduce a unique subcategory of noncoding RNAs that are known as long noncoding RNAs (lncRNAs).

3 The Emergence of Long Noncoding RNAs (lncRNAs)

When the sequencing results materialized from the Human Genome Project, it quickly became evident that the majority of our genome ($>$ 98%) is comprised of nonprotein-coding DNA, while less than 2% of the total genomic sequence is represented by protein-coding regions (amounts to ~20,000 protein-coding genes) (Human Genome Sequencing Consortium 2004; Birney et al. 2007). Furthermore, as

Table 1 Regulatory capacities of lncRNAs

Type of mechanism	Description of mechanism
Scaffold	LncRNAs can serve as adaptors that tether pertinent protein subunits into distinct complexes (Zappulla and Cech 2006)
Guide	Once bound to a protein partner (i.e., chromatin-modifying enzymes), lncRNAs can direct enzymes to select regions of the genome (Rinn et al. 2007)
Enhancer	Certain lncRNAs can directly enhance the activation of neighboring genes (Ørom et al. 2010)
Decoy/sponge	LncRNAs can sequester proteins (i.e., transcription factors and alternative splicing factors) and small regulatory RNAs (i.e., miRNAs) in order to affect their regulation of target genes (Kino et al. 2010; Wang et al. 2018)

demonstrated through genomic tiling arrays and large-scale cDNA cloning projects (Birney et al. 2007; Bertone et al. 2004), the process of transcription is pervasive throughout the mammalian genome and is not only restricted to protein-coding regions. In fact, more than 90% of our genome is transcribed, and the resulting output of transcription produces a dynamic network of transcripts that includes thousands of nonprotein-coding RNAs (Birney et al. 2007). Although several classes of noncoding RNAs (ncRNAs) have been annotated over the years and continue to be uncovered with the advancements in genomic technologies (Kapranov et al. 2007), long noncoding RNAs (lncRNAs) are a fundamental class of RNA molecules that are defined by their length (greater than 200 nucleotides) and limited protein-coding capacities. LncRNAs have putative roles in both biological (i.e., embryonic development (Perry and Ulitsky 2016)) and pathological processes (Biswas et al. 2019) and can take on unique regulatory capacities when governing the expression of genes (Table 1). Moreover, certain lncRNAs may execute more than one mechanism and present with specific functionalities depending on their subcellular localization. For instance, lncRNAs that are localized in the cytoplasm are generally involved in posttranscriptional modifications that govern the stability and translation potential of mRNAs (Yoon et al. 2013), whereas lncRNAs primarily residing in the nucleus have implications in organizing nuclear architecture (Lai et al. 2013), alternative splicing (Hutchinson et al. 2007), and transcriptional regulation (Vance and Ponting 2014). Interestingly, certain lncRNAs can also be found in both the nucleus and cytoplasm (Miao et al. 2019) or other cellular compartments (such as the mitochondria (Mercer et al. 2011)), where these RNA molecules have versatile roles in shaping the epigenome, influencing biological processes (such as transcription and translation), and regulating organelle formation and function (Krause 2018). Aside from their subcellular localization, the site of biogenesis can also classify lncRNAs (Table 2). Namely, lncRNAs can be broadly categorized as either intergenic (not overlapping with any protein-coding loci) or intragenic/genic (overlapping with protein-coding genes), where intragenic lncRNAs are further classified as "antisense," "bidirectional," "intronic," or "sense" depending on their transcriptional orientation on the protein-coding loci. When compared to intragenic lncRNAs, long intergenic ncRNAs (lincRNAs), which arise from the intergenic

Table 2 Classification of lncRNAs based on their site of biogenesis

LncRNA classification	Subcategory	Description of lncRNA subcategory
Intergenic (does not intersect with protein-coding genes)	Long intervening/intergenic lncRNA	Transcribed from the intergenic DNA regions (regardless of orientation) and generally possess greater evolutionary conserved regions compared to intragenic lncRNAs
Intragenic/genic (overlaps with protein-coding genes)	Antisense	Transcribed from the antisense/opposite strand of a protein-coding gene and may overlap with coding exons
	Bidirectional	Transcribed in the opposite direction from the promoter of a protein-coding gene (generally less than 1 kb away)
	Intronic	Transcribed entirely within the intronic regions of a protein-coding gene and does not overlap with coding exons
	Sense	Transcribed from the sense/coding strand of a protein-coding gene and may overlap coding exons

regions of the human genome, oftentimes possess greater evolutionary conservation at both the sequence and RNA secondary structure level (Guttman et al. 2009; Ransohoff et al. 2018). Although particular differences in biogenesis exist between intragenic and intergenic lncRNAs, it is plausible that a majority of these lncRNAs share similar modes of action, through *cis*- or *trans*-regulatory mechanisms, in order to govern fundamental biochemical and cellular processes. Undoubtedly, long noncoding RNAs (lncRNAs) continue to evolve our understanding of the genomic landscape. The versatility of these nonprotein-coding molecules warrants serious consideration for in-depth investigations of their roles in pathophysiological mechanisms, as novel information on these critical mediators will not only add to the existing molecular paradigms, but such knowledge will also contribute to the development of better-targeted diagnostics and therapies.

4 LncRNAs and Diabetic Complications

Aberrant expressions of lncRNAs have been well documented in diabetic complications. Beginning with DR, microarray profiling of diabetic retinal tissues from STZ-induced mice at 2 months revealed that differential expressions of 303 lncRNAs existed between nondiabetic and diabetic retinas (downregulation of 214 lncRNAs and upregulation of 89 lncRNAs) (Yan et al. 2014). Subsequent bioinformatics analyses of the microarray findings further revealed that a dynamic co-expression network exists between mRNAs and lncRNAs and such a network is linked to critical signaling pathways (including MAPK and chemokine signaling) that are tightly associated with several DR-related pathological processes, such as neovascularization, neurodegeneration, and inflammation. Furthermore, extending to clinical samples, another study has also reported the presence of differential expression profiles of lncRNAs (263 upregulated and 164 downregulated) and mRNAs (192 upregulated and 379 downregulated) between the fibrovascular membranes of PDR patients who did or did not receive anti-VEGF (conbercept) treatment (Wang et al. 2019), which allude to the dynamic transcriptomic changes that ensue within the diabetic milieu.

When examining the heart, RNA sequencing has shown that 18,480 lncRNAs can be detected in the left ventricular tissues of humans and the profiles of these lncRNAs are dynamically regulated in advanced heart failure and after the support of a left ventricular assist device—these differential expressions can produce a highly unique lncRNA expression signature that can discriminate between cardiomyopathies of ischemic and nonischemic origins (Yang et al. 2014). In the context of diabetes, however, limited studies have examined the cardiac transcriptome (including lncRNAs) using a DCM model. In fact, presently, one study has profiled the global transcriptome in the diabetic heart, where cardiac tissues from 3-month-old male diabetic Akita mice (Insulin2 heterozygous, a T1D model) were obtained and subjected to transcriptomic profiling using both next-generation sequencing (NGS) and microarray technologies. Namely, using stringent filtering, NGS analyses revealed 12 differentially expressed unannotated lncRNAs in the Akita diabetic heart, while the microarray results determined two lncRNAs with differential expressions (*H19* was upregulated, while *NEAT1* was downregulated) (Kesherwani et al. 2017). Accompanying their ncRNA data, the researchers also found several coding transcripts (including angiopoietin-like 4 and CTGF) that were commonly upregulated across both NGS and microarray analyses in the diabetic Akita heart, and ingenuity pathway analysis further implicated these molecules in critical DCM signaling pathways, such as cardiac metabolism. Nevertheless, these findings, along with the work from others (Wilson et al. 2008), clearly demonstrate that alterations in the transcriptional network can profoundly impact other signaling cascades and in-depth investigations into such cellular cross-talks are warranted for future studies examining DCM.

Compared to the profiling work done in DR and DCM, extensive transcriptomic analyses have been carried out in animal renal tissues using a DN model (Tang et al. 2017; Chen et al. 2017; Zhang et al. 2018b; Reichelt-Wurm et al. 2019). RNA sequencing has identified that the transcriptome profiles of diabetic animal kidneys are well separated during the early development of DN; time points in the study included 0, 2, 4, and 8 weeks (Tang et al. 2017). Compared to the sorted proximal tubular cells (PTCs) obtained from kidneys (STZ-induced) of diabetic DBA/2 J mice at 0 weeks, the largest differential expressions of lncRNAs (~3426) and mRNAs (~3068) were observed at the 2-month time point—suggesting that a more altered global transcriptome is associated with the progression of DN. Interestingly, using kidney-specific ChIP-Seq data obtained from the ENCODE database, the same authors also demonstrated that 39.7% of transcriptional start sites in differentially expressed lncRNA genes overlapped with significant peaks for promoter-associated (H3K4me3, trimethylation of lysine 4 in histone 3) or enhancer-associated (H3K4me1) histone modifications, which infers that other epigenetic mechanisms are intricately involved in the regulation of ncRNAs during the pathogenesis of disease. Furthermore, gene ontology enrichment analysis (GO) of the sequencing data revealed that protein-coding genes that shared *cis* correlations with lncRNA genes had significantly enriched genomic products involved in transmembrane transport, protein kinase cascades, cell death, and cytokine stimulus. Undoubtedly, the collective findings from other transcriptomic studies (Chen et al. 2017; Zhang et al. 2018b; Reichelt-Wurm et al. 2019) also show that the dysregulation of both mRNAs and lncRNAs is evident during the pathogenesis of DN. Using this information, future mechanistic-based studies are required to further characterize the roles of lncRNAs in such a dynamic cellular network, as doing so will provide unique insights into existing molecular paradigms. Throughout the following sections, we will examine the individual contributions of well-studied lncRNAs in each of the major pathological processes related to diabetic complications.

5 LncRNAs as Regulators of Inflammation and Oxidative Stress in Diabetic Complications

Numerous lncRNA-based studies on diabetic complications are beginning to emerge, which are providing novel insights into the elaborate epigenetic networks. As noted in the previous subsection, several lncRNAs are dynamically altered in chronic diabetic complications, and such changes can influence distinct signaling cascades that ultimately initiate damaging cellular processes including oxidative stress and inflammation. It is important to note that the total number of lncRNAs continues to grow on a day-to-day basis and, as such, covering all of the pertinent lncRNAs across diabetic complications in one chapter will be quite difficult. Therefore, in the subsection below, we will take an in-depth look at one of the popular lncRNAs reported to influence such inflammatory processes in DR, DCM, and DN: *MALAT1*.

5.1 MALAT1 (Metastasis-Associated Lung Adenocarcinoma Transcript 1)

MALAT1 is a heavily studied intergenic lncRNA that was originally identified in non-small cell lung carcinoma (Ji et al. 2003). Since its initial discovery, *MALAT1* has been shown to be involved in numerous diseases, ranging from cancers (Hirata et al. 2015) to neural development (Bernard et al. 2010), and is highly conserved across evolution and ubiquitously expressed in tissues (Eißmann et al. 2012). In the context of DR, aberrant *MALAT1* expressions have been documented in the retinal tissues of diabetic animals, high glucose (HG)-treated RF/6A (choroid-retinal endothelial) cells, and HRECs (human retinal microvascular endothelial cells) and in the aqueous humors, vitreous humors, and fibrovascular membranes of DR patients (Yan et al. 2014; Liu et al. 2014; Biswas et al. 2018b). Accordingly, the knockdown of *MALAT1* (via intraocular administration of a *MALAT1* shRNA) alleviates retinal inflammation in diabetic rats by preventing the induction of inflammatory and vasoactive proteins: ICAM-1, VEGF, and TNF-α (Liu et al. 2014). Accompanying the reduction in retinal inflammation, the knockdown of *MALAT1* also decreases vascular leakage, electroretinogram abnormalities, impairments in the retinal vessels (i.e., pericyte loss and acellular capillaries), and retinal cell apoptosis in the diabetic animal eye—confirming the involvement of *MALAT1* in microvascular dysfunction during DR. Likewise, in vitro siRNA-mediated knockdown experiments of *MALAT1* revealed that this lncRNA could additionally regulate the viability, migration, and tube formation potential of RF/6A cells in diabetic environments potentially through the p38 MAPK signaling pathway. In particular, *MALAT1* knockdown can substantially reduce phosphorylated p38 levels in HG environments, and the pretreatment of *MALAT1*-overexpressing RF/6A cells with a specific p38 MAPK pathway inhibitor (SB203580) can significantly prevent the hyper-proliferative effects of *MALAT1* in these cells (Liu et al. 2014). Similarly, other researchers have indicated that HG-induced upregulations of *MALAT1* can promote apoptosis and oxidative stress in human lens epithelial cells through heightened p38 MAPK signaling and the subsequent knockdown of *MALAT1* in such hyperglycemic environments can dramatically reverse HG-induced upregulations of p38 and oxidative stress (Gong et al. 2018). Based on these findings and the known implications of p38 MAPK signaling in diabetes-induced retinal inflammation (Du et al. 2010), it becomes apparent that other key regulatory molecules, such as lncRNAs, exist in the cross-talk between inflammation, oxidative stress, and the progression of DR. Further extending the inflammatory capabilities of *MALAT1*, a recent study from our laboratory demonstrated that *MALAT1* regulates several inflammatory markers (TNF-α, IL-6, MCP-1, and IL-1β) through its association with other epigenetic mediators, such as histone and DNA methyltransferases (Biswas et al. 2018b). For example, when we examined the direct relationships between *MALAT1*, inflammation, and polycomb repressive complex 2 (PRC2, a histone methyltransferase complex responsible for the repressive chromatin mark, H3K27me3 (Holoch and Margueron 2017)), it became evident that the knockdown or knockout of *MALAT1* in HG-treated HRECs or diabetic animal retinas, respectively, directly prevented glucose-induced increases in both inflammatory cytokines and PRC2 components (*EZH2*, *SUZ12*, and *EED*). Interestingly, in HRECs, the knockdown of *MALAT1* can markedly reduce the protein expressions of EZH2, and RNA immunoprecipitation (RIP) experiments also confirmed that *MALAT1* shares a strong binding association with EZH2 (the main catalytic subunit of PRC2) under hyperglycemic environments. Furthermore, the in vitro disruption of histone or DNA methylation, through the administration of pharmacological inhibitors, greatly alters the expressions of *MALAT1* and its inflammatory targets in HG-treated HRECs, which allude to the perplexing interactions shared between lncRNAs and chromatin-modifying enzymes in diabetic environments.

Not only does *MALAT1* have a role in DR, but studies have also implicated this lncRNA in the pathogenesis of DCM (Zhang et al. 2016a, b; Bacci et al. 2018; Gordon et al. 2018). For example, *MALAT1* levels were found to be significantly elevated in the cardiac tissues of diabetic Sprague-Dawley rats (Zhang et al. 2016a). Accompanying the elevated *MALAT1* levels in diabetic rats, hemodynamic measurements of the diabetic hearts also revealed increases in left ventricular end-diastolic pressures and reductions in the left ventricular systolic pressures and in the maximal ascending and descending rates of left ventricular pressure. Remarkably, intracoronary injections of *MALAT1*-shRNA could significantly improve left ventricular diastolic and systolic functions in diabetic hearts, as well as reduce diabetes-induced apoptosis of cardiomyocytes compared to control diabetic rats administered a scrambled shRNA. In a separate study by the same researchers, *MALAT1* knockdown in diabetic rats was shown to improve diabetes-induced cardiac dysfunction and reduce myocardial inflammation, where significant decreases in IL-6, IL-1β, and TNF-α were noted (Zhang et al. 2016b). Further extending the molecular mechanisms for *MALAT1* in DCM, hyperglycemia augments *MALAT1* expressions in cardiomyocytes through the heightened binding of the transcription factor CREM in the *MALAT1* promoter. Accordingly, ChIP analyses demonstrated that HG causes significant recruitment of CREM to the *MALAT1* promoter in H9C2 cells and subsequent silencing of CREM abrogates the HG-mediated inductions of *MALAT1* (Bacci et al. 2018). In addition to CREM, the authors also found that the treatment with a nitric oxide donor or a phosphodiesterase type 5 inhibitor dramatically counteracted the diabetes-induced upregulations of *MALAT1* in the hearts of diabetic mice—demonstrating the potential involvement of nitric oxide signaling in the transcriptional regulation of lncRNAs in DCM. As for the work done from our laboratory, cardiac tissues from *Malat1* knockout diabetic mice demonstrated significantly lower levels of pro-inflammatory cytokines (TNF-α, IL-1β, IL-6, and IFN-γ) compared to wild-type diabetic animals at both 1- and 2-month time points (Gordon et al. 2018). Intriguingly, our echocardiographic analyses also confirmed that the knockout of *MALAT1* improved cardiac/diastolic function in diabetic mice compared to wild-type diabetic controls, further confirming *MALAT1*'s ability to mediate diabetes-induced tissue damage.

Current research also provides direct evidence for *MALAT1* in the pathogenesis of DN (Gordon et al. 2018; Li et al. 2017; Hu et al. 2017; Puthanveetil et al. 2015). For instance, one study has reported that *MALAT1* can regulate renal tubular epithelial pyroptosis (pro-inflammatory programmed cell death) in DN. More specifically, siRNA-mediated knockdown of *MALAT1* in HG-treated human proximal tubular epithelial cells (HK-2) significantly downregulated the expressions of pyroptosis-related proteins, which included ELAVL1, NLRP3, caspase-1, and IL-1β (Li et al. 2017). The authors also shed unique insights into the perplexing interactive relationships between miRNA (miR)-23c and *MALAT1*, where *MALAT1* targets miR-23c and miR-23c can directly repress ELAVL1 expressions, which in turn reduces the downstream expressions of NLRP3, caspase-1 (a pro-pyroptotic protein), and IL-1β (a pro-inflammatory cytokine). Accordingly, at the in vivo level, *MALAT1* expressions are increased and further accompanied with upregulations of the aforementioned pyroptosis-related proteins, while miR-23c expressions were reduced in renal tubular epithelial tissues of diabetic rats. Conversely, when miR-23c mimics were administered to HG-treated HK-2 cells, *MALAT1* RNA expressions were unaffected, but ELAVL1, NLRP3, caspase-1, and IL-1β protein levels were markedly reduced, further implicating miR-23c's direct ability in inhibiting HG-induced pyroptosis in HK-2 cells. Furthermore, sequence alignment between miR-23c and the 3′ untranslated region (UTR) of *MALAT1* revealed that a putative miR-23c target site existed in the *MALAT1* gene and luciferase assays further confirmed the specific interaction between these two ncRNAs. Based on these findings, the authors proposed that *MALAT1* may be sequestering miR-23c, as a molecular sponge, to ultimately downregulate miR-23c levels, which do not effectively suppress ELAVL1 expressions, ultimately leading to pyroptosis. Moreover, in another study, *MALAT1* was shown to induce podocyte damage in early DN by heightening the nuclear translocation of β-catenin (a key mediator in the WNT signaling pathway) and expressions of SRSF1 (a pre-mRNA splicing factor which binds to *MALAT1*). Namely, knockdown of *MALAT1* in HG-treated podocytes can partially restore podocyte function and prevent the nuclear accumulation of β-catenin and overexpression of SRSF1, which confirms the pivotal role of *MALAT1* in early HG-induced podocyte damage (Hu et al. 2017). Further supporting the inflammatory profile of *MALAT1* in DN, a previous report from our laboratory demonstrated that *MALAT1* is upregulated in the renal tissues of diabetic animals and is accompanied with elevated IL-6 and TNF-α levels (Puthanveetil et al. 2015). Mechanistic experiments further revealed that *MALAT1* mediates the expressions of inflammatory markers through the activation of serum amyloid antigen 3 (SAA3, an inflammatory ligand) in HG-treated endothelial cells and *MALAT1* silencing can subsequently dampen both glucose-induced oxidative stress and inflammatory marker expressions. We have also confirmed that the knockout of *MALAT1* in diabetic animals protects the kidneys against diabetes-induced upregulations of inflammatory markers (TNF-α, IL-1β, IL-6, and IFN-λ) and 24-h albumin-to-creatinine ratios compared to wild-type diabetic animals (Gordon et al. 2018). Taken together, it is evident that the increased production of *MALAT1* exerts an inflammatory phenotype in diabetic complications. Nevertheless, it is important to note that *MALAT1* is not only restricted to inflammation, but this lncRNA has also been shown to regulate angiogenesis in HG-treated HRECs (Liu et al. 2019) and the neonatal retina (Michalik et al. 2014), which suggests that future investigations should consider holistic approaches when assessing lncRNAs in the pathobiology of diabetes.

6 LncRNAs as Angiogenic Mediators

Although limited research has examined angiogenic lncRNAs in DCM and DN, several studies have explored the angiogenic capabilities of lncRNAs in DR. Among these lncRNAs, *ANRIL* and *MIAT* are critical angiogenic regulators in the pathogenesis of DR and will be mainly discussed in the following subsections.

6.1 ANRIL (Antisense RNA to INK4 Locus, Also Known as CDKN2B-AS1)

Being antisense to the *CDKN2B* ($p15^{INK4B}$) locus, which is critically involved in cell growth regulation, *ANRIL* is a 3.8-kb lncRNA that arises from the human chromosome 9p21.3 region (Zhuang et al. 2012). Once RNA polymerase II transcribes *ANRIL* (which initially spans nearly 126.3 kb and consists of at least 19 exons), this massive preliminary transcript then undergoes alternative splicing mechanisms that ultimately produce numerous circular and linear isoforms in a tissue-specific manner (Holdt et al. 2016). Interestingly, when *ANRIL* was first identified in a melanoma-neural system tumor family (Pasmant et al. 2007), subsequent studies have unraveled its implications in glaucoma (Burdon et al. 2011), various cancers (Li et al. 2016b), and atherosclerosis/coronary artery disease (Holdt et al. 2010; Zhuang et al. 2012), among many other conditions. In the context of diabetes, however, very few studies have actually examined the role of *ANRIL* using diabetes-related experimental models. As a matter of fact, only one report (Zhang et al. 2017b) and reports from our laboratory have alluded to *ANRIL*'s pathogenetic role in diabetic complications (Thomas et al. 2017, 2018). Notably, in the cortical tissues of diabetic rats complicated with cerebral infarction, *ANRIL* levels were elevated and accompanied increases in VEGF, NF-κB, FLT-1 (also known as VEGFR1), and p-IκB/IκB proteins, when compared to control rats (Zhang et al. 2017b). Remarkably, when *ANRIL* expressions were knocked out in the cerebral infarcted diabetic rats, significant reductions of VEGF, FLT-1, NF-κB, and p-IκB/IκB protein expressions were observed in the brain tissues. Conversely, further overexpression of *ANRIL* augmented VEGF expressions and improved angiogenesis through the heightened activation of NF-κB signaling in diabetic rats with cerebral infarction (Zhang et al. 2017b). Extending *ANRIL*'s angiogenic capabilities, our laboratory has shown that a hyperglycemic environment upregulates *ANRIL* expression in both HRECs and diabetic mice retinas and the inhibition of *ANRIL* markedly lowers glucose-induced retinal angiogenesis in both in vitro and in vivo models (Thomas et al. 2017). Specifically, siRNA-mediated knockdown of *ANRIL* in HG-treated HRECs suppressed glucose-induced increases in cellular proliferation, endothelial cell tube formation, and VEGF expressions, while retinal tissues of *ANRIL* knockout diabetic animals similarly revealed drastic reductions in VEGF expressions and retinal microvascular permeability (as shown by diminished IgG leakage) compared to wild-type diabetic retinas. When examining the mechanistic role of *ANRIL* in DR, this lncRNA was shown to exert its pro-angiogenic functionalities through possible interactions with important epigenetic mediators, which include PRC2, miR-200b, and P300 (a prominent histone acetyltransferase). In fact, RIP analyses confirmed that *ANRIL* shares strong binding associations with PRC2 (specifically EZH2) and P300 during hyperglycemia and the silencing or knockout of *ANRIL* dramatically repressed glucose-induced increases in *P300* and *PRC2* (*EZH2* and *EED* but not *SUZ12*) RNA levels in DR. In parallel with these findings, we also found that silencing of *ANRIL* prevents glucose-induced downregulations of miR-200b (Thomas et al. 2017), which is a direct target of VEGF (Ruiz et al. 2015; McArthur et al. 2011). On a similar note, we have additionally reported that *ANRIL* regulates the heightened production of ECM proteins (Col1α4 and FN) and VEGF in DCM and DN, which is possibly driven by *ANRIL*'s association with PRC2 and P300. Remarkably, such diabetes-induced alterations were prevented in the heart and kidneys of *ANRIL* knockout diabetic mice (Thomas et al. 2018). *ANRIL* knockdown has also been shown to prevent glucose-induced upregulations of endothelin-1 (Biswas et al. 2018a), a potent vasoconstrictor that is abnormally regulated in diabetic organs (Khan et al. 2006). Taken together, our reported findings do not only allude to the highly interactive networks that exist between epigenetic molecules, but we also provide novel evidence on shared lncRNA mechanisms that may exist across diabetic complications.

6.2 *MIAT (Myocardial Infarction-Associated Transcript, Also Known as Gomafu, RNCR2, or AK028326)*

Since its initial identification as a susceptible locus for myocardial infarction (Ishii et al. 2006), the lncRNA *MIAT* has been profoundly implicated in a multitude of biological and pathological processes including brain and retinal development (Aprea et al. 2013; Blackshaw et al. 2004), cancers (Luan et al. 2017), schizophrenia (Barry et al. 2014), post-infarct cardiac fibrosis (Qu et al. 2017), and cataract formation (Shen et al. 2016). Although *MIAT* expressions can also mediate cardiomyocyte apoptosis in DCM (Zhou et al. 2017) and glucose-induced renal tubular epithelial injury in DN (Zhou et al. 2015), its angiogenic capabilities are strongly demonstrated in DR (Yan et al. 2015). In particular, significant upregulations of *MIAT* were detected in several HG-treated endothelial cell lines (ranging from RF/6A cells to HUVECs), in retinal tissues of type 1 diabetic rats (STZ-induced) and type 2 diabetic mice (db/db), and in the fibrovascular membranes of diabetic patients. Furthermore, in order to determine the mechanistic insights of *MIAT* in diabetic conditions, Yan et al. directly administered intravitreal injections of *MIAT* shRNA into the eyes of diabetic rats. Remarkably, *MIAT* knockdown dramatically alleviated diabetes-induced electroretinogram aberrations, retinal vascular leakage, apoptosis of retinal cells, capillary degeneration, and retinal inflammation;

in fact, direct reductions of *MIAT* parallel reductions in VEGF, ICAM-1, and TNF-α proteins. Moreover, both in silico analyses and in vitro functional assays further established that *MIAT* could act as a molecular sponge for specific miRNAs in endothelial cells. In particular, StarBase prediction software indicated that four specific miRNA binding sites (miR-150-5p, miR-29a-3p, miR-29b-3p, and miR-29c-3p) existed in the *MIAT* sequence, and using this information, the researchers cloned the *MIAT* cDNA into a luciferase vector and subsequently transfected this vector into RF/6A cells prior to adding different miRNA mimics. Based on their luciferase findings, only miR-150-5p mimics were able to markedly reduce the activity of *MIAT* luciferase vectors. Similarly, at the in vivo level, intravitreal injections of miR-150-5p mimics into the eyes of nondiabetic rats directly reduced *MIAT* levels, while miR-150-5p antagomirs enhanced *MIAT* expressions; similar relationships between miR-150-5p and *MIAT* were also observed in diabetic rat eyes (Yan et al. 2015). With *MIAT* being confirmed as a direct target of miR-150-5p, the researchers further revealed that *MIAT* could influence *VEGF* gene expressions by sponging miR-150-5p, which is a negative regulator of VEGF. Interestingly, using different concentrations of miR-150-5p mimics with the presence or absence of *MIAT* in RF/6A cells, it was found that *MIAT* overexpression can significantly upregulate VEGF levels and such upregulations can be tapered by increments of miR-150-5p—alluding to the unique interplay shared between lncRNAs, miRNAs, and mRNAs. *MIAT* can exert similar sponging effects on another miRNA, miR-29b, in HG-treated Müller cells, which subsequently promotes cellular apoptosis in DR (Zhang et al. 2017a). Nevertheless, many other lncRNAs are also surfacing as critical angiogenic mediators in DR (i.e., *MEG3* (Zhang et al. 2018a)), and as a result, we encourage researchers to continue exploring the vast capabilities of lncRNAs across all diabetic complications, as such studies will bring novel insights into the impact of diabetes on organ-specific transcriptomes.

7 Fibrogenic Potential of LncRNAs in Diabetic Complications

In addition to the inflammatory and angiogenic profiles of lncRNAs, emerging studies are beginning to emphasize the importance of lncRNAs in fibrosis-related processes. While studies have identified numerous fibrogenic lncRNAs in DN, very few studies have actually examined such lncRNAs using DCM and DR models. In the subsections below, we look at two fibrosis-related lncRNAs that are unique to diabetic complications.

7.1 *H19*

Nearly 2.3 kb in length after splicing, this conserved lncRNA arises from the maternally imprinted *H19* gene and is a crucial lncRNA that is implicated in growth and development (Gabory et al. 2010). In addition to *H19*'s role in embryonic development, emerging studies in recent years have documented the involvement of *H19* in corneal neovascularization (Sun et al. 2019), cancers (Jiang et al. 2016), atherosclerosis (Pan 2017), muscle insulin sensitivity (Geng et al. 2018), ischemic stroke (Wang et al. 2017), and many other processes. When considering diabetic complications, downregulated *H19* levels have been reported in the diabetic hearts of STZ-treated Sprague-Dawley rats (Li et al. 2016a; Zhuo et al. 2017). Notably, in both studies, the overexpression of *H19* exerted considerable protective effects in the hearts of diabetic rats, where diabetes-induced inflammation, oxidative stress, apoptosis of cardiomyocytes, and cardiomyocyte autophagy were significantly attenuated and left ventricular functions were dramatically improved. Mechanistic examinations further demonstrated in cardiomyocytes that the *H19*/miR675/VDAC1 (voltage-dependent anion channel 1) axis contributed to HG-induced apoptosis in DCM (Li et al. 2016a). As for the process of autophagy in DCM, *H19* can directly bind to EZH2, and the subsequent knockdown of *H19* could reduce EZH2 occupancy and H3K27me3 binding in the promoter region of *DIRAS3*, ultimately heightening DIRAS3 expressions and autophagy in cardiomyocytes (Zhuo et al. 2017). Moreover, a recent study from our laboratory has demonstrated the ability of *H19* to mediate the phenotypic switch of endothelial cells in diabetic environments (this process is known as endothelial-to-mesenchymal transition, EndMT, where endothelial cells lose their endothelial markers and develop a more mesenchymal phenotype) (Thomas et al. 2019). More specifically, HG-treated HRECs showed decreases in *H19* and endothelial cell markers (VE-cadherin and CD-31), while mesenchymal markers were significantly upregulated (SM22-α, α-SMA, and FSP-1). Remarkably, the overexpression of *H19* in HRECs drastically prevented the trends elicited by HG, which suggests that *H19* has a protective role in impeding EndMT in DR. In parallel with our in vitro findings, significant reductions of *H19* RNA levels were also observed in the vitreous humors of PDR patients. As well, retinal tissues from *H19* knockout nondiabetic mice revealed an EndMT phenotype (decreased endothelial markers and increased mesenchymal markers) that was comparable to retinal tissues from diabetic wild-type and diabetic *H19* knockout mice. Further expanding our findings, *H19* was found to suppress glucose-induced EndMT via the MAPK-ERK1/2 pathway of TGF-β signaling. Collectively, our study has identified a novel role for *H19* in EndMT (a critical mechanism which can contribute toward the progression of organ fibrosis), and future research should continue to explore the implications of *H19* in diabetes-related processes (neovascularization, inflammation, and ECM protein alterations) across other diabetic complications.

7.2 *Erbb4-IR*

Initially identified in a kidney fibrosis mouse model by RNA sequencing, *Erbb4-IR* (np_5318) is a novel lncRNA that is significantly upregulated in the unilateral ureteral obstructed kidneys of mice and is a critical mediator of TGF-β/Smad3-mediated renal fibrosis (Zhou et al. 2014; Feng et al. 2018). Uniquely, *Erbb4-IR* is situated within an intronic region that lies between the first and second exons of the *Erbb4* gene on mouse chromosome 1, and previous ChIP analyses confirmed that the Smad3 protein directly binds to the promoter region of *Erbb4-IR* (Feng et al. 2018). In the context of DN, a recent report has demonstrated significant upregulations of *Erbb4-IR* in the diabetic kidneys of *db/db* mice, and such upregulations were associated with the activation of TGF-β/Smad3 signaling and progressive renal fibrosis (Sun et al. 2018). Interestingly, shRNA-mediated knockdown of *Erbb4-IR* in the kidneys of *db/db* mice improved renal injury and decreased serum creatinine and microalbuminuria excretion. In terms of renal fibrosis, glucose-induced increases in collagen I, collagen IV, TGF-β1 expression, and Smad3 phosphorylation were markedly suppressed following the knockdown of *Erbb4-IR*, which also paralleled the findings from in vitro experiments. Furthermore, in vitro mechanistic experiments revealed that AGEs could specifically induce *Erbb4-IR* expressions through Smad3-dependent mechanisms, since *Smad3* knockout (and not *Smad2*) was able to prevent AGE-induced *Erbb4-IR* expressions. Extending the mechanisms behind *Erbb4-IR*, the researchers also identified that *Erbb4-IR* shares an interactive relationship with the anti-fibrotic miRNA, miR-29. In fact, as evident by their in vitro and in vivo findings, *Erbb4-IR* knockdown markedly increased miR-29 levels, and transfection of miR-29b mimics significantly reduced *Erbb4-IR*-mediated collagen I expression in mouse tubular epithelial cells (Sun et al. 2018). Nevertheless, although the fibrogenic roles of *Erbb4-IR* have not yet been documented in DCM and DR, the continual discovery and characterization of novel ECM mediators will improve our understanding of the cellular and molecular mechanisms underlying fibrosis in diabetic organs.

8 Concluding Remarks

In recent years, the rapid advent of genomic technologies has drastically improved our capacity to survey the intricate complexities of the genomic landscape. What were once considered "junk DNA" or "dark matter," lncRNAs (and other ncRNAs) are now proving to be dynamic regulators of gene expression, as these versatile RNA molecules can interact with different layers of regulation: transcriptional, post-transcriptional, and epigenetic. In the present chapter, we have discussed some of the major pathological mechanisms that commonly occur in diabetic complications and the implications of lncRNAs in such processes (Fig. 1).

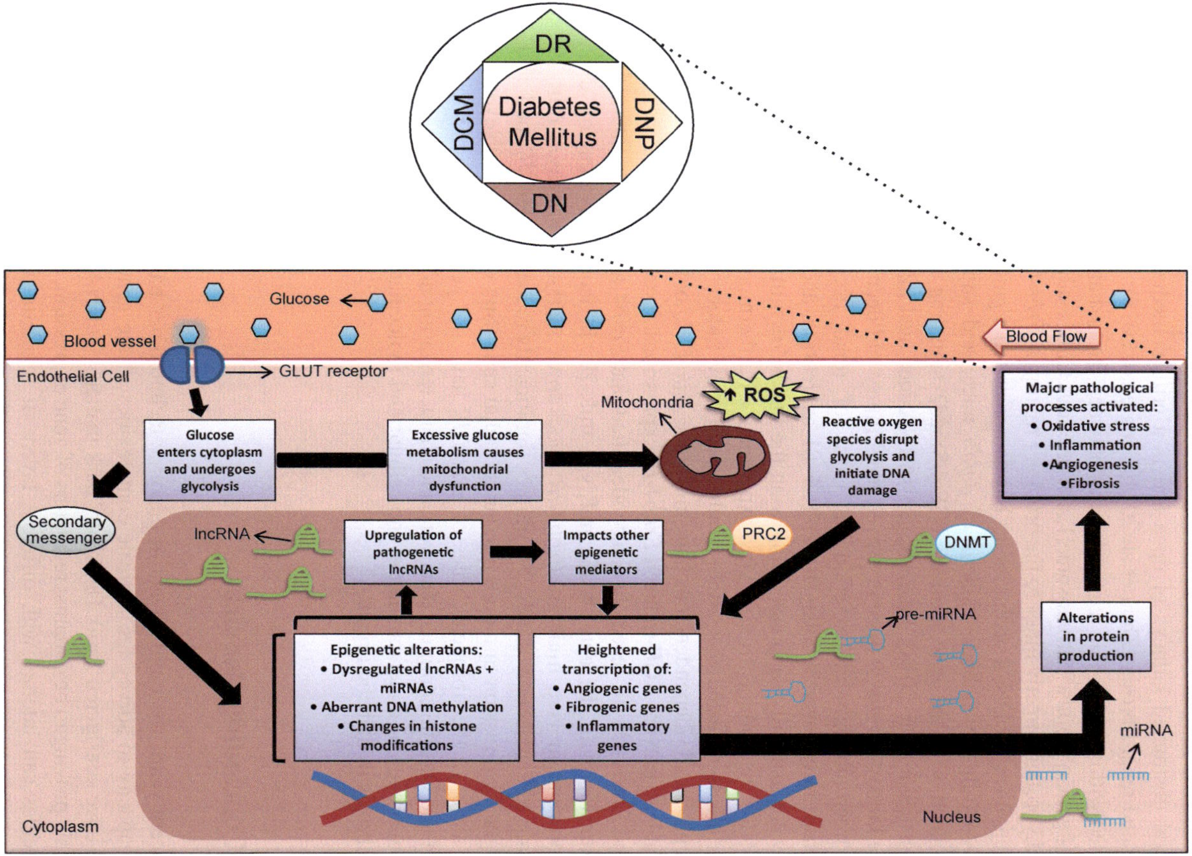

Fig. 1 The major pathogenetic processes underlying diabetic complications. In a diabetic environment, endothelial cells (ECs) are constantly exposed to hyperglycemia. The increased influx of glucose into the cell ultimately evokes mitochondrial dysfunction, which promotes the generation of reactive oxygen

species (ROS) and subsequent DNA damage. Following DNA damage, DNA repair enzymes (such as poly(ADP-ribose) polymerase) are activated and further inhibit the activity of certain glycolytic enzymes (GAPDH), which leads to the buildup of metabolites during glycolysis. These metabolites then activate other signaling cascades that heighten the transcription of pertinent diabetes-related genes and induce various epigenetic alterations. During diabetes, pathogenetic long noncoding RNAs (lncRNAs) are generally upregulated, and such upregulations impact other epigenetic mediators (histone methyltransferases, PRC2; DNA methyltransferases, DNMT; microRNAs, miRNA) and diabetes-related molecules through a variety of mechanisms. Nevertheless, as a result of aberrant molecular signaling and the upregulation of diabetes-related molecules, major pathological processes, which include angiogenesis, inflammation, oxidative stress, and fibrosis, are activated and contribute to the development of microvascular complications: diabetic retinopathy (DR), cardiomyopathy (DCM), nephropathy (DN), and neuropathy (DNP)

Although new, unannotated lncRNAs are frequently being discovered, added efforts must be taken to functionally characterize these lncRNAs in the field of diabetes. Additional mechanistic understanding on lncRNAs will allow us to appreciate the dynamic regulatory networks behind these RNA molecules in pathophysiological processes, which will further create novel avenues for research and drug discovery in diabetes. As well, from the lncRNAs we have examined in our chapter, it is evident that lncRNAs coexist in a very coordinated molecular network. Therefore, going forward, future studies should consider implementing integrated experimental approaches. Such an approach, and with the help of appropriate computational and database-driven tools, will help pinpoint potential lncRNA regions in disease-specific contexts. Once the target lncRNA is identified, subsequent gain-of-function and loss-of-function experiments will further provide more in-depth understanding on the lncRNA functionalities. Nevertheless, given the pervasive nature of transcription throughout the mammalian genome, it is not surprising that other new ncRNA groups (circular RNAs, snoRNAs, and piwi-interacting RNAs) are also beginning to emerge as critical players in diabetes (Shang et al. 2018; Lee et al. 2016; Henaoui et al. 2017), which truly adds new dimensions to this perplexing transcriptional paradigm and requires further consideration as well.

Acknowledgments The research presented in this chapter was supported by the Canadian Institute of Health Research and the Heart and Stroke Foundation of Ontario. The authors would also like to acknowledge all past and present members in the Chakrabarti Laboratory for their ongoing support in the advancement of diabetes research.

References

Angelo LS, Kurzrock R (2007) Vascular endothelial growth factor and its relationship to inflammatory mediators. Clin Cancer Res 13:2825–2830

Aprea J, Prenninger S, Dori M et al (2013) Transcriptome sequencing during mouse brain development identifies long non-coding RNAs functionally involved in neurogenic commitment. EMBO J 32:3145–3160

Arroyo AG, Iruela-Arispe ML (2010) Extracellular matrix, inflammation, and the angiogenic response. Cardiovasc Res 86:226–235

Asbun J, Villarreal FJ (2006) The pathogenesis of myocardial fibrosis in the setting of diabetic cardiomyopathy. J Am Coll Cardiol 47:693–700

Bacci L, Barbati S, Colussi C et al (2018) Sildenafil normalizes MALAT1 level in diabetic cardiomyopathy. Endocrine 62:259–262

Barry G, Briggs J, Vanichkina D et al (2014) The long non-coding RNA Gomafu is acutely regulated in response to neuronal activation and involved in schizophrenia-associated alternative splicing. Mol Psychiatry 19:486–494

Bernard D, Prasanth K, Tripathi V et al (2010) A long nuclear-retained non-coding RNA regulates synaptogenesis by modulating gene expression. EMBO J 29:3082–3093

Bertone P, Stolc V, Royce T et al (2004) Global identification of human transcribed sequences with genome tiling arrays. Science 306:2242–2246

Birney E, Stamatoyannopoulos J, Dutta A et al (2007) Identification and analysis of functional elements in 1% of the human genome by the ENCODE pilot project. Nature 447:799–816
Biswas S, Chakrabarti S (2017) Pathogenetic mechanisms in diabetic retinopathy: from molecules to cells to tissues. In: Kartha CC, Ramachandran S, Pillai R (eds) Mechanisms of vascular defects in diabetes mellitus, vol 17. Springer, Cham, pp 209–247
Biswas S, Chakrabarti S (2019) Increased extracellular matrix protein production in chronic diabetic complications: implications of non-coding RNAs. Non-Coding RNA 5:1–25
Biswas S, Feng B, Thomas A et al (2018a) Endothelin-1 regulation is entangled in a complex web of epigenetic mechanisms in diabetes. Physiol Res 67:S115–S125
Biswas S, Thomas AA, Chen S et al (2018b) MALAT1: an epigenetic regulator of inflammation in diabetic retinopathy. Sci Rep 8:1–15
Biswas S, Thomas AA, Chakrabarti S (2018c) LncRNAs: proverbial genomic "junk" or key epigenetic regulators during cardiac fibrosis in diabetes? Front Cardiovasc Med 5:1–13
Biswas S, Sarabusky M, Chakrabarti S (2019) Diabetic retinopathy, lncRNAs, and inflammation: a dynamic, interconnected network. J Clin Med 8:1–26
Blackshaw S, Harpavat S, Trimarchi J et al (2004) Genomic analysis of mouse retinal development. PLoS Biol 2:1–21
Bommer C, Sagalova V, Heesemann E et al (2018) Global economic burden of diabetes in adults: projections from 2015 to 2030. Diabetes Care 41:963–970
Burdon KP, MacGregor S, Hewitt A et al (2011) Genome-wide association study identifies susceptibility loci for open angle glaucoma at TMCO1 and CDKN2B-AS1. Nat Genet 43:574–578
Buse MG (2006) Hexosamines, insulin resistance and the complications of diabetes: current status. Am J Physiol Endocrinol Metab 290:E1–E8
Chakrabarti S, Cukiernik M, Hileeto D et al (2000) Role of vasoactive factors in the pathogenesis of early changes in diabetic retinopathy. Diabetes Metab Res Rev 16:393–407
Chen S, Puthanveetil P, Feng B et al (2014) Cardiac miR-133a overexpression prevents early cardiac fibrosis in diabetes. J Cell Mol Med 18:415–421
Chen S, Dong C, Qian X et al (2017) Microarray analysis of long noncoding RNA expression patterns in diabetic nephropathy. J Diabetes Complicat 31:569–576
Cho NH, Shaw J, Karuranga S et al (2018) IDF diabetes atlas: global estimates of diabetes prevalence for 2017 and projections for 2045. Diabetes Res Clin Pract 138:271–281
Cukiernik M, Hileeto D, Evans T et al (2004) Vascular endothelial growth factor in diabetes induced early retinal abnormalities. Diabetes Res Clin Pract 65:197–208
De Vriese AS, Tilton R, Elger M et al (2001) Antibodies against vascular endothelial growth factor improve early renal dysfunction in experimental diabetes. J Am Soc Nephrol 12:993–1000
Devaraj S, Jialal I (2000) Low-density lipoprotein postsecretory modification, monocyte function, and circulating adhesion molecules in type 2 diabetic patients with and without macrovascular complications: the effect of α-tocopherol supplementation. Circulation 102:191–196
Devaraj S, Glaser N, Griffen S et al (2006) Increased monocytic activity and biomarkers of inflammation in patients with type 1 diabetes. Diabetes 55:774–779
Devaraj S, Cheung A, Jialal I et al (2007) Evidence of increased inflammation and microcirculatory abnormalities in patients with type 1 diabetes and their role in microvascular complications. Diabetes 56:2790–2796
Du Y, Tang J, Li G et al (2010) Effects of p38 MAPK inhibition on early stages of diabetic retinopathy and sensory nerve function. Invest Ophthalmol Vis Sci 51:2158–2164
Eißmann M, Gutschner T, Hämmerle M et al (2012) Loss of the abundant nuclear non-coding RNA MALAT1 is compatible with life and development. RNA Biol 9:1076–1087
El-Osta A, Brasacchio D, Yao D et al (2008) Transient high glucose causes persistent epigenetic changes and altered gene expression during subsequent normoglycemia. J Exp Med 205:2409–2417
Felsenfeld G (2014) A brief history of epigenetics. Cold Spring Harb Perspect Biol 6:1–10

Feng B, Cao Y, Chen S et al (2016) MiR-200b mediates endothelial-to-mesenchymal transition in diabetic cardiomyopathy. Diabetes 65:768–779

Feng B, Chen S, Gordon A et al (2017) miR-146a mediates inflammatory changes and fibrosis in the heart in diabetes. J Mol Cell Cardiol 105:70–76

Feng M, Tang P, Huang X et al (2018) TGF-β mediates renal fibrosis via the Smad3-Erbb4-IR long noncoding RNA axis. Mol Ther 26:148–161

Fioretto P, Mauer M (2007) Histopathology of diabetic nephropathy. Semin Nephrol 27:195–207

Gabory A, Jammes H, Dandolo L (2010) The H19 locus: role of an imprinted non-coding RNA in growth and development. BioEssays 32:473–480

Galiano RD, Tepper O, Pelo C et al (2004) Topical vascular endothelial growth factor accelerates diabetic wound healing through increased angiogenesis and by mobilizing and recruiting bone marrow-derived cells. Am J Pathol 164:1935–1947

Geng T, Liu Y, Xu Y et al (2018) H19 lncRNA promotes skeletal muscle insulin sensitivity in part by targeting AMPK. Diabetes 67:2183–2198

George B, Chen S, Chaudhary V et al (2009) Extracellular matrix proteins in epiretinal membranes and in diabetic retinopathy. Curr Eye Res 34:134–144

Ghosh AK, Vaughan DE (2012) PAI-1 in tissue fibrosis. J Cell Physiol 227:493–507

Giacco F, Brownlee M (2010) Oxidative stress and diabetic complications. Circ Res 107:1058–1070

Giulietti A, van Etten E, Overbergh L et al (2007) Monocytes from type 2 diabetic patients have a pro-inflammatory profile. 1,25-Dihydroxyvitamin D3 works as anti-inflammatory. Diabetes Res Clin Pract 77:47–57

Gong W, Zhu G, Li J et al (2018) LncRNA MALAT1 promotes the apoptosis and oxidative stress of human lens epithelial cells via p38MAPK pathway in diabetic cataract. Diabetes Res Clin Pract 144:314–321

Gordon AD, Biswas S, Feng B et al (2018) MALAT1: a regulator of inflammatory cytokines in diabetic complications. Endocrinol Diabetes Metab 1:1–11

Guttman M, Amit I, Garber M et al (2009) Chromatin signature reveals over a thousand highly conserved large non-coding RNAs in mammals. Nature 458:223–227

Hayden MR, Sowers JR, Tyagi SC (2005) The central role of vascular extracellular matrix and basement membrane remodeling in metabolic syndrome and type 2 diabetes: the matrix preloaded. Cardiovasc Diabetol 4:1–20

Henaoui I, Jacovetti C, Guerra Mollet I et al (2017) PIWI-interacting RNAs as novel regulators of pancreatic beta cell function. Diabetologia 60:1977–1986

Heng E, Huang Y, Black S et al (2006) CCN2, connective tissue growth factor, stimulates collagen deposition by gingival fibroblasts via module 3 and α6- and β1 integrins. J Cell Biochem 98:409–420

Hirata H, Hinoda Y, Shahryari V et al (2015) Long noncoding RNA MALAT1 promotes aggressive renal cell carcinoma through Ezh2 and interacts with miR-205. Cancer Res 75:1322–1331

Holdt LM, Beutner F, Scholz M et al (2010) ANRIL expression is associated with atherosclerosis risk at chromosome 9p21. Arterioscler Thromb Vasc Biol 30:620–627

Holdt LM, Stahringer A, Sass K et al (2016) Circular non-coding RNA ANRIL modulates ribosomal RNA maturation and atherosclerosis in humans. Nat Commun 7:1–14

Holman R, Paul S, Bethel M et al (2008) 10-year follow-up of intensive glucose control in type 2 diabetes. N Engl J Med 359:1577–1589

Holoch D, Margueron R (2017) Mechanisms regulating PRC2 recruitment and enzymatic activity. Trends Biochem Sci 42:531–542

Hu M, Wang R, Li X et al (2017) LncRNA MALAT1 is dysregulated in diabetic nephropathy and involved in high glucose-induced podocyte injury via its interplay with β-catenin. J Cell Mol Med 21:1–16

Human Genome Sequencing Consortium I (2004) Finishing the euchromatic sequence of the human genome international human genome sequencing consortium*. Nature 431:931–945

Hutchinson JN, Ensminger A, Clemson C et al (2007) A screen for nuclear transcripts identifies two linked noncoding RNAs associated with SC35 splicing domains. BMC Genomics 8:1–16

Ihnat MA, Thorpe J, Kamat C et al (2007) Reactive oxygen species mediate a cellular "memory" of high glucose stress signalling. Diabetologia 50:1523–1531

Ingber DE, Folkman J (1989) Mechanochemical switching between growth and differentiation during fibroblast growth factor-stimulated angiogenesis in vitro: role of extracellular matrix. J Cell Biol 109:317–330

Ishii N, Ozaki K, Sato H et al (2006) Identification of a novel non-coding RNA, MIAT, that confers risk of myocardial infarction. J Hum Genet 51:1087–1099

Ji P, Diederichs S, Wang W et al (2003) MALAT-1, a novel noncoding RNA, and thymosin b4 predict metastasis and survival in early-stage non-small cell lung cancer. Oncogene 22:8031–8041

Jiang X, Yan Y, Hu M et al (2016) Increased level of H19 long noncoding RNA promotes invasion, angiogenesis, and stemness of glioblastoma cells. J Neurosurg 124:129–136

Kanasaki K, Taduri G, Koya D (2013) Diabetic nephropathy: the role of inflammation in fibroblast activation and kidney fibrosis. Front Endocrinol 4:1–15

Kapranov P, Cheng J, Dike S et al (2007) RNA maps reveal new RMA classes and a possible function for pervasive transcription. Science 316:1486–1488

Kesherwani V, Shahshahan HR, Mishra PK (2017) Cardiac transcriptome profiling of diabetic Akita mice using microarray and next generation sequencing. PLoS One 12:1–17

Khan ZA, Chakrabarti S (2012) Glucose-induced cellular signaling in diabetic retinopathy. In: Tombran-Tink J, Barnstable CJ, Gardner TW (eds) Visual dysfunction in diabetes: the science of patient impairment and health care, vol 1. Springer, New York, pp 211–232

Khan ZA, Cukiernik M, Gonder J et al (2004) Oncofetal fibronectin in diabetic retinopathy. Invest Ophthalmol Vis Sci 45:287–295

Khan ZA, Farhangkhoee H, Mahon J et al (2006) Endothelins: regulators of extracellular matrix protein production in diabetes. Exp Biol Med 231:1022–1029

Kino T, Hurt D, Ichijo T et al (2010) Noncoding RNA Gas5 is a growth arrest- and starvation-associated repressor of the glucocorticoid receptor. Sci Signal 3:1–33

Kowluru RA (2003) Effect of reinstitution of good glycemic control on retinal oxidative stress and nitrative stress in diabetic rats. Diabetes 52:818–823

Kowluru RA, Chakrabarti S, Chen S (2004) Re-institution of good metabolic control in diabetic rats and activation of caspase-3 and nuclear transcriptional factor (NF-kB) in the retina. Acta Diabetol 41:194–199

Krause HM (2018) New and prospective roles for lncRNAs in organelle formation and function. Trends Genet 34:736–745

Kuiper EJ, Hughes J, Van Geest R et al (2007) Effect of VEGF-A on expression of profibrotic growth factor and extracellular matrix genes in the retina. Invest Ophthalmol Vis Sci 48:4267–4276

Lai F, Orom U, Cesaroni M et al (2013) Activating RNAs associate with mediator to enhance chromatin architecture and transcription. Nature 494:497–501

Lee J, Harris A, Holley C et al (2016) Rpl13a small nucleolar RNAs regulate systemic glucose metabolism. J Clin Invest 126:4616–4625

Levy AP, Levy N, Wegner S et al (1995) Transcriptional regulation of the rat vascular endothelial growth factor gene by hypoxia. J Biol Chem 270:13333–13340

Li JH, Huang X, Zhu H et al (2003) Role of TGF-β signaling in extracellular matrix production under high glucose conditions. Kidney Int 63:2010–2019

Li X, Wang H, Yao B et al (2016a) LncRNA H19/miR-675 axis regulates cardiomyocyte apoptosis by targeting VDAC1 in diabetic cardiomyopathy. Sci Rep 6:1–9

Li Z, Yu X, Shen J (2016b) ANRIL: a pivotal tumor suppressor long non-coding RNA in human cancers. Tumour Biol 37:5657–5661

Li X, Zeng L, Cao C et al (2017) Long noncoding RNA MALAT1 regulates renal tubular epithelial pyroptosis by modulated miR-23c targeting of ELAVL1 in diabetic nephropathy. Exp Cell Res 350:327–335
Ling C, Groop L (2009) Epigenetics: a molecular link between environmental factors and type 2 diabetes. Diabetes 58:2718–2725
Liu Y, Wang Z, Yin W et al (2007) Severe insulin resistance and moderate glomerulosclerosis in a minipig model induced by high-fat/high-sucrose/high-cholesterol diet. Exp Anim 56:11–20
Liu JY, Yao J, Li X et al (2014) Pathogenic role of lncRNA-MALAT1 in endothelial cell dysfunction in diabetes mellitus. Cell Death Dis 5:1–10
Liu P, Jia S, Shi J et al (2019) LncRNA-MALAT1 promotes neovascularization in diabetic retinopathy through regulating MIR-125b/VE-cadherin axis. Biosci Rep 39:1–10
Ljubimov A, Burgeson R, Butkowski R et al (1996) Basement membrane abnormalities in human eyes with diabetic retinopathy. J Histochem Cytochem 44:1469–1479
Lorenzi M (2007) The polyol pathway as a mechanism for diabetic retinopathy: attractive, elusive, and resilient. Exp Diabetes Res 2007:1–10
Lu P, Takai K, Weaver V et al (2011) Extracellular matrix degradation and remodeling in development and disease. Cold Spring Harb Perspect Biol 3:1–24
Luan T, Zhang X, Wang S et al (2017) Long non-coding RNA MIAT promotes breast cancer progression and functions as ceRNA to regulate DUSP7 expression by sponging miR-155-5p. Oncotarget 8:76153–76164
Luttenberger T, Schmid-Kotsas A, Menke A et al (2000) Platelet-derived growth factors stimulate proliferation and extracellular matrix synthesis of pancreatic stellate cells: implications in pathogenesis of pancreas fibrosis. Lab Investig 80:47–55
Mason RM, Wahab NA (2003) Extracellular matrix metabolism in diabetic nephropathy. J Am Soc Nephrol 14:1358–1373
McArthur K, Feng B, Wu Y et al (2011) MicroRNA-200b regulates vascular endothelial growth factor-mediated alterations in diabetic retinopathy. Diabetes 60:1314–1323
McLennan SV, Yue DK, Turtle JR (1998) Effect of glucose on matrix metalloproteinase activity in mesangial cells. Nephron 79:293–298
Mercer TR, Neph S, Dinger M et al (2011) The human mitochondrial transcriptome. Cell 146:645–658
Miao H, Wang L, Zhan H et al (2019) A long noncoding RNA distributed in both nucleus and cytoplasm operates in the PYCARD-regulated apoptosis by coordinating the epigenetic and translational regulation. PLoS Genet 15:1–24
Michalik K, You X, Manavski Y et al (2014) Long noncoding RNA MALAT1 regulates endothelial cell function and vessel growth. Circ Res 114:1389–1397
Nathan DM (2014) The diabetes control and complications trial/epidemiology of diabetes interventions and complications study at 30 years: overview. Diabetes Care 37:9–16
Nerlich A, Schleicher E (1991) Immunohistochemical localization of extracellular matrix components in human diabetic glomerular lesions. Am J Pathol 139:889–899
Okamoto T, Yamagishi S, Inagaki Y et al (2002) Angiogenesis induced by advanced glycation end products and its prevention by cerivastatin. FASEB J 16:1928–1930
Ørom UA, Derrien T, Beringer M et al (2010) Long noncoding RNAs with enhancer-like function in human cells. Cell 143:46–58
Pacher P, Szabó C (2005) Role of poly(ADP-ribose) polymerase-1 activation in the pathogenesis of diabetic complications: endothelial dysfunction, as a common underlying theme. Antioxid Redox Signal 7:1568–1580
Pan JX (2017) LncRNA H19 promotes atherosclerosis by regulating MAPK and NF-kB signaling pathway. Eur Rev Med Pharmacol Sci 21:322–328
Pasmant E, Laurendeau I, Héron D et al (2007) Characterization of a germ-line deletion, including the entire INK4/ARF locus, in a melanoma-neural system tumor family: identification of ANRIL, an antisense noncoding RNA whose expression coclusters with ARF. Cancer Res 67:3963–3969

Perry RBT, Ulitsky I (2016) The functions of long noncoding RNAs in development and stem cells. Development 143:3882–3894

Potente M, Gerhardt H, Carmeliet P (2011) Basic and therapeutic aspects of angiogenesis. Cell 146:873–887

Pradhan AD, Manson JE, Rifai N et al (2001) C-reactive protein, interleukin 6, and risk of developing type 2 diabetes mellitus. JAMA 286:327–334

Puthanveetil P, Chen S, Feng B et al (2015) Long non-coding RNA MALAT1 regulates hyperglycaemia induced inflammatory process in the endothelial cells. J Cell Mol Med 19:1418–1425

Qu X, Du Y, Shu Y et al (2017) MIAT is a pro-fibrotic long non-coding RNA governing cardiac fibrosis in post-infarct myocardium. Sci Rep 7:1–11

Ransohoff JD, Wei Y, Khavari PA (2018) The functions and unique features of long intergenic non-coding RNA. Nat Rev Mol Cell Biol 19:143–157

Reichelt-Wurm S, Wirtz T, Chittka D et al (2019) Glomerular expression pattern of long non-coding RNAs in the type 2 diabetes mellitus BTBR mouse model. Sci Rep 9:1–13

Rinn JL, Kertesz M, Wang J et al (2007) Functional demarcation of active and silent chromatin domains in human HOX loci by noncoding RNAs. Cell 129:1311–1323

Risau W (1997) Mechanisms of angiogenesis. Nature 386:671–674

Roy S, Sala R, Cagliero E et al (1990) Overexpression of fibronectin induced by diabetes or high glucose: phenomenon with a memory. Proc Natl Acad Sci U S A 87:404–408

Roy S, Amin S, Roy S (2016) Retinal fibrosis in diabetic retinopathy. Exp Eye Res 142:71–75

Ruiz MA, Feng B, Chakrabarti S (2015) Polycomb repressive complex 2 regulates MiR-200b in retinal endothelial cells: potential relevance in diabetic retinopathy. PLoS One 10:1–21

Russo I, Frangogiannis NG (2016) Diabetes-associated cardiac fibrosis: cellular effectors, molecular mechanisms and therapeutic opportunities. J Mol Cell Cardiol 90:84–93

Schratzberger P, Walter D, Rittig K et al (2001) Reversal of experimental diabetic neuropathy by VEGF gene transfer. J Clin Invest 107:1083–1092

Shang FF, Luo S, Liang X et al (2018) Alterations of circular RNAs in hyperglycemic human endothelial cells. Biochem Biophys Res Commun 499:551–555

Shen Y, Dong L, Zhou R et al (2016) Role of long non-coding RNA MIAT in proliferation, apoptosis and migration of lens epithelial cells: a clinical and in vitro study. J Cell Mol Med 20:537–548

Shimizu M, Umeda K, Sugihara N et al (1993) Collagen remodelling in myocardia of patients with diabetes. J Clin Pathol 46:32–36

Shoelson SE, Lee J, Goldfine AB (2006) Inflammation and insulin resistance. J Clin Invest 116:1793–1801

Spirin KS, Saghizadeh M, Lewin S et al (1999) Basement membrane and growth factor gene expression in normal and diabetic human retinas. Curr Eye Res 18:490–499

Stamenkovic I (2003) Extracellular matrix remodelling: the role of matrix metalloproteinases. J Pathol 200:448–464

Sun SF, Tang P, Feng M et al (2018) Novel lncRNA Erbb4-IR promotes diabetic kidney injury in db/db mice by targeting miR-29b. Diabetes 67:731–744

Sun B, Ding Y, Jin X et al (2019) Long non-coding RNA H19 promotes corneal neovascularization by targeting microRNA-29c. Biosci Rep 39:1–12

Tang W, Zhang D, Ma X (2017) RNA-sequencing reveals genome-wide long non-coding RNAs profiling associated with early development of diabetic nephropathy. Oncotarget 8:105832–105847

Thomas AA, Feng B, Chakrabarti S (2017) ANRIL: a regulator of VEGF in diabetic retinopathy. Invest Ophthalmol Vis Sci 58:470–480

Thomas AA, Feng B, Chakrabarti S (2018) Anril regulates production of extracellular matrix proteins and vasoactive factors in diabetic complications. Am J Physiol Endocrinol Metab 314:E191–E200

Thomas AA, Biswas S, Feng B et al (2019) lncRNA H19 prevents endothelial–mesenchymal transition in diabetic retinopathy. Diabetologia 62:517–530
Vance KW, Ponting CP (2014) Transcriptional regulatory functions of nuclear long noncoding RNAs. Trends Genet 30:348–355
Wang Y, Zang Q, Liu Z et al (2011) Regulation of vegf-induced endothelial cell migration by mitochondrial reactive oxygen species. Am J Phys Cell Physiol 301:C695–C704
Wang J, Zhao H, Fan Z et al (2017) Long noncoding RNA H19 promotes neuroinflammation in ischemic stroke by driving histone deacetylase 1-dependent M1 microglial polarization. Stroke 48:2211–2221
Wang Y, Sun L, Wang L et al (2018) Long non-coding RNA DSCR8 acts as a molecular sponge for miR-485-5p to activate Wnt/β-catenin signal pathway in hepatocellular carcinoma. Cell Death Dis 9:1–13
Wang J, Gao X, Liu J et al (2019) Effect of intravitreal conbercept treatment on the expression of long noncoding RNAs and mRNAs in proliferative diabetic retinopathy patients. Acta Ophthalmol 97:1–11
Wilson KD, Li Z, Wagner R et al (2008) Transcriptome alteration in the diabetic heart by rosiglitazone: implications for cardiovascular mortality. PLoS One 3(7):1–11
Yan B, Tao Z, Li X et al (2014) Aberrant expression of long noncoding RNAs in early diabetic retinopathy. Invest Ophthalmol Vis Sci 55:941–951
Yan B, Yao J, Liu J et al (2015) LncRNA-MIAT regulates microvascular dysfunction by functioning as a competing endogenous RNA. Circ Res 116:1143–1156
Yang KC, Yamada K, Patel A et al (2014) Deep RNA sequencing reveals dynamic regulation of myocardial noncoding RNAs in failing human heart and remodeling with mechanical circulatory support. Circulation 129:1009–1021
Yoon Y, Uchida S, Masuo O et al (2005) Progressive attenuation of myocardial vascular endothelial growth factor expression is a seminal event in diabetic cardiomyopathy. Circulation 111:2073–2085
Yoon JH, Abdelmohsen K, Gorospe M (2013) Posttranscriptional gene regulation by long noncoding RNA. J Mol Biol 425:3723–3730
Zappulla DC, Cech TR (2006) RNA as a flexible scaffold for proteins: yeast telomerase and beyond. Cold Spring Harb Symp Quant Biol 71:217–224
Zhang M, Gu H, Xu W et al (2016a) Down-regulation of lncRNA MALAT1 reduces cardiomyocyte apoptosis and improves left ventricular function in diabetic rats. Int J Cardiol 203:214–216
Zhang M, Gu H, Chen J et al (2016b) Involvement of long noncoding RNA MALAT1 in the pathogenesis of diabetic cardiomyopathy. Int J Cardiol 202:753–755
Zhang J, Chen M, Chen J et al (2017a) Long non-coding RNA MIAT acts as a biomarker in diabetic retinopathy by absorbing miR-29b and regulating cell apoptosis. Biosci Rep 37:1–10
Zhang B, Wang D, Ji T et al (2017b) Overexpression of lncRNA ANRIL up-regulates VEGF expression and promotes angiogenesis of diabetes mellitus combined with cerebral infarction by activating NF-κB signaling pathway in a rat model. Oncotarget 8:17347–17359
Zhang D, Qin H, Leng Y et al (2018a) LncRNA MEG3 overexpression inhibits the development of diabetic retinopathy by regulating TGF-β1 and VEGF. Exp Ther Med 16:2337–2342
Zhang X, Shang J, Wang X et al (2018b) Microarray analysis reveals long non-coding RNA SOX2OT as a novel candidate regulator in diabetic nephropathy. Mol Med Rep 18:5058–5068
Zhong Q, Kowluru RA (2011) Epigenetic changes in mitochondrial superoxide dismutase in the retina and the development of diabetic retinopathy. Diabetes 60:1304–1313
Zhou Q, Chung A, Huang X et al (2014) Identification of novel long noncoding rnas associated with TGF-β/Smad3-mediated renal inflammation and fibrosis by RNA sequencing. Am J Pathol 184:409–417
Zhou L, Xu D, Sha W et al (2015) Long non-coding MIAT mediates high glucose-induced renal tubular epithelial injury. Biochem Biophys Res Commun 468:726–732

Zhou X, Zhang W, Jin M et al (2017) lncRNA MIAT functions as a competing endogenous RNA to upregulate DAPK2 by sponging miR-22-3p in diabetic cardiomyopathy. Cell Death Dis 8:1–8
Zhuang J, Peng W, Li H et al (2012) Methylation of p15 INK4b and expression of ANRIL on chromosome 9p21 are associated with coronary artery disease. PLoS One 7:1–9
Zhuo C, Jiang R, Lin X et al (2017) LncRNA H19 inhibits autophagy by epigenetically silencing of DIRAS3 in diabetic cardiomyopathy. Oncotarget 8:1429–1437
Zoghbi HY, Beaudet AL (2016) Epigenetics and human disease. Cold Spring Harb Perspect Biol 8:1–28

Long Noncoding RNAs in Diabetes and β-Cell Regulation

Simranjeet Kaur, Caroline Frørup, Verena Hirschberg Jensen, Aashiq H. Mirza, Joana Mendes Lopes de Melo, Reza Yarani, Anne Julie Overgaard, Joachim Størling, and Flemming Pociot

Contents

Abstract Diabetes is characterized by an insufficient physiological response to increases in blood glucose. There are two major types of diabetes: type 1 diabetes (T1D) and type 2 diabetes (T2D). T1D is the result of an immune-mediated destruction of the pancreatic β cells, whereas T2D is characterized by reduced

S. Kaur · C. Frørup · V. H. Jensen · J. M. L. de Melo · R. Yarani · A. J. Overgaard · J. Størling
Steno Diabetes Center Copenhagen, Gentofte, Denmark
e-mail: simranjeet.kaur@regionh.dk; caroline.froerup.01@regionh.dk; verena.hirschberg.jensen@regionh.dk; joana.mendes.lopes.de.melo@regionh.dk; reza.yarani.01@regionh.dk; anne.julie.overgaard@regionh.dk; joachim.stoerling@regionh.dk

A. H. Mirza
Steno Diabetes Center Copenhagen, Gentofte, Denmark

Department of Pharmacology, Weill Cornell Medicine, New York, NY, USA
e-mail: aah2003@med.cornell.edu

F. Pociot (✉)
Steno Diabetes Center Copenhagen, Gentofte, Denmark

Faculty of Health and Medical Sciences, University of Copenhagen, Copenhagen, Denmark
e-mail: flemming.pociot@regionh.dk

S. Jurga, J. Barciszewski (eds.), *The Chemical Biology of Long Noncoding RNAs*, RNA Technologies 11, https://doi.org/10.1007/978-3-030-44743-4_20

β-cell function and insulin resistance. The incidence of both diabetes forms is increasing worldwide and has doubled since the 1980s. Long-term complications of diabetes include serious micro- and macro-vascular complications and an increased risk of premature death due to dysfunction and failure of various organs caused in part by the toxic effects of high blood glucose levels. In both forms of diabetes, genetic, epigenetic, and environmental factors contribute to the risk of developing the disease. Aberrant epigenetic modifications such as DNA methylation, histone modifications, and noncoding RNAs (ncRNAs) are well-recognized players in both T1D and T2D. Recent advances in chromosome conformation capture technologies have provided novel insights into the spatial arrangement of chromatin and have revealed the importance of β-cell-specific lncRNAs in gene regulation and 3D chromatin folding in the β cells. This chapter provides a comprehensive review covering lncRNAs in diabetes and their role in 3D chromatin architecture and β-cell dysfunction and apoptosis.

Keywords Biomarkers · Diabetes · Liver and muscle metabolism · Long noncoding RNA

1 Introduction

The most common forms of diabetes, type 1 diabetes (T1D) and type 2 diabetes (T2D), currently affect more than 400 million people worldwide, and the disease is predicted to rise beyond 642 million (55% increase; to 1 in 10 adults) in less than 25 years (http://www.diabetesatlas.org). Both T1D and T2D are complex multifactorial metabolic disorders resulting from the loss of glucose homeostasis due to insulin resistance and/or β-cell dysfunction. Central to both types of disease is the islets of Langerhans, which harbor the insulin-producing β cells inside minute endocrine organs dispersed throughout the pancreas.

Onset of T1D is the result of immune-mediated β-cell destruction, leading to absolute or near-absolute insulin deficiency. T cells (autoreactive CD8+ effector cells and CD4+ regulatory cells) together with antigen-presenting cells (dendritic cells and macrophages) play a major pathogenic role in islet-cell infiltration and destruction, which together with cytokines/chemokines (e.g., IL-1β, IFNγ, TNFα) are the constituents of the autoimmune processes (von Herrath et al. 2007). The etiology of T1D is still not fully understood. The onset of clinical diabetes occurs when 80–90% of the β-cell function is lost and manifests itself with insufficient insulin production. T1D most often occurs in childhood or during adolescence.

In T2D, insulin resistance in the skeletal muscle, liver, and adipose tissue leads to an increased demand for insulin. The disease manifests itself when the β cells no longer can compensate for this increased demand due to β-cell exhaustion and loss of β-cell mass and function. Several lifestyle factors such as obesity, physical inactivity, and an unhealthy diet, as well as chronic low-grade inflammation, are the main components leading to T2D (Kahn et al. 2014). T2D is the most common type of diabetes, accounting for around 90% of all diabetes cases, and is most commonly diagnosed in older adults, however it is increasingly seen in children, adolescents, and younger adults due to rising levels of obesity, physical inactivity, and poor diet especially in low- and middle-income countries (Kahn et al. 2014). Both T1D and T2D are caused by failure of the insulin-producing β cells caused by either immune- and/or metabolic-mediated functional impairment or destruction. Despite differences in etiology, genetics, epigenetics, and environmental factors contribute to the risk of both forms of diabetes, and compiling evidence suggests that many of these contributory factors converge at the β-cell level to mutually facilitate β-cell impairment.

2 LncRNAs and Diabetes

A large number of studies have highlighted the potential role of long noncoding RNAs (lncRNAs) in β-cell identity and function in human pancreatic islets (Singer et al. 2015). LncRNAs are characterized as RNA molecules longer than 200 nucleotides that map to the nonprotein-coding regions of the genome, previously thought not to be transcribed and referred to as "junk DNA." Over the past decade, however, lncRNAs have emerged as one of the most highly abundant RNA species in the human genome. Based on their genomic location, lncRNAs are categorized into five main classes: long intergenic RNAs (lincRNAs), intronic lncRNAs, antisense lncRNAs, divergent lncRNAs, and enhancer-derived lncRNAs (Ulitsky and Bartel 2013). The functional mechanisms of the lncRNAs are still to be fully elucidated. However, they have been studied for their roles in regulation of allelic expression, e.g., in genomic imprinting and X chromosomal inactivation, in embryonic and neuronal development, differentiation, and epigenetic changes, and act through interactions with protein, DNA, and RNA molecules (Ulitsky and Bartel 2013). LncRNAs can exert their functions in *cis* or *trans*, i.e., acting on genes from the chromosomes from which they are transcribed or acting on genes from other chromosomes than the ones they are transcribed from (Ulitsky and Bartel 2013).

LncRNAs have been implicated in more than 300 diseases (http://www.cuilab.cn/lncrnadisease). Their roles in diabetes etiology have not been fully deciphered yet; however, recent studies have shown that dysregulated lncRNAs play vital roles in inflammation, β-cell dysfunction, accelerated senescence, insulin secretion, insulin resistance (Morán et al. 2012; Arnes et al. 2016; Mirza et al. 2017; Motterle et al. 2017), and related complications (Leti and DiStefano 2017; Leung and Natarajan 2018). In recent studies, around a thousand islet-enriched lncRNAs have been identified. Many of these are specific for β-cell function or associated with disease susceptibility (Singer and Sussel 2018; DiStefano 2015). *MEG3* (You et al. 2016), *HI-LNC25* (Morán et al. 2012), *βlinc1* (Arnes et al. 2016), *TUNAR* (Akerman et al. 2017), and *PLUTO* (Akerman et al. 2017) are examples of lncRNAs associated to diabetes and β-cell (dys)function. Based on the recent genome-wide association studies (GWAS), many of these lncRNAs are located within the T1D/T2D risk loci or associated with islet/β-cell-specific chromatin domains (Mirza et al. 2014; Singer et al. 2015). In T1D, immune cells during the initial phase of the disease processes affect the islets and β-cell gene expression and result in progressive damage to these cells. The changes in transcriptome signature of both immune cells and target cells show a wide dysregulation of lncRNAs in both T1D and T2D (Morán et al. 2012; Fadista et al. 2014; Akerman et al. 2017; Sathishkumar et al. 2018). However, a better understanding of the precise mechanisms by which dysregulated lncRNAs affect the disease phenotype to advise novel treatment strategies is needed.

2.1 LncRNAs and Metabolic Control

LncRNAs execute metabolic control in tissues such as fat, liver, and muscle on various levels, from epigenetic regulation over direct transcriptional control to protein stabilization. Thereby, they can affect cell proliferation, lipid accumulation, or glucose sensitivity (Ji et al. 2018).

2.1.1 LncRNAs and Adipose Tissue Metabolism

LncRNAs play an important role in the maintenance of adipose tissue homeostasis, as well as in adipocyte differentiation. Several lncRNAs, including *PARAL1*, *ADINR*, and *U90926*, have been described to modify adipose tissue differentiation through interaction with the transcriptional regulators and through histone modifications (Xiao et al. 2015; Ji et al. 2018).

LncRNAs can also regulate adipose tissue function by modulating the secretion of adipokines, e.g., leptin and adiponectin. *Lnc-leptin* is transcribed from a locus upstream of the leptin gene and acts as an enhancer for the transcription of the adipose hormone leptin. *Lnc-leptin* expression is sensitive to insulin and upregulated by high-fat diet, and it can therefore act as a metabolic sensor, facilitating adipogenesis (Lo et al. 2018).

In the metabolically active brown adipose tissue (BAT), lncRNAs have been demonstrated to play a role in tissue development as well as tissue activation. For instance, *H19* is upregulated by cold in BAT and promotes adipogenesis (Schmidt et al. 2018), whereas *UC.417* impairs the activation of BAT in rodents and acts as an inhibitor of p38 phosphorylation (Ji et al. 2018). *βLNC1, GM13133*, and *BATE1* are involved in the thermogenic gene program and required for establishing BAT identity and thermogenic capacity (Ji et al. 2018). *GAS5* is a negative regulator of adipogenesis through the interaction with miR-18a (Li et al. 2018a) and miR-21a-5p (Liu et al. 2018).

2.1.2 LncRNAs and Liver Metabolism

In the liver, several lncRNAs are expressed depending on nutritional status and help to adjust liver metabolism accordingly. *LSTR* and *LGR* are regulated by fasting/refeeding and in turn modify bile acid production to regulate lipid homeostasis or switching from glycogen synthesis to gluconeogenesis during fasting (Zhao et al. 2017; Ji et al. 2018). *SRA* acts as a coactivator of nuclear receptors in hepatocytes and leads to reduced mobilization of fatty acids and increased lipid accumulation (Zhao et al. 2017). *APOA1-AS* modulates histone methylation patterns to reduce the expression of *APOA1*, the main component of HDL in plasma (Zhao et al. 2017). The expression of *H19* temporally increases during fasting, leading to increased hepatic glucose production. *H19* is chronically elevated in the liver of diet-induced obese mice and individuals with T2D and can lead to hyperglycemia and insulin resistance (Zhao et al. 2017).

MALAT1 plays a role in liver fibrosis and hepatic steatosis (Zhao et al. 2017). *MEG3* downregulation favors liver fibrosis, whereas its overexpression reduces fibrosis (He et al. 2014). *MEG3* competes with mir-let-7c-5p and increases *NLRC5* expression by preventing the binding of the miRNA to the *NLRC5* 3′UTR. Increased *NLRC5* expression promotes alcohol-induced liver fibrosis (Wang et al. 2018b). *MEG3* is also upregulated by high-fat diet and in ob/ob mice, and the overexpression of *MEG3* leads to increased hepatic gluconeogenesis and impaired hepatic glycogen accumulation (Zhao et al. 2017). *LncSHGL* is highly expressed in the liver and regulates both glucose and lipid metabolism in normal conditions (Ji et al. 2018). *LncSHGL* is reduced in the livers of obese mice, and overexpression of *LncSHGL* leads to improved glucose regulation and lipogenesis. *GOMAFU* has been shown to play a role in hepatic insulin resistance (Yan et al. 2018). *GOMAFU* expression is increased in the livers of ob/ob mice, which in turn leads to upregulated hepatic glucose production and impaired insulin sensitivity. It acts as a miR-139 sponge, which leads to a derepression of the target gene *Foxo1* (Yan et al. 2018). Overexpression of *RISA* in primary mouse hepatocytes leads to an attenuated phosphorylation of the insulin receptor and reduces the activation of AKT and GSK3B. On the contrary, knockdown of *RISA* alleviates insulin resistance (Wang et al. 2016).

2.1.3 LncRNAs and Skeletal Muscle Metabolism

The expression of *H19* is reduced in skeletal muscle of individuals with T2D, which is connected to impaired insulin signaling and reduced glucose uptake (Gao et al. 2014). *H19* acts as a sponge for miRNA let-7 and therefore leads to reduced expression of let-7 targets, such as the insulin receptor and lipoprotein lipase. Acute hyperinsulinemia, however, reduces *H19* expression (Gao et al. 2014). In a myogenic cell line, *H19* depletion, as well as let-7 overexpression, improves differentiation (Gao et al. 2014).

Muscle differentiation is also subject to lncRNA control. For example, *DUM* is expressed during myogenesis and promotes myoblast differentiation, as well as muscle regeneration. *Dum* acts via silencing its neighboring gene *Dppa2* through the recruitment of DNA methylases (Wang et al. 2015). *LncMD1* is a muscle-specific lncRNA, which sponges miR-133 and miR-135, leading to the activation of muscle-specific gene expression (Cesana et al. 2011).

2.2 *Circulating LncRNAs as Potential Biomarkers in Diabetes*

Several studies have explored the potential of circulating lncRNAs as potential biomarkers in T2D. Most studies have taken advantage of the recent developments in sequencing techniques to profile lncRNAs using peripheral blood mononuclear cells (PBMCs), serum, plasma, or whole blood.

Recently, it has been shown that significantly increased levels of *HOTAIR*, *MEG3*, *MALAT1*, *MIAT*, *CDKN2B-AS1/ANRIL*, *XIST*, *PANDA*, *GAS5*, *Linc-p21*, *ENST00000550337.1*, *PLUTO*, and *NBR2* are found in PBMCs from individuals with T2D compared to healthy controls (Sathishkumar et al. 2018). *THRIL* and *SALRNA1* are significantly downregulated in PBMCs of T2D individuals (Sathishkumar et al. 2018). Several of these lncRNAs have previously been associated with diabetes and diabetic complications (You et al. 2016; Akerman et al. 2017; Li et al. 2017). *PANDA* is linked to cellular senescence and cell cycle regulation, which correlate with poor glycemic control, insulin resistance, accelerated senescence, and increased inflammation (Puvvula et al. 2014). *THRIL* has been reported to regulate TNFα expression and negatively correlates with inflammation (Li et al. 2014). Inhibition of *SALRNA1* induces aging in fibroblasts by upregulating p53 (Abdelmohsen et al. 2013).

GAS5 expression levels was found in a study to be decreased in serum of male veterans diagnosed with T2D (Carter et al. 2015). Here, results indicated that individuals with *GAS5* levels lower than 10 ng/μl have almost 12 times higher likelihood of having T2D. Also, *GAS5* is decreased in individuals with hemoglobin A1c (HbA1c) levels between 41 and 46 mmol/mol, which are considered prediabetic levels (American Diabetes Association 2018), highlighting the possibility of using *GAS5* as a potential diagnostic biomarker of T2D. Circulating lncRNAs in serum of individuals with T2D and diabetic nephropathy have recently been identified as potential novel biomarkers (Yang et al. 2019). *ARAP1-AS1* is downregulated and *ARAP1-AS2* is upregulated in diabetes and in individuals with diabetic nephropathy. Also, *ARAP1* mRNA expression, which is regulated by *ARAP1-AS2*, is upregulated in the aforementioned groups. GWAS have shown that the *ARAP1* gene maps to a T2D susceptibility locus; thus, its upregulation could lead to increased predisposition to T2D.

Three lncRNAs (n342533, n335556, and n336109) are upregulated in whole blood from newly diagnosed T2D individuals and correlate with fasting plasma glucose and HbA1c (Wang et al. 2017). Another lncRNA *p3134* which was found to be upregulated in serum exosomes and whole blood from T2D individuals showed significant positive association with fasting blood glucose, fasting C-peptide, and HOMA-IR levels (Ruan et al. 2018).

Until now, none of the studies have investigated the potential of circulating lncRNAs as biomarkers for T1D.

3 Islet-Enriched LncRNAs

Islet-cell dysfunction is central to the pathophysiology of both T1D and T2D. It has been hypothesized that a reduced number of β cells rather than impaired β-cell function is a major factor for development of T2D (Butler et al. 2003). Over the last decade, an increasing number of studies have focused on the role of lncRNAs as novel regulators of β-cell differentiation and function (Singer et al. 2015; Mirza et al. 2017; Wong et al. 2018).

More than 1100 islet-specific lncRNAs have been identified in human pancreatic islets and purified β cells as well as in mouse pancreatic islet cells (Ku et al. 2012; Morán et al. 2012; Benner et al. 2014). The very first catalog of lncRNAs expressed in human islets and β cells from three human islet donors and two fluorescence-activated cell sorting (FACS)-purified β-cell preparations was created in 2012 (Morán et al. 2012). By combining ChIP-seq (to identify sites of active transcription) and RNA-seq (for expression profiling of transcripts), islet-specific lncRNA transcription was found to be linked with clusters of open chromatin. Several of these islet lncRNAs are β-cell specific and activated during β-cell differentiation. Furthermore, β-cell-specific lncRNAs have been shown to be regulated by extracellular glucose concentrations, which is suggestive of their role in functional adaptation of β cells to increased insulin secretory demands (Morán et al. 2012).

Several islet-specific lncRNAs have been identified based on (a) enriched expression in human islets and FACS-purified β cells relative to the exocrine pancreas and a panel of non-pancreatic tissues, (b) expression in the EndoC-βH1 cell line, and (c) chromatin profile in human islets consistent with active promoters (Akerman et al. 2017). Knockdown of these islet lncRNAs caused significant transcriptional changes on the expression of neighboring protein-coding genes, suggesting both *cis* and *trans* regulatory mechanisms.

3.1 β-Cell LncRNAs

The true relevance of the β-cell lncRNAs in pathogenesis of both T1D and T2D still remains to be addressed systematically. This section describes the role of β-cell

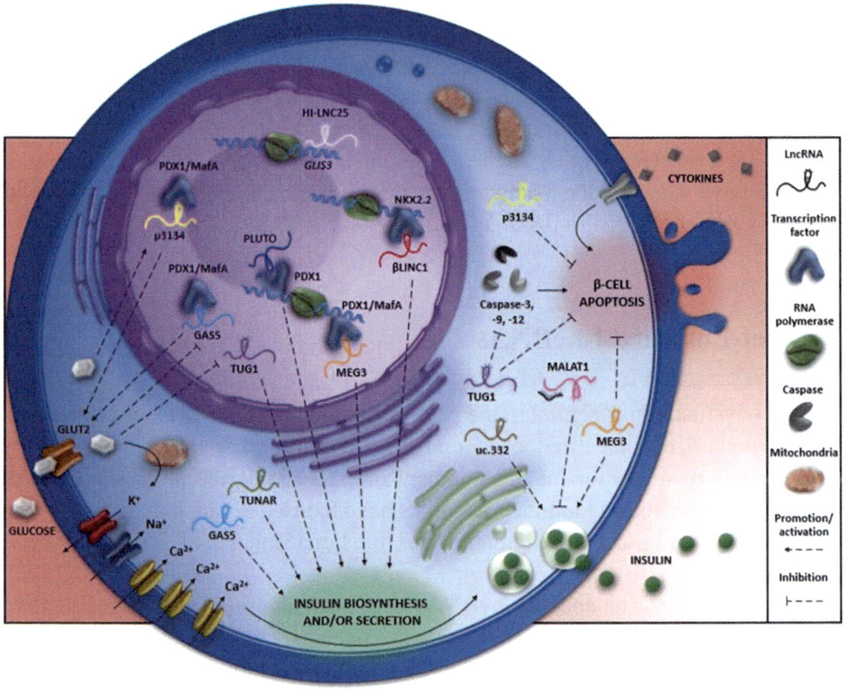

Fig. 1 LncRNAs associated with β-cell function. LncRNAs regulate multiple functions in the β cell, e.g., insulin expression, processing and secretion, as well as programmed cell death, apoptosis. This regulation is mediated partly by modulation of transcription factors involved in the expression of key genes to these mechanisms. This figure highlights ten lncRNAs reported to be involved in β-cell function: *MEG3*, *PLUTO*, *βLINC1*, *TUNAR*, *TUG1*, *HI-LNC25*, *MALAT1*, *uc.322*, *p3134*, and *GAS5*. Promotion/activation and inhibition are visualized by dotted arrows and represent most probable mechanisms by which the lncRNAs regulate β-cell function and apoptosis

lncRNAs in regulating β-cell gene expression programs and β-cell development and function. Figure 1 shows an overview of simplified mechanisms underlying the effects of lncRNAs in β-cell function.

HI-LNC25

The regulatory function of an islet- and β-cell-specific lncRNA *HI-LNC25* has been vastly described in human β cells (Morán et al. 2012). *HI-LNC25* is transcribed from the islet-specific active chromatin domain and regulates the expression of *GLIS3* mRNA, which encodes an islet transcription factor. *GLIS3* is mutated in a form of monogenic diabetes and contains T1D and T2D risk variants and is also involved in β-cell growth and proliferation in response to insulin resistance (Nogueira et al. 2013; Onengut-Gumuscu et al. 2015). Overexpression of *HI-LNC25* in human β cells promotes the expression of *GLIS3,* while knockdown of *HI-LNC25* reduces *GLIS3* expression in EndoC-βH1 cells. *HI-LNC25* and *GLIS3* are located on separate chromosomes (chr20 and chr9, respectively), indicating a *trans*-regulatory effect of

HI-LNC25 on *GLIS3*. These results highlight that *HI-LNC25* is an important regulator of *GLIS3* and might have a role in development of both T1D and T2D.

TUNAR

HI-LNC78 (*TUNAR*) is a highly β-cell-enriched lncRNA with >200-fold enrichment in β cells (Akerman et al. 2017). *TUNAR* regulates genes in pathways controlling insulin secretion and β-cell transcription factors. *TUNAR* is also one of 12 β-cell-enriched lncRNAs selected for functional studies based on its proximity to known genes important for β-cell function and presence of chromatin marks indicating active chromatin in β cells (Akerman et al. 2017). Knockdown of *TUNAR* reduces insulin content and consequently leads to impaired glucose-stimulated insulin secretion in human EndoC-βH3 cells. The transcriptional changes that occur after *TUNAR* knockdown significantly correlate with those observed after inhibition of islet-specific transcription factors *HNF1A* and *MAFB*. Interestingly, *TUNAR* is also upregulated by glucose treatment in human islets. These results indicate that *TUNAR* and islet transcription factors regulate common gene expression programs (Akerman et al. 2017; Wong et al. 2018).

GAS5

Growth arrest-specific transcript 5 (*GAS5*) is a well-characterized lncRNA, known to regulate cell proliferation and growth. *GAS5* is also the most highly expressed lncRNA in human islets and normal pancreas, while it is significantly downregulated in db/db mice (which is a widely used mouse model for T2D) (Fadista et al. 2014). Knockdown of *Gas5* in mouse islets reduces the expression of insulin gene and islet transcription factors including Pdx1 and MafA (Jin et al. 2017). In the mouse pancreatic β-cell line MIN6, *Gas5* knockdown promotes cell cycle arrest at G1 and decreases insulin biosynthesis and secretion (Jin et al. 2017). *GAS5* has also been shown to interact with the glucocorticoid receptor (GR) and acts as its repressor (Esguerra et al. 2020). *GAS5* expression is significantly reduced in human islets and EndoC βH1 cells treated with dexamethasone. Knockdown of *GAS5* in EndoC-βH1 cells significantly reduces glucose-stimulated insulin secretion and increases apoptosis; however, insulin secretion is recovered from dexamethasone-treated cells upon introduction of active segment of *GAS5* (Esguerra et al. 2020). Furthermore, both *GAS5* knockdown and dexamethasone lead to significant reduction of NKX6-1, PDX1, and GR. The effect of dexamethasone on these genes is reversed in both human islets and EndoC-βH1 cells upon addition of a GR inhibitor. These data indicate that *GAS5* might be an important regulator of β-cell identity and function by affecting insulin synthesis and secretion.

MALAT1

Metastasis-associated lung adenocarcinoma transcript 1 (*MALAT1*) is one of the very well-characterized lncRNA and has recently been shown to be a predictor of islet isolation quality (Wong et al. 2019). *MALAT1* is upregulated in serum of individuals with diabetes who smoke compared to nonsmokers (Sun et al. 2018). *MALAT1* has predicted binding sites for miR-17, which showed to be expressed at lower levels in the sera of people with diabetes who smoke compared to nonsmokers. In MIN6 cells, cigarette smoke extract (CSE) upregulates thioredoxin-interacting protein (*TXNIP*) and *MALAT1* and downregulates MafA and miR-17 (Sun et al. 2018). *MALAT1* knockdown induces an increase in miR-17, which suppresses TXNIP and promotes the production of insulin. These results highlight that CSE suppresses the production of insulin by increasing *MALAT1* and decreasing *miR-17*, indicating that *MALAT1* is a potential biomarker of pancreatic β-cell dysfunction caused by cigarette smoke.

TUG1

Taurine-upregulated gene 1 (*TUG1*) is a highly conserved lncRNA in mammals and highly expressed in pancreas compared to other tissues. *TUG1* has been shown to affect apoptosis and insulin secretion in pancreatic β cells. Both in vitro and in vivo downregulation of *Tug1* expression increased apoptosis and decreased insulin synthesis and secretion in mouse β cells (Yin et al. 2015). Islet-specific transcription factors, Pdx1, NeuroD1, MafA, and *Glut2* mRNA levels, are decreased after downregulation of *Tug1* in MIN6 cells. Based on these data, it can be suggested that *TUG1* might contribute to the impairment of β-cell function and therefore be implicated in diabetes pathogenesis.

MEG3

Mouse maternal expressed gene 3 (*Meg3*) is an imprinted intranuclear lncRNA essential for development. Expression of *Meg3* in islets is decreased in T1D (NOD female mice) and T2D (db/db mice) models. In MIN6 cells and isolated mouse islets, *Meg3* expression is modulated dynamically by glucose. Knockdown of *Meg3* results in decreased expression of Pdx-1 and MafA expression at both mRNA and protein levels, impairs insulin synthesis and secretion, and increases β-cell apoptosis (You et al. 2016). Furthermore, knockdown of *Meg3* in vivo leads to impaired glucose tolerance and decreases insulin secretion. Recently, it has been shown that *Meg3* regulates MafA expression in mouse β cells by binding *Ezh2*, a methyltransferase belonging to the polycomb repressive complex-2 (Wang et al. 2018a). Knockdown of *Meg3* and *Ezh2* inhibits the expression of MafA and *Ins2* in mouse β cells. These findings indicate that *Meg3* might be a novel regulator of maintaining β-cell identity via affecting insulin production and cell apoptosis.

p3134

LncRNA *p3134* is a newly discovered lncRNA, which is upregulated in serum exosomes and whole blood from T2D individuals (Ruan et al. 2018). *p3132* negatively correlates with HOMA-β, which is a marker of β-cell function, suggesting its potential role in the regulation of β-cell function. Overexpression of *p3134* in MIN6 cells increases glucose-stimulated insulin secretion and also upregulates key insulin transcription factors, including Pdx-1, MafA, Glut2, and Tcf7l2. Additionally, *p3134* overexpression in islets of db/db mice shows increased mRNA and protein levels of Pdx-1, MafA, and Tcf7l2 when compared with db/m (nondiabetic littermate controls) mice. These results indicate that *p3134* is dynamically modulated by glucose in both MIN6 cells and mouse islets, and might have a role in enhancing insulin synthesis and secretion by promoting key β-cell regulators (Pdx-1, MafA, Glut2, and Tcf7l2). In MIN6 cells, *p3134* overexpression significantly reduces caspase 3 and caspase 9 expression and increases Bcl-2 expression in MIN6 cells. This further confirms the overall protective effect of *p3132* on β-cell proliferation and apoptosis.

uc.322

LncRNA *uc.322* is located within the exonic region of the *SOX6* (SRY-related HMG-box 6) gene and is highly expressed in pancreatic islets compared to other tissues. Recently, it was shown that overexpression of *uc.322* upregulates islet transcription factors PDX1 and FOXO1, promotes insulin secretion, and increases ATP concentration in MIN6 cells (Zhao et al. 2018). Furthermore, knockdown of uc.*322* shows opposite effects and leads to reduced insulin secretion and ATP concentration in MIN6 cells. These results highlight that *uc.322* modulates islet transcription factors and thereby could serve as a therapeutic target for diabetes.

βLINC1

β-cell long intergenic noncoding RNA 1 (*βLINC1*) is a highly conserved, nuclear-enriched, islet-specific lncRNA located 20 kb upstream of the islet transcription factor NKX2.2. The promoter region of *βLINC1* is specifically enriched for islet transcription factor binding sites including FOXA2, NEUROD1, and PDX1 (Arnes et al. 2016). Knockdown of *βlinc1* in mouse β cells reduces the expression of Nkx2.2, which is an important β-cell transcriptional regulator and involved in β-cell maturation and function. *βlinc1* knockout mice show abnormal insulin secretion and glucose intolerance in low-glucose conditions. In MIN6 cells, *βlinc1* silencing downregulates the expression of another islet transcription factor Pax6 and upregulates the expression of the growth hormone-inhibiting hormone somatostatin. Deletion of *βlinc1* results in defective islet development and disrupts glucose homeostasis in the adult mice. These results highlight that *βlinc1* is a novel islet-specific transcriptional regulator of β cells.

Two other mouse orthologue lncRNAs, *βlinc2* and *βlinc3*, are reported to be differentially expressed in the islets of diet-induced obese mice (Motterle et al. 2017). *βLINC3* expression is downregulated in T2D patients compared to healthy controls. Interestingly, expression of *βlinc3* negatively correlates with body weight, insulinemia and glycemia, whereas *βlinc2* has an opposite effect. Upregulation of *βlinc2* or downregulation of *βlinc3* promotes apoptosis in MIN6 cells without affecting the insulin secretion, which is suggestive of their role in the development of T2D.

4 Regulatory Landscape of β Cells Based on 3D Chromatin Architecture

Recent T1D and T2D GWAS have revealed more than 60 and 200 susceptibility loci, respectively (Onengut-Gumuscu et al. 2015; Mahajan et al. 2018). Most of these disease-associated regions harbor noncoding variants that affect both protein-coding and noncoding genes. These noncoding variants impact the regulatory elements such as promoters, enhancers, silencers, and lncRNAs. Interestingly, some of the islet lncRNAs map within the T2D susceptibility loci and are dysregulated in islets from donors with T2D (Morán et al. 2012; Fadista et al. 2014). Patterns of chromatin accessibility (ATAC-seq) and DNA methylation (whole-genome bisulfite sequencing) in 18 human islet preparations generated high-resolution islet chromatin state maps through integration with established ChIP-seq marks (Thurner et al. 2018). Enrichment of T2D GWAS signals was observed within islet enhancers characterized by open chromatin and hypomethylation and within islet expression quantitative trait loci (eQTLs). Further fine mapping analysis using larger numbers of human islet samples from healthy and diabetic islets is required for providing complete description of the causal interactions between DNA methylation, chromatin states, and T1D and T2D susceptibility.

4.1 Human Pancreatic Islet 3D Chromatin Interactions

The first genome-wide map of long-range chromosomal interactions between gene promoters and distant regulatory elements, such as enhancers in human pancreatic islets, by capture Hi-C technology was recently generated (Miguel-Escalada et al. 2019). Capture Hi-C (CHi-C) is a method for profiling chromosomal interactions involving targeted regions of interest, such as gene promoters, globally and at high resolution. An open-source R package CHiCAGO ("Capture Hi-C Analysis of Genomic Organization") was recently described to identify such promoter-interacting regions and 3D chromatin interactions (Cairns et al. 2016). Using CHi-C, 175,784 high-confidence interactions (CHiCAGO score $>$ 5) between annotated promoters and distal genomic regions in human islets were identified (Miguel-Escalada et al. 2019). Further integration of islet ChIP-seq and ATAC-seq datasets with 3D chromatin maps resulted in refined human islet epigenome annotations. Interestingly, islet-selective interacting regions are enriched in active enhancers and islet-specific promoters. This genome-scale map of human islet regulatory interactome is accessible at http://isletregulome.org/. Islet Regulome DB (Mularoni et al. 2017) is a comprehensive, interactive, and annotated database of human islet 3D chromatin interactions that provides an interactive exploration of pancreatic islet genomic data. Figure 2 shows an example of complex regulatory interplay between islet transcription factors, long-range chromosomal interactions, and active enhancers at the *GLIS3* locus, which is a shared risk locus for both T1D and T2D. *GLIS3-AS1* lncRNA, which is antisense to *GLIS3*, is also highly expressed in human islets.

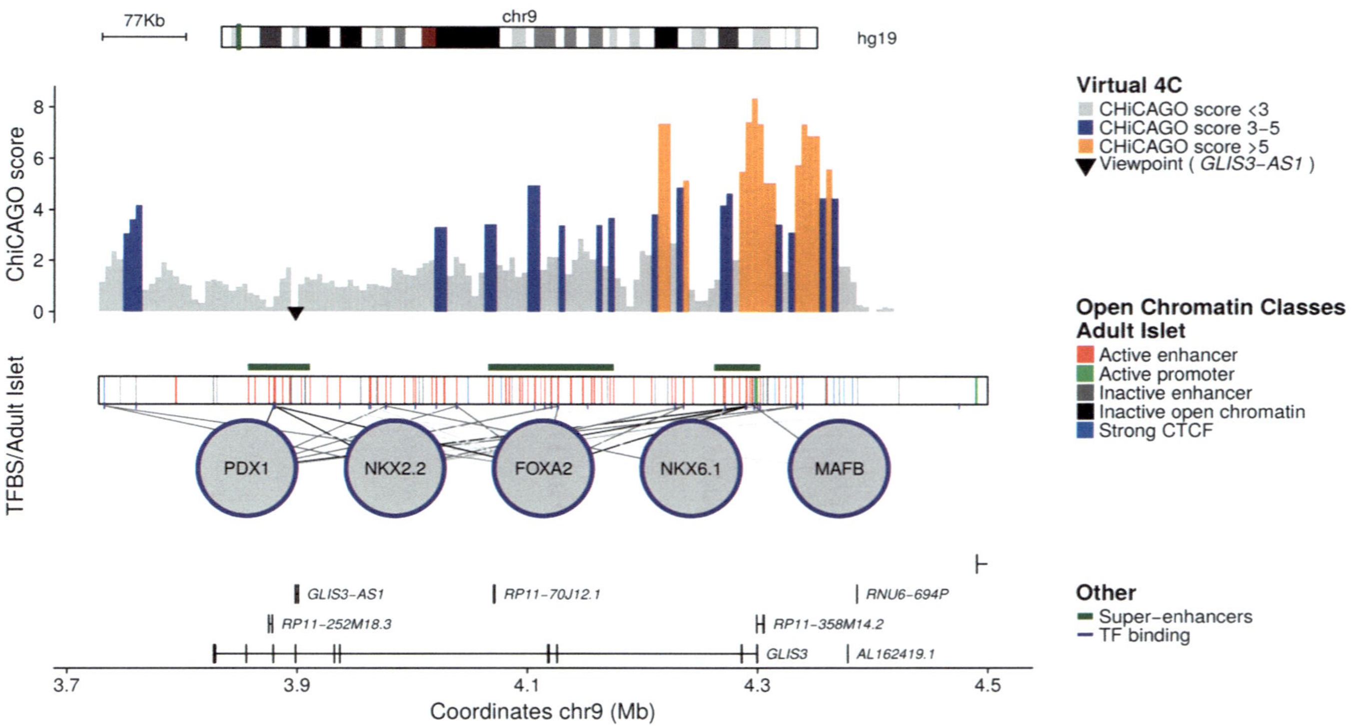

Fig. 2 A snapshot of complex regulatory interplay between islet enhancers and islet transcription factors at the *GLIS3* locus. *GLIS3* harbors both T1D and T2D risk variants and is an important islet transcription factor. The figure illustrates the genomic complexity at the *GLIS3* locus, which includes active islet enhancer clusters, promoters, islet transcription factor binding sites, and long-range chromosomal interactions. The chromosomal interactions as detected by CHiCAGO are shown centered on the viewpoint of antisense lncRNA *GLIS3-AS1*

4.2 LncRNAs Regulate β-cell Identity and Function via Disruption of Local 3D Chromatin Structure

A unique pattern of islet transcription factors (e.g., PDX1, PAX6, *HNF1A*, *GLIS3*, *MAFB*, MAFA, NKX2.2) is responsible for the sustained expression of β-cell selective genes and for maintaining β-cell identity. Among these, PDX1, an essential transcriptional regulator of pancreas development and β-cell function, is significantly downregulated in islets from donors with T2D and impaired glucose tolerance (Akerman et al. 2017). *HI-LNC71* (*PLUTO*), a nuclear-enriched β-cell lncRNA located upstream of the PDX1 locus, is one of the most downregulated lncRNA in islets from donors with T2D and impaired glucose tolerance (Akerman et al. 2017). *PLUTO* knockdown in EndoC-βH3 cells reduces insulin content and, consequently, impairs glucose-stimulated insulin secretion. *PLUTO* encompasses a cluster of enhancers that make 3D contacts with the *PDX1* promoter in human islets and in EndoC-βH1 cells. Knockdown of *PLUTO* reduces *PDX1* expression at both mRNA and protein levels in human islets and in EndoC-βH1 cells. *PLUTO* knockdown also impairs 3D contacts between the *PDX1* promoter and its adjacent enhancer cluster, resulting in reduced PDX1 activity. These findings suggest key analogous roles of β-cell lncRNAs in regulating β-cell networks similar to β-cell transcription factors. It is plausible that defects in β-cell lncRNAs might contribute to pathogenesis of diabetes, and further studies should investigate their role in β-cell programming strategies.

5 LncRNAs Associated with Diabetes-Related Complications

A growing number of lncRNAs have been implicated in various diabetic complications (Table 1). *MALAT1*, *ANRIL*, *linc-MIAT*, and *MEG3* have been studied for their role in diabetic retinopathy (Leti and DiStefano 2017; Leung and Natarajan 2018). Under hyperglycemia, diabetic retinas from STZ-induced rats and retinal endothelial cells show elevated *MALAT1*, *ANRIL*, and *MIAT* expression, suggesting that lncRNAs promote retinopathy in diabetes (Yan et al. 2014; Qiu et al. 2016; Thomas et al. 2017; Li et al. 2018b). A number of lncRNAs including *PVT1, MALAT1, lnc-MGC*, and *TUG1* have been implicated in diabetic nephropathy (Alvarez and DiStefano 2011; Long et al. 2016; Kato et al. 2016; Li et al. 2017). *Lnc-MGC* and *PVT1* are upregulated in human renal mesangial cells with high-glucose treatment, leading to increased fibrosis in renal cells due to the accumulation of extracellular matrix (Alvarez and DiStefano 2011; Kato et al. 2016). *TUG1* plays a protective role in diabetic nephropathy by modulating mitochondrial function in podocytes (Long et al. 2016). Under hyperglycemia, *TUG1* is downregulated which further decreases autoregulation of PGC-1a (a transcription factor important for mitochondrial biogenesis). In mice podocytes, overexpression of *TUG1* improves mitochondrial

Table 1 LncRNAs associated with β-cell dysfunction and diabetic complications

LncRNA	Sample source	Expression	Major findings	References
β-Cell dysfunction				
PLUTO (HI-LNC71)	Islets of T2D donors, EndoC-βH1 cells	Down	Affects 3D chromatin structure and transcription of PDX1	Akerman et al. (2017)
βLINC1	MIN6 cells, genetically mixed C57BL6/129SV adult mice		Controls islet β-cell formation and function by regulating key islet-specific transcription factors (NKx2.2, Pax6, and MafB)	Arnes et al. (2016)
TUG1	MIN6 cells		In vitro and in vivo downregulation of *Tug1* expression induce increased apoptosis and decreased insulin synthesis and secretion in mouse β cells	Yin et al. (2015)
GAS5	db/db mice/mouse islets	Down	Affects the expression of insulin gene and islet transcription factors Pdx1 and MafA and regulates cell cycle arrest and insulin secretion	Jin et al. (2017)
TUNAR (*HI-LNC78*)	Human pancreatic EndoC-βH3 cells		Regulates insulin content and glucose-stimulated insulin secretion	Akerman et al. (2017)
HI-LNC25	Human islets		Regulates the expression of GLIS3, an islet transcription factor	Morán et al. (2012)
MEG3	T1D (NOD female mice), T2D (db/db mice), and MIN6 cells	Down	siRNA-mediated knockdown of *MEG3* impairs insulin synthesis and secretion in vitro; in vivo knockdown impairs glucose tolerance, decreases insulin secretion, and leads to decreased Pdx-1 and MafA at both mRNA and protein levels	You et al. (2016)
MALAT1	MIN6 cells		*MALAT1* knockdown induces an increase in miR-17, which suppresses TXNIP and promotes the production of insulin	Sun et al. (2018)
uc.322	MIN6 cells		Overexpression of *uc.322* upregulates PDX1 and FOXO1, promotes insulin secretion, and increases	Zhao et al. (2018)

(continued)

Table 1 (continued)

LncRNA	Sample source	Expression	Major findings	References
			ATP concentration; on the contrary, knockdown of uc.*322* leads to reduced insulin secretion and ATP concentration	
p3134	MIN6 cells		Overexpression of *p3134* increases glucose-stimulated insulin secretion; upregulates Pdx-1, MafA, GLUT2, and TCF7L2; and reduces apoptosis	Ruan et al. (2018)
Diabetic nephropathy				
MALAT1	STZ rats, HK-2 cells	Up	Upregulates miR-23c, represses expression of ELAVL1, NLRP3, Caspase-1, and IL-1β; prevents hyperglycemia-induced cell pyroptosis	Li et al. (2017)
PVT1	Human renal mesangial cells	Up	Increased fibrosis in renal cells due to the accumulation of extracellular matrix	Alvarez and DiStefano (2011)
lnc-MGC	Human renal mesangial cells, diabetic mice	Up	Knockdown of lnc-MGC inhibits glomerular extracellular matrix accumulation and hypertrophy in diabetic mice	Kato et al. (2016)
TUG1	Podocytes, diabetic mice	Down	Overexpression of *Tug1* improves mitochondrial bioenergetics and reverses several key processes associated with diabetic nephropathy	Long et al. (2016)
Diabetic cardiomyopathy				
MALAT1	STZ rats	Up	Involved in inflammatory response, likely through TNFα, IL-1β, and IL-6	Zhang et al. (2016)
ANRIL	Wt and ANRILKO-STZ mice	Up	Controls expression of extracellular matrix proteins and VEGF through histone acetylator p300 and EZH2 component of PRC2 complex	Thomas et al. (2018)
MIAT	Cardiomyocytes, diabetic rats		Upregulates DAPK2 expression through miR-22-3p, leading to apoptosis	Zhou et al. (2017)

(continued)

Table 1 (continued)

LncRNA	Sample source	Expression	Major findings	References
H19	STZ rats, cardiomyocytes	Down	Regulates DIRAS3 and mTOR signaling and consequently autophagy; regulates miR-675, which targets VDAC1, regulating high-glucose-induced apoptosis	Li et al. (2016)
Diabetic retinopathy				
MEG3	Retinas of STZ-induced diabetic mice and endothelial cells	Down	*MEG3* knockdown regulates retinal endothelial cell proliferation, migration, and tube formation in vitro	Qiu et al. (2016)
ANRIL	Human retinal endothelial cells, ANRILKO-STZ mice	Up	Regulates VEGF expression and function, mediated by p300, miR200b, and EZH2 of the PRC2 complex	Thomas et al. (2017)
linc-MIAT	Plasma from patients with diabetic retinopathy, ARPE-19 adult retinal pigment epithelial cells	Up	Promotes diabetic retinopathy by upregulating TGF-β1 signaling	Li et al. (2018b)
MALAT1	Retinas of STZ-induced rats and db/db mice	Up	Knockdown of *MALAT1* attenuates retinopathy in STZ-induced rats	Yan et al. (2014)

bioenergetics and reverses several key processes associated with diabetic nephropathy. *MALAT1*, *ANRIL*, *H19*, and *MIAT* have been shown to affect multiple pathological processes associated with heart failure and have been associated with diabetic cardiomyopathy (Li et al. 2016; Zhang et al. 2016; Zhou et al. 2017; Thomas et al. 2018).

6 LncRNAs as Biomarkers and Therapeutic Targets for Diabetes

Exploiting lncRNAs for therapeutic advantages requires development of new and targeted interventions, which can selectively silence the expression of or inhibit the functions of the lncRNAs of interest (Ozcan et al. 2015). Alternatively, it may in some cases be beneficial to increase the amount of certain lncRNAs by inhibition of natural antisense transcripts or by introducing lncRNA mimics. For instance, increasing the level of some of the lncRNAs shown to be associated with β-cell survival and function, such as *βLINC1*, could be explored to protect β cells from immune-mediated killing and/or to boost β-cell function.

Currently, several preclinical and clinical siRNA-based therapeutics are in clinical trials for other diseases (Ozcan et al. 2015), which can be inspirational for diabetes treatments. Up until now, only a few studies have attempted to target lncRNAs in diabetes, and these studies have mainly focused on diabetic complications (Leti and DiStefano 2017; Leung and Natarajan 2018). The reason for the limited number of studies in the field so far is probably due to our lack of knowledge about the causative roles of lncRNAs in diabetes development. Moreover, low bioavailability, short half-life, and lack of studies addressing systemic side effects and toxicities currently hamper in vivo applications with oligonucleotides.

Challenges in targeting lncRNAs to block their functions or exploiting the lncRNAs themselves for therapeutic use include obtaining of appropriate stability and affinity of the molecules. As modulation of selected lncRNAs are often tissue or cell-type specific, delivery and target specificity may also be a great obstacle when considering lncRNAs therapeutically.

Besides their possible pathogenic role, lncRNAs may also serve as reliable disease status markers. Ideal biomarkers have high clinical specificity and sensitivity and are quantifiable and cost-effective. They should further be reliable, quick to test, and easy to obtain, e.g., from a blood sample. Thus, noninvasive body fluid-based biomarkers such as lncRNAs offer an excellent window of opportunity to improve our ability for risk prediction, early diagnosis, and monitoring of diabetes progression and prognosis. Since the first lncRNA, *PCA3*, that has been approved by the Food and Drug Administration (FDA) as a biomarker for prostate cancer (Filella et al. 2018), there have been tremendous attempts to find lncRNAs as biomarkers for other diseases including diabetes and its complications (Zhang et al. 2019). The differential expression of lncRNAs in diabetes makes them potential candidates as diagnostic biomarkers. For example, the expression level of *MALAT1* is consistently higher in patients when compared to healthy individuals in several studies, making *MALAT1* an interesting and valid diabetes biomarker candidate. Interestingly, a recent study also identified *MALAT1* as a biomarker for human pancreatic islet isolation quality (Wong et al. 2019).

Undoubtedly, further studies are needed to increase our understanding of lncRNAs in health and disease and their potential as novel antidiabetic therapeutics. Perhaps the greatest potential of this specific class of RNA molecules, at least in the nearest future, is to serve as biomarkers of diabetes for early prediction, better prognostics, and risk of complications, thereby facilitating the possibility of personalized treatment regimens. However, lncRNAs perhaps also hold a key to the future development of novel therapies designed to avoid β-cell loss and failure in diabetes. A mandatory step toward these promising perspectives of lncRNAs is to obtain more detailed knowledge about the causative roles of lncRNAs in the diseased tissues including the pancreatic β cells. More studies addressing this is therefore warranted to advance this exciting research field.

References

Abdelmohsen K, Panda A, Kang M-J et al (2013) Senescence-associated lncRNAs: senescence-associated long noncoding RNAs. Aging Cell 12:890–900

Akerman I, Tu Z, Beucher A et al (2017) Human pancreatic β cell lncRNAs control cell-specific regulatory networks. Cell Metab 25:400–411

Alvarez ML, DiStefano JK (2011) Functional characterization of the plasmacytoma variant translocation 1 gene (PVT1) in diabetic nephropathy. PLoS One 6:e18671

American Diabetes Association (2018) 2. Classification and diagnosis of diabetes: standards of medical care in diabetes-2018. Diabetes Care 41:S13–S27

Arnes L, Akerman I, Balderes DA et al (2016) βlinc1 encodes a long noncoding RNA that regulates islet β-cell formation and function. Genes Dev 30:502–507

Benner C, van der Meulen T, Cacéres E et al (2014) The transcriptional landscape of mouse beta cells compared to human beta cells reveals notable species differences in long non-coding RNA and protein-coding gene expression. BMC Genomics 15:620

Butler AE, Janson J, Bonner-Weir S et al (2003) Beta-cell deficit and increased beta-cell apoptosis in humans with type 2 diabetes. Diabetes 52:102–110

Cairns J, Freire-Pritchett P, Wingett SW et al (2016) CHiCAGO: robust detection of DNA looping interactions in capture Hi-C data. Genome Biol 17:127

Carter G, Miladinovic B, Patel A et al (2015) Circulating long noncoding RNA GAS5 levels are correlated to prevalence of type 2 diabetes mellitus. BBA Clin 4:102–107

Cesana M, Cacchiarelli D, Legnini I et al (2011) A long noncoding RNA controls muscle differentiation by functioning as a competing endogenous RNA. Cell 147:358–369

DiStefano JK (2015) Beyond the protein-coding sequence: noncoding RNAs in the pathogenesis of type 2 diabetes. Rev Diabet Stud RDS 12:260–276

Esguerra JLS, Ofori JK, Nagao M et al (2020) Glucocorticoid induces human beta cell dysfunction by involving riborepressor GAS5 LincRNA. Mol Metab 32:160–167

Fadista J, Vikman P, Laakso EO et al (2014) Global genomic and transcriptomic analysis of human pancreatic islets reveals novel genes influencing glucose metabolism. Proc Natl Acad Sci U S A 111:13924–13929

Filella X, Fernández-Galan E, Fernández Bonifacio R et al (2018) Emerging biomarkers in the diagnosis of prostate cancer. Pharmacogenomics Pers Med 11:83–94

Gao Y, Wu F, Zhou J et al (2014) The H19/let-7 double-negative feedback loop contributes to glucose metabolism in muscle cells. Nucleic Acids Res 42:13799–13811

He Y, Wu Y-T, Huang C et al (2014) Inhibitory effects of long noncoding RNA MEG3 on hepatic stellate cells activation and liver fibrogenesis. Biochim Biophys Acta 1842:2204–2215

Ji E, Kim C, Kim W et al (2018) Role of long non-coding RNAs in metabolic control. Biochim Biophys Acta Gene Regul Mech

Jin F, Wang N, Zhu Y et al (2017) Downregulation of long noncoding RNA Gas5 affects cell cycle and insulin secretion in mouse pancreatic β cells. Cell Physiol Biochem Int J Exp Cell Physiol Biochem Pharmacol 43:2062–2073

Kahn SE, Cooper ME, Del Prato S (2014) Pathophysiology and treatment of type 2 diabetes: perspectives on the past, present, and future. Lancet Lond Engl 383:1068–1083

Kato M, Wang M, Chen Z et al (2016) An endoplasmic reticulum stress-regulated lncRNA hosting a microRNA megacluster induces early features of diabetic nephropathy. Nat Commun 7:12864

Ku GM, Kim H, Vaughn IW et al (2012) Research resource: RNA-Seq reveals unique features of the pancreatic β-cell transcriptome. Mol Endocrinol Baltim Md 26:1783–1792

Leti F, DiStefano JK (2017) Long noncoding RNAs as diagnostic and therapeutic targets in type 2 diabetes and related complications. Genes 8:207

Leung A, Natarajan R (2018) Long noncoding RNAs in diabetes and diabetic complications. Antioxid Redox Signal 29:1064–1073

Li Z, Chao T-C, Chang K-Y et al (2014) The long noncoding RNA THRIL regulates TNFα expression through its interaction with hnRNPL. Proc Natl Acad Sci U S A 111:1002–1007

Li X, Wang H, Yao B et al (2016) lncRNA H19/miR-675 axis regulates cardiomyocyte apoptosis by targeting VDAC1 in diabetic cardiomyopathy. Sci Rep 6:36340

Li X, Zeng L, Cao C et al (2017) Long noncoding RNA MALAT1 regulates renal tubular epithelial pyroptosis by modulated miR-23c targeting of ELAVL1 in diabetic nephropathy. Exp Cell Res 350:327–335

Li M, Xie Z, Wang P et al (2018a) The long noncoding RNA GAS5 negatively regulates the adipogenic differentiation of MSCs by modulating the miR-18a/CTGF axis as a ceRNA. Cell Death Dis 9:554

Li Q, Pang L, Yang W et al (2018b) Long Non-Coding RNA of Myocardial Infarction Associated Transcript (LncRNA-MIAT) promotes diabetic retinopathy by upregulating Transforming Growth Factor-β1 (TGF-β1) Signaling. Med Sci Monit 24:9497–9503

Liu H, Li H, Jin L et al (2018) Long noncoding RNA GAS5 suppresses 3T3-L1 cells adipogenesis through miR-21a-5p/PTEN signal pathway. DNA Cell Biol 37:767–777

Lo KA, Huang S, Walet ACE et al (2018) Adipocyte long-noncoding RNA transcriptome analysis of obese mice identified Lnc-Leptin, which regulates leptin. Diabetes 67:1045–1056

Long J, Badal SS, Ye Z et al (2016) Long noncoding RNA Tug1 regulates mitochondrial bioenergetics in diabetic nephropathy. J Clin Invest 126:4205–4218

Mahajan A, Taliun D, Thurner M et al (2018) Fine-mapping type 2 diabetes loci to single-variant resolution using high-density imputation and islet-specific epigenome maps. Nat Genet 50:1505–1513

Miguel-Escalada I, Bonàs-Guarch S, Cebola I et al (2019) Human pancreatic islet three-dimensional chromatin architecture provides insights into the genetics of type 2 diabetes. Nat Genet 51:1137–1148

Mirza AH, Kaur S, Brorsson CA et al (2014) Effects of GWAS-associated genetic variants on lncRNAs within IBD and T1D candidate loci. PLoS One 9:e105723

Mirza AH, Kaur S, Pociot F (2017) Long non-coding RNAs as novel players in β cell function and type 1 diabetes. Hum Genomics 11:17

Morán I, Akerman I, van de Bunt M et al (2012) Human β cell transcriptome analysis uncovers lncRNAs that are tissue-specific, dynamically regulated, and abnormally expressed in type 2 diabetes. Cell Metab 16:435–448

Motterle A, Gattesco S, Peyot M-L et al (2017) Identification of islet-enriched long non-coding RNAs contributing to β-cell failure in type 2 diabetes. Mol Metab 6:1407–1418

Mularoni L, Ramos-Rodríguez M, Pasquali L (2017) The pancreatic islet regulome browser. Front Genet 8:13

Nogueira TC, Paula FM, Villate O et al (2013) GLIS3, a susceptibility gene for type 1 and type 2 diabetes, modulates pancreatic beta cell apoptosis via regulation of a splice variant of the BH3-only protein Bim. PLoS Genet 9:e1003532

Onengut-Gumuscu S, Chen W-M, Burren O et al (2015) Fine mapping of type 1 diabetes susceptibility loci and evidence for colocalization of causal variants with lymphoid gene enhancers. Nat Genet 47:381–386

Ozcan G, Ozpolat B, Coleman RL et al (2015) Preclinical and clinical development of siRNA-based therapeutics. Adv Drug Deliv Rev 87:108–119

Puvvula PK, Desetty RD, Pineau P et al (2014) Long noncoding RNA PANDA and scaffold-attachment-factor SAFA control senescence entry and exit. Nat Commun 5:5323

Qiu G-Z, Tian W, Fu H-T et al (2016) Long noncoding RNA-MEG3 is involved in diabetes mellitus-related microvascular dysfunction. Biochem Biophys Res Commun 471:135–141

Ruan Y, Lin N, Ma Q et al (2018) Circulating LncRNAs analysis in patients with type 2 diabetes reveals novel genes influencing glucose metabolism and islet β-cell function. Cell Physiol Biochem 46:335–350

Sathishkumar C, Prabu P, Mohan V et al (2018) Linking a role of lncRNAs (long non-coding RNAs) with insulin resistance, accelerated senescence, and inflammation in patients with type 2 diabetes. Hum Genomics 12:41

Schmidt E, Dhaouadi I, Gaziano I et al (2018) LincRNA H19 protects from dietary obesity by constraining expression of monoallelic genes in brown fat. Nat Commun 9:3622
Singer RA, Sussel L (2018) Islet Long noncoding RNAs: a playbook for discovery and characterization. Diabetes 67:1461–1470
Singer RA, Arnes L, Sussel L (2015) Noncoding RNAs in β cell biology. Curr Opin Endocrinol Diabetes Obes 22:77–85
Sun Q, Xu H, Xue J et al (2018) MALAT1 via microRNA-17 regulation of insulin transcription is involved in the dysfunction of pancreatic β-cells induced by cigarette smoke extract. J Cell Physiol 233:8862–8873
Thomas AA, Feng B, Chakrabarti S (2017) ANRIL: a regulator of VEGF in diabetic retinopathy. Invest Ophthalmol Vis Sci 58:470–480
Thomas AA, Feng B, Chakrabarti S (2018) ANRIL regulates production of extracellular matrix proteins and vasoactive factors in diabetic complications. Am J Physiol Endocrinol Metab 314: E191–E200
Thurner M, van de Bunt M, Torres JM et al (2018) Integration of human pancreatic islet genomic data refines regulatory mechanisms at type 2 diabetes susceptibility loci. elife 7:7
Ulitsky I, Bartel DP (2013) lincRNAs: genomics, evolution, and mechanisms. Cell 154:26–46
von Herrath M, Sanda S, Herold K (2007) Type 1 diabetes as a relapsing-remitting disease? Nat Rev Immunol 7:988–994
Wang L, Zhao Y, Bao X et al (2015) LncRNA Dum interacts with Dnmts to regulate Dppa2 expression during myogenic differentiation and muscle regeneration. Cell Res 25:335–350
Wang Y, Hu Y, Sun C et al (2016) Down-regulation of Risa improves insulin sensitivity by enhancing autophagy. FASEB J Off Publ Fed Am Soc Exp Biol 30:3133–3145
Wang X, Chang X, Zhang P et al (2017) Aberrant expression of Long non-coding RNAs in newly diagnosed type 2 diabetes indicates potential roles in chronic inflammation and insulin resistance. Cell Physiol Biochem 43:2367–2378
Wang N, Zhu Y, Xie M et al (2018a) Long noncoding RNA Meg3 regulates Mafa expression in mouse beta cells by inactivating Rad21, Smc3 or Sin3α. Cell Physiol Biochem 45:2031–2043
Wang Q, Li M, Shen Z et al (2018b) The long non-coding RNA MEG3/miR-let-7c-5p Axis regulates ethanol-induced hepatic steatosis and apoptosis by targeting NLRC5. Front Pharmacol 9:302
Wong WKM, Sørensen AE, Joglekar MV et al (2018) Non-coding RNA in pancreas and β-cell development. Non-Coding RNA 4:41
Wong WK, Jiang G, Sørensen AE et al (2019) The long noncoding RNA MALAT1 predicts human pancreatic islet isolation quality. JCI Insight:5
Xiao T, Liu L, Li H et al (2015) Long noncoding RNA ADINR regulates adipogenesis by transcriptionally activating C/EBPα. Stem Cell Rep 5:856–865
Yan B, Tao Z-F, Li X-M et al (2014) Aberrant expression of long noncoding RNAs in early diabetic retinopathy. Invest Ophthalmol Vis Sci 55:941–951
Yan C, Li J, Feng S et al (2018) Long noncoding RNA Gomafu upregulates Foxo1 expression to promote hepatic insulin resistance by sponging miR-139-5p. Cell Death Dis 9:289
Yang Y, Lv X, Fan Q et al (2019) Analysis of circulating lncRNA expression profiles in patients with diabetes mellitus and diabetic nephropathy: differential expression profile of circulating lncRNA. Clin Nephrol 92:25–35
Yin D, Zhang E, You L et al (2015) Downregulation of lncRNA TUG1 affects apoptosis and insulin secretion in mouse pancreatic β cells. Cell Physiol Biochem 35:1892–1904
You L, Wang N, Yin D et al (2016) Downregulation of long noncoding RNA Meg3 affects insulin synthesis and secretion in mouse pancreatic Beta cells. J Cell Physiol 231:852–862
Zhang M, Gu H, Chen J et al (2016) Involvement of long noncoding RNA MALAT1 in the pathogenesis of diabetic cardiomyopathy. Int J Cardiol 202:753–755
Zhang W, Zheng J, Hu X et al (2019) Dysregulated expression of long noncoding RNAs serves as diagnostic biomarkers of type 2 diabetes mellitus. Endocrine 65:494–503

Zhao Y, Wu J, Liangpunsakul S et al (2017) Long non-coding RNA in liver metabolism and disease: current status. Liver Res 1:163–167

Zhao X, Rong C, Pan F et al (2018) Expression characteristics of long noncoding RNA uc.322 and its effects on pancreatic islet function. J Cell Biochem 119:9239–9248

Zhou X, Zhang W, Jin M et al (2017) lncRNA MIAT functions as a competing endogenous RNA to upregulate DAPK2 by sponging miR-22-3p in diabetic cardiomyopathy. Cell Death Dis 8: e2929

Correction to: Long Noncoding RNAs as Players in Breast Tumorigenesis

Cyrinne Achour and Francesca Aguilo

Correction to:
Chapter 15 in: S. Jurga, J. Barciszewski (eds.), *The Chemical Biology of Long Noncoding RNAs*, RNA Technologies 11, https://doi.org/10.1007/978-3-030-44743-4_15

The original version of chapter 15 was inadvertently published with an error in the chapter title. The correct chapter title is mentioned below and the chapter has now been corrected.

Chapter title: 'Long Noncoding RNAs as Players in Breast Tumorigenesis'

The updated online version of this chapter can be found at
https://doi.org/10.1007/978-3-030-44743-4_15

S. Jurga, J. Barciszewski (eds.), *The Chemical Biology of Long Noncoding RNAs*,
RNA Technologies 11, https://doi.org/10.1007/978-3-030-44743-4_21

GPSR Compliance

The European Union's (EU) General Product Safety Regulation (GPSR) is a set of rules that requires consumer products to be safe and our obligations to ensure this.

If you have any concerns about our products, you can contact us on ProductSafety@springernature.com

In case Publisher is established outside the EU, the EU authorized representative is:

Springer Nature Customer Service Center GmbH
Europaplatz 3
69115 Heidelberg, Germany

Batch number: 10726058

Printed by Printforce, the Netherlands